APS Blasting 4

NEW DEVELOPMENT ON ENGINEERING BLASTING

Editor in Chief
Prof. WANG Xuguang

METALLURGICAL INDUSTRY PRESS

Copyright © 2014 by Metallurgical Industry Press, China

Published and distributed by
Metallurgical Industry Press
39 Songzhuyuan North Alley, Beiheyan St
Beijing 100009, P.R.China

All rights reserved. No part of this publication may be reproduced, stored in a retrieval system, or transmitted in any form or by any means, electronic, mechanical, photocopying, recording or otherwise, without the prior written permission of the copyright owner.

图书在版编目（CIP）数据

工程爆破新进展. 4：New development on engineering blasting. 4：英文／汪旭光主编. —北京：冶金工业出版社，2014.10
ISBN 978-7-5024-6767-8

Ⅰ．①工… Ⅱ．①汪… Ⅲ．①爆破技术—文集—英文 Ⅳ．①TB41-53

中国版本图书馆 CIP 数据核字（2014）第 240967 号

出 版 人　谭学余
地　　址　北京市东城区嵩祝院北巷 39 号　邮编　100009　电话　(010)64027926
网　　址　www.cnmip.com.cn　电子信箱　yjcbs@cnmip.com.cn
责任编辑　廖 丹　程志宏　美术编辑　彭子赫　版式设计　孙跃红
责任校对　王永欣　责任印制　牛晓波
ISBN 978-7-5024-6767-8
冶金工业出版社出版发行；各地新华书店经销；三河市双峰印刷装订有限公司印刷
2014 年 10 月第 1 版，2014 年 10 月第 1 次印刷
210mm×297mm；34 印张；1269 千字；530 页
150.00 元

冶金工业出版社　投稿电话　(010)64027932　投稿信箱　tougao@cnmip.com.cn
冶金工业出版社营销中心　电话　(010)64044283　传真　(010)64027893
冶金书店　地址　北京市东四西大街 46 号（100010）　电话　(010)65289081（兼传真）
冶金工业出版社天猫旗舰店　yjgy.tmall.com

（本书如有印装质量问题，本社营销中心负责退换）

4th Asia-Pacific Symposium on Blasting Techniques (2014)

ORGANIZER

China Society of Engineering Blasting

CO-ORGANIZER

Shenzhen Society of Engineering Blasting

ORGANIZING COMMITTEE

CHAIRMAN

Prof. Xuguang Wang
Academician of Chinese Academy of Engineering
President of China Society of Engineering Blasting

VICE-CHAIRMAN

Xiwu Liang
Professorate Senior Engineer
President of Shenzhen Society of Engineering Blasting

ORGANIZING COMMITTEE MEMBERS (in alphabetical order)

Debasis Deb	(India)	Shiro Kubota	(Japan)
Yintong Gao	(China)	Jiankang Li	(China)
Kiyoshi Hashidume	(Japan)	Xiaojie Li	(China)
Yongsheng Jia	(China)	Zhanjun Li	(China)
Jamiyan Jijig	(Mongolia)	Qian Liu	(Canada)

Vladimir S. Litvinenko	(Russia)	Kliment N. Trubetskoy	(Russia)
Wenbo Lu	(China)	Sergei D. Victorov	(Russia)
Cameron Mckenzie	(Australia)	Hao Wang	(China)
Bibhu Mohanty	(Canada)	Xianqi Xie	(China)
Mike Osborne	(America)	Chaoyang Yang	(China)
Amar Pagva	(Mongolia)	Jun Yang	(China)
G. P. Paramonov	(Russia)	Ruilin Yang	(America)
Bayan Rakishev	(Kazakhstan)	Bingxu Zheng	(China)
Alan B. Richards	(Australia)	Jiayu Shi	(China)
Chang-Ha Ryu	(Korea)	Yayu Shi	(China)
Fuqiang Shi	(China)	Valerii N. Zakharov	(Russia)
Alexey V. Shlyapin	(Russia)	Beilong Zhang	(China)
Pradeep Singh	(India)	Dingxiang Zou	(China)

EDITORIAL BOARD

EDITOR IN CHIEF
Xuguang Wang

DEPUTY EDITOR IN CHIEF
Xiwu Liang

EDITORIAL BOARD MEMBERS (in alphabetical order)

Shaopan Chen	(China)	Xianqi Xie	(China)
Yunzhang Fu	(China)	Peixing Xue	(China)
Yintong Gao	(China)	Jun Yang	(China)
Wenxue Gao	(China)	Nianhua Yang	(China)
Yicheng Gu	(China)	Yalun Yu	(China)
Zhiqiang Guan	(China)	Beilong Zhang	(China)
Jiankang Li	(China)	Jingjing Zhang	(China)
Xiaojie Li	(China)	Yongzhe Zhang	(China)
Zhanjun Li	(China)	Zhengzhong Zhang	(China)
Dianshu Liu	(China)	Zhengyu Zhang	(China)
Jinquan Song	(China)	Bingxu Zheng	(China)
Hao Wang	(China)	Changqing Zheng	(China)
Yinjun Wang	(China)	Jiahan Zhou	(China)
Zhongqian Wang	(China)		

SECRETARIES
Wenfeng Song

Pingcheng He

Foreword

The Asia-Pacific Symposium on Blasting Techniques (APS BLASTING) is a symposium on blasting techniques and explosives in the Asia-Pacific region, initiated by the China Society of Engineering Blasting. The International Conference on Physical Problems of Rock Destruction (ICPPRD) is an academic conference about the latest achievements of rock fragmentation theory, technology, engineering and so on, initiated by the Institute of Comprehensive Exploitation of Mineral Resources, Russian Academy of Sciences. To expand conference scale and share conference resources, the organizing committees of two conferences decided to hold APS BLASTING and ICPPRD together from APS BLASTING 2 (ICPPRD 6) in 2009. Since then, the two conferences have been successfully held together for two times (APS BLASTING 2 & ICPPRD 6, 2009 and APS BLASTING 3 & ICPPRD 7, 2011) and have become the most important academic symposium on aspects of blasting techniques and rock fragmentation in the Asia-Pacific region, Russia and the Commonwealth of Independent States.

APS BLASTING 4 & ICPPRD 8 are held in Shenzhen as scheduled, aiming at strengthening the academic exchange and technological cooperation among countries in Asia-Pacific region, Russia and the CIS enhancing the inter-industry and inter-disciplinary penetration, dealing with the opportunities and challenges met by blasting and rock destruction and forecasting the application prospects of blasting technology in various fields to jointly promote the development of blasting technology and rock destruction in the world. I hope these two conferences could lead the communication and discussion to a great climax.

The organizing committees have received 111 abstracts from China, Russia, Canada, Japan, Korea, Sweden, India, Mongolia, Australia, Kazakhstan and the United States. 95 papers are accepted by the proceedings, of which 39 papers are from the countries except China. The Symposium proceedings cover a wide range of subjects and present the leading technological innovations and achievements in industrial explosives, detonating facilities, rock fragmentation theory, physical problems of rock destruction, blasting vibration effect, blasting numerical simulation, blasting excavation, blasting demolition and so on.

Thanks to the great attention and contribution of members of the organizing committees in publicizing the conference themes and calling for more papers and the great support of Shenzhen Society of Engineering Blasting, the two conferences have been well organized. Here I would like to show my sincere gratitude to the all committee members home and abroad and all the conference staff.

I wish the APS BLASTING & ICPPRD community gathered here a fruitful and memorable conference, and I hope you could support APS BLASTING & ICPPRD as ever.

Prof. Xuguang Wang

Academician, China Academy of Engineering

President of China Society of Engineering Blasting

Chairman of the 4th Asia-Pacific Symposium on Blasting Techniques

Honorary Chairman of the 8th International Conference on Physical Problems of Rock Destruction

Contents

1 Blasting Theory

Scientific and Technological Aspects of Blast Action Control in Mineral Deposit Mining
   K.N. Trubetskoy, S.D. Viktorov, V.M. Zakalinsky (Russia) ... 3

Study on the Mechanism of Linear Shaped Jet Splitting Target Based upon High-Speed 3D DIC
   XU Zhenyang, YANG Jun, YU Qi (China) ... 10

Beyond Environmental Vibration Compliance
   Cameron McKenzie, Mike Osborne (USA) ... 16

Study on Crush Progress of Rock in Delay Blasting
   SHI Fuqiang, LIAO Xueyan, JIANG Yaogang, GONG Zhigang(China) ... 25

Camouflet Blasting of a Finite-Length Borehole Charge
   S.D. Viktorov, N.N. Kazakov, A.V. Shlyapin, I.N. Lapikov(Russia) ... 28

Multibody-Discretebody Dynamics to Control Building Demolished by Blasting
   WEI Xiaolin(China) ... 32

Mathematical Model of Seismic Impact on Operating Underground Gas Pipelines
   A.P. Gospodarikov(Russia) ... 44

Numerical Analysis for Blasting Demolition of the Smaller Height-Width Ratio Framed Structure
   LIU Wei, YAN Shilong, HU Kunlun, LI Hongwei(China) ... 48

Explosive Cratering to Finite Size of Earth Structure
   Yumin Li, Ettore Contestabile, Abass Braimah, Bert von Rosen(Canada) ... 52

Numerical Simulation for Effect of Stemming in Blasting
   WANG Huxin, WU Chunping, CUI Xinnan, Chen He(China) ... 57

Estimation of Quasi-Static Action of Explosion Products in the Directed Destruction of Rock
   V.N. Kowalewski, Z.G. Dambaev(Russia) ... 64

Damage Characteristics of Surrounding Rock Subjected to VCR Mining Blasting
   JIANG Nan, ZHOU Chuanbo, LUO Xuedong, LU Shiwei(China) ... 68

Dynamic Failure Mechanism of Rock Mass Using Smoothed Particle Hydrodynamics
   Ranjan Pramanik, Debasis Deb(India) ... 75

A Supernal Security Regression Method of Blast Vibration
   JIANG Dongping, LIU Weizhou, ZHANG Xiliang(China) ... 80

Numerical Estimation of Crater Size of the RC Wall Caused by Blasting
   Shiro Kubota, Kana Nishino, Yuji Wada, and Yuji Ogata(Japan) ... 83

Research on Energy Consumption Characteristics and Energy Density Per Unit Time of Rock Crushing by Impact Load
   GUO Lianjun, SHAO Anlin, ZHANG Daning, MA Xufeng, YANG Yuehui, ZHANG Guojian,
   SUN Houguang(China) ... 87

Evaluation of Progressive Collapse of Blasting Demolition for Culvert Structures
   Young-Hun Ko, Jung-Gyu Kim, Yousong Noh, Myeong-Jin Shim, Hyung-Sik Yang(South Korea) ... 93

Vertical Shaft Blasting Parameters Design Software based on Visual Basic
   ZHANG Jiaming, LI Pingfeng, ZHOU Ming'an(China) ... 96

Distribution of Blasting-Induced Dynamic Events in Rock Mass
   A.A. Eremenko, S.V. Fefelov, V.A. Eremenko(Russia) ... 101

Study on Numerical Simulation of Rock Blasting Based on SPH

LIAO Xueyan, SHI Fuqiang, JIANG Yaogang, GONG Zhigang(China) ········ 104

Numerical Simulation Analysis of Hydraulic Pressurized Blasting Effect
LI Ming, CHEN Nengge, LIU Weizhou(China) ········ 108

2 Rock Fragmentation

Characteristic Response of Rock and Rock-like Materials to Explosive Loading in Controlled Experiments
B. Mohanty, R. Raghavaraju(Canada) ········ 115

Control of Yield Rate of Fines of Different Specification of Stone Quarrying with High Intensity
ZHENG Bingxu, LI Zhanjun, SONG Jinquan(China) ········ 121

Improving the Methodology of Rock Breakage by Blast for Underground Mineral Mining Technologies
S. D. Viktorov, V. M. Zakalinsky, A. A. Osokin(Russia) ········ 125

Research on Key Technology of Digital Rotary Drilling in Surface Mines
DUAN Yun, XIONG Daiyu, YAO Lu, WANG Fengjun, ZHA Zhengqing, GONG Bing(China) ········ 130

Controlled Blasting for Safe and Efficient Mining Operations at Rampura Agucha Mine in India
Pradeep K Singh, M. P. Roy, Amalendu Sinha(India) ········ 137

Study on Technology of Cableway Station Tunnel Demolition Blasting in Scenic Area
SHI Liansong, GAO Wenxue, LIU Dong, CHEN Gui, YAO Shaomin, ZHU Xuyang(China) ········ 152

Study on Open-Pit Precision Control Blasting of Easily Weathered Rock and Its Application
YI Haibao, YANG Haitao, LI Ming, HAN Bin, ZHENG Lujing(China) ········ 157

Influence of Power Characteristics of Explosives on Strength Properties of Pieces of the Blown-up Mountain Weight
G.P. Paramonov, V.A. Isheysky(Russia) ········ 161

Load Analysis of Hammer Crusher
NI Jiaying, ZHAO Sihai(China) ········ 166

Controlled Underwater Blasting near Sensitive Structures at Mithi River Extension Project, Mumbai, India
M. Ramulu, M. Gurharikar(India) ········ 171

Study on the Engineering Method of Excavating Shaft by One-Step Smooth Blasting
WANG Li, GAO Yan, SUN Ning(China) ········ 177

Kinematic Characteristics of Movement of the Explosive Cavity Wall in Different Rocks
B.R.Rakishev, Z.B.Rakisheva(Kazakhstan) ········ 180

Control Blasting Technique for Rock Cutting Expansion of Existing Expressway
CAI Xu, QIU Zhou, GUO Tiantian, ZHU Jinhua(China) ········ 185

Effect of Confining Pressure on the Failure of Model for Reinforced Concrete Slab Member
Yousong Noh, Jung-Gyu Kim, Young-Hun Ko, Myeong-Jin Shim, Hyung-Sik Yang(South Korea) ········ 191

Successful Practice of Ultra 1 km Wells Increasing Water Blasting
GONG Wenxin, ZHANG Zhongyi, JIANG Guixiang, YU Haibin, LI Peng(China) ········ 195

Regularities of the Formation of Submicron Particles at the Explosive Destruction of Rocks
S.D.Victorov, A.N. Kochanov(Russia) ········ 199

Research on Blasting Scheme Matching Semi-continuous Mining Process Used in East Open Pit
LIU Xiaoming, QI Maofu, CHENG Fei(China) ········ 202

Relationship between the Fragmentation Effect and the Blast Wave on the Controlled Blasting for Breaching
Kana Nishino, Shiro Kubota, Yuji Wada, Yuji Ogata, Norio Ito, Masayuki Nagano, Atsuya Fukuda, Mieko Kumasaki(Japan) ········ 207

The Application of Pre-splitting Blasting near Slope Technique in Yuanjiacun Iron Mine
JIA Chuanpeng(China) ········ 211

Study of Energy Characteristics Gasifier Cartridges Onpreservation of Raw Materials Division Monoliths of Rock Mass
G.P. Paramonov(Russia) ········ 216

Application of Multi Luntai Millisecond Blasting Technique on Limestone Mine
OUYANG Hairong, SUN Guang, BAI Hongfeng, ZHANG Jinbang, XIAO Chun(China) ········ 222

The Specification of Granulometric Composition of Natural Jointing in the Rock Massif by Their Average Size
    B. R. Rakishev, Z. B. Rakisheva, A. M. Auezova(Kazakhstan) ⋯⋯⋯⋯ 225

Examples of Blasting Using Emulsion Explosive in Mining in the 200 Meters Ultra Deep Water
    ZHOU Ming'an, XIA Jun, OU Liming, REN Caiqing(China) ⋯⋯⋯⋯ 229

The Town of Rock Blasting Engineering under Complex Environment
    JIN Huishi, YUAN Maoyu, LIU Chongyao, TAO Yongsheng(China) ⋯⋯⋯⋯ 233

3 Explosives and Initiation Techniques

Sensitivity to Impact of Binary Mixes on a Basis Fluoropolymer F-2M
    A.V. Dubovik, A.A. Matveev(Russia) ⋯⋯⋯⋯ 239

Research on Non-primary Explosive Slapper Detonator System
    SHEN Zhaowu, MA Honghao, CHEN Zhijun(China) ⋯⋯⋯⋯ 244

Methodological Aspects of Properties and Blast Energy Kinetics Control of Industrial of Explosives
    N.N. Efremovtsev, S.I. Kvitko(Russia) ⋯⋯⋯⋯ 256

Research of Manufacturing Process Improvement of Expanded Ammonium Nitrate Explosive
    ZHENG Bingxu, GAN Dehuai, SONG Jinquan, LI Zhanjun(China) ⋯⋯⋯⋯ 263

Some Behavior Characteristics of Condensed Composite Materials in Explosive Processes
    A.V. Starshinov, S.S.Kostylev, I. Y.Kupriyanov, N.Y.Yargina, JijigJamiyan(Russia) ⋯⋯⋯⋯ 267

The Comparison of the Commercial Explosives in United States to that in China and Discussion on Development of Chinese Commercial Explosives
    YAN Shilong, GUO Ziru, SHEN Zhaowu, WU Hongbo, WANG Quan(China) ⋯⋯⋯⋯ 270

Nonel Detonator Delay's Error and One by One Detonate Network
    LIAO Xiaolin, CAI Jianhua(China) ⋯⋯⋯⋯ 275

Production of Nitroester-Based Explosives in Russia
    A.S. Zharkov, E.A. Petrov, N.E. Dochilov, R.N. Piterkin(Russia) ⋯⋯⋯⋯ 281

Preliminary Discussion on Greening Design and Assesment of Industrial Explosives
    ZHANG Daozhen, LI Yunxi, SONG Zhiwei, JIANG Tiansheng(China) ⋯⋯⋯⋯ 284

Properties of Emulsion Matrices of Explosives Based on the Best Russian, Kazakh and Chinese Emulsifiers
    E.A. Petrov, P.G. Tambiev, P.I. Savin(Russia) ⋯⋯⋯⋯ 288

Anti-dynamic Pressure Overloads Performance and Detonation Characteristics of Emulsion Explosives Sensitized by MgH_2
    CHENG Yangfan, MA Honghao, SHEN Zhaowu(China) ⋯⋯⋯⋯ 291

Explosive Energy Distribution in Explosive Column Case Study of Some Indian Mines
    G.K.Pradhan(India) ⋯⋯⋯⋯ 297

Application of Digital Electronic Detonators in Blasting Safety in Construction Process
    CHENG Guangxiang, KANG Quanyu(China) ⋯⋯⋯⋯ 306

4 Special Blasting and Demolition Blasting

Blasting Demolition of "L" Shape Frame Structure Building in Downtown District
    XIE Xianqi, YAO Yingkang, JIA Yongsheng, LIU Changbang, WANG Honggang(China) ⋯⋯⋯⋯ 313

Design and Construction Technology of Large Dock Cofferdam Demolition Blasting
    GUAN Zhiqiang, ZHANG Zhonglei, LI Qiang, WANG Lingui, YE Jihong(China) ⋯⋯⋯⋯ 319

Experimental Research on the New Technology of Explosive Cladding
    MIAO Guanghong, MA Honghao, SHEN Zhaowu, YU Yong, LI Xuejiao, REN Lijie(China) ⋯⋯⋯⋯ 329

Study on the Blasting Technology of the Emergency Ice Breaking Used in the Ningxia-Inner Mongolia Reach of the Yellow River
    YAN Junwei, YANG Xusheng, LI Xiukun, DONG Dekun(China) ⋯⋯⋯⋯ 335

Research and Practice of New Technique for Blasting Demolition of Hyperbolic Cooling Tower
　　ZHU Jinhua, SUN Xiangyang, GUO Tiantian, WANG Qingtao(China) ········· 340
Removal of the Exhaust Tube Blasting in Complicated Environment
　　YU Hui, LIU Guixin, ZHANG Chao, FANG Limin(China) ········· 345
Trial Blasting Technology in Demolition Blasting of Building and Its Analysis
　　FEI Honglu, ZHANG Longfei, HE Wenbin, YANG Zhiguang(China) ········· 348
Directional Blasting Demolition of 60m Brick Chimney
　　YAO Jinjie, ZHAO Rundong, XIONG Yazhou, LUO Zhenhua(China) ········· 353
The Irregular Structure of Meteorological Office Building Demolition Blasting
　　LI Wei, WANG Qun, ZHAO Jingsen, LI Dan(China) ········· 356
Blasting Demolition of Urban Viaduct 3.5 km (2.175Mile) in Length
　　XIE Xianqi, JIA Yongsheng, YAO Yingkang, LIU Changbang, HAN Chuanwei(China) ········· 361
Water Pressure Blasting in Demolition of Domestic Architecture Analysis and Practice
　　YAO Jinjie, XIONG Yazhou, ZHAO Rundong, LUO Zhenhua(China) ········· 370

5　Blasting Vibration and Safety

Blast Vibration Impact Analyses for Waste Dump Stability
　　P.K. Singh, A. Sinha(India) ········· 377
Measurement and Analysis of Environmental Vibration Caused by Blasting Excavation of Foundation Pit in Town
　　TIAN Yunsheng, ZUO Jinku, LIU Weihua, WANG Ning(China) ········· 389
State of the Art Review of R&D on the Blast Vibration at KIGAM
　　Chang-Ha Ryu, Byung-Hee Choi, Hyung-Su Jang, Myoung-Soo Kang(South Korea) ········· 393
Engineering Blasting Safety Monitoring System Based on Internet and Cloud Computing
　　YANG Min, YANG Xun, XU Qiang(China) ········· 399
Blasting Harmonics
　　Adrian J. Moore, Alan B. Richards(Australia) ········· 404
Application and Research on Slight Vibration Controlled Blasting Technology for Metro Tunnel Adjacent Intermediate High Pressure Gas Pipe
　　ZHANG Junbing, ZHENG Baocai, LI Zihua, HU Yunfeng(China) ········· 416
Gas-Dynamic Hazard in Presentday Highly Productive Mines—Prediction Problems and Solutions
　　Zakharov V.N., Malinnikova O.N., Feit G.N. (Russia) ········· 423
The Implementation and Application of Demolition Data's Informationalized Managerial System
　　OU Liming, XU Cheng, ZHOU Ming'an(China) ········· 428
Effect of Blast Induced Ground Vibrations on Green Concrete
　　M. Gurharikar, M. Ramulu(India) ········· 433
Achievement of Safe Distance of Air Shock Wave in Tunnel Blasting
　　YU Haihua, YANG Haitao(China) ········· 438
Technology for Safety and Environmental Control in Blasting Operations
　　Sushil Bhandari, Reema Bhandari(Australia) ········· 442
Study on Blasting Operations Hazard Identification and Risk Assessment Based on Improved Evaluation Method LEC in Xiaocishan Mine
　　HUANG Kaihe, ZHANG Xiliang, XIE Liangbo(China) ········· 450
Monitoring of Coal and Rock Mass Conditions, Coal Mine Air and Extraction Equipment State
　　S.S. Kubrin(Russia) ········· 454
Effect Mechanism of Excavation Blasting in an Under-Cross Tunnel on Airport Runway
　　LU Shiwei, ZHOU Chuanbo, JIANG Nan, XU Xing(China) ········· 461
Spreading out Dust and Gases, Formed by Blasting Operations at Erdenetiin-Ovoo Ore Open Pit Mine and Application of Remote Control (Firing) System for Blasting
　　B.Laikhansuren, G.Tulga, D.Nyamdorj, N.Davaakhuu(Mongolia) ········· 467

Mechanism of the Short-delay Controlled Blasting in the Air and Its Application in Anti-terrorist
 YAN Honghao, ZHAO Tiejun, LI Xiaojie, WANG Xiaohong(China) ······ 473

Development of Damage Criteria for Underground Mine Structures Subjected to Blast Vibration from Neighbouring Surface Mine
 A. K. Jha, D. Deb(India) ······ 481

Blasting Open Pit Mining Impact on the Surrounding Forest Ecological Benefits
 ZHOU Xiaoguang, JIANG Zhou(China) ······ 490

Load on the Atmosphere with Carrying out Explosive Works in a Quarry
 V.I. Papichev, E.S. Chechneva(Russia) ······ 497

Test and Study on the Low Dust Blasting Charge Structure
 YANG Haitao, LIU Weizhou, ZHANG Xiliang, YI Haibao(China) ······ 501

Vulnerability Assessment of Pressurized Pipeline by Surface Explosives
 Yumin Li, Ettore Contestabile(Canada) ······ 505

SVM Prediction Model of Tunnel Blasting Excavation for the Destruction of the Ancient Great Wall
 LI Longfu, CHEN Nengge, ZHANG Xiliang, JIANG Dongping(China) ······ 511

Study on Control Measures for Blast-Induced Ground Vibration in Open Pit Metal Mine by Means of Numerical Simulation
 Kento Fukui, Hirokuni Inoue, Takashi Sasaoka, Hideki Shimada, Akihiro Hamanaka, Kikuo Matsui,
  Shiro Kubota, Tei Saburi(Japan) ······ 516

Analysis of Vibration Effects of Shallow Highway Tunnel under Blasting Seismic Waves
 YAO Jinjie, LUO Zhenhua, XIANG Aiguo, ZHAO Rundong, XIONG Yazhou(China) ······ 522

The Various Reasons and the Preventive Measures of Misfire in the Nonel Detonation Circuit
 LI Shuming, LUO Wei, ZHANG Fuyang(China) ······ 525

Using Sublevel Pre-splitting Blasting Effect to Reduce the Seismic Wave Velocity
 Sun Jing, Yan Jianhua(China) ······ 528

Blasting Theory

Scientific and Technological Aspects of Blast Action Control in Mineral Deposit Mining

K.N. Trubetskoy, S.D. Viktorov, V.M. Zakalinsky

(*Institute of Comprehensive Exploitation of Mineral Resources (IPKON), Russian Academy of Sciences, Russia*)

ABSTRACT: The article describes the available reserves of blasting process efficiency enhancement in mineral deposit mining in unfavorable conditions on the basis of further development of methods of large-scale bulk blasting of the rock mass.

KEYWORDS: breakage by blast; explosive; hollow charges; breakage scale; fragmentation; closely-spaced charges; mining methods; selective mining; tight face

Blast energy application still remains the only universal and most efficient method of hard rock breaking by bulk blasts and serves a basis for the development of different technologies for mineral deposit mining [1]. Intensive development of mineral deposits in relatively favorable geological conditions results in the depletion of mineral reserves and necessity of involving in production the reserves of deposits with unfavorable conditions, particularly those occurring at great depths, liable to rock bumps, and characterized by morphology diversity, requiring selective and directed blasting of ore and rock boreholes, with the use of novel geotechnologies.

New approaches and models are required for consistent description of physical processes in mining and providing timely results including fast assessment opportunities. At that, the most important physical-technical process in mining is determined by the perfection of mining methods for complicated and multi-component solid ore bodies and rock mass in surface and underground operations. Actually, explosive energy and scientifically grounded blast action control provide the only "tool" (facility) for the purpose.

On the one hand, the problem can be reduced to the classification of methods and blast action control facilities with respect to their fitness for particular conditions, i.e., analysis of the quality of their correlation. On the other hand, it can be reduced to the expansion of the scope of facilities and methods of blast action control with the purpose of meeting the mining process requirements in a more flexible and "delicate" manner. There is also the third aspect, namely, testing and refinement of the calculation technique for estimation of parameters of drilling and blasting processes associated with novel drilling and blasting equipment, and a broad range of explosives offered by the market.

As the final objective of all innovations and solutions is the perfection of a mining method and its main process – ore breakage by blast, the objective and task are determined by the most specific and important physical-technical and technological-economic aspects for current conditions. As it is proved by the analysis, one of them is on the way towards the increase of the direction component of blast action. This has resulted in the development of the technique for the selection of reasonable types of breakage by blast based on the integral criterion of the assessment of breakage scale in complicated geological and morphological conditions. Particularly, in underground mining of ore deposits the perfection of mining methods goes within the framework of expanding the application area of a known research and engineering area, such as large-scale breakage of rock mass by blast.

The technique is based on the approach, which is aimed at the development of technology for the formation of a blast wave of any shape and intensity due to borehole drilling patterns for closely-spaced charges of conventional design, larger diameter boreholes and special-type charges effecting the directed action of a large-scale blast. As a result, broad opportunities are opened for the application of large-scale blasting in different geological and mining conditions. For instance, it can be implemented with methods of induced level caving by slices with parallel closely-spaced explosive charges of larger diameter in tight-face conditions and ground control due to reasonable effect of blast on the rock mass. In surface mining operations it is rather promising for selective mineral mining in complicated geological conditions for reasonable fragmentation in block sections differing in quality. Besides, conditions of large-scale blasting operations in confined and stress-strained conditions at large depths and solution of the problem of mining method class selection and providing the desired quality of rock material fragmentation also require blasting technologies with directed blasting elements.

Corresponding author E-mail: krasavin_08@mail.ru

The main idea is in the novel methodological approach, which envisages a solution as the selection of one among variants differing in scale of drilling and blasting processes using a certain generalized criterion, which takes into account geological, geomechanical, technological and some other factors related directly to explosive charge blast action and charge design that on the whole provides the minimization of the quantity of variants [2].

In specific variants a spectrum of estimated data, mining method class are presented and the integrated effect of the process conditions on blast results is shown. This has made possible the formulation of integrated characteristics (criterion) of the rock mass breakage by blast as a factor of breakage scale, with which the quantity of variants was reduced to three types of breakage by blast: small-scale, medium-scale and large-scale breakage.

A type of breakage scale reflects the integrated effect of blasting conditions as main characteristics and indices of drilling processes generalized by blasting methods and reduced to three variants that characterize it as an integrated one. It was used for the solution of a problem of rock material fragmentation quality and selection of a mining method class for the implementation of large bulk (large-scale) blasts.

In the first instance, mineral mining including blasting processes performed by any method, in particular those with controlled large-scale blast action, requires the solution of a fundamental problem of reasonable fragmentation of the medium being broken and maintaining of the rock mass stability in different mining conditions and for different mining methods. As the analysis of the estimated data shows, application of the available technique in the task under consideration is hardly acceptable for some fundamental reasons of a theoretical nature.

A new approach was required to the technique of charge size calculation for ore breakage with large directional explosive charges in underground conditions and based on the application and development of the scale effect [2]. This approach is based on the development of the principle of scale effect described by the known formula:

$$Q = qV \qquad (1)$$

and based on its analysis and transformation with the purpose of estimation of equal degree of fragmentation for any volumes and conditions of large-scale blasting. Its main shortcoming was found – it does not take into account rock material fragmentation, as historically it was derived from the task of earth mass (volume) moving. It has been perfected in such a way that q value is presented as a dependent variable of V value that results in the following differential equation:

$$dQ = q(V)dV \qquad (2)$$

which takes into account the quality of rock material fragmentation depending on the blasting volume (scale). With known dependences of explosive unit consumption on the line of least resistance for rocks with varying fissuring and hardness the required charge size can be calculated with due account for the desired fragmentation degree by the following formula

$$Q = c \int_{w_0}^{w_1} q(w) dw \qquad (3)$$

It reflects both in a "traditional" way the volume of the rock slice being broken (q = const, formula(1)) and in addition, "integrally" its fragmentation ($q = f(w)$, formula(3), a new factor). In practice, q to a varying degree depends on w, but it can also be independent, for instance, in case of breakage in intensively fractured ores with a size limit at mine being shorter than the distance between fractures. In this case, the formula (3) is identical to the formula (1).

Based on the analysis of test data the formula $q = f(w)$ in the formula (3) can be written as the following dependence:

$$q = q_0 + kw^\varphi \qquad (4)$$

where, q_0 is the explosive unit consumption with w_0, corresponding to the lower limit of the integral in the formula (3); q is the same, but with w_1 in the upper limit of the integral; term kw^φ is the scale added component corresponding to the increase of the line of least resistance; k, φ are proportionality and scale factors accordingly.

The formula was considered for all types of ore: from intensively fractured to relatively solid varieties, with condition of nearly identical fragmentation. Strong dependence on the intensity of rock mass fracturing (~ 70%), slight dependence on rock hardness f (~ 20%) and ~10% dependence on other factors were discovered. With the decrease of fracturing the nature of dependence (6) varies, but not in an obvious way. The greatest dependence is typical for crack-free or slightly fractured ores, and it decreases with the gradual increase of fracturing, and if matching the size limit it reaches the relative minimum in moderately fractured ores. The formula determines the "dynamic" way of its application, readily (promptly) influencing on the process of calculation of the amount of explosives required for bulk blast, notably, at the very beginning of the process, as all methods of calculation of the amount of explosives are based on the dependence (1) altering its structure to meet the local blasting conditions.

The dependence is controllable, as it is characterized by the "dynamic" nature of its application as compared with the classic "static" formula (1). This allows promptly, at the beginning of the calculation of the required amount of explosives for a large-scale bulk blast to differentially take into account the geological structure of the whole block.

With the available technique this volume of the rock mass being broken would correspond with some other overestimated amount of explosives meeting the calculation by formula (1).

Using the generalized law of similarity with the condition of the fragmentation quality being unchanged, for comparative blasting in identical conditions from formulas (3) and (4) one can formulate an expression, which includes in compliance with energy principle such values as explosive charge energy margin, borehole pattern and line of least resistance:

$$\frac{Q_n}{Q_o} = \frac{E_n}{E_o} = \frac{S_n}{S_o}\left(\frac{W_n}{W_o}\right)^\varphi \quad (5)$$

where, n and o indices correspond to new and old values.

In this formula the added component to the main charge in the formula (4) is seen, which is required for the maintaining of fragmentation quality with the variation of the volume of the rock mass being broken (scale correction). Basically, it is possible in advance, based on the geological data, to design large-scale blasts for stoping in any underground conditions, with no need of test blasting, with the same fragmentation quality as can be achieved with small-scale parameters of blasts. This methodological aspect can be used also with different standard techniques for drilling and blasting operations by way of comparing and correcting the charge and rock mass mining unit parameters for a particular borehole charge. This methodological approach for charge size calculation will acquire great importance for the development and design of new technologies with large-scale blasting.

Below goes the discussion of some large-scale blasting technologies with the use of explosive charges and implementing the effect of controllable blast action, including the directed blast action, in various conditions [3].

In mineral deposit mining such methods as tight-face ore blasting have become a worldwide practice. Application of level induced caving with ore body pillarless extraction at iron-ore deposits liable to rock bumps in Western Siberia provides for ground control due to filling of mined-out areas with caved enclosing rock (Fig.1). With a widely used variant of ore breakage with ellipse-shaped compensation chamber and tight-face conditions blasting operations are effected with bunches of parallel closely-spaced explosive charges of 105 mm diameter providing for efficient mining in unfavorable geological and geomechanical conditions [4].

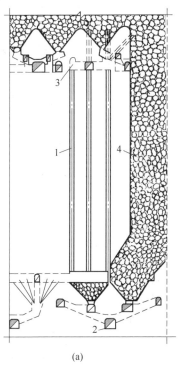
(a)

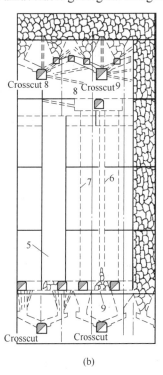
(b)

Fig.1 Level induced caving with ore tight-face breakage with parallel closely-spaced borehole charges

1—closely-spaced charges; 2—crosscut; 3—drill workings; 4—tight face; 5—compensation chamber; 6—vertical concentrated charge;
7—larger-diameter boreholes of 250 mm diameter; 8—closely-spaced boreholes; 9—block undercutting level

In this case, due to the effect of geomechanical situation the greatest influence on the rock mass is produced by blasting operations. The mass of an explosive charge for process (0.7~20 t as an average) and bulk blasts (120~370 t) ranges from 0.5 to 700 t, and the volume of the caved rock mass ranges from 30 to 250 thousand cub. m. During in-block development and after ore bulk caving the repair and maintenance costs increase in workings of haulage and drill levels, and block bottom exposed to dynamic effects.

Disadvantages of 105 mm diameter borehole explosive

charges include the loss of 30% to 50% of boreholes during block development, lack of opportunity of enhancing output per man in drilling operation, as well as no opportunity of charging all boreholes in a block in unfavorable conditions that results in the increased yield of oversize material, lower output per man in ore drawing, etc.

For safety and mining efficiency enhancement, and minimization of the negative effect of massive blasts on the rock mass a method has been developed for steeply dipping bodies of great and medium thickness and varying stability of ore, which contains a novel element for blast action control. It involves minimizing the quantity of borehole per bunch with the same specific consumption of explosives for primary breaking due to a greater borehole diameter and an opportunity of using some other patterns of borehole arrangement in a bunch, which generate a blast wave more directionally, with due account for geomechanical situation (Fig.2).

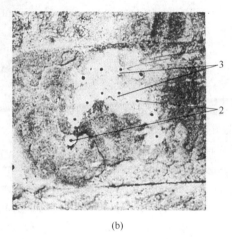

Fig.2　Cumulative pattern of closely-spaced charge arrangement generating a directed blast wave

(a) block before charging; (b) block after charging

1, 2—boreholes; 3—charge hollow

Main stress concentrators in a mining structure are compensation chambers and a great number of breakoffs at undercutting and drill levels. For their elimination the parameters have been estimated of a single-stage method of level slicing and caving with ore tight-face breakage with 250 mm diameter boreholes.

Areas of certain geological structures are mined, for instances, intensively fractured ore and rock of medium stability, or on the contrary, stable solid slightly fractured rock mass. Negative effect of geodynamic situation is controlled due to design and process characteristics of elements of the mining method. In these conditions the effect of blast action directionality can be neglected, and the quantity of 250 mm diameter boreholes in a bunch can be reduced to three-one.

The proposed method has made possible the involvement of additional production reserves with 4-6-fold growth of output per man in borehole drilling. The seismic effect of blast on the adjoining rock mass was reduced due to the decrease of explosive charge mass to 10 t for breaking every slice of ore in a block. If compared with the minimum mass of an explosive charge of at least 120 t for bulk caving of blocks, the blast effect on the rock mass is several times lower.

Application of bunch borehole charges differing in quantity, diameter and arrangement patterns, with an opportunity of their directed blasting has improved performance and economics of the mining method, reduced the specific consumption of explosives both for primary and secondary fragmentation, enhanced output per man, borehole length and drilling accuracy. In particular, for the formation of the central vertical concentrated charge of large mass and special design in an ore block with a shape similar to cylindrical use is made of delineating closely-spaced bunch charges. They are placed at a certain distance around the central charge and fired first [5,6]. Compensation chambers, newly formed free faces, controlled blasting with direction of a blast wave towards the earlier caved area provide the protection against a seismic effect of charges of large mass and the desired fragmentation of the rock mass by blast.

Development of the processes of the known mining methods and techniques in terms of physical-chemical and combined geotechnologies is based on the elaboration of the rock mass blasting methods providing the efficiency of the leaching process. Combined mining methods of physical-engineering and physical-chemical geotechnologies were considered for the development of hard rock and physical-chemical conditions for the selection of explosive for intensification of hard ore leaching with the involvement of elements of directed controlled blasting[7, 8].

Drilling and blasting in surface mining operations with bulk large-scale breakage of the rock mass are characterized by certain peculiarities, which must be taken into account in directed blast action control. The matter is that borehole diameter in surface mining operations for some reasons has reached its limit of 300 mm thus preventing from the application of some technologies (conditions similar to camouflet), dramatically narrowing the opportunities of multi-row short-delay blasting and lowering the acceptable quality of rock material fragmentation. In this case, application of parallel closely-spaced borehole charges is actually the only technologically acceptable method of increasing explosives energy, except for flame-jet drilling and some other exclusive methods of vertical hole formation.

Complex faces in ore and barren blocks of mineral deposits, which parts differ significantly in the geological structure and mineral quality are worked by bulk and selective methods. Bulk mining is effected with the use of the simplest technology, including its simple organizational patterns, and accordingly, with high-performance mine hauling machinery. In this case, the technology differs insignificantly from that used for ore or stone face development. However, conventional bulk blasting in complex faces results in significant blending of the useful component that significantly decreases the efficiency of loading and hauling processes and worsen the conditions of ROM material processing. Ore output is to a great extent diluted with barren rock, and a relatively large part of mineral is lost once and for all.

Selective working of complex hard-rock faces with the use of drilling and blasting processes and selective extraction is effected either at the stage of rock mass loosening, or at the stage of loading and hauling, with or without material grading in faces. Grading may be simple, as selective extraction of mineral and barren rock at the bench front without grading over shotpile height, and it may be also complex, with selective extraction of different type of ore and rock both along the width and height of faces.

In development of such faces the conditions are created for simple grading by shovels. In this case, the more intensive is mineral blending with barren rock, the more time is spent for grading, and thus, the lower is th loading rate.

In open-face blasting bulk and selective methods of complicated deposit mining have a number of important shortcomings, which in some cases put in question the opportunity (the idea) of selective mineral mining.

Therefore, the main requirement to blasting operation is that a site for ROM material with mineral and barren rock disposal after blast were most suitable for selective mining.

Selective breakage of rock is conventionally based on physical and technological aspects of blast action control. Methods and facilities of the first aspect may include charges of various design and explosive compositions characterized by directiveness of blast action, for instances, closely-spaced and special-purpose charges. The second aspect includes methods and equipment for controlled directional blasting with blast wave shielding and localizing, as well as their various combinations.

According to a bulk blast design the conventional technology for selective mining envisages drilling of blasthole rows, charging them with explosives, wiring-up patterns, consecutive short-delay blasting on ROM material left in place. The quantity of rows to be fired is minimized; blast capacity (energy) of a single-borehole charge of specified diameter is limited. Specific features of blasting technology used for complex and multi-component ore and rock mass mining are not reduced to the above mentioned elements of large-scale breakage with larger-size charges proportional to the quantity of tight-face blasting rows, i.e., breakage scale. Methods and technology of complex face decomposition require differentiation of useful component- and barren-rock areas, in ideal case, clearly on "ore-rock" contacts and interface, in contrast to fracturing in case of conventional breakage by blast. This requires further research and involvement of new blasting methods, equipment and facilities.

In addition to closely-spaced charges of diverse geometry such methods and facilities may also include explosive charges with cumulative effect, as well as defense-related special explosive charges with blast wave shielding and localizing. These technological innovations give rise to new variants of blasting technologies for mineral deposits with complicated conditions of occurrence.

ICEMR RAS researchers and engineers have developed such blasting technology (method) for selective mining of deposits with a complicated structure in open pits (Fig.3). It takes into account the characteristic aspects and actually does not envisage any energy limit as such for subsequent rows of borehole charges depending on drilling equipment capabilities [9].

This method is characterized by the feature envisaging that for selective fragmentation of areas with diverse rock types, in end boreholes of bunch closely-spaced explosive charges the hollow charges with plane symmetry of linear shape are installed against each other. Along the contact of areas with diverse rock types the hollow charges are installed with plane symmetry of ring shape and concavity axis oriented towards the contact (Fig.4).

Application of bunches of closely-spaced high explosive charges, which quantity in every subsequent row is greater by 1~2 beginning from the 2nd~3rd row provides for an actually unlimited multiple-row pattern of bulk blast. Borehole ends are charged with conversion or some other high explosives, with conventional explosives charged above them. Combinations of delay circuits, including instantaneous camouflet blasting, provide invariance (shift) of geometry of the location of large volumes of the rock mass with a complicated structure before and after large-scale bulk blast, up to zero. This allows selective ore mining in parallel with barren rock excavation, or rock excavation is effected as a second stage after ore extraction. Besides, selective fragmentation of ore and rock areas in a variety of tight-face conditions, production of large volumes of non-blended ROM material left at the place of blast open opportunities for long-term and sustainable selective removal of complex-structure face by shovels. The quantity of bulk blasts and the number of runs of loading and hauling machinery decreases thus enhancing the efficiency of the novel technology.

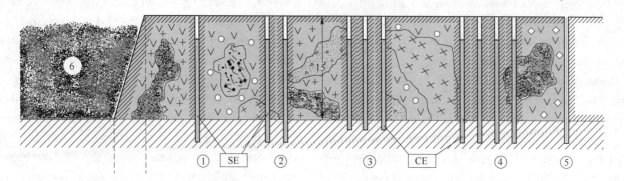

Fig.3 Schematic flow chart of large-scale blasting method for a rock mass block of complicated structure

1—row of single borehole charges; 2—bunch charge in two boreholes; 3—bunch charge in three boreholes; 4—bunch charge in five boreholes; 5—contour charge; 6—rock material broken at the earlier stage; SE—standard explosive; CE—conversion explosive

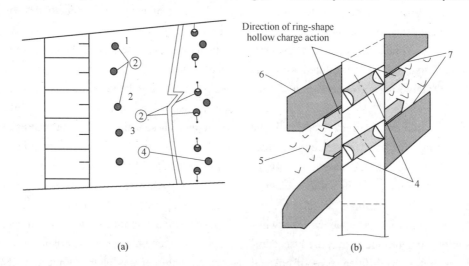

Fig.4 Rows of single conventional and closely-spaced borehole hollow charges

(a) general view;(b) charge hollow fragment

1—row of single borehole charges; 2—row of closely-spaced charges with plane symmetry of linear shape; 3—opposite hollow charges with plane symmetry; 4—hollow charges with plane symmetry of ring shape; 5—ore; 6—rock; 7—interface

Thus, scientific and technological aspects of large-scale blast action control determine the mineral deposit mining geotechnology for complicated conditions developing one of the most promising areas of blasting processes in Russia.

REFERENCES

[1] Trubetskoy K N, Malyshev Yu N, Puchkov L A et al. Mining Sciences. Development and conservation of the mineral wealth of the Earth. // RAS, AMS, RAES, IEA; Under Editorship of K.N. Trubetskoy: - M.: Publishing House of the Academy of Mining Sciences 1997:478. (in Russian)

[2] Viktorov S D Disintegration of rocks with closely-spaced charges Viktorov S D, Galchenko Yu P, Zakalinsky V M, Rubtsov S K. Under the Editorship of Acad. K.N. Trubetskoy. M. Nauchtekhlitizdat Publishers OOO, 2006: 276. (in Russian)

[3] Trubetskoy K N, Viktorov S D, Zakalinsky V M. A new conception of drilling and blasting process perfection at underground ore mines // Gorny Zhurnal. – 2002. – No 9. (in Russian)

[4] Eremenko V A, Karpov V N, Filatov A P, Kotlyarov A A. Perfection of the mining method with ore tight-face blasting in the development of mineral deposits liable to rock bumps// Gorny Zhurnal. – 2014. – No 1. (in Russian)

[5] Eremenko V A. Perfection of drilling and blasting technology for iron-ore deposits of Western Siberia/ A.A. Eremenko – Novosibirsk: Nauka Publishers, 2013:192. (in Russian)

[6] Viktorov S D. Technology for large-scale blasting at Siberian ore deposits liable to rock bumps / Viktorov S D, Eeremenko A A, Zakalinsky V M, Mashukov I V. – Novosibirsk: Nauka Publishers, 2005: 212. (in Russian)

[7] Zakalinsky V M, Frantov A E. Physical and chemical preconditions of explosive selection for hard ore leaching // Vestnik, Russian Academy of Natural Sciences. – 2013/6. – Vol. 13. – p. 97-102. (in Russian)

[8] Frantov A E. Revisiting: Substantiation of the properties of concersion explosives with special account for characteristics of blasting operations in geotechnologies // Marksheideriyai Nedropolzovaniye. 2013. – No 6 – p. 11-15. (in Russian)

[9] Viktorov S D, Zakalinsky V M, Frantov A E, Galchenko Yu P. Method of large-scale breakage of the rock mass of a complicated structure for selective mineral mining in surface operations. RF Patent No 251330, Invention priority of 16.07.2012, Application No 2012129943. (in Russian)

Study on the Mechanism of Linear Shaped Jet Splitting Target Based upon High-Speed 3D DIC

XU Zhenyang, YANG Jun, YU Qi

(*State Key Laboratory of Explosion Science and Technology of Beijing Institute of Technology, Beijing, China*)

ABSTRACT: The mechanism of linear shaped charge configuration explosions metal jet splitting target is studied based upon High-Speed 3D DIC, the development of cracks in the side split from the cylindrical concrete target was high-speed photographed, and three-dimensional deformation data is analyzed. The conclusions can be drawn: it is a reliable method to study splitting mechanism of target under the action of linear shaped charge structure explosion using High-Speed 3D DIC. The average speed of crack propagation is 84.47 m/s, side cracks of the target develop downward in "s" shape round the axis. The impact of linear shaped charge jet has significant guidance to the splitting of the target. Crack propagation zone is the main strain concentration zone which presents obvious open-type characteristics, and its fracture morphology is tensile fracture.

KEYWORDS: DIC-3D; crack; speed; strain

1 INTRODUCTION

Secondary crushing operations play an important role in blasting engineering, especially in disaster relief. Using shaped charge to realize the chunk splitting morphology control improve work efficiency, and have practical significance in controlling blasting harmful effects range. To symmetrical cylindrical concrete specimens, when linear length of charge base on bottom diameter of 1/10, the penetration depth is less than the height of 1/4, we can achieve the purpose of controlling the splitting form. Building High-Speed 3D DIC experimental test platform, using high-speed photography technology can carry out splitting test non-contact strain measurement under shaped explosion effect. Focus on the splitting phenomenon of the penetration of linear jet to axial single primary plane, calculated the speed of splitting crack propagation, and analyzed the relationship between the maximum principal strain field and crack of split.

2 LINEAR SHAPED SPLITTING TARGET BODY

2.1 Splitting Principle

Concrete is a brittle material, it will fracture when the strain rate is very small, and under the sufficient strength uniaxial compressive load will be brittle fractured. Using the jet of linear shaped charge structure explosion cylindrical specimen, the advantages of the axial length of linear jet splitting will have a significant role in guiding to direction of the target body, so that the specimen split into two parts.

Splitting cracks extension along the Linear Shaped axial length of the agents, when target body subjected to linear shaped charge jet, target body under tensile load effects appeared only one main jet penetration axial plane cleavage, splitting crack unbranched substantially and does not appear other parallel to the fracture surface, enabling precise control of the splitting effect. The diagram of test as showns in Fig.1(a), the effect of specimen' splitting as showns in Fig.1(b).

Linear charge length and width of the wedge-shaped cover ratio greater than 1, allows penetration axial end form higher intensity stress concentration effect than the radial endpoints, then will be formed along the axial penetration of a single main plane splitting, showed significant directional splitting effect.

2.2 Linear Shaped Charge Structure and Concrete Parameters

According to the empirical formula design the linear shaped charge parameters, using the center of the top surface detonation mode. The linear shaped charge parameter was shown in Tab.1. The schematic diagram of the linear shaped charge as shown in Tab.1.

Concrete target select 30 cm of bottom diameter 30 cm of high cylinder. Refer to "ordinary concrete mechanical test

---

Corresponding author E-mail: xuzhenyang10@foxmail.com

method standards" with the ratio of water ∶ cement ∶ sand ∶ gravel = 0.38 ∶ 1 ∶ 1.11 ∶ 2.72, elastic modulus of 40 MPa, the conservation of 28 days to reach strength.

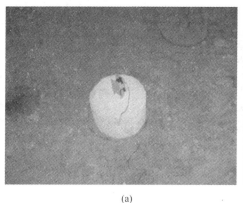
(a)

(b)

Fig.1 Picture of experiment

(a) diagram of the test; (b) effect of the specimen' splitting

Tab.1 Linear shaped charge parameter table

Shaped charge liner		Charge height /mm	The axial length shaped charge liner /mm	Loading dose /g	Shaped charge liner quality /g	Drug cover wall thickness /mm	Target body parameters		Height of burst /mm
Caliber /mm	Wedge angle /(°)						Bottom radius /mm	High /mm	
35	45	30	30	45.3	2.64	0.2	150	300	40

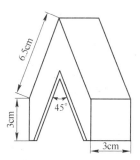

Fig.2 Schematic of linear shaped charge

3 DIC-3D EXPERIMENTAL DESIGN

3.1 DIC-3D Principle

High-Speed 3D DIC non-contact audience response measurement systems rely on two high speed camera which have angle to shoot the specimen at a common site, by deformation of speckle at each image the three-dimensional seat standard (x, y, z) of specimen surface can be obtained, and thus obtain the distortion components (U, V, W) in the xyz each direction, three-dimensional displacement data obtained by testing, the specimen crack growth and strain data can be calculated. Fig.3 is a schematic diagram of High-Speed 3D DIC.

3.2 Experimental Design

The test conducted in Beijing Institute of Technology's Explosion Holes, the angle of two high speed cameras is 12°, target body from the photographic instrument lens 2.5 m, photographic instrument lens and shoot the target body surface placed in the same horizontal plane. Use two 1000 W lights at a distance of 1 m at the front of the specimen shooting.

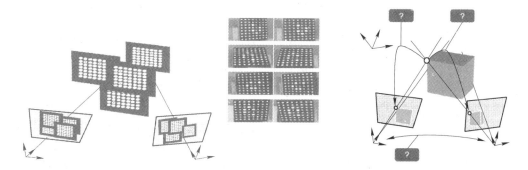

Fig.3 Schematic diagram of High-Speed 3D DIC

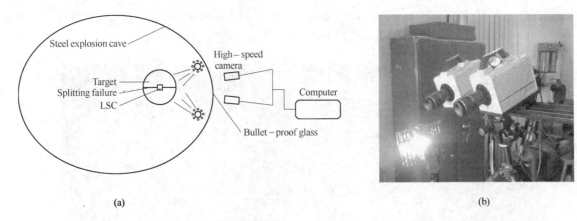

Fig.4 Test systems and equipment

(a) High-Speed 3D DIC test schematic; (b) speed cameras

3.3 Instrument Parameter Settings

Shooting speed is an important parameter of high speed camera shooting, a reasonable choose has direct impact on whether or not to get a splitting deformation data. In this experiment, estimated the crack growth rate under explosion shaped was 200~800 m/s, selected shooting speed of 50.000/s. Taking into account the effects of explosive bring light and detonation products, shaped charge part is completely covered with steel pipe, and specimen wrapped with rubber, leaving only a 200 mm×120 mm range for shooting, after several tests, the development of cracks can through the selected range. Pointing black speckles in the shooting surface of the specimen, in order to calculate the displacement, Fig.5(a) is shooting calibration graph, Fig.5(b) is the specimen protection and reserved shooting area graph.

Fig.5 Laboratory equipment commissioning

(a) shooting calibration; (b) specimen protection and reserved shooting area

4 CRACK GROWTH ANALYSIS

4.1 Field of View Three-Dimensional Reconstruction

High-Speed 3D DIC is the analysis method which based on dynamic speckle, the date at the site of fracture or crushing will be lost or distorted, data before breaking reliability is guaranteed. Select the square area in central of field of view shooting, select the red zone area for the data analysis, after the reconstruction the three-dimensional parameter is high 90 mm×wide 70 mm×thickness 52 mm. The red area is the area of data analysis as shown in Fig.6(a), three-dimensional reconstruction of analysis region as shown in Fig.6(b).

4.2 Crack Growth Rate

At the crack tip pick 16 measuring points marked c_1, c_2, ..., c_{16}, time interval is 0.04 ms, starting time is 3.76 ms, the end point is 4.36 ms, as shown in Fig.7. Mark crack tip coordinates at the moment ti is (x_i, y_i, z_i), crack tip propagation path is irregular arc, since during the unit time Δt the arc is very small, the crack can approximate that extend along a straight line. Define 3.8 ms as t_1, 4.36 ms as t_{16}. During the time $\Delta t_i = t_i - t_{i-1}$, crack growth rate is v_i, v_i can be approximated considered the instantaneous velocity when the crack to measure point c_i.

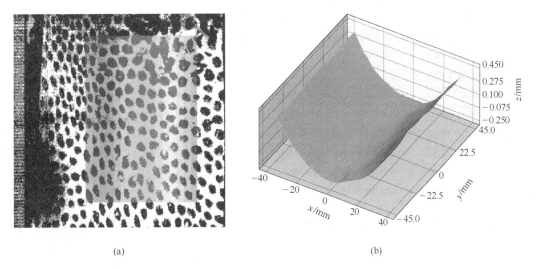

(a) (b)

Fig.6 Experimental analysis area

(a) the red area is the area of data analysis; (b) three-dimensional reconstruction of analysis region

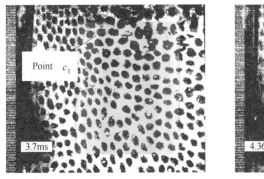
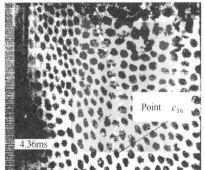

Fig.7 The results observed

Displacement of the crack tip

$$|S_i| = \|S_i| - |S_{i-1}\| = \sqrt{(x_i - x_{i-1})^2 + (y_i - y_{i-1})^2 + (z_i - z_{i-1})^2} \quad (1)$$

Crack speed during Δt

$$v_i = \frac{\partial S_i}{\partial t} \quad (2)$$

Average speed of crack

$$v_{avg} = \frac{\sum_{i=2} v_i}{i - 1} \quad (3)$$

Can be obtained by calculating the crack growth rate is an average of 84.47 m/s. Specimens cracks, damage and cumulative damage of the crack tip material combined effect of crack propagation.

Shown in Fig.8, the crack growth rate curve showed periodic non-stationary, crack growth rate only in a shorter period of time show linear low. At 4.2 ms, the peak velocity is 221.81 m/s, 3 times close to the average velocity value. At this time the crack develop to the specimen's central, because the central part of the shear stress concentration, dislocation in both sides will cause cracking pace sudden acceleration.

Distribution of crack growth rate is stepped, can be divided into four stages: the 3.76~3.85 ms, speed after the first increased and then decreased, the maximum speed is lower than 45 m/s. Within 3.85~4.12 ms, the speed quickly rose to the level of around 150 m/s, and showed periodic variation. Within 4.12~4.21 ms, the speed reaches the maximum value on the entire curve, after the maximum value, then the linear attenuation to 34.27 m/s, then the periodic variation. Within 4.21~4.36 ms, the speed variation is relatively stable, the average speed is lower than the before one.

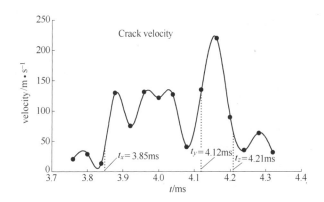

Fig.8 Crack growth rate curve

Analysis combined the experimental data, we can see splitting crack propagation law: presented stepped cycle; during each stage, the peak wills reduction after increase; after the peak of each stage, it will quickly reduce, to the minimum the next stage. When the crack growth rate rapidly decreased to 1 m/s, the crack will stop expansion; if the crack growth rate is maintained, then before the crack stop, it will run through the test piece.

Power of crack propagation comes from the elastic energy release rate gover the resistance R, changes in the magnitude of the resistance R is quite small, so that the crack propagation speed depends on the size of the crack tip stress intensity. The crack growth rate increases suddenly bound to the elastic energy release rate increased instantly consumed before breakage of the system stored elastic energy, the remaining kinetic energy is suddenly decreased. In the crack propagation process, excessive fragmentation partial produce new surface much larger than the surface of cracks, when broken to a certain extent, crack growth will certainly stop.

In this test, the crack propagation rate is not high, cracks are low, only in the zone near penetration cracks speed is high. Rapid expansion cracks more easily bifurcation, visibly, control initial penetration is more critical to the cleavage direction.

4.3 Crack Morphological Characteristics

Brittle materials such as concrete or rock under the impact, mainly characterize is tensile damage. When the compression stress wave propagation to the free surface, the compression wave reflect formation stretching wave, stretching and compression wave generated superposition, the near the free surface synthesis tensile stress, result a net tensile stress near the free surface. Before macroscopic damage phenomenon, the tensile stress has made the deformation of the specimen showed some regularity, and the distribution of pull stress concentration determines the failure modes of the specimen. Under penetration effect, incremental damage caused by tensile stress is far greater than by compressive stress.

Shown in Fig.9, the specimen appears multiple injuries destruction, deterioration of surface roughness, damage and failure in near the splitting crack area and the upper area of the specimen is more obvious. Original defect and injury by the tensile stress some will develop into micro cracks. As the main cleavage plane is always in a strong field of stress concentration, some extent led to the micro-cracks in the near splitting crack field at extended process gradually converge toward splitting crack, making the micro-cracks cannot affect the stability of splitting crack extension. Splitting crack tip appear three significant local crushing, crushing changed the laws of crack tip local force, resulting in a significant direction shift of crack propagation. With partial crushing appear, three branch cracks have begun to expand from here, the branch cracks propagation path are short and developed independently, and does not appear forked again. Stress concentration distribution had a fundamental impact to crack propagation path, stress concentration of splitting crack competition defects cracks always dominate, making splitting crack to stabilize expansion.

Fig.9　Damage rupture morphology

4.4 The Maximum Principal Strain

In the field, the three strain concentration regions will are labeled zone 1 and zone 2, zone 3, as shown in Fig.10.

Show as Fig.10(a), at 3.8 ms time, the crack first into zone 2, where the highest strain. Zone 2 shapes change with time, and gradually develop to form strain concentration. In zone 1 exist two oval strain concentration areas; comparison with test photos can be found the site where on the specimens right side is the fragments produced, features can be more accurate identification of specimens broken by the maximum principal strain variation.

Fig.10(b), the development to 3.9 ms, two oval-shaped in zone 1 has basically disappeared. At this point, zone 1 formed a weaker intensity horizontal strain concentration band, increases the possibility of transverse fracture within this site. Zone 2 and zone 3 formed a connected strain concentration band which perpendicular to the strain concentration band at zone 1, zone 2 where these two bands interchange has the biggest strain, splitting cracks first appeared here.

Figs.10(c), 10(d) shows, along with the development of splitting cracks, zone 2 and zone 3 separate red strain concentration zone is gradually elongated and connected to form a strain concentration at crack propagation path. When cracks appear only in zone 2, the zone 3 strain concentration has been formed, changes in the shape of three zones can be considered as the development of the internal morphology of the specimen splitting external performance.

Splitting crack begin at the point of jet departure continues to expand; the expansion will cause the primary cleavage planes within the specimen. For a "scheduled" part at specimen side splitting crack propagation path, the internal split the crack tip will arrive before surface cracks, when the surface crack arrive here, they will be convergence and co-expansion.

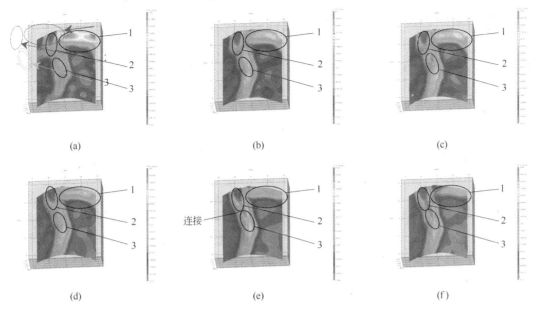

Fig.10 Maximum principal strain field

(a) 3.8 ms; (b) 3.9 ms; (c) 4 ms; (d) 4.1 ms; (e) 4.2 ms; (f) 4.3 ms

5 CONCLUSIONS

(1) Based on High-Speed 3D DIC design the experiment testing non-contact audience dynamic deformation of specimen splitting under shaped process to ensure the accuracy of the data to prove the feasibility study methods.

(2) The average speed of crack propagation is 84.47 m/s, "s" type crack development down around the sides of the axis.

(3) The impact of linear shaped charge jet form stress concentration at the ends of the long axis of the jet open pit has obvious guide to the splitting of overall target body.

(4) The maximum principal strain always appear in crack propagation area, fracture morphology of tensile fracture, splitting performance obviously open-type fracture characteristics.

REFERENCES

[1] Tate A. Further results in the theory of long rod penetration[J]. *Journal of the Mechanics and Physics of Solids,* 1969, 17(3): 141~150.

[2] Alekseevskii V P. Penetration of a rod into a target at high velocity[J]. *Combustion Explosion and Shock Waves,* 1966, 2(2): 63~66.

[3] Sternberg J. Material properties determining the resistance of ceramics to high velocity penetration[J]. *Journal of applied physics,* 1989, 65(9): 3417~3424.

[4] Forrestal M J, Longcope D B. Target strength of ceramic materials for high-velocity penetration[J]. *Journal of Applied Physics,* 1990, 67(8): 3669~3672.

[5] Kirugulige M S, Tippur H V. Measurement of Fracture Parameters for a Mixed-Mode Crack Driven by Stress Waves using Image Correlation Technique and High-Speed Digital Photography[J]. *Strain,* 2009, 45(2): 108~122.

[6] Chevalier L, Calloch S, Hild F, et al. Digital image correlation used to analyze the multiaxial behavior of rubber-ike materials[J]. *European Journal of Mechanics-A/Solids,* 2001, 20(2): 169~187.

[7] Toupin R A. Saint-Venant's principle[J]. *Archive for Rational Mechanics and Analysis,* 1965, 18(2): 83~96.

[8] Inglis C E. Stresses in a plate due to the presence of cracks and sharp corners[J]. *Spiemilestone Seriesms,* 1997, 137: 3~17.

[9] Walker J D, Anderson C E. A time-dependent model for long-rod penetration[J]. *International journal of impact engineering,* 1995, 16(1): 19~48.

[10] Erdogan F, Sih G C. On the crack extension in plates under plane loading and transverse shear[J]. *Journal of basic engineering,* 1963, 85(4): 519~525.

[11] Evans W M, Ubbelohde A R. Formation of Munroe jets and their action on massive targets[J]. *Research a journal of science and its applications,* 1950, 3(7):331~336.

[12] Seokbin Lim. Steady state equation of motion of a linear shaped charges liner Meyers[J]. *International Journal of Impact Engineering,* 2012 (44): 10~16.

Beyond Environmental Vibration Compliance

Cameron McKenzie[1], Mike Osborne[2]

(1. Blastechnology, Brisbane, Australia; 2. Austin Powder Company, USA)

ABSTRACT: More and more, mining operations are required to increase focus on controlling environmental impacts and complying with stricter and stricter limits of acceptability as regards blast-induced vibrations and overpressure. The introduction and extensive use of electronic initiation enhances the ability of operations to comply with limits, though systems to permit optimisation are not readily or widely available, and are seldom used in any routine manner in the mining industry. Environmental vibration impacts, either in the medium-field as an impact on vibration-sensitive slopes or mine sectors, or in the medium to far-field as an impact on nearby occupied structures, can be minimised by careful selection of delay timing. While seed wave modelling provides a unique opportunity to identify optimum timing from a vibration control perspective, the method also provides the opportunity to go beyond compliance and start to address the complex issue of complaint and human perception – an opportunity which has received little attention. This paper presents the use of seed wave modelling as an everyday tool which is being used at the blaster level in the USA to identify delay timing which can be used either to minimise peak vibration levels, or to minimise the perception of vibration to humans occupying residential structures. The paper presents the methodology of the Seed Wave Model, how it is used to identify optimum delay timing, and how a Perception Index is defined and used to rank different delay timing options. The paper reproduces, in part, information presented by the same author (McKenzie, 2012).

KEYWORDS: environmental vibration; human perception; perception index; frequency index; seed wave modelling

1 INTRODUCTION

The now-common application of electronic initiation of blasts in open pit mining appears to be delivering the promised benefits of precise and programmable firing times, judging by the plethora of papers extolling the wide range of benefits. However, it also appears that from the user perspective, there remains a technology gap to assist site personnel in choosing the optimum timing to provide best results in terms of fragmentation (fines control and oversize control), muckpile profile, excavator productivity, damage, or environmental impact control. While electronic initiation enables a much greater flexibility as regards delay timing and blast sequencing, the shot-firer is facing a dilemma of choice as regards the selection of that particular timing which will deliver the desired results. Furthermore, it seems unlikely that there exists a timing and sequencing which will simultaneously optimize the process from the perspective of any two of the above factors, meaning that a compromise solution will usually be required.

In the experience of the author, Seed Wave Modelling provides very useful assistance in identifying initiation timing which takes maximum advantage of either constructive or destructive interference of the vibration waves generated by individual charges as they detonate. By enhancing the constructive interference of waves within the body of the blast, the state of induced tension is maximised and will likely yield an increase in both macro-and micro-fracturing, as various studies have reported, e.g. Fribla, 2006, Hamdi et al. 2006, Katsabanis et al, 2006, Katsabanis et al, 2008, Kojavic et al, 1995, Vanbrabant & Escobar, 2006, Paley 2010. Modelling can assist in identifying those delay times which maximise the constructive interference of waves from individual charges, thereby intensifying induced stress fields and micro-fracture intensity, with such studies requiring the use of near-field seed waves in the modelling process. By maximising the destructive interference of waves at a particular location outside the blast area, the levels of induced vibration may be reduced, with likely benefits in terms of reduced environmental impact (including reduced impact on pit walls and nearby rock slopes).

In the context of blast vibration impacts, the Seed Wave Model can be applied to either ground-borne vibration, or air-borne vibration, to either maximize or minimize impacts. Ground vibration applications can involve either velocity-based vibration waves or acceleration-based waves.

2 TECHNICAL BACKGROUND TO THE SEED WAVE MODEL

Seed Wave Modelling is based on the principle of linear

Corresponding author E-mail: cameron@blastechnology.com

wave superposition, i.e. that the net result from the detonation of multiple charges can be found by simple linear summation of the waves generated by each individual charge. The concept and application are well explained by Hinzen (1988). Commonly, it is assumed that each charge produces the same shape of vibration wave, and that the amplitude of each wave is dependent on the weight of explosive associated with each charge. In some applications, individual waves have unique shapes derived from a single measured "parent" wave and related via "parent-sibling" Fourier manipulation, while in most applications the seed wave shape is maintained constant and only the amplitude is varied. Some landmark studies of the concept of Seed Wave Modelling are worth mention, including those by Blair (1999), Spathis (2006), Yang (2007), Yang & Scovira (2010).

In its purest form, the seed or signature wave from a single charge is fired in the same location as the ensuing multi-hole blast, and its impact (vibration or overpressure) is measured at the point of interest. Modelling then becomes very specific to that geometry, since the shape and characteristics of the recorded seed wave are very dependent on the seismic source-receiver path. Further, the weight of explosive in the single charge should be equal to that of each of the charges in the multi-hole blast so that no further amplitude adjustment need be performed. In reality, however, the seed wave requires adjustment, sometimes in both shape and amplitude, since the distances between individual holes (or deck charges) and the monitoring point are not constant and not equal to the distance at which the seed waveform was recorded, and the charge weight commonly varies from hole to hole. The angle between the charge and the monitoring point may also vary for long or wide blasts, affecting the relative amplitudes of the horizontal components of a triaxial monitoring system, and requiring a coordinate transformation, or blending of the three orthogonal components of vibration.

Blair (1999) presents particularly comprehensive details on methods of adjustment of seed wave amplitude and shape, drawing heavily on Monte Carlo methods to reflect actual variability in both shape and amplitude of seed waves both as they are generated at the blasthole and as they are received at the monitoring location. Yang (2007) also presents methods to adjust seed wave shape and amplitude for both differences in charge weight and propagation distance for both close-in monitoring and far-off monitoring.

It is worth special mention that adjustment of the seed wave amplitude to account for charge weight variability between different holes or deck charges should also have a degree of randomness associated with it to reflect the actual variability observed in traditional square root scaling regressions involving single charges measured over varying distances. Effectively, the K term of the charge weight scaling Equation (1) should be considered to have a log-normal distribution, with a standard deviation derived from field measurements of single charges.

$$PPV = KSD^{-n} \qquad (1)$$

where, PPV represents the estimated vibration amplitude; SD represents scaled distance (distance divided by the square root of the charge weight); K, n are regression parameters characteristic of the rock mass in which the measurements are conducted.

In its simplest form, the Seed Wave Model can be represented mathematically, for a blast containing N blastholes with individual delay firing times of d_i, by Equation (2).

$$R(t) = \sum_{i=1}^{N} A_i S(t - d_i - T_i) \qquad (2)$$

where, $R(t)$ is the resultant time history waveform representing the sum of the separate seed waves $S(t)$, A_i represent the amplitudes of each seed wave, and T_i represents the arrival time of the vibration waves at the receiver. calculated using Equation (1) according to the explosive weight and the distance of propagation for each of the individual charges. It is convenient to include the firing time delay and the travel time into one term D_i representing the arrival time of each seed vibration wave at the receptor.

As Hinzen (1988), Blair (1999), and Anderson (2008) point out, the same calculation can be undertaken as a convolution in either the frequency or time domain. The calculation can be made significantly more complicated, and probably more realistic, by adding a degree of stochastic behaviour to the shape of the different seed waves, and to the amplitude of each wave, as explained by Blair (1999).

Equation (2) is applied to each of the three geophone components which comprise a standard vibration monitoring system, so that Equation (2) can be considered to apply to each of the triaxial geophone/accelerometer components, i.e.:

$$R_L(t) = \sum_{i=1}^{N} A_i S_L(t - D_i)$$
$$R_T(t) = \sum_{i=1}^{N} A_i S_T(t - D_i) \qquad (3)$$
$$R_V(t) = \sum_{i=1}^{N} A_i S_V(t - D_i)$$

where, the subscripts L, T and V refer to the Longitudinal, Transverse, and Vertical components of a triaxial configuration of geophones or accelerometers, and where the longitudinal geophone axis is aligned in the direction of the single hole used to generate the seed wave.

The greatest complication, however, comes from the requirement to undertake a coordinate transformation to account for the different angles of incidence of the waves from different blastholes, according to the alignment of the triaxial monitoring transducer, and the lateral spread in blasthole coordinates. This is also explained in considerable detail by Blair (1999), who also points out the special case of the transverse component of the triaxial geophone or accelerometer system. Interestingly, no other worker has described the same irregular behaviour of the transverse component.

The question also arises as to the test conditions which provide the best estimate of the seed wave shape. While the ideal conditions would include a charge weight for the seed wave equal to that typically used in a production hole, and confinement conditions equal to those applicable for the production holes (i.e. a normal free face), it is frequently difficult to achieve such conditions, especially since in large blasts, different holes can contain quite different charges. For example, if the modelling exercise is trying to decide whether to use 1 deck or 2 decks per hole, what charge weight should the single hole use? Many quarry operators do not want to fire a single hole to a free face for fear of creating later problems due to an irregular face, forcing the use of a seed wave obtained under fully confined conditions, away from a free face. Finally, how close to the production blast should the single charge be fired? Ideally, within the same volume, but does that mean the Seed Wave Model requires a new seed wave prior to every blast?

3 EXPERIMENTAL AND MONITORING CONSIDERATIONS

The answer to the questions above depends quite strongly on the level of precision required of the modelling. If the intention of the modelling is to estimate the vibration characteristics (peak amplitude, shape, frequency and duration) as precisely as possible, then it probably is necessary to obtain a seed wave within a few tens of metres of each production blast. If the objective is to comply with statutory limits (of either vibration or overpressure), then it may be sufficient to simply estimate the levels within a known margin of error such as plus or minus 20%, applying appropriate percentiles to the predictions to ensure statutory compliance. The author's approach is to use the same seed wave until the accuracy of prediction becomes unacceptable, at which time a new seed wave is required.

Bernard (2012) notes that through a process of deconvolution (using the measured production blast waveform and the known time-delayed Dirac delta function) he is able to extract a new seed wave from the previous production blast, obviating the need for repetitive single hole firing. Bernard presents statistics in terms of accuracy of prediction after around 600 applications (an unknown number of which utilised the seed wave extraction method), with half of predictions being within 10% of the measured value, and on 28% of occasions, the actual measurements exceed the predicted levels by more than 10%. Further, approximately 70% of the outcomes represented a vibration reduction of at least 30% relative to previous measurements.

Anderson (2008) presented the same deconvolution idea proposed by Bernard (2012), referring to the time-delayed Dirac delta function as a "comb function", though he was much more guarded as regards the success of the process. Personal experience suggests that the deconvolution process to extract a seed wave from a measured seismic record usually produces a wave bearing little resemblance to a single charge wavelet.

As regards the configuration of the charge used to record the seed wave, it is again considered ideal if the charge can as closely approximate a normal production charge as possible, in terms of size, diameter, length, explosive density, and confinement conditions. Anderson (2008) implies the expectation that charge confinement (i.e. a charge fired to a free face with a normal burden or fully confined with infinite burden) will affect vibration levels – an expectation supported by most text books. Interestingly, though, Brent et al (2001) shows that burden confinement appeared to have no effect on peak vibration levels. Ramulu et al (2002) show similar results (i.e. no influence of burden) for the far field (scaled distance $>\sim30$ m/kg$^{0.5}$), but note very different behaviour in the near-field (scaled distance < 20 m/kg$^{0.5}$). The author's own experience is that seed waves obtained under conditions of infinite confinement (very large stemming columns and no free face) do not necessarily generate higher vibration levels (Fig.1), or show a greater tendency to over-predict vibration impacts, though that experience is limited to scaled distances in excess of 15 m/kg$^{0.5}$.

Fig.1 is not presented to imply that free face vibration levels are higher than those from fully confined charges, but rather that charge confinement in this test did not appear to increase peak vibration level. When considering the differences in amplitude between waves produced by charges of equal weight, it should be remembered that normal amplitude variability appears to have a coefficient of variance (standard deviation divided by mean) of the order of 10%. The source of the variation is unknown, but may relate to differences in explosive coupling, explosive detonation characteristics, or reproducibility of the vibration measurement system.

The author's experience also suggests that the charge weight used to obtain the seed wave can be as small as one third to one half of the size of the production charges being modelled, while still providing reliable vibration predictions. Despite this finding, it is still considered preferable that the seed wave be generated by a charge which closely approximates the size of the production charges (in diameter, length, and weight).

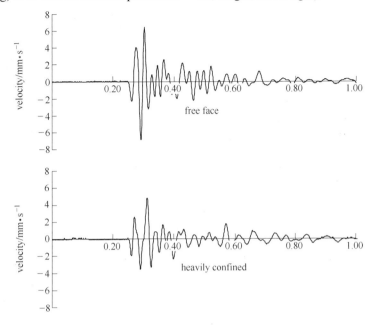

Fig.1 Seed waves from 118 kg charges fired to free face (upper) and fully confined (lower), amplitudes scaled for constant scaled distance of 20 m/kg$^{0.5}$

In the author's experience, the more detailed that a seed wave collection campaign can be, the more reliable will be the modelling. An example is cited for the case of modelling conducted for every overburden blast over a period of more than 12 months in an Australian open cut coal mine, based on extensive single hole firing and vibration monitoring. The weight of charge used in the various single holes varied from 80 kg to 200 kg, while production charge weights varied over the range 120 kg to 300 kg. The seed wave collection campaign enabled accurate determination of the vibration propagation conditions for single charges, as well as the variability in amplitude for the same scaled distance as previously discussed, as shown in Fig.2.

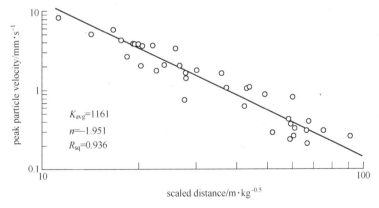

Fig.2 Scaled distance regression for single charges of weight varying from 80 kg to 200 kg

A complication to the Seed Wave Model seems to lie in the use of permanent monitoring systems. Such systems are installed with a fixed alignment of the geophones, which will generally not be consistent with the assumed convention of the radial (or longitudinal) geophone pointing towards the location of the seed wave charge.

As Fig.3 shows, the use of a permanent monitor which is not well aligned with a blast being modelled, in which the transverse component will be better aligned with the single hole charge than the radial component. Considering the comments made by Blair (1999) with respect to amplitude adjustment of the transverse geophone/accelerometer signal, this misalignment may cause difficulties in modelling. When using seed wave modelling, it is considered advisable that monitoring always be conducted with the radial geophone oriented with its axis pointing to the centre of the production blast being modelled and monitored. This may mean that seed wave modelling should utilise a well-aligned roving monitor rather than a permanent, randomly aligned (with respect to a constantly variable blast location) monitor.

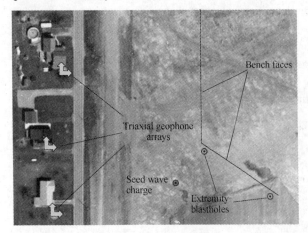

Fig.3 Permanent monitors poorly aligned (arrows show orientation of radial and transverse geophones) for seed wave modelling of single-row blast (extremity holes shown)

4 APPLICATIONS – FAR FIELD VIBRATION

While the seed wave model can be applied to great value in the near-field for the purposes of minimizing impacts on the pit wall, or maximizing stress concentrations within the blast volume, this paper has environmental vibration impact control as its principal focus. Environmental applications focus on determination of inter-deck, inter-hole and inter-row delay times to either minimise peak particle vibration (*PPV*) levels, or alternatively to adjust vibration frequencies.

4.1 Controlling *PPV*

The application of seed wave modelling for the prediction of vibration levels from even large and complicated blasts has been well demonstrated (Hinzen et al 1987, Blair 1999, Yang 2007, Yang & Scovira, 2010). This demonstrated ability then becomes the basis for adding a search routine to the models, so that the modelling can be repeated for many different timing combinations. Bernard (2012) is probably the first to publish the results of such search routines, along with measures of the accuracy of prediction, for a statistically large number of blasts. Bernard's data suggest that, through careful selection of delay times (inter-deck, inter-hole and inter-row), vibration levels can probably be lowered by an average of 30% relative to those achieved with conventional non-electric initiation, and in some case by much greater amounts.

In the author's experience, using a proprietary Monte Carlo based model, vibration levels for different timing combinations can vary over a factor of at least 4, as illustrated in Fig.4 for 1053 different delay timing combinations (single row blast, 5 decks per hole, inter-deck delays from 8 to 20 ms, inter-hole delays from 20 to 100 ms) within the same blast and modelled at the same receptor.

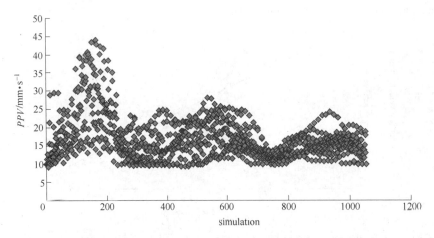

Fig.4 Variability in induced vibration levels for 1053 different delay combinations, for a single-row blast fired within 100 metres of housing

Perhaps the most interesting feature in Fig.4 is the large number of delay combinations which will produce relatively low levels of vibration (9~10 mm/s), suggesting that there is no unique delay timing combination to produce low induced vibration levels, and that the blaster will have quite a range of delay times from which to choose a timing appropriate to the other goals of fragmentation, displacement, and overpressure. The wide range of delay timing which generate low vibration levels also raises the possibility of choosing timing based on factors other than just the *PPV*. The data of Fig.4 also highlight the penalty in terms of imprudent or random timing selection, and the need for an engineering tool to assist in selection of times which deliver control over blasting outcomes – a tool desperately needed in the light of the newly-won flexibility offered by programmable electronic initiation. Adoption of the same electronic timing that was previously used with non-electric initiation may well lead to very unsatisfactory results in terms of vibration impacts.

4.2 Adjusting Frequency

Statutory vibration limits in many countries are frequency dependent, as highlighted in the various vibration standards such as DIN 4150, RI 8507, and UNE 22 381 193. The "frequency" of a vibration wave as reported by commercial seismographs relates to the time between successive zero-cross points which lie on either side of the peak particle velocity on each of the triaxial gauges. While the validity of this measurement as a meaningful measure of frequency can be debated, it is the system adopted by all known commercial blasting seismographs, and it forms the basis for determining compliance in many countries.

In addition to the frequency of the peak vibration, commercial blasting seismographs also display many other amplitude/frequency points determined by finding local maxima/minima between successive zero cross points on each seismograph channel. Fig.5, shows two different possible outcomes for the same blast, depending on assigned inter-hole and inter-row delay timing, and Fig.6 presents the RI 8507 spectra for the same waves. The units for waveforms of Fig.5 and the frequency spectrum of Fig.6 have been kept in the original (US) units for which the study results were obtained.

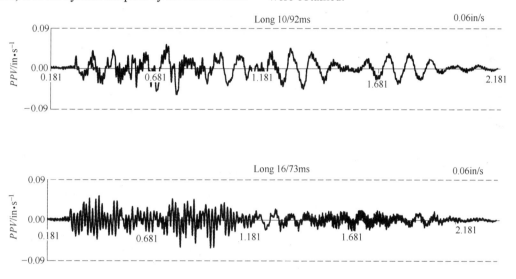

Fig.5 Modelled waveforms obtained from the same blast at the same receptor with different inter-hole and inter-row delay times, but similar values of PPV. Numbers above each waveform show inter-hole / inter-row delay times

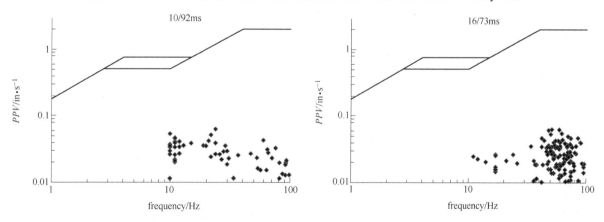

Fig.6 Two different frequency spectra for the modelled waveforms of Fig.5

In spite of the low amplitudes being recorded (< 2.5 mm/s), blasting at this site is generating strong complaint, possibly as a result of the low frequencies of vibration tending to induce building resonances, noting that complaints invariable come from building occupants rather than people working outdoors at the time of the vibration disturbance. In order to reduce community perceptions of blasting vibration impacts (and structural resonances), it is possible to search for delay timing combinations which shift the mass of points in the zero cross frequency spectrum towards the right, towards higher frequencies. As long as this can be done while still remaining compliant with respect to peak vibration amplitudes, blasters are able to choose delay timing which addresses both regulatory compliance as well as community perception.

Searching for a delay combination which maximises frequency requires that the mass of points in the RI 8507 plot be characterised by a single index such as the frequency of the centre of mass. Because of the significance of low frequencies in evaluating the potential for exciting house resonance, the author prefers an index which focuses on the low end of the distribution of points, and is using the 10% point on the cumulative amplitude spectrum, as illustrated in Fig.7. The cumulative spectrum can be estimated either from Zero Cross spectra using average vibration amplitudes within discrete frequency bins, or from the Fourier spectrum via the Cumulative Power Spectral Density function.

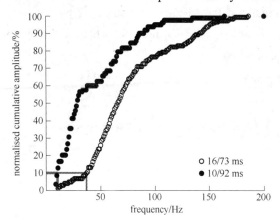

Fig.7   Cumulative amplitude spectra for the blasts of Fig.5, derived from the frequency spectra of Fig.6, also shown are the 10% values (red lines), representing the frequency indices for each event

To calculate the Frequency Index, each point in the cluster of points in the RI 8507 graph is assigned to a frequency bin, typically with a 1 Hz bin width. Where multiple points fall into the same bin, the most appropriate action is to use the maximum value, since summing or averaging of these points has no particular significance. The "cumulative spectrum" is then obtained by summing of the individual bin values to create a typical "S" shaped, cumulative distribution curve. Normalising the curve (peak value = 100%) means that the frequency at which the cumulative amplitude is 10% of the total amplitude is easily determined. The method assigns the same frequency index to a wave, irrespective of its amplitude, with frequency indices less than about 14 Hz considered to indicate the potential to excite resonance in a typical residential structure, and larger multistory structures tend to have even lower resonant frequencies.

The 10% point has no particular significance other than representing the onset of a significant component of the total impact. The cumulative power spectral density obtained from a Fourier analysis would be preferred, and more easily understood, but suffers the disadvantage of not being relevant to legislative code or regulatory bodies.

Using the cumulative Zero Cross spectrum method de-scribed above, the two blasts of Fig.5 are assigned frequency indices of approximately 10 Hz (10 / 92 ms) and 35 Hz (16 / 73 ms). While delay timing can go some way towards changing personal perception, it must be remembered that the frequency spectrum of vibrations from blasting is controlled principally by the rock mass. Major shifts in frequency by changing delay timing will rarely be possible. In the example above, there may be very significant benefit to the occupant of a building, if the selection of blast timing were based not only on low PPV values, but also on highest possible frequency index.

Accepting that human perception of vibration depends on a combination of amplitude, duration and frequency, the issue of personal perception can be taken a step further, by proposing a Perception Index (PI), which takes into account the peak particle velocity (PPV), vibration duration (T_v) and the derived frequency index (I_f) as:

$$PI = PPV^a \times T_v^b \times I_f^c \qquad (4)$$

where, PPV can be in either US or metric units; T_v is in seconds; I_f is in Hertz.

In seeking appropriate values for the exponents a, b, and c, reference is made to both international standards and broadly accepted trends. Since all international standards focus principally on peak particle velocity, the exponent for this term in the perception index is assigned a value of 1.0. In deciding an appropriate value for the exponent of the vibration duration term, reference is made to the British Standard BS 6472: 1992 Evaluation of Human Exposure to Vibration (it is also presented in the 2008 version of the same standard but has an error in the sign of the N exponent), which proposes the following adjustment factor for acceptable vibration levels from blasting based on the number of blasts per day and the duration of the induced vibration:

$$F = 1.7 \times N^{-0.5} \times T^{-d}$$

where, N is the number of blast events per day; T is the duration of the vibration in seconds.

The adjustment factor for the duration of vibration, T^d, shows that as vibration duration increases, acceptable limits are lowered, and suggests values of "d" of 0.32 for buildings with wooden floors, or 1.22 for buildings with concrete floors. The value of the exponent used to derive the perception index could there be adjusted according to the predominant type of house construction, or perhaps an average value can be used where both types of houses are present.

In deciding an appropriate value for the exponent of the vibration frequency term, I_f, reference is made to the USBM RI 8507 curve since this appears to be the most widely used guide, and refers principally to the expected response of the structure which is believed to drive complaint from building occupants.

In RI 8507 the overall frequency trend seems to be quite well defined by the dashed line in Fig.8, which has an overall slope of +0.5. Since the RI 8507 graph indicates acceptable vibration levels, perception levels can be considered to be the inverse of acceptability, i.e. low levels of acceptable vibration suggest high levels of human perception (inside the building) and vice versa. The exponent, c, for the frequency term in the perception equation (equation 4) is therefore assigned a value of -0.5.

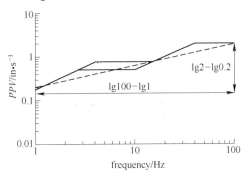

Fig.8 Frequency trend defined by USBM RI 8507, based on building response

In Equation 4, each term can be derived from analysis of either the modelled or measured seismic waveforms. The exponents a, b, and c in the equation can be considered to reflect the relative sensitivities of the different components of the index, and can be adjusted from the values proposed in this paper according to community response and feedback, and field experience. Clearly, the exponent c will always be negative, since personal perception of vibration by building occupants is expected to increase with decreasing frequency. Based on the above proposal, peak particle velocity therefore has the strongest influence on perception, followed by the vibration duration, with the frequency having the weakest influence, and the sensitivity analysis suggesting that people may be twice as sensitive to changes in peak vibration amplitude as they are to the same percentage changes in vibration frequency. These relative sensitivities can easily be changed by reassignment of the exponents a, b and c in equation 4.

Search algorithms can then find the delay times that minimise the peak particle velocity, maximise the frequency index, or minimise the perception index. The perception index is dimensionless, and its absolute value has no particular significance, allowing it to readily be implemented in either US or metric units. For the two blasts of Fig.5, the perception indices were calculated to be 0.034 for the upper waveform, and 0.022 for the lower waveform. While the two impacts would appear to be the same in terms of the PPV induced, the perception indices suggest that the delay timing of 10 ms inter-hole and 92 ms inter-row may generate vibrations which are significantly ($\sim$50%) more perceptible, and bothersome to building occupants than those produced with delay times of 16 ms and 73 ms. These values were derived for the exponent values of $a = 1$, $b = 0.77$, $c = -0.5$.

5 CONCLUSIONS

Complaint and compliance are clearly unrelated issues when it comes to the impact of blast-induced vibrations on surrounding communities, and while compliance with statutory regulations offers some consolation to mining and quarrying operations, incessant complaints are driving operators to look for opportunities for further reducing environmental impacts. Precise initiation timing, currently attainable only through electronic initiation systems, offers the possibility not only of reductions in peak vibration levels, but also of significant adjustments to the frequency content of the vibrations and the probability that they will invoke complaint from building occupants. While the model described here has been used quite widely and successfully in the US quarry industry, and frequently with a focus on minimizing perception rather than minimizing PPV, results in terms of reduced personal discomfort and complaint have been very promising though also purely anecdotal, so it remains a challenge to quantify the extent to which human perception to blast-induced vibrations can be measured and quantified. The ability to focus on perception as well as compliance with a statutory peak vibration level is a strong incentive to use models such as the seed wave model which predict the complete vibration waveform, rather than models which focus purely on peak vibration amplitudes.

REFERENCES

[1] Anderson D A. Signature hole blast vibration control – 20 years hence and beyond., *34th Annual Conference on Explosives and Blasting Technique, International Society of Explosives Engineers*, Volume 2, Nashville, TN. 2008.

[2] Bernard T. The truth about signature hole method, *38th Annual Conference on Explosives and Blasting Technique, International Society of Explosives Engineers*, Volume 2, Orlando, FL. 2012.

[3] Blair D P. Statistical models for ground vibration and air blast, *Int. J. Blasting and Fragmentation*, 1999,3:335~364.

[4] Brent G F, Smith G E, Lye G N. Studies on the Effect of Burden on Blast Damage and the Implementation of New Blasting Practices to Improve Productivity at KCGM's Fimiston Mine, *AusIMM, Explo 2001 Conference*, Hunter Valley, NSW, 2001: 28~31.

[5] Hamdi E, du Mouza J, Le Cleac'h J-M. Micro fragmentation

energy evaluation in rock blasting, Proceedings *8th International Symposium on Fragmentation by Blasting - Fragblast 8*, Santiago, Chile, 2006: 134~139.

[6] Hinzen K G, Ludeling R, Heinemeyer F, Roh P, Steiner U. A new approach to predict and reduce blast vibration by modelling of seismograms and using a new electronic initiation system, *13th Annual Conference on Explosives and Blasting Technique, International Society of Explosives Engineers*, Miami, FL, 1987: 144~161.

[7] Hinzen K G. Modelling of blast vibrations, *Int. J. Rock Mech. Min. Sci, & Geomech Abstr.*, , 1988, 25(6): 439~445.

[8] Katsabanis P D, Tawadrous A, Braun C, Kennedy C. Timing effects on fragmentation, Proceedings of the *32nd Annual Conference on Explosive and Blasting Technique*, 2006, 2, 29 Jan – 01 Feb, Grapevine, Texas.

[9] Katsabanis P D, Tawadrous A, Sigler J. Effect of powder factor on the impact breakage of rocks, Proceedings of the *34th Annual Conference on Explosive and Blasting Technique*, 2008, 2, Jan 27~30, New Orleans, Louisiana.

[10] Kojavic T, Michaux S, McKenzie C. Impact of blast fragmentation on crushing and screening operations in quarrying, *AusIMM Explo 95 Conference*, Brisbane 4-7 September, 1995: 427~436.

[11] McKenzie C. Seed Wave Modelling Applications for Fragmentation, Damage, and Environmental Impact Control, *10th International Symposium on Rock Fragmentation by Blasting, Fragblast 10*, India, November, 2012.

[12] Paley N. Testing electronic detonators to increase SAG mill throughput at the Red Dog Mine, *36th Annual Conference on Explosives and Blasting Technique, International Society of Explosives Engineers*, 2010,2, Orlando, Florida.

[13] Ramulu M, Chakraborty A K, Raina A K, Reddy A H, Jethwa J L. in *Proceedings 7th International Symposium on Rock Fragmentation by Blasting - Fragblast 7*, Beijing (Ed: Prof WANG Xuguang) 2002:617~624 (Beijing Metallurgical Industry Press).

[14] Spathis A T. A scaled charge weight superposition model for rapid vibration estimation, *Int. J. Blasting and Fragmentation*, 2006,10(1~2): 9~31.

[15] Vanbrabant F, Espinosa A. Impact of short delays sequence on fragmentation by means of electronic detonators: theoretical concepts and field validation, in *Proceedings 8th International Symposium on Fragmentation by Blasting - Fragblast 8*, Santiago, Chile, 2006: 326~331.

[16] Yang R. Near-field blast vibration monitoring, analysis and modelling, *33rd Annual Conference on Explosives and Blasting Technique, International Society of Explosives Engineers*, 2007, 2, Nashville, TN.

[17] Yang R. Scovira D S. A model for near and far-field blast vibration based on multiple seed waveforms and transfer functions, *36th Annual Conference on Explosives and Blasting Technique, International Society of Explosives Engineers*, 2010, 2, Orlando, FL.

Study on Crush Progress of Rock in Delay Blasting

SHI Fuqiang, LIAO Xueyan, JIANG Yaogang, GONG Zhigang

(*Sichuan Province Academy of Safety Science and Technology, Chengdu, Sichuan, China*)

ABSTRACT: With the development of blasting technology and improvement of safety requirement, hole by hole initiation technique is commonly used in rock blasting engineering for lowering blasting vibration. And as the formation of new free surface resulted from the former hole initiated, thus its blasting effect is better than holes initiated simultaneously. But hole by hole initiation may cause high boulder yield, much bedrock, and so on in rock blasting engineering. For the solution of above problems, this paper study on explosion shock wave and explosion product taking action on rock, proposes scientific blasting design idea and basis for effective and safe blasting design blasting scheme by combination of theory, numerical simulation and practice.

KEYWORDS: blasting; crush progress; rock

1 INTRODUCTION

Rock blasting process involve explosion, dynamic response, propagation of stress wave in rock and so on, additionally very short duration of blasting, high stress and great deformation of rock, strain rate sensitivity of rock and so on, thus it is difficult to study the rock blasting. Another aspect, rock is anisotropic material and has cracks & damage, so it has greater difficult to study rock's blasting than other homogeneous materials[1]. The heterogeneity exists in rock may lead to three main effects: first lowering the strength of rock, resistance of deformation and breaking of rock; second weakening the action and propagation of stress wave; third varying the distribution of energy and the action of blasting. Zhai Yue et al[2] drew a conclusion that concrete in delay blasting, early blasting rock form crush zone and fracture zone. The crush zone and fracture zone could influence latter rock blasting, so considering the early blasting rock as free surface or complete crush zone couldn't reveal practical situation. For exact study of influence of early blasting rock to latter blasting rock, this paper studied the influence of early blasting rock to latter blasting rock by theory analysis and numerical simulation using damage model.

2 THEORY ANALYSIS OF CRUSH PROGRESS OF ROCK IN DELAY BLASTING

For analyzing influence of early blasting rock to latter blasting rock, rock model containing damage parameters was established first and then a theory analysis of crush progress in delay blasting by using damage model was made.

2.1 Damage Model of Rock

Considering influence of rock damage to rock strength, a factor damage degree D was introduced. Assuming damage occurred when rock was in plastic strain, damage degree D was described as follows:

$$D = \sum_{i=1}^{n} \frac{\Delta \varepsilon_p}{\varepsilon_p^{\text{failure}}} \quad (1)$$

$$\varepsilon_p^{\text{failure}} = D_1 \left[\frac{P}{f_c}\left(1 - \frac{f_t}{f_c}\right) \right]^{D_2} \quad (2)$$

in the formula, $\Delta \varepsilon_p$ was plastic strain increment, $\varepsilon_p^{\text{failure}}$ was failure strain, f_c was uniaxial compression strength, f_t was uniaxial tensile strength, P was stress, D_1, D_2 were damage constants.

Once damage occurred in rock（$0 \leqslant D \leqslant 1$）, breaking strength of rock was described as follows[3]:

$$Y_{\text{fractured}} = (1-D)Y_{\text{failure}} + DY_{\text{residual}} \quad (0 \leqslant D \leqslant 1) \quad (3)$$

in the formula, $Y_{\text{fractured}}$ was breaking strength of damage material, Y_{failure} was failure strength, Y_{residual} was residual strength.

Y_{failure} was the function of P, similarity angle θ and strain rate $\dot{\varepsilon}$:

$$Y_{\text{failure}}(P, \theta, \dot{\varepsilon}) = Y_{\text{TXC}}(P) \times F_{\text{rate}}(\dot{\varepsilon}) \times R_3(\theta) \quad (4)$$

$$Y_{\text{TXC}}^*(P) = \frac{Y_{\text{TXC}}(P)}{f_c} = A(P^* - P_{\text{spall}}^* F_{\text{rate}})^N \quad (5)$$

$$P^* = P/f_c \quad (6)$$

$$P_{\text{spall}}^* = P^*\left(\frac{f_t}{f_c}\right) \quad (7)$$

$$F_{\text{rate}}(\dot{\varepsilon}) = \left(\frac{\dot{\varepsilon}}{\dot{\varepsilon}_0}\right)^\alpha \quad \text{with} \quad \dot{\varepsilon}_0 = 30 \times 10^{-6} \text{s}^{-1} \quad (8)$$

Corresponding author E-mail: sfq@swjtu.cn

$$R_3(\theta) = \frac{2(1-Q_2^2)\cos\theta + (2Q_2-1)\left[4(1-Q_2^2)\cos^2\theta + 5Q_2^2 - 4Q_2\right]^{1/2}}{4(1-Q_2^2)\cos^2\theta + (2Q_2-1)^2} \quad (9)$$

$$Q_2 = Q_{2,0} + BP^* \quad (10)$$

In the above formulas, $Y^*_{TXC}(P)$ was compression meridian, $F_{rate}(\dot{\varepsilon})$ was strain rate reinforcement parameter, $R_3(\theta)$ was angle function, A was failure surface constant, N was failure surface exponent, $\dot{\varepsilon}$ was equivalent strain rate, $\dot{\varepsilon}_0$ was reference strain rate, α was compression strain rate exponent, δ was tensile strain rate exponent, $Q_{2,0}$ was tensile-to-compression meridian ratio.

When $D=1$, the material was in complete failure situation and bore a certain degree of compression rather than tension, and the residual compression strength was:

$$Y_{residual} = b(f_c)^M \quad (11)$$

In the above formula, b was residual failure constant, M was residual failure exponent.

2.2 Theory Analysis of Crush Progress of Rock

In loose blasting, the load acting on rock cause by explosive shock wave and explosive gas raised sharply. Rock was damaged when the load was greater than limit of elastic strength. At this moment, D was equal to 1 and rock was crushed. Crushed rock could bear compression rather than tension, so it is not correct to consider crashed rock as free surface, but to take it as discontinuity surface for rock's mechanical property was saltatorial. According to theory of stress wave, stress wave should reflect on discontinuity surface, and tension stress would formed when it propagated from higher strength material to lower strength material. Rock's tension strength was far less than compression strength, thus rock near discontinuity surface was fractured by tension, so in this situation the discontinuity surface formed by early blasting rock was conducive to crush of latter blasting rock. Another aspect, crack was formed on rock caused by explosive stress wave. Stress wave reflected on the cracks leading to asymmetrical stress around hole and escaping of gas from cracks, thus rock blocks were asymmetrical. In delay blasting, the influence of early blasting rock to latter blasting rock was dual. One aspect, discontinuity surface benefit crush of latter blasting rock; another aspect, cracks led to asymmetrical rock blocks.

3 NUMERICAL SIMULATION OF CRUSH PROGRESS OF ROCK IN DELAY BLASTING

3.1 Simulation of Blasting Crater

Simulation could analyze crash progress of rock intuitively in delay blasting. Rock material used damage strength model and explosive used JWL equation[4] in simulation. Crash zone was formed 1ms after initiation from simulation of typical blasting, shown as Fig.1. Stripped damage zone existed at the bottom and two sides of crush zone. So in rock blasting not only crushable blasting crater but also damage zone and irregular cracks around crushable blasting crater formed.

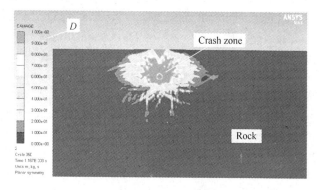

Fig.1 Damage cloud of one hole blast for rock

3.2 Influence of Rock Cracks to Blasting

Rock blasting progress was simulated by cracks placed in rock previously for influence simulation of cracks to blasting. From simulation, blasting crater and cracks were formed 10ms after initiation leading to asymmetric stress of rock and large rock block, shown as Fig.2.

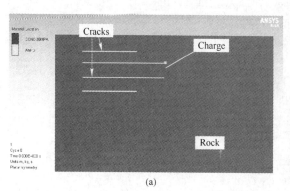
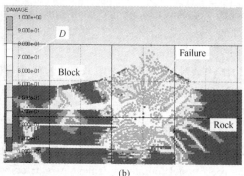

(a)　　　　　　　　　　　　(b)

Fig.2 Damage cloud of rock

(a) 0ms; (b) 10 ms

3.3 Simulation of Delay Blasting

As shown in simulation, rock damage zone tended to be stale 4ms after initiation. The second blasting cartridge initiated 5ms after the initiation of first blasting cartridge in simulation for avoidance of termination of calculation as great deformation of mesh. The calculation result showed that irregular damage crack formed around damage zone and crush zone after initiation of first blasting cartridge, shown as Fig.3. Then after initiation of second blasting cartridge, surrounding rock was crushed and stress wave concentrated on damage crack leading to extension of crush zone. With the further action of explosive load, rock damage zone close to early blasting rock was greater than opposite side. In delay blasting, latter explosive energy concentrated on damage crack caused by early blasting load and lead to crush of rock. Stress wave reflected and penetrated on discontinuity surface when propagating to it and formed tension stress leading to fracture of rock. Thus rock damage zone and damage degree close to early blasting rock were greater than opposite side.

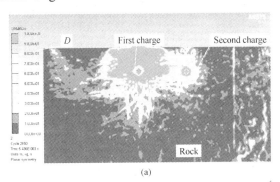

(a)　　　　　　　　　　　　　　(b)

Fig.3　Damage cloud of delay blast

(a) the scend hole detonate 0.4ms later; (b) the scend hole detonate 0.7ms later

4　CONCLUSIONS

(1) The influence of early blasting rock damage to latter blasting rock was analysed by rock damage strength model. The result showed that discontinuity surface was conducive to crush of latter blasting rock, and the same time irregular damage crack may led to asymmetrical rock blocks. It is critical to control early blasting rock zone as reflection and penetration surface and damage crack not to extend to further blasting zone for lowering rock blocks.

(2) In delay blasting, significant blownout and bedrock may happened as rock around block up segment moved diagonally for discontinuity surface weakened explosive load.

(3) In line by line initiation, the holes initiated simultaneously in the same line, so there was no formation of discontinuity surface, thus blownout was decreased significantly. When hole by hole initiation was adopted, the situation of blownout and bedrock were improved significantly by taking step of enlarging hole distance and decreasing line distance.

(4) The simulation of crush progress of rock in delay blasting was made through considering rock damage strength model. The results were agreeable to the engineering practice and reveal that penetration and reflection on discontinuity surface was key factor of altering rock's blasting property. The results may also provide scientific basis of understanding rock blasting principle and optimizing design scheme.

REFERENCES

[1] Wang Xuguang. Handbook of Blasting[M]. Beijing: Metallurgy Industry Press, 2010.

[2] Zhai Yue, Ma Guowei, Zhao Junhai, et al.Study on Comparison of Dynamic Properties of Granite and Concrete Subjected to Impulse Load[J]. Chinese Journal of Rock Mechanic and Engineering, 2007, 26(4): 762～768.

[3] AUTODYN materials library version 6.1.

[4] Sun C W, Wei Y Z, Zhou Z K. Applied Detonation Physics[M]. Beijing: National Defence Industry Press, 2000(in Chinese).

Camouflet Blasting of a Finite-Length Borehole Charge

S.D. Viktorov, N.N. Kazakov,
A.V. Shlyapin, I.N. Lapikov
(*Institute of Comprehensive Exploitation of*
Mineral Resources (*IPKON*), *Russian Academy of Sciences, Russia*)

ABSTRACT: The article presents the description of the calculation technique proposed by the authors with due account for end effects, main parameters of the process of camouflet blast action of a finite-length borehole charge in the rock mass. The technique can be used for research into the processes of rock fragmentation by blast of borehole charges in open pits.

KEYWORDS: charge model; hollow model; zone model; cylindrical charge; 3D setting; camouflet zone; end effects; main parameters

Borehole charges are widely used in mineral mining. Nearly in all blasting conditions the camouflet phase of a blast is effected as a process phase, which development accounts for a significant part of charge energy consumption. Per se and completely it is implemented in the blast of a finite-length charge in unconfined space.

Objective laws governing the camouflet phase of a concentrated charge blast are not thoroughly investigated and described in the technical literature [1~3]. Numerous studies and published papers are available on the investigation, 2D setting, objective laws governing the development of an infinite-length cylindrical charge [3, 4].

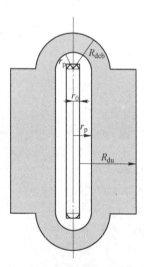

Fig.1 Models of a charge shape, hollow and zone of blast camouflet action

Insufficient attention was attached to the research into processes of spherical symmetry at ends of a charge being fired. These end processes have different effects on the development of the cylindrical part of the process. In terms of technology they can participate in some other task solving. The end effect of the upper part of a charge pertains to fly rock travel. It also influences the specific consumption of explosives. The end effect of the lower part of a charge pertains to bench floor working and shovel performance.

The camouflet phase of a blast is the first phase of the process of blast effect on the rock mass. Its coverage area borders closely a charge. Outside its boundaries a wave phase of a blast is implemented. Fig. 1 shows the models of a cylindrical charge shape, gas hole shape and blast camouflet action zone shape.

The model of a charge shape is a cylinder with hemispheres at the charge ends. It is assumed that a hemisphere radius r_0 is equal to the radius of the borehole charge. It is assumed that the model of the camouflet gas hole shape is a cylinder with hemispheres at the ends. The radii r_p of cylindrical and spherical parts of the gas hole are assumed to be equal. External boundaries of the blast camouflet action zone are calculated from the condition of equal maximum stresses in a strain wave and rock ultimate compression strength.

Borehole charge energy

$$E_{ch} = 41868 MQ$$

where, Q is heat of explosive blasting, kcal/kg; M is charge mass, kg.

The volume of a camouflet hollow is calculated by the formula [1, 2]

$$V_p = \frac{38 E_c}{\rho c^2} \left(\frac{\rho c^2}{250 \sigma_{comp}} \right)^{2/3}$$

where, ρ is rock density; c is P-wave velocity in rock; σ_{comp} is rock ultimate compression strength; V_p is the volume of camouflet hollow.

The radius of a camouflet hollow is calculated from simultaneous solution of the following equations

$$V_{pc} = \pi r_p^2 (L_3 - 2r_0) ; \quad V_{psph} = \frac{4 \pi r_p^3}{3} ; \quad V_p = V_{pc} + V_{psph}$$

Corresponding author E-mail: victorov_s@mail.ru

where, V_{pc} is volume of the cylindrical part of a hollow; V_{psph} is volume of the spherical part of a hollow.

The radii of external boundaries of cylindrical and spherical parts of the camouflet zone are radii of cylindrical and spherical strain wave emission.

The radius of cylindrical strain wave emission is calculated by the formula [4]

$$R_{dc} = r_{pc}\sqrt{\frac{\rho c^2}{4\sigma_{comp}}}$$

The radius of spherical strain wave emission is calculated by the formula [1,2]

$$R_{dsph} = r_{psph}\sqrt[3]{\frac{\rho c^2}{4\sigma_{comp}}}$$

The radius of the cylindrical zone of the blast camouflet phase action R_{dc} is greater than the radius of the spherical part of the camouflet zone R_{dsph} due to different conditions of wave spreading in cylindrical and spherical zones.

The maximum mass velocity at the boundary of emission is proposed to calculate by the formula [3]

$$U = \frac{c\sigma_{comp}(1-\mu^2)}{E(1-0.36\mu)}$$

where, E is rock modulus of elasticity; μ is rock Poisson's ratio.

This maximum mass velocity in a strain wave is the criterion of the external boundary of the blast action camouflet zone.

At the external boundaries of the camouflet hollow the maximum mass velocities are equal in cylindrical and spherical parts, but this velocity is formed at different distances from the seat of blast.

With the use of these formulas the software has been developed in Mathcad shell, which provides the calculation of geometry parameters of the camouflet zone for specific conditions of finite-length borehole charge blasting at open pits.

Tab.1 Geometry parameters of the camouflet zone

No.	Parameter	In the cylindrical part of the zone	In the spherical part of the zone
1	Charge radius r_0 =0.125 mm	—	—
2	Charge length L_{ch} = 11 m	—	—
3	Charge mass M= 486 kg	—	—
4	Charge energy E_{ch}= 2.034 MJ	—	—
5	Radius of camouflet hollow/m	0.232	0.232
6	Volume of camouflet hollow/(m³/%)	1.81/97.3	0.05/2.7
7	Radius of camouflet zone/m	1.90	0.94
8	Volume of camouflet zone/(m³/%)	112.3/97.3	3.5/2.7

Numerical values of geometry parameters of the camouflet zone were calculateв for the following blasting conditions: charge radius r_0 =0.125 mm; borehole length L_{bh} =17 m; stemming length L_{st} = 6 m; explosive blast heat Q = 1000 kcal/kg; charge density ρ_0 =900 kg/m³; rock ultimate compression strength σ_{comp} = 160 MPa; P-wave velocity in rock c = 4000 m/s; granite elasticity modulus E = 56000 MPa; Poison's ratio μ = 0.23; granite density ρ = 2700 kg/m³.

Tab.1 contains calculated results for blasting conditions of a borehole TNT charge of 250 mm diameter in granite. From the Table it can be seen that the volume of the camouflet hollow and the volume of the blast camouflet phase action zone accounted for by spherical zones by many times less than those accounted for by the cylindrical zone.

Energy parameters for the blast camouflet zone can be calculated as follows.

The process of gas hole expansion can be described with sufficiently high accuracy for the practice by two conjugate isentropes with different fixed isentrope parameters.

The hole volume in the point of isentrope conjugation can be calculated by the following formula

$$V_c = V_{ch}\sqrt{\frac{P_{mean}}{0.35\rho_0\left(427Qg - \frac{D^2}{16}\right)}}$$

where, V_{ch} is charge volume; P_{mean} is mean pressure in a charge hollow; g is acceleration of gravity.

Pressure in the point of isentrope conjugation

$$P_c = \left(\frac{V_{ch}}{V_c}\right)^3$$

With the known values of the charge volume, pressure in the hole and the hole volume in the point of isentrope conjugation it is possible to calculate the pressure of detonation products in the finite camouflet hollow. Pressure in the finite camouflet hollow is calculated from the solution of the following equations

$$P_p = P_{mean}\left(\frac{V_{ch}}{V_c}\right)^3 ; \quad P_p = P_c\left(\frac{V_c}{V_p}\right)^{1.3}, \text{ with } V_p > V_c$$

$$P_p = P_{mean}\left(\frac{V_{ch}}{V_c}\right)^3, \text{ with } V_p < V_c$$

Energy in the camouflet hollow remaining in detonation

products is calculated from the solution of the following equations

$$Э_o = \frac{P_p V_c}{2} - \frac{P_c V_c}{2} + \frac{P_c V_c}{0.3}, \text{ with } V_p < V_c$$

$$E_o = \frac{P_p V_p}{0.3}, \text{ with } V_p < V_c$$

Energy already spent by the moment of blast was consumed for excessive fragmentation of rock in the camouflet zone and partially transferred to cylindrical and spherical strain waves.

Energy transferred to a cylindrical strain wave is calculated as follows:

The wave length at the boundary of the cylindrical zone is calculated by the formula

$$\Lambda = 3R_{dc}$$

For the estimation of energy transferred to cylindrical and spherical waves it is necessary to accurately calculate a tensor of unit strain variation in strain waves at the boundary of wave emission.

Variation of radial and tangential unit strain over wave length is calculated by the formulas [3]

$$\varepsilon_r = \frac{-U}{0.18c\Lambda}\left(2\lambda - \frac{3\lambda^2}{0.54\Lambda}\right), \text{ with } 0 \leqslant \lambda \leqslant 0.36\Lambda$$

$$\varepsilon_r = \frac{-2.222U}{c\Lambda}\left[\left(\frac{3\lambda^2}{0.54\Lambda} - 6\lambda\right) + 1.44\Lambda\right]$$

with $0.36\Lambda \leqslant \lambda \leqslant 0.54\Lambda$

$$\varepsilon_r = \frac{-0.4U}{c\Lambda}\left(\frac{2\lambda}{0.92} - 2.17\Lambda\right), \text{ with } 0.54\Lambda \leqslant \lambda \leqslant \Lambda$$

$$\varepsilon_\Theta = \frac{U}{0.18c\Lambda R_{dc}}\left(\lambda^2 - \frac{\lambda^3}{0.54\Lambda}\right), \text{ with } 0 \leqslant \lambda \leqslant 0.36\Lambda$$

$$\varepsilon_\Theta = \frac{0.4U}{0.18c\Lambda R_{dc}}\left[\left(\frac{\lambda^3}{0.54\Lambda} - 3\lambda^2\right) + 1.44\lambda\Lambda - 0.108\Lambda^2\right]$$

with $0.36\Lambda \leqslant \lambda \leqslant 0.54\Lambda$

$$\varepsilon_\Theta = \frac{0.4U}{c\Lambda R_{dc}}\left(\frac{\lambda^2}{0.92} - 2.17\lambda\Lambda + 1.335\Lambda^2\right)$$

with $0.54\Lambda \leqslant \lambda \leqslant \Lambda$

where, ε_r is radial unit strain; ε_Θ is tangential unit strain; λ is the phase of wave length; U is the maximum mass velocity of particles at the boundary of wave emission; c is P-wave velocity in rock.

Variation of the stress tensor components over the wave length at the boundary of emission is calculated by the known dependences [4]

$$\sigma_r = \frac{E}{1-\mu^2}(\varepsilon_r + \mu\varepsilon_\Theta); \quad \sigma_\Theta = \frac{E}{1-\mu^2}(\varepsilon_\Theta + \mu\varepsilon_r)$$

$$\sigma_z = \mu(\sigma_r + \sigma_\Theta)$$

where σ_r, σ_Θ, σ_z are stress tensor components; E is elasticity modulus; μ is Poisson's ratio.

Strain wave energy carried by the wave through the cylindrical surface with a radius R_{dc} is calculated by the formula

$$E_{wc} = \frac{\pi(L-2r_0)R_{dc}}{E}\left[\int_0^\Lambda (\sigma_r^2 + \sigma_\Theta^2 + \sigma_z^2) - 2\mu(\sigma_r\sigma_\Theta + \sigma_\Theta\sigma_z + \sigma_z\sigma_r)d\lambda\right]$$

where, L is a borehole charge length.

Unit strains in a spherical wave are calculated by the same formulas and replacement of R_{dc} with R_{dsph}, assumption $\Lambda = 3R_{dsph}$ and assumption $\sigma_z = \sigma_\Theta$

Strain wave energy carried by the wave through the spherical surface with a radius R_{dsph} is calculated by the formula

$$E_{wsph} = \frac{2\pi R_{dsph}^2}{E}\left[\int_0^\Lambda (\sigma_r^2 + \sigma_\Theta^2 + \sigma_z^2) - 2\mu(\sigma_r\sigma_\Theta + \sigma_\Theta\sigma_z + \sigma_z\sigma_r)d\lambda\right]$$

Energy transferred to cylindrical and spherical strain waves is calculated by the formula

$$E_w = E_{wc} + E_{wsph}$$

Energy spent for rock breakage in the blast camouflet zone is calculated by the following formula

$$E_{bc} = E_{ch} - (E_o + E_w)$$

With the use of these formulas the software has been developed in Mathcad shell, which provides the calculation of energy parameters of the camouflet zone for specific conditions of finite-length borehole charge blasting at open pits. The article presents the calculated results for blasting conditions of a borehole TNT charge of 250 mm diameter in granite. The calculation results for the given variant are shown in Tab.2.

In the illustrative example provided in the Table the borehole charge energy is 2034.3 MJ. Energy amounting to 71.1 MJ was transferred to a cylindrical strain wave. Energy of 2.4 MJ was transferred to a spherical strain wave. For excessive rock fragmentation in the camouflet zone 701.9 MJ were spent.

The share of energy spent for rock fragmentation in the camouflet zone accounts for 35% of the total charge energy. Therefore, any methods of minimizing energy consumption for rock fragmentation in the camouflet zone will have a favorable effect on the quality of fragmentation of all material being broken and on the minimization of drilling and blasting costs.

Tab.2 Energy distribution in the camouflet zone by energy consumption type

Energy distribution by consumption type	Energy	
	MJ	%
Borehole charge energy	2034.6	100
Energy remaining in detonation products by the end of blast camouflet phase development	1259.3	61.9
Energy transferred to a cylindrical strain wave	71.1	3.5
Energy transferred to a spherical strain wave	2.4	0.1
Energy spent for rock fragmentation in the blast camouflet zone	701.9	34.5

With the variation of blasting conditions the ratio of energy consumption types will also vary. However, energy transferred to a cylindrical or spherical strain wave will probably account for at most 5%~6% of the charge energy. Energy consumption for excessive fragmentation of rock in the camouflet zone will consistently remain high.

CONCLUSIONS

(1) A model of a finite-length cylindrical charge and a model of a camouflet hollow as a cylinder with hemispheres at charge ends are proposed.

(2) A model of the shape of the zone of finite-length borehole charge blast camouflet action is proposed. The external boundary of the camouflet zone is the cylinder surface with hemispheric surfaces at ends. In the proposed model the cylinder radius is greater than that of hemispheres.

(3) A calculation method and software have been developed for the estimation of main parameters of the camouflet zone of a finite-length cylindrical charge. Main parameters of the camouflet zone of a finite-length cylindrical charge have been calculated for one variant of borehole charge blasting conditions.

REFERENCES

[1] Adushkin V V. Simulation studies of rock breakage by blast. "Physical problems of rock mass breakage by blast". – M.: IPKON RAS, 1999: 18~29. (in Russian)

[2] Adushkin V V, Spivak A A. Geomechanics of large-scale blasts. – M.: Nedra Publishers, 1999: 52. (in Russian)

[3] Viktorov S D, Kazakov N N. Wave parameters in the zone of rock fragmentation by blast. Vestnik, Kremenchug Polytechnic University. Issue 5/2005: 141~144. (in Russian)

[4] Kazakov N N. Particle mass velocity in a wave at the emission boundary. Collected works "Vzryvnoe Delo", Issue No106/63. – M.: MVK Po Vzryvnomu Delu ZAO, Academy of Mining Sciences, 2011: 27~32. (in Russian)

Multibody-Discretebody Dynamics to Control Building Demolished by Blasting

WEI Xiaolin

(Guangdong Hongda Blasting Co., Ltd., Guangzhou, Guangdong, China)

ABSTRACT: The course of the development of science and technology of building demolished by blasting is described. The Building Toppling Dynamics, which includes to Multibody-discretebody Dynamics, is defined. It is stands out that the dynamic characteristic is difference from traditional multi-body system. The dynamic equations are erected. The examples of its representative equations of demolition are enumerated. Their analytic and approximate solutions are obtained. The dynamic equations similarity and applications of dimensionless of standard and regularity are advanced. The complete chessboard emulating of variable topological Multibody-Discretebody Dynamics is realized. The parameters about disrepair material mechanics and impact dynamics of concrete component concerned for equation compute are clarified. It has been demonstrated by field observation and engineering examples that Building Toppling Dynamics, which includes to Multibody-Discretebody Dynamics, is correct. Applied the Multi-body Dynamics of the author, the dimensionless expression of cutting size of incision, cheap of exploded heap(its front wide and height), recoil, sitting down(example: single rear post), burst order and blasting time difference (with the aid of single incision closure) of many collapse manners of many kinds buildings have been simply educed by the demolition model. In order to select reasonable toppling manners, demolition measurements and incision parameters, new comprehensive theory and simple practical algorithm for general are provided and it has been proven by engineering practical examples that exact control of demolition with blasting can be achieved by MBDC. However, the Multibody-Discretebody Dynamics is new scientific developing stage of demolition by blasting.

KEYWORDS: demolition by blasting; building; Multibody-Discretebody Dynamics; exact control

1 INTRODUCTION

Demolition by blasting is an important technique in China engineer blasting. In large numbers of engineering practices abundant experiences were accumulated. In industrialization and civic architecture building configurations were in the gross adopted from laying bricks and stones to reinforced concrete, so that buildings are urged to develop high storey, super high storey and great structure and that forms tend to multiplicity and complexity, and buildings are firm and are difficult to topple down since unsteadying. Therefore, mechanics and technology to be China characteristic about blasting demolition of building are urged to create.

Fore century 90's the tidal wave of new technology revolution around centre of modern communication technique swept over world. Under promote of fifth scientific revolution, in blasting demolition domain the digitalization technology for judgment and reading to close range photogrammetry long camera, multi monitor photographs controlled by computer and multi strain measurement had separately been introduced, which were made up of comprehend sive observation technology for blasting demolition. Therefore, the movement gesture of building framework, disrepair material mechanics and impact characters of elastic and brittle body were deeply understood. In breeding of computing numerical technique, the Multibody-discretebody dynamics for building collapse and topple of reinforced concrete configuration is occurred.

2 SCIETIFIC SEEDTIME OF BUILDING DEMOLITION BY BLASTING

2.1 Press Pole Unstability

Traditional principle of press pole is that after blasting the bare part of steel bar is looked upon signal pole, the minimum height by blasting is calculated by method used to unstabilized critical stress. In 1992 unstable model[1] of small steel frame was advanced by LU Wenbo, who had confirmed the minimum height of blasting pole. In 2000 the finite element method of variable stiffness of building frame and cutting steel bars were advanced by ZHANG Qi, who computed arrangement of plastic joints of configure

---

Corresponding author E-mail: wxl_40@163.com

tion and judged its initial unstability[2].

2.2 Mass Centre Move Forward and Static Unstability

All appearance, since configuration unsteady when incision closes building may or may not topple. Therewithal, in 2003 JIN Jiliang presented, when mass centre of building did forward and did exceed front toe closed to incision the building must be overturn and toppled by statics, so that incision parameter[3] was also deduced. But, actually under that smaller incision a lot of buildings can also be overturned and toppled. Therefore, model is not used to overturning kinetic energy.

2.3 Kinetic Energy Overturn and Dynamic Numerical Simulation

From engineering examples of a great of building demolished to overturn, in 2007 Multibody-Discretebody Dynamics[4] for building demolished by blasting was presented by the author, and from the principle of overturning kinetic energy and its destroy the model of Building Toppling Dynamics was established, as shown in Fig.2. And according to its toppling gesture the collapse vibration, splash and dust can also be to assessed. The dynamics has been demonstrated correct by field observations and engineering practical examples, and has been affirmed by academician QIAN Qihu and academician WANG Xuguang[5, 6].

However, as building configuration is developed to firm and high storey, blasting demolition is promoted from technology to science and from careless to exact control. In night before high wave of new scientific revolution clue of building toppling dynamics has been appeared.

3 BLASTING DEMOLITION IS NEEDED BY CHINA

Recently the design theory founded by structure mechanics and single body mechanics can not be satisfied for blasting demolition design. China is the most large county of producing cement and steel bar in the world, a lot of multilayer and high storey building is made of reinforced concrete. Therefore, building toppling dynamics of configuration of reinforced concrete demolished by blasting and its corresponding technology are needed in China.

4 BUILDING TOPPLING DYNAMICS OF REINFORCED CONCRETE

While configuration of reinforced concrete destroys, necessarily that comes through concrete ruptured and steel bar drawn out to escape from mucilaginous. As the plastic angular θ_u >2%~6%, it has been considered by reinforced concrete structural mechanics to form plastic joint in destroying point. The building framework of toppling movement is taken for multi-body system to be connected with moving joints. Therefore, since configuration is initially unsteadied, necessarily it comes through multi-body system movement, then multi-body maybe dispersed to be non fully discrete[5] only by the traction of steel bars, until or direct damage to fully discrete, that latest collapses to on ground and piles to blasting heap. However, in order to reflect that whole(or all of) process and view the building structure all demolition failure types, the limit analysis of initial unsteadying, variable topological Multi-body System Dynamics, dynamic analysis from multi-body to discretebody and dynamic analysis of complete discrete-bodies are made up of unified dynamics by the author, so that whole toppling process of building framework is described, the dynamics of that whole (or all of) process is call as Multibody-Discretebody Dynamics. It is combined by dynamic analysis of impact component to be made up of Buildings Toppling Dynamics. The dynamics has been correspondingly proven correct by 7 engineering practical examples in author's book[5]. Thus it is can be seen, among them the variable topological Multi-body System Dynamics is all absolutely and necessarily the most impotent process in simulating each building toppling.

4.1 Multi-body System Combining with Building Demolition

The ideology of Multi-body System is applied to building demolition, that building mechanism is combined by natural-born joint, so that component is come to rigid-plastic body. Therefore, multi-body dynamic equations can be established, so that it is found on theory foundation to control natural key movement of building toppling demolished by blasting.

But building toppling dynamics must be made basic principle of Multi-body Dynamics to combine realism of blasting demolition and to form Multibody-discretebody Dynamics. The characters of building mechanism are that, First, in toppling some support components and poles can be possibly pressed to burst greatly, so that the support length of component is made short and its mass is reduced tremendously. Those components already can not be abstracted to rigid, but also separately make deformable body and the body to reduce mass, are all called anamorphosis, but another majority of components is also maintained to abstract rigid. Second, the joints of traditional Multi-body System are processed beforehand and body number and dynamic topology are prepared to stipulate, but the joints of building mechanism are all so nature-bore according to dynamic condition and corresponding joints made new bodies. The dynamic topological changing point is decided naturally by condition of load and intension, that process is natural topology. Third, the plastic joints are decided by contact shape between bodies, material and dynamic characters, that turning joint around neutral axis is easily formed by entity bended to break, the press joint around center axis on contact shape is formed by content of cockle surface of canister of thick wall, the joint on surface of forging thick head is formed by impact of elastic and plastic body, the joint on surface of forging head dispersed to thin is formed by impact of fragile body. Mechanism remain bending moment of plastic joint is decided by uninstall property of concrete damnified. The equivalent strength of break up work of pole compacted and pressed to burst is considered by the model of destroy cycle "uninstall dispersed and whole load" of pole head. Forth, configuration of building is made up of a lot of girder, poles and walls. When that is changed to mechanism, the plastic joints formed by girders and poles can be abstracted to system without tree structure by many multi-body. As these same span beams and same layer columns are paralleled to move each other and tediously surplus constraints are existed, so multi-body without tree structure of parallel beam and column can be simplified to dynamic body of equal effect with one freedom degree. Therefore, building mechanism can be great simplified to process number bodies without tree structure, till ultimate to 1~3 bodies of equal effect and configuration bodies. However, the establishment of differential equations and its numerical integration are great simplified. While number bodies are decreased and relation between bodies is simplified, a part of dynamics equations of building toppling can be got approximate solution and even resolve solution, so that formula express of toppling process is convenient. Building mechanism multi-body is made up of parallel beams and columns, which is also a character. Finally, building multi-body excessively constrained is dispersed to non-complete discrete-body and before to completely dispersed or directionally is broken to complete discrete-body, which is another character. By this taken, building mechanism multi-body is combined by characteristic of building mechanism from universal and rifeness principles, that increases variable mass body, natural body, natural joint, natural dynamic topology, different characteristic joints and the system without tree structure made up of parallel beam and column are simplified to single opening chain multi-body, so as Multi-body System Dynamics has been abundant and developed.

4.2 Building Multi-body Dynamics Equations

Roberson-Wittenburg method is one of the common methods to establish the multi rigid body system dynamics equation of 3 dimensional. Building bodies system with tree structure is mostly plane single opened chain system. The high-rise buildings such as chimney, shear walls and frame and cylinder structure are single opened chain system, which dynamics equation of rooted system is [5, 8]

$$\{[B]^T \text{diag}[m][B]+[C]^T \text{diag}[J][C]\}[\ddot{q}]+$$
$$\{[B]^T \text{diag}[m][\dot{B}]+[C]^T \text{diag}[J][\dot{C}]\}[\dot{q}]-$$
$$\{[B]^T [F]+[C]^T [M]\} = 0 \qquad (1)$$

where $[q] = [q_1, q_2, ..., q_f]^T$ ——the independent generalized coordinates of single opened chain body system as the f degrees of freedom for the n body, in the system (component) the $r_{\mu s}$ position vector and the φ_μ angular position of μ mass centre (or any point) of bodies are $[q]$ function.

$$\left. \begin{array}{l} r_{su} = r_{su}(q_1, q_2, ..., q_f) \\ \varphi_u = \varphi_u(q_1, q_2, ..., q_f) \end{array} \right\} \quad \mu = (1, 2, ..., n) \qquad (2)$$

From the type (2) velocity and angular velocity matrix form can be obtained by time derivative.

$$\left. \begin{array}{l} [\dot{r}_s] = [B][\dot{q}] \\ [\dot{\varphi}] = [C][\dot{q}] \end{array} \right\} \qquad (3)$$

where $[\dot{q}] = [\dot{q}_1, \dot{q}_2, ..., \dot{q}_f]^T$;
$[\dot{r}_s] = [\dot{r}_{s1}, \dot{r}_{s2}, ..., \dot{r}_{sn}]^T$;
$[\dot{\varphi}] = [\dot{\varphi}_1, \dot{\varphi}_2, ..., \dot{\varphi}_n]^T$;
$[B] = \text{jacobian}(r_{su}, q)$;
$[C] = \text{jacobian}(\varphi, q)$;
n ——the system in number;
jacobian ——q Jacobi matrix.

From type (3) to be time derivative the matrix form acceleration and angular acceleration can be obtained.

$$\left. \begin{array}{l} [\ddot{r}_s] = [B][\ddot{q}]+[\dot{B}][\dot{q}] = [\ddot{r}_s(\ddot{q})]+[\ddot{r}_s(\dot{q})] \\ [\ddot{\varphi}] = [C][\ddot{q}]+[\dot{C}][\dot{q}] = [\ddot{\varphi}(\ddot{q})]+[\ddot{\varphi}(\dot{q})] \end{array} \right\} \qquad (4)$$

where $[\ddot{r}_s] = [\ddot{r}_{s1}, \ddot{r}_{s2}, ..., \ddot{r}_{sn}]^T$;
$[\ddot{\varphi}] = [\ddot{\varphi}_1, \ddot{\varphi}_2, ..., \ddot{\varphi}_n]^T$;
$[\ddot{q}] = [\ddot{q}_1, \ddot{q}_2, ..., \ddot{q}_f]^T$;
$[\dot{B}] = \dfrac{\mathrm{d}}{\mathrm{d}t}(\text{jacobian}(r_{su}, q))$;
$[\dot{C}] = \dfrac{\mathrm{d}}{\mathrm{d}t}(\text{jacobian}(\varphi, q))$.

where diag$[m]$ ——the body mass diagonal matrix ;
diag$[J]$ ——diagonal matrix of inertia principal moment of the bodies;
$[F]$ ——the main vector matrix of the out force of bodies, in building collapse mechanism body the out force is of only gravity in the gravitational field;
$[M]$ ——the out main force moment and principal moment matrix of resistance,

that is all the resistance moment matrix of out or inner the plastic joints of apiece ends of building collapse mechanism bodies, the plastic joint moment of out and inner joints should be added.

Computed positive expression (1) kinetic equation, numerical simulation of the motion of n bodies posture of apiece topologies is gotten. The dynamic equation is inversely calculated, that calculates interactional force between of bodies to determine dispersing body. Therefore, the analytic theory of arbitrary multi fold chimney, shear wall and frame demolished by controlled blasting is founded. With different n, f and specific $[B]$, $[C]$ in the type (1), the computer will make the symbolic computation, automatic modeling, the concrete dynamics equation of different topology. The general formula (1) can only be generally solved numerically. Because the type (1) for the implicit two order ordinary differential equations, it is translated into the explicit two order equations and is available to solve by the numerical order 4 and 5 stage Runge-Kutta method.

4.3 Solution of Kinetic Equation

In general, building collapse dynamics (multi body dynamics) equation is two order ordinary differential equations. So far, the demolition area has only individual incomplete analytical solution in the explosion, but no approximate solution, but has only the numerical solutions. The blasting demolition collapse of building is essentially in the finite field gravity field (less than $\pi/2$), with the body movement of variable topologies, angle finite domain of topology collapse movement and sometimes is only up to 0.3, follows the mechanical laws of the gravity field. Therefore, in this section integrable power series main item will be instead of angle function, so that the approximate dynamic equations can be formed, by which analytical solution can be gotten, or the approximate solution can be obtained from the summed up numerical solution, Taking a typical demolition parameters, case-based reasoning is implemented, the application domain of approximate solutions is constructed and the corresponding error is determined, in order to solve and simulate the motion of body. The following case, list the main kinetic equation, others are shown in the literature [5].

4.3.1 Single Span, Multi Span Cantilever Frame Beam and Continuous Beam Inclining with Rotation

This is a classic problem of blasting demolition China circles, which is called moment method by cross disintegration [9], also known as "implosion" in foreign. Used the building itself gravity bending moment and shear force by the delay of successive initiation on cross, the beams in the horizontal direction by cross will be fractured. The first cross is first initiated, and then successive is initiated after cross, and because of fault movement before the cross the initial speed and initial displacement of the cross are obtained. The approximate solution of the kinetics equation is simple, shown in literature [5].

4.3.2 The Toppling of High-Rise Building One-Way

4.3.2.1 The Initial One-Way Toppling

While chimney, shear wall, frame and frame shear wall structure of high-rise buildings are initially directional falling, as shown in Fig.1, from the formula (1) multi-body system with $n=1$, $f=1$ dynamic equations of vertical layout of the bottom end of the plastic joint axis rooted system can be obtained for

$$J_b \frac{d^2 q}{dt^2} = Pr_c \sin q - M_b \cos(q/2) \quad (5)$$

where P——the monomer weight force, kN, $P = mg$;

m——mass of monomer, 10^3 kg;

r_c——distance from mass center to bottom end joint, m;

J_b——the inertia moment of monomers on the bottom joint axis, 10^3 kg·m^2;

M_b——resistance moment for the bottom of the plastic joint (residual moment of mechanism), kN·m;

q——angle between line from mass center to bottom joint and a vertical line, (°).

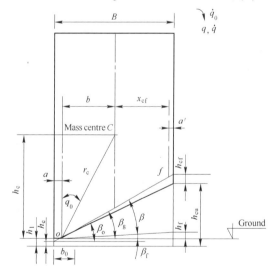

Fig.1 Gesture of shearing building toppling down to ground

When blasting demolition the initial condition is

$$t = 0, \quad q = q_0, \quad \dot{q} = \dot{q}_0 \quad (6)$$

It can get the numerical solution and the analytical solution $\dot{q}$, t of the type (5) [4]:

$$\dot{q} = \sqrt{\frac{2Pr_c(\cos q_0 - \cos q)}{J_b} + \frac{4M_b[\sin(q_0/2) - \sin(q/2)]}{J_b} + \dot{q}_0^2}$$

(7)

$$t = \sqrt{J_b/mgr_c}(\ln(q - M_b/mgr_c +$$
$$\sqrt{2(m_0-1) - 2M_b q/mgr_c + \dot{q}^2}) - \ln(q_0 - M_b/mgr_c +$$
$$\sqrt{2(m_0-1) - 2M_b q_0/mgr_c + \dot{q}_0^2}))$$
(8)

where $m_0 = \dot{q}_0^2(J_b/2mgr_c) + \cos q_0 + 2M_b \sin(q_0/2)/mgr_c$

4.3.2.2 The Incision Closed into the Turning of Collapse on Ground

After the incision of frame, frame shear wall and silo structure are blasted and support columns as lower body fall backwards, building framework of whole structure up the incision layer as upper body along the column end hinge b subsequently is toppled forward, to form its movement of folding mechanism of the body $n=2$ and of $f=2$ degrees of freedom, as shown in Fig.2. When the incision is closed, the building is hit on ground and rotated (including Fig.1). According to the principle of momentum conservation, the speed after hitting on ground [5]

$$\dot{q}_f = [\dot{q}_c J_c + m_2 r_f(v_{cx} \cos \dot{q}_{rf} + v_{cy} \sin q_{rf})]/J_f \quad (9)$$

where r_f and q_{rf} —— respectively for the distance of building mass center C to the front toe f and angle between line of mass center C to the front toe f and a vertical line;

v_{cx} or v_{2cx} —— the horizontal velocity of building mass center;

v_{cy} or v_{2cy} —— vertical velocity of building mass center;

J_f —— the inertial moment at point f of building;

$\dot{q}_c$ —— otational speed of buildings before hitting on ground;

J_c —— nertia principal moment.

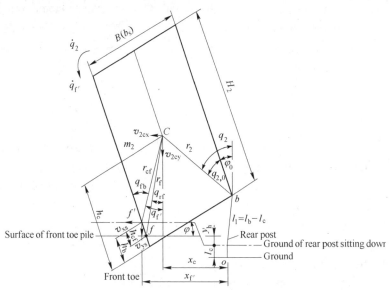

Fig.2 Gesture of frame front post toppling to ground (v_{cy} up for+)

Since building incision is closed, the toppling of monomers around front toe f of incline buildings or silo can be simplified as a body $n=1$ movement of single opened chain of a single degree q of freedom $f=1$, which dynamic equation is

$$J_f(d\dot{q}/dt) = mgr_f \sin q \quad (10)$$

Initial conditions: $t=0, q=q_f, \dot{q}=\dot{q}_f$ (11)

where q —— angle between r_f and a vertical line, (°).

Rotated the building, which mass center is raised and its potential energy is increased. If you do not take into account the anterior column collision damage, when the centroid distance do not exceed toe hitting location f distance, that is $x_c \leqslant x_f$.

Impact kinetic energy $T_f = J_f \dot{q}_f^2/2$;

$$T_f \geqslant w_f = r_f m_2 g(1 - \cos q_f)$$

Or energy ratio turning to guarantee rate

$$K_{to} = T_f / w_f \quad (12)$$

where w_f —— the centroid potential forward rotation energy increased.

Considered building overcome rolling resistance, reinforced bars of behind support, leave guarantee surplus overturned and calculation errors, when $K_{to} \geqslant 1.4 \sim 1.5$, architecture mechanism can be ensured to overturn. After incision of the frame and frame shear wall structure are closed, the drift in storey or collapse in span is satisfied by static and dynamic conditions, respectively according to the kinetic equations, and is collapsed [5, 10].

4.3.3 Building Layers Collapse with Mass Loss and Toppling

Crushed pillar of high-rise building incision, if sliding of building bottom is prevented by the basement, high-rise mass is lost and building is toppled with sitting downstairs from the fixed axis d on ground, that can be regarded as a single body model with variable mass [5], as shown in Fig.3.

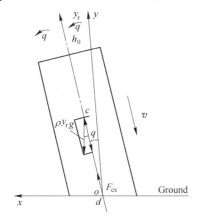

Fig.3 Toppling and mass dissipate of building

That dynamic equation [5] is
$$\sum Y_r = 0$$
$$d(\rho y_r v)/dt = -\rho y_r g \cos q + F_{cs} + y_r^2 \rho \dot{q}^2 /2 \quad (13)$$
$$\sum M_d = 0$$
$$d(J_b \dot{q})/dt = (\rho y_r^2 g \sin q)/2 + M \quad (14)$$

where ρ ——the line mass of building along the height, 10^3 kg/m;

F_{cs} ——bottom radial average resistance as the variable mass while sitting down, kN;

y_r ——the radial height of the building to radial coordinate origin, m;

J_b ——the inertia moment of the building for the point d, which is the function y_r, 10^3 kg·m²;

v ——radial velocity of building sitting, m/s;

M ——the bending moment at the bottom, $M = 0$.

Type (13), (14) initial conditions:
$t = 0$, $v = v_0 = v_{y0} \cos q_0$, $q = q_0$, $\dot{q} = \dot{q}_0$, $y_r = h_0$,
$$J_b = J_{d0} \quad (15)$$

where h_0 ——the building high above incision, m;

J_{d0} ——the inertia moment of building to the medial axis of the upper margin on incision, 10^3 kg·m²;

v_{y0} ——the speed of building to hit on ground when incision closed highness h_p.

Consolidated the differential equation to have integral factor, its analytical solution [5] is
$$v = -[F_{cs}(1-h_0^2/y_r^2)/\rho + (2/3)g(h_0^3 \cos q_0 / y_r^2 - y_r \cos q) + (y^2\dot{q}^2 - h_0^4 \dot{q}_0^2 / y_r^2)/4 + v_{0s}^2 h_0^2 / y_r^2]^{\frac{1}{2}} \quad (16)$$

When buildings collapse in situ, i.e. $q_0 = 0$, $q = 0$, $\dot{q}_0 = 0$, $\dot{q} = 0$, $v_{0s} = v_{y0}$.

Type (16) is as the analytical solution, when the collapse and toppling, by which, $q_r = q_0 (q_0 \neq 0)$, $\dot{q} = \dot{q}_0$, the error caused is aligned by v_{0s}, formula (16) is the approximate solution [5] according to v_{0s} table adjustment.

4.3.4 Building Binary Toppling

A considerable number of buildings collapsed could be included toppling in the same and double direction of double bodies, such as a whole section shear wall double bodies with bidirectional folding toppling. The dynamic equations are obtained by multi-body dynamics equations of single opened chain (1) - when $n = 2$, $f = 2$, while the toppling direction φ_1 and φ_2 are corresponding to the opposite or the same direction of double bodies [5].

4.4 Variable Topologcal Multibody-discretebody Global Simulation

The contact of each object of multi body system is called topological configuration, referred to as the topology [8]. Collapsed building structure by blasting, forming mechanism is the system of topology changes, collectively referred to as the variable topology Multibody-Discretebody system. Programmed the topology according to the time sequence, motion results as before topology is as the initial conditions of adjacent topological latter [5, 11~14], the whole process of building collapse can be calculated and simulated.

4.5 Similar Properties of the Kinetic Equation

In order to avoid repeated observations of blasting demolition, repeated experiments for models and computational duplicate values of engineering case, we can apply cases, that is shown the correct numerical results proven by experiments, according to the similar properties of the kinetic equation, since the solutions and the derived quantity with non dimensional regularization the criterion formula or rendering are established, as to popularize into practice. The Dimensionless matrix

$$[B_n] = [B]/B$$
$$[\dot{B}_n] = [\dot{B}]/[\dot{q}]^T$$
$$[F_n] = [F]/m_2$$

where m_2 ——the mass of main structure.

$$\text{diag}[m_n] = \text{diag}[m]/m_2$$
$$\text{diag}[J_n] = \text{diag}[J]/(m_2 B^2)$$
$$[\dot{C}_n] = [\dot{C}]/[\dot{q}]^T$$

introduced into equation (1).

where B (no B in $[B]$)——the width of main frame body;

H ——highness sitting down of main frame body;

$[M] \approx 0$;

$[\ddot{q}] = [\dot{q} d\dot{q}/dq]$.

introduced into equation (1), the implicit express $[\dot{q}]$ have been obtained by integration, when $[m]/m_2$ is constant, obviously $[\dot{q}]$ is independent of m_2, and $[\dot{q}B^{0.5}]$ is independent of B. Thus, overturning guarantee rate K_{to} in type (12) has nothing to do with m_2 and B, but only is mainly dependent on $\eta_h = H/B$, $\eta_c = h_c/B$ and $\lambda = h_{cud}/B$. In the majority of demolition collapsed type K_{to} is dependent on $k_j = J_c/J_{cs}$ less, than that before.

where h_c, h_{cud} —— respectively highness of gravity center and highness of incision of main frame body;

J_c —— the principal inertia of main body;

J_{cs} —— the principal inertia computed by main solid of graphics mass with uniformly distributed.

$$J_{cs} = (H^2 + B^2)m_2/12$$

Thereby removed independent variables, raced slightly less off variables by error, protruded main variables, to reduce the number of equational variables, guidelines chart $\lambda - \eta_c, \lambda - \eta_h$ with universal significance can be established, shown in section 5.1. Building k_j is about $0.75 \sim 1.28$. With application of criteria rendering the individual collapse pattern has errors larger, than ordinary. The computing modeling with method "reality transformed from picture" can be simply applied to compute k_j, then use the k_j interpolation on standard nomogram. When $[M] \neq 0$, dimensionless parameters $M/(m_2 g l_o)$ can be introduce, l_o as the span length, and establish guidelines for rendering, shown in section 5.1. In addition, because $[\dot{q}B^{0.5}]$ is nothing to do with B, but similarity ratio of incision closure time is reasoning to $B^{0.5}$, shown in section 5.4. The building blasting pile and recoil can be determined according to the corresponding standard nomograms.

4.6 Mechanics of Materials of Component Damage

In force balance the equations on normal section of reinforced concrete member, the probability theory of material strength is adopted, inclusively limit resistance (bending moment) of structural safety is calculated by the material standard strength, structural instability (bending moment) is calculated by the random strength mean [5] of concrete and the residual resistance (bending moment) of mechanism is calculated by the damage strength. Residual resistance should be considered by pull-out strength of reinforced bar from adhesion among concrete and reinforced bar bending coefficient[5] $\alpha_t = \cos(\theta_p/2)$, in type θ_p of plastic joint rotation angle, protective layer of concrete in compressive zone is fallen off and the equivalent rectangular stress coefficient α_c [5] is adopted in damaged compression zone. Among that the ductile failure of reinforced concrete column designed by "Ductile (tensile) failure", $\alpha_c = 0.8 \sim 0.9$[5]; sup-

port of incision of reinforced concrete chimney for more "fragile (compression) failure", $\alpha_c = 0.25 \sim 0.4$[5]. The calculative method above has been demonstrated proper by field observation and engineering practice in literature [5].

4.7 Impact Dynamics of Concrete Member

Due to hit the impact stress wave on the ground and floor beams is constantly reflected, stress superposition can be nearly doubled, first the crushing failure zone and the plastic joints are formed at both ends of the column [5] in layer. As frame column after blasting of a single row holes is hit, the plastic zone is not generally formed in the middle part of post.

When the reinforced concrete wall and column support member are shocked, the asymptotic steady state compression, non steady state compression and combination compression will be experienced. Failure work of non steady compression per m³ column should be graphics area surrounded by reinforced concrete stress-strain curve. The crushing work of asymptotic steady state shock compression, according to the cycle "full pressure loading - escape unloading"[5] formed by the crushing mechanism on the column ends in broken pieces, can be deduced the relational formula of fixed mass with steady state of impact crushing. The principle above has been demonstrated correct by field engineering observation in literature [5]. The equivalent strength [5] of the total compression process is

$$\sigma_e = k_d k_{sc} k_l \sigma_{cs} \quad (17)$$

where k_d —— dynamic load coefficient of concrete strength, take 1.1;

k_{sc} —— equivalent coefficient of contact section area of collapse column with column section, measured 0.74[5];

k_l —— collapse column ratio k_l = (column high-residual column length) / column height, measured 0.58[5]; When the back column hits the ground, it sits down less than highness in storey, only stable compression asymptotical in one storey, $k_l = 1$.

Because building mass above incision is larger $25 \sim 50$ times than column own mass, when building hit, high of column failure must be considered to increase the potential energy consumption of the whereabouts of building compression process. The principle above has been demonstrated correct by experiment in literature [15]. To crush height of bottom column h_f

$$h_f = m_2 v_{2,0}^2 / [2(s_1 \eta - m_2 g)] \quad (18)$$

where η —— work ratio required to crush unit volume m³ of reinforced concrete, 10^3 kJ; $\eta = \sigma_e$. Supporting strength of collapse of high-rise building layer is directly calculated[5] by measure value and the type (16).

5 DEMOLITION TECHNOLOGY OF INCISION CONTROL OF MULTI-BODY DYNAMICS(MBDC)

5.1 Building Cutting Parameters of Incision

In different demolition of buildings, after blasting of incision, according to equation (1) or (13) toppling, incision closing, architecture model and similar nature of the dynamics equation, buildings collapse can be judged by the dimensionless criterion, shown in Fig.4.

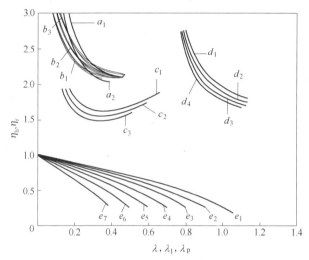

Fig.4 Relation of incision parameter ($\lambda(\lambda_1,\lambda_p) - \eta_h(\eta_r)$)

Standard curve $\lambda - \eta_h$ in the figure the a family for the single incision of shear wall and frame shear wall structure of buildings turn over in overall,

$$\lambda = h_{cud} / B$$
$$\eta_h = H / B$$

where h_{cud} ——the incision highness since sitting down for building;

B ——building width, shown in Fig.1;

H ——building highness, after building sitting down;

a ——distance of the neutral axis joint O on support bottom from behind side of wall.

where K_{to} =1.5, a_1 in order to the line of assure, but a_2 is high risk lines, thus for K_{to} =1.1.

The calculative method above has been proven right by engineering example in literature [5].

Lines the b family are criterion curve [16] $\lambda_1 - \eta_h$ of lower incision while toppling of building two incision to same direction, when the upper incision is first closed the combinatorial monomer to overturn is formed and K_{to} =1.5. Among them

$$\lambda_1 = h_{cud1} / B$$

$$\lambda_2 = h_{cud2} / B$$

where h_{cud1} ——lower incision highness since sitting down;

h_{cud2} ——upper incision highness; λ_2 =0.22;

l_1 ——lower body highness since sitting down;

l_j ——the dimensionless highness of lower body, $l_j = l_1 / B$, l_j of b_1, b_2 and b_3 are respectively 0.92, 1.14 and 1.36.

The upper, centre and lower dimensionless curve within the family are a/B =0.066, 0.0825 and 0.11. The calculative method has been proven accurate by engineering example in literature [17].

Lines the c family are criterion curve [10] for span collapse of frame and wall frame buildings. Among them $n_c = 4$, K_{to} =1.9, l_o of curve c_1, c_2 and c_3 are respectively 3.2 m, 3.5 m and 3.8 m, thus K_t is correspondently 1.5352, 1.4036 and 1.2928. The method has been proven correct and exact by 4 practical engineering examples in literature [10].

where n_c ——cross number of framework, $n_c = 4$;

l_o ——span length (or the average span), l_o of line C_1, line C_2 and line C_3 is respectively for 3.2 m, 3.5 m and 3.8 m;

K_t ——dimensionless moment, K_t of line C_1, line C_2 and line C_3 is respectively for 1.5352, 1.4036 and 1.2928;

$$K_t = K_{to} M_{dh} / (mgl_o)$$
$$M_{dh} = M_f + M_r + M_q$$

M_f ——mechanism residual moment initial value of front-end[5] on each span;

M_r ——mechanism residual moment initial value of back-end[5] on each span;

M_q ——mechanism shear moment initial value [5] of the wall;

m ——mass of frame building, 10^3 kg.

Line the d family shows, that after blasting of the frame building incision upper head of behind single column is formed to plastic joint b, so that 2 bodies and 2 degree of freedom motion are formed, see Fig.2. Each of line d_1, d_2 line d_3 and line d_4 is as criterion curve $\lambda - \eta_h$ overturned after incision closure of the frame, the corresponding principal inertia ratio $k_j = J_{c2} / J_{co}$ is 0.75, 1, 1.25, 1.5, among that K_{to} =1.3~1.5. Its calculative method has been demonstrated correct by 3 engineering practice examples in literature [5].

Line the e family is for criterion curve sitting down of high-rise building collapse in multi storey with mass loss (shown in Fig.3), the dimensionless parameters:

$$\lambda_p = h_p / h_o$$

$$\eta_r = y_r / h_o$$

where h_p ——incision highness;

y_r ——building highness on the heap by blasting;

h_o ——building highness above incision.

As formula (16) $q=0$, $\dot{q}=0$, $v=0$, the dimensionless equation can be derived.

$$\lambda_p = F_P(1-\eta_r^2)/2 - (1-\eta_r^3)/3 \quad (19)$$

$$F_p = K_{to}F_{sp}, F_{sp} = S_1\sigma_{cg}/(gh_o\rho)$$

where S_1 ——support section of building lower floor;

σ_{cg} ——the equivalent support dynamic strength[5], C30 concrete column is $\sigma_{cg} = 13.645$ MPa;

K_{to} ——the guarantee rate of collapse, 1.1.

From graph, line e_1, line e_2, line e_3, line e_4, line e_5, line e_6 and line e_7 are respectively as the criterion equation curve of F_{sp}=2.6, 2.4, 2.2, 2, 1.8, 1.6, 1.4. The reader can be actually happening and interpolated from criterion curve of Fig.4. The model in 4.3.3 part has been demonstrated correct by engineering observation in literature [5], so that the calculative method above can also be right.

If the coordinate point $(\lambda(\lambda_1, \lambda_p) \cdot \eta_h(\eta_r))$ of demolition project is upper right from standard curve, the building will be collapsed. Standard curve of building of reverse double incision toppling is shown in the literature [18] and has been proven accurate by 3 practical examples in literature [18]. The η_h of toppling of one-way incision composed by raise incision, down incision and shallow incision is less than the standard curve a about 0.25~0.35, that has been proven right by engineering, as shown in the literature[19]. The demolition permitted high risk can be reduced K_{to}. With the cases of demolition increased, it can ensure success, according to actuality the guarantee rate can be reduced, correspondingly can decrease η_h and increase η_r.

5.2 Front Width and Height of Blasting Pile

The mutual relation among bodies and its ground connection in blasting pile can be defined as accumulation. Original deposit can be instantly formed by hitting of structure on ground. With structure fell, splash off stone, individual member forward and flip the secondary accumulation can be formed. Primary accumulation can be divided into five types, class Ⅰ—the overall turnover accumulation, class Ⅱ—accumulation by collapse in span, class Ⅲ—accumulation by drift in storey, class Ⅳ—granular deposit, class Ⅴ—integral incline rather than collapse.

According to building structure, the blasting demolition method and incision criterion the type of blasting pile can be basically determined. And the deposit shape formula is deduced by dynamics equation and its similarity. Such as shear wall structure, frame shear wall, frame building to be not generated collapse in span and drift in storey, single incision blasting with toppling class Ⅰ accumulation is formed. The front width of pile is

$$L_{gf1} = dx_s + H_1 - h_{cu} - h_{cf} \quad (20)$$

Highness of pile is

$$h_{gf1} = B \quad (21)$$

where dx_s ——distance between detonation before and space hit to ground after blasting of front column; dx_s is positive, hit to the front place of anterior column. But dx_s is negative, hit to place behind anterior column;

H_1 ——building highness, m;

h_{cu} ——front cutting-edge highness of building incision, m;

h_{cf} ——breaking highness of anterior column when building hit to ground, it can be set from cutting top of incision to floor beam bottom of upper storey;

B ——building width, m.

$$dx_s = dx'_s B$$

where dx'_s ——the dimensionless distance between front column before detonation and site of hit of front column to ground after blasting, that of framework is selected by interpolation from the standard rendering with mass height width ratio η_c of building and λ_1 about incision, correspondingly shown in Fig.5.

Tab.1 The η_c in Fig.5

Curve name	c_1	c_2	c_3	c_4	c_5	c_6	c_7
η_c	0.85	0.8	0.75	0.7	0.65	0.6	0.55

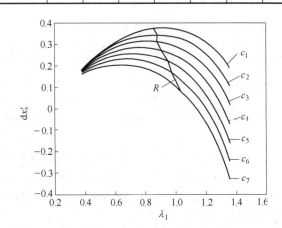

Fig.5 Relation between $\lambda_1 - dx'_s$ ($k_j=1$; R: the whole turn for right, $K_{to}=1.4\sim1.6$)

When $\lambda_1=1$, $\eta_c=0.85\sim0.55$ (the building height to width ratio $\eta_h = 2\eta_c + \lambda_1$, λ_1 is ratio between height of post column joint b and floor width since sitting down), corresponding $dx'_s=0.37\sim0.1$.

That of shear wall and frame shear wall structure is by [5]

$$dx_s = (B-a)(1/\cos\beta - 1) \quad (22)$$

where β —— incision angle, (°).

The accumulation of class III, class II blasting pile are formed by collapse of frame shear wall and frame building with drift side in storey and collapse in span, shown in the literature[5], dx_s is with the above. The calculation principle of blasting pile of building with multiple incisions and of building sitting down is similar above [20], as shown in the literature [5]. The part calculative method has been proven usable by 18 engineering examples in literature[5].

5.3 The Recoil of Building

In the directional toppling, rear pillars support with upper structure forward at the same time is also accompanied by some backward movement, that maximum is recoil value. Obviously, the recoil is prerequisite to rear column support. When the supporting column loses support capacity, its upper structure is fallen, which is referred to as the sitting down. From its formation mechanism the recoil is divided into mechanism recoil, backward of column root and columns support inverted [5]. Back width along blasting pile is the end result of recoil.

When building forward toppling, bottom joint O of shear wall and frame shear wall structure is due to backward, and under the radial stress and tangential thrust force, followed out the dynamic equation (5), building overcomes the friction and slides backwards along ground, shown in Fig.1. Because wall or column after blasting are linked by reinforced bars in column and larger friction force, sliding distance of rear wall or column will not be larger than the height of blasting height h_e and impact crushing h_p of wall (column). When the incision of frame and frame shear structure is cut by blasting, the folding mechanism motion with 2 degrees of freedom formed body, and mechanism recoil of joint b at the same time is formed[5], shown in Fig.2.

Recoil value of mechanism is
$$dx_b = dx_b' B$$

where dx_b' —— the dimensionless mechanism recoil value, this dx_b' of framework is selected by interpolation according to the ratio η_c between building height to width of mass centre from the criterion chart(figure) 6 (Corresponding $\eta_h = 2\eta_c + \lambda_1$);

λ_1 —— the ratio(l_1/B), which is of b joint height to building width after the column sitting down.

Standard rendering is decided by dynamic equations, derived quantity and similarity. Calculating principle of factory bent back is similar above, shown in literature [5]. The part calculative method has been proven usable by engineering example[5].

Tab.2 The η_c value of curve in Fig.6

Curve name	c_1	c_2	c_3	c_4	c_5	c_6	c_7
η_c	0.85	0.8	0.75	0.7	0.65	0.6	0.55

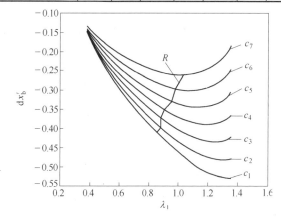

Fig.6 Relation between λ_1 - dx_b' (k_j=1; R: the whole turn for correspondingly the right, K_{to}=1.4~1.6)

5.4 Detonating Sequence and Time

Computed equation (1) forward and inverse, as multi incisions it can be demonstrated that cut incision upstream initiation sequence is sitting down more easily than to cut incision by downstream initiation[5]. Incision initiation interval can be referenced single incision closure time t as type (7).

$$t = t'\sqrt{B(1+k_j/3)} \quad (23)$$

where t' —— similar time function, $t' = t''/\sqrt{g}$, s/m$^{0.5}$;

$t''(\lambda, \eta_c)$ —— dimensionless time.

From formula (7) to chart (figure) 7, and after front cross breaking down of building the t' can be selected by interpolation of the average ratio η_c of height to width of building mass centre and λ. The part calculative method has been proven right by engineering example[5].

5.5 The Building Sits Down

When rear column in incision blasting is only remained, the pseudo static load of frame column with toppling is reduced, but heavy load of frame column more than 3 spans has be surpassed single post instable strength in multilayer, single slender column is easy to be broken and building can be sat down[5]. After blasting the bottom column, buildings sitting down and rear column crushing, its crushing height increases as long as the blasted and is accelerated growingly. As framework not more 3 spans, these rear impact collapse height is calculated by formula (18) and the theorem of momentum, but dimensionless sitting down λ_{hp} and $h_{pf}/(h_e - 0.2)$ of the single post is not correlation in order to analysis of the kinetic equation.

where h_{pf} —— the larger highness sitting down ($k_j =$ 1.25), m;
h_e —— blasting highness, m.

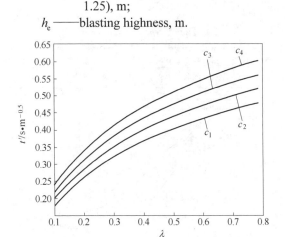

Fig.7　Relation between $\lambda - t'$ (line c_1, c_2, c_3 and c_4 are whole turn η_c 1.0, 1.1, 1.2 and 1.3)

As $h_e \leqslant 0.7$ m, $h_{pf} \approx \lambda_{hp}(h_e - 0.2)$ is for the quasi linear in order to analysis of the kinetic equation, thus

$$\lambda_{hp} = C_p F_p + C_h \quad (24)$$
$$F_p = S_1 \sigma_e / P$$

where σ_e —— the equivalent strength of rear column in layer, shown in type (17), C20 concrete 15.6 MPa; among them $k_1 = 1$;

P —— building the weight force, 10^6 N;

S_1 —— single post cross-sectional area, m².

C_p, C_h are selected by interpolation B, l_1 in Fig.8. The conditions are for the principal inertia ratio $k_j = 1.25$ with maximum sitting down h_{pf} of rear column, initiation delay of rear column root as 0.5 s and joint b of reinforced concrete column to be had bending moment. As $h_e > 0.7$ m, h_{pf} is nonlinear, formula (24) does not be suited. The shallow incision blasting can be adopted in structure of frame and the frame shear wall more than 2 spans, with double rear column, the height of collapse is basically linear relationship with high of blasting, sitting down is decreased and recoil are rarely.

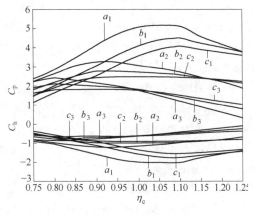

Fig.8　Relation among η_c, C_p and C_h

Tab.3　The value of B and l_1 in Fig.8

Curve name	a_1	a_2	a_3	b_1	b_2	b_3	c_1	c_2	c_3
l_1/m	6.1	6.1	6.1	7.6	7.6	7.6	9.1	9.1	9.1
B/m	10.7	12.0	13.3	10.7	12.0	13.3	10.7	12.0	13.3

5.6　Chimney

The chimney incision parameters of highness below 150 m according to toppling static mechanism of incision meshing are decided[5] and are proven usable by observational engineering examples[5]. According to dynamics above its parameters related motion are determined. The collapse mechanism of cockle surface of canister and incision parameters of reinforced concrete chimney of the thin wall and tall above highness 210 m are proven right by observational engineering example and are shown as the literature [21].

6　CONCLUSIONS

Building demolition by blasting is described by the Buildings Toppling Dynamics contained the Multibody-discretebody Dynamics, that mechanism is clear, correct and consistent with the practice. By its variable topology dynamic equations the analytical solution, the approximate solution and numerical solution are obtained. The toppling global process of simulation for demolition can be made by its topology combination. With similarity and its regularization of dynamic equation, the incision size, collapse shape of blasting pile, recoil, sitting down, detonating sequence and segmentation time of its each demolition model in various architectural structure can be easily expressed by dimensionless. In order to select demolition reasonable toppling mode, demolition measurements and cutting parameters of incision of high and large architectural structure the new complete theory and widely suitable, simple and practical algorithm are provided, which applies the foundation for accurate control of the blasting demolition. Although some parameters need also to measure, but, in the range of professional disciplines, the key technology of building of the civic high and large architectural structure demolished by blasting is basically completed, that is the Multi Body Dynamics Incision Control Demolition Technology (MBDC) [5]. It has been proven by field observations and engineering practical examples that Buildings Toppling Dynamics, which includes to the Multibody-discretebody Dynamics, is correct and MBDC is usable and accurate. Thus, Multibody-discretebody Dynamics is new technology of blasting demolition and the imminent new strategic opportunities of technological revolution of demolition blasting, which shows the new stage in the development of science and technology.

REFERENCES

[1] Lu Wenbo. Model lost steady of steel bars framework demolished by blasting[J]. *Blasting*, 1992, 19(2): 31~35 (in Chinese).

[2] Zhang Qi, Wu Feng, Wang Xiaolin. Model of finite element in frame demolished by blasting to lose stability[J]. *China Engineering Science*, 2005, 10(3): 22~28 (in Chinese).

[3] Jin Jiliang. The parameters of blasting cut for directional collapsing of high rise buildings and towering structure[J]. *Engineering Blasting*, 2003, 9(3):1~6 (in Chinese).

[4] Wei Xiaolin, Fu Jianqu, Li Zhanjun. Analysis of multibody-discretebody dynamics and its applying to building demolition by blasting// Collectanea of discourse abstract of CCTAM2007 (Down).Beijing: *China Mechanics Academy Office*, 2007: 690 (in Chinese).

[5] Wei Xiaolin. *Building toppling dynamics (Multibody-discretebody dynamics) and technology of blasting demolition exact control[M]*. Guangzhou: Zhong Shan University Press, 2011 (in Chinese).

[6] Wang Xuguang. Foreword//Corpus of China 125 field science and engineering technology forum "New materiel composed by explosion and key science and engineering technology of high effective and safe blasting". Beijing: Metallurgical Industry Press, 2011 (in Chinese).

[7] Hong Jiazhen. *Computational dynamics of multibody systems* [M]. Beijing: Higher Education Press, 1999 (in Chinese).

[8] Yang Tingli. *Basic theory of mechanical system-structure, kinematics, dynamics*[M]. Beijing: Mechanical Industry Press, 1996 (in Chinese).

[9] Yang Renguang, Shi Jiayu. *Blasting demolition of buildings* [M]. Beijing: China Architecture Industry Press, 1985 (in Chinese).

[10] Wei Xiaolin. Cutting parameters of toppling frame building demolished with collapse in beam span by blasting[J]. *Engineering Blasting*, 2013, 19(5):1~7 (in Chinese).

[11] Hong Jiazhen, Ni Chunbi. Whole simulation of varying topological multi-body dynamics[M]. *Mechanics Transaction*, 1996, 28(5): 633~636 (in Chinese).

[12] Wei Xiaolin, Zheng Bingxu, Fu Jianqiu. Mechanical analysis and numerical simulation of folding dumping of reinforced concrete chimney[C]// China blasting technology. Beijing: *Metallurgical Industry Press*, 2004: 564~471 (in Chinese).

[13] Wei Xiaolin, Fu Jianqiu, Wang Xuguang. Numerical modeling of demolition blasting of frame structure by varying-topological multibody dynamics[C]//New Development on Engineering Blasting. Beijing: *Metallurgical Industry Press*, 2007: 333~339(in Chinese).

[14] Zheng Bingxu, Wei Xiaolin. Modeling studies of high-rise structure demolition blasting with multi-folding sequences [C]// New Development on Engineering Blasting. Beijing: *Metallurgical Industry Press*, 2007: 236~332(in Chinese).

[15] Du Xingwen, Song Hongwei. Cylindrical shell shock dynamics and crashworthiness design[M]. Beijing: *Science Press*, 2005 (in Chinese).

[16] Wei Xiaolin. Cutting parameter of building demolished by blasting with two cutting[C]//New Techniques in China III. Beijing: *Metallurgical Industry Press*, 2012: 576~580 (in Chinese).

[17] Li Chao, Wu Jianfeng, Zhu Yanhui, Li Zhongfei. Directional folded explosive demolition of 10-storey framed building [J]. *Blasting*, 2013, 30(1):79~81, 89(in Chinese).

[18] Wei Xiaolin. Cutting coefficient of building demolished by blasting with two cutting in reverse direction[J]. *Blasting*, 2013, 30(4) :99~103 (in Chinese).

[19] Zhao Hongyu, Wang Shouxiang, Liu Yunjian, Xie Zenglin. Demolition of a high framed shear-wall structure by controlled blasting[J]. *Blasting*, 2008, 25(2):53~56(in Chinese).

[20] Wei Xiaolin. Muckpile of building explosive demolition with many cutting[J]. *Blasting*, 2012, 29(3):15~19 (in Chinese).

[21] Wei Xiaolin. Mechanism of press burst with cockle of supporting and cut parameters of high and thin wall chimney[J]. *Blasting*, 2012, 29(3):75~78 (in Chinese).

Mathematical Model of Seismic Impact on Operating Underground Gas Pipelines

A.P. Gospodarikov

(*National mineral resources university "University of mines", Russia*)

ABSTRACT: Conducting explosive works near the operating underground gas pipeline is closely connected with a problem of seismosafe parameters determination for the drilling-and-blasting works allowing both their effective maintaining, and ensuring safety of the pipeline.

In existing normative documents still there are no direct instructions by seismosafe mass of a charge determination when conducting explosive works near the operating underground gas pipeline.

The mathematical model for joint dynamic deformation of system soil-pipeline taking into account their contact interaction nonlinear effects is offered on the basis of researches of seismoblast waves with underground gas pipelines interaction processes. Within the offered mathematical model the complex of computing programs in the Matlab environment, based on effective numerical methods application using perfectly matched layer (pml) as absorbing boundary conditions is developed.

On the basis of the developed algorithms and computing programs the calculation method for a charge's seismosafe mass while conducting explosive works near operating underground gas pipelines is offered.

KEYWORDS: underground gas pipeline; seismic influences; mathematical model; numerical algorithm; calculation procedure

Blasting operations near existing underground pipeline is closely related to the problem of determining shown seismosafe blasting parameters, compelling stakeholders as their effective management and preservation of the pipeline.

Existing regulations are still no direct guidance on the definition of the seismic safety of the charge mass for blasting operations near existing underground pipeline.

In this paper, based on studies of interaction processes of seismic waves from underground pipelines, a mathematical model of joint dynamic deformation of the soil-pipe, taking into account the nonlinear effects of the contact interaction. Under the proposed mathematical model developed complex computational programs in Matlab, based on the use of efficient numerical methods involving perfectly coordinated layers (pml) as absorbing boundary conditions.

On the basis of the developed algorithms and computer programs proposed method of calculating the weight of the seismic safety of charge for blasting operations near existing underground pipelines.

Different formulations of research problems of dynamic deformation of mechanical systems using the finite element method (FEM), as experience has shown, the most effective is the use of the variational principle of virtual displacements in conjunction with the Lagrange d'Alembert principle.

To derive the equations of motion allowing the soil mass used functional Π - the total energy of the mechanical system:

$$\Pi = \int_V \frac{1}{2}\sigma_{ij}\varepsilon_{ij}dV - \int_V \left(F_{Vi}u_i - \rho\frac{\partial^2 u_i}{\partial t^2}u_i\right)dV - \int_S F_{Si}u_i dS \quad (1)$$

where
- σ_{ij} —— the components of the stress tensor, Pa;
- ε_{ij} —— components of the strain tensor;
- u_i —— elastic displacement of the particle in the direction of a deformable body coordinate axis x_i, m;
- F_{Vi} —— mass forces, H;
- F_{Si} —— surface forces defined on the outer surface S, the volume of ambient space V, H;
- ρ —— density of the deformed ones, kg/m³;
- t —— time, s.

Model describing the dynamic deformation of the pipeline uses the theory of thin shells involving Kirchhoff-Liave hypotheses.

Due to their fundamental differences in the nature of the deformation, the soils can be described by two different equations of state: the equation of state for soft soil (sand) and the equation of state for rocky soils (granite). For soft soils known equation of state adopted academician S.S. Grigorian (Fig.1).

Compression diagram has an initial linear-elastic portion

Corresponding author E-mail: Gospodarikov@spmi.ru

($p < p_e$), where the soil deformations are purely elastic; p_e and ρ_e – pressure limit values p and density ρ on the elastic deformation of the medium section;. The pressure and density in the range of pressure variation $-p_e$ to p_e is linear form:

$$p - p_0 = C^2(\rho - \rho_0), \quad \text{at } p_0 \leqslant p \leqslant p_e \qquad (2)$$
$$C = (p_e - p_0)/(\rho_e - \rho_0)$$

where p_0 —— the initial pressure in the soil, Pa;
ρ_0 —— the initial density in the soil, kg/m³;
C —— numerical constant.

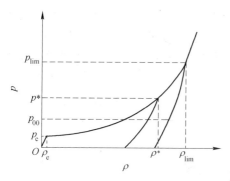

Fig.1 Diagram of compression in the model academician S.S

In the field of elastoplastic shock adiabatic approximation is used, based on the additive approximation:

$$\frac{\rho}{\rho_0} = \varepsilon_e + \left\{\sum_{i=1}^{3} \alpha_i [1 + \gamma_i(p - p_e)/E_i]^{-\frac{1}{\gamma_i}}\right\}^{-1} \qquad (3)$$

where $\alpha_1, \alpha_2, \alpha_3$ —— bulk concentration of free porosity (entrapped air) of liquid (water) and the solid component;
E_1, E_2, E_3 —— bulk modulus of the components, Pa;
$\gamma_1, \gamma_2, \gamma_3$ —— numerical coefficients.

Unloading medium from reached in the process of loading the maximum density of the soil ρ^* describes a two-tier sloping:

$$p = \begin{cases} p^* + C_1^2(\rho - \rho^*) & p > p_{00} \\ p_{00} + C_2^2(\rho - p_{00}) & p < p_{00} \end{cases} \qquad (4)$$

where, C_1 and C_2 are the speed of sound, determined by the slope to the axis ρ first and second links in a broken line, respectively; $p_{00} = p^*/\gamma_c$ characterizes the ratio of lengths of polylines; (ρ_{00}, p_{00}) is the break point in the coordinates of the unloading curve (ρ, p); parameter γ_c sets ratio C_1 to C_2 at $\rho^* = \rho_{\lim}$; $\rho_{\lim}$ is density limit ρ, after which unloading curve coincides with the load curve. Shear deformation elastoplastic soil environment in the state described by the model of exact linear elastic medium. In the area of elastoplastic deformation dependence $p(\rho)$ described by the scheme Prandtl-Reuss with Mises plasticity condition [1,2].

As the equations of state for rocky soils accepted model of linear viscoelastic medium:

$$\begin{cases} \frac{1}{2}S_{ij} = G\varepsilon'_{ij} + \eta\dot{\varepsilon}'_{ij} \\ \sigma_S = 3K\theta \end{cases} \qquad (5)$$

where S_{ij} —— components of the stress tensor deviator, Pa;
G —— shear modulus, Pa;
ε'_{ij} —— deviatoric components of the strain tensor;
$\dot{\varepsilon}'_{ij}$ —— deviatoric components of the strain rate tensor;
σ_S —— mean stress, $\sigma_S = \frac{1}{3}\sigma_{ii}$, Pa;
θ —— Average strain, $\theta = \frac{1}{3}\varepsilon_{ii}$;
K —— compression modulus, Pa;
η —— coefficient of dynamic viscosity, Pa·s.

As for solving the system of governing equations describing the mathematical model adopted to be effective, as a rule, numerical methods, then this system of equations must be converted to pre-form corresponding to the applicable numerical method. In this paper, for the direct integration of the system governing equations, the following numerical methods: sampling for spatial variables used finite element method; for sampling time variable used finite-difference schemes.

As a result, numerical implementation of FEM computational domain V is partitioned into N finite element, volume V_e each. Inside the required elements of the stress function σ_{ij} and displacements u_i approximated by linear interpolation polynomials, matrices recorded in the shape functions for stress $[N_G]$ and displacements $[N_U]$ relative stress values $\{G_e\}$ and displacements $\{U_e\}$ elements at the nodes. Next, for each finite element composed discrete analogs of the original differential equations for the nodal values. After summing these equations over all the elements in the grid finite element equations of the dynamics of the mechanical system are transferred to the matrix differential equation with respect to the global vector of unknowns $\{G\}$, $\{U\}$. The result is a finite-element analogues of all the necessary governing equations of the mathematical model of the joint dynamic deformation of the soil and pipeline [1,2].

To exclude the reflection of waves from a conditional loop is used kind of absorbing boundary conditions - a method of introducing perfectly coordinated layers (pml) on fictitious boundaries of the computational domain. Pml-layers are the specific absorption band which catch and loosen them in the incoming wave, thus preventing reflection in the computational domain. Constitutive equations describing the motion of stress waves in the pml in the plane formulation, have the form:

$$\left([F_2^e]\frac{\partial}{\partial x_1}+[F_1^e]\frac{\partial}{\partial x_2}\right)\{\sigma\}+\left([F_2^p]\frac{\partial}{\partial x_1}+[F_1^p]\frac{\partial}{\partial x_1}\right)\int_0^t\{\sigma\}\mathrm{d}t$$
$$=\rho f_m\{\ddot{u}\}+\rho c_s f_c\{\dot{u}\}+\rho c_s^2 f_k\{u\} \tag{6}$$

$$f_m\{\dot{\sigma}\}+c_s f_c\{\sigma\}+c_s^2 f_k\int_0^t\{\sigma\}\mathrm{d}t=[D]\left\{\frac{2\mu}{\omega_0}\left([F_1^e]^T\frac{\partial}{\partial x_2}+\right.\right.$$
$$\left.[F_2^e]^T\frac{\partial}{\partial x_1}\right)\{\ddot{u}\}+\left[\left([F_1^e]^T+\frac{2\mu}{\omega_0}[F_1^p]^T\right)\frac{\partial}{\partial x_2}+\right.$$
$$\left.\left.\left([F_2^e]^T+\frac{2\mu}{\omega_0}[F_2^p]^T\right)\frac{\partial}{\partial x_1}\right]\{\dot{u}\}+\left([F_1^p]^T\frac{\partial}{\partial x_2}+[F_2^p]^T\frac{\partial}{\partial x_1}\right)\{u\}\right\} \tag{7}$$

where $\{\sigma\}$ —— the stress vector, Pa;

$\{u\}$ —— the displacement vector, m;

$[D]$ —— matrix of elastic characteristics of the environment;

μ —— damping coefficient, Pa·s;

ω_0 —— the reference frequency, c^{-1};

c_s —— shear wave velocity, m/s; points are marked on top of the time derivatives.

matrix $[F_1^e],[F_2^e],[F_1^p],[F_2^p]$ have the form:

$$[F_1^e]=\begin{bmatrix}0 & 0 & 1+f_1^e \\ 0 & 1+f_1^e & 0\end{bmatrix};\ [F_1^p]=\begin{bmatrix}0 & 0 & c_s f_1^p \\ c_s f_1^p & 0 & 0\end{bmatrix}$$
$$[F_2^e]=\begin{bmatrix}1+f_2^e & 0 & 0 \\ 0 & 0 & 1+f_2^e\end{bmatrix};\ [F_2^p]=\begin{bmatrix}c_s f_2^p & 0 & 0 \\ 0 & 0 & c_s f_2^p\end{bmatrix} \tag{8}$$

where f_i^e —— damping profile for waves propagating perpendicular to coordinate extended $\tilde{x}_i$;

f_i^p —— damping profile for waves propagating parallel to coordinate extended $\tilde{x}_i$.

Expressions f_m, f_c and f_k determined by the formulas:

$$\begin{aligned}f_m &= (1+f_1^e)(1+f_2^e)\\ f_c &= (1+f_1^e)f_2^p+(1+f_2^e)f_1^p\\ f_k &= f_1^p f_2^p\end{aligned} \tag{9}$$

The developed numerical model [2, 3] obtained solving the problem of propagation of seismic waves in an infinite medium without first pml-layers and then using the latter (Fig.2).

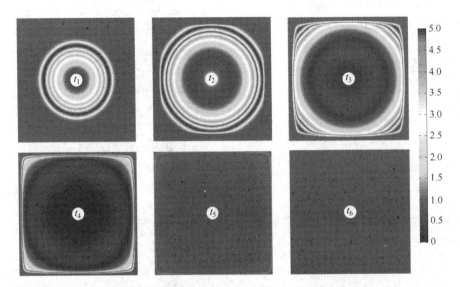

Fig.2 Distribution of spherical waves in an infinite medium using pml-layers: t_i—time, s
($t_1 = 5$ ms; $t_2 = 10$ ms; $t_3 = 15$ ms; $t_4 = 20$ ms; $t_5 = 25$ ms; $t_6 = 30$ ms)

As seen from the results of numerical experiment, application technology pml-layers can effectively absorb the incident waves. To jointly solve equations of the dynamics of soil and pipeline necessary to ensure that the conjugation of the contacting surfaces between the pipe and the surrounding soil. In this paper we used for this purpose finite element contact interface designs (Fig.3).

The magnitude of the energy derivative of the contact interaction, which determines schaya a load of contact forces, including forces prior conjugation is used in the formation of a global force vector. At the same time, the formation of

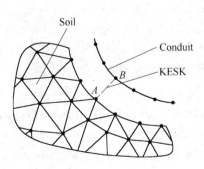

Fig.3 Schematic diagram KESK

A, B—the contextual nodes finite element

this vector is accompanied by the necessary modification in the global stiffness matrix caused by the use of the penalty function method. Iterative process for solving the nonlinear problem concerning organized minimize residual value (inequality) contact displacement vector in the global system considered deformed configuration.

For the numerical integration of the equations of dynamics finite element model in this paper the method of direct time integration (scheme of Habolt).

REFERENCES

[1] Berenger J. A perfectly matched layer for the absorption of electromagnetic waves, Journal of Computational Physics, 1994, 114 (2): 185~200, C.52.

[2] Gospodarikov A P, Gorokhov N L Evaluation of the strength of the pipeline located in the ground for blasting operations // Notes Mining Institute. ie 195. SPb, 2012. - S. 89~94.

[3] Gospodarikov A P, Gorokhov N L. Dynamic analysis of seismic effects on pipelines // Notes Mining Institute. ie 193. SPb, 2011.-S. 318~321.

Numerical Analysis for Blasting Demolition of the Smaller Height-Width Ratio Framed Structure

LIU Wei, YAN Shilong, HU Kunlun, LI Hongwei

(*School of Chemistry Engineering, Anhui University of Science and Technology*, *Huainan, Anhui, China*)

ABSTRACT: The blasting cut parameters has important influence on building blasting demotion. Combined with project, the numerical model of a small height-width ration and 12 layers framed structure demolition due to blasting was built by ANSYS/LS-DYNA. The structure was built with BEAM and SHELL element, the reinforced concrete was simplified as homogeneous isotropic and modeled with piecewise linear plasticity material. The ground model was built with solid element and rigid material. The demolition effect of different blasting cuts were simulated and analyzed, the simulation and the real results showed that: the height of blasting cut is only necessary but not sufficient condition for building collapse, the reasonable pre-demolition and time-delay should be combined with the blasting height to achieve ideal blasting effects.

KEYWORDS: framed structure; blasting demolition; cut parameters; numerical analysis

1 PREFACE

Directional collapse is one of the most common controlled demolition methods of buildings. The principle is to use triangle blasting cut and control detonating sequence to let the building rotate about a certain axis and then collapse[1~4]. For the buildings with little height-width ratio, its center of gravity is not easy to move out after the blasting cut formed and collapse with difficulty, so the blasting cut should design accurately and reasonably for this type of building. The demolition project was discussed in terms of cut height, delay time and preprocessing by means of simulation combined with engineering project.

2 INTRODUCTION OF DIRECTIONAL BLASTING DEMOLITION

Directional demolition project usually use the blasting cuts such as Fig.1, where the height of blasting cuts should satisfy two criterias: (1) the bearing columns were unstable; (2) center of gravity was moved out. For the first criteria, depth study were conducted by scholars at home and abroad such as the classical theory of bucking of compression bar and small-scale steel frame model while empirical formula was usually adopted in the practice. For the second, some calculation was needed and Jin ji-liang made it out by geometry theory, the frame structure was assumed to be plane and the height of center of gravity is H, height of blasting cut is h, span is L, as shown in fig1. From which, when the height of blasting cut equal $H/2$ the structure get the maximum eccentricity, the height of blasting cut is in the range as follows:

$$(H - \sqrt{H^2 - 2L^2})/2 \leqslant h \leqslant H/2 \qquad (1)$$

The formula above was true only when $H \geqslant \sqrt{2}L$.

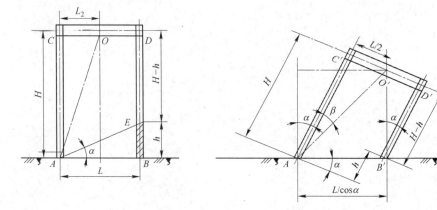

Fig.1 Sketch of directional blasting project

Corresponding author E-mail: wliu@aust.edu.cn

But buildings with smaller height-width ratio cannot satisfy $H \geqslant \sqrt{2}L$, terms of height of blasting cut, sequence and time-delay of detonation should be overall considered when carrying out a directional blasting demolition, means of "cut beams and columns" should be adopted to achieve ideal collapse effect if necessary.

3 PROJECT CASE

3.1 General Situation

A twelve-storied frame building with 52 m in length, 32 m in width and 39 m in height was to be demolished by blasting, the first floor was 6 m in height and others were 3 m, there were 48 columns in every floor with dimensions of 1.4 m×0.8 m, 1.2 m×1.2 m, the beam with dimension of 0.8 m×0.4 m, the floors are 110mm thick and the concrete grade in C30, shown as Fig. 2.

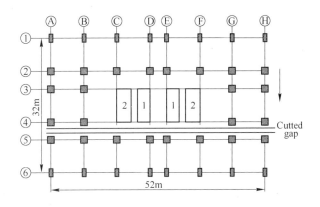

Fig.2   Plane sketch

3.2 Design of Blasting Project

The building wouldn't move the center of gravity out when the cut formed for its smaller ratio of height to width, according to analysis above. Higher blasting cut and long time-delay were employed, the details as follows:

(1) The height of blasting cut is up to 23 m, including 7 floors.

(2) Half-second detonators were adopted and reasonably allocated in the blasting cut.

(3) Beams along with collapse direction were blasted from 1F to 7F.

(4) Cut 20 cm width gaps from 1F to 12F, between column 4 and column 5, details were shown in Fig.2 and Fig.3.

3.3 Finite Element Modeling

ANSYS/LS-DYNA was adopted to build 3D model of the frame building[5~8], beams and columns were modeled with BEAM elements and floors were modeled with SHELL elements. There were no walls in this building. The reinforced concrete was simplified as homogeneous isotropic and modeled with piecewise linear plasticity material. The ground was built with SOLID element and rigid material.

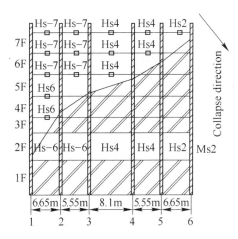

Fig.3   Sketch of blasting cut

4 ANALYSE ON BLASTING DEMOLITION AND SIMULATION

4.1 Effect of Blasting Demolition and Simulation

There were two different behavior of the building along the cutted gap under the gravity and time-delay, the front part fall as free and the back part rotated and inclined, the duration took about 6s and eventually disjointed well. The simulation presented the whole process from the cuts formation to the whole building collapse and the results were in good accordance with that of reality in terms of collapse time, collapse length and stack height, the effect of blast and simulation were shown in Fig.4.

4.2 Height of Blasting Cut on the Effect of Demolition

Another three blasting demolition projects were simulated and compared to explore the height on the effect of demolition. The height of the triangle cuts were 9m(up to 2F), 15m(up to 4F) and 23m(up to 7F) respectively   and the cut was detonated without time-delay ,columns in the cut were destroyed at the same time and parts outside the cut were not blasted. The results showed that:

(1) When blasting with the height of 23m, the stack height is minimum and the recoil distance is longer.

(2) When blasting with the height of 15m, the center of gravity of the building can not move out, but the dropping behavior cause failure in the joints of beams and columns, the building was transformed from determinate structure to statically indeterminate structure and leading to collapse by gravity and free falling.

(3) Smaller blasting height (9 m) would not cause the building collapse but incline.

(4) Measures of employing Time-delay detonating net work, cutting gap and increase detonating position can reduce collapse length and recoil distance.

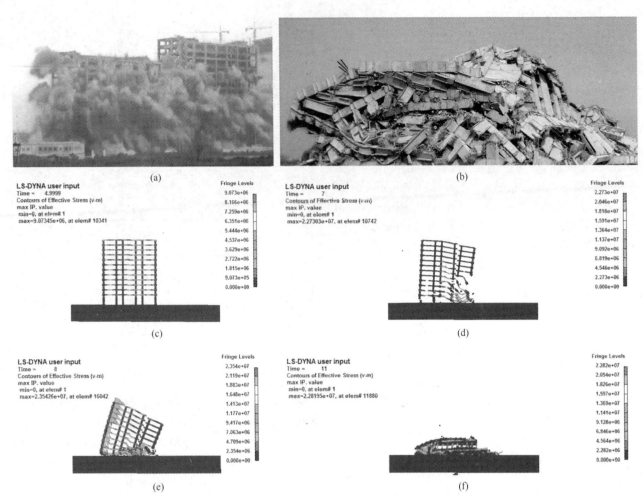

Fig.4　Effect of demolition and simulation

(a) moment of the collapse; (b) demolition effect; (c) *t*=0s; (d) *t*=2s; (e) *t*=4s; (f) *t*=6s

Tab.1　Result of blasting and simulation

	Blasting height/m	Stack height/m	Collapse length/m	Recoil distance/m	Remarks
Project case	23	13.0	45.0	4.5	
	23	11.5	46.0	4.5	Time-delay detonating
	23	11.5	59.6	7.2	No time-delay
Simulation	15	11.2	56.5	2.5	Time-delay detonating
	15	21.0	56.0	0.0	No time-delay
	9		Didn't collapse but incline		Time-delay detonating
	9		Didn't collapse but incline		No time-delay

5　CONCLUSIONS

The result of the project case and simulation showed that:

(1) In demolition buildings of smaller height-width ratio, height of blasting cut and time-delay should be designed reasonably, certain height is necessary for building falling, inclining and impact to ground; reasonable time-delay not only contribute the powerful bending but the more efficient in large deformation and disjointing of the parts so as to make the building collapse even thoroughly.

(2) Blasting height and time-delay have significance influence on the building'blasting and safety, the interaction mechanism of them should be further investigate.

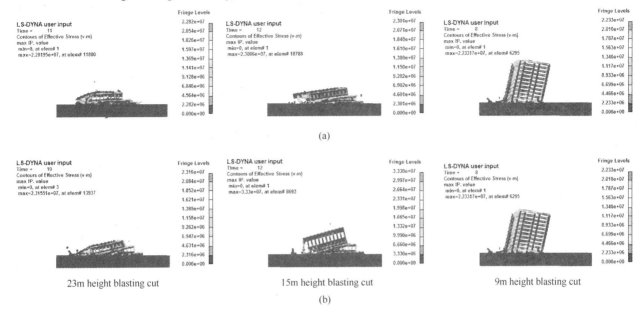

Fig.5 Simulation result of different blasting project
(a) detonating with time-delay; (b) detonating without time-delay

REFERENCES

[1] Jin jiliang. The parameters of blasting cut for directional collapsing of highrise buildings and towering structures[J]. *Engineering blasting*, 2003, 9(4): 1～6.

[2] Peng jianhua, Chen shouru, Wang Weiwei. Study on blasting cut height for directional collapse of frame buildings[J]. *Engineering Blasting*, 2005, 11(2): 44～46.

[3] Zhou Fengyi, Zhu Jinhua, Qiu Jinfen. Discussion on height of blasting cut in blasting demolition of buildings[J]. Ming Technology, 2007, 7(3): 91～92.

[4] Lou Jianwu, Zhang Weixin, Fang Xiang. Application of low-height blasting cut in directional blasting demolition of frame buildings[J]. *Engineering blasting*, 2002, 8(3): 21～24.

[5] Liu Wei, Liu Lilei, Simulation of frame structure demolition due to blasting[J]. *Engineering blasting*, 2008, 14(1):12～15.

[6] Georgios Michaloudis, Steffen Mattern, Karl Schweizerhof. Computer Simulation for Building Implosion Using LS-DYNA. High Performance Computing in Science and Engineering '10, 2011: 519～528.

[7] Steffen Mattern, Gunther Blankenhorn, Karl Schweizerhof. Computer-Aided Destruction of Complex Structures by Blasting. High Performance Computing in Science and Engineering'06, 2007: 449～457.

[8] Gunther Blankenhorn, Steffen Mattern, Karl Schweizerhof. Controlled building collapse analysis and validation. LS-DYNA Anwenderforum, Frankenthal, 2007.

Explosive Cratering to Finite Size of Earth Structure

Yumin Li[1], Ettore Contestabile[2], Abass Braimah[3], Bert von Rosen[2]

(1. 9132-0663 Quebec Inc., Montreal, Canada; 2. Canadian Explosives Research Laboratory, Ottawa, Canada; 3. Ottawa University, Ottawa, Canada)

ABSTRACT: Earthen structures such as embankment dams are vulnerable to explosive cratering. Software codes such as ConWep are often used to calculate expected crater sizes based on soil type and explosive mass. However most of these codes, including ConWep, have been developed for explosives placed on or buried in soil that is typically considered an infinitely large flat surface, in contrast with the geometry of finite size of earth structures. Therefore, some error in the estimation of crater dimensions of finite earth structure by ConWep is expected. In an effort to understand the boundary effect of finite earth structure on explosively formed crater parameters such as crater size, diameter and depth, comparative experiments were performed on a scale model of a concerned earth structure (embankment dam) and on a infinitely flat surface of the same soil. Due to presence of two side slopes, the average crater diameter caused by surface explosions on the dam crest is greater than that of the craters formed on the infinitely flat ground. The effect of the slope was reflected in the disproportionately large diameter of the craters in the direction perpendicular to the length of the dam. These experiments on infinitely flat surface show that ConWep overestimates both the crater diameter and to a greater extent the crater depth. Therefore, ConWep's ability to correctly predict crater sizes for the range of explosive masses used in these experiments is questionable. The suggestion is that an alternative to using ConWep to calculate the crater depth would be to use the 1990 version of ConWep to calculate the crater radius and then a crater index of 4 to determine the crater depth. This approach results in a crater depth within 10% of the experimental results.

KEYWORDS: explosion crater; embankment dam; earth structure; surface explosion; experiment

1 INTRODUCTION

Earthen embankment dams are vulnerable to cratering, either internal cratering due to charges that have been sunk into the dam using existing vertical or horizontal shafts, or surface cratering which would occur when a charge is placed on the crest or slope of the dam. Internal cratering can result in internal seepage and internal erosion, which may eventually lead to dam failure (there is a similar concern with animal burrows and sink holes). Surface cratering can lead to overtopping and subsequent dam failure (Nistor et al, 2005). Software codes such as ConWep code (Hyde, 1990 & 1997) and Holisapple code (Halisapple, 1980 & 2003) can be used to calculate expected crater sizes based on soil type and explosive mass. However most of these codes have been developed for explosives placed on or buried in soil that is typically considered an infinitely large flat surface, in contrast with the geometry of embankment dams. Therefore, some error in the estimation of crater dimensions of embankment dams by these codes is expected, where the burden is reduced by the presence of the embankment slopes. Fig.1 is a sketch showing such a scenario where ConWep would overestimate the burden.

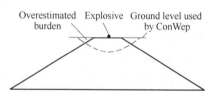

Fig.1 Reduction in burden due to the sloping sides of the dam as compared to the burden used to generate crater sizes in ConWep

In an effort to understand the effect of the sloped sides of an embankment dam on explosively formed crater parameters such as crater shape, diameter and depth, comparative experiments were performed on a scale model of a typical embankment dam and on a flat surface of the same soil. The test results provide an insight into the damage arising from the detonation of explosive charges and allow the estimation of the crater size due to the detonation of explosives on the crest of a full-scale dam.

2 EMBANKMENT DAM MODEL

Surveying equipment was used to mark the boundaries of

Corresponding author E-mail: yuminli205@gmail.com

the model embankment dam on a flat grassy area to obtain the profile shown in Fig.2. Such a profile provides opportunities for tests to be performed on the embankment dam (indicated on the left) and for comparison, a semi-infinite, undisturbed flat surface (indicated on the right).

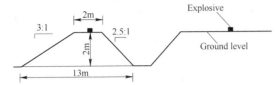

Fig.2 Cross-section of model dam and infinite flat surface

The model dam was constructed by digging the sandy native soil of the test site away from ground level, leaving the native undisturbed topsoil to form the crest of the model dam while the slopes of the model were formed by digging away soil. Wooden jigs were constructed and used in conjunction with a level to assist the excavator operator to accurately form the required slopes. The model dam, which represents a one-fifth scale of a 10 m high embankment dam with 3:1 upstream slope and 2.5:1 downstream slope, is 33 m long and 2 m wide at the crest and 13 m wide at the base.

3 EXPERIMENT

ANFO (Ammonium Nitrate and Fuel Oil) was used as the surface explosive charge (the charge placed on the crest of the dam) because it was assumed to be the explosive of choice for a saboteur or terrorist since it is relatively inexpensive and readily available. As listed in Tab.1, a total of ten tests were conducted to investigate the effect of explosions on the model dam. Four tests were performed with the charge placed on the centreline of the crest. These tests were compared to four similar tests that were performed on the native flat land (with a little vegetation). In order to determine the effect of the vegetation on the crater size, the tests performed on flat undisturbed native soil were compared to two additional tests where 50 kg explosive charges were placed on flat ground from which either the vegetation or both the vegetation and the topsoil had been removed.

Tab.1 Test matrix

Test No.	Explosive/kg	Explosive Type	Location	Vegetation
1	10	ANFO	Flat ground	yes
2	10	ANFO	Dam crest centre	yes
3	25	ANFO	Flat ground	yes
4	25	ANFO	Dam crest centre	yes
5	50	ANFO	Flat ground	no
6	50	ANFO	Flat ground	yes
7	50	ANFO	Dam crest centre	yes
8	50	ANFO	Flat surface without top soil	no
9	100	ANFO	Flat ground	yes
10	100	ANFO	Dam crest centre	yes

4 CRATER MEASUREMENT

The resulting craters were comprised of two concentric craters, a small, deep inner crater concentric with a large, shallow outer crater as shown in Fig.3. Measurements were taken across the diameter along four axes at 45°, while the depth profile was determined by measuring the depths at five different locations along each axis as shown in Fig.3. Section 1-1' (North-South) in Fig.3 refers to the orientation along the crest of the dam while Section 3'-3 (West-East) refers to the orientation perpendicular to the crest of the dam. Point D designates the centre of the crater, while points C and E are at the edge of the inner crater and points A and G are at the edge of larger outer crater. Two additional measurements were taken at points B (equidistant between A and C) and F (equidistant between E and G) to better characterize the crater profile.

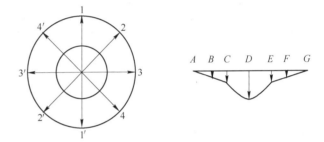

Fig.3 Sketch showing measurement locations of crater diameter and depth

5 RESULTS AND DISCUSSION

5.1 Effect of the Embankment Slopes on Crater Profiles

Fig.4 shows the average crater diameters from surface explosions on the native flat ground and on the dam crest. The average crater diameter was determined from the average of the measured diameters in the four orientations. The average crater diameter on the crest of the dam for the tests with the larger explosive charges was typically larger than that of the craters formed with equal masses of explosive on the infinite flat ground. In such situations, the crater diameter on the crest of the dam was larger than the width of dam crest. Fig.4 indicates that the presence of the dam slope

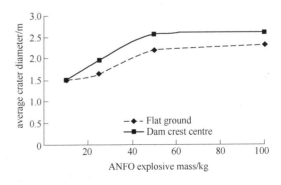

Fig.4 Average crater diameter

increases the overall blasting efficiency. However, when the crater diameter is substantially smaller than the dam width, as in the case with the 10 kg of ANFO, the overall effect of the dam profile is diminished.

From blasting theory (Persson, 1994), the throwing of fragments is predominantly in the direction of the minimum burden. The presence of the dam slope reduces the burden and allows the blast energy to discharge the lower mass of burden toward the upstream and downstream sides of the dam. Therefore, it is expected that an elliptical crater will be formed by surface explosions on the dam crest, where the long axis of elliptical crater is across the dam crest while the short axis is along the dam crest centreline. Fig.5 shows a comparison of crater diameters along the length of and normal to the model dam. In general, the crater diameter toward the upstream and downstream faces of the dam is significantly larger than that along the length of the dam. On infinitely flat ground the explosion energy is distributed evenly in all directions parallel to the ground plane resulting in a roughly round crater as indicated by the results plotted in Fig.6. Within the measurement error, the magnitude of the crater diameters in two orientations shows no significant difference.

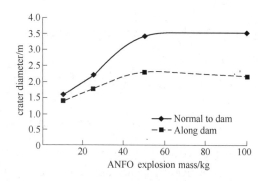

Fig.5 Crater size in two orientations for surface explosions on the centre of the dam crest

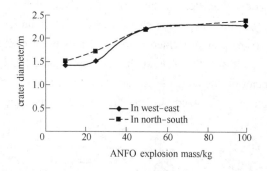

Fig.6 Crater sizes for two diameter orientations for surface explosion on native flat ground

To further demonstrate the effect of the dam slope on the crater dimensions, a comparison of crater diameters in two directions for surface explosion on the dam crest and on infinitely flat native ground is shown in Figs.7 and 8. Fig.7 shows that the crater diameters across the dam crest (i.e. in West-East orientation) are significantly larger than the corresponding diameters of the craters on the infinite flat native ground. On the other hand, Fig.8 indicates that the crater diameters along the length of the dam crest (i.e. in North-South orientation) are practically the same as the corresponding diameters of the craters on the infinite flat native ground.

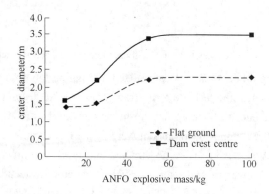

Fig.7 Comparison of normal-to-dam crater diameter with crater diameter on flat ground

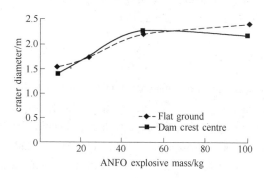

Fig.8 Comparison of crater diameter along the length of the dam with crater diameter on flat ground

Fig.9 shows the crater depths for the surface charge detonations on flat ground and on the dam crest. There is no significant difference in these two sets of crater depths. Therefore it can be concluded that the profile of the dam does not affect the explosion crater depth for this dam model size and profile and the explosive charge sizes used. This can be expected from the blasting theory since the presence of the dam slope does not alter the burden in vertical direction.

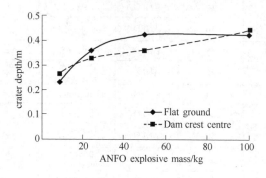

Fig.9 Crater depth for surface explosions on flat ground and dam crest

5.2 Comparison of Experimental Results with Codes

Fig.10 and Fig.11 show the crater depths and diameters resulting from the experiments and those predicted by ConWep code and Holsapple code in dry sand. From Fig.10, the crater depths predicted by Holsapple code in dry sand is in a good agreement with experimental data, while both ConWep codes greatly over-predict the crate depth by ANFO surface explosive.

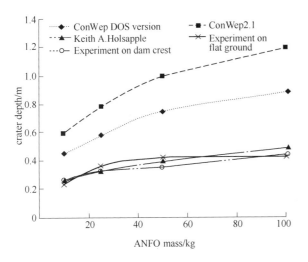

Fig.10 Crater depth of dry sand by surface explosion

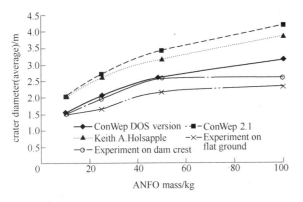

Fig.11 Crater diameter of dry sand by surface explosion

On the other hand, ConWep DOS version (Hyder, 1990) is better than ConWep 2.1 version (Hyder, 1997) and Holsapple code in predicting the crater diameter in dry sand as shown in Fig.11. The current windows version of ConWep gives the worst result in predicting both crater diameter and depth. This may be due to inappropriate assumption in this version that TNT equivalency of ANFO is 1.72 for cratering. The DOS version of ConWep uses a TNT equivalency of 0.82 and is more consistent with the test results than ConWep 2.1 version, particularly with respect to the crater diameter results. However, it should be kept in mind that this analysis is based on only a few tests with only small masses of explosives in dry sand.

5.3 Effect of Vegetation

Tab.2 lists the experimental results of the crater sizes for three topographies. No single test dominates the others, in either crater diameter or depth. This indicates that in these tests the vegetation did not play significant role in the formation of the resulting craters.

Tab.2 Crater size in different geographies

ANFO/kg	Topography	Crater Diameter/m	Crater Depth/m	Inner Crater Diameter/m
50	Natural flat ground with vegetation	2.17	0.423	1.08
50	Flat ground without top soil	2.22	0.33	0.97
50	Natural flat ground without vegetation	2.11	0.37	0.90

6 CONCLUSIONS

The objective of this test program was to determine the effect of the dam embankment on crater formation by surface charges and to make a comparison with the predicted values from software codes. Due to presence of dam slope, the average crater diameter caused by surface explosions on the dam crest is greater than that of the craters formed on the infinitely flat ground. The effect of the slope was reflected in the disproportionately large diameter of the craters in the direction perpendicular to the length of the dam.

These experiments show that ConWep 2.1 version overestimates both the crater diameter and to a greater extent the crater depth. ConWep DOS version is much better than ConWep 2.1 version in predicting crater diameter although it overpredicts the crater depth as well. Holsapple code is relatively good for predicting the crater depth.

Due to the fact that the dam material used in this preliminary model was an uncompacted native sandy soil which did not truly represent a typical embankment construction, further tests on a properly constructed dam with, possibly, a head of water, may be required.

REFERENCES

[1] Nistor I, Rennie C D, Contestabile E. 2005. Earth fill dam

breach failure due to explosion, *CERL report* 2005(20).

[2] Hyde D W. 1990. ConWep (DOS version), *U.S. Army Corps of Engineers*.

[3] Hyde D W. 1997. ConWep (window 2.1 version). *USAE Engineering Research & Development Centre*.

[4] Holsapple K A. 2003. Users Manual: Crater sizes from explosions or impact, *http://keith.aa.washington.edu/craterdata/scaling/usermanual.html*.

[5] Holsapple K A, Schmidt RM. 1980. On the scaling of crater dimensions 1. explosion processes. *J. Geophys. Res.* 85 (B12): 7247~7256.

[6] Persson P A, Holmberg R, Lee J. 1994. Rock Blasting and Explosives Engineering. *CRC Press*.

Numerical Simulation for Effect of Stemming in Blasting

WANG Huxin[1], WU Chunping[1], CUI Xinnan[2], Chen He[1]

(*1. Beijing General Research Institute of Mining & Metallurgy, Beijing, China;*
2. Beijing GXAK Technology Co., Ltd., Beijing, China)

ABSTRACT: Good stemming can improve the utilization ratio of blasting energy reduce explosive consumption. But the importance of stemming is usually neglected in operation. In this article 3 kinds of stemming including of non-stemming, stemming with tamping plug and full charging, were simulated to analyze the blasting effect, by using of the software ANSYS. It is shown that as to the others, blasting with tamping plug can get better result.

KEYWORDS: blasting; numerical simulation; stemming

1 INTRODUCTION

After charging explosives, the remaining space of blasthole usually be filled with a certain material. The materials are called tamping plug or stemming materials.

1.1 The Role of Stemming

1.1.1 Enhance the Blasting Effect

As well known, in blasting, rock is broken by the combined action of detonation wave and detonation gas. After initiation, the rock near the hole is crushed by detonation wave, meanwhile, the rock in the distance is fragmented by it. Then, detonation gas will expand the cracks of rocks and will throw the rocks out. Stemming can prevent detonation gas diffusing out of the hole soon, prolong the affecting time of detonation gas and transfers more energy of explosives into mechanical energy to crack rocks. Be better to blasting safety.

Good stemming can contribute to full reaction of explosive, reduce noxious gases, decrease fly-rocks in open pit and protect facilities from destroying. Stemming is benefit for preventing gas explosion and coal dust explosion in coal mining.

Furthermore, a certain additive added into ordinary water tamping plug can improves the atomization effect of explosive and speeds up sedimentation velocity of the blasting dust [1].

1.1.2 Misunderstanding to Stemming

Stemming is often ignored in blasting operation. To simplify charging, blastholes are usually charged fully or not stemmed. A wrong stemming will not only waste explosives but also do harmfully to blasting.

1.2 About ANSYS/LS-DYNA software

The finite element method (FEM) is a kind of simulation one which can simulate a certain loading condition and its response [2,3]. Many blocks called "elements" are used to simulate a real object. The model response is reflected by through all the elements response. The FEM is based on the solid flow variation principle. The elements will be added some given boundary conditions, loading and material properties. The displacement, stress, strain, energy and other results will be computed by solving linear or nonlinear equations.

ANSYS/LS-DYNA software is used widely to analyze nonlinear problems such as impact, explosion and shock. It is powerful in preprocessor and postprocessor as same as the ANSYS software. It is also formidable in dynamic solution as same as the LS-DYNA Software.

2 SIMULATION FOR STEMMING

2.1 Modeling and Solving Control

In order to compare the different kind of charge, the full charging was equivalent to stemming with explosives. As shown in Fig.1, three kinds of models such as stemming with tamping plug, stemming with explosives (i.e. full charging) and stemming with air (i.e. non-stemming), were simulated.

The size of the 3 models are length×width×height = 1m × 0.03m × 1.1m. The diameter of the blastholes is 45mm. The length of the blastholes is 1m. The length of the stemming is 0.5m. All the explosives were detonated from the bottom of the blastholes.

To simplify simulation, the 1/4 of each model was established. The back, bottom and left side were set to be non reflecting boundary as an infinite rock mass. The front side was the symmetry along the x axis. The right side was the symmetry along the y axis. The top side is free face.

It was 140964 elements in total. The size of elements was increased gradually along the radial direction. The size of explosives elements and blastholes one is 5mm. The calculation time was set for 2ms. The total calculation was set for 100 steps. Each step is 20 μs.

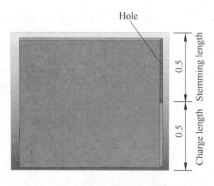

Fig.1 Model of the rock and blasthole

2.2 Parameters of Materials

In LS-DYNA module of ANSYS software, elastic-plastic material constitutive models are often used to analysis rock material. For example, MAT_PLASTIC_KINEMATIC kinematic hardening plastic material, is suitable for the strength and strain rate related materials, while exhibiting plastic materials properties. The physical and mechanical parameters of rock are listed in Fig.1. In order to simplify simulation, the parameters of tamping plug are same to rock.

Tab.1 Physical and mechanical parameters of rock

Rock	Density /g·cm^{-3}	Elastic modulus /GPa	Compressive strength /MPa	Tensile strength /MPa	Poisson ratio	Shear modulus /GPa
Silicalite	2.80	1.18	81.30	5.49	0.18	0.50

The high explosive model, *MAT_HIGH_EXPLOSIVE_BURN, was used with JWL equation. The performance parameters and constants of JWL equation of the 2# rock explosive are listed in Tab.2 and Tab.3.

Tab.2 Performance parameters of the explosive

Explosives	Density/g·cm^{-3}	Detonating velocity/m·s^{-1}	Transmission distance/cm	Brisance/mm
2# rock explosive	1.2	3200	5	14

Tab.3 Constants of JWL equation

Explosives	A/GPa	B/GPa	R_1	R_2	ω
2# rock explosive	47.6	0.53	4.50	0.90	0.30

2.3 Comparison with Simulation Results

2.3.1 Central Elements of Models

On the basis of the distance to blastholes, 5 elements were chosen from the middle of each model. The x axis of the 5 elements increase progressively. The y and z axis keep the same value. But the number of elements of each model was different. The models were shown in Fig.2. The coordinates of central elements in each model were shown in Tab.4.

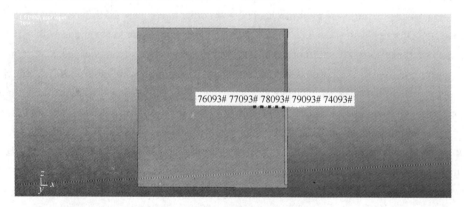

Fig.2 Central elements of the models

Tab.4 Coordinates of central elements of the models

Number of elements tamping plug	x-coordinate	Number of elements full charging	x-coordinate	Number of elements non-stemming	x-coordinate	Distance to blasthole/m
74093	−0.022	39893	−0.022	69393	−0.022	0.022
79093	−0.072	49893	−0.072	74293	−0.072	0.072
78093	−0.122	47893	−0.122	73293	−0.122	0.122
77093	−0.172	45893	−0.172	72293	−0.172	0.172
76093	−0.222	43893	−0.222	71293	−0.222	0.222

The pressure-time curves of elements in the middle of models, which are stemmed by tamping plug, full charging and non-stemming, are shown in Fig.3.

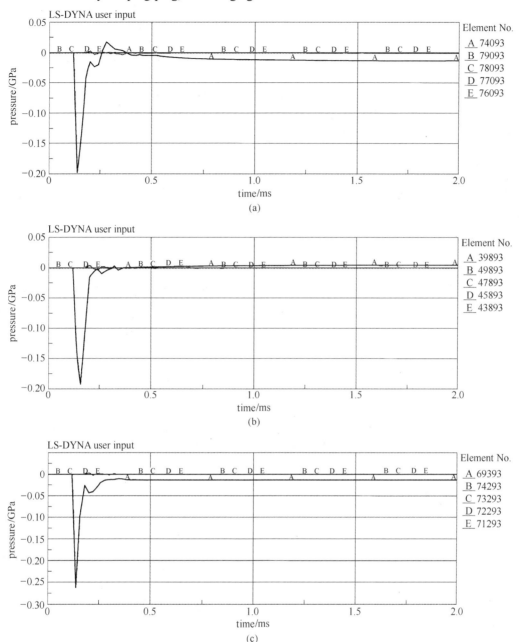

Fig.3 Pressure-time curves of elements in the middle of models
(a) tamping plug; (b) full charging; (c) non-stemming

It can be seen from Fig.3, the 3 models' pressures of wall elements which were far from blatholes 0.022m, are −0.2GPa, −0.19GPa and −0.26GPa respectively. Due to the shock wave's transmission to blasthole wall, the rocks surrounding the blastholes were crushing. Then, a crushing zone formatted with a lot of energy depleting. The peak pressure decreases sharply along the distance increasing.

The 4 pressure-time amplified curves of elements which were in middle-far zone are presented in Fig.4.

Seeing from Fig.4, along the distance increasing from 0.072m to 0.222m, the peak pressure of each element diminished exponentially generally. Because of the 3 curves are basically same, the middle stress of 3 models are almost same.

The peak pressure values are presented in Tab.5. A concept of "Relative distance" is conducted. It means a ratio between the distance of element to blasthole axis and the radius of the blasthole.

According to the Tab.5, peak pressures change slowly with the increase of relative distance. The pressure wave propagates in almost stable velocity and reached a same distance within almost same time.

The pressure wave within the same distance of propagation with time is almost as same, change propagation velocity.

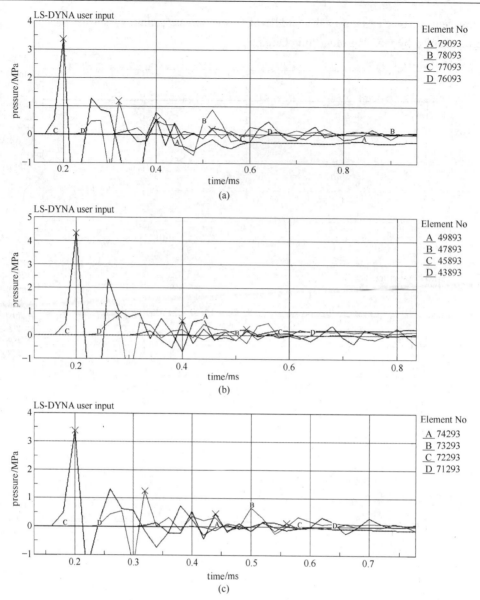

Fig.4 Pressure-time amplified curves of elements

(a) tamping plug; (b) full charging; (c) non-stemming

Tab.5 Peak pressures of elements

Number of elements tamping plug	Time /ms	Peak pressure /MPa	Number of elements full charging	Time /ms	Peak pressure /MPa	Number of elements non-stemming	Time /s	Peak pressure /MPa	Relative distance /r·R^{-1}
79093	0.2	3.4	49893	0.2	4.3	74293	0.2	3.3	3.4
78093	0.31	1.2	47893	0.28	0.9	73293	0.32	1.3	5.5
77093	0.4	0.5	45893	0.4	0.6	72293	0.44	0.5	7.9
76093	0.52	0.1	43893	0.52	0.2	71293	0.56	0.2	10.1

2.3.2 Elements on the Top of Models

The top of models are free face. Some elements were selected in accordance with relative distance from short to long. The Fig.5 shows the model which is stemmed by tamping plug. Another 2 models are same to Fig.5 except for the numbers of elements.

The coordinates of elements which are selected on the top of models, are listed in Tab.6.

The pressure-time curves of elements on the top of the models are presented in Fig.6.

Fig.5 Elements selected on the top of the models

Tab.6 Coordinates of elements on the top of models

Number of elements tamping plug	X-coordinate	Number of elements full charging	X-coordinate	Number of elements non-stemming	X-coordinate	Distance to blasthole/m
9892	−0.022	39992	−0.022	5092	−0.022	0.022
13892	−0.122	47992	−0.122	8992	−0.122	0.122
12892	−0.172	45992	−0.172	7992	−0.172	0.172
11892	−0.222	43992	−0.222	6992	−0.222	0.222
53692	−0.495	39593	−0.495	48992	−0.495	0.495

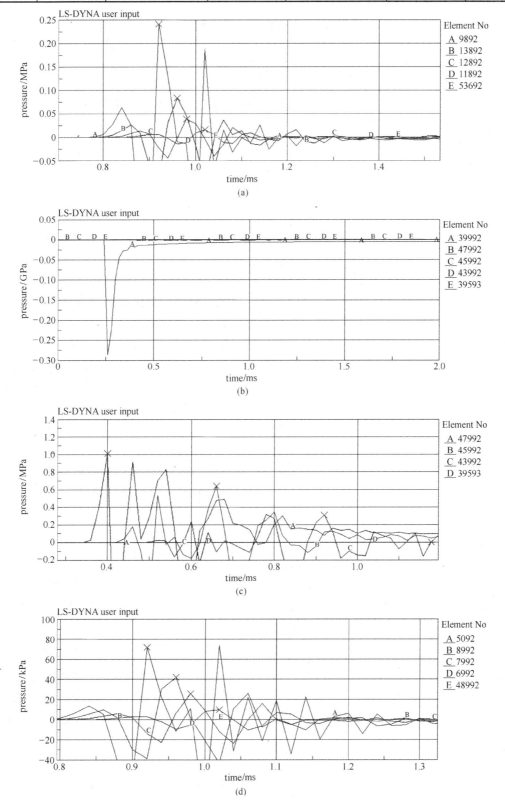

Fig.6 Pressure-time curves of elements on the top of the models

(a) tamping plug; (b) full charging No.1; (c) full charging No.2; (d) non-stemming

According to Fig.6, the variation is wide in the peak pressures of elements on the top of the models. The pressure of blasthole wall in each model is most high. Shown in Fig.6(a), the peak pressure of element on the top of the model which stemmed by tamping plug, is 0.24MPa. Shown in Fig.6(b), The peak pressure of element on the top of the model which is full charging, is 0.28GPa. In order to observe conveniently, the pressure-time curve of other element was presented in Fig.6(c) except for the one of the blasthole wall. The peak pressure has been reduced to 1.0MPa. The peak pressure of model which was stemmed by air is 71.9kPa. The peak pressures of distant elements took as a same law.

The peak pressures of all curves are showed in Tab.7 according to relative distances.

Tab.7 Peak pressures of elements

Number of elements tamping plug	Time/ms	Peak pressure /MPa	Number of elements full charging	Time/ms	Peak pressure /MPa	Number of elements non-stemming	Time/s	Peak pressure /MPa	Relative distance/r·R⁻¹
9892	0.92	0.24	39992	0.26	285.5	5092	0.92	0.0719	1
13892	0.94	0.16	47992	0.4	1.0	8992	0.97	0.0421	5.5
12892	0.96	0.08	45992	0.66	0.6	7992	0.98	0.0258	7.9
11892	0.98	0.04	43992	0.92	0.3	6992	1.02	0.0098	10.1
53692	1.2	0.01	39593	1.4	0.02	48992	0.56	0.002	22.5

Presented in Tab.7, the peak pressures of the elements which are at the same relative distance will present a rule: the full charging model > the tamping plug model > the non-stemming model. At the full charging model, the rocks around top of blasthole were broken by explosives entirely. The pressures of the elements on the top of model with the relative distance 5.5, 7.9, 10.1, are far greater than the one of the tamping plug model. In such condition, the rocks around the top of blasthole will be broken too shattered. At the relative distance 22.5, pressures of the two models are almost same. It is shown that the advantage of full charging model is more and more small along the increasing of relative distance. Comparing with another models, the peak pressure of top element of non-charging model is nearly 3 orders of magnitude in difference.

The reason is that the pressure wave of top element of full charging model was conducted by the explosion of top explosives. But the pressures of another two models was conducted by explosion from bottom explosives to top one. In non-stemming model, the top pressure will descend since of gas born in blast was escaped along blastholes.

3 CONCLUSIONS

Some conclusions are drawn as follows by through analyzing the 3 kinds of stemming with tamping plug, full charging and non-stemming.

(1) The 3 pressure-time curves of central elements are almost same. The propagation time and peak pressure are small in difference. It indicates that the stress state of central rocks in 3 models are basically same.

(2) In full charging model, the rocks around blasthole were crushed. In non-stemming model, the rocks on the top of blasthole cannot be crushed enough because of gas born in blast has been overflowed from orifice and pressure of top element was not enough.

(3) Compared with the 3 models, the non-stemming model cannot meet blasting requirements in some times because of gas born in blast escapes prematurely and acts in rocks too short. It can only form a blasting cavity inside a rock. The pressure around free surface cannot break rocks.

Compared with full charging model, tamping plug can enhance the efficiency of explosives by though saving explosives and bringing gas born in blast into full play.

ACKNOWLEDGEMENT

This work was financially supported by the National Science and Technology Supporting Plan of China(No. 2012 BAB08B01). We wish to express our sincere thanks to the persons who work for the investigation.

REFERENCES

[1] JIN Longzhe, YU Meng, LIU Jieyou, et al.Experimental

study on reducing the dust of explosion by the new water stemming[J]. JOURNAL OF CHINA COAL SOCIETY, 2007,32(3):253~257(in Chinese).

[2] BAI Jinze. LS-DYNA3D Theory Basis and Example Analysis[M]. Beijing: Science Press, 2005(in Chinese).

[3] ZHAO Haiou. LS-DYNA Dynamical Analysis Manual[M]. Beijing:Weapon Industry Press,2003(in Chinese).

[4] LI Yuchun, SHI Dangyong, ZHAO Yuan. ANSYS10.0/LS-DYNA theoretic basis and engineering practice[M].Beijing: China Water & Power Press, 2005(in Chinese).

[5] SHANG Xiaojiang, SU Jianyu. ANSYS/LS-DYNA dynamical analysis method and application at project[M].Beijing: China Water & Power Press, 2006(in Chinese).

Estimation of Quasi-Static Action of Explosion Products in the Directed Destruction of Rock

V.N. Kowalewski[1], Z.G. Dambaev[2]
(*1. National Mineral Resources University "University of Mines", St. Petersburg, Russia;*
2. Buryat State University, Ulan-Ude, Russia)

ABSTRACT: The article deals with the mathematical calculation of explosion products movement in the blast-hole for justification of quasistatic process of rock destruction. The results of model experiment of elongated cylindrical charge explosion to check theoretical calculations using the hypothetical accounting of tension waves interaction between adjacent cavities are described, and estimated time of crack formation process is defined. For studies of numerical calculations of the model experiments were carried out on an optically transparent models by high-speed photography. As a result of the shooting was received sequence of frames of the process gas exhaust and stress wave propagation, as well as the process of crack development. When considering the interference pattern observed the development of major cracks in the plane of division and the remaining visible gaseous combustion products in the hole. For modeling the interaction of stress waves from adjacent charges to the two side faces of the model were applied smooth rigid plates, thereby allowing the interference of stress waves from the rigid boundaries (symmetry axes). As the result the resource opportunities of explosion quasistatic action for the directed rock destruction between the adjacent elongated charges is demonstrated.

KEYWORDS: model experiment; the quasistatic effect of the explosion; an elongated cylindrical charge stress wave; rocks

During elongated cylindrical charge explosion in the blast-hole fast-proceeding difficult gas dynamics processes occur which depend on type of a charge, initiation method, explosive transformation kinetics and the further expiration of explosion gaseous products in the atmosphere. Thus, it is necessary to provide dynamic impact of explosive loading on the rock for the crack formation process between the adjacent elongated charges.

Mathematical calculation of dynamic loads determination in a charging cavity is reduced to consideration of the corresponding gas dynamics problem about distribution of explosion gaseous products pressure and, on this basis, an explosion impulse along an internal surface of the blast-hole. Thus, it is necessary to consider two mechanical problems: (1) the problem of explosion gaseous products movement in a charging cavity; (2) the problem of the rock movement as a result of explosion. Movement is meant as large-scale media movement. For justification it is required to compare speeds of movement of various media: explosion gaseous products movement and large-scale movement of rock particles. It is necessary to consider joint movement of both media (explosion products media and a large-scale movement of rock) which determines the development of the main crack between adjacent charges.

Division of process into phases is conventional and is made according to existence of boundary conditions in mechanical process of elongated charge explosion.

The theory description of gas dynamics processes in the blast-hole which is based on the assumption that the process caused by detonation wave distribution on charge length is one-dimensional non-stationary movement of gases. Legitimacy of such statement is based on the estimates which derived from the initial conditions: the cavity diameter is much less than its length; rock compressibility is much more than explosion products compressibility; speeds of gas dynamics indignations much more than a speed of sound in air. It is known that solid rocks have big acoustic rigidity and insignificant expansion of a camouflage cavity. In this case, alignment of explosion products parameters distributions on the section of the blast-hole occurs in a time comparable with the time of all length charge explosive movement. As a result, by process consideration "as a whole", i.e. on the scale of charging cavity length, time of pressure alignment on section can be neglected and we can consider as of this alignment is instant.

The main functions of gaseous products condition and movement depend on the spatial (axial) coordinate of Z and t time. Gas is considered as ideal. The condition of gas on length of the blast-hole is described by system of the equations of the gas dynamics which is based on the accounting of mechanics fundamental laws – preservations of weight, an impulse, energy, in a form corresponding to one-dimensional model of a non-stationary current. At a detonation of a charge there is a process of explosive transformation into gaseous products which occupy volume equal at an instant detonation to the section which is earlier occupied by a charge. We believe that gases enter work on generating of a quasistatic field of tension in the course of expansion of the explosive camera to the final sizes. The non-stationary expiration is caused by disintegration of an initial gap (the 1st sort) between these parameters of gaseous substances and parameters not indignant air.

Appearing gases flow is automodel, that is develops like itself uniformly in time, movement of the explosion products (EP) is accompanied by the emergence of a shock wave which is forming in the course of movement of gases on length of the charging camera. Values of parameters of mass speed, pressure and density at the front a shock wave are determined by formulas:

$$U_1 = \frac{2}{\kappa+1} U_S \left(1 - \frac{1}{M^2}\right)$$

$$\frac{P_1}{P_0} = \frac{2\kappa}{\kappa+1}\left(M^2 - \frac{\kappa-1}{2\kappa}\right)$$

$$\frac{\rho_1}{\rho_0} = \frac{(\kappa+1)M^2}{2+(\kappa-1)}$$

where U_S —— the speed of the shock wave extending on not indignant gas;

P_0, ρ_0 —— respectively pressure and density of not indignant gas;

κ —— adiabatic curve indicator for gas;

M —— mach number $M = \frac{U_S}{a_0}$;

a_0 —— sound speed in air.

The ratio defining Mach number for a shock wave, arisen at gap disintegration $\frac{P_m}{P_0}$, registers in a look:

$$\frac{P_m}{P_0} = \left(\frac{2\kappa_0}{\kappa_0+1}M^2 - \frac{\kappa_0-1}{\kappa_0+1}\right)\left(1 - \frac{\kappa_1-1}{\kappa_0+1}\left(1-\frac{1}{M^2}\right)\frac{Ma_0}{a_1}\right)^{-\frac{2\kappa_1}{\kappa_1-1}}$$

where, P_m is EP pressure after expansion to well section; local speed of a sound in EP; indicator of an adiabatic curve of EP [1].

The arisen flow on the charging camera length contains a shock wave, a contact surface and a depression wave. Further, with the expiration of products of explosion from the blast-hole mouth, process changes therefore the area of applicability of this decision is limited.

Let's consider on a practical example of production of a block stone with a blast-hole H depth = 3 m and length of a charge of equal 2/3 depths. Considering the linear speed of a detonating cord equal 6.5×10^3 m/s and the speed of movement of gaseous products in the m/s blast-hole 0.7×10^3 m/s that established pressure in average part of a column of the extented charge makes 2000×10^{-6} s.

It is necessary to compare the received values of pressure with estimates of duration of explosion impact in average part of a column of a charge at distance between blast-holes of 0.5 m. The wave of tension covers this site approximately for 50×10^{-6} s, that is during 35~40 times it is less, than time of established pressure in the considered section of the blast-hole. In this case, it is possible to consider that in the field of the mass if between charging cavities the quasistatic field of tension is established. It is also known that the speed of distribution of cracks in rocks make (0.2÷0.4) cp speeds of a longitudinal wave. Therefore, for this case in rocky breeds crack formation process between adjacent charging cavities is carried out for 125×10^{-6} with, that is a crack formation process of rock happens in time significantly more, than time of action of a wave of tension and is much less, than time of action of established quasistatic pressure of products of explosion [2].

In Fig.1 it is shown epure distributions of pressure of EP on length of the blast-hole various timepoints. Thus, a numerical assessment of effect of quasistatic pressure, on the basis of the executed calculations also it is possible to claim that process of the directed destruction happens at the expense of quasistatic action of products of explosion.

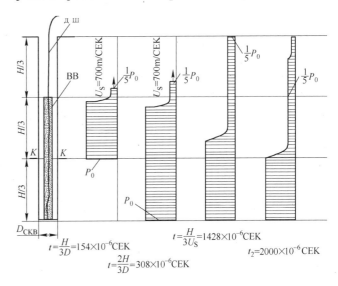

Fig.1 Distribution of pressure of EP in the blast-hole

For justification of numerical calculations and receiving more reliable picture model experiments on optically transparent models were made by a method of high-speed photographing. The record of process of the gases expiration and tension waves distribution and also development of crack formation process is presented on Fig.2 with an interval 18 ms. By results of an interferential picture development of the main crack in the planes of split and the remained gaseous products of explosion in the blast-hole is visible.

Models represented cubes, $100 \times 100 \times 100$ in size in which center there were cylindrical cavities with a diameter of 7 mm and 75 mm in depth in which the charge with a diameter of 2 mm, 50 mm high and tetranitropentaerytrite weight of 1.4 g coaxially took place.

Fig.2 The record of explosion of the extended cylindrical charge

At interaction of tension waves there is a development of the main crack between the extended cylindrical cavities that leads to increase in growth rate of the main crack which size makes (0.3~0.4) cp where cp is the speed of a longitudinal wave in optical glass. At further movement of this plane there is a merge in the uniform plane of a crack formation process. On the basis of an evident picture of distribution of tension waves in the course of a crack formation process it is possible to present destruction of the firm environment as follows: in a vicinity of a charging cavity the stretching tension which exceed strength on stretching are formed and by that there is a start of radial cracks in the area of an arrangement of charges. Thus process can be considered as quasistatic in which the arrangement of tension defines development of a crack [3].

As a result of a crack formation process the step surface of split of model is formed of optical glass that testifies to germination of a crack of spasmodic character. It is explained by pulsing nature of movement of a crack at the expense of intermediate relaxation processes of tension its tops. The pulsation is caused by that the size of tension intensity in the crack top, necessary for further growth and is defined by time of feed of necessary level of intensity of tension.

For modeling of tension waves interaction process from adjacent charges smooth rigid plates were put to two lateral sides of model. It is visible from the record (Fig.2) that the speed of charge detonation is 6000 m/s, and the speed of the gases expiration is 700~1000 m/s. Tension waves in model extended with a speed of a longitudinal wave of this environment. As a result of an interference of tension waves from rigid borders (symmetry axes) the directed radial cracks which determined model split on two equal parts were formed. Thus the average speed of development of the main crack made about 2000 m/s, and process of destruction of a sample happened for 25×10^{-6}c. Therefore, it is possible to claim that the defining role in development of the main crack is played by interaction of tension waves though thus in a cavity of the charging camera gaseous products of explosion still continue to remain that testifies to quasistatic process of destruction.

Further footage shows a split-second, that cracks developing, cover model of a single visible plane. We should also note the influence of the upper free surface on the distribution of stress fracture. In this case, the distance from the free of charge to the upper limit is 25 mm.

After reaching the top of the stress wave svoodnoy border, she reflected on her back as a rarefaction wave. Interaction of direct and reflected waves begins with 4.6 microseconds, and a split-second frames visible meeting these waves.

At a meeting of the reflected wave with negative phase of the incident wave form on the line dual amplitudes, ie lines that will likely split plane. Subsequent frames show the influence of the reflected wave on the plane split, ie plane split the top of the plane ahead of the bottom of the division and also moving, and subsequently merged into a single plane split.

On the basis of theoretical research, analysis, visual picture of wave propagation and the nature of the stress fracture can imagine the destruction of the solid medium as follows.

After the explosion, the elongated cylindrical charge in a solid under the pressure of the explosion products are formed radial and tangential stresses. In the vicinity of the charge voltage exceeds the tensile limit of the tensile strength and because of formation of radial cracks, buckling cavity forming part of the explosive occurs because the pressure of the explosion products is less than the compressive strength.

Most brittle materials compressive strength is much greater than the tensile strength. Interaction of stress waves increases the probability of failure of the solid medium by a superposition of waves of tensile stresses. The largest area of the tensile stress is the line between the explosive cavity in which a crack. Further development of a crack occurs superposition of tensile stresses, which are perpendicular to the line split. Final destruction occurs mainly under the influence of the quasi-static pressure of the explosion products.

Under the influence of the quasi-static pressure of the explosion products is the development of the main crack, which Accompanying stress concentration at the tip of the crack. This process is explosive action can be regarded as quasi-static process in which the voltage is carried distribute cracking process.

Thus, the results of laboratory tests confirmed the theoretical conclusions about the impact of the interaction of waves on the process priority development magistranoy fracture line location of holes.

REFERENCES

[1] Zhigalko E F. Dynamics of shock waves. Publishing house of St. Petersburg State University, 1987.

[2] Dambaev Zh G, Kowalewski V N. Mathematical model of movement of products of explosion in the blast-hole for ensuring process of the directed destruction of rocks. Messenger of the Buryat State University, 2011(9): 249~252.

[3] Kowalewski V N, Dambaev Zh G. Optimization of dynamic loading at explosion of adjacent charges for crack formation process between them. Messenger of the Buryat State University, 2012(SB): 203~206.

Damage Characteristics of Surrounding Rock Subjected to VCR Mining Blasting

JIANG Nan, ZHOU Chuanbo, LUO Xuedong, LU Shiwei
(Faulty of Engineering, China University of Geosciences, Wuhan, Hubei, China)

ABSTRACT: For limiting the damage range caused by explosive impact loads in vertical crater retreat (VCR) mining, the blasting damage characteristics of surrounding rock were studied by two methods-numerical simulation and ultrasonic testing. Combined with the mining blasting in Dongguashan Copper Mine, the VCR blasting characteristics under different conditions are obtained by using LSDYNA. Based on statistical fracture mechanics and damage mechanics theories, a damage constitutive model for rock mass subjected to blasting load was established. Then by using the fast Lagrange analysis codes FLAC3D, the blasting damage characteristics of surrounding rock were analyzed by applying the blasting loads obtained from the VCR mining and the damage zone is obtained. At last, the relationship between the amount of explosives and the radius of damaged surrounding rock mass was discussed, and its formula was also derived. The research provides a theoretical basis for rationally controlling stope boundaries and optimizing mining blasting parameters.

KEYWORDS: VCR; blasting loads; surrounding rock; damage; numerical simulation

1 INTRODUCTION

VCR (vertical crater retreat) mining technology is widely used in mine engineering because it possesses many better features, such as higher efficiency and more simple operation. In the mining process, the blasting vibration originating simultaneously from the rock-fracturing blasting load also can damage surrounding rock. For better control of the stope boundary, it is a key prospect in engineering application to ascertain damage characteristics of surrounding rock under mining blasting load and to optimize blasting parameters.

Damage effects of rock mass under blasting load were extensively studied at home and abroad, but these researches were mostly based on in-site tests and laboratory experiments (Palmström and Singh, 2001; Singh, 2002; Zhang et al., 2003; Cai et al., 2004). In recent years, with the development of computer technology, numerical simulation was increasingly adopted to study blasting damage of rock mass (Wu and Hao, 2006; Ma and An, 2008; Wu et al., 2004; Zhu et al., 2008). Hao et al analyzed the rock damage under stress wave of blasting based on an anisotropic damage constitutive model (Hao et al., 2002a, 2002b). Wei et al studied the damage of rock mass subjected to underground explosion (Wei X Y, 2009). Wang et al analyzed the tensile damage of brittle rock mass subjected to underground explosion and the evolution characteristics of craters (Wu C et al., 2007).

Based on the theories in statistical fracture mechanics and damage mechanics, a damage model for rock mass subjected to blasting load was established. At the same time, damage characteristics of surrounding rock subjected to mining blasting load in VCR mining and conventional blasting load were analyzed by adopting numerical simulations.

2 VCR MINING BLASTING LOAD

2.1 Numerical Model

Dongguashan Copper Mine is located in Tongling City, Anhui Province, PR China. It has the ability to produce around 4.3 million tons of copper ore annually, and its service life is 28 years. As the largest copper mine of downhole pit mining in Asia, its mining depth is more than 1000 meters, which ranks first among the nonferrous metal mines in Asia. The deposit is as long as more than 1800 m in trend, more than 500 meters wide and 20~70 m thick. It is divided into panels every 100 m, and there is an 18 m wide barrier pillar in each pair of adjacent panels. A panel is 100 m wide, whose length and height equal to the width and the thickness of the deposit respectively. Every panel consists of 20 stopes, which are arranged along the trend of the deposit and 18 m wide. The room stope and pillar stope are 82 m long and 78 m long respectively.

VCR mining method was adopted for underground mining in Dongguashan Copper Mine. According to the reality, large-

---

Corresponding author E-mail: happy john @ foxmail.com

diameter deep-hole blasting is introduced. The blasthole diameter is 165 mm, the charge length is 1.5 m to 10.5 m, the stemming length is 1.2 m to 2.0 m. In this study, a three-dimensional model is established using the software LS-DYNA, as shown in Fig.1, which is measuring 20 m in x-direction, 20 m in y-direction and 10m in z-direction. 10 blasting holes are equally divided into 2 rows. The distance between two rows is 3 m, and the distance between 2 adjacent holes in the same row is 2.8 m. Only the bottom surface is free, and the rest surfaces are applied non-reflecting boundary.

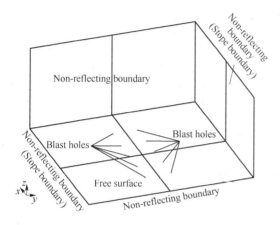

Fig.1 LSDYNA model

2.2 Calculation Parameters

In LSDYNA simulation, Mohr-Coulomb (M-C) model is selected as the rock material's model (Jiang N and Zhou C, 2012). According to the test results, parameters of stope rock mass are listed in Tab.1.

Tab.1 Parameters of the rock material

Density /g·cm^{-3}	Elastic modulus/GPa	Poisson's ratio	Cohesion/MPa	Internal friction/ (°)
3.22	69.00	0.31	21.43	56.21

The JWL state equation can simulate the relationship between pressure and specific volume in the explosion process (Hallquist, J. O., 2007). The equation is as follows:

$$p = A\left(1 - \frac{w}{R_1 V}\right)e^{-R_1 V} + B\left(1 - \frac{w}{R_2 V}\right)e^{-R_2 V} + \frac{\omega E_0}{V} \quad (1)$$

where, A, B, R_1, R_2, w are material constants; p is pressure; V is relative volume and E_0 is specific internal energy. The physical and mechanical parameters of the dynamite are the same with that of the field test and are listed in Tab.2.

Tab.2 Parameters of the explosive

Density /g·cm^{-3}	Detonation velocity /cm·μs^{-1}	A/GPa	B/GPa	R_1	R_2	ω	E_0/GPa
1.09	0.4	214.4	18.2	4.2	0.9	0.15	4.192

2.3 Blasting Loading

Due to the model which is built symmetrically, a quarter of the model is calculated to reduce the size of the research object. So the model was simplified as a 10 m cube, and the charge length is 3 m, 3.5 m, 4.0 m, 4.5 m, 5.0 m, 5.5 m and 6.0 m respectively. The charging length of 6.0 m is shown in Fig.2. The bottom surface is free, the front surface and the right surface are applied with normal displacement constraint. The rest surfaces are applied with non-reflecting boundary, and the left sides are stope boundaries (Chen M et al., 2007).

As illustrated in the rock blasting theory, the crushed zone radius is 2~3 times larger than the blasthole radius (Persson P A et al., 1993). In order to obtain the blasting load applied on the crushed zone boundary, the monitoring element was selected at a horizontal distance of 0.33~0.50 m away from the center of the blasthole. For example, when the charge length is 6 m, element H54050 was selected as the monitoring element, as shown in Fig.3.

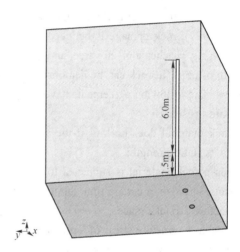

Fig.2 Numerical models with different charging lengths of 6.0 m

Fig.4 shows the pressure curve with time at element H54050. Shock wave pressure applied on the boundary of crushed zone under blasting load increases to its maximum within 0.5 ms, up to 1.74 GPa. In the same way, when the charge length is 3 m, 3.5 m, 4 m, 4.5 m, 5 m and 5.5 m, the maximum shock wave pressure applied on the boundary of crushed zone under blasting load is 1.4 GPa, 1.43 GPa, 1.46 GPa, 1.52 GPa, 1.57 GPa and 1.61 GPa respectively.

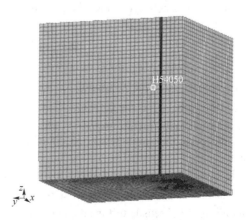

Fig.3 Monitoring element in the model with 6 m charging length

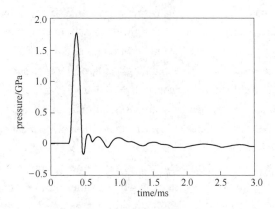

Fig.4 Pressure time history of Element H54050

3 ROCK DAMAGE MODEL UNDER BLASTING LOAD

According to statistical fractured mechanics (Liu L, Katsabanis P D, 1997), rock damage is believed to occur when the crack density reaches a given value, and the cumulation and growth of crack in rock can be demonstrated in probability terms. The following agreements have been achieved by the related research:

(1) a rock material does not fail if the applied stress is lower than its static strength;

(2) when a rock material is subjected to a stress higher than its static strength, a certain time duration is needed so that the fracture can take place;

(3) the dynamic fracture stress of a rock material is higher than its static strength and approximately cube root dependent on the strain rate.

So damage due to blasting loading can be defined as the probability of fracture, written as follows:

$$D_i = p_f = 1 - e^{-C_{di}^2} \quad i = 1,2,3 \quad (2)$$

where, p_f is the fail probability of damaged rock mass; C_{di} is the crack density in i direction and D_i is damage value. Obviously D_i has a value between 0 and 1, to respond to the stiffness of the intact, undamaged rock with crack density $C_{di} = 0$ and the fully fragmented rock with an infinite C_{di}, respectively. Crack density is defines as follows:

$$C_{di} = \begin{cases} 0 & (\varepsilon_i \leqslant \varepsilon_{cri}) \\ \alpha_i(\varepsilon_i - \varepsilon_{cri})^{\beta_i} t & (\varepsilon_i > \varepsilon_{cri}) \end{cases} \quad (3)$$

where, α_i, β_i are material constants; ε_i is the principle strain in i direction; ε_{cri} is the corresponding critical strain; t is the time to reach fracture stress. In terms of isotropic damage, the principle strain would be equivalent volume modulus.

Tensile strain is a quite important index to evaluate whether rock mass is damaged or not (Kuszmaul J S, 1987), so generally uniaxial tensile test is adopted to approximately simulate rock damage. critical strain can be obtained based on uniaxial tensile test as follows:

$$\varepsilon_{cri} = \frac{1-2\nu}{E_i}\sigma_{sti} \quad i=1,2,3 \quad (4)$$

where, σ_{sti} is static tensile strength in i direction, E_i is equivalent elastic modulus. We define ε_{fi} and C_{dfi} as the tensile strain and crack density in i direction when rock mass is cracked. Then C_{dfi} can be expressed as follows:

$$C_{dfi} = \alpha_i(\varepsilon_{fi} - \varepsilon_{cri})^{\beta_i}(t_i - t_{ci}) \quad i=1,2,3 \quad (5)$$

t_i is the total time when the rock mas reach the fracture stress, t_{ci} is the time duration when the tensile strain reaches the critical value. Both of them can be obtained by:

$$t_i = \frac{\varepsilon_{fi}}{\dot{\varepsilon}_i} \quad i=1,2,3 \quad (6)$$

$$t_{ci} = \frac{\varepsilon_{cri}}{\dot{\varepsilon}_i} \quad i=1,2,3 \quad (7)$$

where, $\dot{\varepsilon}_i$ is tensile strain rate. Accordingly, Eq. (5) can further be written as:

$$C_{dfi} = \alpha_i(t_i\dot{\varepsilon}_i - t_{ci}\dot{\varepsilon}_i)^{\beta_i}(t_i - t_{ci}) \quad i=1,2,3 \quad (8)$$

The time interval between the critical damage to the fracture of rock mass is given by

$$t = t_i - t_{ci} = \left(\frac{C_{dfi}}{\alpha_i}\right)^{\frac{1}{1+\beta_i}} \dot{\varepsilon}^{\frac{-\beta_i}{1+\beta_i}} \quad i=1,2,3 \quad (9)$$

The relationship between the fracture stress and the corresponding strain is

$$\sigma_{fi} = \frac{E_i(1-D_{fi})}{1-2\nu}\varepsilon_{fi} \quad i=1,2,3 \quad (10)$$

Submitting Eqs. (4) and (9) to Eq.(10), the strain rate dependent constitutive relation is obtained that

$$\sigma_{fi} = (1-D_{fi})\sigma_{sti} + \frac{E_i(1-D_{fi})}{1-2\nu}\left(\frac{C_{dfi}}{\alpha_i}\right)^{\frac{1}{1+\beta_i}} \dot{\varepsilon}^{\frac{1}{1+\beta_i}} \quad i=1,2,3$$

(11)

where, D_{fi} is the damage variable under the fracture state, other variables are the same as previous meanings. Because the dynamic tensile strength of a rock material is higher than its static strength and is approximately cube root dependent on the strain rate. So the value of β can be given as 2. For given D_{min} and C_{df}, the material constant α_i can be calculated based on the blasting crater test results.

According to *the construction technical specification on rock foundation excavation engineering of hydraulic structure (SL 47—1994)*, D_{min} should be 0.2 (Chuanbo Z, et al., 2012). At the same time, based on the in-site tests, α_i is equal to 3.16×10^6, and other parameters are listed in Tab.1.

4 NUMERICAL ANALYSIS OF ROCK DAMAGE ZONE

4.1 Numerical Model

Like the blast model above, a 3D model was established

using FLAC3D. Due to the model which is built symmetrically, a quarter of the model is calculated to reduce the size of the research object. For example, when the charge length is 6 m, a 10 m long, 14 m wide, 14 m high model is established, as shown in Fig.5. The left side is the stope boundary, and the hollow part is the boundary of crushed zone of 2 m× 6 m×7.5 m.

The boundary conditions are the same with that in Chapter 2.3. Because the maximum pressure varies depend on the charge, the blasting damage zone depending on the charge can be simulated by applying different pressure on the boundary of crushed zone. Mohr-Coulomb (M-C) model is considered as the material model for numerical simulation (Itasca F, 2009), and the parameters are listed in Tab.1. The applied dynamic loads are equal to the results calculated above.

4.2 Analysis of the Damage Characteristics

For example, when the charge length is 6 m, the damage growth on the free surface and along the height direction is shown in Fig.6 and Fig.7.

Fig.6 and Fig.7 show that the rock mass are not damaged yet in 0.1 ms after detonation whether on the free surface or along the height direction. In the time interval between 0.1 ms to 1.5 ms, the damage zone radius is approximate 6.785 m and it reaches about 8.142 m in the time interval between 1.5 ms to 3.0 ms.

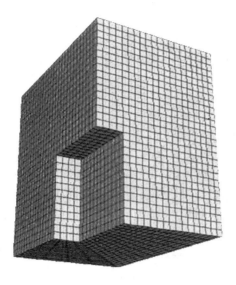

Fig.5 Numerical model with the charging length of 6 m

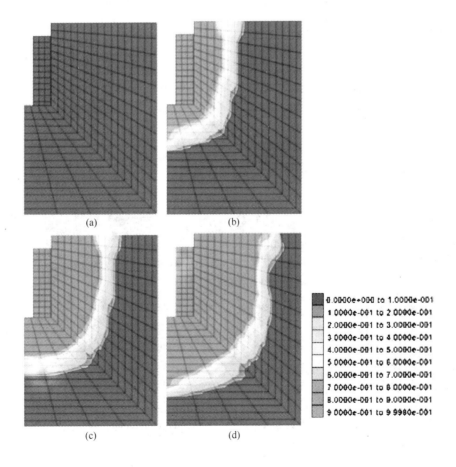

Fig.6 Damage growth on the bottom free surface
(a) t=0.1 ms; (b) t=1.0 ms; (c) t=1.5 ms; (d) t=3.0 ms

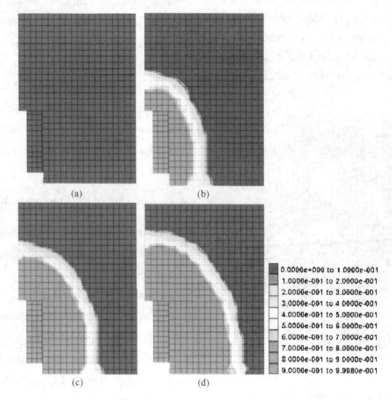

Fig.7 Damage growth along the height direction

(a) t=0.1 ms; (b) t=1.0 ms (c) t=1.5 ms; (d) t=3.0 ms

Using the same way, the damage zone with the charge length of 3 m, 3.5 m, 4.0 m, 4.5 m, 5.0 m, 5.5 m and 6.0 m is shown in Fig.8. Damage under blasting loads grows faster within the first 1.5 ms after denotation than that in the latter 1.5 ms. Combined with the above analysis, the calculation results agree well with the first two agreements (1) and (2) in section 3.

Fig.8 shows the crushed zone with the damage value of 0.2. The damage zone radiuses are listed in Tab.3.

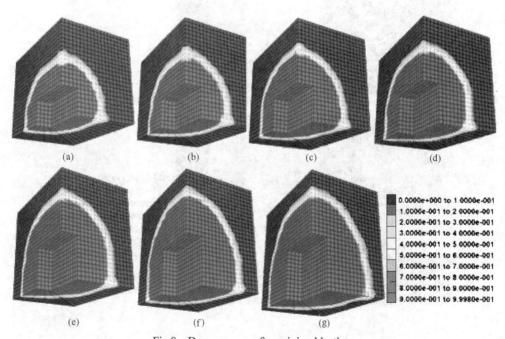

Fig.8 Damage zone after mining blasting

(a) charge length is 3.0 m; (b) charge length is 3.5 m; (c) charge length is 4.0 m; (d) charge length is 4.5 m;
(e) charge length is 5.0 m; (f) charge length is 5.5 m; (g) charge length is 6.0 m

Tab.3 Damage zone radiuses with different charge weights

Charging length/m	Maximum one-stage charge weight/kg	Damage zone radius/m
3	696.75	5.504
3.5	812.875	5.942
4	929	6.478
4.5	1045.125	6.763
5	1161.25	7.065
5.5	1277.375	7.684
6	1393.5	8.141

Based on the in-site parameters, the statistical relationship between charge amount and damage zone radius is established.

As shown in Fig.9, the function relation can be expressed by Eqn. (12).

$$r = 3.693\ln Q - 18.778 \quad (12)$$

where, r is damage zone radius of surrounding rock; Q is one-stage charge.

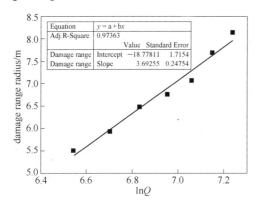

Fig.9 Fitting curve between damage zone radius and charge weight

The rooms and barrier pillars are 18 m wide. As illustrated in Eqn. (12), when $r=9$ m, the maximum one-stage charge is 1848.26 kg. So the in-site maximum one-stage charge should be less than 1848.26 kg. The maximum one-stage charge can be pre-calculated by empirical formula when to blast at different locations, so that the stope boundary can be under control, the surrounding rock mass are stable, and the safe production can be ensured.

5 CONCLUSIONS

(1) In LSDYNA, when the charge length is 3 m, 3.5 m, 4 m, 4.5 m, 5 m and 5.5 m, the maximum shock wave pressure applied on the boundary of crushed zone under blasting load is 1.4 GPa, 1.43 GPa, 1.46 GPa, 1.52 GPa, 1.57 GPa and 1.61 GPa respectively.

(2) Based on statistical fracture mechanics, damage due to blasting loading can be defined in the probability form. The damage model was established. And the damage characteristics were obtained by numerical simulation.

(3) According to the results in different blasting conditions, the statistical relationship between charge amount and damage zone radius is established. Furthermore, the maximum one-stage charge was proposed to be less than 1848.26 kg on the purpose of controlling the stope damage boundary.

ACKNOWLEDGMENT

The study was sponsored by the National Natural Science Foundation of China (Grant No. 41372312 and No. 51379194), the Fundamental Research Funds for the Central Universities, China University of Geosciences (Wuhan) (Grant No. CUGL140817) and the China Postdoctoral Science Foundation (Grant No. 2014M552113). We are also grateful to the China Scholarship Council (CSC) for supporting.

REFERENCES

[1] Cai M, Kaiser P K, Tasaka Y, et al. Generalized crack initiation and crack damage stress thresholds of brittle rock masses near underground excavations[J]. International Journal of Rock Mechanics and Mining Sciences, 2004, 41(5): 833~847.

[2] Chen M, Lu W, Yi C. Blasting vibration criterion for a rock-anchored beam in an underground powerhouse[J]. Tunnelling and underground space technology, 2007, 22(1): 69~79.

[3] Chuanbo Z, Nan J, Gang L. Study on blasting vibration cumulative damage effect of medium-length hole mining[J]. DISASTER ADVANCES, 2012, 5(4): 468~473.

[4] Fourie A B, Green R W. Damage to underground coal mines caused by surface blasting[J]. International Journal of Surface Mining and Reclamation, 1993, 7(1): 11~16.

[5] Haibo L, Xiang X, Jianchun L, et al. Rock damage control in bedrock blasting excavation for a nuclear power plant[J]. International Journal of Rock Mechanics and Mining Sciences, 2011, 48(2): 210~218.

[6] Hallquist J O. LS-DYNA keyword user's manual. Livermore Software Technology Corporation, 2007.

[7] Hao H, Wu C, Seah C C. Numerical analysis of blast-induced stress waves in a rock mass with anisotropic continuum damage models Part 2: Stochastic approach[J]. Rock mechanics and rock engineering, 2002, 35(2): 95~108.

[8] Hao H, Wu C, Zhou Y. Numerical analysis of blast-induced stress waves in a rock mass with anisotropic continuum damage models part 1: equivalent material property approach[J]. Rock Mechanics and Rock Engineering, 2002, 35(2): 79~94.

[9] Itasca F. Fast Lagrangian Analysis of Continua in 3 Dimensions, Version 4.0. Minneapolis, Minnesota, Itasca Consulting Group, 2009: 438.

[10] Jiang N, Zhou C. Blasting vibration safety criterion for a tunnel liner structure[J]. Tunnelling and Underground Space Technology, 2012, 32: 52~57.

[11] King M S, Myer L R, Rezowalli J J. Experimental studies of elastic-wave propagation in a columnar-jointed rock mass[J]. Geophysical Prospecting, 1986, 34(8): 1185~1199.

[12] Kuszmaul J S. A new constitutive model for fragmentation of rock under dynamic loading[R]. Sandia National Labs., Albuquerque, NM (USA), 1987.

[13] Liu L, Katsabanis P D. Development of a continuum damage model for blasting analysis[J]. International Journal of Rock Mechanics and Mining Sciences, 1997, 34(2): 217~231.

[14] Ma G W, An X M. Numerical simulation of blasting-induced rock fractures[J]. International Journal of Rock Mechanics and Mining Sciences, 2008, 45(6): 966~975.

[15] Palmström A, Singh R. The deformation modulus of rock masses—comparisons between in situ tests and indirect estimates[J]. Tunnelling and Underground Space Technology, 2001, 16(2): 115~131.

[16] Persson P A, Holmberg R, Lee J. Rock blasting and explosives engineering[M]. CRC press, 1993.

[17] Singh P K. Blast vibration damage to underground coal mines from adjacent open-pit blasting[J]. International Journal of Rock Mechanics and Mining Sciences, 2002, 39(8): 959~973.

[18] Wang Z L, Li Y C, Shen R F. Numerical simulation of tensile damage and blast crater in brittle rock due to underground explosion[J]. International Journal of Rock Mechanics and Mining Sciences, 2007, 44(5): 730~738.

[19] Wei X Y, Zhao Z Y, Gu J. Numerical simulations of rock mass damage induced by underground explosion[J]. International Journal of Rock Mechanics and Mining Sciences, 2009, 46(7): 1206~1213.

[20] Wu C, Hao H. Numerical prediction of rock mass damage due to accidental explosions in an underground ammunition storage chamber[J]. Shock Waves, 2006, 15(1): 43~54.

[21] Wu C, Lu Y, Hao H. Numerical prediction of blast induced stress wave from large-scale underground explosion[J]. International journal for numerical and analytical methods in geomechanics, 2004, 28(1): 93~109.

[22] Yang R, Bawden W F, Katsabanis P D. A new constitutive model for blast damage[C]//International journal of rock mechanics and mining sciences & geomechanics abstracts. Pergamon, 1996, 33(3): 245~254.

[23] Zhang Y Q, Hao H, Lu Y. Anisotropic dynamic damage and fragmentation of rock materials under explosive loading[J]. International Journal of Engineering Science, 2003, 41(9): 917~929.

[24] Zhu Z, Xie H, Mohanty B. Numerical investigation of blasting-induced damage in cylindrical rocks[J]. International Journal of Rock Mechanics and Mining Sciences, 2008, 45(2): 111~121.

Dynamic Failure Mechanism of Rock Mass Using Smoothed Particle Hydrodynamics

Ranjan Pramanik, Debasis Deb

(*Department of Mining Engineering, Indian Institute of Technology Kharagpur, India*)

ABSTRACT: The paper presents a numerical procedure based on SPH framework to analyze the fracture and fragmentation process of rock medium under dynamic stress wave followed by gas expansion. To analyze the dynamic fracture mechanism related to blast-induced shock wave and gas expansion, a rectangular rock mass containing multiple blast holes were numerically blasted in the proposed SPH framework. The damage pattern around the boreholes and formations are mainly generated due to tensile cracks and are simulated using the developed numerical tool. It is found that the developed procedure has the potential to provide valuable information to understand the physical phenomena those occur in the failure process of rock mass under blast induced dynamic loads.

KEYWORDS: blast; damage; SPH; explosive; stress wave

1 INTRODUCTION

Fragmentation size is an important parameter in wide variety of applications such as rock blasting in mines, tunnel excavation, structural demolition etc., in which solid masses are damaged and fragmented by detonating explosive material confined within the space of solid medium. It is well known that rock damage and fragmentation by blasting occur due to detonation induced stress wave and product gas driven fracture propagation. The efficiency of such operations greatly depends on the understanding of detonation mechanism of explosive and subsequent failure process of surrounding brittle material.

In order to understand rock failure process under blast load, knowledge of failure due to initiation of microcracks and propagation of moderate cracks are essential. Mechanical properties and failure characteristics of rock are sensitive to dynamic loading rate and rock sample undergoes through multiple physical processes which occur at different time and length scales. Continuum modelling based on grid base methods does not always work well for the simulation of large deformation, fracture and fragmentation, especially if discontinuities occur in the failure process. Recent development of meshless or meshfree methods, Smoothed Particle Hydrodynamics (SPH) have the advantages for simulating large deformation, fracture propagation and fragmentation those may occur during failure process by including elasto-plastic and damage theories in the framework (Das & Cleary, 2010; Deb & Pramanik, 2013; Pramanik & Deb, 2013). SPH was first developed to simulate nonaxisymmetric phenomena in astrophysical dynamics, in particular polytropes (Benz et al., 1989; Gingold & Monaghan, 1977).

In this paper, a numerical procedure based on SPH framework is developed to investigate the key physical phenomena of fracture and fragmentation processes of rock medium under dynamic stress wave followed by gas expansion. For detonation dynamics of explosive, SPH formulation of the Euler equation is implemented with Jones-Wilkins-Lee (JWL) equation of state with a reaction zone. For non-linear behavior of rock material, modified Grady-Kipp continuum damage model is used to analyze dynamic fracture behavior of rock mass in tension due to blast loading. For shear failure, Drucker-Prager yield criterion with associative rule is employed in the constitutive model for plastic deformation. A SPH model is developed by considering a square shape rock medium with centrally located emulsion explosive in a borehole. The damage pattern around the blasthole and the formation of tensile cracks near free surfaces were subsequently simulated using the developed numerical tool. The model is simulated by assuming all four sides of the rock medium as free surfaces. The results of peak pressure are also validated with those obtained from AUTODYN and are found to be in good agreement. To show the efficacy of the proposed method, a rectangular rock mass containing multiple

---
Corresponding author E-mail: deb.kgp@gmail.com

blast holes are also numerically blasted in the proposed SPH framework.

2 PROBLEM DESCRIPTIONS

2.1 Explosive Burn

A traditional programmed burn model has been implemented in many hydrocodes (Bdzil et al., 2001) of explosive engineering to describe a detonation process. In this case, each SPH particle is pre-assigned its burn time, t_b based on the detonation velocity, D_{CJ} and its distance from the detonating point. If the current time is below the particle burn time, $t < t_b$, then the particle will not be allowed to burn in this time and the burn fraction Y for this particle is assigned to be zero. If $t > t_b$, then the burn fraction is updated according to the rule

$$Y(x) = \begin{cases} 0 & \text{for} \quad t \leq t_b \\ \dfrac{t(x)-t_d(x_d)}{\Delta t} & \text{for} \quad t_b < t < t_b + \Delta t \\ 1 & \text{for} \quad t \geq t_b + \Delta t \end{cases} \quad (1)$$

where, $t_b = \|x - x_d\|/D_{CJ}$ is the burn time, x_d is the position of the detonation point and Δt is the burning interval, given by

$$\Delta t = \begin{cases} n_b \dfrac{\Delta x}{D_{CJ}} & \text{for} \quad \text{1-D} \\ n_b \dfrac{\Delta x \Delta y}{D_{CJ}\sqrt{\Delta x^2 + \Delta y^2}} & \text{for} \quad \text{2-D} \end{cases} \quad (2)$$

where, $n_b = 3 \sim 6$, Δx and Δy are the initial spacing of the explosive particles in x and y directions respectively.

2.2 Equation of State

The Jones-Wilkins-Lee (JWL) equation of state is adopted to describe the pressure of gaseous products of the explosive.

$$p_{JWL} = A\left(1 - \dfrac{\omega\theta}{R_1}\right)e^{-R_1/\theta} + B\left(1 - \dfrac{\omega\theta}{R_2}\right)e^{-R_2/\theta} + \omega\theta\rho_0 E \quad (3)$$

where, A, B, and C are constants in GPa pressure units; R_1, R_2 and ω are real numbers; $\theta = \rho/\rho_0$; ρ_0 is the initial density of the explosive; E is initial internal energy per unit volume in J/kg. The JWL parameters of emulsion explosives used in this paper are $\rho_0 = 1310$, $E = 3.2 \times 10^6$, $D_{CJ} = 5500$, $A = 214.36$, $B = 0.182$, $R_1 = 4.2$, $R_2 = 0.9$, $\omega = 0.15$.

By incorporating the above program burn model in Eq.(3), the modified equation of state is given by

$$p(\rho, E, Y) = p_0 + [p]Y \quad (4)$$

where, p_0 is the initial pressure of the unreacted explosive and $[p] = p_{JWL}(\rho, E) - p_0$. The above scheme replaces the pressure with partial pressure in the reactive zone according to the burn fraction.

2.3 Governing Equations

The interaction between product gas and rock medium is a multi-physics phenomenon in which high pressure of the detonation induced gas causes a considerable deformation in rock medium. In this paper, detonation induced gas is considered as compressible fluid that interact with surrounding brittle rock material when explosion occurs in confined conditions. As shown in Fig.1, the domain occupied by gas is denoted by Ω_t^g and the rock by Ω_t^r at time $t \in [0,T]$.

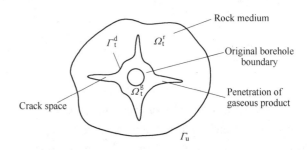

Fig.1 Schematic of the domains occupied by gas and rock in rock blasting phenomenon

The governing equations of motion of gas and rock in combined form can be written as

$$\dfrac{d\rho}{dt} = -\rho \dfrac{\partial v^\alpha}{\partial x^\alpha} \quad (5)$$

$$\dfrac{dv^\alpha}{dt} = \dfrac{1}{\rho}\dfrac{\partial \sigma^{\alpha\beta}}{\partial x^\beta} \quad (6)$$

$$\dfrac{de}{dt} = \dfrac{\sigma^{\alpha\beta}}{\rho}\dfrac{\partial v^\alpha}{\partial x^\beta} \quad (7)$$

where, α and β denote the cartesian components; v^α is the velocity component; e is the internal energy; d/dt denotes the material derivative following the motion. The stress components, $\sigma^{\alpha\beta}$ for gas and elastic solid are defined by

$$\sigma^{\alpha\beta} = \begin{cases} -p\delta^{\alpha\beta} & \forall x \in \Omega_t^g \\ 2G\varepsilon_d^{\alpha\beta} + K\varepsilon_v\delta^{\alpha\beta} & \forall x \in \Omega_t^r \end{cases} \quad (8)$$

where, the pressure of detonation gas, p is evaluated from Eq. (4); G and K are shear and bulk modulus respectively; ε_d and ε_v are the deviatoric and volumetric strain of rock material. It should be noted that as the detonation induced gas has high pressure and high dispersion speed, the product gas is treated as invicid in adiabatic process.

2.4 Non-linear Behavior of Rock Medium

It is well known that rock fragmentation by blasting is one of the dynamic failure process. The explosive loading plays an important role for propagation, reflection and interaction of the stress waves that result in crushing, spalling and fragmentation of rock materials. This inelastic response is due to principally stress induced pre-existing microcracks and defects. The growth of these discontinuities render portion of the rock volume unable to carry load. This is then reflected in the increase of crack density and decrease of strength and stiffness of the material. It is neither straight forward process nor computationally cost effective to simulate the dynamic failure of individual crack or defect in rock mass using classical theories of fracture mechanics. Therefore, with the development of advance computational methods, efforts have been directed towards an effective continuum descriptions of fracture, fragmentation and wave propagation by taking into account the rock constitutive model, strength and failure characteristics. During the process of stress wave propagation, tensile stresses or shear stresses do occur and cause rock material to fail in tension or in shear respectively. In this study, generalised Grady and Kipp (1980) damage model for higher dimension is incorporated to determine damage variable, $D(0 \leqslant D \leqslant 1)$ if it is in tension. After evaluation of damage variable, the common approach is to scale the entire stress tensor by a factor $(1-D)$. However, this approach equally modifies compressive components along with tensile components of the total stress tensor which should not be the case if the failure occurs due to tensile stress. In this paper, a rational approach similar to that of Das and Cleary (2010) is then adopted to treat only the tensile stress components keeping the compressive part unaltered.

Under compressive loads, pressure-dependent inelastic response is also observed for rock medium. Therefore, the material is treated with the context of Drucker-Prager strength theory for elaso-plastic response (Deb and Pramanik, 2013).

3 SPH APPROXIMATION OF THE GOVERNING EQUATIONS

SPH methodology overcomes the disadvantages of traditional mesh based numerical methods in treating large deformations, large inhomogeneities, tracing free surfaces and moving boundaries in transient analysis under explosive induced stress wave. SPH approximation is presented for solving the equations of motion Eq. (5)~Eq. (7). For more comprehensive details on SPH method one can refer to Monaghan (1992).

In SPH, the state of particles is represented by a set of points with fixed volume, which possess material properties interact with the all neighboring particles by a weight function or smoothing function or smoothing kernel (Gingold & Monaghan, 1977). This function required to be continuous and differentiable and satisfy the normalization, delta function and compactness properties. Each particle has a support domain, $\Lambda_a, \forall a \in \Omega$, specified by a smoothing length, h_a. The value of a function at a typical particle is obtained by interpolating values of that function at all particles in his support domain weighted by smoothing function. Gradients that appear in the flow equation are obtained via analytic differentiation of the smoothing kernel.

The transformation of the set of governing equation Eq.(5)~Eq.(7) into particle approximation yields the following set of SPH equation

$$\frac{d\rho_a}{dt} = \sum_{b \in \Lambda_a} m_b (v_a^\alpha - v_b^\alpha) \frac{\partial W_{ab}}{\partial x_a^\alpha} \qquad (9)$$

$$\frac{dv_a^\alpha}{dt} = \sum_{b \in \Lambda_a} m_b \left(\frac{\sigma_a^{\alpha\beta}}{\rho_b^2} + \frac{\sigma_b^{\alpha\beta}}{\rho_b^2} + \Pi_{ab} \delta^{\alpha\beta} \right) \frac{\partial W_{ab}}{\partial x_a^\beta} \qquad (10)$$

$$\frac{de_a}{dt} = \frac{1}{2} \sum_{b \in \Lambda_a} m_b \left(\frac{\sigma_a^{\alpha\beta}}{\rho_b^2} + \frac{\sigma_b^{\alpha\beta}}{\rho_b^2} + \Pi_{ab} \delta^{\alpha\beta} \right) (v_b^\alpha - v_a^\alpha) \frac{\partial W_{ab}}{\partial x_a^\beta} \qquad (11)$$

4 NUMERICAL RESULTS AND DISCUSSION

Two numerical examples are presented below to show the efficacy of the proposed methodology. First example explain and verifies with dynamic fracture mechanism related to blast-induced borehole breakdown and crack propagation considering a rectangular rock mass with a single centrally located blast hole. The initial pressure of unreacted SPH explosive particle is assumed to be zero. The second example deals with dynamic failure mechanism for five blast holes exploded at the same time.

4.1 Single Explosion in a Rectangular Rock Mass

This example is to understand how the damaged and fractured zones are developed under dynamic stress wave and latter, due to expansion of high pressure gas.

A square rock medium of dimension 0.1 m×0.1 m is considered with emulsion explosive at center having a diameter of 0.01 m as shown in Fig.2. Simulation is performed in 2D plain strain condition. The rock medium has density of 2261 kg/m³, elastic modulus of 17.83 GPa, Poisson's ratio of 0.271. A total of 62500 square particles with spacing 0.4 mm are used to discretize the rectangular domain with an initial smoothing length of 0.48 mm. The particles inside the borehole represent emulsion explosive and the rest represents rock particles that surround the explosive. Free surface boundary conditions are assumed for the outside surfaces of the rock medium.

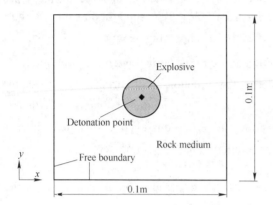

Fig.2 Schematic of rectangular rock medium containing centrally located explosive

A similar model is developed in AUTODYN in order to validate the results of the proposed method in term of pressure distribution. Fig.3 depicts the distribution of peak pressure from detonation point to free surface at a spacing of 2 mm. It is found that results predicted by SPH method agree well with those determined from AUTODYN. The devolvement of damage in medium at different time step is shown in Fig.4. The formation of radial cracks and spalling near the free surface can be observed in Fig.4(a). It is worth mentioning that major cracks are developed in the radial directions however spalling zone is parallel to the free surface. The subsequent fragmentation in the rock medium occurs due to expansion and penetration of high pressure product gases into void space generated by radial cracks. This is evident from the Fig.4(b) that gases penetrate into the radial cracks near the borehole and causes further displacement of rock fragments.

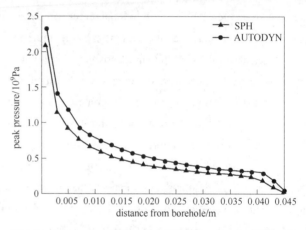

Fig.3 Comparison of generated peak pressure in SPH and AUTODYN from the detonation point to free surface

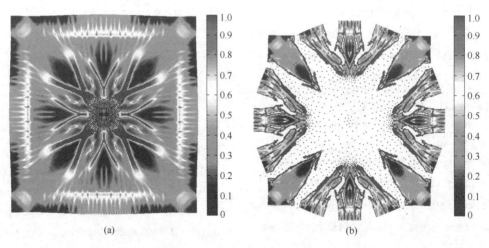

Fig.4 Accumulation of damage in the rock medium at two different time step
(a) time= 23.48 μs; (b) time= 351.56 μs

4.2 Multiple Explosions in a Rock Medium

This example is similar to previous example but with multiple blast-holes of diameter 0.009 m each (Fig.5). The properties of rock medium and boundary conditions are retained identical to those of previous example. All the blast holes are detonated in the same time level. The development of the damage in the rock medium after detonation of explosive can be observed in Fig.6 at two different time steps. The volume of explosive in previous example is same as total volume of distributed explosive in five holes. From Fig.4(b) and Fig.6(b), it can be observed that damage and development of cracks in the medium are much higher in the case distributed explosive in five blast-holes.

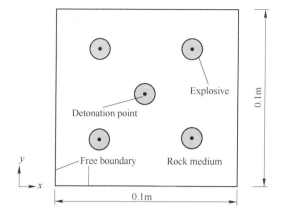

Fig.5 Schematic of rectangular rock medium containing multiple sources of explosive

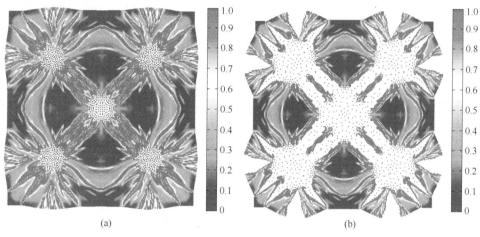

Fig.6 Damage in the rock medium for multiple explosions at two different time step
(a) time= 67.74 μs; (b) time = 209.15 μs

5 CONCLUSIONS

In this paper, the interaction of stress wave and product gases with the surrounding brittle rock material is presented in SPH framework. Explosion in single hole and multiple bore holes in rock medium emphasizes the role of stress wave loading on crack initiation and propagation. The major radial cracks in the medium are formed due to principal stress (tensile). The reflected stress wave from the free surface is responsible for formation of spalling zone. The developed procedure has shown the potential to provide valuable information to understand the physical phenomena those occur in the failure process of homogeneous rock mass under blast induced dynamic loads. However, issues relating to the applicability and performance of the procedure to even more complicated physical problems are yet to be investigated. This will be the subject of additional research work in near future.

REFERENCES

[1] Bdzil J, Stewart D, Jackson T. Program burn algorithms based on detonation shock dynamics: discrete approximations of detonation shock with discontinuous front models. *Journal of Computational Physics*, 2001, 174: 870~902.

[2] Benz W, Cameron A, Melosh H. The origin of the Moon and the single-impact hypothesis III. *Icarus* 81, 1989: 113~131.

[3] Das R, Cleary P. Effect of rock shapes on brittle fracture using smoothed particle hydrodynamics, *Theoretical and Applied Fracture Mechanics*, 2010, 53: 47~60.

[4] Deb D, Pramanik R. Failure process of brittle rock using smoothed particle hydrodynamics, *Journal of Engineering Mechanics*, 2013, 139: 1551~1565.

[5] Gingold R, Monaghan J. Smoothed particle hydrodynamics-theory and application to nonspherical stars. *Monthly Notices of the Royal Astronomical Society*, 1977, 181: 375~389.

[6] Grady D, Kipp M. Continuum modelling of explosive fracture in oil shale, International *Journal of Rock Mechanics and Mining Sciences & Geomechanics Abstracts*, 1980, 17(3): 147~157.

[7] Monaghan J. Smoothed particle hydrodynamics, *Annu, Rev, Astron. Astrophys*, 1992, 30: 543~574.

[8] Pramanik R, Deb D. Rock failure analysis using smoothed-particle hydrodynamics, *Geosystem Engineering*, 2013, 16: 92~99.

A Supernal Security Regression Method of Blast Vibration

JIANG Dongping[1,2], LIU Weizhou[1,2], ZHANG Xiliang[1,2]

(1.Sinosteel Maanshan Institute of Mining Research Co.,Ltd., Maanshan, Anhui,China;
2.State Key Laboratory of Safety and Health for Metal Mine, Maanshan, Anhui,China)

ABSTRACT: Least squares fitting is usually adopted in regression of blast vibration. On the basis of Sadov's formula, which is invented by a former Soviet Union scholar named Steve Sadove, the regression elements are $\ln(v)$ and $\ln(\sqrt[3]{Q}/R)$ and they linearly increase each other. Actually, if $\ln(v)$ and $\ln(\sqrt[3]{Q}/R)$ are analyzed directly, there are 50% data points over the regression line and 50% data points below. So 50% actual measured data of blast vibration velocity are unsafe for the regression equation. So we introduce regression safety coefficient $f(0<f<1)$ in the regression analysis in this paper and find a new regression method which is safer. The f is different in different projects, such as $f=0.9$ means 90% measured sample points is below the regression line and 10% is up. So the veracity of blasting safety is improved. We can see, the new regression method contract with common method, the topography and geological conditions coefficient K from blasting points to fixed point is increase by f, and the damping coefficient α is all the same. That means slope of regression line is invariant while intercept increase along with f.

KEYWORDS: blast vibration; analysis of regression; numerical analysis; blasting safety

1 INTRODUCITON

Vibration velocity is usually adopted to show vibration intensity for its better pertinence and stability. For the attenuation law of blast vibration, former Soviet Union scholar Steve Sadove generalize experience formula blast vibration intensity with experiments. the vibration intensity relates to maximum charging weight per delay Q and distance of blasting center to measuring point R. He also come up with the topography and geological conditions coefficient K and damping coefficient α. In addition, USMBE, P.B.Attwell(Europe) and Japan Mining and so on come up with attenuation formula that relate to K, α, Q and R. Generally, we adopt least square method to achieve K and α, then K and α substitute into the attenuation formula. With this method, there are 50% actual measured points up the regression line. So the regression formula obtained by common regression method is unsafe. This paper put forward a supernal security regression method which based on Sadov's formula.

2 COMMON REGRESSION METHOD OF BLAST VIBRATION VELOCITY

The blast vibration attenuation formula of Sadov's is

$$v = K\left(\frac{\sqrt[3]{Q}}{R}\right)^{\alpha} \quad (1)$$

where v ——the highest vibration velocity of ground particle, cm/s;

Q ——maximum charging weight per delay, kg;

R ——distance of blasting center to measuring point, m;

K ——the topography and geological conditions coefficient, (dimensionless);

α ——K and damping coefficient, (dimensionalss).

If $\hat{y} = \ln v$, $\hat{x} = \ln\left(\frac{\sqrt[3]{Q}}{R}\right)$, $a = \alpha$, $b = \ln K$, with Least square fitting,

$$a \approx a_0 = \frac{\sum_{i=1}^{n}\left(x_i - \frac{1}{n}\sum_{i=1}^{n}x_i\right)\left(y_i - \frac{1}{n}\sum_{i=1}^{n}y_i\right)}{\sum_{i=1}^{n}\left(x_i - \frac{1}{n}\sum_{i=1}^{n}x_i\right)^2} \quad (2)$$

$$b \approx b_0 = \frac{1}{n}\sum_{i=1}^{n}y_i - a_0 \cdot \frac{1}{n}\sum_{i=1}^{n}x_i$$

$\hat{y} = a_0\hat{x} + b_0$ is the regression line of all sample data and it situates center of sample points. Their location is showed in Fig.1.

Corresponding author E-mail: yht 03121 @ 163. com

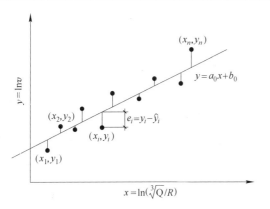

Fig.1 The regression line and sample points

3 A SUPERAL SECURITY REGRESSION METHOD

Lots of sample points are over the regression line which we can see from Fig.1. On the basis of mathematical statistics theory and the assumption that sample data obey normal distribution when applying least square fitting, there are 50% sample points meet $y_i > a_0 x_i + b_0$. It is unsafe. So this paper introduce the safety coefficient of regression fitting and named it $f(0 < f < 1)$. f means ratio of sample points which below the regression line. For example, $f=0.9$ means 90% sample points are below the regression line.

If $e_i = y_i - \hat{y}_i$, $e_1, e_2, ..., e_n$ obey normal distribution $N(0, \sigma^2)$,

$$e_x = \frac{1}{\sqrt{2\pi}\sigma} e^{-\frac{x^2}{2\sigma^2}} \tag{3}$$

where σ —— standard deviation of $e_1, e_2, ..., e_n$.

Discrete distribution of y_i and e_i are the same. If the regression line of (x_i, y_i) meets f, the regression line of (x_i, e_i) meets f, and the regression line of (x_i, y_i) is

$$y = a_0 x + b_0 + e_0 \tag{4}$$

where e_0 —— ratio of $e_1, e_2, ..., e_n$ less than e_0 is f, it can be seen details in Fig.2.

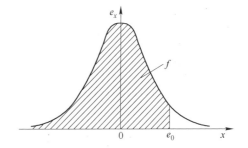

Fig.2 Relationship of e_0 and f

According to Fig.2,

$$\int_{-\infty}^{e_0} \frac{1}{\sqrt{2\pi}\sigma} e^{-\frac{x^2}{2\sigma^2}} dx = f \tag{5}$$

If $U = x/\sigma$, formula (5) can be converted into form of standard normal distribution.

$$\int_{-\infty}^{e_0/\sigma} \frac{1}{\sqrt{2\pi}} e^{-\frac{U^2}{2}} dU = f \tag{6}$$

Each e_0 of corresponding f can be achieved by referring to normal distribution function table. Tab.1 shows some e_0 corresponding to f. So

$$K = \exp\left[\frac{1}{n}\sum_{i=1}^{n} y_i - a_0 \cdot \frac{1}{n}\sum_{i=1}^{n} x_i + e_0(f)\right] \tag{7}$$

where $e_0(f)$ —— e_0 corresponding to f.

Tab.1 e_0 corresponding to f

f	0.7	0.75	0.8	0.85	0.9	0.95
e_0	0.5244 σ	0.6645 σ	0.8416 σ	1.0364 σ	1.2816 σ	1.6449 σ

4 COMPARISON OF TWO REGRESSION METHODS IN SOME IRON MINE

To compare the common regression method and the supernal security regression method, we measured underground blast vibration velocity in some iron mine and acquired 13 groups data in Tab.2.

Tab.2 Measured blast vibration data of some iron mine

Serial number	Distance R/m	Maximum charging weight per delay Q/kg	Maximum vibration velocity by measured v/cm·s^{-1}
1	86.4	11.7	0.5157
2	94.5	11.7	0.801
3	157.3	11.7	0.2113
4	83.9	12.9	0.5139
5	83.9	12.9	0.5721
6	151.4	12.9	0.393
7	151.4	12.9	0.3756
8	650.2	12.9	0.0413

Continues Tab.2

Serial number	Distance R/m	Maximum charging weight per delay Q/kg	Maximum vibration velocity by measured v/cm·s^{-1}
9	650.2	12.9	0.0528
10	114.8	13.35	0.4203
11	287.8	13.35	0.0918
12	287.8	13.35	0.0716
13	298.7	13.35	0.0556

It can be calculated $a_0 = 1.4507$, $b_0 = 4.7466$, $r=0.9376$ which is relative coefficient, with common regression method. Regression graph can be seen in Fig.3. Corresponding, $\alpha = 1.4507$, $K = 115.193$ and attenuation formula of blast vibration that calculated by common regression method is $v = 115.193 \left(\dfrac{\sqrt[3]{Q}}{R} \right)^{1.4507}$.

By calculating, the standard deviation σ obtained by common regression is 0.1028, and each e_0, α, K corresponding to different f appear in Tab.3. The relationship between f and α, K show in Fig.4.

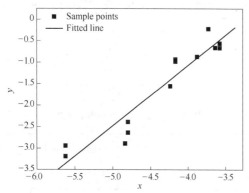

Fig.3 The fitted line and sample points

Tab.3 Each e_0, α, K corresponding to different f

f	0.7	0.75	0.8	0.85	0.9	0.95
e_0	0.05391	0.06831	0.08652	0.10654	0.13175	0.16910
α	1.4507	1.4507	1.4507	1.4507	1.4507	1.4507
K	121.572	123.336	125.602	128.142	131.413	136.414

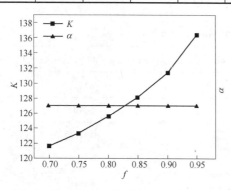

Fig.4 The relationship between f and α, K

5 CONLUSIONS

(1) This paper aims at improving the safety of regression formula. There are 50% data points over the regression line and 50% data points below in common regression method, so this paper introduce safety coefficient f ($0<f<1$) and quantifying safety in f.

(2) On the basis of common regression method and assumption that blast vibration sample data obey normal distribution, we can obtain each adaptive value e_0 of residual error corresponding to different f. Then obtain value of α and K by formula (5) and (7).

(3) By derived regression equation and living example regression, α is all the same and K increases when f increases.

REFERENCES

[1] Zong Q, Wang H B, Zhou S B. Research on monitoring and controlling techniques considering effects of seismic shock[J]. *Chinese Journal of Rock Mechanics and Engineering*, 2008, 27(5):938~945.

[2] He X Q. *Practical regression analysis*[M]. Beijing: Higher Education Press, 2008.

[3] The National Standards Compilation Group of People's Republic of China, GB 6722—2003 *Safety regulations for blasting*[S]. Beijing: Standards Press of China, 2004.

[4] Wang X G, Yu Y L, Liu D Z. *Enforceable handbook of safety regulations for blasting*[M]. Beijing: China Communications Press, 2004.

[5] The National Standards Compilation Group of People's Republic of China. GB 4086.1—1983 *Tables for statistical distributions normal distribution*. Beijing, 1984.

[6] Liu, J Q, Tang T Y. Selection of formulas for calculation of blasting vibration[J]. *Mining Engineering*, 2005, 3(6):43~45.

Numerical Estimation of Crater Size of the RC Wall Caused by Blasting

Shiro Kubota, Kana Nishino, Yuji Wada, and Yuji Ogata
(*National Institute of Advanced Industrial Science and Technology, Tsukuba, Japan*)

ABSTRACT: When we apply the demolition blasting partially to the reinforced concrete wall, the stress waves reflected from the free surfaces are interacted with each other. The fragmentation interact those stress waves and the phenomena become very complicated. Four types of the situations of the fragmentations can be considered. One is the no crater, second case has the crater of the borehole side, the third case is the crater of the opposite side, and last case is the crater of both sides. If the diameter and the angle of the borehole, and the kind of the stemming are fixed, those situations may be determined by the amount of explosive and depth of borehole. In this paper, we investigate the estimation method for crate size using numerical simulations. In this paper, the numerical results are shown.

KEYWORDS: numerical estimation; reinforced concrete (RC) wall; blasting; crater

1 INTRODUCTION

The breaching breaks down a part of a reinforced concrete wall, in order to rescue survivors in need of help in collapsed building at the natural disaster such as big earthquake. Because the road is severed and the situation is chaotic, the development of the breaching technology with mobility and swiftness are necessary. For such technology, we considered the utilization of the high energetic materials, i.e. blasting (Kubota, 2013, Nishino, 2014). The reinforced concrete wall was selected as the subject at the breaching. Because the thickness of the concrete wall is thin compared to the common subject of the demolition by blasting, the conventional design cannot be applied for this technique. The crater size caused by blasting is important parameters to estimate the effectiveness of the blasting conditions. In addition, we have tried to extend the application of the blasting technology for the partial demolition or repair the RC structures. In this paper, the numerical estimation of the blasting efficiency for the RC wall has been conducted by very simple model using ANSYS Autodyn.

2 NUMERICAL PROCEDURES

Main purpose of the simulations is to examine the influence of the free surface. Axisymmetric coordinate system was employed as simple model. To ignore the reflection wave from wall edges, the diameter of the calculation field was set to 3000 mm. The thicknesses of the walls were 150 mm and 300 mm. The depths of the boreholes were varied as 75 mm and 90 mm, respectively.

The porous model and the Drucker-Prager model were used as the equation of state (EOS) and constitutive equation as the concrete. Those parameters were selected from the material data base in ANSYS Autodyn. The failure model was selected hydro tensile limit. Because this report concentrates the influence of the free surfaces, the influence of the parameters for material model was not discussed here. The most important model for the estimation of the fragmentation is the failure model. Since the strain rate in the simulations relatively fast, the tensile limit was set to 7 MPa. Fig.1 shows the schematic illustration of the calculation filed for an axisymmetric model. The water was used as the stemming material. The composition C4 explosive was set in to 2×11 cells to attain the total weight of 3.02 g. Fig.2 shows the calculation field for the 3-D model. One quarter was calculated by using two symmetry planes. The composition C4 was set in to 2×9 cells to attain the total weight of 3.15 g. The Eulerian-Lagrangian interaction algorism was employed for both axisymmetric and 3-D model. The mesh sizes for Eulerian and Lagrangian were 2.5 mm and 5 mm, respectively.

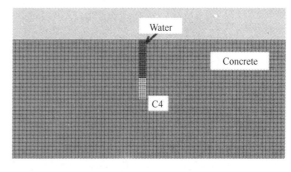

Fig.1 The calculation filed for an axisymmetric model

Correspoding author E-mail: kubota.46@aist.go.jp

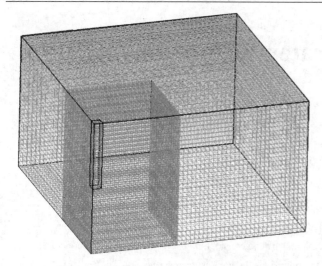

Fig.2 The calculation filed for the 3-D model

3 RESULTS AND DISCUSSIN

Fig.3 shows the bulk failure distributions at 0.7 ms after the initiation of the C4 explosive obtained by the axisymmetric models with the 75 mm borehole depth.The part of the bulk failure corresponds to 'red color. Although the crater size could not be predicted only the bulk failure distributions, the influence of the difference of the thickness of the walls could be examined qualitatively.The bulk failure distributions at the borehole side of the 150 and 300 mm thicknesses of the walls were almost the same each other, so the sizes of the craters were independent of the difference of the thickness in this consideration. As the quantitative information, the histories of the free surface velocities were dumped during numerical estimation. The points of the gauges were set 15, 65, and 115 mm from the axis at both the borehole sides and the bock sides. Fig.4 shows the histories of the free surface velocities at the borehole side in the case of the 75 mm borehole depth.The influence of the difference of the thickness can be confirmed at the gauge 1 only. The bulk failure distributions obtained by the axisymmetric models with the 90 mm borehole depth is shown in Fig.5. The examination related to the prediction of the crater size was same as the 75 mm borehole case.

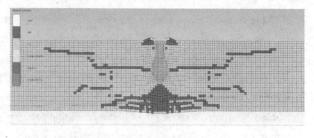

Fig.3 The bulk failure distributions at 0.7 ms after the initiation of the C4 explosive obtained by the axisymmetric models with the 75 mm borehole depth

(a) thickness of wall: 150 mm; (b) thickness of wall: 300 mm

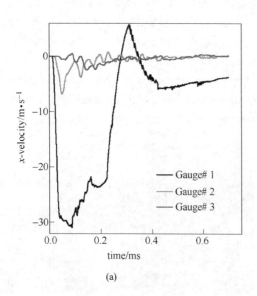

(a)

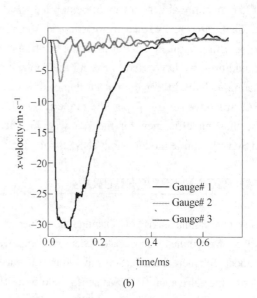

(b)

Fig.4 The free surface velocities at the borehole side obtained by the axisymmetric models with the 75 mm borehole depth

(a) thickness of wall: 150 mm; (b) thickness of wall: 300 mm

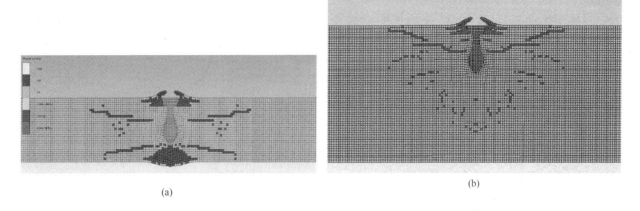

Fig.5 The bulk failure distributions at 0.7 ms after the initiation of the C4 explosive obtained by an axisymmetric model with the 90 mm borehole depth

(a) thickness of wall: 150 mm; (b) thickness of wall: 300 mm

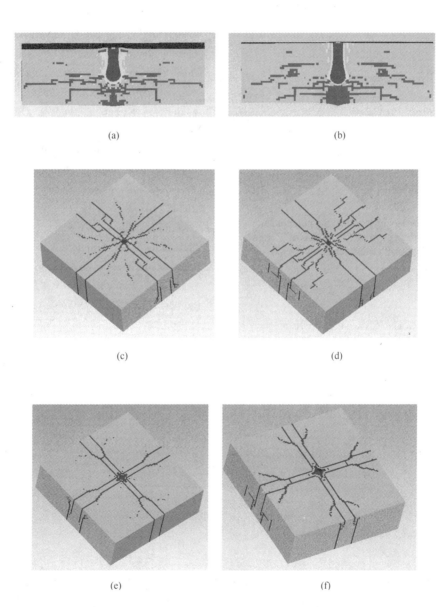

Fig.6 The bulk failure distributions at 0.7 ms after the initiation of the C4 explosive obtained by an axisymmetric model with the 90 mm borehole depth

(a) Cross section (borehole depth 75 mm); (b) Cross section (borehole depth 90 mm);
(c) borehole side (borehole depth 75 mm); (d) borehole side (borehole depth 90 mm);
(e) back side (borehole depth 75 mm); (f) back side (borehole depth 90 mm)

Fig.6 shows the bulk failure distributions obtained by 3-D models. The bulk failure distribution for 75 mm borehole depth was compared with that for 90 mm borehole depth. From only these distributions only, the judgment is difficult.

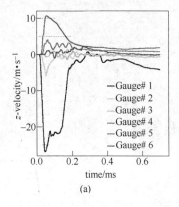
(a)

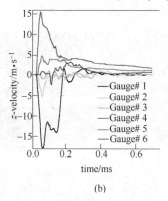
(b)

Fig. 7　The free surface velocities both at the borehole and back sides obtained by the axisymmetric models
(a) 75 mm borehole depth; (b) 90 mm borehole depth

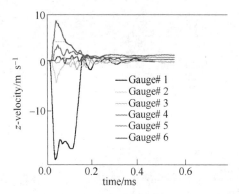

Fig.8　The free surface velocities both at the borehole and back sides obtained by the axisymmetric models of 75 mm borehole depth with reinforcing bar

REFERENCES

[1] Kubota S, Nishino K, Wada Y, Ogata Y, Ito N, Nagano M. zNakamura S, Taguchi T, Fukuda A. Application of micro-blasting for rescue work[C]//*Proceeding of the 7th International Conference on Explosives and Blasting. Beijing*: Metallurgical Industry Press, 2013: 235~240.

[2] Nishino K, Kubota S, Wada Y, Ogata Y, Ito N, Nagano M, Nakamura S, Taguchi T, Fukuda A. Estimation of crater in reinforced concrete wall caused by mini-blasting. *Explosion [C]//Shock Wave and High-Energy Reaction Phenomena 2,Materials Science Forum*767, 2014: 154~159.

[3] Nishino K, Kubota S, Wada Y, Ogata Y, Ito N, Nagano M, Nakamura S, Taguchi T, Fukuda A. Partial destruction of reinforced concrete wall by small blasting[C]// *Proceedings of the 8-th International Symposium on Impact Engineering, Applied Mechanics and Materials* 566, 2014: 268~273.

Research on Energy Consumption Characteristics and Energy Density Per Unit Time of Rock Crushing by Impact Load

GUO Lianjun [1], SHAO Anlin [2], ZHANG Daning [1], MA Xufeng [2], YANG Yuehui [1], ZHANG Guojian [1], SUN Houguang [2]

(1. University of Science and Technology Liaoning, Anshan, Liaoning, China; 2. Anstell Minging, Anshan, Liaoning, China)

ABSTRACT: The HSPB impact test was used to simulate the fragmentation process of rock blasting. Granite, phyllite and magnetite quartzite were selected to be tested by changing the impact velocity. By analyzing the input waves, reflect waves, transmit waves, and the crushing size features of samples, the stress-strain relationships, stresses and strains rates time related features were discussed, and the relation models between coefficient of energy utilization and the degree of crushing, as well as the models between the coefficient and input energy were set up. Based on this, the new concept and expression of energy density per unit time were put forward which can effectively describe the energy output structures and explored a new way to research on explosives and rocks coupling.

KEYWORDS: impact load test; rock fragmentation; coefficient of energy utilization; energy density per unit time

1 INTRODUCTION

The best coupling of explosives and rocks was the key factor of increasing the coefficient of energy utilization and improving rock blasting effect. Currently, the crush mechanism of different explosives used in different rocks is not clearly understood, moreover, there is not suitable correlation index to describe explosive energy feature and energy consumption of rock fragmentation. Hence, to find a new correlation index between explosive energy output and rock dynamic broken is the key issue of optimizing coupling of explosives and rocks.

As rocks dynamic respond changes with different explosion energy[1,2], the index of rock impedance cannot reveal the whole features of dynamic respond of rocks. Meanwhile, rocks efficient fragmentation only occur when the energy transmitted to rocks matches the rocks dynamic respond, otherwise, the energy will transform into elastic wave and dissipate. Therefore, it is necessary to research rocks dynamic respond and broken characteristics, and find out the reliable indexes for establishing more suitable coupling principle between explosives and rocks[3].

2 PREPARATION OF ROCKS IMPACT LOAD TEST

2.1 Apparatus

The rocks dynamic characteristics were measured by Split Hopkisson Press Bar(SHPB) system which consists of test, loaded, measure apparatus and computer. The test apparatus includes input bar, transmit bar and absorb bar which are made by steel 40Cr. The steel's density is 7800kg/m^3, modulus of elasticity is 200GPa, and the Poisson's ratio is 0.3. Both the input and transmit bar's size are ϕ50mm×1800mm, and the impact bar's is ϕ50mm×400mm. The loaded apparatus uses high pressure bottle of liquid nitrogen with up to 15MPa pressure. The velocity of impact bar was measured by BC-202 detonation velocity meter, and the strains were detected by SDY2107B super dynamic strain instrument. The strain gages were stuck on the middle along the long axis of input and transmit bars. The reliability test must be done before actual test. Copper sheet with 1mm thick and 1mm diameter was selected as pulse shaper, and the software of matlab was selected as programming platform for analyzing and calculating of stress-strain, modulus of elasticity and energy consumption.

2.2 Basic Principles

According to the hypothesis of one dimension stress, the samples' stress $\sigma(t)$, strain $\varepsilon(t)$, strain ratio $\dot{\varepsilon}(t)$ can be calculated by[4]:

$$\sigma(t) = \frac{EA_b}{2A_s}[\varepsilon_i(t) - \varepsilon_r(t) - \varepsilon_t(t)] \quad (1)$$

$$\varepsilon(t) = \frac{c_0}{B}\int_0^t [\varepsilon_i(t) - \varepsilon_r(t) - \varepsilon_t(t)]dt \quad (2)$$

$$\dot{\varepsilon}(t) = \frac{c_0}{B}[\varepsilon_i(t) - \varepsilon_r(t) - \varepsilon_t(t)] \quad (3)$$

Corresponding author E-mail: yht 03121 @ 163. com

From the hypothesis of homogenization, there is

$$\varepsilon_i(t) + \varepsilon_r(t) = \varepsilon_t(t) \quad (4)$$

Then, the expression of (1),(2),(3) can be simplified as follow:

$$\sigma(t) = \frac{A_b}{A_s} E \varepsilon_t(t) \quad (5)$$

$$\varepsilon(t) = \frac{2c_0}{B} \int_0^t \varepsilon_r(t) \quad (6)$$

$$\dot{\varepsilon}(t) = \frac{2c_0}{B} \varepsilon_r \quad (7)$$

where A_s ——cross sectional area of sample, mm^2;
 B ——thick of sample, mm;
 A_b ——cross sectional area of press bar, mm^2;
 E ——elasticity modulus of press bar, GPa;
 c_0 ——velocity of longitudinal wave in press bar, m/s.

2.3 Samples Processed

Granite, phyllite and magnetite quartzite were chosen as research object. The suitable size rocks were selected carefully from one open pit mine blasted muck pile and took to lab. First, the core samples were drilled by drilling machine of model ZS-100, then the core samples were cut into small samples according to the designed size by cutter. After this, the cut column samples were put on the two end-plane grinding machine of model SHM-200 to process into final samples. The samples size was designed as 50mm in diameter and 25mm in thickness, and 0.2mm of error in diameter and end-plane.

3 IMPACT PRESS TEST

3.1 Dynamic Stress and Strain

Five sets samples for every type of rocks were tested under different impact velocities. There were three qualified samples in every set. The test results were shown in Tab.1.

Tab.1 Data of impact test

Rock type	No.	Thickness/diameter/mm	Impact velocity/m·s^{-1}	Average strain ratio/s^{-1}	Peak stress/MPa
Granite	1-1	25/50	6.03	56.60	40.2
	1-2	25/50	5.89	57.56	23.3
	1-3	25/50	5.94	58.32	35.5
	2-1	25/50	8.79	84.96	96.0
	2-2	25/50	9.13	87.38	116
	2-3	25/50	8.89	80.35	120
	3-1	25/50	11.48	83.20	131
	3-2	25/50	12.10	68.20	118
	3-3	25/50	11.72	74.38	123
	4-1	25/50	14.33	135.57	129
	4-2	25/50	14.10	130.12	120
	4-3	25/50	13.88	139.07	127
	5-1	25/50	17.60	196.26	179
	5-2	25/50	18.14	213.07	118
	5-3	25/50	17.84	207.90	130
Phyllite	1-1	25/50	6.77	46.56	51.3
	1-2	25/50	6.39	46.68	31.7
	1-3	25/50	6.35	48.05	38.0
	2-1	25/50	10.90	58.35	26.9
	2-2	25/50	10.52	59.40	25.0
	2-3	25/50	10.78	50.37	23.5
	3-1	25/50	12.80	99.93	39.3
	3-2	25/50	12.95	86.19	45.7
	3-3	25/50	12.64	89.25	41.2
	4-1	25/50	14.36	61.69	58.9
	4-2	25/50	14.46	74.38	73.4
	4-3	25/50	14.67	85.04	63.0
	5-1	25/50	16.88	294.01	82.9
	5-2	25/50	16.41	271.68	75.4
	5-3	25/50	16.57	280.35	70.3

Continues Tab.1

Rock type	No.	Thickness/diameter/ mm	Impact velocity/m·s^{-1}	Average strain ratio/s^{-1}	Peak stress/MPa
Quartzite	1-1	25/50	7.69	110.76	44.5
	1-2	25/50	7.21	102.28	55.6
	1-3	25/50	7.31	95.56	43.4
	2-1	25/50	9.39	110.28	49.9
	2-2	25/50	9.60	105.88	54.7
	2-3	25/50	9.50	137.57	60.9
	3-1	25/50	12.22	114.91	144
	3-2	25/50	12.03	137.90	120
	3-3	25/50	12.35	125.88	116
	4-1	25/50	14.18	155.37	113
	4-2	25/50	14.36	135.06	137
	4-3	25/50	14.45	140.72	154
	5-1	25/50	16.91	181.53	183
	5-2	25/50	17.27	171.26	190
	5-3	25/50	17.08	175.36	185

From Tab.1,the stress-strain curves, stress-time curves and strain ratio-time curves can be obtained. They shown as Fig.1,Fig.2 and Fig.3. Fig.1 shows that all the stress-strain curves have similarly style with statics. But every peak stress and modulus of elasticity changed with the different impact load.

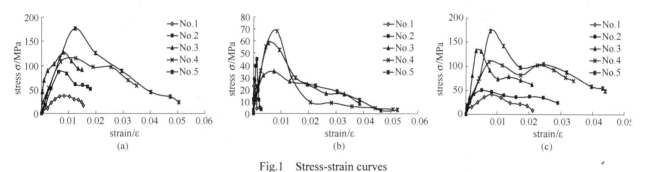

Fig.1 Stress-strain curves
(a) granite ;(b) phyllite; (c) quartzite

Fig.2 dedicated that the dynamics stress increased with the impact velocity and the peak stress occurred earlier. These phenomena could be explained as that there was not enough time for the samples' inside original micro-flaw to dehisce or link up and then the deform delayed. Furthermore, the delay became more obviously along with the impact velocity speed up.

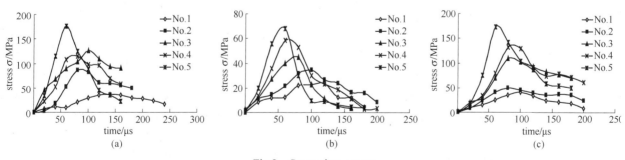

Fig.2 Stress-time curves
(a) granite ;(b) phyllite; (c) quartzite

It shows from Fig.3 that the strain ratios of three types samples are well correlated with loading rate. The slower the loading rate, the smaller the changing of strain ratio, and the curves were gently. When the loading rate was faster, the arrange of strain ratio changing became wider, and there exist minus at the end of curve because of the rebounded action of samples load-off.

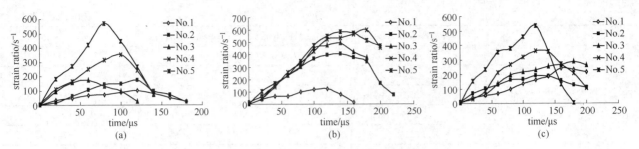

Fig.3 Strain ratio-time curves

(a) granite ;(b) phyllite; (c) quartzite

The above data analysis shows that the same type of rock has different deformation and strength features under different loading rate. The dynamics features of different types of rock especially the time depend features have obviously difference with similar impact.

3.2 Impact Velocity and Strain Ratio

It revealed from the dynamics test that loading rates have obviously influence on average strain ratio of samples. The relationship between loading rates and average strain ratios can be described as Fig 4. The impact velocity in the test was controlled in the range of 5m/s to 17m/s, thus, the average ratio changed from $50s^{-1}$ to $300s^{-1}$. It is believed that every time of the test had same initial position and distance of run of the impact bar. The average ratio shows nonlinear changing along with different impact, and the dynamics responds of three types of sample differ distinctively. The curves in Fig.4 could be described in expressions by fitting method among which polynomial format had good fit.

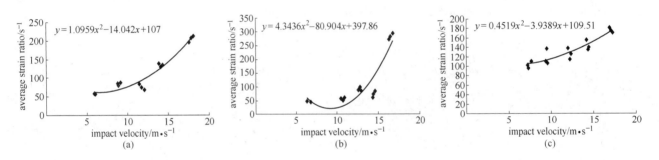

Fig.4 Impact velocity and average strain ratio curves

(a) granite ;(b) phyllite; (c) quartzite

Fig.4 also revealed that the dense particles and little of joint and crack reduced small deformation during the original structure being closed for granite and quartzite. The most energy of impact was used to produce new cracks, thus, the strength impact deduced obviously crushing. On the other hand, the original joints and cracks grew well in phyllite, the impact load made the original joints and cracks closed first and reduced a lager deformation. Furthermore, the strain ratio reduced when the moment of the joints and cracks closed, and soon, the strain ratio would further increase by the new structures growing. This process told that unsuitable higher energy input would deduce the original structure efficiency in crushing because the energy was not to guide the original structure broken but to produce new structures. Therefore, the strength and time features decided the utility efficiency of energy in crushing different types of rocks.

4 STUDY ON ENERGY CONSUMPTION OF ROCK CRUSHING

4.1 Particle Size and Energy Consumption

The strength and deformation characteristics of the three types of samples have been analyzed above, and the energy consumption will be discussed below. The energy distribution in the test for the three types of samples is shown in a table(omit). The fitting method is used here to find out the rule between the average crushing particle size of samples and the efficient energy using, and the curves were drew in Fig.5, the expressions could be extracted as follows:

Granite:
$$\eta_1 = 0.2305f^3 - 4.9123f^2 + 28.306f + 3.7278 \tag{8}$$

Phyllite:
$$\eta_2 = 55.374f^2 - 316.1f + 456.1 \tag{9}$$

Quartzite:
$$\eta_3 = 0.0206f^4 - 0.7661f^3 + 9.5175f^2 - 44.917f + 109 \tag{10}$$

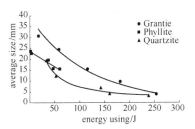

Fig.5 Particle size and energy using curves

For the difference of the dynamics characteristics of the three types of rocks, the output structure of energy has important influence on the utilization of impact energy.Fig.6 shows the different relationship of samples crushing between utilization of energy and crushing rate. The energy utilization means the ratio of useful energy to input energy, whereas the crushing rate means the ratio of longest size to average crushing size of samples. It can be found from fig.6 that the granite and phyllite samples' useful energy utilization are undulate with the changing of crushing rate. but for phyllite samples, the energy utilization is becoming lower with the crushing rate increase, and most of energy turned into reflection. This told that different rocks when being crushed into different particles need different output parameters of energy to get higher energy utilization.

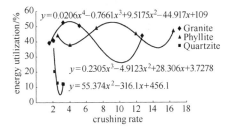

Fig.6 Relation of energy utilization and crushing rate of samples

Fig.7 gave out the relationship of energy utilization with energy input amount for the three types of samples. In the range of this test, there existed an area which the energy utilization was higher for granite and quartzite, however, for phyllite,the energy utilization had a steep decrease and then remained relatively steady.

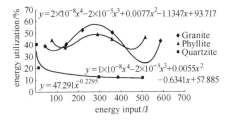

Fig.7 Energy input and utilization curves

The test showed that the energy utilization of rocks crushing is closely related to crushing rate of rocks first, and is also closely related to input structure of energy. Therefore, it is important to analyze the characteristics of energy input so as to increase the energy utilization of rock crushing.

4.2 Energy Density Per Unit Time Analysis

As is discussed above, both the dynamics features of rocks and energy input structures decided the energy utilization, the rocks' properties have been discussed and the suitable indexes of energy structures will be dealt with bellow.

By the principle of dynamics impact test, if the energy dissipation of impact heat and particles fly off were not considered, the useful energy consumption in the impact test can be described as:

$$W_e = W_I - W_r - W_T \quad (11)$$

$$W_e = \int_0^{\tau_1} \varepsilon_I^2(t)dt - \int_0^{\tau_2} \varepsilon_r^2(t)dt - \int_0^{\tau_3} \varepsilon_T^2(t)dt \quad (12)$$

where W_e——effective energy consumption, J;
W_I——input energy, J;
W_r——reflected energy, J;
W_T——transmitted energy, J.

The expression (12) can be divided into two parts: the first part is input energy which includes energy strength and working time and is decided by impact bar's quality and velocity in the test. The second part includes item two and item three which were decided by rocks'features and the amount controlled by strength and working time of reflective and transmitted waves. Both energy output explosion and machine impact have time depended strength features, and the energy absorbed in rock crushing has the same feature. So the concept of energy density per unit time is selected as index of describing energy structure and used to reveal the features of energy released and absorbed. The concept means the energy input unit volume of samples from impact bar in unit time. The output energy density per unit time was expressed as:

$$K_I = \frac{1}{\tau_1 V \rho} \int_0^{\tau_1} \varepsilon^2(t) dt \quad (13)$$

Meanwhile, the effective energy density per unit time was expressed as

$$K_e = \frac{c_0}{BV\rho}(W_I - W_r - W_T) \quad (14)$$

where ρ——sample density, kg/cm^3;
V——volume of sample,cm^3.

In bench blasting of open pit mine, the energy density per unit time of explosive transmit to rock mass was expressed as:

$$K = \frac{\pi D_0^2 Q \rho_0 D}{4ab\rho} \quad (15)$$

where D_0——holes diameter, m;
Q——heat of explosive released, J/kg;
ρ_0——density of charge, kg/m^3;
D——explosion velocity, m/s;

ab——blast hole burden area, m^2;
ρ——density of rock, kg/m^3.

5 CONCLUSIONS

Rocks dynamics characteristics and crushing energy consumption were analyzed by SHPB system in the test and useful data were obtained. By analyzing, the rules of samples strength, deformation and time effect were discussed. For different rocks, the energy utilization was strongly influenced by energy density per unit time and crushing size, and there existed obvious relationship between rocks dynamics respond, energy output structure and crushing size. The concept, combining explosive energy with fragmentation energy utilization of energy density per unit time, set up an effective path to research the optimal coupling of explosives and rocks.

REFERENCES

[1] Guo L J, Yang Y H, Hua Y H. Test and Analysis on Distortion and Damage of Rock under Impact Loading[J]. *Journal of Water Resources Architectural Engineering*, 2013, 11(6):31~34.

[2] Guo L J, Yang Y H, Hua Y H, Li L. Experimental Study of Dynamic Characteristics of Grantie under Impact Loading[J]. *Engineering Blasting*, 2014, 20(1):1~4.

[3] Wang X G, Zheng B X, Song J Q. Blast technology present situation and development in China[C]//*New Blast Technology in China* III, Beijing: *Metallurgical Industry Press*, 2012.

[4] Fu Zh L. Test Guide of Rock Mechanics[M]. Beijing: *Chemical Industrial Press*, 2011.

Evaluation of Progressive Collapse of Blasting Demolition for Culvert Structures

Young-Hun Ko, Jung-Gyu Kim, Yousong Noh, Myeong-Jin Shim, Hyung-Sik Yang
(*Department of Energy and Resource Engineering Chonnam National University, Gwangju, Korea*)

ABSTRACT: In this study, objects of demolition are underground culvert structures to be removed for entire width of road. In order to evaluate the effectiveness and verification of collapse, numerical analysis and scaled model test were performed. Collapse were evaluated by DCR (demand-capacity ratio) method for linear static analysis based on the GSA(General Service Administration) guideline. Scaled model tests were carried out to verify the numerical analysis results. As a result of model experiment, it was confirmed that roof slab and soil cover were collapsed due to internal blast pressure when the center column was eliminated by explosive.

KEYWORDS: culvert structures; GSA; DCR; scaled model test

1 INTRODUCTION

The object in this study, underground structure for blasting is applied in areas under relatively wide road and high mobility as Fig.1. Object of the demolition is an artificial structure located under the wide municipal road where traffic is heavy(Fig.1).In this structure, explosive is charged on center columns and slabs for instantaneous detonation but progressive collapsing of the structure. But in this case, serious reservations about effectiveness and possibility of demolition were raised on account that the design of this structure was not built considering demolition nor blasting theory. It was built simply designed by general consideration for structures. In this study, numerical analysis and model experiment were implemented to verify effectiveness of the structure.

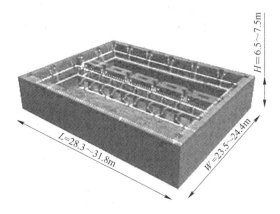

Fig.1 Underground structure for demolition
(ROK Army Headquarters, 2006)

2 PROGRESSIVE COLLAPSE ANALYSIS METHOD

In June 2003, the United States General Services Administration (GSA) released their latest version for progressive collapse mitigation guidelines to be applied for all federal buildings in the USA(The U.S. General Services Administration, 2003).

Applied load of progressive collapse analysis is as Fig2. For static analysis loading, 2.0(1.0DL + 0.25LL) was imposed on RC beam around removed column and for near beams, 1.0DL + 0.25LL was imposed as gravity load. For dynamic analysis loading, load combination of 1.0DL + 0.25LL was applied on the model that column is removed.

The limiting criterion of efficiency of progressive collapse mitigation is drawn by DCR(demand-capacity ratio) for linear analysis, angle of rotation(θ) of structure member for non-linear analysis.

The term, "demand", for linear-static analysis is defined as the stress value of the structure according to vertical loading due to column removal, and "capacity" is defined as the momentary ultimate strength of each component of building analyzed from consideration working load and horizontal load in steady-state before column removal. *DCR*, which is the criterion for efficiency of progressive collapse mitigation for linear-static analysis, is value of demand divided by capacity.If *DCR* is below 2.0 for typical buildings, 1.5 for a typical buildings, respectively, then, it is considered that additional collapse by bending will not be happen.

Corresponding author E-mail: hsyang@chonnam.ac.kr

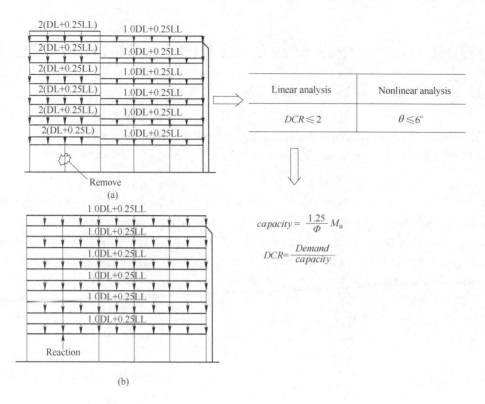

Fig.2　Applied load of progressive collapse analysis and allowable value for linear and nonlinear analysis of RC beam (GSA)

(a) static load GSA; (b) dynamic load GSA

3　NUMERICAL ANALYSIS

For progressive collapse analysis of this structure, MIDAS CIVIL, which was finite-element analyzing program, was used to evaluate collapse of roof of the structure by comparison the result with *DCR* value from linear-static analysis from *GSA* guidelines. Fig.3 is the result of modeling and numerical analysis.

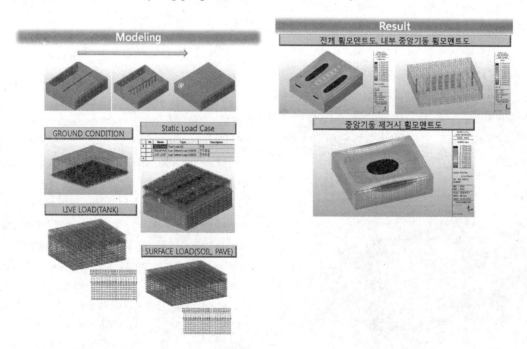

Fig.3　Model and result of numerical analysis

The result of analysis is compared with *DCR* from linear-static analysis, which is one of progressive collapse analysis methods, and if *DCR* value was above 2.0 then it was assumed that additional collapse would occur.

The moment of slab and end of slab due to vertical loading by center column removal was substituted for demand value, and the momentary ultimate strength of slab structure was substituted for capacity value, thereby *DCR* was calculated to 6.89 for slab, 4.38 for slab end, respectively, which are vastly above the limit figure of *DCR* of typical building, and this result shows that demolition of the slab by vertical and live load according to removal of center column by blasting is convinced to occur. This was indicated on Tab.1.

Tab.1 DCR(demand-capacity ratio) of structures

non Removal column $Mu/\mathrm{kN} \cdot \mathrm{m}^{-1}$		Removal column $Mu/\mathrm{kN} \cdot \mathrm{m}^{-1}$		Strength Reduction Factor φ_f
Slab	slab end	slab	slab end	
597	1296	6049	8346	0.85
DCR		6.89	4.38	

4 SCALED MODEL TEST

To examine collapse aspect of the structure after removing center column of the target, model was produced and experimented as Fig.4. The model has dimensions of 5 m length, 4.1 m width and height of 1.9 m, designed and constructed as a reinforced concrete box structure and was covered up with soil.

Charged quantity of explosive on the center column was 44kg which was calculated by corresponding military formula, and aspect of destruction of roof slab due to internal blast pressure and collapsing phase of covered soil was examined while blasting.

(a)

(b)

(c)

(d)

Fig.4 Demolition blasting of scaled model
(a) construction, curing; (b) after covering by soil; (c) TNT arrangement; (d) after blasting

5 CONCLUSIONS

The result from the numerical analysis and the model experiment on the underground structure for blasting is as in the following. Due to GSA(General Service Administration) standards, DCR(Demand-capacity ratio) of the slab and slab end were 6.89 and 4.38 respectively, which are above the limit figure of *DCR*, 2.0. And this result shows that demolition of the slab by vertical and live load according to removal of center column by blasting is convinced to occur. To verify the result on numerical analysis, model experiment was implemented. It was supposed that center column of the structure was removed by blasting, TNT was loaded on the midst bottom and blasted. The result showed demolition of roof slab and soil cover.

Hereafter, effect on roof slab and road on earth according to internal interference of blasting pressure in underground structure should be considered and measured additionally.

REFERENCES

[1] ROK Army Headquarters, 2006. Military Engineering Training Manual: 272.

[2] GSA, 2003.Progressive Collapse Analysis and Design Guidelines for New Federal Office Buildings and Major Modernization Projects, The U.S. General Services Administration: 4~10.

[3] NIST, 2007, Best Practices for Reducing the Potential for Progressive Collapse in Buildings: 91~107.

Vertical Shaft Blasting Parameters Design Software based on Visual Basic

ZHANG Jiaming[1], LI Pingfeng[2], ZHOU Ming'an[1]

(*1. College of Basic Education for Commanding Officers, NUDT, Changsha, Hunan, China;
2. Hunan Lianshao Construction Engineering Limited Liability Company, Guangdong Hongda Blasting Company, Loudi, Hunan, China*)

ABSTRACT: The design of blasting parameters, the most important first step in vertical shaft blasting engineering, which mainly adopts the empirical method at present, affects the safety and quality of the follow-up process directly. Visual Basic, one of the most widely used at present, is based on a visual programming tool Windows environment programming tool. It can be used in the development of database management systems, multimedia applications and graphics applications. Based on Visual Basic platform, software design of vertical shaft blasting parameters is developed, and it is based on the theory of blasting and a large amount of domestic and international experience. As long as we input the parameters of surrounding rock and necessary data of section sizes of vertical shaft, the software can quickly call the database related algorithm and optimized design. It can automatically generate the holes' distance, aperture and single holes' charge and other parameters of the cut holes, auxiliary holes and Peripheral holes. At the same time, according to the design results of optimal parameter generation blast holes' layout, it will provide the designer with image, intuitive, accurate data and image, leading to vertical shaft blasting design more systematic, fast and precise. It is of great significance to control the blasting effect, shorten the construction period and improve economic efficiency.

KEYWORDS: vertical shaft blasting; Visual Basic; parameters design; software development

1 FOREWORD

The design of blasting parameters is the most important first step in vertical shaft blasting engineering. Though people have gathered a great deal of experience in the long engineering practices and researches, we mainly adopt the empirical method at present because of the complexity and variety of vertical shaft blasting engineering. The design results of blasting parameters varies from person to person frequently, and it can't meet the demand of vertical shaft blasting engineering, which affects the safety, schedule and quality of the process directly.

Based on Visual Basic platform, software design of vertical shaft blasting parameters is developed, and it is based on the scientific theory of blasting as well as a large amount of experts' experience and engineering. The software can quickly call the database related algorithm to design and optimize the parameter. At the same time, it will provide the image of design results. As a result, the current situation of many design errors from empirical method will be solved. Besides, the software simplifies the operation, leading to vertical shaft blasting design more systematic, precise and quantitative.

2 THEORETICAL PRINCIPLE AND SOFTWARE SUPPORT

2.1 Classification of Wall Rock

The software design of vertical shaft blasting parameters mainly adopts the classification of wall rock, which takes the wall rock's characteristic condition and stabilization into account synthetically.

2.2 Theoretical Principle of Vertical Shafting Blasting

The database of software is based on the theory of blasting and a large amount of domestic and international experience, including shock wave and explosive gas interaction theory, the rock stress wave theory, dynamic strength theory of rock under loading and so on.

2.3 Visual Basic

Visual Basic, one of the most widely used at present, is based on a visual programming tool Windows environment programming tool. The concept of control is draw into Visual Basic, which provides a large number of interface elements for the software designers, such as forms, button,

Corresponding author E-mail: 441230756@qq.com

menu, textbox and so on. Visual basic achieves the modularity design of these controls. The software designers only need to set these controls on the appropriate place freely instead of programming, which simplifies the software development greatly. It can be used in the development of database management systems, multimedia applications and graphics applications.

2.4 Feedback Design Method

Due to the uncertainty of blasting parameters, blasting of rock in the process of high complexity and the influence factors of diversity, it is difficult to realize the optimization of parameters. Using feedback design in this software, feedback design software of vertical shaft blasting parameters design is developed based on theoretical study and experimental study. It can feedback the real-time data from the test gun to the software and further parameters, which realizes the optimization of vertical shaft blasting parameters furthest and improves the accuracy of parameter design.

3 THE PRINCIPLE OF SOFTWARE DESIGN

3.1 The Simplification of Human Computer Dialogue Interface

The human computer dialogue interface, being used to input the parameters of surrounding rock and output the results of parameter design, adopts the concise and convenient dialog box or drop-down menu. Only need simple button orders and data input, it can finish the design of vertical shaft blasting parameters. It also provides prompt message for example, the classification of wall rock.

3.2 Automated and Visualized Design

As long as we input the parameters of surrounding rock and necessary data of section sizes of vertical shaft, the software can quickly call the database related algorithm and optimized design. It can automatically generate the holes' distance, aperture and single holes' charge and other parameters of the cut holes, auxiliary holes and Peripheral holes. At the same time, according to the design results of optimal parameter generation blast holes' layout, it will provide the designer with image, intuitive, accurate data and image.

3.3 The Design of Blasting Parameters Precise

Based on the input parameters and database of blasting theory, we can finish the blasting engineering experiment by the initial parameters design form software. Then, the software is designed precisely depending on the experiment data, as well as designs and optimizes the parameters, ensuring the blasting parameters accurate and leading to vertical shaft blasting design more precise.

4 MAIN FUNCTIONS OF SOFTWARE

Owing to Visual basic is an object oriented computer language, object can data and operate singly. Thus, according to separate the definition and application object by encapsulating, it can achieve the expansion and extension of the meaning of things by reducing the inheritance mechanism.

The entire shaft blasting parameters design software includes the following seven parts:

4.1 Original Data Input

It includes the parameters of surrounding rock, necessary data of section sizes of vertical shaft and so on.

4.2 Program Run Part

Based on the original data and the database of software, it can design a serious of blasting parameters.

4.3 Preserve and Output the Result

According to the blasting design, it can output a serious of parameters like the holes' distance, aperture and single holes' charge and other parameters of the cut holes, auxiliary holes and Peripheral holes. The software also can provide the blasting instruction automatically.

4.4 Image of Design Results

According to the design results of optimal parameter generation blast holes' layout, it will provide the designer with image, intuitive, accurate data and image.

4.5 Optimal Design of Feedback Function

According to the real-time test gun and obtained data, designers can feedback the data to software, modifying and optimizing the blasting parameters, leading to blasting parameters more accurate and feasible results.

4.6 Dynamic Design

During the engineering, the software can design the blasting parameters dynamically with the changing of wall rock's characteristic condition.

4.7 Prompt Files

The software prompt files of wall rock classification standard simplify the design and convenient designer operating the computer.

5 PROGRAM DESIGN

5.1 Introduction of the Main Program

The main program mainly includes 4 basic forms and the 2 dialog windows, and as shown in Fig.1. 4 basic forms are landing interface, classification file, selection of prompt interface and the output interface. The dialogue windows are mainly used for feedback input original data input and data of the test gun.

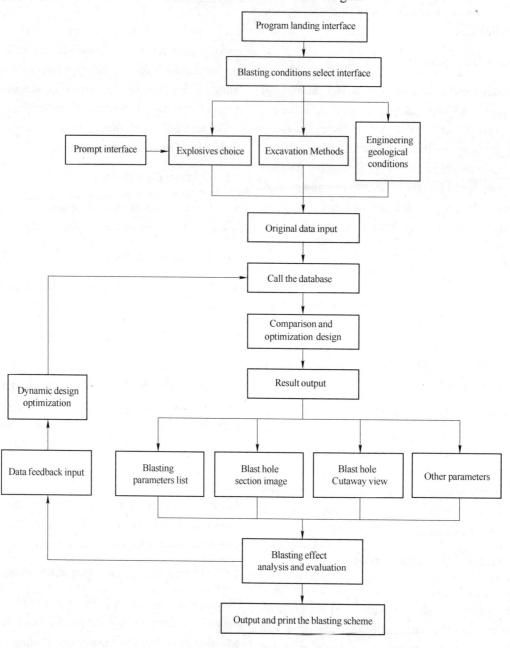

Fig.1　The main program structure

5.2　The Program Algorithm Example

The unit consumption of explosives Q for example, explosive consumption experience formula, due to engineering change limit, only as a reference.

Unit explosive consumption calculation formula examples are as follows:

(1) Lange FLS type

$$q = 1.5 \times 10^{-3} \times \left(\frac{a_1}{d_1}\right) \times (a_1 - 0.5d_1)$$

(2) Lange FLS simplification

$$q = q_n(a_1 - 0.5d_1)$$

(3) the modified formula of the PU

$$q = 1.1k_0\sqrt{f/S}$$

(4) according to the rock solid coefficient and basal area estimation formula

$$q = 1.4\sqrt{f/S}$$

(5) according to the empirical formula of surrounding rock classification and section size

$$q = KB$$

The theoretical formula and empirical formula of a series

of, have certain error with the practical construction. Based on the empirical formula of rate, function setting unit explosive consumption Q and rock solid coefficient F, basal area of S:

$$q = a + b\sqrt{f/S}$$

Engineering live data calls in database, using the linear fitting coefficient optimization calculation.

No.	f_i	S_i	$x_i = \sqrt{f/S}$	q_i	$x_i q_i$	x_i^2
1	8	26.4	0.550	2.15	1.183	0.303
2	12	24.6	0.698	1.79	1.249	0.487
3	6	15	0.632	1.20	0.758	0.399
4	13	12	1.041	2.72	2.832	1.084
5	16	28.4	0.751	2.33	1.750	0.564
⋮	⋮	⋮	⋮	⋮	⋮	⋮
m	6	19.3	0.558	1.08	0.603	0.311
Σ			21.807	59.90	43.443	16.035

Method of fitting curve equation, a solution using Cramer's rule:

$$a = \left(\sum_{i=1}^{m} q_i \sum_{i=1}^{m} x_i^2 - \sum_{i=1}^{m} x_i \sum_{i=1}^{m} x_i q_i\right) \bigg/ \left[m\sum_{i=1}^{m} x_i^2 - \left(\sum_{i=1}^{m} x_i\right)^2\right]$$

$$b = \left(m\sum_{i=1}^{m} x_i q_i - \sum_{i=1}^{m} x_i \sum_{i=1}^{m} q_i\right) \bigg/ \left[m\sum_{i=1}^{m} x_i^2 - \left(\sum_{i=1}^{m} x_i\right)^2\right]$$

The solution

$a=0.62296 \qquad b=1.87995$

The fitting equation

$$q = 0.62296 + 1.87995\sqrt{f/S}$$

Optimization function relations unit explosive consumption Q and rock solid coefficient F, S basal area, refer to table were compared, and the national standard unit explosive consumption by checking, error is less than 5%. The theoretical formula and empirical formula in the database as the foundation, then call a lot of engineering live data calculation and optimization, database of engineering data has been added, so that the software optimization results become more accurate.

5.3 Program Demo

5.3.1 Program Landing (as shown in Fig.2)

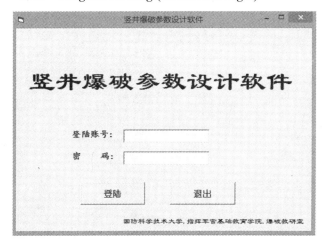

Fig.2 Program landing interface

5.3.2 The Blasting Conditions and Initial Data Input (as shown in Fig.3)

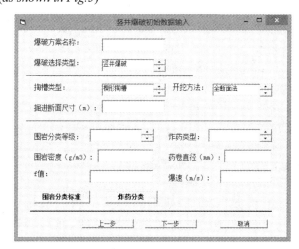

Fig.3 Original data input interface

5.3.3 Shaft Blasting Parameters Design (as shown in Fig.4)

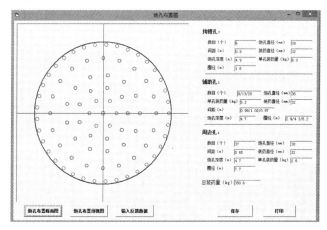

Fig.4 Blast hole section image

Background: well blasting case design, shaft diameter 7.7 m, platts hardness coefficient of rock is 10, saturated compressive ultimate strength of more than 30 MPa, belonging to the V grade hard rock. The rock geological structure effect is slight, joint dysplasia, without interlayer, blocky structure. The parallel cut blasting method, the excavation depth 4.7 m. The emulsion explosive, charge diameter is 32 mm, the hole diameter 38 mm.

Along with the construction of a deep, well depth increasing, rock hardness enhancement, platts hardness coefficient is 14, the saturated compressive ultimate strength of more than 60 MPa, which belongs to the class of hard rock. The original blasting parameters are not suitable for the construction of. On the basis of the original blasting parameters design, using software algorithm, gradually improve the hardness coefficient of rock, the dynamic optimization design of parameters, the design of blasting parameters to meet the construction requirements that live, dynamic design.

6 CONCLUSIONS

(1) Based on Visual Basic, the shaft blasting parameters design software is developed. The concise operation interface makes the design more convenient, leading to vertical shaft blasting design more systematic and fast.

(2) For the different wall rock's characteristic condition, performance of explosive, section sizes of vertical shaft and so on, the software has the corresponding humanized prompt table files, which clears the classification standard and prompts the performance of parameters.

(3) It is applied to the design of blasting parameters for different environments, enabling designers to accurately design scheme. Meanwhile, the image of design results, including section and Cutaway view, provide the designer with image, intuitive, accurate data and image.

(4) Using feedback design method and dynamic design method, designers can modify and optimize the blasting parameters on the basis of the real-time test gun and obtained data, leading to blasting parameters more accurate and feasible results.

(5) Software database including theory formula, empirical formula and a lot of engineering live data, every construction are added to the data in the database, the software optimization design is more and more precise.

REFERENCES

[1] Wang J Q, Xu W Y. Development and Application of computer aided design software of open pit blasting[J]. *Metal World*, 2009 (z1): 56~60.

[2] Wang H M, Dong Z X. Tunnel excavation blasting chart of computer aided design[J]. *Engineering Blasting*, 2003, 9(3): 38~41.

[3] Xie Q Z, Li Z L. Blasting rock chart computer aided design[J]. *Blasting*, 2002, 19(1):21~23.

[4] Fu G M, Zhou M A. *Military Demolition Project*[M]. Changsha: National University of Defense Technology press, 2011.

[5] Liu D S, Li S L. *Demolition Project*[M]. Beijing: Science Press, 2011.

[6] Gao C Y, Liu B B. *Visual Basic development of actual collection*[M]. Beijing: Tsinghua University press, 2010.

Distribution of Blasting-Induced Dynamic Events in Rock Mass

A.A. Eremenko[1], S.V. Fefelov[1], V.A. Eremenko[2]
(1. Chinakal Institute of Mining SB RAS, Russia;
2. Institute of Comprehensive Exploitation of Mineral Resources (IPKON), Russian Academy of Sciences, Russia)

ABSTRACT: The authors have developed multiple-row blasting patterns for beam cetiguous charges 105～250 mm in diameter and vertical concentrated charges arranged across the width of rock blocks to be blasted.

The article describes the geomechanical condition of rocks under blasting. It has been found that the quality of blasting fragmentation and the seismic energy of tremors are influenced by weight distribution of explosive per delay intervals.

KEYWORDS: seismic events; explosion; rock mass; energy; rock block

Mining in highly rockburst-hazardous rock masses requires adequate ground control, for instance, by means of optimized geometrics of stopes and goafs, or parameters of geotechnology that affect rock mass behavior. Out of all technology processes, blasting exerts the strongest impact on enclosing rocks. Different scale blasting invokes various dynamic events in mines. The connection of blasting and dynamic events implies possibility to use blast energy to control energy of seismic ad dynamic events in rock masses and to reach quality fragmentation of rocks by blasting. Intensity of a dynamic event is influenced by location of a blast source, weight of explosive charges and sequence of blasting. Block caving of ore is simultaneously carried out toward balance chambers (specially driven slot raises to accommodate expanded volume of shattered loose rock) and toward confined environment (muck) [1]. With the accepted two-row arrangement of borehole fans, the line of least resistance was 7.5～10 m on the side of the slot raise and 6～9 m on the side of the confined environment. The fans were composed of 20 to 30 contiguous blastholes. The powder factor ranged from 0.45 to 0.67 kg/t; yield of oversize was high—from 6% to 8%.

Improvement of the blasting performance was reached by arranging the fans of blastholes 105～250 mm in diameter in three–five acrwise rows the width of a block to be blasted; the rational number of advance charges in an arc was 0.3～0.5 of their number in the other arc, which allowed reduction in the powder factor from 0.17 to 0.03～0.05 kg/t for the secondary fragmentation and decreased the number of charges in the fans down to 5～10 charges.

Owing to the optimized arrangement of fans of contiguous and vertical charges of increased diameter (150～250 mm and 1000 mm), with the advance blasting toward four–six balance chambers and then toward the confined environment [2], the blasting powder factor was decreased 1.2 times, the secondary fragmentation powder factor was lowered to 0.03～0.04 kg/t, and the blasting performance was increased by 15%～20%.

In level mining, upward and ring holes 105 mm in diameter are drilled in a block roof. The blocks are mainly 40～110 m long, 22～80 m wide and 60～150 m high. Usually in such blocks 5～8 rows down fans of contiguous and parallel-contiguous holes 105, 150 and 250 mm in diameter are drilled. The spacing of the rows is 5.0～5.5 m. The blasting powder factor is 0.5～0.65 kg/t; the secondary fragmentation powder factor is 0.03～0.05 kg/t.

Geomechanical assessment of rock mass condition during blasting. Blasting of block 20～21 at the center of the ore body induced 209 shocks with their total energy not higher than the blast energy of 6.6×10^6 J. Concrete lining in crossdrifts in block 20～21 was broken. Caving of block 7 at the north edge of the ore body provoked no large seismic events (Fig. 1).

It is found that distribution of explosive weight by the delay intervals influences the rock fragmentation quality and affects the seismic energy of shocks. Caving of blocks 19～13 and 14 on the East site and blocks 2～4 on the South-East site with the delay intervals of 225, 250, 300 and 350 ms used explosive charges 28～29 t in weight, which caused higher total seismic energy of dynamic events—5×10^8 J (Fig. 2). With the delay intervals of 150 and 300 ms in blocks 2 and 3 on the South-East site 2); the seismic energy of shocks was 5.7×10^6～1.62×10^8 J; the energy class index Ke (the ratio of the seismic energy of blast to the weight of explosive) varied between 2×10^{-4} and 7×10^{-5} (Fig. 2). The decrease in the weight of explosive charges to 10～15 (20) t with delay intervals from 0 to 350 ms reduced the seismic energy of the dynamic events.

Corresponding author E-mail: eremenko@ngs.ru

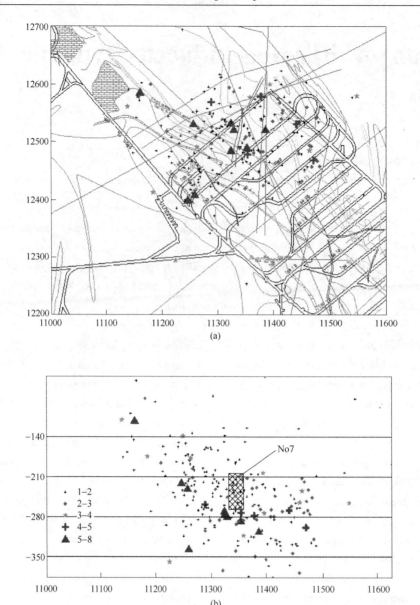

Fig. 1 Dynamic activity in mine after large-scale blasting of block 7 in (a) plan view and (b) elevational view

1~8 mean the energy; 11000~11600, 12200~12700— the coordinates x and y; −140~−350 m are the mine levels

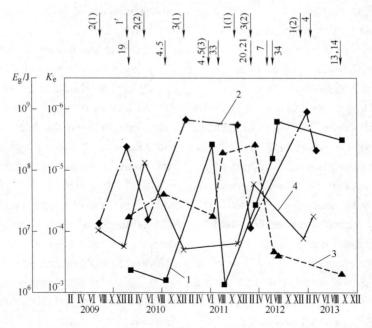

Fig.2 The variation of the energy class index Ke in blasting of blocks on the East (curve 1) and South-East (curve 2) sites and the change of the total seismic energy of shocks, Eg (curves 3 and 4): 19–13 are the East site blocks; 2(1)–4 are the South-East site block

Thus, the distribution and intensity of dynamic events in mines are influenced, aside from rock mass stresses, by the mining system, explosive weight and location of a blast source.

REFERENCES

[1] Eremenko A A. Improvement of blasting technology for iron ore mines in West Siberia. Novosibirsk: Nauka, 2013 (in Russian).

[2] Viktorov S D, Eremenko A A, Zakalinsky V M, Mashukov I V. Large-scale blasting technology for rockburst-hazardous orebodies in Siberia. Novosibirsk: Nauka, 2005 (in Russian).

Study on Numerical Simulation of Rock Blasting Based on SPH

LIAO Xueyan, SHI Fuqiang, JIANG Yaogang, GONG Zhigang

(*Sichuan Province Academy of Safety Science and Technology, Chengdu, Sichuan, China*)

ABSTRACT: The Lagrange, Euler and ALE are often used in simulation of explosion, but each has both advantages and disadvantages in simulation of propagation of explosion shock wave, deformation and velocity of medium. And all of them could not simulate the progress of crack propagation, so this paper proposes a new method——SPH. SPH could simulate the whole rock explosion from explosion、rock crack formation and propagation to rock fragment.

KEYWORDS: blasting; SPH; numerical simulation

1 INTRODUCTION

The rock blasting is a large compressive deformation, the numerical simulation based on Lagrange or Euler can not continue, because the deformation of grid is too large. The finite element method usually use erosion or remeshing to get over the difficulty. The smoothed particle hydrodynamics named SPH, which was invented by Lucy, Gingoldd and Monaghan in 1977, was used for celestial successfully [2,3]. Then the SPH was applied to numerical simulation of underwater explosion[4~6]. Compared with the finite element and finite difference grid method, SPH method can solve the problem of large deformation and breakage. This paper uses the SPH method to simulate the process of rock blasting. The results show that, SPH method can the whole process simulation of rock blasting, the blasting explosive, explosive effect on rock, rock stress and strain, cracks and expansion, rock crushing and throwing. Therefore, the SPH method for the simulation of rock blasting and blasting scheme optimization provides a new method.

2 PRINCIPLE OF SPH

The core of the basic idea of SPH is a kind of interpolation theory, which is based on the kernel function, the continuum is discretized into a series of mass of a particle, the kernel approximation to equation, the basic process is as follows:

(1) The continuum is discretized into a series of quality SPH particle, no task connection between particles, so the SPH method without grid.

(2) Field function with integral representation to approximate, in the SPH method called kernel approximation method.

(3) Application support neighboring particles within the domain of the corresponding value sum to replace the integral expression of field function of field function of particle approximation.

(4) The particle approximation method is applied to all the partial differential equations of the field function related items, the partial differential equation.

(5) The particles are attached to the quality, it means that with the material properties of the particles of these particles; field variables using explicit integration method to get all the particles changes with time value.

3 SIMULATION OF ROCK BLASTING

The SPH method was used for the typical blasting crater and throwing blasting process simulation. In order to study the feasibility by simulation of rock blasting process of the SPH method. The RHT material model is used on the two calculation model, the uniaxial compressive strength is 35MPa; for ANFO explosives, the density is 930 kg/m^3, the equation of state is JWL.

3.1 Simulation of Blasting Crater

In order to verify the feasibility of SPH method. The simulation of typical blasting crater is be done. The physical dimension of model for rock is 6m×8m; the geometry size of charge is 0.08m×0.08m; the charge from the left boundary and the right boundary distance both is 4m, from the upper boundary 1m; the particle spacing is 40mm, as shown in Fig.1.

The blasting crater simulated forming process based on SPH is shown in Fig.2. As can be seen from Fig.2: first the rock which surround the charge is crushed by the explosive

Corresponding author E-mail: stagger @ mail.ustc.edu.cn

stress wave after the charge explosion; then the free surface of the rock is cracked by reflective tensile stress; at the same time inside the rock cracks and gradually expand the formation fracture zone, fracture and crack mainly concentrated in the upper part, a funnel is shaped; the upper part of the rock fly out, then the explosive funnel is formed eventually.

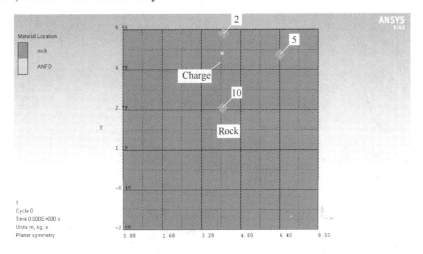

Fig.1　Model of blasting crater

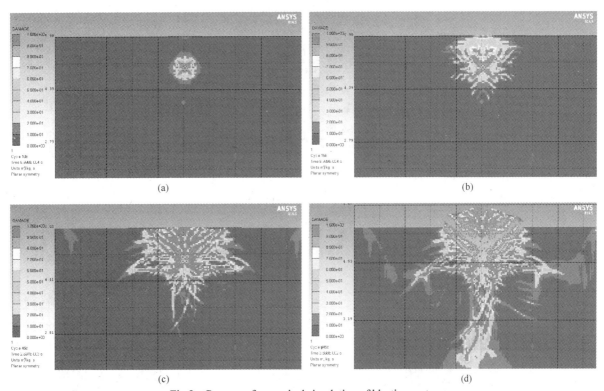

Fig.2　Process of numerical simulation of blasting crater
(a) 0.6 ms; (b) 0.9 ms; (c) 2.7 ms; (d) 3.9 ms

The SPH method can simulate the whole process of the formation of blasting crater, including the explosive. The process of the rock is compressed by explosive impact. The process of rock crack, propagation, fractured, the whole simulation process is consistent with the classic theory of blasting crater. In addition the stress, speed and other important parameters of any point in the model can be calculated as shown in Fig.3 and Fig.4.

3.2　Simulation of Blasting Crater

In order to verify the SPH method applicability in blasting engineering, two free surface casting blast is established. The physical dimension of model for rock is 8m×8m; the physical dimension of charge is 0.08m×0.08m; the charge from the left boundary distance is 2.6m, from the upper boundary 2m; Particle spacing is 40mm, as shown in Fig.5.

Using SPH method to simulate the rock throwing blasting process is shown in Fig.6. From the Fig.6 can be seen clearly the process of the charge explosion and surrounding rock crushed and stress reflection that reach the boundary is the same as blasting crater; but the difference: the stress wave in the left occur reflection, then crack upward extended also to the left, so the upper left part break and crack, upper left part rock throw out finally. The blasting crater process simulated by SPH and engineering actual match.

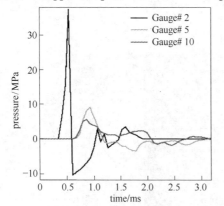

Fig.3　Stress time-history curves of particle

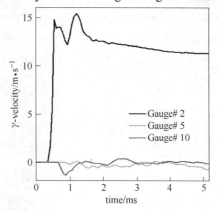

Fig.4　Velocity time-history curves of particle

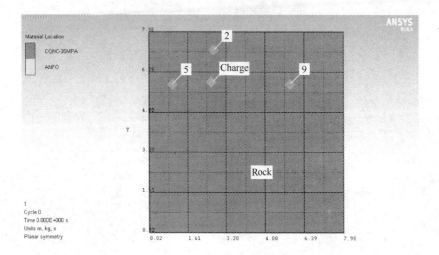

Fig.5　Model of casting blast

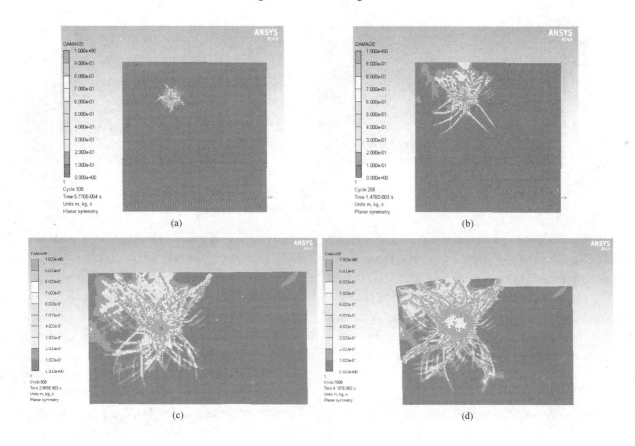

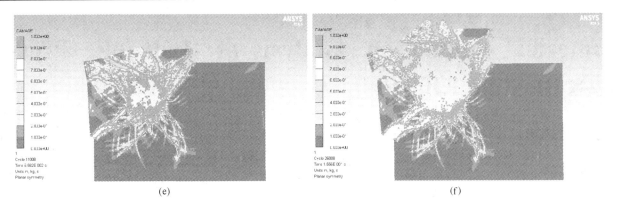

Fig.6 Damage cloud of casting blast
(a) 0.6 ms; (b) 1.5 ms; (c) 3.0 ms; (d) 41.9 ms; (e) 65.8 ms; (f) 155.6 ms

4 CONCLUSIONS

(1) The explosion crater model is simulated by SPH method. The result of numerical simulation shows that the SPH method can simulate the whole process of blasting that include the rock breaking and the formation of explosive crater. The result is consistent with the classical theory of explosion crater.

(2) The SPH method is used to simulate the casting blast. The results indicate that SPH method can simulate the whole process of casting blasting. The simulation of rock blasting throwing process and engineering actual match.

(3) The whole process of rock blasting can be simulated by SPH methods, especially the process of rock crushing and throwing which traditional Lagrange method and Euler method can't be effective simulation. Therefore, SPH method for rock blasting fragmentation and blasting engineering research provides a new method.

REFERENCES

[1] Wang Xuguang. Handbook of blasting[M]. Beijing: Metallurgical Industry Press, 2010.
[2] Lucy L B.Numerical approach to testing the fission hypothesis[J]. Astronomical Journal, 1977(82):1013~1024.
[3] Gingold R A,Monaghan J J.Smoothed Particle Hydrodynamics:Theory and Application to Non-spherical stars[C]//Monthly Notices of the Royal Astronomical Society, 1977(181): 375~389.
[4] Johnson G R, Stryk R A,Beissel S R.SPH for high velocity impact computations[J].Computer Methods in Applied Mechanics and Engineering,1996, 139:347~373.
[5] Libersky L D,petschek A G,et al. High strain lagrangianhydrodynamics:A three-dimensional SPH code for dynamic material response[J].J Comput Phys., 1993, 109: 67~75.
[6] Johnson G R,Beissel S R.Normalized smoothing functions for SPH impact computations[J].International Journal for Numerical Methods in Engineering, 1996, 39: 2725~2741.
[7] AUTODYN materials library version 6.1.

Numerical Simulation Analysis of Hydraulic Pressurized Blasting Effect

LI Ming[1,3], CHEN Nengge[2], LIU Weizhou[1,3,4]

(1.Maanshan Kuangyuan Blasting Engineering Co., Ltd., China; 2.Maanshan Iron and Steel Group Mining Co., Ltd., China; 3.Sinosteel Maanshan Institute of Mining Research, China; 4.State Key Laboratory of Safety and Health for Metal Mines, China)

ABSTRACT: During the deep-hole blasting in open pit mine, ores and rock chunk mainly appeared in the upper bench. In order to improve the broken quality of ores and rock in the upper level, the hydraulic pressurized blasting was put forward. In this paper, on the basis of mechanism analysis of pressurized hydraulic blasting, the mathematical models of different water column heights were established by LS-DYNA dynamics analysis software. Based on the calculation the stress distribution states of different water column heights were obtained. In the same time, when the water column height was 2.5m, the peak stress increased 6.66% compared with no water column. While the water column height was 4m, it increased 9.38%. Under the same conditions, the peak stress of 4m water column height increased 2.55% than that of 2.5m. The hydraulic pressurized effect of 4m water column height was better. And the hydraulic pressurized effect was in conformity with the theoretical analysis.

KEYWORDS: numerical simulation; hydraulic pressurized blasting; mathematic model; water column

1 INTRODUCTION

During the deep-hole blasting in open pit mine, ores and rock chunk mainly appeared in the upper bench. In order to improve the broken quality of ores and rock in the upper level, the hydraulic pressurized blasting technology was put forward. The hydraulic pressurized blasting uses water as the working medium of transfering energy. The hydraulic pressurized blasting can more fully use the explosive energy in underwater blasting. Under high pressurized, the compressibility of water is larger than rocks, therefore water can be used as the buffer layer between explosive product and rock mass. The existence of buffer layer can not only extend the time of explosion shock wave in rock mass, but also can reduce or eliminate the energy loss in ore-bearing rock of the plastic deformation zone. After the explosion under water, it produces shock waves, and will make a lot of water evaporation, explosive gas in the form of the bubble. And they continue to swell, compressed, forming multiple fluctuation. Pressurized wave formed by the first pulse (secondary pressurized wave) effect time is longer, impulse is bigger. It can be compared with the energy of the blasting shock wave. Accordingly, it makes the water contacting with rock acquire larger pressurized in a long time, and makes the rock mass by blasting destroyed evenly. In this paper, on the basis of mechanism analysis of the hydraulic pressurized blasting technology, the ANSYS LS-DYNA simulation software was used to explore the hydraulic pressurized blasting effect.

2 HYDRAULIC PRESSURIZATION MECHANISM AND CHARGE STRUCTURE

2.1 Action Mechanism

In the hydraulic pressurized blasting charge structure of blast hole, after the main charge initiation of explosion, the wavefront of blasting shock wave propagation spreads along the hole axial pressurized to the water column. After the water column is compressed by the wavefront of shock wave, and the water column absorbs the surface wave peak pressurized. When the wavefront pressurized decays gradually, the water column no longer absorbs pressurized and turns to release energy gradually, so as to prolong the affect time in rock mass. At the same time water absorbs the wavefront peak pressurized in the compression zone, not only reduces the energy loss in the compression zone, but also plays a important role of energy transmission, making rock mass in the upper bench under a greater pressurized. Accordingly it improves the broken quality of rock mass in the upper bench.

Corresponding author E-mail: yht 03121 @ 163. com

2.2 Charge Structure

When choosing a charging structure, types of mine geological conditions, hydrologic conditions, ore-bearing rock properties and application of explosive type and other factors should be considered. In the model test, comparison test of dynamite package position in launching intervals, and the radial water charge structure have been carried out. There are four types of charge structure and their application condition and effect are shown as followed.

2.2.1 Radical Water Coupling

Water was the radial arrangement. The explosive was in the middle of blast holes, and it was surrounded by water. Under the condition of the same blasting effect, the charge structure of radial water coupling blasting used less dynamite quantity, and rock broken effect was better with dropped powdered mine rate significantly. This kind of charge structure was suitable for easily explosive or medium hard ore and rock. Generally, the water resistant explosives should be used as well.

2.2.2 Central Water Interval

The water resistance explosive was in the bottom of blast hole, while the ordinary explosive was in the upper. The water was in the middle of the two kinds of explosives. Under the condition of the same blasting effect, reducing the range of using explosives could get a better rock broken effect in different kinds of ore and rock. The water resistant explosives should be used in lower while ordinary explosives used in upper.

2.2.3 Lower Water Interval

Water was in the bottom of the blast hole and explosive was located in the upper of the water. Under the condition of the same blasting effect, reducing the range of using explosives could get a better rock broken effect in different kinds of ore and rock. It could be applied to the ore-bearing of exceeding medium hard rock. In a lot of open pit mines, water existing in the bottom of blast holes is common. At this time water resistant explosives should be used in the blast hole bottom. When the charge structure was used, it could make full use of the water medium, then common explosives were loaded in the upper.

2.2.4 Upper Water Interval

Water was in the upper of the explosive. It could make more apparent broken quality especially in upper bench. It could be applied to different properties rocks. For difficult blasting rock, the effect was also good. It had better use emulsion explosive.

In this paper, the hydraulic pressurized blasting simulation used the interval charge structure of upper water, as shown in Fig.1.

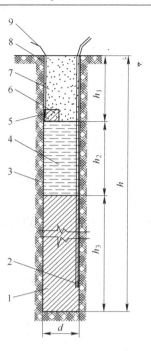

Fig.1 Charging structure of hydraulic pressurized blasting

1—main explosive column; 2—detonator of main explosive column;
3—water bag; 4—spout; 5—advanced tamponade cartridge primer detonator;
6—advanced tamponade explosive package; 7—packing; 8—blast hole; 9—detonator

2.3 The Role of Hydraulic Pressurized

When hydraulic pressurized blasting was used, it could appropriately reduce the amount of explosive. According to the spreading laws of shock wave, the total energy of explosive E_w spreading into water could be calculated as well as the explosive energy E_e that water medium got and passed to rock mass.

$$E_w = \frac{lh_w}{\rho_w C_{p1}} \int_p^t [\overline{p}_w f(t)]^2 dt \tag{1}$$

$$E_e = \frac{lh_e}{\rho_0 C_{p2}} \int_o^t [p_0 f(t)]^2 dt \tag{2}$$

Compared to (1) and (2) type

$$\frac{E_w}{E_e} = \frac{h_w \overline{p}_w^2}{h_e p_0^2} \tag{3}$$

where
l——blast holes section of the perimeter;
h_w——the water column height of blast hole water injection;
$\rho_0 C_{p2}$——rock wave impedance;
$\rho_w C_{p1}$——water wave impedance;
$\overline{p}_w$——the average initial pressurized rock sound water medium;
P_0——the initial stress in the rock;
h_e——explosive column height;
$f(t)$——the stress wave attenuation coefficient.

After dynamite explosion in rock mass, broken zone generated around the dynamite. The energy loss of blasting shock waves was big as it passed through the plastic deformation zone.

$$\text{Suppose}: E_e = 0.65 E_w \quad (4)$$

Type to large size hydraulic pressurized blasting energy rate is:

$$\xi = \frac{h_w \overline{p_w^2}}{0.65 h_e p_0^2} \quad (5)$$

The amount of explosives reducing could roughly estimate by this equation in hydraulic pressurized blasting. It could improve the utilization rate of explosives.

3 NUMERICAL SIMULATION ANALYSIS

In the condition of the same hole depth, ore and explosives parameters, three mathematical models of different water column height and packing height were designed in this paper. And the numerical simulation was conducted to analysis and verify the influence of water column height on blasting effect.

3.1 Water Column Height Calculation

According to the current popular theory of hydraulic pressurized blasting both at home and abroad, reasonable water column height calculation formula was shown as followed.

$$h_w = \left[\frac{25(\rho_0 C_{p2} + P_e D) P_{iw}}{9(\rho_0 C_{p2} + P_w C_{p1}) P_o} - 1\right]^2 \quad (6)$$

where
$\rho_0 C_{p2}$ ——ore-bearing rock wave impedance;
$P_w C_{p1}$ ——reasonable water highly water wave impedance;
$P_e D$ ——explosive wave impedance;
P_o ——wavefront pressure of explosive;
P_{iw} ——water on the surface of the medium wave initial pressure.

The calculated results of water height were shown in Tab.1.

Tab.1 Water height of the same ore and rock

Category	Ore-bearing rock wave impedance /10^6kg·m^{-2}·s^{-1}	Wave impedance /10^6kg·m^{-2}·s^{-1}	Explosive wave impedance /10^6kg·m^{-2}·s^{-1}	Wavefront pressurized of explosive /GPa	Water on the surface of the medium wave initial pressurized /GPa	Water column height/m
Ore	16	3.0	5.5	5.26	5.1	4
Diorite	7.5	3.0	5.5	5.26	5.1	5

When the blast hole depth was 12m in ore blasting, the calculated reasonable water column height was 4m. Model 1 was no water blasting, while water column height 2.5m in model 2 and 4m in model 3.

3.2 The Minimum Packing Height

In order to ensure the packing effect of hydraulic pressurized blasting, referring to conventional blasting, it determined that the minimum packing height was 5.5m (including water column), and the upper stuffing was 2.5~2.5m. An advancing tamponade cartridge (25ms initiation in advance) 3kg in weight was set between water column and stuffing, acting as strengthening measures.

3.3 Numerical Simulation Parameters Selection

3.3.1 The Performance of Explosive Index

The rock emulsion explosive was used, the main performance indexes as shown in Tab.2.

Tab.2 Main performance indicators of rock emulsion explosive

The density of explosive/g·cm^{-3}	Gap distance/cm	Power capability/mL	Brisance /mm	Detonation velocity/m·s^{-1}	Storage and Shelf life/d	Content of poisonous gas/L·kg^{-1}
1.10~1.35	≥5	≥260	≥12	≥3.2×10^3	180	≤80

3.3.2 Performance Parameters of the Ore

Lithology and physical and mechanical parameters of ores and rocks were shown in Tab.3.

Tab.3 Lithology and physical and mechanical parameters of ores and rocks

Ore-bearing rock category	f	Proportion /kg·cm^{-3}	Compressive strength/MPa	Strength of extension/MPa	Velocity of longitudinal wave/m·s^{-1}
Lump ore	12~16	4.04	116.0	8.27	3800~4200
Diorite	8~10	2.5	87.24	8.29	2800~3200

3.3.3 Stuffing Material Parameters (Dust)

Stuffing material was rock powder, related parameters shown in Tab.4.

Tab.4 The material parameters of rock powder

Density/g·cm^{-3}	Elasticity modulus（E）/Mbar	Poisson ratio	Yield stress/Mbar	Tangent modulus/Mbar
1.80	2×10^{-2}	0.3	2×10^{-4}	2×10^{-3}

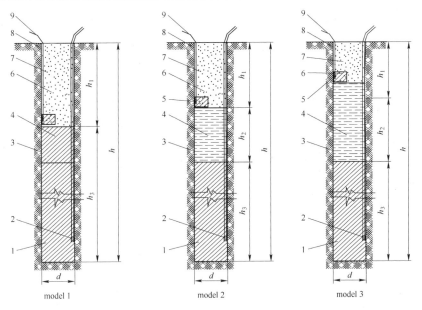

Fig.2 Diagram of numerical simulation geometric model

1—main explosive column; 2—detonator of main explosive column; 3—water bag; 4—spout; 5—advanced tamponade cartridge primer detonator; 6—advanced tamponade explosive package; 7—packing; 8—blast hole; 9—detonator

3.4 The Simulation Results Analysis

The related parameters of material properties and state equation were set in ANSYS LS-DYNA. Then the adjust K file was submitted to LS-DYNA program to solve. When the calculation was over, the results were processed by LS-PREPOST post-processing software, eventually the rock pressurized profile of different moments were shown in Figs.3(a)~3(c) and the stress attenuation figures of monitoring point were shown in Figs.4(a)~4(c). As can be seen from Figs.3 (a)~3 (c), in the same time, the peak stress of non-water column was 62.88 MPa. When the water column height was 2.5m, the peak stress was 67.07MPa, 6.66% larger than that of non-water column. When the water column height was 4m, the peak stress was 68.78MPa, 9.38% larger than compared with that of non-water column. The peak stress of 4m column height increased 2.55% relative to that of 2.5m. The hydraulic pressurized effect of the former was better. According to the stress attenuation curves of monitoring points, as shown in Fig.4 (a) ~ Fig.4 (c), the peak stress of monitoring points increased 8%~17% due to the presence of water column, and the blasting time prolonged 5%~13%, which indicated that the hydraulic pressurized effect was more obvious.

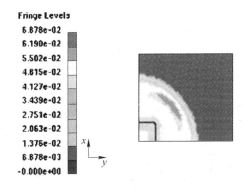

Fig.3(b) t=25ms model 2 pressurized profile

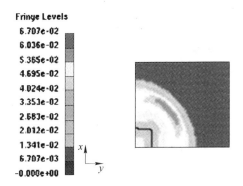

Fig.3(c) t=25ms model 3 pressurized profile

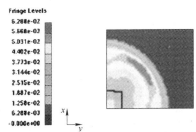

Fig.3(a) t=25ms model 1 pressurized profile

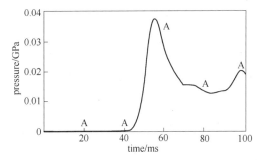

Fig.4(a) The stress attenuation figure of monitoring point in model 1

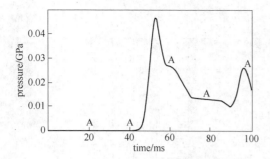

Fig.4(b) The stress attenuation figure of monitoring point in model 2

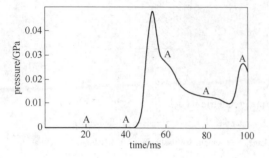

Fig.4(c) The stress attenuation figure of monitoring point in model 3

4 CONCLUSIONS

In this paper, the main conclusions were shown as follows:

(1) Hydraulic pressurization mechanism analysis showed that, under the condition of equal quantity, hydraulic pressurized blasting was superior to the ordinary dry hole blasting. The former could more fully use of the energy of explosive in underwater explosion, extend the time of blasting effect, improve the quality of ore-bearing rock crushing. After calculating it was determined that the most appropriate water column height was 4m.

(2) The numerical analysis showed that in the same time, when the water column height was 2.5m, the peak stress increased 6.66% compared with non-water column. While the water column height was 4m, it increased 9.38%. Under the same conditions, the peak stress of 4m water column height increased 2.55% than that of 2.5m. The hydraulic pressurized effect of the former was better. And the hydraulic pressurized effect was in conformity with the theoretical analysis.

(3) The peak stress of monitoring points increased 8%~17% due to the presence of water column, and the blasting time prolonged 5%~13%, which indicated that the hydraulic pressurized effect was more obvious.

REFERENCES

[1] LIU W ZH, YUAN Y J, etc. Hydraulic boosting blasting test[J]. *Engineering Blasting* 2014(2):10~13.

[2] CHEN X H. Deep hole hydraulic blasting charge structure and applied research[J]. *Journal of Coal Society*. 2000, 25(S1): 112~116.

[3] ZHANG X L. Water pressurized blasting broken characteristics research[M]. Beijing: *Beijing University of Tron and Steel Technology*, 1986.

[4] LIU W ZH, ZHANG X L, YUAN Y J, etc. Large size hydraulic booster blasting environmental protection technology research[Z]. MaAnShan: *Ma'anshan Kuangyuan blasting engineering Co., Ltd.; Nanshan Mining Company of Magang (Group) Holding Co.,Ltd.;*2012.

[5] SHI SH Q, KANG J G, etc. ANSYS/LS-DYNA in explosion and impact in the field of engineering application[M]. Beijing: China Architecture & Building Press, 2011.

[6] SHANG X J, SU J N, etc. ANSYS/LS-DYNA dynamic analysis method and the engineering examples[M]. Beijing: China Water & Power Press, 2006.

Rock Fragmentation

Characteristic Response of Rock and Rock-like Materials to Explosive Loading in Controlled Experiments

B. Mohanty[1,2], R. Raghavaraju[1]

(1. Department of Civil Engineering and Lassonde Institute of Mining University of Toronto, Toronto, Canada;
2. Department of Mining Engineering Indian Institute of Technology, Kharagpur, India)

ABSTRACT: This study describes the results of single-hole blasting experiments in three well-characterized materials, a granite, a limestone, and a grout-based synthetic rock. The parameters studied are the nature of the dynamic stress field around the borehole and its decay in the near-field (i.e. 0.5 to 5 borehole diameters), and as a function of the dynamic properties of the target rocks. Detonating cord of varying strengths (3 g/m, 5.3 g/m, and 10 g/m of PETN) with air coupling in 6.4 mm and 9.5 mm diameter define the explosive source conditions. A thin-walled copper tube was installed in the borehole to isolate the explosion gas, so that only shock stress-induced fractures and their distribution could be studied. The cylindrical test samples measured 15 cm high and 15 cm in diameter. The resulting fracture patterns were mapped by a dye penetration technique, and fragment size distribution by sieve analysis. The observed near-field pressure decay, and fracture and fragmentation characteristics will serve to calibrate numerical models for prediction of blast results.

KEYWORDS: blasting; detonating cord; single hole; fracture pattern; fragment size; shock pressure; dynamic stress-induced fracture

1 INTRODUCTION

Accurate prediction of blast results, both in small-and large-scale operations continues to present a continuing and significant challenge, despite much recent advance in understanding of the detonation process, easy availability of monitoring equipment, improved explosives products, and numerical tools. It represents a particularly difficult task because it involves three different but inter-linked components: explosive energy source, blast geometry and target rock. Of these, only the blast geometry is easily controlled; the rest can be highly variable. Each of these exhibits a range of properties that is often neither controlled nor fully understood. Therefore, systematic study, under controlled blasting conditions in small scale, is one step towards achieving the required degree of quantification. The results from such studies can then be utilized in calibrating existing or emerging numerical models for accurate prediction of blast results, in both small-scale and large-scale operations.

The respective roles of the shock pressure and the expanding explosion gas pressure form detonation of an explosive column is now well established in the literature (Clark, 1987; Hustrulid, 1999; Olsson et al., 1999; Anon., 2011). However, there continues to be some debate regarding the respective contributions of each of these phases towards the overall fragmentation and rock movement. Various researchers in the past have ascribed different fractions of the total explosive energy in the shock phase through a number of experimental investigations, with the shock contribution varying from merely 2% to as high as 25% for varying blast conditions (Langefors, 1978; Brinkmann, 1990; Lownds et al., 2000). However, these estimates were obtained through either indirect measurements or theoretical postulates, and often at distances far larger than the typical spacing to diameter ratio in a blast. This study is part of a continuing work in trying to correlate direct measurement of dynamic pressure close to the borehole (< 10 borehole diameters) with resulting fracture network and fragment size, restricted to the initial shock phase only, and to elucidate the role of the latter in controlling the overall fragmentation (Paventi and Mohanty, 2002; Mohanty and Dehghan-Banadaki, 2007; Raghavaraju and Mohanty, 2014).

2 EXPERIMENTAL PROCEDURE

Three types of rocks were selected for single-hole laboratory scale blasting experiments; Laurentian granite, Flamboro limestone and a synthetic rock/concrete grout sample prepared from readily mixed sand and cement (Sika Grout212) (Raghavaraju and Mohanty, 2014). These target rocks were selected on the basis of their wide ranging uniaxial com

---

Corresponding author E-mail: bibhu.mohanty@utoronto.ca

pressive strength values (259 MPa) for Laurentian, 130 MPa for Flamboro, and 64 MPa for the Synthetic rock. Laurentian granite is a fine grained rock, and Flamboro limestone is a very fine grained homogeneous rock.

The laboratory-scale cylindrical samples measured 15 cm in diameter and 15 cm in height, with a central borehole running through their length. The detonating cord (PETN) of varying strength (3 g/m, 5.3 g/m, and 10 g/m) was centrally located in the boreholes measuring 6.4 mm and 9.5 mm in diameter. Use of rubber plugs with a central hole to accommodate the cord at both ends of the borehole assured accurate centering of the explosive charge. The outer diameters of the three types of cord measured 3.7 mm, 3.1 mm, and 5.1 mm. Two types of decoupling media were employed, air and water. The detonation velocity of the cord was 6800 m/s, which was of course much higher then the P-wave velocity in the test rocks. The target rocks represent a very wide range in terms of their strength and related properties (Tab.1).

Tab.1 Properties of the target rocks

Rock type	P-wave velocity /km·s^{-1}	S-wave velocity /km·s^{-1}	Density /g·cm^{-3}	Young's modulus /GPa	Unconfined compressive strength/MPa	Tensile (brazilian) strength/MPa
Laurentian granite	4.39	2.79	2.66	47.5	259	12.5
Flamboro limestone	6.24	3.02	2.65	64.3	130	6.0
Synthetic rock	4.24	2.65	2.27	37.6	64	5.5

In addition to providing variable coupling conditions, each borehole was lined with a tight fitting thin-walled copper tube serving as a borehole liner cemented to the borehole wall with high strength epoxy in order to prevent the explosion gases entering the surrounding rock mass, so that only the effect of the dynamic stress induced fractures could be studied. The wall thickness of the copper tube was 0.6 mm for the 6.4 mm diameter borehole, and 0.8 mm for the 9.5 mm diameter borehole. The cross section of the borehole in this experimental set-up is shown in Fig.1. No stemming of any kind was used in the test. The detonating cord extended a short distance beyond the top of the test rock, and was initiated by means of a shock tube detonator, so as to provide the necessary trigger signal to the data acquisition system and at the same time, minimize any extraneous noise for the pressure sensors. The latter consisted of 510 ohm carbon composite resistors (Ginsburg and Assay, 1991; Rosenberg et al., 2007). These measured 1.9 mm in diameter and 6 mm in length, and were embedded in the rock by means of epoxy in 4 mm diameter holes placed at distances ranging from 3 mm to 50 mm from the borehole wall (i.e. 0.5 to 5 borehole diameters). The experimental arrangement is shown schematically in Fig.2.

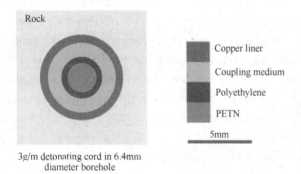

Fig.1 Cross-sectional view of the test sample with central detonating cord with PETN charge and its Polyethylene wrapping, coupling medium (air or water), and Copper tube borehole liner

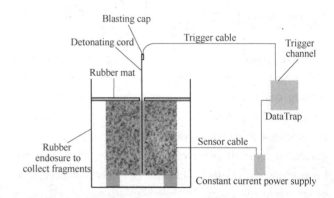

Fig.2 Schematic diagram of the experimental set-up for measurement of transmitted pressure from single-hole blasts and subsequent collection of rock fragments

The carbon composite resistors (CCR) were connected to a constant current power supply (PCB Model 482A16), and their output connected to a multi-channel high-speed data acquisition system with a sampling speed of 10 MHz (DataTrap II). The change in resistance value of the CCR can then be converted to the corresponding pressure value in the pressure range of interest (Dehghan-Banadaki, 2010). A typical output from a CCR gauge in these experiments is shown in Fig.3 for the transmitted pressure. The precursor signal in this record is attributed to minor electrical noise. However, the arrival of the actual shock pressure at the sensor is unambiguous and noise-free.

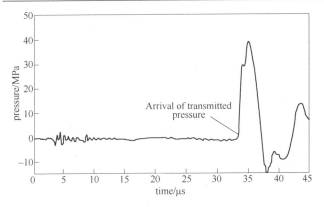

Fig.3 Typical dynamic pressure signature recorded by the CCR sensor at a distance of 10 mm from the borehole wall (6.4 mm dia.) from a single-hole blast (5.3 g/m of PETN) in Flamboro limestone

3 NATURE OF DYNAMIC PRESSURE TRANSMISSION CLOSE TO BLASTHOLE

The effect of decoupling on the transmitted pressure and its decay in Laurentian granite is shown in Fig.4(a). Decoupling is varied in this case by increasing the borehole diameter from 6.4 mm to 9.5 mm, while keeping the same charge (5.3 g/m of PETN). For both cases, the amplitude of the transmitted pressure drops off rapidly close to the borehole wall, but decays more slowly with increasing distance from the borehole. However, the attenuation factor is found to be slightly less in the higher decoupling case than the lower decoupling one, as expected. The estimated pressures at the borehole wall for the two cases are 472 MPa and 193 MPa respectively.

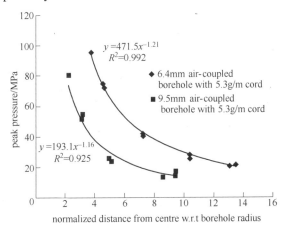

Fig.4(a) Transmitted peak pressure decay with distance in Laurentian granite from air-coupled borehole of 6.4 mm and 9.5 mm diameter with Cu liner of 0.6 mm and 0.8 mm thickness charged with 5.3 g/m detonating cord

The effect of rock types on the dynamic pressure transmission for identical blasting parameters is shown in Fig.4(b), for the air-coupled case with 5.3 g/m detonating cord in 9.5 mm diameter borehole. As expected, the synthetic rock being the weakest transmits much lower pressures than either Flamboro limestone or Laurentian granite. Although there is some scatter in the recorded data for both limestone and grout samples, the trend in pressure decay is quite obvious. The estimated blast pressures at the borehole wall are 74 MPa, 140 MPa, and 193 MPa for the synthetic rock (grout), limestone and granite respectively. These values nearly mirror the respective UCS values for these rocks shown in Tab.1. The attenuation constant for pressure decay in synthetic rock is lower than for either limestone or the granite rocks, due to its lower transmitted pressure values. Due to higher decoupling (i.e. 9.5 mm dia. vs. 6.4 mm borehole) the estimated pressure at the borehole wall is considerably lower (i.e. 140 MPa and 74 MPa for the limestone and the synthetic rock respectively, compared to 193 MPa in granite).

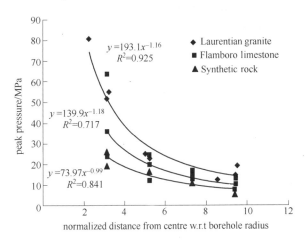

Fig.4(b) Transmitted peak pressure decay with distance in the three rock types from air-coupled borehole of 9.5 mm diameter with Cu liner of 0.8 thickness charged with 5.3 g/m detonating cord

4 PRODUCTION OF CRACK NETWORK AND FRAGMENTATION

Since the transmitted pressure is considerably higher than the strength of the target rocks, even for decoupled charges in this investigation, a network of cracks will result due detonation of the explosive in the borehole. Visualization of the resulting cracks are examined by employing the dye-impregnation technique, whereby a fluorochromatic dye is mixed in with an epoxy resin-hardener mixture and applied to the rock surface of interest (Dehghan-Banadaki, 2010). When the latter is illuminated by an ultraviolet source of light on the surface, the cracks become easily visible for mapping by photography (Fig.5(a)). The fractures are then manually traced, and the resulting crack density (i.e. total length of all the cracks over the area of a specific zone; mm/mm^2) determined through the use of GIMP 2.8, an open source image editing software.

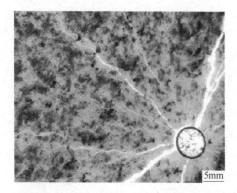

Fig.5(a) Blast-induced crack network around borehole at bottom surface of a Laurentian granite sample, with 6.4 mm diameter air-coupled borehole with 0.6 mm Cu liner charged with 3 g/m cord

Since the explosive charge in the borehole is of finite length detonating at a velocity of 6800 m/s, the transmitted pressure field will change with depth from the top of the test sample, i.e. the point of initiation. Thus the resulting crack density at any particular radial plane will depend on its depth from the top of the test sample. Therefore, the test sample after each test is sliced into three sections as shown in Fig.5(b), and the crack density measured along each radial plane to study the variation of cracks with depth from the point of initiation. An example of this variation for Laurentian granite (3.5 g/m air-coupled charge in 9.5 mm diameter copper lined borehole) is shown in Fig.6. The minimum crack length considered in this study was 2.5 mm.

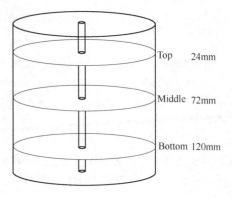

Fig.5(b) Schematic diagram showing cross-sectional surfaces considered for crack density analysis (top section 24 mm, middle section 72 mm, and bottom section 120 mm from top of test sample)

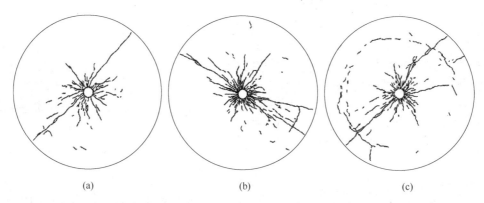

Fig.6 Crack mapped surfaces of Laurentian granite charged with 3 g/m cord in air-coupled borehole (9.5 mm dia.) with copper liner at different sections along the borehole

(a) top; (b) middle; (c) bottom

In order to further quantify the nature of crack network, each radial plane is divided into three equal zones, and the crack density measured in each to yield information on its distribution as a function of distance from the borehole. This is shown for a typical test in Laurentian granite.

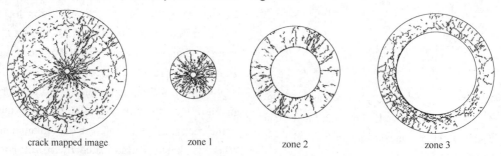

Fig.7 Typical crack mapped image and the three different zones to study variation of damage with radial distance in a 144 mm diameter granite sample with air-coupled borehole (6.4 mm dia. with 0.6 mm thick Cu liner) with 3 g/m PETN charge

The effect of varying charge weight in the borehole for variation in crack density across the three zones for identical coupling condition is shown in Fig.8. As expected, Zone 1 yields the highest crack density across all sections. The crack density decreases as one moves away from the hole, except near the outer boundary, where additional cracks appear due to reflection of the stress wave from the boundary. However, the highest crack density occurs across the bottom section due to relection of stress waves from the bottom of the rock sample. Increased decoupling by replacing 6.4 mm diameter hole with 9.5 mm diameter borehole reduces the crack density across all three sections, resulting in a 30%~40% reduction across the mid-section in all three zones. Fig.9 shows a comparison of the observed crack density for the three rock types (i.e. Laurentian granite, Flamboro limestone, and Synthetic rock) across the mid-section for a 3 g/m cord detonating in an air-coupled 9.5 mm diameter borehole with Copper liner. The crack density for the Synthetic rock is observed to be the highest in all three zones in the radial plane as well as in the three sections. This is to be expected because of its much lower strength. However, Laurentian granite despite having much higher strength both in compression and tension (Tab.1), exhibits nearly the same crack density as the Synthetic rock. However, with increasing distance from point of initiation at the top, the crack density for both Laurentian granite and Flamboro limestone appears similar, and both considerably lower than that in the Synthetic rock.

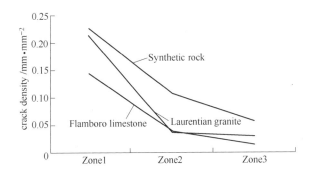

Fig.9 Average crack density in the mid-section for three rock types (granite, limestone and synthetic rock), with 3 g/m charge in air-coupled 9.5 mm diameter borehole with Cu liner

5 FRAGMENT SIZE ANALYSIS

A detailed fragment size analysis was carried out in this investigation to correlate crack density with resulting frgment size distribution for the three rock types. The size distribution was obtained through sieve analysis. To achieve actual fragmentation of the test samples in contrast to producing just crack network, it was necessary to employ a stronger explosive charge (i.e. 10 g/m detonating cord) in the borehole in these experiments. The results are shown in Fig.10. The results show that Laurentian granite yielded the coarsest fragments, whereas, both Flamboro limestone and the Synthetic rock yielded nearly equivalent size distribution below 10 mm. However, two majot problems were encountered in the size distribution analysis: (1) there was nearly 7% loss of extra fines in the grout samples and only 2%~3% loss in the rest; (2) the cylindrical nature of the test samples tended to produce wedge-shaped fragments, whose proper analysis in terms of assigning appropriate size distribution functions (e.g. Rosin-Rammler, Swebrec, etc.) requires further study (Ouchterlony, 2009). The data shown in the graph do not take this loss of extra fines into account, incorporation of which would have shown the Synthetic rock to produce finer fragments than the other two rock types.

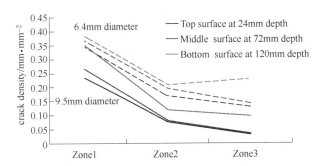

Fig.8 Crack density in Laurentian granite sample at three sections charged with 5.3 g/m cord in air-coupled in 9.5 mm (solid line) and 6.4 mm (dashed line) diameter borehole with Cu liner

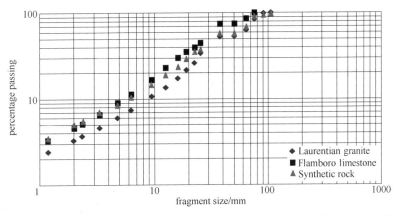

Fig.10 Fragment size distribution in target rocks with 10 g/m detonating cord in air-coupled borehole of 6.4 mm diameter with Cu liner of 0.6 mm thickness

6 CONCLUSIONS

The transmited pressures from exploding charges in a borehole have been successfully measured in laboratory-scale studies, involving three homogeneous rock types but exhibiting a very wide range of strength properties. These laboratory-scale studies involved single-hole blasts with detonating cords of varying strengths serving as the explosive charge. The objective was to define the dynamic pressure field very close to a blast hole with the aim of correlating its role in creating dynamic stress induced fracture and fragmentation in selected rocks. The effect of the role of the subsequent explosion gases in extending the stress-induced cracks was eliminated by suitable Copper lining of the borehole wall. The measurement distances ranged from 0.5 to 5.0 borehole diameters, which are directly relevant to determining the role of stress waves in creating the initial fracture network in a blast. The study showed that a given rock strength does not automatically translate into the expected fracture network or the fragment size distribution. The study also provides excellent data base for calibration of vaious numerical models curently under development for prediction of blast results in the field.

ACKNOWLEDGEMENT

The authors are grateful to the Center for Excellence Mining Innovation (CEMI) and Natural Sciences and Engineering Research Council of Canada (NSERC) for financial support. Technical help from Professor Karl Petersen of the University of Toronto is also gratefully acknowledged.

REFERENCES

[1] Anon, 2011, *"Blasters' Handbook"*; 18th Ed. Int. Soc. Explosives Engrs., Cleveland, USA, p.1~1030.

[2] Brinkmann J R. 1990. An Experimental study of the effects of Shock and Gas penetration in Blasting', *Proc. 3rd Int. Symp. on Rock Fragmentation by Blasting – FRAGBLAST 3*, Brisbane, Australia, p.55~66.

[3] Clark G B. 1987; *"Principles of Rock Fragmentation"*; Wiley & Sons, USA, p.1~610.

[4] Dehghan-Banadaki M D. 2010. Stress-wave induced Fracture in Rock due to Explosive Action, Ph.D. Thesis, University of Toronto, Toronto, Canada.

[5] Ginsberg M J. Asay B W. 1991."Commercial carbon composite resistors as dynamic stress gauges in difficult environments", *Review of Scientific Instruments*, 62-9, p.2218~2227.

[6] Lownds M. 2000, Measurement of shock pressures in splitting of dimensional stone. *Proc. 1st World Conf. on Explosives and Blasting Techniques*,(ed. Holmberg R), Balkema, Rotterdam, p.241~246.

[7] Langefors U, Kihlstrom B. 1978, *"Rock Blasting"*, 3rd Ed., Wiley, New York.

[8] Mohanty B, Dehghan Banadaki M M. 2009, "Characteristics of stress-wave-induced fractures in controlled laboratory-scale blasting experiments," *Proc. 2nd. Asia-Pacific Symp. on Blasting Techniques*, Wang X. (ed), Metallurgical Industry Press, Beijing, p.43~49.

[9] Olsson M, Nie S, Bergqvist I, Ouchterlony F. 1999. "What causes cracks in rock blasting? " *Proc. 6th Int. Symp. on Rock Fragmentation by Blasting- FRAGBLAST 6*, Johannesburg, SA.

[10] Ouchterlony F. 2009. "Fragmentation Characterization; the Swebrec function and its use in blast engineering", *Proc. 9th Int. Symp. on Rock Fragmentation by Blasting- FRAGBLAST 9* (ed:Sanchidrian J.A), p.3~22.

[11] Paventi M, Mohanty B. 2002. "Mapping of blast-induced fractures in rock"; *Proc. 7th Int. Symp. on Rock Fragmentation by Balsting - FRAGBLAST-7*(ed. Xuguang, W,) Metallurgical Industry Press; China, p.166~172.

[12] Srirajaraghavaraju R, Mohanty B. 2014, "Dynamic Stress Field around a Blast Hole – A Laboratory Study", *Proc. 40th Ann. Conf. on Explosives and Blasting Techniques*, Int. Soc. Explosives Engrs. Fort Worth, USA, p.1~10.

Control of Yield Rate of Fines of Different Specification of Stone Quarrying with High Intensity

ZHENG Bingxu, LI Zhanjun, SONG Jinquan
(*Guangdong Hongda Blasting Co., Ltd., Guangzhou, Guangdong, China*)

ABSTRACT: According to the characteristics of deep-hole bench blasting and the geological conditions, the size requirement of finished stone and so on, the factors influencing the yield rate of fines were studied, and some new achievements have been made, and are of practical guiding significance to mining in super-large scale quarries and mines.

KEYWORDS: bench blasting; yield rate of fine stone; charge structure

1 INTRODUCTION

In southern china a large quarry produce 3.5 ten thousand cubic meter a day and 80 ten thousand cubic meter blasting stones monthly average. The height of mine area is 180 m and covers an area of 40 ten thousand square meters. The total blasting volume is about ten million cubic meters. The blasting production of stone is separated into 15 varieties according to the blocks specifications 10 to 2000 kg and its grading is quite strict, the stone which above 10 kg shall not be greater than 10 percent.

The rock in the blasting areas is diorite granite, f-value is $8 \sim 14$, joints and fissures of weathering groove crush zone are well-developed and the fissure water is rich. There have vein invasion in multiple places in the same platform, even in the same working face of blasting operation there have several different geological conditions of rock, and it will be very bad for blasting and sorting. The mining area of rock is separated into 3 varieties according to the complexity of blasting, which is first categories, relatively hard blasting rock; second categories, secondary explosive rock; third, explosive rock. According to the request of sorting stone, the mining machinery choose 1.0 to 1.6 cubic meter hydraulic backhoe, generally the efficiency of excavator can reach average 800 cubic meter per machine-team, because of the need to choose the qualified stone fromblasting heap, the efficiency of drivers is reduced to an average of 400 cubic meter per machine-team, it needs 90 backhoe to work, each backhole need 20m wire length to work, the normal operation needs 1800m critical pile length, it need more than 600 m length of working face for drilling charging and so on. So, in order to achieve continuous balanced work, the total length of lines needs more than 2400 m.

The quarries are united to 15m high bench, blasted $3\sim 4$ bench every day. The number of blasting row is not more than 3 rows; width is $100\sim 150$ m, after blasting it will dig clean in one week. The volume of borehole can reach 2000m every day, the secondary blasting need more than 2000 detonator every day, using 10 DTH drill, the equipment of rock drilling are Ingersoll-Rand VHP750 DTH drill, the blast hole is $\phi 140$ mm vertical hole. The charge structure of full hole is which the transverse uncoupling loading coefficient of charging at the bottom is small(using $\phi 110$ mm cartridges) and the toptransverse uncoupling loading coefficient of charging is big (using $\phi 70$ mm $\phi 80$ mm $\phi 90$ mm $\phi 100$ mm cartridges), we use waterproof emulsion explosive, the density of explosive is 1.16g per cubic meter. When meeting with the hard broken location of rock, we usually don't charge in vacuum space at the bottom of blasting hole.

2 DIFFERENT CHARACTER OF FINES IN BLASTING MUCK PILE

In the process of cleaning in muck pile, the amount of stone and spoils of dailydelivery in the muck pile can be seen in the follow figure. From the figure we can see that it delivered rock materials most in a few days after blasting, then the spoils has increasing exponentially. The fines are mainly come from the central of muck pile, and it corresponds to the position of the coupling charging (bottom charging). The next is the weakness of interlayer and crush zone. The fines are less in the corresponding parts of stemming (blasting stone is distributed in the surface of muck piles), the yield rate of fines in the part of top charging (the part of uncoupling charging) is much less than that in the part of bottom charging. It should explain that the fines can be quarried after accumulating to a certain amount, and this is one of the reasons that the fines are less in preliminary stage.

Corresponding author E-mail: 274991025@qq.com

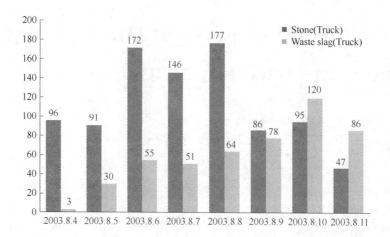

Fig.1 Qualified stone and waste slag quantity statistics

The parameters of drilling and blasting on site: the aperture is 140mm, the vertical hole, the bench height is 15m, the sub drilling is 1.3m, and the bottom burden is 3.8m, the distance of hole is 6.5m.

3 RESEARCH OF THE INFLUENCE OF YIELD RATE OF FINES

According to the characteristics of engineering, the block which is less than 10kg are designated as fines, the yield rate of fines are defined as the percentage which the block weight of less than 10kg in the total weight of blasting hole.

The range of the crushing zone which near the blast hole is depending on the conditions of the characteristics of rock and explosive drilling and blasting parameters charge structure and so on.

(1) On site the yield rate of fines after changing the unit volume consumption of dynamite, the parameter of the blasting hole in different explicability of rock. It can be seen in the follow table.

Tab.1 Statistics of yield rate of fines

Number	Name of platform	Types of rock (explicability of rock)	Unit consumption	Row spacing/m	Distance of hole	Spacing	Yield rate of fines/%
1	+70 m	I	0.32	3.8	6.5	24.7	9.28
2	+70 m	I	0.36	3.5	6.5	22.8	13.93
3	+40 m	I	0.43	3.5	6.2	21.7	15.21
4	+40 m	I	0.46	3.5	6.5	22.8	15.74
5	+40 m	I	0.48	3.5	6.0	21.0	28.76
6	+130 m	II	0.31	4.2	6.0	25.2	16.93
7	+40 m	II	0.32	3.8	6.5	24.7	21.19
8	+85 m	II	0.32	4.2	5.8	24.4	21.49
9	+100 m	II	0.36	3.5	6.5	24.7	41.30
10	+85 m	II	0.39	4.2	5.5	23.1	48.40
11	+130 m	II	0.44	3.5	6.5	22.8	53.09
12	+100 m	III	0.29	3.8	6.5	24.7	126.12
13	+100 m	III	0.30	3.8	6.5	24.7	90.24
14	+100 m	III	0.30	3.8	6.5	24.7	298.94
15	+130 m	III	0.34	4.2	6.0	25.2	50.05
16	+115 m	III	0.35	3.5	6.5	22.8	64.33
17	+115 m	III	0.37	3.5	6.5	22.8	102.21
18	+115 m	III	0.40	3.6	6.2	22.3	95.51

(2) In the same situation of unit volume consumption of dynamite, the yield rate of fines is decreased with the increase of the resistance line; the yield rate of II rock have more influence than that of I rock by the bottom burden. The bottom burden is reduced from 4.2m to 3.0m, the yield rate of II rock increase 1.63 times, while the yield rate of I rock only increase 0.79 times.

(3) In the range of same unit volume consumption of dynamite and a certain distance of hole, the yield rate of fine stone are decreased with the increase of the distance of

hole, furthermore, the yield rate of fine stone of Ⅱ rock have more influence than that of Ⅰ rock on the distance of hole; the bench bottom burden have more effect than the distance of hole on the yield rate of fine stone; the broad pore technique has been test before in the field, but it produced the powder ore fines more easily, this project is to reduce the powder ore fines, the distance of hole should not be more than 6.0m and less than 5.0m, the value of spacing burden ratio is less than 1.5 and more than 1.2.

(4) Charge structure has much more influence on yield rate of fine stone than the properties of rock distance of hole and the bench bottom burden. In order to reach the requirements of the design of rock grading, we use the charge structure which is the transverse uncoupling loading coefficient of charging at the bottom is small and the upside transverse uncoupling loading coefficient of charging is big. Generally the density of bottom charge Δ_{bottom} is 11kg/m, the density of the top part Δ_{top} is no less than 4.5kg/m, in order to meet the need of density of charge line and uncoupling loading coefficient(1.2～2.0), we use the cartridges which the bottom lifting diameter is 110mm and thetop are 70mm 80mm 90mm and 100mm.

In the same situation of borehole net parameters, the yield rate of fines increase in power characteristic with the increase of density of top charge, and it has a significant reduction with the decrease of density of bottom charge; Ⅱ rock have much more influences than Ⅰ rock on the yield rate of fines affected by density of top and bottom charge; The lower parts have much more influences on yield rate of fines with other factors such as bottom burden properties of rock and pith of holes; the smaller the density of charge, the larger uncoupling loading coefficient, with the decrease of maximum stress of the inner wall of blast hole, the broken circle is getting smaller, along with the decrease of powder fines.

(5) The influence on the property of explosive to yield rate of fines is great. This project adopt strip emulsion explosive in several factories, the detonation velocity is 3300～5600m/s, the greater the explosive detonation velocity, the larger range of hole damage is, and the yield rate of fines are high naturally. To reduce the fines, we should adopt low detonation velocity explosive.

4 TECHNICAL MEASURES AND ENGINEERING EFFECT OF CONTROLLING YIELD RATE OF FINES

(1) To reduce the ore fines, do not adopt the wide-space patterns of blasting holes, the coefficient density of hole should be in 1.2～1.5.

(2) The charge structure has closely relationship with ore fines, thebottom charge need the bottom rocks which has large pushing clamping force, the uncoupling loading coefficient should be small, the ore fines produced is difficult to avoid, so, the top part is the key point to control ore fines, the upper charge should be controlled strictly, to make the top rock fall down simply.

(3) Themore the rock broken badly, the more valuable it is to control ore fines.

(4) By adopting comprehensive control technical measures, theyield rate of fines of this quarry is reduced from 50% to 18%.

(5) According to the influence degree that all technical measures to yield rate of fines, it is sorted in order of size: charge structure, lines charge density, unit volume consumption of dynamite, explosive detonation velocity, bottom burden, distance of holes.

REFERENCES

[1] Rose Manis H P. Proceedings of the Fourth International Symposium On Rock Fragmentation By Blasting[M].Beijing: Metallurgical Industry Press, 1995.

[2] LIU Dianzhong, YANG Shichun. A practical handbook of engineering blasting[M]. Beijing: Metallurgical Industry Press, 2003.

[3] WAN Yuanlin, WANG Shuren. Analysis of calculating borehole pressure under decoupled charging[J]. Blasting, 2001, (1): 13～15.

[4] ZHANG Zhengyu.Modern water power engineering blasting[M]. Beijing: China WaterPower Press, 2003.

[5] LUO Xiuhao.Science and technology novelty search report[R]. Guangdong: Institute of Science and Technology Information, 2008.

[6] ZHENG Bingxu and so on.Appraisal documents of research achievement on high-intensive exploitation technique of different dimensions stones[R]. Guangzhou: Guangdong Hongda Blasting Engineering Co.Ltd., 2008.

[7] LIU Dianzhong. A practical handbook of engineering blasting[M]. Beijing: Metallurgical Industry Press, 2003.

[8] YU Yalun. Theory and technique of engineering blasting[M]. Beijing: Metallurgical Industry Press, 2004.

[9] WANG Xuguang, YU Yalun. Present situation and development of engineering blasting [C]//Engineering blasting corpus (the sixth period), Shenzhen: Haitian Press, 1997.

[10] WU Congshi, QI Baojun, LIU Yufeng and so on. Press change and blasting effect of shot hole in slope sections[J]. Journal of Liaoning technical university (natural science edition), 2001,20(4).

[11] XIAO Fuguo, LI Qiyueand so on. Summary of joint rock blasting block prediction model[J]. Mining Research and Development, 2001, 20(4).

[12] TAN Zheng, LI Guangyue, LI Changshan and so on. Gary correlation analysis of blasting parameters influencing size performance of block blasted[J]. Mineral engineering, 2003, (6).

[13] LI Xiangdong, ZHANG Qiang. Study on optimization design of medium-length-hole blasting parameters[J]. Mining Research and Development, 1995,(1):35~39.

[14] WU Zijun, GAO Shantang, HU Renxing and so on. Research on rock fragmentation by blasting design parameters[J]. Mining Research and Development, 1983, 3(2).

Improving the Methodology of Rock Breakage by Blast for Underground Mineral Mining Technologies

S. D. Viktorov, V. M. Zakalinsky, A. A. Osokin

(*Institute of Comprehensive Exploitation of Mineral Resources (IPKON), Russian Academy of Sciences, Russia*)

ABSTRACT: The authors propose a new approach to the methodology of rock mass breakage by blast and present a physical model and classification of rock breakage by blast to choose the right type for different conditions thus enhancing the efficiency of blasting technology in mineral mining.

KEYWORDS: blast; mining method; model; blasting method; intensively fractured; moderately fractured; solid ore in place; classification; small-scale; medium-scale; large-scale rock breakage by blast

Ever-growing requirements to the efficiency and safety-in-mining in worsening conditions of mineral mining determine the necessity of improving the existing mining methods and development of novel drilling and blasting technologies. With that, the efficiency of blast energy utilization does not always matches well with the process objectives. A direct consequence of this is the intensive degradation of quality characteristics of newly developed mineral deposits with extremely complicated geological structure. This problem requires the development of novel geotechnologies meeting the variability level of deposit geological characteristics, right choice of drilling and blasting methods and parameters, with due account for the growing toughness of economic and environmental constraints.

It is important to classify a variety of factors for the analysis and estimation of the complexity of deposit development conditions for the elaboration of a new methodological approach. The prerequisites are based on the analysis and application of the known results in mining theory and drilling and blasting practice.

ICEMR RAS researchers have investigated a new area of rock mass breakage by blast, which main point is based on the estimation of the compliance of the rock blasting scale (parameters and design of explosive charges) with the conditions of efficient mining [1]. A key moment in this case is the development of an algorithm of classification and criterion in the selection of methods for the process control, which take into account the internal structure of various conditions of deposit development. This has resulted in the idea of getting the result as a criteria – integral characteristics of breakage by blast including geological, geomechanical, technological factors and those directly relating to explosive charge blast action and charge design. The criterion is expressed as a selection of one among three variants of the criterion types of blasting scale generalized on the basis of all known drilling and blasting methods and parameters. Thus, a type of blasting scale reflects an integrated effect of blasting conditions of any complexity, including the growth of mining depth, that allows its characterization as integrated. Further step is the tie-in of a particular variant to particular mining conditions.

This is based on the principle of the estimation of a blast scale effect of one borehole charge, which is calculated from the correlation (comparing) of values and parameters of a charge and a mining block of the rock mass for this charge. It can be used with different standard techniques for drilling and blasting operations by way of comparing the scale fitness of the above mentioned values and their correction. The necessity of this approach application for the calculation of a charge amount arises at the stage of design and development of novel technologies for large-scale breakage by blast [2]. It is possible on the basis of mining and geological data to make preliminary calculations of drilling and blasting parameters for stoping of any scale with the desired quality of fragmentation and no need for test blasting.

A reasonable extent of fragmentation of the media being broken and maintaining the stability of the rock mass are provided by the selection of parameters of explosive energy release kinetics and design of detonation systems (group of charges) for different methods of breakage by blast. Technically, for the solution of the task of blast action control in complicated conditions numerous methods are available of commercial explosion control. Besides, even the classification of such methods by process factor is available [3]. However, it cannot sort out a number of contradictions between the requirements of geotechnologies in complicated geomechanical conditions, including great mining depth, and limited opportunities of controlling the processes of blast energy release and its distribution in the rock mass. The fact

---

Corresponding author E-mail: victorov_s@mail.ru

is in the lack of a mechanism of selection and classification of methods of blasting process control by characteristic, which inner structure contains a group of initial (explosive charge, its shape, location in the rock mass, design, energy, wiring-up, etc.) and limiting (type and shape of a rock mass block to be broken, its physical properties, geological and mining characteristics, open air) conditions. Selection of mining methods includes these conditions categorizing them by blasting process classification characteristic (generalized criterion of the scale of breakage by blast).

In terms of science and technology the new approach can be reduced to the solution of the task of development and substantiation of a physical model of intensive advance blasting of the rock mass depending on the diversity (complexity) of geological conditions. The idea is to get an express method of assessment and selection of a blasting technique based on the integral characteristics of rock mass breakage by blast.

The model consists of two interrelated parts: its structure (algorithm, block diagram) and a test bench, which is a physical part of the general model, see Fig.1. The main point of the model structure part is a block diagram for the description and visualizing of the process of blasting intensification in different conditions.

A particular type of breakage by blast is integrally determined by its scale and data spectrum, while the integrated effect of blasting operation conditions with depth on the selection of the class of mining methods and type of breakage can be substantiated by the following.

Ore deposit underground mining methods for stoping at thick ore deposits [4,5] in terms of "blasting" can be arbitrarily divided based on such an assumed "system forming" parameter as "stability" into three main classes:

Ⅰ—mining methods with open stope, including tight-face blasting, ore and enclosing rock block, slice and panel caving with large-scale parameters of rock mass sections being broken;

Ⅱ—mining methods with backfilling, ore shrinkage, support of underground workings;

Ⅲ—mining methods with enclosing rock caving and small-scale parameters of rock mass sections being broken.

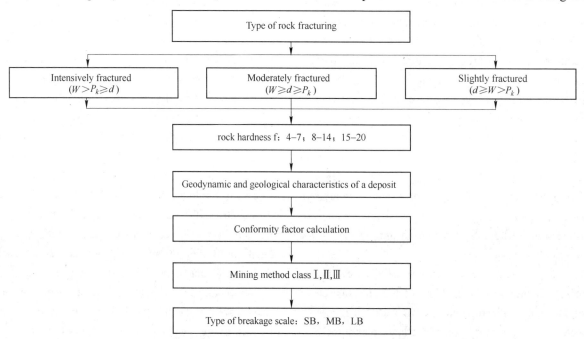

Fig.1 Block diagram of the physical model for the selection of rock mass high intensity blasting

W—line of least resistance; d—mean distance between fractures in the rock mass section being broken; P_k—size limit at an ore mine

The scale of breakage by blast proposed as a criterion for the selection of its integral type is the failure of (degree) proportionality of the volume being broken and the charge energy with the condition of achieving an equal fragmentation ratio. This effect known as a "scale factor" at earlier stages was implicitly present in methods of breakage parameters calculation, however, there was no impulse for its system-wide development. Small-scale breakage (SB) is the blasting of sections (layers) of the rock mass with blasthole or borehole charges of small, arbitrarily of up to 50 mm diameter, and a line of least resistance W of up to 1.2~1.5 m.

Large-scale breakage (LB) is the blasting with larger (large) parameters of the rock mass sections being broken. It is performed with borehole charges of large diameter ($100 < d \leqslant 1000$ mm), chamber charges and some charges differing in geometry, which are equivalent in terms of the scale of blast action (W reaches 10 m and even more).

Medium-scale breakage (MB) is an intermediate type in terms of its parameters.

Fig.1 shows rock types in terms of correlation of main

parameters of a blasting process. Conventional classes of mining methods are used as arguments of function, which values are given in the last row of its structure based on the classification of breakage by type selection for different conditions described in the table below.

Tab.1 Classification of breakage by type selection for different conditions

Mining depth	Rock types	Rock characteristics	Correlation of blast parameters, W, d, P_k	Rock hardness, f	Integrated effect of blasting operation conditions with depth on the selection of	
					mining method class, No	type of breakage scale (SB, MB, LB) and its efficiency ($\geqslant$)
1	2	3	4	5	6	7
200~500	1. Intensively fractured	Fractures of small to medium size, where the d distance between fractures of all systems does not exceed the size limit P_K. At least 1-2 fractures in any direction per one meter.	$W > P_k \geqslant d$	4~7	II, III	MB
				8~14	I	LB
				15~20	I	LB
500~700				4~7	II	MB = LB
				8~14	I	LB
				15~20	I	LB
>700				4~7	II, III	MB = SB
				8~14	II, III	MB
				15~20	I, II, III	MB $\geqslant$ LB
200~500	2. Moderately (irregularly) fractured	Medium block-size rocks. The d distance between fractures: from the size limit to a distance between boreholes or W line of least resistance	$W \geqslant d \geqslant P_k$	4~7	II, III	SB $\geqslant$ MB
				8~14	II, III	SB = MB
				15~20	II, III	SB
500~700				4~7	II, III	SB $\geqslant$ MB
				8~14	II, III	SB
				15~20	II, III	SB
>700				4~7	II, III	SB
				8~14	II, III	SB
				15~20	II, III	SB
200~500	3. Slightly fractured (solid)	Rarely encountered fractures, the distance between which is close to the parameters of a layer being broken; the presence of rock loosening, microfractures and more dense fractures, which are well wave-conductive	$d \geqslant W > P_k$	4~7	I, II, III	MB
				8~14	I, III	MB = LB
				15~20	I, III	MB = LB
500~700				4~7	II, III	MB
				8~14	I, III	MB $\geqslant$ LB
				15~20	I, III	MB = LB
>700				4~7	II, III	SB
				8~14	II, III	SB $\geqslant$ MB
				15~20	II, III	SB = MB

These values are shown as a resultant criterion type of rock mass breakage (blasting) and their comparative efficiency with due account for the integrated effect of main conditions and characteristics of the surrounding medium of a charge and expressed as follows: SB – small-scale breakage; MB – medium-scale breakage; LB – large-scale breakage; "$\geqslant$" – more efficient or equivalent; "=" – competitive.

As it is known, a significant effect on the results of rock fragmentation, its intensity in particular, is produced by the stress state of the rock mass depending on the "stability" and determined by the mining depth and geological conditions that is also reflected in the fifth row of the physical model.

A hypothesis of the correlative relationship between classes of mining methods and types of breakage scale was checked with samples of various materials.

The results of these experimental studies were used for the development of the above mentioned classification.

Before mining starts the rock mass is in its natural (original) stress state governed by gravity and tectonic fields.

Stresses in the vertical plane for ideally elastic, homogeneous and isotropic rock mass in conditions of impossibility of horizontal strains depend on the weight of overlying rock γH, where γ is the volumetric weight of rock; H is the depth of the point under consideration from the surface. By results of experimental research into vertical stresses in many points of the globe at depths of down to 3 km the following dependence was found: $\sigma_3=0.026H$. In contrast to vertical stresses, horizontal stresses are rather complicated, as parameters of the tectonic component may vary significantly both in space and with time. This peculiarity can be explained by irregular distribution of the velocities of tectonic movements in space, and the earth's crust deformation rates. In rocks of crystalline and folded basement horizontal stresses are greater than vertical in 60% of all cases, for sedimentary formations – the number of such cases accounts for 10%~15%. In some cases the difference may reach 5~10 times. Such measurements were taken in numerous ore mines and underground structures worldwide, in particular, in ore mines and tunnels of the Kola Peninsula, Gornaya Shoriya, Donets Basin, Ural, Altai, as well as in Paleozoic folded sands of Norway, Ireland, Africa, Southeastern Australia, Portugal, etc. As a result, nearly in all these cases horizontal stresses by several times exceeding vertical stresses were registered.

Some cases are known, when in ore mines at relatively shallow depth (100~150 m) intensive dynamic phenomena, such as, rock flaking and "bursting" as thin plates from roadway walls accompanied by a strong sound effect were registered. As a rule, the scale of these phenomena depends on geological and mining conditions including the increase of mining depth.

Physical modeling of some of these phenomena was performed in laboratory conditions. ICEMR RAS researchers developed a test bench (physical part of the model) for the purpose, where rock samples were exposed to uniaxial compression, and parameters were estimated of the particles chipped-off from the surface of a cylinder-shaped hole specially made in the center of the samples. In this case, a cylinder-shaped hole is in the role of an underground working. Uniaxial compression of a sample results in non-uniform distribution of stresses on the hole contour, and according to the elasticity theory, zones are formed with maximal compressing and tensile stresses.

The general arrangement of the laboratory plant for experiments is shown in Fig.2.

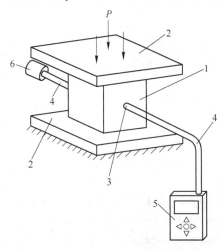

Fig.2　General arrangement of the laboratory plant for experimental studies

According to the proposed research methods and procedures in the center of the test sample 1 the test volume was created as a through cylinder-shaped hole 3 of 6 mm diameter. Sampling tubes 4 were attached on two sides to this hole. One sampling tube was connected with an aerosol particle counter 5, and another with an air filter 6. The sample was placed between support plates of the press 2 and exposed to uniaxial compression ranging from zero to a breakdown point. Over the total time of experiment the parameters of emission of submicron particles ranging from 0.3 to over 5.0 μm were registered. During uniaxial compression of different rock type samples the formation of submicron particles from the cylinder-shaped hole surface and their quantity depending on the size and position of the cylinder-shaped hole were recorded[7]. The quantity of newly-formed particles was explained by rock instability and, as a consequence, by the values of underground working exposure determining the selection of a class of mining methods.

Fig.3 reflects the results of experimental studying of a dolomite sample, where the x-axis shows the compression stress σ, MPa, and the y-axis shows the specific quantity of particles, $1/(m^2 \cdot s)$.

Registration of the quantity of particles and analysis of particle-size distribution were performed at 60-second intervals. The sample uniaxial compression strength was 56 MPa. From the Figure below it can be seen that with compression stress nearing 45 MPa the emission of particles increases in all ranges, while with the values nearing the critical state (50 MPa) a fast growth of particle emission was observed that is an evidence of a forthcoming macro-failure of the test sample. Similar dependences were found for urtite, sandstone, granite, ferruginous quartzite, etc. [8]

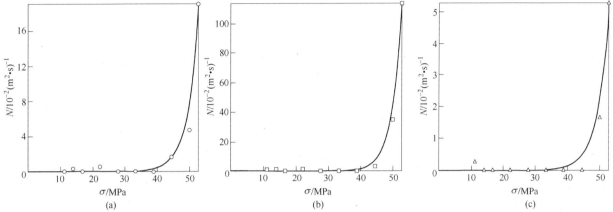

Fig.3　Dependence of the quantity of particles on compression stress for dolomite exposure to uniaxial compression in different ranges

(a) 0.3~0.5 μm; (b) 0.5~5.0 μm; (c) >5.0 μm

By results of these studies, the conformity factors were calculated, which numerical values characterize the rock mass stability via its stress state value. The analysis of these results shows the correlative relationship of the classes of mining methods and types of breakage scale. This provided and opportunity of identifying the final matching of breakage scale types and specified conditions of their efficient implementation. In particular, if breakage efficiency types are equal, the preference is determined by the value of conformity factor.

The algorithm of the physical model application is rather simple. Based on the information contained in cells of horizontal same-name rows, the class of underground mining methods conditionally generalized in terms of breakage by blast is shown in the last but one bottom row. Then, as a final result, in the last row the type of breakage scale matching this data is identified. Identification of the latter makes it possible in advance of drilling and blasting operations to select in the spectrum of known design options the one, which matches the mining process and drilling and blasting parameters of a block or stoping area of an ore mine.

Thus, the improvement of methodology of breakage by blast including the elaboration of a physical model of the process of the rock mass intensive blasting, and based on this model the classification of breakage by its type selection for different conditions is a novel approach in this area of mining science.

REFERENCES

[1] Rock blasting with closely-spaced charges/Viktorov SD, Galchenko YuP, Zakalinsky VM, Rubtsov SK. – M., Nauchtekhlitizdat Publishers OOO, 2006: 276 (in Russian).

[2] Technology of large-scale rock breakage by blast for the Siberian ore deposits liable to rock bumps. // Viktorov SD, Eremenko AA, Zakalinsky VM, Mashukov IV. – Novosibirsk: Nauka Publishers, 2005: 212 (in Russian).

[3] Kazakov NN. Classification of methods of commercial explosion control in surface mining operations/ Votrosy Razrusheniya Gornykh Porod. – M.: IPKON RAS, 1994: 5~15 (in Russian).

[4] Agoshkov MI, Malakhov GM. Underground mining of ore deposits. – M.: Nedra Publishers, 1966 (in Russian).

[5] Kaplunov RP, Cheremushentsev IA. Underground mining of ore and placer deposits. – M.: Vysshaya Shkola Publishers, 1966 (in Russian).

[6] Baikonurov OA. Classification and selection of underground mining methods for mineral deposits/Alma-Ata, Nauka Publishers, 1969: 606 (in Russian).

[7] Viktorov SD. Formation of submicron particles in mining, and a new method of assessment of catastrophic phenomena / Vestnik, Russian Academy of Sciences, 2013, 83(4): 300~306 (in Russian)

[8] Viktorov SD, Kochanov AN, Osokin AA. Objective laws governing the generation of micro- and nanoparticles in the course of rock breaking and deformation. Mining information and analysis bulletin. Transactions of Symposium: Miner's Week-2011. – M.: Gornaya Kniga Publishers, 2011 (OB1): 185~191 (in Russian).

Research on Key Technology of Digital Rotary Drilling in Surface Mines

DUAN Yun, XIONG Daiyu, YAO Lu, WANG Fengjun, ZHA Zhengqing, GONG Bing
(*Beijing General Research Institute of Mining & Metallurgy, Beijing, China*)

ABSTRACT: In order to achieve the digitization and refinement of rotary drilling in surface mines, we develop the Blasthole Design Software using AutoCAD ActiveX (C#) technologies, the computer-assisted positioning terminal system based on GPS OEM board, the Navigation Software of locating blast holes using LabView and C++. These technologies realize the functions of boundary lines measurement, the blasthole layouts design, digital drilling according to assistant software and drilling data transmission, and also provide the reliable basis for blast refinement in surface mines.

KEYWORDS: blasthole layouts; GPS OEM; rotary drills; surface mines; digital mining

1 INTRODUCTION

Rotary drill is the main drilling machine in open pit mine, with many advantages such as a high rotary speed. The main process of blasting combines the blasthole layouts design, digital drilling, filling explosives, detonating network design and evaluation of blasting effects. However, blasting design and digital drilling decide the blast effect.

At present, in China's large-scale open pit mine, blasting design, work face preparation, blasthole acceptance, filling explosives and detonating network design rely entirely on the ground blast technicians and blasting operations staff manually is ubiquitous, which cannot guarantee the accuracy. Especially, in the aspect of layout parameters determine in the deep hole blasting, the majority of mining enterprises usually use a simple method with rulers and tapes to measure blasting site situations. The process is relatively rough, and cannot realize the refinement and digitalization. Meanwhile, drilling is the primary part of the open-pit mine production. For a long time, due to lack of effective technical means, parameters of drilling and blasting cannot be accurately measured, safe and efficient blasting is difficult to comprehensively conduct, and blast hole drilling quality is in the low level, seriously affects the blasting effect.

From the information above, it shows that the key technology of blasting effect is digital blasting design and drilling process. Domestic mining enterprises have not implemented automated drilling rigs and intelligent technology. On the contrary, foreign research instituti-ons have developed advanced rigs detection system, but its technology is secret and its price is relatively expensive. As an consequence, the subsequent blast design process is difficult to achieve real-time interaction between each other, and applying widely is not easy in our mines. Therefore, the study of blasthole affordable GPS system and suitable blasting design software is significant to improve blasting effect, achieve digital and intelligent drilling and reduce blasting mining costs.

2 CORE FUNCTIONS OF DIGITAL DRILLING SYSTEM

2.1 Utilizing GPS RTK Technologies to Position Precisely

GPS reference station is established in the mine dispatch center, GPS rover equipment is installed on the rig. The antenna of GPS rover station is mounted on the roof of the rig operating room, GPS rover receiver module is installed in the rig intelligent monitoring host internally, communicating through the RS232 interface with the host, and the whole system is shown in Fig. 1. With this module, the monitoring host obtains the location coordinates of this point. This product is easy to be installed and used, with high precision and real-time access to satellite signals to locate, and can meet rig measurement requirements of explosion area brow line and the boundary line.

After the installation of GPS base station is completed, we uses a handheld GPS on site to measure a variety of location information such as brow lines, blast hole and so

---

Corresponding author E-mail: yduan-nmg@126.com

on. With the advantages of using simple and quickly, only one person can complete all measurements. After the measurement, the data can be imported to a computer for analysis and processing, you can use a dedicated blasting design software to display accurately and analysis efficiently for a specific function.

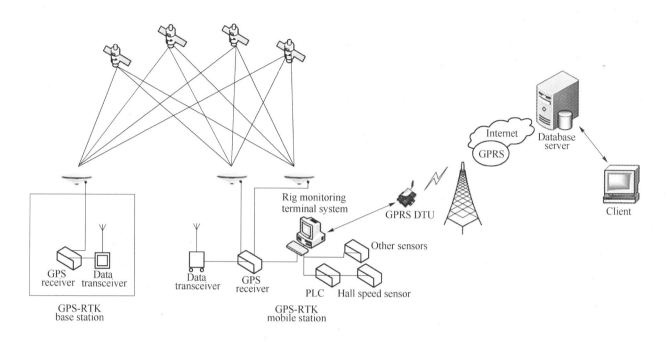

Fig.1　Structure of the whole digital drilling system

2.2　The Development of Blast Design Software

Blast design software system includes blasthole mesh parameter analysis and design, communicating with server, blasthole display, lithology identification and printing reports. This software enables to download and upload blasting parameters. It can also provide visualization display and editing functions combining with real geological mapping system, using the secondary development of AutoCAD software. According to the brow line on site, perforated hole information and parameters set by the user, the software automatically optimizes holes distribution. In addition, based on the data collected in field, the software can display terrain of blasting area, drilling conditions and lithology identification information. The blast design parameters will be adjusted by feedback actual perforation information to achieve optimum design goal of mine blasting.

Blast design software system includes the following seven sections:

(1) Installation this design software to the local computer, and make it run as administrator.

(2) Through a remote connection to the server, login server database, the software reads the blasting parameters from severs to local, mainly providing brow line and on-site drilling rig operating information. Users can select parameters corresponding to the time and the rig number, and arrange periods of drilling and blasting reasonably.

(3) In the design process of holes network, click the Show Brow Line button, and then the site measured brow line data is shown to topographic maps. In order to ensure the accuracy and application of automatic holes layout, users can draw standard line by themselves, and then configure parameters in the module of automatic holes layout, including holes orientation, interval, row spacing, and number of holes and regional shape of blast area information. Trough software calculation, we get holes location information automatically, and it can be displayed on the geological mapping of CAD drawings, enabling designers to see holes information clearly. Designers can modify location according to the actual situation, in particular of drawing the front holes.

Client interface of this function is shown below in Fig. 2, click anywhere in the window, except for the position occupied by the control having recommended data,can be automatically refreshed. Based on changes in terrain blasting area, software system can generate a different holes layout area.

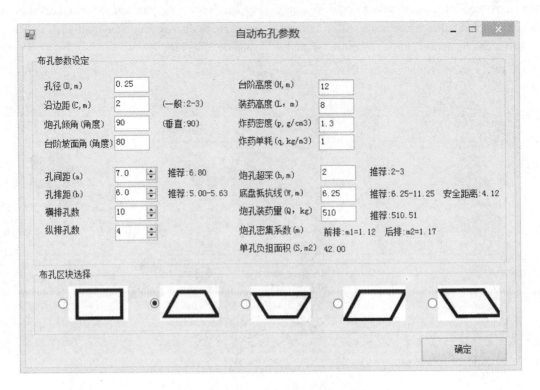

Fig.2　The automatic holes layout module

(4) After completing the design, designers can make design information (shown in Fig. 3)stored in the local computer, or upload it to the database which is available for drilling monitoring terminal reads perforation tasks.

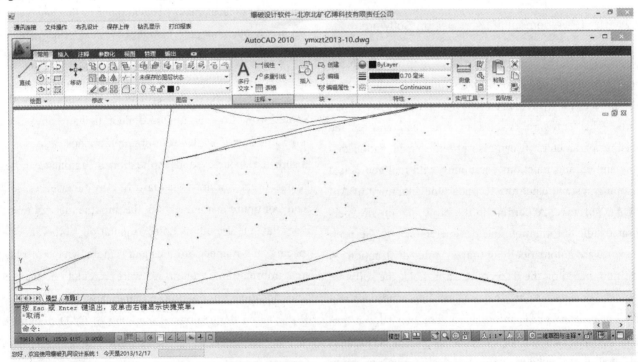

Fig.3　Diagram of automatic holes layout

(5) Three-dimensional visualization function is mainly used to display the terrain undulation of blasting area, the specific drilling status and query holes properties, providing a reference for blast designers, and as shown in Fig.4. Parameters are obtained by reading the server database.

This feature provides actionable tools modules: display selection module (optional display topographic maps and rig drilling chart), view Angle adjustment module (azimuth and viewing angle can be adjusted separately for three-dimensional graphics comprehensive view), zoom module (click + / - zoom adjustment), blasthole properties query module. In addition, users can drag mouse to move the 3D picture.

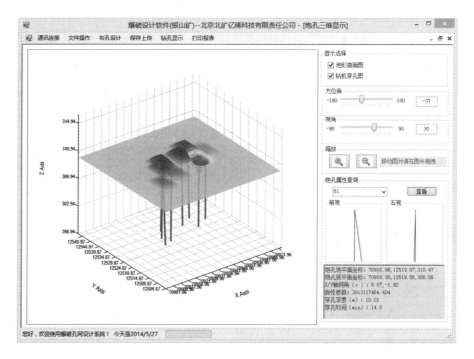

Fig.4　Three-dimensional visualization function

(6) Lithologic identification and its display is mainly applied after drilling is completed. According to the parameters collected during the drilling process, we use blasting design software to analyze and process lithological parameters in this blasting area and display it. The main function is providing blast designers charging reference, and they can adjust depending on different lithological information of blastholes to achieve optimal explosion. Lithology information is calculated by blast design software utilizing drilling machine operating parameters, and lithological map is shown in Fig.5.

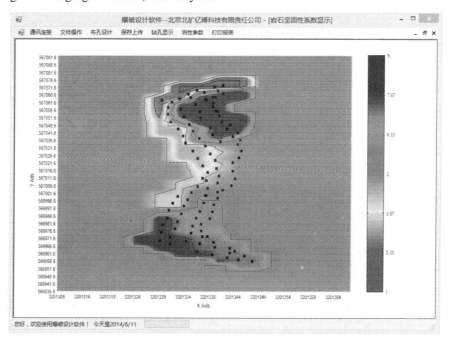

Fig.5　Lithological map in a blasting area

(7) In order to facilitate the management, the software provides printing reports. Reports record holes coordinates, design time, designers, and other relevant information, and print it out for easy management and archiving. When everything is done, you can use the file operation function, save design drawings before exiting the system.

2.3　Digital Blasting Integrated Management Platform

We built up a digital blasting integrated management platform which is shown in Fig. 6 and set up a central server with database service. We also deployed a monitoring system on this server to record client side activities. With the help of the monitoring system, we were able to communicate with client side system in real-time as well as establish communications between multiple drill machines. In additional, engineers worked in control center were able to monitor the state of drilling machines in real-time which providing data support for effective resource management. What is more, this administrative system collected important scientific data about punching location, rotating speed, rotary rate and lithology information for future optimum design. It constructed a high performance administrative platform for mine explosion and implemented information sharing and data synchronization via computer network.

序号	眉线点X	眉线点Y	眉线点Z	时间
1.1	70663.321	12566.6899999999	323.325	2013-11-22 12:52:17
1.2	70655.9889999999	12568.6899999999	323.978	2013-11-22 12:52:17
1.3	70648.149	12569.6099999999	323.87	2013-11-22 12:52:18
1.4	70640.115	12570.3620000002	323.912	2013-11-22 12:52:18
1.5	70631.589	12570.9720000001	323.557	2013-11-22 12:52:18
1.6	70624.373	12572.7489999998	323.169	2013-11-22 12:52:18
1.7	70619.54	12566.6290000002	322.446	2013-11-22 12:52:18
1.8	70626.4449999999	12565.969	323.056	2013-11-22 12:52:19
1.9	70634.421	12564.878	323.372	2013-11-22 12:52:19
1.10	70641.485	12564.273	323.507	2013-11-22 12:52:19
1.11	70651.992	12563.3229999999	323.332	2013-11-22 12:52:19
1.12	70657.26	12561.8289999999	323.384	2013-11-22 12:52:19
1.13	70661.573	12561.4679999999	323.19	2013-11-22 12:52:20
1.14	70658.841	12541.034	322.92	2013-11-22 12:52:21
1.15	70653.519	12537.6719999998	323.059	2013-11-22 12:52:21
1.16	70647.397	12536.5	323.264	2013-11-22 12:52:21
1.17	70640.682	12537.4109999998	323.4	2013-11-22 12:52:21
1.18	70634.606	12539.5290000001	323.433	2013-11-22 12:52:21
1.19	70629.6629999999	12542.264	323.366	2013-11-22 12:52:22
1.20	70625.638	12546.3569999998	323.131	2013-11-22 12:52:22
1.1	70663.321	12566.6899999999	323.325	2013-11-22 21:04:49
1.2	70655.9889999999	12568.6899999999	323.978	2013-11-22 21:04:50
1.3	70648.149	12569.6099999999	323.87	2013-11-22 21:04:50
1.4	70640.115	12570.3620000002	323.912	2013-11-22 21:04:50

12345678910…>>

Copyright (C)2007| All right Reserved | 北京矿冶研究总院 版权所有
隐私保护 北京北矿亿博科技(TEL:010-63299307, FAX:010-88370780) 版权所有
©2011 北京北矿亿博科技有限责任公司

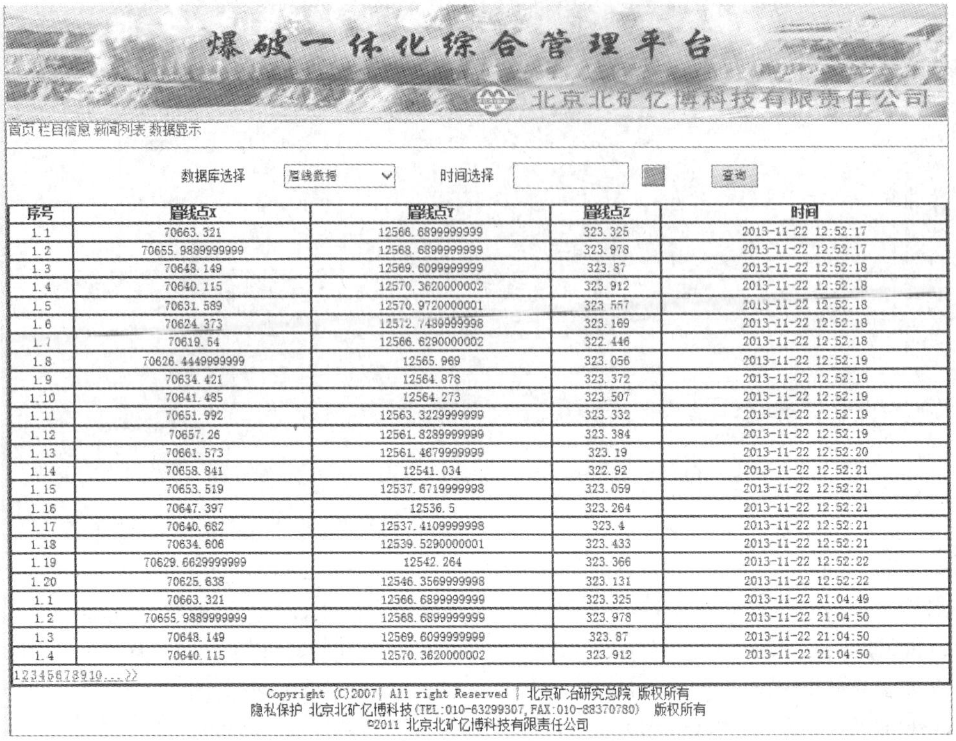

Fig.6　Digital blasting integrated management platform

2.4　Drilling Machine Monitoring Terminal System

In this study, we develop a drilling machine monitoring terminal system based on LabVIEW to position accurately and navigate automatically. Users can enter design drilling depth according to lithology and blasting demand. The operation state monitoring page can display drilling depth visualization, rigs running parameters and GPS positioning information, and this page is shown in Fig.7.

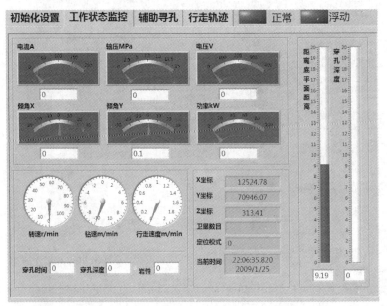

Fig.7　The operation state monitoring page

Blast holes navigation page mainly assists rig drivers to look for holes, and holes on the map, designed by blasting design software, are transfer to the display terminal through the task module. Users can use the keyboard to select upcoming perforated blast hole numbers, and can also change the navigation mode. This interface is shown in Fig. 8.

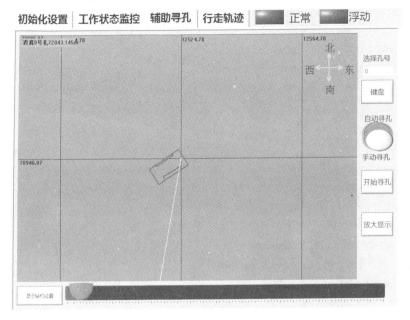

Fig.8 Blast holes navigation page

Specific navigation process is divided into manual and automatic modes, and the navigation process is shown in Fig.9. In manual mode, select the aperture number, and then the software prompts rig navigation information to reach the current location of the hole, and besides, drivers can zoom, pan the map and make other operations for viewing; In automatic mode, based on the relative position of blast holes and the drilling rig, and they can be automatically displayed in the center of the screen. When the distance between the blast hole and drilling rig is within a certain range, the color of the hole turns yellow, and drivers can work.

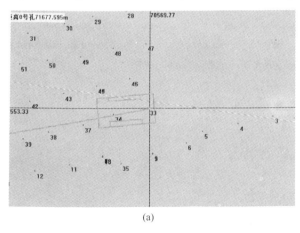
(a)

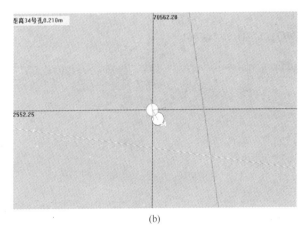
(b)

Fig.9 Navigation process
(a) manual mode; (b) automatic mode

3 APPLICATION OF RESEARCH RESULTS

The digital rotary drilling system is applied in Dexing Copper Mine of Jiangxi Copper Company Limited. The client side system was deployed on computers of blasting designers, and the server was set in the central computer-room of Beijing General Research Institute of Mining and Metallurgy. The entire system provided a visualized and digital solution for copper mine explosion. The implementation details were described as follows.

First of all, we deployed the entire system based on Fig.1. This including database service installation, explosion design software installation, drill machine monitoring system installation as well as GPRS network set up. Then we did hardware installation for GPS RTK and sensors on drill machines and conducted system communication testing to make sure we could collect data from the sensors and locate drill machines from GPS. In addition, we used drilling machines to collect real time data and send them back to explosion design engineers. The engineers computed the data based on lithology information and store them into database. At last, the on-site workers would check the computed data in drilling machine monitoring system and conduct accurate drilling based on those pre-calculated parameters. At the same time, drilling data would be collected and uploaded to the central database. In this way, explosion expects were able to analyze the real-time data and send instructions back to job site.

4 CONCLUSIONS

In this study, we propose some major researches on digital mining technology based on rotary drilling machine. It achieves intelligent precise explosion control with GPS positioning and blasting design software. It implements the automation process for perforation parameters, computation and visualization for drilling machine perforation. The research not only improves the accuracy of explosion work but also reduces the labour intensity. The combination of accurate perforation positioning and real-time monitoring system result in better explosion effects, much looser muck piles and flat bench outlines. This research eliminates the potential unpredictable factors in traditional explosion work which affecting the explosion quality. It improves the production efficiency and reduces equipment wearing, so reduce the total cost for mining production.

ACKNOWLEDGEMENT

The authors gratefully acknowledge financial support from the National Natural Science Foundations of China (No. 51104018) and Beijing Xicheng District Outstanding Personnel Training Project (No. 20120071).

REFERENCES

[1] John H, Datavis P, John V. Drill Monitoring Systems and the Integration with Drill and Blast Software. ISEE Annual Conference. Nashville, Tennessee, 2011.

[2] Chen Q Y, Dang Y Y, YI F. Design of GPS receiver based on GPS-OEM board[J]. Journal of Xi'an Polytechnic University, 2008, 22(6): 775~778.

[3] Zhao X D. Study on drilling construction technologies in open-pit mine[J]. Non-Ferrous Mining and Metallurgy, 2002, 18(6): 5~7.

[4] Dong Y D. CAD secondary development theory and technology[M]. Hefei: Hefei University of Technology Press, 2009.

[5] Zhang Y T, Gu Y L. ASP.NET From Entry to Master. Beijing: TSINGHUA University Press, 2008.

[6] Shen L J, Liu Y, Wang X G. Bench blasting technique in open pit mines at home and abroad. Engineering Blasting, 2004. 10(2): 54~58.

[7] Duan Y. Research on blast holes automatic design and development of GPS positioning technology and equipment. Beijing General Research Institute of Mining and Metallurgy Research Paper, 1~110.

[8] O M, Pang B Y. Research on drilling and blasting method of rock crushing in open pit mines. Second International Conference on Fragblast, 1987: 381~392.

Controlled Blasting for Safe and Efficient Mining Operations at Rampura Agucha Mine in India

Pradeep K Singh, M. P. Roy, Amalendu Sinha

(*CSIR- Central Institute of Mining & Fuel Research, Dhanbad, India*)

ABSTRACT: Rampura Agucha open pit mine is a lead zinc mine producing 6 mtpa of ore and has started its underground part which is slated to produce 4.5 mtpa of ore in near future. The mine is, currently working at 300 m depth and is designed to reach to a depth of 421 m. The open-pit as well as underground operations are in full swing. This leads a greater emphasis as its footwall is achieving its final configuration while hang wall is on dynamic stage. During blasting, back break was the main concern as despite of carrying out pre-splitting; the benches experienced back breaks adversely affecting the stability. Normal production blast consists of firing 7~16 rows involving 150~450 holes in a blast round. A master plan of the pre-split holes position at spacing of 1.2 m was prepared. The available explosive was in the cartridge diameter of 25 mm which provided decoupling by factor of 3.6. Every bench of footwall and of all stages of hang wall is pre-splitted to minimize damage to the rock mass. Pre-split holes of about 8 00 000 m are being drilled annually to make pit-wall stable. In the experimental trials the mouths of the pre-split holes were left without explosives from 1 to 2.7 m. In the latter stage the inclined holes of 80°, 70° and 60° with 115 mm diameter were experimented. The pre-split at a spacing of 1.2 m and inclined at 60° on footwall and 70° on hang wall yielded desired results.

KEYWORDS: pre-split; wall control; damage to wall; blast optimisation; controlled blasting

1 INTRODUCTION

Drilling and blasting continues to be an important method for rock excavation and rock breaking. Drill and blast technique has a disadvantage that sometimes it produces cracks in an uncontrolled manner and generates micro cracks within the post blast opening geometry. One way of achieving controlled crack growth along specific directions and inhibit growth along other directions is to generate stress concentrations along those preferred directions. Several researchers have suggested a number of methods for achieving fracture plane control by means of blasting. There are four methods of controlled blasting, and the one selected depends on the rock characteristic and the feasibility under the existing conditions. These methods are line drilling, cushion blasting, smooth-wall blasting and pre-splitting. The paper deals with challenges faced during standardisation of pre-split blast design at Rampura Agucha mine in India for safety and long term stability of pit-wall. The control of fractures in undamaged brittle materials is of considerable interest in several practical applications including rock fragmentation and over-break control in mining (Fourney et al., 1975; Fourney, 1993; Kaneko et al., 1995). One way of achieving controlled crack growth along specific directions and inhibit growth along other directions is to generate stress concentrations along those preferred directions. Several researchers have suggested a number of methods for achieving fracture plane control by means of blasting. Fourney, et al.(1978)suggested a blasting method which utilizes a ligamented split-tube charge holder. Nakagawa et al.(1982), examined the effectiveness of the guide hole technique by model experiments using acrylic resin plates and concrete blocks having a charge hole and circular guide holes. Katsuyama et al.(1983) suggested a controlled blasting method using a sleeve with slits in a borehole.

There are four methods of controlled blasting, and the one selected depends on the rock characteristic and the feasibility under the existing conditions. These methods are line drilling, cushion blasting, smooth-wall blasting and pre-splitting (Lopez et al., 1995). If the rock is incompetent, smooth-wall blasting may not be satisfactory (Rossmanith et al., 2008). Cushion and pre-splitting blasting are the most commonly used methods, with the main difference between the two beings that in cushion blasting the final row of holes is detonated last in the sequence, while in pre-splitting the final line holes are detonated first in the sequence. Pre-splitting consists of creating a plane of shear in solid rock on the desired line of break(Sharafisafa et al., 2011). It is somewhat similar to other methods of obtaining

Correspoding author E-mail: pradeep.cimfr@yahoo.com

a smoothly finished excavation, but the main difference is that pre-splitting is carried out before any production blasting and even in some cases before production drilling (Fourney et al., 1978; Mohanty et al., 1990). Pre-splitting utilizes lightly loaded, closely spaced drill holes, fired before the production blast. The purpose of pre-splitting is to form a fracture plane across which the radial cracks from the production blast cannot travel. The fracture plane formed may be cosmetically appealing and allow the use of steeper slopes with less maintenance. Pre-splitting should be thought of as a protective measure to keep the final wall from being damaged by the production blasting(Rossmanith et al., 2008).

The theory of pre-splitting is that when shock waves from simultaneously detonating charges in adjoining blast holes collide, tension occurs in the rock, forming a crack in the web between the holes. For that reason it is important that charges are detonated simultaneously or as close as possible. Firouzadi et al. (2006) concluded that small diameter holes in pre-splitting row such as 102 mm using decoupled charge is difficult whereas continuous charging of pre-split row in such hole yields better result. Among different approaches of continuous charging (decoupling, explosive mixing), decoupling due to its operation difficulties of small diameter holes (102 mm) was rejected by Lopez et al. (1995). The measured effects of decoupling, explosives etc. on the length of radial cracks behind the half casts have been reported by Ouchterlony et al. (2000). The detonation behavior of the explosive is also responsible for the difference in their damage potential. Velocity of detonation (VOD) is an important indicator of the performance of an explosive. It controls the rate of release of explosive energy and also influences the energy partitioning with respect to shock and gas energies.

If back break is not controlled, it ultimately decreases the overall slope angle, with major economic consequences such as decreased recoverable ore reserves and increased ore to waste ratios. The best approach is to control the effects of blasting so that the inherent strength of the walls is not destroyed. The purpose of pre-splitting is that to isolate the blasting area from the surrounding rock mass by forming an artificial plane to limit gas and stress wave penetration into the remaining nearby rock formation (Scott et al., 1996; Olofsson et al., 1998; Cho et al., 2004; Singh et al., 2009; Adamson, 2012; Gopinath et al., 2012).

Although some relevant work has been done on the theoretical study of stress waves from blasting in cylindrical boreholes (Heelan P A, 1953; Tubman et al., 1984; Meredith et al., 1993; Nakamura Y, 1999; Blair et al., 2006; Blair D P, 2007), and attempt has been made to use numerical methods to estimate blast-induced damage from stress waves (Trivino et al., 2009). There is a significant shortage of field-scale experimental studies on blasting. Field scale measurements of blast-induced damage are not common either, and are limited to specific rock types and conditions. Ouchterlony et al. (1999; 2001) and Olsson et al. (2002), for example, have contributed with the study of blast-induced damage in rock by conducting crack measurements in granite blocks and bench blasting. Other authors have used pressure sensors to estimate damage extent by measuring rock swelling (Brent et al., 1996; 2000) and gas penetration into cracks (Yamin G A, 2005).In particular, little experimental work has been done to properly study blast-induced stress waves. Amongst the few examples of proper wave measurement and analysis are the works of White & Sengbush (1963), Vanbrabant et al. (2002), Trivino & Mohanty (2009), Singh et al. (2009)and Trivino & Mohanty (2012).

The field study has been conducted at Rampura Agucha mine which is the India's largest Lead-Zinc open-pit mine of M/s Hindustan Zinc Limited. The present depth of the pit is 300 m and the ultimate pit will be 421 m. Pre-split blasting was implemented in all the benches in an attempt to make them safer and more competent. Pre-split blasts were carried out at all the benches in foot wall (FW) side in existing cut back. One of the objectives in implementation of pre-splitting was to maintain high-wall stability for access roads, ramps, entry points etc. which must remain for extended periods during the mine life. This also reduces clean up times and wears & tears on haulage equipment as a result of reduced material falling-off of the high-wall and onto the haulage road.

2 GEOLOGICAL INFORMATION ABOUT THE MINE

Rampura Agucha mine is located at 225 km north-northeast of Udaipur and is the largest & richest Lead Zinc deposit in India containing ore reserves and resources of 107.33 million tonnes grade of 13.9% zinc and 2% lead. Rampura Agucha is a stratiform, sediment-hosted Lead Zinc deposit, occurs in Pre-Cambrian Banded Gneissic complex and forms a part of Mangalwar complex of Bhilwara geological cycle (3.2~2.5 billion years) of Archean age and comprising of magmatites, gneisses, graphite mica schist, pegmatite and impure marble. The rocks have been subjected to polyphase deformations and high-grade metamorphism.

The deposit is a plunging isoclinal synform with rock units showing NE–SW strike with steep dips(75°~80°) in hanging wall and moderate dips (60°~65°) in footwall towards SE. The host rock occupies the core of the synform and plunge in southwestern limit is 65°~70° due NE. The Rampura Agucha mixed sulphide deposit is a massive lens shaped ore body with a NE-SW strike length of 1500 m and a width varying from a few meters in the NE direction widening to as much as 120 m in the central to SW section with average of 58 m. The ore minerals are mainly sphalerite and galena. The ore body dips from 50° to 70° towards SE. The host rock for mineralization is Graphite-mica-sillimanite Gneiss/Schist (GMS) and consists of mica (white, green and brown varieties), feldspar, quartz and an appreciable amount of graphite. Walls are composed of Garnet Biotite-Sillemanite Gneiss (GBSG) and intrusions of Pegmatite and Amphibolite and Mylonite (on footwall only) while GBSG forms the major chunk amounting around 70%~80% of the mass.

There are 3 major joint sets on footwall. Foliation is the most prevalent discontinuity 60°~80° /N130° affecting the blast damage. Rocks at the mine are moderately competent. The physic-mechanical properties of the rocks are presented in Tab.1. Foliation is the main plane of discontinuity which governs the bench profile. The pit strike was divided in 100m wide zones viz. N100~N200, N200~N300, S100~S200, etc. (Tab.2). The area is geologically complex. The majority of the country rock is a gneiss of moderate strength, with a pervasive foliation and lenses of minor rock units (intrusions of stronger amphibolite and pegmatite) uniformly dipping at approximately 70° to 90° to the east at upper elevations and, in the northern part of the pit, down to about 50° to the east at lower elevations. The working benches and overview of the mine is shown in Fig.1.

Tab.1 Physico-mechanical properties of rocks at Rampura Agucha Mine

Rock properties	Lithology			
	Ore	Amphibolite	GBSG	Pegmatite
Density /g·cm^{-3}	3.00	2.99~3.11	2.67~2.98	2.69~2.74
Young's modulus /GPa	57.54~84.63	97.82~131.91	26.47~88.83	65.72~90.57
Uniaxial compressive strength /MPa	18.1~77.5	99.45~257.35	40.45~74.23	136.76~148.92
Tensile strength /MPa	1.97~12.5	14.50~21.59	3.12~11.27	10.27~13.33

Tab.2 Trend of foliation on footwall (FW) of the pit (S8/S7 means area between S800 & 700 & 80/141 means joint is dipping at 80° due N141°)

FW Foliation						
S8S7	S7S6	S6S5	S5S4	S4S3	S3S2	S2S1
81/141	80/130	86/137	76/135	78/141	76/135	79/153
NS0N1	N1N2	N2N3	N3N4	N4N5	N5N6	N6N7
82/142	80/139	83/147	85/144	77/149	72/145	81/137

Fig.1 Overview of the benches of Rampura Agucha Mine

3 BACKDROP OF MINING OPERATIONS

The extent, size and shape of ore body at Rampura Agucha Mine, ideally suited to workout upper part by open-pit technique up to techno-economic pit bottom depth and below it by underground mining (Fig.2). Ore body has been proved down to 1065 m and upper part of ore body is being worked with open pit. Firming of open-pit depth is dynamic, influenced by updated geotechnical and financial parameters. However, based on updated input parameters, present consideration for ultimate pit depth is 372 m containing 48.7 Mt ore and 528 Mt of waste. The open pit will be continued with shovel dumper combination delivering ore production of 6.15 Mt and would be supplemented by underground working from June 2013 onwards.

The mine works with shovel–dumper combination deploying large sized HEMM like 34 m^3 shovels and 240 tonne trucks navigation with Truck Dispatch System (TDS). The excavation is carried out by a drill fleet of 165 mm for production blasts and 115 mm diameter drills for trim and presplitting purposes. The footwall has reached to its final configuration while its hang wall is under four staged cut-backs. Operation is being carried out with bench height of 10 m, berm width of 9.5 m with batter angle of 60° and 70° on FW and HW respectively. Two access ramps of 25 m wide each in F/W and H/W, with gradient of 1 in 12, provides the entry to the pit. The ultimate pit has been designed in DATAMINE software using slope angles of 35° & 42° in FW and HW respectively as outcome of the geo technical study.

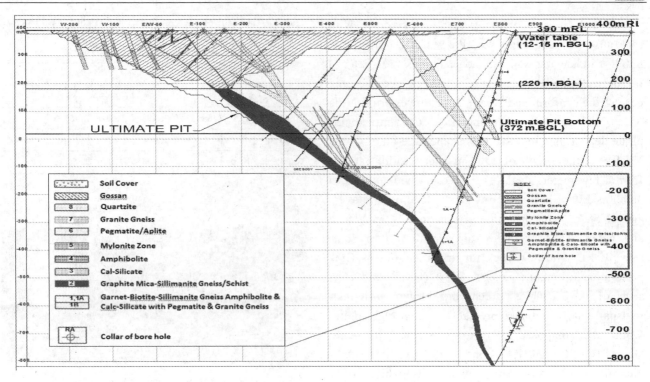

Fig.2 Section showing ultimate pit and present pit of Rampura Agucha Mine

4 PRODUCTION BLASTS DETAIL

Blasting with varying designs and charging patterns were conducted at different benches of the mine. The depth of holes varied between 10 and 12 m. Generally in the waste the burden is kept 4 m and spacing of 5 m is being practiced but in the ore, the burden is 3 m and spacing is 3.5 m. SME explosives were used and were detonated with Unitronic (electronic delay detonator), or Nonel shock tubes. The recorded in-the-hole VOD of explosives were in the range of 5200 to 5600 m/s (Fig.3) and fragmentation were good (Fig.4) in most of the blasts without any back breaks in competent rock. Nevertheless, back breaks were noticed when the blasting faces are geologically disturbed. The explosive weight per delay ranged between 140 and 540 kg. Total explosive weight detonated in a blast round varied between 2792 to 34170 kg. Different drill pattern geometries were tested to determine the best design that would result in good rock breakage with lower costs and greater safety. Fig.5 shows a typical blast design followed at the mine for a bench having bench height of 11 m, hole depth of 11.5 m and 165 mm hole diameter. Burden and spacing are 4 by 5. Utmost care was taken in production blasts near the pre-split area but sometimes the foliations planes and geological discontinuities played in opening of foliation planes as shown in Figures 6 and 7.

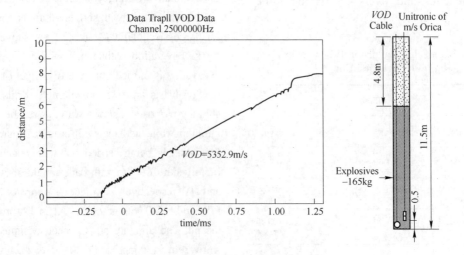

Fig.3 The traces of recorded in-the-hole *VOD* SME explosives

Fig.4 Negligible back-breaks were noticed in compact rock

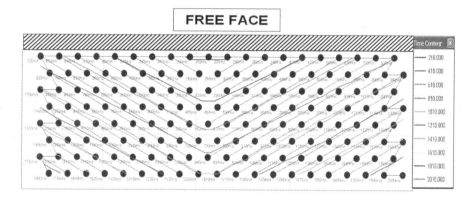

Fig.5 Blast design and charging pattern of holes for medium hard rock patch at Rampura Agucha Mine

Fig.6 Opening of foliation planes due to production blast

Fig.7 Geologically disturbed faces which are strengthen by bolting and wire netting

Ground vibrations data recorded at different locations due to blasting of production blasts were grouped together for statistical analysis correlating the maximum explosive weight per delay (Q_{max} in kg), distance of vibration measuring transducers from the blasting face (R in m) and recorded peak particle velocity (v in mm/s). The established equation for the mine is given as equation 1 and the regression plot is presented at Fig.8.

$$v = 179.56 \left(\frac{R}{\sqrt{Q_{max}}} \right)^{-1.197} \quad (1)$$

Correlation co-efficient = 84.1%.

where v ——peak particle velocity, mm/s;
  R ——distance between vibration monitoring point and blasting face, m;
  Q_{max}——maximum explosive weight per delay, kg.

5 BOREHOLE PRESSURE CALCULATION FOR PRE-SPLIT HOLES

Borehole detonation pressure is considered to be a good criterion to describe the intensity of an explosive. It has very important role in success of pre-split blast. Dynamic in-situ compressive strength of the rock could be two times or more than its static value. The range of compressive strength of most of the rocks is 40～80 MPa and the range of tensile strength is 5～9.25 MPa, so the dynamic compressive strength of the rock is considered as 80 ～160 MPa. Decoupling factor is defined as the ratio of explosives cartridge diameter (32 mm) and blast hole diameter (115 mm) and it is 0.278. The recorded in the hole *VOD* of 32 mm cartridge explosives was 3739 m/s (Fig.9).

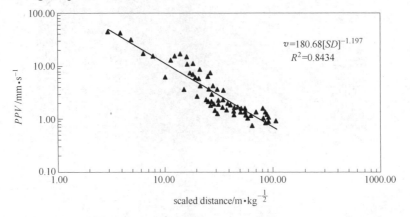

Fig.8 Regression plot of recorded *PPV* at their respective scaled distances.

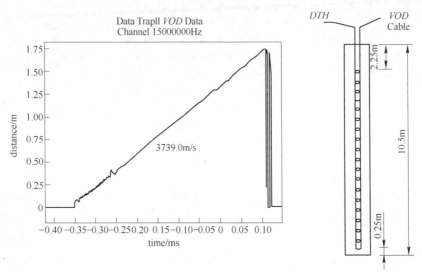

Fig.9 Recorded in-the-hole *VOD* of cartridge slurry explosives of 32 mm diameter

The effective borehole pressure in pre-split holes was calculated by equation 2 (Chiappetta RF, 2001). The recorded *VOD* of explosives was 3739 m/s and specific gravity was 1.1 g/cm³, so the calculated borehole pressure comes to:

$$P_b = 1.25 \times 10^{-4} \rho (VOD)^2 (r_e / r_b)^{2.6} \quad (2)$$
$$= 1.25 \times 10^{-4} \times 1.1 \times 3739^2 \times 0.2783^{2.6}$$
$$= 69.2 \text{ MPa}$$

where P_b——borehole pressure in MPa;
ρ——specific gravity of explosives;
VOD——detonation velocity of explosives in m/s;
r_e——radius of explosives charge in mm;
r_b——radius of borehole in mm.

The lower value of the borehole pressure (40%～60%) to the dynamic compressive strength of the rock is required to get the optimum pre-split result. So, the required borehole pressure should be in the range of 32～92.4 MPa. Latter on the two cartridges of explosives were tagged together and were loaded by providing air-gaps which resulted with good pre-split result (Fig.10).

Fig.10 Loading of cartridge explosives in pre-split holes

6 EXPERIMENTS WITH VERTICAL PRE-SPLIT HOLES AND RESULTS

The workings were nearing the final pit limits in FW, pre-split work started in FW. The designed angle of individual bench slope is 70° which are because of the dip angle of most of the foliation planes, as such the most appropriate method is to do pre-split drilling at this angle. However, since drilling inclined holes requires too much of accuracy, so work stated with vertical hole drilling which was subsequently shifted to inclined hole drilling.

Prior to start of the pre-split a master plan of the pre-split hole position at a spacing of 1.2 m was prepared. The pre-split line was designed at 1.0 m away from the final crest line of 370 bench away from the berm to be left. The available bench width at 370 bench from pre-split line was between 25~40 m. It was planned to cut the bench in two slices so as to have the last row of blast holes at least 20 m away from the pre-split line. The production drilling and blasting started from Northern end. The first blast was taken from N-200 to S-175 for production holes. No damage was observed on the bench as already 20 m width was left. The second blast from S-175 to S-250 was taken which was having only two rows. In that production blast, a crack was observed up to 18-19 m. It was then decided to not to take any production blast in critical area on FW bench unless pre-split blast is taken in that area.

A pre-split blast was taken on 370/360 FW-S. The 62 vertical holes were drilled at a spacing of 1.2 m. The depths of holes were 10.5 m. The charge density per length kept was 0.53 and 0.625 kg/m^2. The blast was taken with detonating cord initiation system. The top portion of 2.0 m of drill hole was kept open without explosives and was fired instantaneously. The crack developed along pre-split line was very good (Fig.11). There was absence of any type of back-breaks but at few places excessive crater were noticed where the charge factor was 0.625 kg/m^2 (Fig.11).

Fig.11 Pre-split view of the blast face at 370/360 FW-S

Similar results were encountered at number of blasts but wherever the rock formation was not favorable excessive back breaks were encountered. A pre-split blast was taken at 370/360 FW-S. The coordinate of blast was N-20 to N-96. The total number of hole blasted was 80 (from N-19 to N-48 from N-50 to N-100). The explosive used in the blast was Super-Dyne having diameter of 32 mm. The weight of a single cartridge explosive was 400 g. The average charge density per length used was 0.53 and 0.65 kg/m^2. The top 2.5 m of drill hole was kept uncharged and the holes were kept open during the blasting process. The result of the blast was not good as cracks developed beyond the pre-split blast and excessive crater was recorded where charge factor was 0.65 kg/m^2 (Fig.12).

Fig.12 View of pre-split blast taken at 370/360 FW-S

Another pre-split blast was taken at S-436 to S-503 with 72 holes. The diameter was 115 mm and was drilled vertically. As the rock formation was having clay intrusion so the spacing was 0.9 m and the charge density was 0.44 kg/m^2. The slurry explosive was used. The top portion of 3 m was kept open without any explosives. The excellent pre-split line was observed (Fig.13).

Fig.13 View of pre-split result at 360-350 FW at S-436 to S-503 holes

A number of trial blasts for pre-split line was conducted at different rock types having *RQD* of 50 to 70 and the charge factors used were between 0.44 to 0.9 kg/m^2 depending upon the rock type. It was decided to take a few production blasts along with the pre-split holes at 20~25 m away from the final pit wall line to explore the possibility of detonating pre-split holes together with the production blast holes.

The blast was conducted at 370/60 south FW in hard rock

formation. The production holes were drilled with 165 m diameter at burden and spacing of 4 and 4.5 m. The depths of the hole were 10.5 m and number of rows was three excluding pre-split line holes. The details of the blast design and charging pattern is presented in Fig.14. The blast resulted with mixed result. The one third of the blast portion from where the blast was initiated resulted with poor fragmentation and excessive back breaks were noticed whereas the two third of the blast patch resulted with excellent result (Fig.15). The half cast factors were noticed. A major joint plane was passing in the centre of the blast patch, which was controlled by blast design and initiation sequences and no back break was encountered in that portion too.

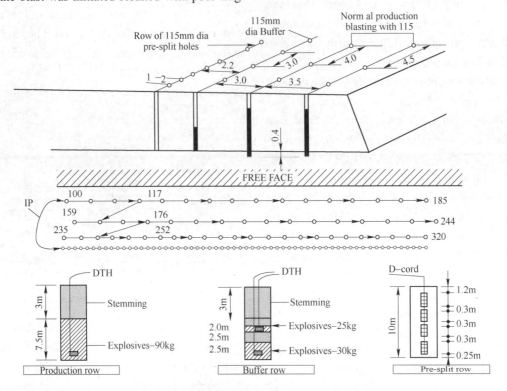

Fig.14 Blast design of production holes along with pre-split holes

Fig.15 The two third portion of the blast with excellent breakage and depression of required muck profile and stable pit walls

7 EXPERIMENTS WITH INCLINED PRE-SPLIT HOLES AND RESULTS

The pre-split planes are sub-parallel to persistent, smooth, undulating joints that dip into the pit at about 30° to 70° at the Rampura Agucha mine. The average dip of these joints is about 50°. Where the pre-split plane intersects one of these joints, some of the explosion gases produced by the pre-split charges jet into and wedge open the joint and, therefore, reduce the joint's shear strength. Therefore, there is an incentive to minimize the probability of pre-splits intersecting these joints. This probability decreases as the inclination of the pre-split holes decreases from 90° to 50°, the latter being the average inclination of these joints. So, 70° pre-splits are better than vertical pre-splits and 60° pre-splits would be even better than 70° pre-splits in the rock mass encountered at the mine.

Pre-splitting holes were drilled at angle from horizontal to conform to steep major joint system present. This is essential for success since vertical pre-split will not work where joint sets steeper than 70° are present. (As J1 and J2 sets at Rampura Agucha Mine are steeply dipping). One of the most important factors determines success of a pre-splitting operation is the manner in which the final row of holes in the production blast (buffer row) is charged. The pre-split holes with 80°, 70° and 60° were experimented. The charge distribution and depth of holes for one of the pre-split blasts with 60° inclined holes of 115 mm and loaded with 32 mm cartridges of slurry explosives is shown in Tab.3. Experimented blast design along with the pre-split line using 115 mm diameter in all the blast holes and drilled with 70° inclinations and charged with bulk explosive at 10 m high bench is presented in Tab.4.

Tab.3 Charge distribution in 60° inclined pre-split holes of 115 mm loaded with 32 mm cartridges of slurry explosives

Pre-split hole ID	Length of hole /m	Spacing /m	Required explosives /kg·m⁻²	Length of explosive column /m
P-136	11.0	1.2	6.45	7.9
P-137	10.9	1.2	6.40	7.7
P-138	10.8	1.2	6.35	7.5
P-139	10.8	1.2	6.35	7.5
P-140	10.7	1.2	6.30	7.4
P-141	10.7	1.2	6.30	7.4
P-142	10.7	1.2	6.30	7.5
P-143	10.7	1.2	6.30	7.3
P-144	10.7	1.2	6.30	7.4
P-145	10.7	1.2	6.30	7.2
P-146	10.7	1.2	6.30	7.2
P-147	10.7	1.2	6.30	7.3
P-148	10.7	1.2	6.30	7.4
P-149	10.7	1.2	6.30	7.3
P-150	10.7	1.2	6.30	7.3
P-151	10.7	1.2	6.30	7.2
P-152	10.7	1.2	6.30	7.2
P-153	10.7	1.2	6.30	7.3
P-154	10.7	1.2	6.30	7.3
P-155	10.7	1.2	6.30	7.3
P-156	10.7	1.2	6.30	7.2

Tab.4 Experimented blast design along with the pre-split line using 115 mm diameter in all the blastholes and drilled with 70° inclinations and charged with bulk explosive at 10 m high bench

	1ˢᵗ row production blast holes	2ⁿᵈ row production blast holes	3ʳᵈ Row production blast holes	Buffer row blast holes	Pre-split blast holes
Length of blasthole /m	10.5	10.5	10.5	10.5	11.3
Unavoidable fallback /m	0.2	0.2	0.2	0.2	0.2
Effective length of blasthole /m	10.3	10.3	10.3	10.3	10.9
Effective sub-drilling /m	0.3	0.3	0.3	0.3	0.5
Stemming length /m	2.5	2.5	3.4	3.2	2.5
Deck length /m	0.0	0.0	2.1	3.3	—
Charge length /m	8.6	8.6	5.1	4.1	8.6
Explosive density /g·cm⁻³	1.1	1.1	1.1	1.1	—
Charge weight /kg	98	98	53	43	6.7
Drilled burden distance /m	3.1	3.1	2.7	2.2	2.7①
Drilled blast hole spacing /m	3.6	3.6	2.5	2.5	1.2

① Really the stand-off distance rather than the burden.

The charge distribution and other parameters were kept constant in the successive sets of blasts whereas the inclinations of holes were changed i.e. 80°, 70° and 60°. A number of trial blasts were conducted at FW with 80°, 70° and 60° inclinations. The pre-split blast results at FW with 80°, 70° and 60° inclinations in one set of experiments are presented in Figures 16, 17 and 18 respectively. The pre-split holes with 60° inclinations gave acceptable result.

Further it was decided to increase the bench height from 10 m to 12 m in view of long term stability of the pit wall. The FW rock is not competent enough having compressive strength between 40~80 MPa in most of the areas. The experience is that even if sub-grade is not done, no problem is faced in excavation up to required limits and as well as in maintaining the required level. Moreover, the possibilities of getting back breaks are more with the sub-grade in such type of rocks and to prevent the floor or crest area for next bench below from these cracks. So, no sub-grade is done at least in the proposed crest area.

Fig.16　View of pre-split blast result with 80° inclination of pre-split blast holes

Fig.17　View of pre-split blast result with 70° inclination of pre-split blast holes

Fig.18　View of pre-split blast result with 60° inclination of pre-split blast holes

The bench heights were kept as 12 m in place of 10 m. Explosives were charged with 25~28 kg in stab holes whereas in buffer holes the explosives loaded were 75~80 kg. The production holes were charged with 175~180 kg. The charge distribution pattern of the holes is presented in Fig.19. At few instances of the pre-split blasting with the above mentioned design some areas were shattered. It was decided to take the blast in two slices, the first slice of 8 m and the subsequent slice of 4 m but the results of taking 12 m bench blasts with previously taken pre-splitting have given better results than taking 8 and 4 m blasts with 12 m pre-split. The reasons are that when taking 12 m blast, the trim blasts are of width of 18~24 m (considering from pre-split line) or less. The lesser width drilling is possible due to additional 9.6 m berm available on the bench and drill machines can be deployed easily. When taking 8+4 m blasts, in upper 8 m, there is no issue as again additional 9.6 m of berm width is available. However, when taking the bottom 4 m blast, since no berm width is available, for manoeuvring the drill machines, the width of blasts to be kept is 25~30 m. Also since the bench height is only 4 m, the drill pattern has been changed. The production holes are drilled at a closer pattern of burden of 3 m and spacing of also 3 m. This is resulting in more charge concentration to get desired fragmentation for mucking which is not giving required results. The width of blasts in second cut will be reduced on bottom benches. Anyhow, the upper two benches have been successfully excavated by pre-split blasting and have reached at their final limits. The view of the upper two benches is shown in Fig.20.

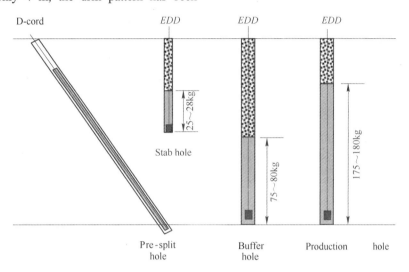

Fig.19 Explosive distribution pattern in the holes

Fig.20 View of the upper two benches after implementation of pre-split blasting

The complete mining plan was redesigned and again the bench height was shifted from 12 m to 10 m. The benches of 10 m are being blasted in one stage and are yielding acceptable results. Till date the blasting is in operation at 260 m from the surface as shown in Fig.1. The annual drilling of pre-split holes are approximately 700000 m. The pre-split blasts have been successfully implemented in all the benches (Figures 21 and 22) and the zones which are having fault, shear are strengthen by cable bolting and wire netting to make it more stable for long term stability (Fig.23). The pre-split design helped enable to excavate 25 benches with stable pit-walls. Further blasting is in operation and ultimate pit depth of 421 m will be achieved in coming 6 years.

Another monitoring and modeling exercise was performed to observe the changes occur in blast vibration data. In approach to this measurement the geophones were placed at 5 m before the pre-split line and 5 m after the presplit line to record the vibration at both the locations in production blasts which were conducted after pre-split blasting. The results show that at the respective scaled distances the vibration recorded after pre-split line was much lower than those recorded before the pre-split line (Fig.24). These results also confirmed that pre-split works as filter in attenuation of vibration. The difference between the amplitude at various distances between the two monitoring locations can be considered to represent the filter performance i.e. pre-splitting.

Fig.21　The close view of few benches with stable pit-walls

Fig.22　View of benches with stable pit-walls

Fig.23　Geologically disturbed final pit-walls strengthen with cable bolts

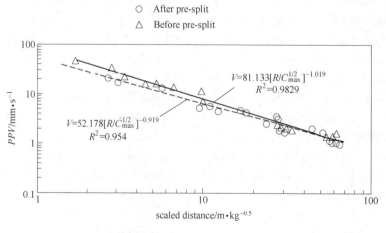

Fig.24　Recorded vibration levels before and after pre-split due to production blasts

8 DISCUSSION AND CONCLUSIONS

Initially pre-split blast results were not sustainable. A number of trials with different combination viz. varying the spacing, hole angle, charge per square metre, use of gas bags, bottom initiation, cartridge diameter, standoff distance between pre-split and buffer row, number of buffer rows, hole diameter of buffer rows, different drill patterns, different blast designs, powder factor, number of rows in a blast, blast size etc. were carried out. With the available data of exploratory drilling, the charge per square meter was calculated. Initially, the actual charge was kept less than that calculated as it was known that not only the dynamic tensile strength of rock but also the overall Rock Mass Rating, the joint sets and foliation planes do also plays important role in success for pre-split blasting. It was a totally practical approach. The charge was gradually increased till an optimum powder factor was obtained.

In achieving satisfactory pre-split results more than thousands pre-split blasts were conducted. This involves, gaining experience based on results achieved and its analysis, now the results are satisfactory and the technique has been established and being improved continuously. There needs to be a "stand-off" distance (measured at bench floor level) between the pre-split plane and the buffer (or nearest production) blast holes. Where, this stand-off distance is too small, the production blast breaks back beyond the pre-split plane. Where the stand-off distance is too large, it will be difficult to dig back to expose the pre-split plane; fillets of unbroken rock will be left adhering to the pre-split plane.

The standard practice till recently involves drilling at $70°$ at designed crest line at a spacing of 1.2 to 1.3 m, charge factor of 0.44 to 0.65 kg/m^2 depending upon rock type, vertical buffer row of small dia (115 mm) at a stand-off distance of 1.0 to 1.5 m, number of production rows of (165 mm) 4 to 5, blast size in a linear length of 70~90 m, and delay timings have been standardized based on linear superposition technique in the case of either electronic delay detonators or Nonel initiation system so as to have proper burden relief and reduced charge per delay to reduce effect of blast induced ground vibrations on the final pit walls.

It has been established that each pre-split blast holes should be fired before the drilling of adjacent production blast holes. Those blasts that were fired along with pre-split blast holes did not yield good results. The cracks were encountered beyond the pre-split line. The radial cracks behind the half casts tend to became longer when an existing fracture disturbs the blast holes. The pre-split holes spacing of 1.2 m was found optimum for 115 mm drill diameter for the bench heights of 10 m and 12 m. The powder factor of 0.44 to 0.60 kg/m^2 yielded desired pre-split results. A number of experiments were carried out with Nonel tubes and detonating cords to initiate the holes in pre-split blasting. The pre-split holes fired with detonating cord gave better results than those fired with Nonel tubes.

The change in bench height from 10 m to 12 m and again back to 10 m, the revised batter angle or slope of benches has been changed to $60°$ from $70°$ and berm width to 9.5 m from 6.5 m in final push back of FW. The results of pre-splitting at top two benches with 12 m height which has reached the final limits found to be excellent and the other benches with 10 m height which has also reached at final limits resulted into excellent pre-split results.

Most of the surface mines operators who use pre-splitting incline their pre-split blast holes at $70°$ to $80°$; $65°$ pre-split blast holes are rare and $60°$ blastholes are rarer. The charging of shallower-dipping pre-split blast holes was difficult, because of the greater frictional/drag force that resists the down-hole movement of pre-assembled charges. The $60°$ pre-splits was successfully introduced at Rampura Agucha mine, final batters were better, largely because smaller volumes of rock was undercut by the pre-split and higher percentages of "half barrels" were visible. Discretization of the pit into different zones based on characterization of ground i.e. Rock Mass Rating (RMR) helped enable in modification in orientation of firing front, firing pattern and firing timings based on rock characteristics including orientation of the major joint/ foliation set in trim blasts facilitated in achieving excellent pre-split results. Special treatment for area ravaged by major geological structures like faults, shears, etc. by wire netting and cable bolting resulted into enhanced stability of the pit walls. The pit walls are monitored in real time by Slope Stability Radar (SSR) and the results shows that pit-walls are stable.

ACKNOWLEDGEMENTS

The authors express their thankfulness to the officials of Rampura Agucha Mine for providing necessary facilities during field investigations. The permission of Director, CSIR-Central Institute of Mining & Fuel Research, Dhanbad, India to publish the paper is thankfully acknowledged.

REFERENCES

[1] Fourney W L, Holloway D C, Dally J W. Fracture initiation and propagation from a center of dilatation. *Int J Fracture*.1975, 11:1011~1029.

[2] Fourney W L. Mechanisms of rock fragmentation in by blasting.In: Hudson J.A, editor. Compressive rock engineering, principles, practice and projects. Oxford: *Pergamon Press*, 1993.

[3] Kaneko K, Matsunaga Y, Yamamoto M. Fracture mechanics analysis of fragmentation process in rock blasting.*J JpnExp-Soc*.1995, 58(3):91~99.

[4] Fourney W L, Dally J W, Holloway D C. Controlled blasting with ligamented charge holes.*Int J Rock Mech Min Sci*.1978, 15:121~129.

[5] Nakagawa K, Sakamoto T,Yoshikai R. Model study of the guide hole effect on the smooth blasting. *J JpnExp Soc*. 1982, 43:75~82.

[6] KatsuyamaK, Kiyokawa H,Sassa K. Control the growth of cracks from a borehole by a new method of smooth blasting. *Mining Safety*.1983, 29:16~23.

[7] Lopez J C, Lopez, J E. Drilling and blasting of rocks. *A.A.Balkema*, Rotterdam, 1995.

[8] Rossmanith H P, Uenishi K. The Cuña problem – reconsidered. In Proc. of 12th International Conference of International Association for Computer Methods and Advances in Geomechanics, (*IACMAG*), Goa, India, 2008.

[9] Sharafisafa M,Mortazavi A. A numerical analysis of the effect of a fault on wave propagation.*In Proc. of 45th US rock mechanics symposium*, San Francisco, USA, 2011.

[10] Mohanty B. Explosive generated fractures in rock and rock like materials. *EngFractMech*.1990: 889~898.

[11] Firouzadj A, Farsangi M A E, Mansouri H,Esfahani S K. Application of controlled blasting (Pre-splitting) in Sarcheshmeh copper mine. *In Proc. of 8th Int. Symp. on Rock Fragmentation by Blasting - Fragblast 8*, Santiago, Chile, 2006: 383~387.

[12] Ouchterlony F, Olsson M,Bavik S O. Perimeter blasting in granite with holes with axial notches and radial bottom slots.*Int. J. Blasting and Frag.*, Netherlands, 2000, 4(1): 55~82.

[13] Scott A, Cocker A, Djordjevic N, Higgins M, La Rosa D, Sarma K S, Wedmaier R. Open pit blasting design analysis and optimization. Published by JKMRC, (Ed. Andrew Scott), Queensland, Australia, 1996: 214~243.

[14] Olofsson S O. Applied explosives technology for construction and mining. *A.A. Balkema*, 1998: 183~186.

[15] Cho S H, Nakamura Y, Kaneko K. Dynamic fracture process of rock subjectedto stress wave and gas pressurization. *Int J Rock Mech Min Sci*.2004, 41:439.

[16] Singh P K, Roy M P, Joshi A,Joshi V P. Controlled blasting (pre-splitting) at an open pit mine in India.In Proc. of 9th*Int. Symp. on Rock Fragmentation by Blasting*,Granada, Spain, 2009: 481~489.

[17] Adamson WR. Reflections on the functionality of pre-split blasting for wall control in surface mining. In Proc. of 10th *Int. Symp. on Rock Fragmentation by Blasting* -Fragblast 10, New Delhi, India, 2012: 697~705.

[18] Gopinath G, Venktesh H S, Balachander R,Theresraj. Presplit blasting for final wall control in a nuclear power project.*In proc. of 10th International symposium on Rock Fragmentation by Blasting – Fragblast 10*, New Delhi, India, 2012: 731~740.

[19] Heelan P A. 1953. Radiation from a cylindrical source of finite length. *Geophysics*.1953, 18: 685~696.

[20] Tubman K M, Cheng C H,Toksoz M N. Synthetic full waveform acoustic logs in cased boreholes.*Geophysics*.1984, 49: 1051~1059.

[21] Meredith J A, Toksoz M N, Cheng C H. Secondary shear waves from source boreholes. *Geophysical Prospecting*, 1993, 41: 287~312.

[22] Nakamura Y. Model experiments on effectiveness of fracture plane control methods in blasting. *Int J Blast Fragment*, 1999, 3:59~78.

[23] Blair D P, Minchinton A. Near Field blast vibration models. *In Proc. 8th Int. Symp.on Rock Fragmentation by Blasting – Fragblast 8*, Santiago, Chile, 2006: 152~159.

[24] Blair D P. A comparison of Heelan and exact solutions for seismic radiation from a short cylindrical charge. *Geophysics*, 2007, *72 (2)*: E33~E41.

[25] Trivino L F, Mohanty B. Seismic radiation from explosive charges in the near-field: results from controlled experiments. *In Proc. of 35th Ann. Conf. on Explosives and Blasting Technique*, Denver, USA 2, 2009: 155~166.

[26] Ouchterlony F, Olsson M, Bavik S O.Bench blasting in granite with holes with axial notches and radial bottom slots.*In Proc. of 6th Int. Symp. on Rock Fragmentation by Blasting - Fragblast 6*, Johannesburg, South Africa, 1999: 229~239.

[27] Ouchterlony F, Olsson M,Bergqvist I. Towards new Swedish recommendations for cautions perimeter blasting. *Proceedings of Explo*.2001: 169~181.

[28] Olsson M, Nie S, Bergqvist I, Ouchterlony F. What causes cracks in rock blasting? In Proc. of *6th Int.Symp.on Rock Fragmentation by Blasting - Fragblast 6*, Johannesburg, South Africa, 2002: 221~233.

[29] Brent G F, Smith G E. Borehole pressure measurements behind blast limits as an aid to determining the extent of rock damage.*In Proc. of 5th Int. Symp. on Fragmentation by Blasting – Fragblast 5*, Montreal, Canada, 1996: 103~112.

[30] Brent G F, Smith G E. The detection of blast damage by borehole pressure measurement.*The Journal of The South African Institute of Mining and Metallurgy*, 2000: 17~21.

[31] Yamin G A. Field measurements of blast-induced damage in rock.MASc Thesis. *University of Toronto*, 2005.

[32] White J E,Sengbush R L. Shear waves from explosive sources. *Geophysics*, 1963, 28 (6): 1001~1019.

[33] Vanbrabant F, Chacon E, Quinones L. P and S mach waves generated by the detonation of a cylindrical explosive charge – experiments and simulations. *In Proc. of 6th Int. Symp.on Rock Fragmentation by Blasting - Fragblast 6*, Jo-

hannesburg, South Africa, 2002: 21~35.

[34] Trivino L F, Mohanty B, Munjiza A. Investigation of seismic radiation patterns from cylindrical explosive charges by analytical and combined finite-discrete element methods. *In Proc. of 9th Int. Symp.on Rock Fragmentation by Blasting - Fragblast 9*, Granada, Spain, 2009: 415~426.

[35] Trivino L F, Mohanty B. Estimation of blast-induced damage through cross-hole seismometry in single-hole blasting ex periments. *In Proc. of 10th Int. Symp. on Rock Fragmentation by Blasting - Fragblast 10*, New Delhi, India, 2012: 685~695.

[36] Chiappetta R F. The importance of pre-splitting and field controls to maintain stable high walls, eliminate coal damage and over break. *Tenth High-tech seminar on State of the art, blasting Technology, instrumentation and explosives application, GI-48*, Nashville, Tennesse, USA; 22~26 July 2001.

Study on Technology of Cableway Station Tunnel Demolition Blasting in Scenic Area

SHI Liansong, GAO Wenxue, LIU Dong, CHEN Gui, YAO Shaomin, ZHU Xuyang
(*The College of Architecture and Civil Engineering, Beijing University of Technology, Beijing, China*)

ABSTRACT: By using reasonable delay of the blasting inside hole and outside hole and deep hole interval charging technology successfully demolished cableway station tunnel, effectively control the influence of blasting vibration on the Badaling Great Wall; use protective materials covered in explosion body, which can effective control of the blast flyrock, to ensures the safety of the equipment and facilities around the Great Wall. The successful experience of the blasting dismantled can be use for future similar tunnel construction, provide a reference implementation of the blasting operations around.

KEYWORDS: demolition of the tunnel; millisecond delay blasting; Badaling Great Wall; protection technology

1 INTRODUCTION

With the mature of demolition blasting technology and it brings significant economic benefit, demolition blasting has become one of the most competitive method in demolition industry. At present, the demolition blasting has been widely used, including, chimney, water tower, silos, buildings, etc. However some demolition blasting environment not only complex, and demolition of building, content itself also has some special structure characteristics, often brought some difficulty to blasting operation, which requires the demolition operations must try to be careful from design to construction. In this paper, the Badaling Great Wall north road station tunnel which will be dismantled have such characteristics, the dismantle the tunnel as part of the thin-walled structure, reinforcement with the internal, and drilling is difficult; in addition, the area is located at the 5A level tourist scenic spot, the latest distance between the tunnel and the Great Wall is only 60m, puts forward high demands on blasting protection work. Based on the blasting design, blasting network and safety aspects, laying the tunnel blasting demolition technology is introduced, From the results, blasting has a complete success.

2 BACKGROUND OF ENGINEERING

In order to fit the requirements of the reconstruction and expansion, the need for clues to the Badaling Great Wall north road station on part of the tunnel. Clues to the Badaling Great Wall north road station tunnel (here in after referred to as the station tunnel), located in the hillside, and the cable car station underground vertical elevation difference of more than 700 meters, and the tunnel which need to dismantle is open-excavated tunnel and it is concrete structure, the tunnel section of straight wall arch, width of 8 m, clear height of 7.1 m, the tunnel length of 17.2 m, tunnel is divided into two segments, the front tunnel length is 8 m, made of thick wall around 55 cm, middle vault 35 cm thick, on both sides to gradually transition to 55 cm; after a period of tunnel length is about 9.2 m, made the wall thickness is 65 cm, middle vault 45 cm thick, vault to on both sides of the tunnel wall thickness gradually transition to the 65 cm; two net section of the tunnel section size is the same, the difference is than the front section of the tunnel wall thickness after 10 cm tunnel wall thickness. Upper tunnel and around 1 m thick slurry block stone layers, sizing block stone layer outside of natural soil, natural soil outside covered with weeds, vegetation, tunnel profile specifications as shown in Fig.1. Disintegration of cableway station on the part of the proposed blasting tunnel on the east side distance about 60 m, Badaling Great Wall slash distance tunnel north and the west to the valley, in addition, no other facilities around the tunnel, station tunnel on the surrounding environment is shown in Fig.2.

Corresponding author E-mail: wxgao@bjut.edu.cn

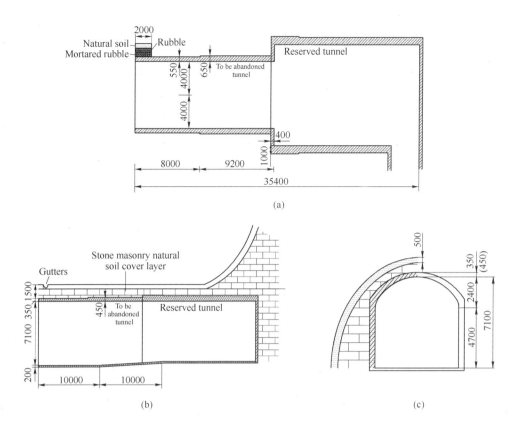

Fig.1 Sectional graphs of the above tunnel of Badaling Great Wall northern cableway

(a) planar graphs; (b) vertical sectional graphs; (c) cross sectional graphs

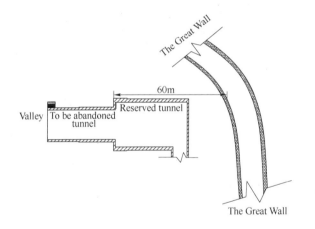

Fig.2 The surrounding environment of the above tunnel of the Badaling Great Wall northern cableway

3 DESIGN OF BLASTING SCHEME

3.1 Demolition Scheme Selection

There are several demolition method according to different structure types: (1) Gravity hammer impact damage removed. (2) Push the lever to dismantle or mechanical traction directional demolition collapsed. (3) Fully mechanized demolition. (4) A static method of expansion agent and broken. (5) Control blasting demolition. Controlled blasting is the control of collapsing direction, scope of damage, broken degree, control blasting harm, and this method is safe, fast, is the main method of building demolition.

This tunnel is located in the steep hillside, can pass from the supremacy of the next stop is only the mountain path, large machinery and equipment cannot be shipped first station; in addition, dismantle period is relatively short. Combining these factors, decided to adopt Controlled Blasting demolition technology to dismantle tunnel.

Before blasting construction, artificial remove arch tunnel and some plasma build by laying bricks or stones on both sides expressed; keep tunnel junction, intends to dismantle and along the small cross section tunnel contour with pneumatic picks out a 20～30 cm wide cutting seam, and cutting off steel bar; blasting design intends to take along the proposed demolition of tunnel wall arrangement of vertical and horizontal hole blasting.

3.2 Design of Blasting Parameters

According to the actual situation of the tunnel, the combination of vertical hole and horizontal hole to hole arrangement, reduce the number of the drill hole, increase the drilling efficiency.

(1) Design along the vertical direction of side wall to cut hole drilling, hole depth of 2.5 m, the remainder is perpendicular to the tunnel wall drilling horizontal hole. Vertical hole, hole spacing a is 50～60 cm, a total of 60, setting spe-

cific charge q is 800 g/m³, puckering take charge $Q=750$ g; filling length of 60 cm, on the filling clay and for air space between the period of the grain. Hole layout as shown in Fig.3 (a), Fig.3(b).

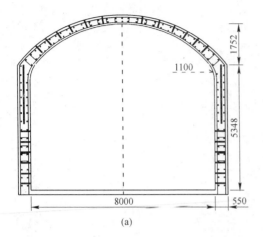

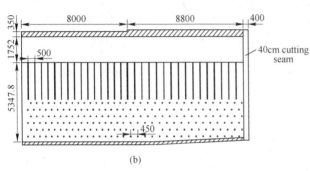

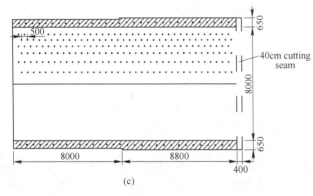

Fig.3 Layout of blast holes of the above tunnel

(a) layout of blast holes of the tunnel section; (b) layout of blast holes of the tunnel sidewall; (c) layout of blast holes of the tunnel sidewall and vault

(2) Head of the tunnel length is 8 m and tunnel wall thickness 55 cm, horizontal hole perpendicular to the wall to decorate, hole-net parameters as follows: a(pitch) $\times b$(row spacing) equals 45cm×40cm, hole depth: L equals $2/3 \times \delta$ (δ is tunnel wall thickness on both sides), take L equals 40 cm; end of the tunnel length is 9.2 m and tunnel wall thickness part 65 cm, horizontal hole is also perpendicular to the wall to decorate, hole-net parameters as follows: $a \times b$ equals 45cm×45cm, hole depth: L equals $2/3 \times \delta$ (δ is tunnel wall thickness on both sides), take $L = 40$ cm; hole layout as shown in Fig.3 (b).

(3) The uneven thickness of tunnel vault from the vault to gradually thickening, middle vault middle thickness of 35 cm and 45 cm, hole depth L equals 25~45 cm, vault gun distributed network parameters as: 60×50cm the main point is to make concrete cracking, hole layout as shown in Fig.3 (c).

Tunnel wall and arch shallow hole, specific charge q equals 700~800 g/m³.

3.3 Detonating Network Design

The tunnel blasting network will be divided into two regions, namely the tunnel roof and shallow hole wall vertical deep hole is divided into an area; wall level is divided into a shallow hole area. Around each large area can be divided into two symmetrical small regions, each small area between hole and row between extensions.

Unilateral wall horizontal shallow hole blasting network as shown in Fig.4, as far as possible to reduce the blasting vibration, the first salvo hole wall in a row of the most close to the ground takes the lead in initiating alone, cut off the top five could spread of blasting vibration to the base. Network latency: vertically delayed twice, the first row after firing, the second to fourth row of initiation, the fifth and sixth row of initiation, meanwhile with 5 section as a hole epitaxial detonator detonating tube detonator; horizontally, the first row of every 5 hole delay time, the second to fourth row every 15 hole (each row of five hole) delay time, meantime with 3 period of detonator detonating tube detonator as aperture extension.

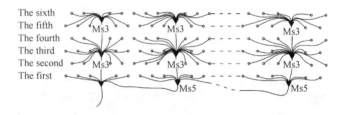

Fig.4 Initiation network of the horizontal shallow blast hole in the unilateral legislation wall

Unilateral vaults and vertical wall vertical deep hole blasting network as shown in Fig.5, the network delay: vertically delayed two times, the vaults, four, five, six after firing number vault in the first, second and third row of

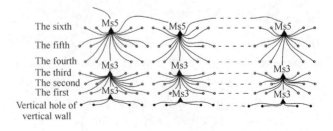

Fig.5 Initiation network of the vertical deep hole in the unilateral vault and legislation wall

initiation, finally establish vertical wall of deep hole blasting, meanwhile with 5 section as a hole epitaxial detonator detonating tube detonator; horizontal and vertical wall vertical deep hole every four hole delay time, the fourth to sixth could vault hole every 12～18 hole (4～6 hole in each row) delay time, the vault in the first, second and third salvo hole every 12～18 hole (4～6 hole in each row) delay time, meantime with 3 period of detonator detonating tube detonator as aperture extension.

In addition, the hole arrangement for internal detonator, vault shallow hole used within 12 period of detonating tube detonators, retaining wall horizontal shallow hole internal use 14 period of detonating tube detonators, vertical deep hole inner wall staggered, 12 tied detonators, 14 tied detonators in deep hole using interval charging at the same time, this way of detonator and charging from a certain extent, reduce the blasting vibration.

4 THE BLASTING SAFETY TECHNOLOGY AND VIBRATION MONITORING AND ANALYSIS

4.1 The Blasting Safety Technology

Reduce the blasting vibration effect and the maximum control blast fly rock, adopted the following methods: (1) to adopt millisecond delay between aperture, between technology, control the largest single segment dose; (2) to determine the rational detonating sequence, from a certain extent, reduce the influence of blasting vibration on the surrounding facilities; (3) the MATS covering the tunnel vault, MATS gland above the sandbags, effective control of the blast fly rock.

4.2 The Blasting Vibration Monitoring and Analysis

The blasting effectively seismic reduction measures include: (1) Using millisecond delay blasting technology, reduce single-stage blasting quantity; the size of the blasting vibration intensity depends on the size of the largest single segment dose [3], this effect is called "single period of independent action principle". So to reduce the effects of vibration effect a blasting must be divided into multiple blasts, therefore, control the largest single segment dose is an effective method to control the blasting vibration. (2) Was carried out on the side near the Great Wall on the side of the hole section charge, because the side wall close to the mountain, therefore, the Great Wall is mainly from the side wall of the blasting vibration; The research proves that segmented propellant and concentration of the blasting vibration effect there exists a significant difference, the ultra shallow hole extremely scattered charge (general building demolition blasting of a hole from a beam, column, wall and other medicines can be seen as ultra shallow hole very scattered charging) compared to the conventional centralized blasting blasting shock can be reduced by more than 90%[4]. Therefore, general demolition blasting of blasting vibration on the surrounding buildings do not constitute a hazard, but should be considered by the landing impact ground buildings collapse caused by seismic effects. (3) Keep tunnel junction, you would waste and be picking a cutting seam, and cutting off steel bar, the maximum extent to reduce the blasting of the influence of the tunnel is preserved.

In order to determine the blasting vibration effect in just 60 m from the blasting point of the Great Wall at the foot of the Great Wall and roof were set up three points, respectively controlled by 3 instruments, vibration plane layout is shown in Fig.6.

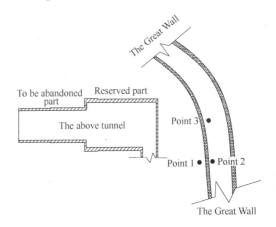

Fig.6 Layout of monitoring points in blasting areas

Point 1 and 2 65 m from blasting area, among them point 1, points 2 at the foot of the Great Wall is located in the place of the Great Wall, 60 m points 3 from the blasting area, and is located in the Great Wall. According to the measured data, the three measuring point vibration velocity and vibration frequency as shown in Tab.1.

Tab.1 The result of vibration measuring

Measuring point	Distance from blasting area/ m	Largest single charge / kg	Vertical		Horizontal of radial		Horizontal of transverse	
			Vibrating velocity /cm·s^{-1}	Frequency /Hz	Vibrating velocity /cm·s^{-1}	Frequency /Hz	Vibrating velocity /cm·s^{-1}	Frequency /Hz
No. 1	The foot of the Great Wall /65	6	0.39	56	0.48	57	0.41	57
No. 2	The Great Wall top /65	6	0.37	57	0.45	50	0.48	50
No. 3	The Great Wall top/60	6	0.49	53	0.38	59	0.40	50

From Tab.1 can be obtained, standing on the blasting area, is apart from the Great Wall recently the vibration velocity of 3 points of a maximum of 0.49 cm/s. No.2 point maximum vibration velocity of 0.48 cm/s, the Great Wall at the foot of 1 point maximum vibration velocity is 0.48 cm/s. After analyzing the frequency spectrum of the vibration wave, standing on the measuring point vibration main frequency in the range of 50~60 Hz, considerably more than the natural frequency of vibration of the building. Point 1 vibration velocity and vibration frequency waveform as shown in Fig.7. According to "The blasting safety regulations "(GB 6722—2003) sites the vibration of a maximum of 0.5 cm/s, the measured results show that the maximum vibration of the Great Wall is less than 0.5 cm/s. Shows that the blasting vibration are controlled effectively, no adverse effects on the Great Wall.

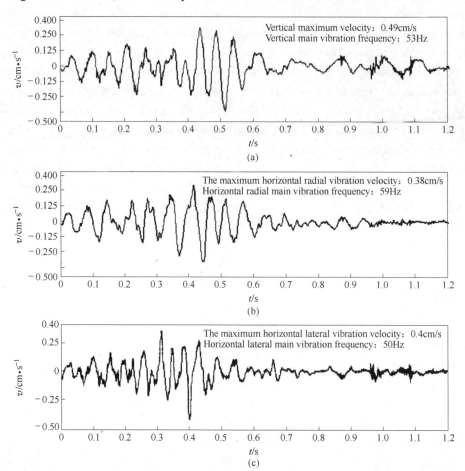

Fig.7 Three-channel vibration velocity waveform of the point 3
(a) vertical vibrating velocity; (b) horizontal radial vibration velocity; (c) horizontal lateral vibration velocity

5 CONCLUSIONS

The demolition blasting work environment is relatively complex, critical area and the minimum distance of the Great Wall is only 60 m, the flying rock and blasting vibration are put forward strict requirements, from the perspective of the results of the blasting, blast fly rock, blasting vibration are controlled in a reasonable range of less than, after blasting is relatively uniform, basic to subsequent slag removal manual working has brought a lot of convenience, the blasting demolition blasting was a successful, for such blasting operations in a complicated environment provides the reference.

REFERENCES

[1] Wang Xuguang, Yu Yalun. Demolition blasting technology in the 21st century [J]. Journal of engineering blasting, 2000, 3.

[2] Wang Xuguang, Zheng Bingxu, Chang Chengjong, Liu Dianshu. Blasting handbook [M]. Beijing: Metallurgical Industry Press, 2010.

[3] Wu Tengfang Wang Kai. Differential blasting technology research [J]. Journal of blasting, 1997, 14 (1): 53~57.

[4] Jin Chungui. Dispersion of the blasting vibration effect [J]. Journal of China mining, 2000, 9 (1): 202~205.

Study on Open-Pit Precision Control Blasting of Easily Weathered Rock and Its Application

YI Haibao [1,2], YANG Haitao [1], LI Ming [1], HAN Bin [2], ZHENG Lujing [3]

(1. Sinosteel Maanshan Institute of Mining Research, Maanshan, Anhui, China; 2. School of Civil & Environmental Engineering, University of Science and Technology Beijing, Beijing, China; 3. College of Resource and Environment Engineering, Guizhou University, Guiyang, Guizhou, China)

ABSTRACT: Based on the characteristics of highly weathered and broken rock in Jinfeng gold mine, the network parameters of bench blasting were optimized by Surpac software, identified as 6m×7m, maintaining the stability of rock slope. By means of the millisecond blasting technology of the hole and outside the hole, the high accuracy initiation by hole was achieved with Orica high-precision detonator. The blasting was safe and reliable, reducing the mine production cost effectively. With the method of ShotPlus simulation, the blasting order and rock movement direction were guaranteed in accordance with the expected direction, which significantly improved the rock blasting effect. It is proved that the slope damage due to blasting vibration was significantly reduced with the technology mentioned above. And the ore loss and dilution rate were controlled effectively, making a good economic and social benefits.

KEYWORDS: open-pit blasting; high precision; initiation by hole; ShotPlus; slope

1 INTRODUCTION

Jinfeng gold mine locates in Zhenfeng County, Guizhou Province. The ore type is fine grained disseminated, with high arsenic, carbon, mercury, antimony and other harmful elements, being a typical "Carlin-type" gold mine, called China's largest difficult mining, metallurgical gold. The mine includes three mineralized zones, Rongban, Lin Tan, Liuchanggou, involving ten mineralized domains. Mineralized domains are strictly controlled by the faults, as constructed epithermal gold deposits.

The mine is excavated by underground and open pit combining mining method, the maximum height of slope being about 300m. As the mineralization controlled by faults, the mine rock mainly are siltstone and claystone, which are easily weathered and disintegrate mudding once encountering water. It results in the poor stability of high and steep slope, slope collapse often occurred, the slope controlling cost remaining high. Meanwhile, because of the large-scale blasting of mining operation, conventional blasting technology is likely to cause the ore and waste rock mixed with each other, increasing the ore losses and dilution.

Therefore, how to use the international advanced blasting technology to minimize the stripping ratio to reduce production costs, effectively reducing the ore dilution and losses, achieving slope stability control, while meeting the requirements of safe and efficient mining, are the key technical problems to be solved.

According to the mine mining characteristics, based on the international advanced Surpac software and ShotPlus working platform, relying on high precision detonators, it realized the high precision deep-hole detonation technology after optimization of blasting parameters. It significantly weakened the blasting vibration damage on the slope, better controlling the ore displacement, reducing the ore loss and dilution, creating good economic and social benefits, having an extensive application value.

2 BLASTING PARAMETERS OPTIMIZATION

Reasonable blasting parameters plays an important role in improving the blasting effect and reducing project costs. Is the selected blasting parameters appropriate or not directly affects the mine productivity and slope stability, and its main factors include: rock types, structure and construction, mine size, slope height and slope angle, rig types and blasthole diameter, etc. The mine rocks are mainly sandstone, mudstone and minor limestone, rock solid coefficient f being between 6 to 8, belonging to a medium hardness rock. The bench height is 20m and the majority of the slope angle is about 65°. The blasthole diameter is 165mm drilled by L8 and 780 Atlas drilling rig.

Corresponding author E-mail: yht 03121 @ 163. com

2.1 Bench Height

Improving blasting bench height could reduce the blasting cycle times and increasing the mine efficiency. To meeting the requirements of mining equipment efficiency and ensuring the safety, the bench height of blasting is the bigger the better. The maximum digging height of PC1250 is about 12m, combining the slope features of 20m, the bench is formed by twice blasting, each blasting height being 10m.

2.2 Blasthole Ultra-Deep

The main function of ultra-deep is to decrease the position of charge center, to increase the dynamite dose in the bottom of blastholes, to overcome the bottom resistance, to avoid or reduce the remaining foundation, forming a flat bench. If the ultra-deep is too small, it appears bedrock easily. While too big it is, it makes severe rock crushing in bench bottom, affecting the efficiency of the next step drilling.

Through field study of different lithologies and ultra-deeps, combining with a large number of on-site engineering practices, the two different ultra-deeps were determined, 0.5m and 0.8m, using 0.5m in mudstone and sandstone and 0.8m in limestone.

2.3 Balsthole Network Parameters and Arrangement Mode

Balsthole network parameters and arrangement mode of open pit blasting are the key factors affecting the blasting effect. By means of the study of optimization of balsthole network parameters in ore blasting area, it was determined that the arrangement mode was plum and network parameters were 6m×7m to control blasting displacement and reduce ore loss and dilution. The front smallest resistance line near free surface was 4~5m. It not only could fully throw rocks in the front rows, creating a better compensation space for the rear holes, but also made the rocks fully squeezed and broken, reducing backward impact and improving ore blasting effect. A horizontal bench partition with surpac software indicated in Fig.1, the blastholes arrangement mode indicated in Fig.2. The network parameters were 6m×7m, the holes strike was perpendicular to that of ore body.

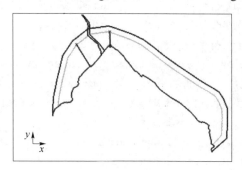

Fig.1 +620m blasting bench

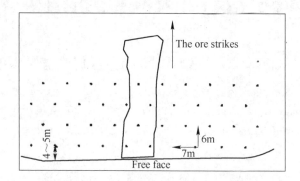

Fig.2 Blastholes arrangement mode

3 PRECISION CONTROL BLASTING TECHNOLOGY OF HOLE BY HOLE INITIATION

The initiation technology of hole by hole is a new blasting process, gradually promoted and used at home and abroad. It can expand blasting scale, reduce blasting hazards, improve blasting effect, improve mining efficiency and decrease the cost of drilling and blasting. It has been widely used in domestic large open pit mine. Compared with millisecond between rows and "V" shape, slash detonation technology, it has the advantages of broken rock uniform and low blasting vibration, etc..

The high-precision detonators produced by Orica company were adopted. According to engineering geological conditions of ore and rock, the 400ms delay detonators were the used in the holes. The connection of surface detonators was that: when master line was 17ms, goose-shaped row was 42ms and when master line was 25ms, goose-shaped row was 42ms or 65ms. Through the combination of delay detonators in the holes and surface, it achieved a high-precision hole by hole initiation, reducing the maximum segment loading dose, controlling the blasting vibration significantly. Practice showed that the detonation technique was safe and reliable, reducing the mine production cost effectively.

Based on the results of mine field tests, the mode of continuous charge was determined, reverse detonation. Loose blasting was used to minimize ore loss and dilution. As can be seen from Fig. 3 the boundary of blasting region and non-blasting area, the blasting rock fragmentation was uniform and the loosening state of muck pile was better, making a good blasting effect.

Fig.3 Open pit effect of loose blasting

4 BLASTING PROCESS SIMULATION BASED ON SHOTPLUS SOFTWARE

The combination of ShotPlus software and precision detonator produced by Orica company can design the firing circuit connection based on characteristics of blasting area. It can simulate the rock movement direction, the single hole initiation time, initiation process, statistics the usage of surface and hole detonators.

When the rock moving direction is consistent with ore strike, it is helpful for controlling ore loss and dilution. In the second phase of slope blasting, rock moves in the opposite direction of the free surface, effectively avoiding the front rock throwing forward. The safe platform is filled by rock, rolled into the first phase pit, which will affect its normal production. In this way it could not determine the boundaries of ore body accurately in ore area, resulting in ore dilution and loss.

By the ShotPlus simulation of blasting process it achieved a precision controlling blasting, ensuring the blasting order and rock movement direction in accordance with the expected direction. It could improve the blasting effect and control the movement direction of muck pile effectively, reaching the purpose of controlling ore loss and dilution. The simulation movement direction of ore and rock was shown in Fig.4.

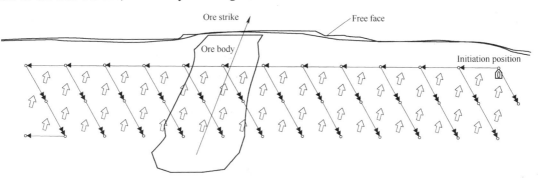

Fig.4 Movement direction of ore and rock

By means of blasting simulating initiation process of ShotPlus software and illustrating initiation orders it could find the firing circuit was linked or not. Meanwhile it was convenient to determine all blastholes with detonate interval less than 8ms (less than 8ms considered simultaneously detonate), as shown in Fig.5. Though adjusting the firing circuit connection, the number of blastholes with detonate interval less than 8ms could be minimized, which would reduce the once blasting charge, control blasting vibration amplitude, weak the slope influence due to blasting vibration and maintain the slope security and stability.

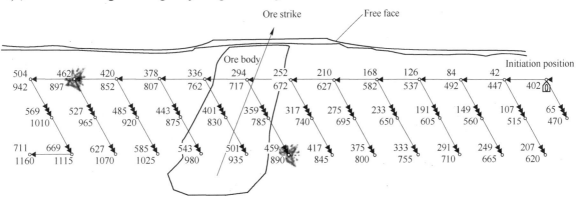

Fig.5 Blasting simulation of detonate interval less than 8ms

Through the optimization of blasting parameters, the unit explosives consumption decreased from 0.4kg/m^3 to 0.33kg/m^3. When calculated based on the annual transport amount of 10.6 million m^3, it could save the costs of explosives and detonators 3.9432 million yuan.

5 CONCLUSIONS

The conclusions from this study are as follows.

(1) Based on the hole by hole blasting technology of high-precision detonator it achieved high-precision controlling blasting, reduced the explosive charge simultaneously, decreased the damage and failure on the slope from blasting vibration and improved the slope stability effectively.

(2) Combining with advanced Surpac and ShotPlus software, by means of firing circuit optimization, ore transport direction could be controlled in line with ore body stike, which effectively controlled the movement direction of muck pile. It achieved the ore and rocks were digged and transported separately, reducing the ore loss and dilution rate.

(3) Through the optimization of blasting parameters, the unit explosives consumption decreased from 0.4kg/m^3 to 0.33kg/m^3. When calculated based on the annual transport amount of 10.6 million m^3, it could save the costs of explosives and detonators 3.9432 million yuan.

(4) It proved that the detonation technology is safe and reliable, making a good blasting effect. The blasting block was uniform and muck pile was centralized and loose, easy to dig and transport. It improved mine productivity, creating good economic benefits, being widely applied.

REFERENCES

[1] Han B, Zheng L J, Wang S Y, etc. Synthetic reinforcement of complicated and broken open pit slope[J]. *Journal of Central South University(Scienceand Technology)*, 2013, 44(2): 772 ~777.

[2] Han B, Yi H B, Zheng L J, etc. Synthetic reinforcement and drainage technology for low-strength easy-weathered broken open pit slope[J]. *Metal Mine*, 2013(3): 14~17.

[3] Chen S, Zhou G S, Zhou Y, etc. Application of hole-by-hole initiation technique in Taihe iron mine[J]. *Engineering Blasting*, 2011, 17(1): 43~45.

[4] Ren X, Guo X B, Guo F, etc. Test research on the blasting ignition borehole sequence in Karst developed limestone quarry[J]. *Metal Mine*, 2008 (11): 12~15.

[5] Luan D D, Pan E B. Application of high-precision hole bottom hole-by-hole blasting technique in broken ore bodies[J]. *Nonferrous Metals*, 2013 65(4): 71~74.

[6] Yin Y J. Application of the hole-by-hole initiation technique in Jinduicheng open-pit mine[J]. *Engineering Blasting*, 2004, 10(3): 71~75.

Influence of Power Characteristics of Explosives on Strength Properties of Pieces of the Blown-up Mountain Weight

G.P. Paramonov, V.A. Isheysky
(*National mineral resources university "University of mines", Russia*)

ABSTRACT: Being an initial link in a technological chain of minerals production and processing, drilling-and-blasting works, in many respects define the subsequent operations efficiency and in particular, crushing and sorting complex functioning. On rubble production pits this communication is traced in an uncontrollable elimination grade yield. Explosion action on the destroyed massif leads to strength characteristics decrease in the destroyed rock pieces which can significantly influence the volume of a sub-standard production yield.

The present work is devoted to research of power characteristics of the explosives, used for rocks destruction, influence on blown-up rock pieces strength characteristics. Model experiment of a single cylindrical charge explosion process in the blast hole for the purpose of destroyed rock pieces in shotpile strength properties dependence establishment taking into account destruction zones from explosive power characteristics is considered.

For numerical calculations justification model experiments are made on special equivalent-material concrete blocks. On the basis of pilot studies the process picture of the block destruction by various power charges is received. Results of the executed model experiments allow to establish connection between the samples' (in shotpile) physicomechanical properties decrease of the blown-up rock from various zones of destruction and power characteristics of applied explosive.

KEYWORDS: model experiment; power parameters; strength characteristics; rocks; explosion; explosives charges; destruction zone

Quality of rock mass blasting preparation in the development of mineral deposits significantly affects the overall economic performance of the enterprise. The blasting technology efficiency has a significant impact on the rock mass blasting preparation. Despite the impressive advances in the field of the blasting technology development, use of emulsion explosives, transition to a reduced diameter wells, the desired rock mass crushing remains unreachable. Therefore, the study of the influence of blasting parameters on grain-size distribution of blasted rock mass seems to be topical for mining companies from scientific and practical point of view.

The purpose of the research is to establish the influence of the energy characteristics of explosives on the distribution of blasted rock fragments in different zones of destruction.

To justify the numerical calculations, modeling experiments were conducted on cylindrical blocks. When implementing a model of zone blasting crushing, a removable mold of hollow thin-walled cylindrical forms of different diameters was originally made. The total number of cylinders is 6. The radius of the cylinder varies from 10 to 100 relative radii of the charge. The height of all cylinders is 80 relative radii. The design parameters of the model are shown in Fig.1. The general model view is given in Fig.2.

The basic condition for choosing the charges and their preparation for the experiment was the same value of linear parameters. Since the common industrial explosives are impossible to use in model conditions because of the large critical diameter, fine-grained pentaerythrite tetranitrate was used as the explosive. The explosive energy characteristics changed by injection of sodium chloride from 0 to 20% in increments of 10%. The change of velocity of detonation from % content of fillers, composed of PETN is shown in Fig.3.

The general planogram of experimental results comparison is shown in Fig.4.

The general view of the models and the blasted mass volume are shown in Fig.5.

To investigate the grain size distribution, the software package WipFrag Granulometry Analysis Software by WipWare Inc. Canada allowing estimation of the fractional composition with the photoplanometric method was applied. The program plots a Rosin-Rammler distribution curve in logarithmic coordinates.

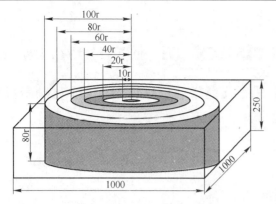

Fig.1 The model design

Fig.2 General view of the model

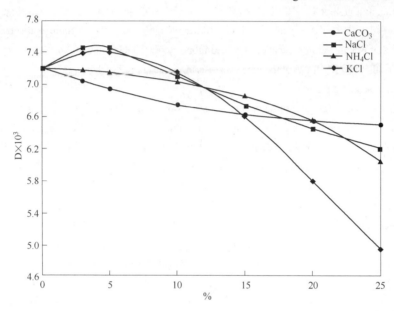

Fig.3 The change of velocity of detonation from % content of fillers, composed of PETN

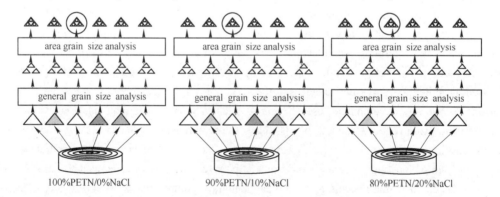

Fig.4 The tests scheme

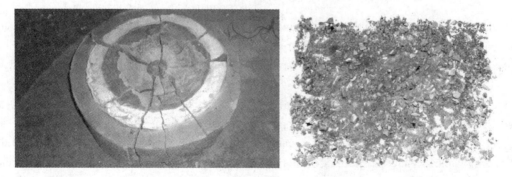

Fig.5 The general view of the models and the blasted mass volume

The blasted mass from different destruction zones was separated by colors. The mass was separated by the color spectrum, and then the grain size distribution of the blasted mass from each zone was determined. Below is an example of data processing in the 100% pentaerythrit.

The total yield of the blasted mass by weight from different destruction zones is given in Tab.1. Details of the grain size distribution for each destruction zone after a series of experiments are summarized in Tab.2.

Tab.1 The total yield of the blasted mass by weight from different destruction zones

Composition of the mixture	Radius of the center of charge					Total mass /gram(%)
	Area 0~10r	Area 11~20r	Area 21~40r	Area 41~60r	Area 61~80r	
PETN/NaCl	mass/(gram/%)					
100%/0%	462(27.07)	383(22.43)	342(20.03)	290(17.0)	230(13.47)	1707(100)
90%/10%	439(30.27)	339(23.37)	308(21.24)	210(14.41)	154(10.62)	1450(100)
80%/20%	440(36.46)	345(28.58)	205(16.98)	152(12.60)	65(5.38)	1207(100)

Tab.2 Yield of the blasted mass by fractions from different destruction zones

fraction	Area 0~10r			Area 11~20r			Area 21~40r			Area 41~60r			Area 61~80r		
	mass/(%/gram)														
25.0	100 / 462	100 / 439	100 / 440	100 / 383	100 / 339	100 / 345	100 / 342	100 / 308	100 / 205	100 / 290	100 / 210	100 / 152	100 / 230	100 / 154	100 / 65
16.0	99.4 / 459.2	99.5 / 436.8	99.5 / 437.8	99.5 / 381	99.1 / 335.9	99.4 / 342.9	90.9 / 310.8	94.5 / 291	93 / 190.6	94.7 / 274.6	94.8 / 199	99.5 / 151.2	93.6 / 215.2	92.1 / 141.8	92.8 / 60.3
12.5	96.3 / 444.9	96.6 / 424	96.5 / 424.6	96.7 / 370.3	94.5 / 320.3	96.6 / 333.2	90.7 / 310.1	89.7 / 276.2	92.8 / 190.2	91.8 / 266.2	91.2 / 191.5	97.1 / 147.5	91.6 / 210.6	91.1 / 140.2	91.7 / 59.6
10.0	93.1 / 430.1	93.7 / 411.3	92.5 / 407	93.9 / 359.6	89.6 / 303.7	96.6 / 333.2	89.2 / 305	86.9 / 267.6	91.8 / 188.1	88.9 / 257.8	89.1 / 187.1	96.4 / 146.5	91.6 / 210.6	90.1 / 138.7	90.1 / 58.6
8.0	92.1 / 425.5	92.7 / 406.9	92.3 / 406.1	92.1 / 352.7	88.8 / 301	95.5 / 329.4	86.6 / 296.1	84.4 / 259.9	88.4 / 181.2	86.6 / 251.1	86.6 / 181.8	94.7 / 143.9	84.7 / 194.8	87.4 / 134.5	86 / 55.9
6.70	90.7 / 419.0	91.5 / 401.6	90.5 / 398.2	89.7 / 343.5	86.3 / 292.5	92.3 / 318.4	84.2 / 287.9	81.6 / 251.3	85.3 / 174.8	84.5 / 245	84.7 / 177.8	91.8 / 139.5	82 / 188.6	83.8 / 129	82 / 53.3
5.60	89.3 / 412.5	90.2 / 395.9	89 / 391.6	86.6 / 331.6	83.9 / 284.4	89.3 / 308	81.1 / 277.3	78.3 / 241.1	81.3 / 166.6	82.1 / 238	82.4 / 173	89 / 135.2	78.8 / 181.2	79.6 / 122.5	77.6 / 50.4
4.75	88 / 406.5	89 / 390.7	88.3 / 388.5	83 / 317.8	82.4 / 279.3	87.3 / 301.1	77.6 / 265.3	74.5 / 229.4	76.8 / 157.4	79.8 / 231.4	79.8 / 167.5	86.7 / 131.7	74.5 / 171.3	75.1 / 115.6	73.3 / 47.6
4.00	87.4 / 403.7	88.5 / 388.5	87.8 / 386.3	79.7 / 305.2	79.9 / 270.8	83 / 286.3	74.1 / 253.4	71.3 / 219.6	72.8 / 149.2	76.3 / 221.2	75.5 / 158.5	82.7 / 125.7	69.5 / 159.8	69.8 / 107.4	67.9 / 44.1
3.35	87 / 401.9	88 / 386.3	87.4 / 384.5	75.6 / 289.5	76.6 / 259.6	78.7 / 271.5	70.5 / 241.1	67.8 / 208.8	68.7 / 140.8	72.6 / 210.5	71.3 / 149.7	78.4 / 119.1	64.8 / 149	65 / 100.1	63.1 / 41
2.00	85.8 / 396.3	86.7 / 380.6	86 / 378.4	62.1 / 237.8	65.2 / 221	65.2 / 224.9	58.6 / 200.4	55.8 / 171.8	55.7 / 114.1	60.9 / 176.6	60.4 / 126.8	65.6 / 99.7	53.1 / 122.1	52.9 / 81.46	51.6 / 33.5
1.40	84.1 / 388.5	84.2 / 369.6	83.3 / 366.5	54.5 / 208.7	57.8 / 195.9	55.7 / 192.1	49.2 / 168.1	46.8 / 144.1	45.5 / 93.2	51.3 / 148.7	49.8 / 104.8	53.5 / 81.3	42.5 / 97.7	41.9 / 64.52	39.7 / 25.8
1.00	75.8 / 350.1	73.9 / 324.4	72.7 / 319.8	45.6 / 174.6	47.8 / 162	44.2 / 152.4	37.6 / 128.5	34.8 / 107.1	31.8 / 65.1	35.6 / 103.2	33.7 / 70.7	35 / 53.2	57.3 / 131.7	26.4 / 40.6	24.2 / 15.7
0.85	64.6 / 298.4	61.8 / 271.3	60 / 264	38.1 / 145.9	39.8 / 134.9	35.7 / 123.1	29.6 / 101.2	26.9 / 82.8	24.2 / 49.6	26.3 / 76.2	24.5 / 54.4	25 / 38	19.9 / 45.7	19.1 / 29.4	17.3 / 11.2
0.60	36.4 / 168.1	33.7 / 147.9	31.9 / 140.3	21.9 / 83.8	20.8 / 70.5	19.5 / 67.2	15.8 / 54	13.9 / 42.8	11.8 / 24.1	11.3 / 32.7	10.3 / 21.6	10.1 / 15.3	8.6 / 19.7	8.3 / 12.7	7.2 / 4.6
	100% PETN 0% NaCl	90% PETN 10% NaCl	80% PETN 20% NaCl	100% PETN 0% NaCl	90% PETN 10% NaCl	80% PETN 20% NaCl	100% PETN 0% NaCl	90% PETN 10% NaCl	80% PETN 20% NaCl	100% PETN 0% NaCl	90% PETN 10% NaCl	80% PETN 20% NaCl	100% PETN 0% NaCl	90% PETN 10% NaCl	80% PETN 20% NaCl

From the analysis of the obtained results, the grain size distribution shows that the largest number of fines was formed in the explosion of a pure pentaerythrite tetranitrate charge. This can be explained by an increase in charge of the blasting action.

Parameters of medium sized samples from each zone after a series of simulation experiments are presented in Fig.6.

0~10r			11~20r			21~40r		
100%/0%	90%/20%	80%/20%	100%/0%	90%/20%	80%/20%	100%/0%	90%/20%	80%/20%
min = 0.000 m max = 0.017 m blocks= 25364 mean = 0.002 m stdev = 0.004 m mode = 0.001 m sph = 0.510 D10 = 0.0004 m D25 = 0.0005 m D50 = 0.00071 m D75 = 0.0010 m D90 = 0.0061 m Xmax= 0.0167 m X50 = 0.0007 m b = 6.3927 Xc = 0.0008 m n = 1.64	min = 0.000 m max = 0.017 m blocks= 26021 mean = 0.002 m stdev = 0.004 m mode = 0.001 m sph = 0.512 D10 = 0.0004 m D25 = 0.0005 m D50 = 0.00072 m D75 = 0.0010 m D90 = 0.0054 m Xmax= 0.0167 m X50 = 0.0007 m b = 6.9457 Xc = 0.0009 m n = 1.69	min = 0.000 m max = 0.017 m blocks= 24027 mean = 0.002 m stdev = 0.004 m mode = 0.001 m sph = 0.512 D10 = 0.0004 m D25 = 0.0005 m D50 = 0.00074 m D75 = 0.0011 m D90 = 0.0064 m Xmax= 0.0167 m X50 = 0.0007 m b = 6.5377 Xc = 0.0009 m n = 1.72	min = 0.000 m max = 0.017 m blocks= 19253 mean = 0.003 m stdev = 0.004 m mode = 0.001 m sph = 0.518 D10 = 0.0004 m D25 = 0.0006 m D50 = 0.0011 m D75 = 0.0033 m D90 = 0.0068 m Xmax= 0.0167 m X50 = 0.0012 m b = 2.2860 Xc = 0.0021 m n = 1.12	min = 0.000 m max = 0.017 m blocks= 20629 mean = 0.003 m stdev = 0.004 m mode = 0.001 m sph = 0.518 D10 = 0.0004 m D25 = 0.0007 m D50 = 0.00117 m D75 = 0.0031 m D90 = 0.0102 m Xmax= 0.0167 m X50 = 0.0011 m b = 2.0260 Xc = 0.0018 m n = 1.17	min = 0.000 m max = 0.017 m blocks= 14877 mean = 0.003 m stdev = 0.003 m mode = 0.001 m sph = 0.520 D10 = 0.0004 m D25 = 0.0007 m D50 = 0.00121 m D75 = 0.0028 m D90 = 0.0059 m Xmax= 0.0167 m X50 = 0.0012 m b = 2.8600 Xc = 0.0019 m n = 1.33	min = 0.000 m max = 0.022 m blocks= 14722 mean = 0.004 m stdev = 0.006 m mode = 0.001 m sph = 0.516 D10 = 0.0005 m D25 = 0.0008 m D50 = 0.00143 m D75 = 0.0042 m D90 = 0.0113 m Xmax= 0.0215 m X50 = 0.0014 m b = 2.3100 Xc = 0.0024 m n = 1.10	min = 0.000 m max = 0.022 m blocks= 13401 mean = 0.004 m stdev = 0.006 m mode = 0.001 m sph = 0.522 D10 = 0.0005 m D25 = 0.0008 m D50 = 0.0016 m D75 = 0.0049 m D90 = 0.0128 m Xmax= 0.0215 m X50 = 0.0016 m b = 2.1630 Xc = 0.0027 m n = 1.11	min = 0.000 m max = 0.022 m blocks= 14854 mean = 0.004 m stdev = 0.005 m mode = 0.001 m sph = 0.517 D10 = 0.0006 m D25 = 0.0009 m D50 = 0.00166 m D75 = 0.0044 m D90 = 0.0089 m Xmax= 0.0215 m X50 = 0.0016 m b = 2.7020 Xc = 0.0027 m n = 1.21

41~60r			61~80r		
100%/0%	90%/20%	80%/20%	100%/0%	90%/20%	80%/20%
min = 0.000 m max = 0.022 m blocks= 21230 mean = 0.004 m stdev = 0.005 m mode = 0.001 m sph = 0.520 D10 = 0.0006 m D25 = 0.0008 m D50 = 0.00154 m D75 = 0.0037 m D90 = 0.0109 m Xmax= 0.0215 m X50 = 0.0013 m b = 2.5000 Xc = 0.0022 m n = 1.19	min = 0.000 m max = 0.022 m blocks= 21039 mean = 0.004 m stdev = 0.005 m mode = 0.001 m sph = 0.518 D10 = 0.0006 m D25 = 0.0009 m D50 = 0.00166 m D75 = 0.0039 m D90 = 0.0111 m Xmax= 0.0215 m X50 = 0.0014 m b = 2.6160 Xc = 0.0023 m n = 1.22	min = 0.000 m max = 0.017 m blocks= 21194 mean = 0.003 m stdev = 0.003 m mode = 0.001 m sph = 0.520 D10 = 0.0006 m D25 = 0.0008 m D50 = 0.00173 m D75 = 0.0029 m D90 = 0.0060 m Xmax= 0.0167 m X50 = 0.0013 m b = 3.5989 Xc = 0.0018 m n = 1.46	min = 0.000 m max = 0.022 m blocks= 20218 mean = 0.004 m stdev = 0.005 m mode = 0.001 m sph = 0.518 D10 = 0.0006 m D25 = 0.0010 m D50 = 0.0017 m D75 = 0.0049 m D90 = 0.0095 m Xmax= 0.0215 m X50 = 0.0017 m b = 2.5930 Xc = 0.0031 m n = 1.17	min = 0.000 m max = 0.022 m blocks= 20206 mean = 0.005 m stdev = 0.006 m mode = 0.001 m sph = 0.517 D10 = 0.0006 m D25 = 0.0010 m D50 = 0.00184 m D75 = 0.0047 m D90 = 0.0094 m Xmax= 0.0215 m X50 = 0.0018 m b = 2.7780 Xc = 0.0031 m n = 1.20	min = 0.000 m max = 0.022 m blocks= 21506 mean = 0.005 m stdev = 0.006 m mode = 0.001 m sph = 0.517 D10 = 0.0007 m D25 = 0.0010 m D50 = 0.00192 m D75 = 0.0051 m D90 = 0.0100 m Xmax= 0.0215 m X50 = 0.0019 m b = 2.7250 Xc = 0.0034 m n = 1.22

Fig.6 Data on grain size distribution parameters

Meaning of the parameters:

$D10, D25$, etc. —— the size and the percent yield in terms of screening, D10 indicates a sieve opening size which will pass through a 10% by weight of the sample;

blocks —— the number of network elements plotted by WipFrag program when analyzing photographs;

max —— the structure maximum size;

min —— the structure minimum size;

mode —— the most common structure size;

n —— the Rosin-Rammler uniformity coefficient equal to the slope of the Rosin-Rammler curve in logarithmic coordinates;

stdev —— the standard deviation of the sample size;

Xc —— characteristic size at which the Rosin-Rammler line in logarithmic coordinates was cut off;

Xmax —— clipping of 100 % passing of the Rosin-Rammler distribution inclined line.

Uniformity coefficient increases depending on the explosive power reduction what indicates a more uniform distribution of grain size for samples of rock according to the Rosin-Rammler law and thus the formation of a more uniform stress field during the detonation of an explosive with lower power settings. Output medium sized piece of each zone increases from the center to remove the charge. With increasing power of the explosive medium sample of the same areas in different model experiments decreases.

For determining the strength characteristics pieces blown up masses of samples from different areas of destruction had been tested on a point of strength performed in accordance with standard ASTM D 5731 apparatus point loads TS706. Strength characteristics of pieces that do not meet the criterion of selection of samples for measurements under ASTM, obtained on the basis of the method of determining the coefficient of fortress on Protodyakonov.

Strength characteristics pieces blown up mountain weight from various areas of destruction to remove the charge after a series of model experiments are given in Tab.3.

Fig.7　The equipment for measurement of strength

Tab.3　Strength characteristics of pieces of blasted rock

composition of the mixture	Average strength in the area tension/compression					
	area 0~10r	area 11~20r	area 21~40r	area 41~60r	area 61~80r	area 81~100r
PETN/NaCl						
100%/0%	—/—	1.32/23.0	1.52/33.3	2.53/44.1	2.88/52.05	2.89/53.04
90%/10%	0.8/15.0	1.8/30.0	2.2/37.0	2.65/47.5	2.87/51.5	2.9/52.5
80%/20%	1.05/25.0	1.98/33.0	2.3/41.0	2.77/48.5	2.86/52.5	2.91/53.5

To determine the power consumption of rock failure in the collapse of the different zones with different levels of explosive loading array technique was developed, which is based in the destruction of a certain amount of band model. Failure occurs at different levels of explosive loading of the array. After a series of experiments, samples were taken of each zone. Further determined by the energy intensity of the destruction of images crushing rocks falling weight impact tester at the shock.

After analyzing the experimental data, Reduced output dropout (20%) after primary crushing stage is achieved by maintaining a 15% strength characteristics of pieces of blasted rock mass based on the dynamic management of the blast on the rock massif.

REFERENCES

[1] Adushkin V V. AN Sukhotin About the destruction of the solid medium explosion. J. Appl N 4, 1961.

[2] Baron L I. Lumpiness and methods of measurement. Moscow, USSR Academy of Sciences Izdvo, 1960.

[3] Biryukov A V, Repin N. The applicability of some of the laws of distribution in the study of lumps mixtures. Kemerovo, Tr. KuzPI, vol. 48, 1973.

[4] Zhurkov S N. Kuksenko V S, Petrov V A. A basis for forecasting the mechanical destruction., Ibid, 1981, 259(6).

[5] Kuznetsov V A. Analytical assessment of the blasted rock grain-size distribution. Blasting work, 91/48, 1998.

Load Analysis of Hammer Crusher

NI Jiaying[1], ZHAO Sihai[2]

(*1.Xinjiang Tianhe Blasting Engineering Co., Ltd., China; 2.China University of Mining and Technology, China*)

ABSTRACT: According to the engine cases of hammer crusher the mechanism of coal breaking by impact is introduced, the simplified dynamic model for performance of hammer crusher is proposed, and the dynamic differential equation is deduced. On that basis the load characteristic of hammer crusher is analyzed in dynamics, which provides a theoretical basis for design and working status analysis of hammer crusher.

KEYWORDS: hammer crusher; load analysis; dynamics; impact load

1 OVERVIEW

Hammer crusher is a kind of common ore crushing machine, it is widely used in coal and cement raw material processing. Into the crusher, the ore will be subjected to the impact of the high-speed rotating hammer and broken. Ordinary hammer head is fixed on the rotary articulated way, in the broken ore block, the hammer head can partial backward, thus avoiding the rotor spindle by the strong impact.

Reproduced machine with hammer crusher is mainly used in the coal mine along the groove, and used with reproduced machine systems to break . Chunks of raw coal. Reproduced machine with hammer crusher is common hammer crusher, the deformation of the main changes are active hammer to fixed hammer head, with the method of impact on coal to coal breakage.

Through the working load of hammer crusher for statics and dynamics analysis. The paper studied the crushing mechanism, and established the corresponding mathematical model, and dynamic simulation, so as to find the optimal design method. For the design of hammer crusher and the analysis of working status, research not only has important theoretical significance, also has important practical application value.

2 DYNAMIC ANALYSIS OF A HAMMER CRUSHER

Reproduced machine with hammer crusher is composed of two parts, motor and reducer are the dynamic parts of the crusher, hammer and hammer body, are shaft and bearing to perform part of the crusher. Motor and reducer, provides torque, and has a great influence on no-load starting, no-load operation of breaker, and the hammer and hammer body part of crusher have a great influence on the load characteristic of normal work. Hammer head and the wheel or hammer body is through the thread to join together rigidly, the impact of hammer head will be passed directly to the hammer body, shaft and bearing, etc.

Reproduced machine with actuator is a hammer, hammer crusher, hammer head and the hammer body is rigid to join together by screw thread, as a result, stress analysis of the reproduced machine use hammer crusher mainly include hammer head and the body's stress analysis, the stress of the shaft and bearing force analysis, etc.

For study convenient, hammer and hammer body, shaft and bearing, including key study as a whole, called hammer axis. Hammer axis is actually a fixed axis rotation of rigid body, is the mainly part of the hammer crusher, load analysis is mainly on hammer axis . Loads on hammer axis mainly include friction load, inertia load and damping load and the working load.

Hammer crusher or hammer axis work normally, the hammer head and coal(or rock), will produce the reaction and a force. The force with the lever is the work load torque. In collision, although the speeds of the coal and (or rock) the hammer axis will change a little, due to the collision time very short, acceleration is very big, there will be a huge impact. Due to the collision is not continuous, but periodically, so the load torque on hammer axis is cyclical, it is equivalent to give hammer axis a periodic excitation, hammer axis will be forced vibration.

3 THE DYNAMIC MODEL OF HAMMER CRUSHER

The hammer axis is made up of bearings, shaft, key hammer, hammer body and top parts, from the dynamic perspective, can be simplified to a degree of freedom of single disc hinge system, schematic diagram below.

The system only around a horizontal axis rotation degrees of freedom, using a rotational angle of theta parameters, its geometry can be completely determined, according to the theory of theoretical mechanics, the general form of differential equation

---

Corresponding author E-mail: ni2789820@sina.com

of movement of the system are as follows:

$$J\frac{d^2\theta}{dt^2}+c\frac{d\theta}{dt}+k\theta=M(t) \quad (1)$$

where J —— system;
  c —— viscous damping coefficient;
  k —— stiffness coefficient of the system.

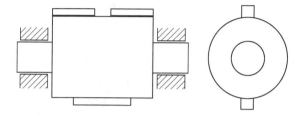

Fig.1 Hammer axis simplified dynamic model

The hammer shaft is an elastomer, excited by periodic load, will cause the forced vibration, when the frequency of periodic external load is equal to the natural frequency of hammer shaft system, will produce intense vibrations, resonance, leading to hammer axis, and even the entire hammer crusher.

Impact coal breaking is a kind of high efficiency breaking coal way, it is completely different from the mechanism of shearer cutting type crushing and milling type broken. It makes good use of the physical and mechanical characteristic of the coal itself. In addition, more energy can be passed in a flash when impact, the common characteristic of all kinds of shock machines is energy concentration, the peak force is several times to dozens times of static loading machine; moreover, the impact is in the form of a pulse output, and jumps in coal breaking process, which can be compatible with each other, so it can achieve high energy utilization; and under the impact of the split type, brittle coal is easy to form relatively stable lump coal, which can greatly improve the lump coal rate. Impact coal breaking includes hammer breaking and reaction coal breaking.

The hammer crusher shaft relies on impact of broken coal (or rock), the impact is discontinuous, and produce great impact in the process of impact, the impact on hammered shaft head, produces load torque. The impact is in the form of rectangular pulse, also is in the form of rectangular pulse load torque.

Sets the speed of hammer axis n, the impact or the frequency of the load torque $f_k=\frac{2n}{60}$ or the duration is $T=\frac{60}{2n}$. Due to the impact in a very short time, the impact or the effect of load torque time is very short, can be thought of a pure pulse form of load. In order to simplify the calculation, the following is still calculated by rectangular pulse. As shown in Fig.2.

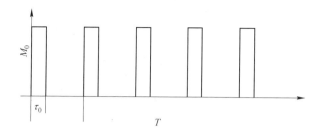

Fig.2 Hammer axis pulse load

Set load torque for $M(t)$, M_0 for peak for the moment, then

$$\begin{aligned}M(t)=&M_0[H(t)-H(t-\tau_0)]+M_0[H(t-T)-\\&H(t-T-\tau_0)]+M_0[H(t-2T)-\\&H(t-2T-\tau_0)]+\ldots\end{aligned} \quad (2)$$

where $H(t)=\begin{cases}1(t>0)\\0(t\leqslant 0)\end{cases}$

Hammer axis motion differential equation of the general form, such as equation 1, is now known to motivate the load torque $M(t)$ as shown in equation 2, because the system damping and load torque, compared to a negligible, time-domain analysis, for the system response can be obtained

$$\theta=\frac{M_0}{k}+\frac{2M_0}{J\pi}\left\{\frac{1}{p^2-\left(\frac{2\pi}{T}\right)^2}\sin\frac{2\pi}{T}t+\frac{1}{3\left[p^2-\left(\frac{6\pi}{T}\right)^2\right]}\sin\frac{6\pi}{T}t+\frac{1}{5\left[p^2-\left(\frac{10\pi}{T}\right)^2\right]}\sin\frac{10\pi}{T}t+\ldots\right\} \quad (3)$$

where M_0 —— as shown in the previous section for maximum impact load torque;
  k —— hammer shaft torsional rigidity coefficient;
  J —— hammer axis of inertia;
  p —— hammer shaft, $p=\sqrt{\frac{k}{J}}$.

Where the power of motor and reducer torque are not discussed, in system analysis, it can be just as dynamic torque superposition in the response of the system.

4 LOAD CHARACTERISTICS OF HAMMER CRUSHER

Detailed analysis on the hammer actuator dynamics characteristic of shaft system, and hammer crusher is composed of motor, reducer shaft main parts, such as analyzing the operating conditions of a hammer crusher, must be integrated into the dynamics of the hammer crusher system and load characteristic.

Studies the load characteristics of hammer crusher, the first analysis of a special case of hammer crusher operation state, without load and no-load characteristics, and then study the hammer crusher operation condition of the load and load characteristic.

4.1 No-load Characterics of Hammer Crusher

When no load, the load includes inertial load, friction load, damping load, the power is provided by the motor. At this point, the running characteristics of hammer crusher system have important relationship with the running characteristics of the motor, the hammer crusher operating characteristics of the system is determined by the operating characteristics of the motor.

Convenient to analyze the moment of inertia of convert of hammer shaft to the rotational inertia of the motor shaft, set the rotational inertia of the hammer shaft for J_1, reducer ratio of i, convert to the motor shaft to the moment of inertia for J_{ml}, then

$$J_{ml} = \frac{J_1}{i^2}$$

Similarly, friction torque to convert to the motor shaft

$$M_{fm} = \frac{M_f}{i\eta}$$

Convert to the damping torque of motor shaft

$$M_{cm} = \frac{M_c}{i\eta}$$

No load, equivalent to the load on the motor shaft torque for conversion to the damping torque of motor shaft

$$M_L = \frac{M_f}{i\eta} + \frac{M_c}{i\eta} + \left(J_m + J_G + \frac{J_1}{i^2}\right)\varepsilon_m$$

where M_f——effect on the hammer shaft friction torque;
M_c——role in the hammer on the shaft of damping torque;
i——from the motor shaft to the hammer on the shaft of the transmission ratio;
η——transmission efficiency;
J_m——the rotational inertia of the motor shaft;
J_G——transmission device convert to the motor shaft of the rotating parts on the moment of inertia;
J_1——hammer axis of inertia;
ε_m——the angular acceleration of the motor.

Therefore, the following discussion no-load characteristic of hammer crusher the no-load characteristic is only need to study, the related parameters for conversion.

4.1.1 The No-Load Staring

When the motor start turning, speed increased gradually, until it reach the steady speed, a process known as the starting process. According to different rotor asynchronous motor, generally have a cage asynchronous motors and two kinds of wound rotor asynchronous motor, there is a difference between their starting way.

4.1.1.1 No-load Starting, Cage Asynchronous Motor

Ordinary squirrel-cage asynchronous motor direct starting curve as shown in Fig.3, corresponding to A specific frequency values have A corresponding t—s curve, different frequency values, have different T—n curve, only when the motor T—n curve is higher than the load torque curve, it can start smoothly, and stable operation in the curve of intersection point (A point). T—n curve is higher than load resistance torque line, the more the starting acceleration stability quickly, on the contrary, if T—n curve under load line, the motor can't start.

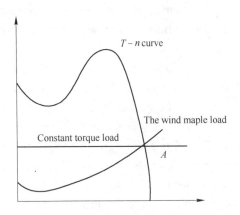

Fig.3 General squirrel-cage asynchronous motor starting curve directly

Cage asynchronous motor direct starting, the starting current is big, and the starting torque is not large, the starting performance is not ideal.

Cage asynchronous motor direct starting at starting moment, as the rotating magnetic field still has a lot to the relative speed of the rotor, so when the rotor induction electromotive force E_{20} is about 20 times the rated speed at the rotor induction electromotive force E_2, such a large current will cause a lot of lines to land, cause power grid voltage is reduced, affect the normal work of the other electrical equipment. So in actual operation, often using motor step-down starting (capacity over 7 kW three-phase asynchronous motor).

Step-down starting principle is to try to not rated voltage during motor starting, and add a lower voltage, when after the start process, plus the impact pressure. Due to reduce the starting voltage, it reduces the starting current. Because the starting torque is directly proportional to the square of the voltage, so the starting torque also significantly reduced, so the step-down starting can only be used for light load or no-load starting.

4.1.1.2 The Wound Rotor Asynchronous Motor Starting

If a wound rotor asynchronous motor rotor loop into the appropriate resistance, which can limit the starting current, and can increase the starting torque, overcome the cage type asynchronous motor starting current is big at the same time, the shortcomings of starting torque is not big, suitable for large capacity of asynchronous motor overload starting. Wound rotor asynchronous motor starting into rotor resistance and the rotor string of frequency-sensitive rheostat two starting method.

Hammer crusher of no-load starting to overcome inertia load and friction load, belongs to the light load or no-load starting, can use directly starting or step-down starting.

4.1.2 No-load Running

Hammer crusher no-load running, you just need to overcome the friction load, and the asynchronous motor no-load running status of basically the same.

Asynchronous motor no-load running of stator current is called the no-load current, With I_0 said. Asynchronous motor no-load running, due to the shaft without mechanical load, the speed is very high, close to the synchronous speed, the stator rotating magnetic field between the rotor and the relative velocity is almost zero, So the rotor induced electromotive force $E_2=0$, Rotor current $I_0=0$, Rotor magnetomotive force $\phi_2=0$.

Hammer crusher no-load running, the asynchronous motor need to overcome the friction load, not entirely without mechanical load state, but because of the friction load is small, mainly is the rolling friction and gear friction, friction torque is small, can be considered to be asynchronous motor no-load running state.

4.2 The Load Characterics of Hammer Crusher

When the normal work of the hammer crusher, asynchronous motor in addition to the need to overcome inertia load and friction load, and the impact of the hammer to break the coal load, the load of shock pulse torque. At this point, the working characteristic of hammer crusher is not only influenced by asynchronous motor, the working condition of hammer crusher hammer axis plays a main role to the characteristics of the system.

When the asynchronous motor with mechanical load, the rotor speed will drop, it increase the relative speed of the stator rotating magnetic field cutting rotor winding and the rotor induction electromotive force E_2 and the rotor current, this will produce electromagnetic torque, and form the motor output torque.

Asynchronous motor mechanical properties parameter expression is as follow:

$$T_{em} = \frac{m_1 p u_1^2 \frac{r_2'}{s}}{2\pi f_1 \left[\left(r_1 + \frac{r_2'}{s} \right)^2 (x_1 + x_2')^2 \right]} \quad (4)$$

where m_1——number of stator phase;
 p——pole logarithmic;
 u_1——stator phase voltage;
 f_1——power frequency;
 r_1——each phase of the stator winding resistance;
 x_1——each phase of the stator winding leakage reactance;
 r_2'——equivalent to the side of the stator rotor resistance;
 x_2'——equivalent to the side of the stator rotor leakage reactance;
 s——slip.

In the equation, only the slip s (or the speed of the motor) is changeful, $s = \frac{n_0 - n}{n_0}$, $n_0 = \frac{60 f_1}{p}$. Or the electromagnetic torque of asynchronous motor is only associated with the rotational speed of asynchronous motor, when the rotating speed is constant, it can be thought of a fixed value.

Now asynchronous motor torque is also reduced to hammer on the shaft, according to the content of the discussion in the previous chapter, you can get the load characteristics of hammer crusher shaft, it can also be considered as the load performance of hammer crusher, the general form of the differential equation is as follow:

$$J \frac{d^2\theta}{dt^2} + c \frac{d\theta}{dt} + k\theta = T_e - M(t) \quad (5)$$

where the T_e for the output torque of the motor converted to hammer crusher shaft torque. $M(t)$ is a pulse torque, and T_e is approximation for a constant, the differential equation for a sum is two linear differential equation.

5 CONCLUSIONS

Reproduced in this paper, the coal mine machine used in the execution part of hammer crusher hammer - axis has carried on the dynamic analysis, the simplified dynamic model is established, and on the basis of deriving its dynamics differential equation, the analytical solutions of the equations is obtained. Hammer crusher is studied under the condition of no load, load of load characteristics, for the design of hammer crusher and working condition analysis provides a theoretical basis.

REFERENCES

[1] Ji Gongying. PCM250 hard thick coal seam working face of hammer crusher[J]. Coal mining machinery, 2003, 8.

[2] Du Erhu, Tian Quzhen. Preliminary discussion on the mechanism of broken coal[J]. Journal of xi 'an institute of mining 1998 (2).

[3] Zhao Sihai, Li Guoping, LuoTieNan. hammer crusher three-dimensional modeling and simulation studies[J]. Coal mining machinery, 2008, 5.

Controlled Underwater Blasting near Sensitive Structures at Mithi River Extension Project, Mumbai, India

M. Ramulu[1], M. Gurharikar[2]

(1. Central Institute of Mining & Fuel Research, Regional Centre, Nagpur, India;
2. Rocktech Systems & Projects, Nagpur, India)

ABSTRACT: Controlled underwater blasting was conducted near sensitive structures in Mumbai city in India for river widening at the mouth of the Arabian Sea. After heavy floods in Mumbai due to narrow passage of Mithi River in 2005, Mumbai Metropolitan Region Development Authority (MMRDA) has under taken deepening and widening of River by underwater blasting. This lead to bed rock blasting by specialised underwater blasting techniques near sensitive structures like flyover, sewerage plant and old Mahim causeway. Control of vibrations and fly rock were important objectives of the project along with desired fragmentation. Underwater drilling was carried out by drill rigs placed on platoons, specially fabricated for the purpose. After the drilling is carried out to the required depth, drill rods with hammer is removed from the hole keeping the casing intact in the hole to avoid hole collapsing and identification. Couplable Plastic Tube (CPT) explosives used for charging in blast holes from the platform called barge. The blasthole diameter was 75 mm and the depth was restricted to 1.75 m to reduce blast induced ground vibrations. Hole-to-hole delay initiation was carried out to get better fragmentation and to reduce maximum charge per delay for reducing intensity of vibrations. Specially manufactured high density seismic explosives were used to break the rock under water. As the water head was just 0.4 m, blasting was carried out during high tide times with a water head of above 4.5 m to reduce the flyrock projectile and avoiding risk to the habitats in the vicinity. The specific charge used was 0.25~0.35 kg/m^3 and specific drilling was 1~1.17 m/m^3 to ensure optimum fragmentation. The vibration intensity in all the blasts was restricted to 5mm/s, which was the threshold limit for the adjacent sensitive structures. The desired optimum fragment size of 0.1 m^3 was achieved in all the blasts was without compromising safety parameters. The specially designed underwater blast patterns could yield desired production to complete the project in stipulated time.

KEYWORDS: controlled blasting; underwater drilling;underwater blasting

1 INTRODUCTION

Underwater blasting technology is going in a rampant way due to great demand of sea and river widening programs for various onshore and offshore applications. As part of river widening and deepening program to avoid choking in monsoon season, controlled underwater blasting was conducted for Mithi river at the mouth of the Arabian Sea in wake of heavy floods in 2005, which submerged some low-lying houses. Mumbai Metropolitan Region Development Authority (MMRDA) has assigned the job of deepening and widening of River by underwater blasting. This lead to bed rock blasting by specialized underwater blasting techniques near sensitive structures like flyover, sewerage plant and old Mahim Causeway bridge and toll buildings as shown in Fig.1.

Therefore, the challenge of conducting successful underwater blasting is doubled due to nearby structures and habitaes. This paper deals with the blast performance and measure taken to contain the side-effects of underwater blasting.

2 UNDERWATER BLASTING

Although underwater blasting is not a new area, extensive research work is not reported to facilitate design and execution for a particular rock mass. The basic work on underwater blasting shock waves carried out by earlier researchers Penney (1941), Roth (1959), Navagin (1960), Rinehart and Pearson (1963), Cole (1965), Shachnazarov and Mittelmen (1966), is very useful for understanding the blasting mechanics.

Underwater blasting is always a challenge because of its complicated nature of execution and application. Unlike surface blasting, the underwater blasting always requires special type of explosives, loading techniques and performance monitoring tools. Underwater blasts differ from blasts in rock because the detonation of the explosives underwater generate an initial shock wave and a gas bubble in

---

Corresponding author E-mail: more.ramulu@gmail.com

water. The gas bubble typically expands and collapses several times before reaching the surface. This expansion and collapse of the gas bubble as it rises to the surface is called a "bubble pulse" (Miller and Yancey, 1999). The bubble pulse generates low frequency vibrations that have the most potential for damaging structures and causing blasting complaints. Fig.2 shows the bubble rupturing the water surface from the detonation of 3kg emulsion explosive.

According to Susanszky, (1976) it is a problem protecting structures when underwater blasting is in their immediate vicinity as the explosives energy is efficiently transferred in the form of shock waves to structures and damage can result even at great distances. Therefore, all the blasts were monitored with seismograph for recording ground vibrations and frequency to ensure the safety of surrounding weak structures. Therefore, the blast induced damage to structures due to underwater blast vibrations cannot be viewed from similar perspective of surface blast vibrations. The nearest structure from the blast site is at a distance of 50m from the blast site.

Fig.1   Existing structures near the underwater blasting site

Fig.2   Bubble rupturing the water surface from the detonation of 3kg emulsion explosive

3 FIELD APPLICATION OF UNDERWATER BLASTING METHOD

As the surroundings are very sensitive, very cautious underwater blasting method was used at Mithi river expansion and deepening works of MMRDA. The details of the site and general blasting practice are explained in the following sections.

3.1 Details of the Test Site

The underwater blast site is well within Mumbai, which is in western India. The site is located at just beside the 100 year old Mahim Causeway bridge towards downstream side at latitude and longitude of 19° 2'53.17"N and 72°50'14.63"E respectively. The Mahim Causeway is a vital link road connecting the city of Mumbai with its northern suburbs. The causeway links the neighborhoods of Mahim to the south with Bandra to the north. The underwater blast site with surrounding important structures is shown in Fig.3.

Fig.3   Underwater blast site with surrounding important structures

3.2 Geotechnical Information

The basic rockmass formation of the blasting site is basalt. These rocks consist primarily of a succession of basaltic lava flows with individual flow thickness varying between few metres to 45 metres with the average thickness being 10 to 15 metres. The two types of flow consist of compact nonvesicular basalt and amygdaloidal basalt with zeolites as cavity filling material. Breccia also forms some intrusions. The compact basalts are relatively more jointed in comparison to amygdaloidal basalts, which generally have one joint set. The strike and dip of the compact basalt joint sets are: 45°～215° and 78° due North-West; 130°～310° and 87° due South-West. The basalt rock is black in colour and resistant to weathering because of their hard and compact nature. Flow of ground water through the jointing is more pronounced in the high-tide period. The joints are by and large tight with aperture < 0.1 mm in general. Some of the random joints are filled with gouge on the wall near the roof. The average joint spacing varies between 0.30 to 0.50 m. The joints are slightly rough, irregular and planer. Since depth is very low and strata is of subsoil type, sometimes clay fillings was noticed in the joints.

The core samples were collected from various blast sites and were subjected to laboratory analysis to determine the uniaxial compressive strength (UCS) and modulus of elasticity (E). The digital Schmidt hammer was used to obtain the *in-situ* strength of the rock mass. The RQD was range was decided based on measurement of joint volume. The uniaxial compressive strength of basalt from Point load testing was 74 MPa and the strength from Schmidt rebound hammer 51 MPa. The RQD was 80% and the elasticity modulus was 14.5 GPa.

3.3 Method of Rock Excavation

The method of rock excavation was by drilling and blasting and material used to be evacuated by excavator of $10m^3$. The rock excavator used to work in low tide period for easy digging and loading purpose. The excavator, while digging and removing blasted material underwater is shown in Fig.4.

Fig.4 Rock excavation and removal after blasting

3.4 Drilling and Blasting

Generally in under water drilling, where water level is more than 3 m, drill rig is generally placed on barge, specially designed to carry out drilling operation. First MS Casing is lowered in water. This casing diameter is more than the diameter of drill bit. One end of casing is rested on rock in water, where hole is to be drilled and another end of casing is held by helpers so that it should not get dislocated during drilling. Once the drilling is carried out to the required depth, rods with hammer is removed from the casing and casing is hold in the same position as was during drilling.

Drilling machine used at Mithi river deepening was tire mounted wagon drill machine with 75 mm bit diameter. A barge is fabricated to facilitate underwater drilling as shown in Fig.1. Generally drilling was carried out in low tide, where water level use to be 0.5-1 m, so that it was feasible to move rig in water and carry our drilling easily. Where water level is to be more than meter, drilling is to be carried out with drill rig placed on barge. The hole is to be plugged by divers by PVC pipe, where depth of water use to be more. As the there was a sand cover of 2~3 m above the bed rock which is to be blasted, there was a problem of hole collapse immediately after drilling. To avoid this problem, holes were inserted with plastic pipes upto the rock bed i.e. in the sand column as shown in Fig.5.

Fig.5 Inserted with plastic pipes after drilling of blast holes

Couplable Plastic Tube (CPT) explosives used for charging in blastholes from the platform called barge. Strength of explosive was 90% and Velocity of Detonation was 5000m/s. Nonel shock tube initiation was used with 25m delay to each hole. Once the hole is charged with CPT explosive, it is tied with nylon wire so that shock tube does not break due to water current and is lowered in hole. The blastholes used to be stemmed with chips of drill cuttings or coarser sand. About 25~30 holes are charged and connection made with 17 ms hole to hole delay and 42 ms row to row. The charge diameter was 65 mm and the total hole depth was 4~5 m and in rock it was varying from 1.75~2 m. Hole-to-hole delay initiation was carried out to get better fragmentation and to reduce maximum charge per delay for reducing intensity of vibrations. Specially manufactured high density seismic explosives were used to break the rock under water.

As the water head during low-tide period was just 0.5~1 m, blasting was carried out during high tide times with a water head of above 4.5 m to reduce the flyrock projectile and avoiding risk to the habitats in the vicinity. The specific charge used was 0.7 kg/m^3 and specific drilling was 1.17 m/m^3 to ensure optimum fragmentation. Blast design and output parameters of underwater blasting are given in Tab.1.

Tab.1 Blast design and output parameters of underwater blasting

Parameter	Value
Diameter of blast hole/mm	75
Total No. of blast holes per round	25~30
Burden/m	1.75
Spacing/m	2
Charge per round/kg	40
Maximum Charge per delay/kg	3
Specific charge/kg·m^{-3}	0.35
Specific drilling/m·m^{-3}	1~1.17

4 UNDERWATER BLAST RESULTS

The underwater blasts could initially gave poor fragmentation due to inadequate specific charge. The trials were started with 1kg per hole, which giving a specific charge of 0.15 kg/m^3 which is very low for the high strength basalt rocks of the site. But this low charge was kept keeping the weak structures in the vicinity. Therefore, the charge quantity is gradually increased from 0.15~0.35 kg/m^3 to get optimum fragmentation and controlled blast vibration. At 0.29 kg/m^3 the rock breakage and fragmentation was manageable, but at 0.35 kg/m^3 the fragmentation was excellent but vibration intensity was on higherside. Therefore, the specific charge was compromised with reduced charge per delay to get less vibrations in the vicinity of structures. The blast results with various charge quantities are given in Tab.2. Therefore, the design parameters varied from point to point to get optimum and safe blast results. The major blast result parameters are explained in the following sections.

Tab.2 Blast results with various charge quantities

Specific Charge/kg·m^{-3}	Maximum charge per delay/kg	Mean fragment size/m	PPV (at 50 m) /mm·s^{-1}	Bubble height/m
0.14	1	1.58	2.5	0
0.21	1.5	1.1	3.2	0.1
0.29	2	0.55	3.8	0.25
0.36	2.5	0.1	4.3	1
0.43	3	0.07	4.8	3

4.1 Blast Fragmentation

Images of fragmentation were captured with high resolution camera and analysed with Wipfrag software to determine mean fragment size, which is considered as the representative fragment size of each blast round. At least 10 representative pictures were captured from each blast round and same procedure is repeated for all the trial blasts. One sample picture taken from the muckpile of blast at low tide period is shown in Fig.6. Fragment size distribution of the image (Fig.6) is shown in Fig.7. The mean fragment size measured by the digital image analysis for very poor fragmentation was 1.5 m and for very good fragmentation it was 0.07 m. It was observed that the finer fragmentation with mean fragment size less than 0.1 m could not be handled by the excavator inside the water. This may be due to buoyancy effect, which was resulting in inefficient muck removal. Therefore, it was considered that the mean fragment size of 0.1m is the optimum fragment size for 10 m^3 bucket size excavator.

Fig.6 Post excavation fragmentation of underwater blasting

4.2 Shovel Loading Performance

The Shovel Loading Performance study was conducted on the blast fragmentation under the water. The shovel loading cycle time observed was 8.5 s/m^3 and 3.5 s/m^3 respectively for poor (1.5 m) and good fragmentation (0.07 m) respectively. There was considerable reduction of shovel loading cycle time by 62% with improved fragmentation.

4.3 Blast Vibrations

The ground vibrations for all the trial blasts were recorded with triaxial geophones of Instantel's Minimate seismograph. Attenuation model for ground vibration levels for various scaled distance is shown in Fig.8. The safe charge per delay was calculated based on the attenuation model indicated in the graph. The high value of correlation coefficient indicates the consistency of the data. The data presented in the Tab.2 indicates the peak particle velocity (PPV) at 50 m distance is less than 5 mm/s. This indicates that the surrounding structures, which are at a distance of 50 m from the blast site are safe.

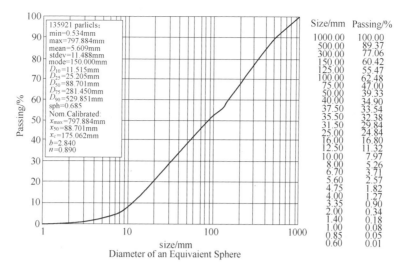

Fig.7 Fragment size distribution of muckpile

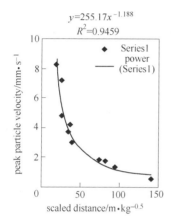

Fig.8 Vibration attenuation model for underwater blasting

4.4 Flyrock

There was practically no flyrock in all the trial blasts. This is due to precise blast design as well as sufficient sand cover and water head of 3~4 m. The water bubble height was varying from 0 to 3 m, but in no cases it ejected flyrock. This helped in conducting blasting operations without interrupting the traffic which was about 80 m from the blast site. Therefore, all the blasts were safer to the surrounding public and habitats. Use of nonel initiation system also helped in containing the flyrock and water bubble.

5 CONCLUSIONS

The underwater trial blasts conducted with optimized blast design parameters at Mithi river extension project were fruitful in containing the side effects of blasting i.e, ground vibrations, air overpressure and flyrock. Blast data was generated on fragmentation, ground vibrations and flyrock analysed for blast performance assessment. All the muck pile images were analysed by image analysis technique using WipFrag software for determination of fragmentation size distribution and mean fragment size. It was observed that the finer fragmentation with mean fragment size less than 0.1 m could not be handled by the excavator inside the water due to buoyancy effect. Therefore, it was considered that the mean fragment size of 0.1 m is the optimum fragment size for 10 m^3 bucket size excavator. The specific charge corresponding to this fragmentation size was 0.35 kg/m^3. The attenuation model developed out of trial blasts gives the vibration intensity of less than 5 mm/s at 50 m distance, where sensitive structures are located. The maximum charge per delay to be used was 2 kg for a distance range of 30~40m from the structures and it was 3 kg for 40~50 m distance. The precise blast design coupled with sand cover of 3 m and water head of 4 m helped in total containment of flyrock. This helped in conducting blasting operations without interrupting the traffic which was about 80 m from the blast site. The application of optimized techniques helped in enhancement of both safety and productivity of underwater blasting.

ACKNOWLEDGEMENTS

The authors thankfully acknowledge the Director, CIMFR, Scientist-in-Charge, CIMFR Regional Centre, Nagpur and M/s MMRDA and NA constructions for their cooperation during the studies. The views expressed in the paper are those of the authors and not necessarily of the organizations they represent.

REFERENCES

[1] Cole R H, (1965), *Underwater Explosions New York Dover Public. Inc.*

[2] Miller D, Yancey B, Matheson G. (1999), *Vibration and Airblast Standards for Underwater Blasting in Virginia, Proceed-*

ings, 25th Annual Conference, February 7~10, Nashville, TN. pp.65~75.

[3] Navagin J S.(1960), *Ispolaovan'ie Energii Podvadnova Vzrueva dl'a Listovoty Shtampovki, D.N.T.P. Leningrad.*

[4] Penney W G (1941), *British Report RC 142.*

[5] Rinehart J S, Pearson J. (1963), *Explosive Working of Metals Pergamon Press.*

[6] Roth J.(1959), *The Forming of Metals by Explosives The Explosive Engineer 37 pp.3~4.*

[7] Shachnazarov A M, Mittelmen I D.(1966), *Visocoscorostnote Deformirovan'ie Mettalov Izd.Mashinostrotentie Moscoww.*

[8] Susanszky Z (1976), *New aspects of shock waves in underwater blasting, General Proceedings, 2nd Annual Conference, , January 28 - 30, Louisville, KY. pp. 364~372.*

Study on the Engineering Method of Excavating Shaft by One-Step Smooth Blasting

WANG Li, GAO Yan, SUN Ning
(*Engineering Scientific Research and Design Institute, Shenyang Military Area Command, China*)

ABSTRACT: The paper discusses technical difficulties in shaft construction and limitations of current constructing methods in the context of actual detonation projects. The paper analyzes a case of shaft excavation by one-step smooth blasting, and then sums up the selection of detonation parameters and items for attention during shaft construction for national defense purpose. Therefore knowledge concerning shaft construction is accumulated for the good of national defense projects.

KEYWORDS: shaft; shaft excavation by one-step explosion; engineering practice

The shaft has a wide range of application in the field of national defense engineering. It is the vertical passing, which is excavated in the stratum all the way to the surface, in various kinds of engineering projects. Shafts with larger cross sections, which have a diameter of more than 1 meter, are usually excavated by the drilling-blasting method, which is divided further into shaft excavation by one-step explosion (SEOES) and shaft excavation by multiple-step explosion (SEMSE). The SEOES is the most economical one. For so many years, however, the SEMSE has been employed as a practice, whereas the SEOES seldom functions in national defense projects, because of few successful cases for reference, high risks of employment, hard-to-control technical parameters, etc. In a national defense project, our unit excavated a shaft of nearly 20 meters depth in a rough geological area by employing the SEOES. The expenditure was saved and process of construction was sped up, in safety and security standards. In the following, the paper discusses, in the context of actual engineering projects, parameters selection, quantity of charged explosive, cutting methods and spacing, time difference of detonating, structure of charging, and sealing of shaft with regard to the decoupling-charge SEOES.

1 COMMON METHODS OF SHAFT CONSTRUCTION

The shaft is widely applied in engineering projects in such fields as national defense, civil air defense, as well as underground and tunnel constructions. Ventilation shaft, cable shaft and communications shaft are just some examples. Shafts vary from centimeters to meters in diameter, and several meters to 10-odd meters in depth. The excavation methods are also divided into SEOES and SEMSE. Although it has advantages of saving time and manpower, the SEOES is seldom employed due to technical difficulties and uncontrollable blasting effects. The SEMSE, which has been employed more often, has the problem of difficult slag removal after every detonation. During the construction of a shaft, slag is removed vertically. Both manual and mechanical methods are not able to solve the problems of small amount of slag each time and danger in operations. Despite of its technical difficulties, SEOES is not wholly beyond control of application. Our unit explores new means in excavating projects, and, based on learned experience, calculation and onsite conditions, we successfully employed the decouple-charge SEOES to make a shaft of 1.5-meter diameter and about-20-meter depth. During the construction, we overcame a series of technical problems such as serious rocks weathering, steeply-sloped position of the shaft, etc., and made a successful one-step explosion by employing the decouple-charge, smooth blasting method. As a result, the project was finished 20 days ahead of the deadline and 500000 RMB was saved.

2 CONSTRUCTING METHODS AND PARAMETERS SELECTION

2.1 Selection of Machinery

The shaft is excavated up-down. The construction site is normally on a slope, which causes problems for mechanical transportation. Tracked excavator can be used to cut an avenue of approach for transporting drilling machines. Because of explosion in the shaft is vertical, the SEOES requires vertical blast holes for run-through explosion, and, as a result, the normal drill carriage cannot meet the demands. After repetitive comparisons and demonstrations, it is concluded that tracked down-the-hole drill is

---
Corresponding author E-mail: sning2008@sina.com

the choice. The down- the-hole drill is capable of cutting 100 meters depth holes. The down- the-hole drill makes the hole which is, though bigger in diameter, applicable for detonation, and ensures the verticality of drilling operation. Before construction, an excavator eliminate the soil on the surface of the construction site and level the ground, which is at least 3 meters long and wide to provide for operation of the down-hole drill. Pay attention to the safety and security during the operation which is usually characterized by employment of big machines on the steep slant.

2.2 Selection of Operation Parameters

During drilling operation, bit of $\phi 115$ mm was first used to make a central hole, and then bit of $\phi 90$mm used to make holes and smoothing blasting holes. To ensure slag can be thoroughly cleaned, the distance between the cutting holes and central hole should be shorter, usually is $2 \sim 5$ times diameter of the central hole. The amount of smooth blasting holes and cutting holes are just as usual. If cross section is smaller and rock friction is larger, when the cutting holes is initiated, the large diameter central hole is no charge and is the free surface of cutting holes, cut holes should be the paralleled small diameters and long depth holes. If the operation is based on the center of the shaft, down-the-hole drill, bit of $\phi 115$mm should be used to make a central hole which is free surface of cut holes and without explosives. Around the central hole are 4 cutting holes which are 0.2 meters away from the central hole, and 8 smooth blasting holes are well distributed along the contour of the shaft. Bit of $\phi 90$mm is used in cutting holes and smooth holes. For loading explosive convenient, $5 \sim 7$ centimeters wide bamboo plates can be used. Explosive cartridges are tied to the plates with scotch tape, and clay is used to fill the space between the cartridges. In this way the decoupleing charge takes shape. When the SEOES is employed, blast holes are drilled to the demanded depth at one time. However, rock swells after detonation. To ensure the enough space to compensate for every explosion, stage blasting should be employed. The height of a layer is determined by rock nature, cross sections, hole diameter, etc. Generally speaking, long height for rocks which are easy to be detonated and short height for rocks which are not. The decoupleing charge requires the tamping, which is 1.2 to 1.5 times of normal parallel detonation. The amount of charged explosive is normal. The amount of explosive charged for smooth hole needs extra attention. According to the learned experience, to ensure the effect of detonation, smooth hole should be filled by 4 to 6 cartridges in every meter, half-second delay, color-changeable, plastic nonel detonator should be employed, and smooth blast holes are detonated following cut holes.

3 CONTROLLING POINTS AND NOTICES

While excavating shaft by means of one-step decouping-charge smooth blasting, it is important to work carefully and heed safety issues. The following points should be highlighted during the construction.

(1) Pay great attention to the camouflage issue during the construction. The shaft construction is often carried out on the slope, with a large excavation work plane and high degree of machinery application, which requires special attention to the camouflage issue during the operation. When the avenue of approach is built in on mountain for mechanical equipment, it is important to cause as little damage to the vegetation as possible. After the construction, the original vegetation should be restored to the utmost to reduce the problems of exposure. When camouflaging the drilling operation, the camouflage net can be applied horizontally to the shaft mouth, and then carry out the operation.

(2) Calibrate the direction of the blasting hole. Deviation of the blasting hole causes an inconsistency resistance of the top and bottom of the hole, and affects on smooth blasting. Therefore, it is very important to assure the precision of the professional survey location. During the operation, surveyors should monitor the vertical degree of the drilling rod based on the designed drilling direction by using the Electronic Total Station. Adjust timely when the offset occurs, in order to make sure of the uprightness of the drilling operation.

(3) Control the precision of the drilling hole, particularly the location of the smooth blasting hole. The distance between the opening of the smooth blasting hole and the designed outline should be no more than 5 cm, and the blasting hole is vertical.

(4) Guarantee the quality of charging explosive. The blasting operation has a high security requirement. Before charging, it is critical to check blasting holes, clean out the sundries and distinguish the charge weight and delay time of detonator. When charging, it is important to handle with care, particularly not to touch the wall of the holes overmuch, with the purpose of avoiding the detritus fallings into and blocking holes, and ensuring the integrity of nonel pipe.

(5) Rigidly implement the onsite management. Irrelevant personnel, Irrelevant machinery and vehicles are prohibited to enter the blasting zone. Personnel who have entered the blasting zone should wear the hard hat and observe all of the rules and regulations relevant to the blasting construction.

(6) Strictly abide by the operation specifications on the electric air compressor and down-the-hole drill, and prohi-

bit the irrelevant personnel to enter the operation zone during the operation.

(7) The blasting personnel are not allowed to enter the site to check the blasting effect until ventilation for half an hour. In addition, slay removal operation should not be carried out until the end of danger examination and disposal.

(8) Officers should be in charge of the operation and the safety personnel should supervise onsite for safety and security purposes, paying attention to the earth and rocks on the slope, preventing slope failure, and guaranteeing the absolute safety of personnel and equipment. The safety warning signs should be set up at the construction site and the safety rules should be rigidly implemented to ensure safety during construction.

(9) Check and seal the holes timely. Blasting holes should be checked timely at the end of blasting and slag discharge. The detected under-excavated parts should be cleaned up timely and substantial deviations should be remedied in time. The construction should proceed to the next stage after appropriate inspection. If the covering operation on the holes is not carried out in short time, the holes can be stuffed with crab sticks and braided fabrics, avoiding the rocks falling into the holes.

(10) Set up distinct signs at the part of the shaft mouth. It is critical to set up distinct signs at the part of the shaft mouth, particularly the upper mouth, after the blasting, and establish separate outpost area, to prevent dangers caused by mistakenly entering of persons. In addition, separate outpost area should also be established at the part of the lower mouth, preventing the falling rocks caused by the mistakenly entering of persons.

4 CONCLUSIONS

The shaft construction has always been difficult in the blasting construction, and particularly in the section construction, the vertical slag discharge is very difficult. The shaft excavation by one-step decouple-charge smooth blasting can reduce the time of the blasting operation, manpower, material and financial resources, and the difficulty of slag discharge. It takes less time, has a high efficiency and reduces the danger of the multiple-step blasting. The field blasting indicates that it is worthwhile to generalize one-step decouple-charge smooth blasting technology in the shaft construction, and that the relatively reliable cutting form is the large empty holes drum cut. This technology adopts decouple charge, which can prevent the rock pieces from "blocking up" or producing "springing" phenomenon, and reduce the explosive consumption at the same time.

REFERENCES

[1] YU Yalun. Engineering Detonation Theories and Technologies[M]. Beijing: Metallurgical Industry Press, 2004.

[2] GU Yicheng. The Operation and Security of Detonating Projects[M]. Beijing: Metallurgical Industry Press, 2004.

[3] TIAN Houjian, etc. Practical Detonating Technologies[M]. Beijing: PLA Publisher, 1999.

Kinematic Characteristics of Movement of the Explosive Cavity Wall in Different Rocks

B.R.Rakishev[1], Z.B.Rakisheva[2]

(*1. K.I.Satpaev KazNTU Almaty, Republic of Kazakhstan; 2. al-Farabi KazNU Almaty, Republic of Kazakhstan*)

ABSTRACT: Based on examination of the first stage of the explosive explosion in the rock massif an expression was proposed for increasing mass of rocks, squeezed out of the explosive cavity. Using the theorem of momentum change of the body with variable mass the dependence was determined of velocity changing of the explosive cavity walls movement from its radius. Giving a certain value of relative movement of the cavity walls, one can easily calculate their velocity during corresponding period according to this dependence. Also this dependence allows to find the expansion time of the explosive cavity during the period between its adjacent positions. According to the proposed formulas the velocities and expansion time were defined of the explosive cavity in rocks with different physical and mechanical properties by using explosives with the different physical-chemical characteristics.

KEYWORDS: explosive explosion; explosive cavity; velocity of the cavity walls movement; time of the cavity expansion

Among the various effects of the explosion in a solid medium the particularly obvious is the cavity formation. It includes integrally all the elements of the explosion process development[1,2]. Kinematic characteristics of the explosive cavity movement in rock massif (velocity of movement, time of the cavity formation) depend on the particular combination of physical and technical properties of the rock mass, chemical and physical characteristics of explosives and conditions of blasting.

Identification of this interaction may be based on the use of regularities in the development of the rock explosion. It is generally accepted that after detonation of the explosive charge in the hole the powerful stress wave extends by the rock, which strongly compresses (destroys) a layer of rock in its path[1~3]. However, after a while the boundaries of the detonation products (DP) begin to lag behind the compression wave front. The gaseous detonation products, developing further already the initiated process of destruction, regrind the medium at the contact zone and expand axially the charging chamber.

By symmetrical development the gas (explosive) cavity reaches its limiting volume, determined by the properties of the medium, the used explosives and conditions of blasting. In other words, the DP will expand axially until their pressure will fall up to the magnitude of P_c, equal to strength resistance of medium in conditions of explosive loading[2,4]. During this time the process of rock massif destruction is continued. Until the wave of compression will reach a free surface, the picture of medium movement and development of destructions occur the same way as by camouflet explosion.

Thus, for definition of the maximal sizes of the gas cavity by explosion of the cylindrical charge in massif with two free surfaces it is enough to restrict consideration of camouflet stage of explosion. In the future, due to the influence of the wave processes and the ledge destruction, axially symmetric development of the cavity is violated. The second stage enters into force - the stage of accelerated movement of the crushed part of massif.

In this formulation the maximal zone of DP expansion can be calculated as the maximal magnitude of moving of the border of contact "DP – medium". Motions of the contact border are connected with medium crushing, its grinding and compaction (Fig.1). Displaced from the cavity volume of rock rising from the minimal to the maximal magnitude and is distributed in certain way at the zone of destruction. Hence, mass of this volume is variable. Its value, falling on unit of the charge length, is determined from the ratio [2]:

$$M = \pi(r^2 - r_H^2)\rho_O \quad (1)$$

where, r is the distance from the charge axis to the border of contact "DP – medium"; r_H is the cavity initial radius; ρ_O is density of the displaced rock.

In equation (1) in the initial moment of movement $M \neq 0$ since before the movement beginning, at least some elementary layer of medium is already compacted or crushed by action of shock wave (SW). Therefore $r_H = r_O(1+\varepsilon)$,

---

Corresponding author E-mail: b.rakishev@mail.ru

where ε is relative thickness of the layer compacted (or crushed) by SW before the beginning of the border movement under action of gaseous products of detonation; r_0 is radius of the charge.

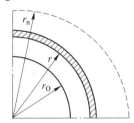

Fig.1 Scheme for definition of the regularities of the explosive cavity movement

Having applied the theorem of movement quantity change of a body of variable mass, for the displaced and involved in movement volume of rock it is possible to write:

$$u\frac{d(Mu)}{dr} = S(P - P_c) \quad (2)$$

where, M is mass of medium, involved in mechanical movement by DP; u is the current velocity of the contact border "DP – medium"; S is the current square of a surface of the contact border; P is current pressure of DP; P_c is the strength characteristic of medium in conditions of explosive loading.

It was found experimentally that the height of the gas cavity by explosion of the cylindrical charge is close to its height. Then the area of surface of the contact border, falling at the unit of the charge height, is equal to:

$$S = 2\pi r \quad (3)$$

Since the strength characteristic of rocky rocks in conditions of the all-round compression is more than $P_k = 200$ MPa, DP expansion at the considered site can be taken as passing entropic according to the law $PV^3 = $ const [2,5], so:

$$P = P_H (r_0/r)^6 \quad (4)$$

where, P_H is initial pressure of DP.

Having substituted values of M, S and P from (1), (3) in equation (2) and having taken a derivative from the left part by r, we obtain:

$$\frac{du^2}{dr} + \frac{4ru^2}{r^2 - r_0^2} = \frac{4r}{\rho_0(r^2 - r_0^2)}\left[P_H\left(\frac{r_0}{r}\right)^6 - P_c\right] \quad (5)$$

After integrating the equation (5) in relevant limits the law of velocity changing of cavity walls depending on its radius was obtained in form:

$$u = \frac{1}{\left(\frac{r}{r_0}\right)^2 - 1}\sqrt{\frac{P_H}{\rho_0}\left(\frac{r_0}{r}\right)^2\left[\left(\frac{r_0}{r}\right)^2 - 2\right] - \frac{P_c}{\rho_0}\left(\frac{r}{r_0}\right)^2\left[\left(\frac{r}{r_0}\right)^2 - 2\right] + \frac{P_H}{\rho_0} - \frac{P_c}{\rho_0}}$$

Denoting r/r_0 as $\bar{r}$, after simple transformations this equation can be reduced to:

$$u = \frac{(\bar{r}^2 - 1)\sqrt{P_H - P_c \bar{r}^4}}{\bar{r}^2(\bar{r}^2 - 1)\sqrt{\rho_0}} = \frac{\sqrt{P_H - P_c \bar{r}^4}}{\bar{r}^2\sqrt{\rho_0}} \quad (6)$$

For use the equation (6) it is necessary to know the magnitudes of values P_H, P_c.

As was mentioned, the cavity expansion under the gaseous products action starts from the separation moment the shock wave front from the contact surface "DP – medium". By this time in the closed volume the same average pressure is set up in the products of detonation. Its magnitude is twice less than pressure at the front of detonation wave[5]. Hence, this pressure is the initial for the considered stage of explosion, i.e.:

$$P_H = 1/8 \rho_{em} D^2 \quad (7)$$

where, ρ_{em} is density of explosives; D is the velocity of explosives detonation.

For maximum pressure in a cavity, formed by the explosion of a cylindrical charge, predetermining the strength characteristic of rock massif in conditions of dynamic loading, we have obtained[2]:

$$P_c = \sigma_{com}\left(\frac{\rho_0 c^2}{\sigma_{com}}\right)^{1/4} \quad (8)$$

where, σ_{com} is the rock strength limit in compression, MPa; ρ_0 is the rock density; c is the sound velocity in rock.

As can be seen from equation (8), the strength characteristic depends both on the compressibility and the rock strength limit in crushing. The expression in parentheses is a dimensionless quantity, which takes into account the condition of all-round dynamic loading of the medium under action of explosives explosion. In other words, this number shows how many times the rock resistance to destruction increases actually at the near zone of the explosion by all-round dynamic load.

Equation (6) expresses the law of velocity changing of the explosive cavity walls from its relative radius. According it one can calculate the velocity of the cavity walls, corresponding to a given value of the relative radius of the cavity.

To illustrate the application capabilities of the proposed methodology for determining the kinematic characteristics of the walls movement of explosive cavity, the strength characteristics of the studied rocks, calculated by the formula (8), are shown in the Tab.1. Initial pressure of DP P_H, calculated by the formula (7), consistently was taken equal to 1500, 2000, 2500, 3000, 3500, 4000, 4500 MPa.

Tab.1 Strength characteristics of the studied rocks

Rocks	P_c/MPa	Rocks	P_c/MPa
Albitophyre (S)	603	Skarned limestone (A)	845
Diorite-porphyrite (S)	843	Coarse-grained limestone (A)	750
Marbleized limestone (S)	520	Marbleized limestone (A)	648
Clayey limestone (S)	614	Diorite-porphyrite (A)	590
Magnetite ore, rich (S)	1103	Clay shales (Zh)	477
Magnetite ore, poor (S)	934	Shales (Zh)	560
Porous martite (S)	401	Diorites (Zh)	672
Quartz sandstone (S)	706	Serpentinites (Zh)	615
Diabase porphyrite (S)	787	Dunites (Zh)	637
Porphyry syenite (S)	706	Quartz diorite (Zh)	850
Albitophyre tufa (S)	699	Diorite peridotites (Zh)	748

Numerical values of the velocities of the cavity walls and the corresponding time of the cavity expansion by the various values of P_H, P_O, $\bar{r}_i$, $\bar{r}_{i+1}$, calculated according to the formula (6), are adduced at the Tab.2.

The expansion time of the explosive cavity at a distance, equal to the segment between adjacent radii of cavity (τ) by the well's radius r_0=0.125m, was determined by the formula:

$$\tau = \frac{0.25(\bar{r}_{i+1} - \bar{r}_i)}{u_i + u_{i+1}} \quad (9)$$

where, $\bar{r}_i$, $\bar{r}_{i+1}$ are the relative radii of the cavity at the considered and the following position of the cavity radius correspondingly (Fig.2); u_i, u_{i+1} are the corresponding to these positions velocities of the cavity walls.

The magnitudes τ for different rocks by the different values of P_H, P_O, $\bar{r}_i$, $\bar{r}_{i+1}$, u_i, u_{i+1} are adduced at the Tab.2. As can be seen from these data, with the cavity expansion its expansion time grows a few. At the same rocks with increasing of P_H the time of cavity formation till the limit position increases.

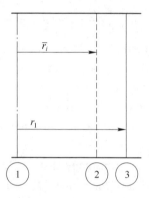

Fig.2 Scheme for definition of the expansion time of the explosive cavity

1—the charge axis; 2—the $\bar{r}_i$ -position of the cavity at the considered site; 3—the $\bar{r}_{i+1}$ - position of the cavity at the following site

Tab.2 Velocities of the cavity walls (u) and its expansion time (τ) at the rocks of Sarbay (S), Akzhal (A), Zheticara (Zh) deposits

$\bar{r}$	u	τ	$\bar{r}$	u	τ	$\bar{r}$	u	τ
1	2	3	4	5	6	7	8	9
	By P_H=1500 MPa at porous martite (S)			By P_H=1500 MPa at clay shales (Zh)			By P_H=2000 MPa at clayey limestone (S)	
1.02	524.48	4.90	1.02	591.78	4.35	1.02	675.96	3.81
1.04	496.13	5.18	1.04	557.31	4.62	1.04	637.01	4.04
1.06	468.90	5.48	1.06	525.01	4.93	1.06	599.44	4.30
1.08	442.68	5.81	1.08	489.06	5.27	1.08	563.08	4.58
1.10	417.33	6.36	1.10	459.45	5.84	1.10	527.76	5.06
1.14	368.87	14.46	1.14	397.38	13.59	1.14	459.53	11.73
1.18	322.65	16.66	1.18	338.21	16.27	1.18	393.27	13.88
1.22	277.71	19.59	1.22	276.42	20.44	1.22	327.05	17.10
1.26	232.86	23.86	1.26	212.91	28.72	1.26	257.84	22.91
1.30	186.25	31.23	1.30	135.22	9.50	1.30	178.74	44.04
1.34	133.95	51.74			$\sum$113.53	1.34	48.35	4.60
1.38	59.31	5.20						$\sum$136.05
		$\sum$190.47						

Continues Tab.2

$\bar{r}$	u	τ	$\bar{r}$	u	τ	$\bar{r}$	u	τ
1	2	3	4	5	6	7	8	9
By P_H=2000 MPa at marbleized limestone (A)			By P_H=2000 MPa at diorite-porphyrite (A)			By P_H=2000 MPa at diorite (Zh)		
1.02	652.24	3.95	1.02	679.99	3.78	1.02	672.88	3.83
1.04	613.56	4.20	1.04	641.58	4.01	1.04	631.70	4.08
1.06	576.18	4.48	1.06	604.57	4.26	1.06	592.75	4.36
1.08	539.92	4.79	1.08	568.82	4.53	1.08	554.11	4.67
1.10	504.60	5.32	1.10	534.14	4.99	1.10	517.36	5.20
1.14	436.04	12.42	1.14	467.38	11.49	1.14	443.65	12.27
1.18	368.81	14.94	1.18	402.92	13.48	1.18	371.67	14.95
1.22	300.59	18.95	1.22	339.12	16.32	1.22	297.03	19.48
1.26	227.15	27.53	1.26	273.53	21.06	1.26	216.29	31.36
1.30	136.13	9.50	1.30	201.33	32.58	1.30	102.54	7.80
		$\sum$106.07	1.34	105.57	7.80			$\sum$108.01
					$\sum$124.31			
By P_H=2500 MPa at quartz sandstone (S)			By P_H=2500 MPa at diabase porphyrite (S)			By P_H=2500 MPa at skarned limestone (A)		
1.02	776.44	3.31	1.02	727.11	3.54	1.02	718.13	3.59
1.04	733.47	3.51	1.04	684.66	3.76	1.04	674.49	3.83
1.06	692.12	3.72	1.06	643.67	4.01	1.06	632.24	4.09
1.08	652.23	3.95	1.08	603.97	4.28	1.08	591.18	4.38
1.14	539.49	9.92	1.14	490.61	11.01	1.10	551.10	4.88
1.18	468.35	11.54	1.18	417.70	13.12	1.14	472.93	11.51
1.22	398.54	13.77	1.22	344.35	16.36	1.18	395.61	14.05
1.26	327.83	17.24	1.26	266.72	22.63	1.22	315.95	18.40
1.30	252.30	24.17	1.30	175.12	11.30	1.26	227.41	30.24
1.34	161.51	10.70			$\sum$94.75	1.30	103.32	7.80
		$\sum$106.15						$\sum$102.77
By P_H=2500 MPa at coarse-grained limestone (A)			By P_H=3000 MPa at poor magnetite ore (S)			By P_H=3500 MPa at rich magnetite ore (S)		
1.02	743.68	3.46	1.02	753.09	3.42	1.02	748.76	3.44
1.04	701.32	3.67	1.04	709.38	3.63	1.04	705.02	3.66
1.06	660.49	3.90	1.06	667.19	3.87	1.06	662.79	3.89
1.08	621.02	4.15	1.08	626.35	4.12	1.08	621.88	4.15
1.10	582.71	4.58	1.10	586.64	4.56	1.10	582.09	4.60
1.14	508.86	10.57	1.14	509.85	10.58	1.14	505.05	10.70
1.18	437.38	12.44	1.18	435.10	12.58	1.18	429.89	12.75
1.22	366.36	15.17	1.22	360.13	15.59	1.22	354.26	15.91
1.26	292.87	19.85	1.26	281.23	21.24	1.26	274.16	22.04
1.30	210.81	32.71	1.30	189.53	11.90	1.30	179.51	11.50
1.34	94.89	7.20			$\sum$91.49			$\sum$92.64
		$\sum$117.71						

Analysis shows that by the given strength and elastic properties of rocks the initial velocities of the cavity walls movement depend strongly on initial pressure of DP. So, by P_H=1500, 2000, 2500, 3000, 4000 and 4400 MPa at albitophyre initial velocities of the cavity walls are correspondingly 543.5; 685.34; 802.50; 904.61; 996.31; 1080.25 and 1142.97 m/s; at diorite-porphyrite are correspondingly 437.94; 595.83; 719.89; 825.51; 919.07; 1003.95 and 1067.0 m/s, at marbleized limestone (S) −565.22; 699.94; 812.63; 911.50; 1000.64; 1082.46; 1143.72 m/s; at clayey limestone (S) −534.64; 675.96; 792.47; 893.92; 984.97; 1068.29; 1130 m/s; at rich magnetite ore (S) − 272.79; 442.69; 563.50; 662.64; 748.76; 825.96; 882.87 m/s; at poor rich magnetite ore (S) −373.41; 531.04; 651.59;

753.09; 842.44; 923.19; 983.02 m/s; at porous martite (S) - 524.48; 635.69; 730.16; 813.73; 889.49; 959.28; 1011.65 m/s and so on.

Thus, for all of the rocks with increasing initial pressure of DP the initial velocities of the cavity walls grow. With expansion of the cavity the velocities of the cavity walls decrease, reaching zero by its limit radius[2].

At the considered rocks the time of maximum cavity formation ranges from 94.75 till 190.47 mcs. There is no any information in literature concerning the kinematic characteristics of the explosive cavity movement in rock massif.

The found values for the movement velocities of the cavity walls and the time for cavity formation of the maximum volume are comparable with the data of laboratory research at the rock samples, recorded by the pulsed X-ray installation of the G.V.Plekhanov Leningrad State Institute[7].

CONCLUSIONS

(1) The performed analysis confirms the possibility of theoretical determining of the movement velocities of the cavity walls by the formula (6) and the time of the cavity expansion between its fixed adjacent positions by the formula (9).

(2) Kinematic characteristics of the movement of the cavity walls are predetermined by elastic and strength properties of rocks and detonation characteristics of the applied type of explosives.

(3) Scientific novelty of the obtained results of study and their practical value consists in their universality, i.e. ability to determine the kinematic characteristics of the explosive cavity walls movement under conditions of each mineral deposit.

REFERENCES

[1] Rodionov V N, Adushkin V V, etc. Mechanical effect of underground explosion[M]. 1971: 200.

[2] Rakishev B R. Power consumption of mechanical destruction of mining rocks. Almaty: Baspager, 1998: 210.

[3] Pokrovsky G I, Fedorov I S. Action of impact and explosion in deformable mediums[M]. 1957: 276.

[4] Rakishev B R. Prediction of technological parameters of the blasted rock at the quarries. Alma-Ata: Nauka, 1983: 240.

[5] Baum F A. Processes of the mining rocks destruction by explosion[C]//Explosive affair. 1963, 52(9): 262~285.

[6] Viktorov S D, Zakalinsky V M. Explosive destruction of mining massifs in Russia // Vzryvnoe delo. – M.:JSC "MVK on explosive affair in AMS", 2012. 107(64): 181~190.

[7] Hanukaev A N. Physical processes by the rocks breaking by explosion [M]. Nedra, 1974: 223.

Control Blasting Technique for Rock Cutting Expansion of Existing Expressway

CAI Xu[1], QIU Zhou[1], GUO Tiantian[2], ZHU Jinhua[2]

(1. Haiwei Engineering Construction Co., Ltd. of FHEC of CCCC, Beijing, China;
2. School of Basic Education for Commanding Officers, NUDT, Changsha, Hunan, China)

ABSTRACT: It is usually impossible to hold up the traffic a long time for a full closed construction during the expressway rebuilding and expanding. Not only the seismic wave and air shock wave, but also the throwing and rolling rocks are the key factors to a unblocked expressway. The control rock cutting expansion blasting technique is researched with the expanding project of G4 Expressway. Firstly, the approach of shallow-hole blasting and thin layer peeling is adopted to control the amplitude of seismic wave and air shock wave by means of strictly control the explosive charge and ensure the stemming length. Then change the direction of minimal resistance line, set up multiple protecting walls and guard nettings by use of the vegetation on the upper side slope and reserved rock wall on the original side slope or revetment to reduce the volume and quantity of throwing and rolling rocks. Followed by the presplitting blasting and multi row millisecond extruding blasting to ensure good smoothness and stability of the side slope and the boulder yield is within 10%. Engineering results show, the proposed methods are economical and practical, vibration, shock wave and the throwing rock have been effectively controlled as well as the quality and construction progress of the rock fill roadbed and has a guiding significance for the construction of similar projects.

KEYWORDS: expressway; rebuilding and expanding; rock cutting expansion; control blasting

1 INTRODUCTION

In the construction process of the rock cutting of expressway expanding project, most attention is paid on the traffic safety and operation, but the control blasting technology is a key factor for the project to reduce the cost, shorten the construction period and ensure the traffic flow. The control rock cutting expansion blasting technique is researched with the ZXTJ-10 Zhumadian to Xinyang section expanding project of G4 Expressway is studied for the safety control and management measures in order to solve the following problems.

(1) To control the direction of blasted muckpile, the throwing and rolling rocks to prevent the damage of the G4 expressway.

(2) To determine the reasonable blast parameters according to the different engineering geology.

(3) To make the throwing and rolling rocks are few and small, to reduce highway interrupt time as far as possible and restore traffic as soon as possible.

(4) To reduce the blasting vibration and noise as far as possible.

(5) To reduce the blasting boulder yield for the mechanical excavation and stone roadbed filling.

2 PROJECT OVERVIEW

The begin and end mileage of ZXTJ-10 section are K105+70 and K115+700 with a distance of 10 km. This section has a complex topography with high hills landform, reservoir area, hilly landform and bird resource in Dongjiazhai national nature reserve. High hills landform is mainly distributed in the mileage of K105+700-K112+810, K113+540-K113+810, terrain undulated roughly, soil erose seriously, gullies developed and strong cut. The top layer are mainly composed of completely/strong weathered pyroclastic rock, cloud-two-long-granulite, quartz schist, amphibolites and plagioclase hornblende schist, the bottom layer are moderately weathered rocks with good engineering geological properties. Shikou reservoir area is between the mileage of K112+810 to K113+540, the silty clay surface is alluvial layer with multiple pits and ponds and the engineering geological properties are poor. The hilly landform is located in K113+810 to K115+700 with undulated roughly terrain, developed gullies and seriously erode shallow layer which is composed of granite and plagioclase hornblende schist with good engineering geological properties.

The width of the original road base is 26 m, the width of

Corresponding author E-mail: ttguo_mail@126.com

the ilateral widening monolithic road base is 41.0 m, the width of the one side widening monolithic road base is 47.25, the one side widening range is 21.25 m. The slope ratio is 1∶1 or 1∶0.75. The width of the separated road base is 20.5 m. The total earthwork and stonework is 454195 m³, earthwork is 89040 m³, stonework is 607737 m³, surface clean area is 482455 m², dredging and filling is 7360 m³, ramming area is 158682 m². Tab.1 shows the basic engineering parameters.

Tab.1 Basic Engineering Parameters

No.	Begin/End mileage	Management mileage	Length/m	Digging height/m	Direction
1	K106+070~K106+140	K979+376~K979+446	70	41.7	Right
2	K106+250~K106+329	K979+556~K979+635	79	41.6	Right
3	K106+390~K106+550	K979+696~K979+856	160	47.9	Left
4	K106+765~K106+960	K980+071~K982+266	195	38.8	Left
5	K108+540~K108+620	K981+846~K981+926	80	34.4	Right
6	K109+045~K109+200	K982+351~K982+506	155	32.1	Right

3 CONSTRUCTION FEATURES AND HAZARD ANALYSIS

3.1 Construction Features

G4 expressway is the North South economic main artery with high traffic density, high traffic speed and strong influence, the blasting operation must ensure the safety and smooth flow of existing expressway.

According to the design requirement, the rock which has been cut off should be used for the road bed, the utilization rate is 100%, 90% and 80% separately for the soft rock, hard rock and solid rock.

The surrounding is very complicated with intensive houses and high-voltage wire which are very close to the expressway, and the construction area passes through the Dongjiazhai national nature reserve.

The area of excavation face is small, the side slope is very high and steep which is bad for the blasting and excavating.

3.2 Hazard Analysis

The main work of this project is rock cutting control blasting of expressway, the surrounding is very complicated, in order to ensure the safety of the existing expressway and do no harm to the environment especially the Dongjiazhai national nature reserve, the following hazards must be strictly controlled.

(1) The distance and direction of throwing and rolling rock and blasted muckpile must be strictly controlled to prevent the existing expressway being damaged.

(2) The blasting noise and shockwave must be strictly controlled to prevent the harm to the birds in Dongjiazhai national nature reserve.

(3) The blasting vibration must be strictly controlled to prevent the damage to the adjacent buildings and structures.

4 BLASTING SCHEME DESIGN

4.1 Blasting Parameters

The engineering environment is complicated and the stonework is a little bit scattered, so the short-hole loose blasting is adopted which cut the side slope to several benches from top to bottom. For example, when the height of the side slope is 8 m, three benches are divided, the height of each bench is 3 m.

Besides the bird resource national nature reserve, there are many buildings and structures adjacent to the expressway need to be protected, the distance of nearest building is 300 m. So the short-hole slice peeling blasting method is adopted, the explosive quantity in a sound is strictly controlled, ensure the stemming length of the blast-hole, detonating cord which is exposed to the earth surface is forbidden, to control the blasting shockwave, vibration and noise to reduce the damage to the birds, building and structures.

G4 expressway is the North South economic main artery, the breaking time cannot be more than 2 hours. In order to control the distance and direction of the throwing and rolling rocks, change the direction of minimal resistance line, set up multiple protecting walls and guard nettings by use of the vegetation on the upper side slope and reserved rock wall on the original side slope or revetment to reduce the volume and quantity of throwing and rolling rocks. The reserved rock wall will be loosened during the blasting and be removed by machinery later. The plan and cut sketches are illustrated in Fig.1 and Fig.2.

The blasting parameters are as follows,

(1) Bench height H=3 m;
(2) Hole depth L=3.75 m;
(3) Hole diameter D=80 mm;

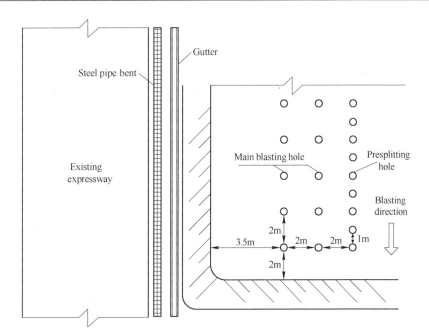

Fig.1　Plan sketch of the short-hole blasting

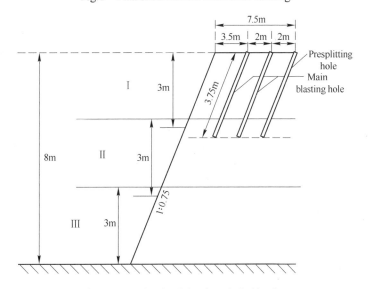

Fig.2　Cut sketch of the short-hole blasting

(4) Minimal resistance line $W=2$ m;

(5) Hole pitch $a=2$ m;

(6) Array pitch $b=2$ m;

(7) Explosive factor $q=0.25$ kg/m³, single hole charge $Q=3$ kg;

(8) Explosive quantity in a sound is 6 kg, the largest total charge of a single blasting is 1100~1600 kg.

Pre splitting blasting parameters are,

(1) Hole depth $L=3.75$ m;

(2) Hole diameter $D=80$ mm;

(3) Hole pitch $a=1$ m;

(4) Linear charge concentration $q'=0.27$ kg/m, single hole charge $Q=1$ kg.

The single cut area is 6000~9000 m², the stonework amount is 18000~27000 m³.

4.2　Charge Structure

4.2.1　Charge Structure of the Main Blasting Hole

The charge of the main blasting hole is continuously $\phi 70$ mm emulsion explosive sticks with one non electric millisecond detonator at the bottom of the hole which is lead to the out of the hole with the plastic detonating tube. The hole is stemmed with loess soil or rock powder, and compacting with bamboo stick. Fig.3 is the charge structure of main blasting hole.

4.2.2　Charge Structure of the Pre-splitting Hole

The charge of the pre splitting hole is separately $\phi 32$ mm emulsion explosive stick which is tied up to the detonating cord and fixed to a bamboo chip. Millisecond detonator is used for initiating and time-delay outside the hole. Fig.4 is the charge structure of the pre splitting hole.

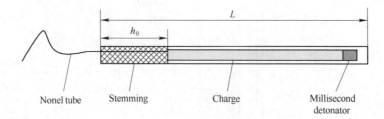

Fig.3 Charge structure of the main blasting hole

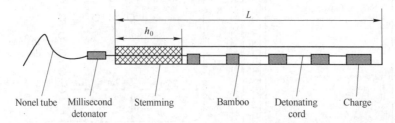

Fig.4 Charge structure of the pre-splitting hole

Both kinds of the hole are covered with sandbag to reduce the throwing rocks and shockwave.

4.3 Design of Detonating Network

In order to reduce the blasting hazards to ensure the safety of the building, expressway and personnel, millisecond blasting is adopted. The detonating network is a non-electric millisecond delay network both inside and outside the hole. Every two holes of the main blasting hole is one sound and every four holes of the pre-splitting hole is one sound. The time delay between neighboring rows is permanent. For the main blasting hole, MS12 nonel detonator is used inside the hole and MS3 nonel detonator is used for the explosion propagation outside the hole. For the pre-splitting hole, MS10 nonel detonator is used inside the hole and MS2 nonel detonator is used for the explosion propagation outside the hole. The pre-splitting hole is initiated 170 ms earlier than the main blasting hole. The detonating network is grouped in clusters and connected by four-way connectors. Fig.5 is the detonating network.

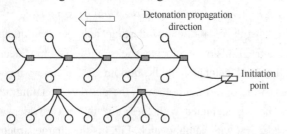

Fig.5 Detonating network

5 SAFETY PROTECTION

5.1 Protection Wall

Two protection walls are set up. The first one is composed of 50 cm wide concrete crash bearers which are painted black and yellow reflective paint and go along the expressway emergency lane. The crash bearers are chained by $\phi 32$ mm steel bars, and each crash bearer is fixed by two expansion bolts to increase the anti-impact ability. The second one is composed of I-steel pole, color steel plate and fence of the expressway. The length of the I-beam is 2.5 m and is fixed in the soil by expansion bolt. From the top of the fence, there are four square steels, the distances to the fence are 30 cm, 50 cm, 50 cm, 50 cm respectively.

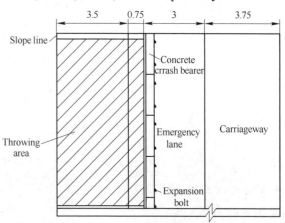

Fig.6 Sketch map of the protection wall

5.2 Protection Netting

Protection netting is set up on the blasting bench by using I-steel pole and U-steel beam. The pole is 5 m $12^{\#}$ I-steel, separation distance is 2 m and is planted into the soil for 1.5 m. The beam is two layers of $10^{\#}$ U-steel, the distance to the earth's surface is 0.25 m, 0.5 m respectively. The pole and the beam are spot welded together as the frame and use the $\phi 8$ mm shaped steel mesh as the protection netting. When the slope protection reach the 3rd grade, two protection nettings are needed, one is set up on the third slope platform, the other goes down with the blasting bench.

Fig.7 Protection wall

Fig.8 Protection netting

5.3 Measures for Traffic Safety

On each side of the expressway there is one liaison personnel with intercom. From 1 km to the blasting point, every 100 m there is a personnel to assist traffic police to guide the traffic on both directions. From 200 m to the blasting point, there are eight clearing personnel. The expressway will be temporarily closed in blasting and will be reopen in 30 minutes after blasting.

(a) (b)

Fig.9 Blasting result
(a) before blasting; (b) after blasting

6 CONCLUSIONS

Cutting blasting of expressway expansion should make fully use of the existing slope, terrain, and space and through the optimal design design to get good loosening effect.

In order to reduce the blasting hazards, various measures are adopted and achieved good results. No damage to the existing expressway, bird, adjacent building and communication cables.

Practice results show that change the direction of minimal resistance line, set up multiple protecting walls and guard nettings by use of the vegetation on the upper side slope and reserved rock wall on the original side slope or revetment can effectively reduce the volume and quantity of throwing and rolling rocks. The traffic can be restored in 30 minutes after blasting.

REFERENCES

[1] Meng H L, Guo Y, Shi J J. Control Blasting on Cutting Expansion nearby Existing Line in the Chongqing-Fuling Railway [C]//*New Technology of Blasting Engineering in China III*, Beijing: Metallurgical Industry Press, 2012:284~290.

[2] Li J B, Lu Q M, Tian A J, Li J S, Li Y Q. Excavation Technology of Overlong-ditch by One Single Blasting in Complicated Surroundings[J]. Engineering Blasting, 2009,15(1):28~30.

[3] Wang X G. *Design and Construction of Blasting*[M]. Beijing: Metallurgical Industry Press, 2011.

[4] Liu D Z. *Practical handbook for engineering blasting*[M]. Beijing: Metallurgical Industry Press, 1999.

[5] Zhang Z Y. *Traffic civil engineering blasting Engineers Handbook*[M]. Beijing: China Communications Press, 2002.

Effect of Confining Pressure on the Failure of Model for Reinforced Concrete Slab Member

Yousong Noh, Jung-Gyu Kim, Young-Hun Ko, Myeong-Jin Shim, Hyung-Sik Yang

(*Department of Energy and Resource Engineering, Chonnam National University, Gwangju, Korea*)

ABSTRACT: Experiments were carried out to examine the effect of confining pressure on the failure behavior of models for slab member using block type reinforced earth retaining wall and reinforced soil. The result showed that the bigger confining pressure was given, the less damage of member, bending of rebar and relaxation of reinforced soil were produced. Thus, coefficient of lateral pressure should be considered for underground structure blasting and increasing confining pressure due to concrete or asphalt should be considered for practical application.

KEYWORDS: slab member; underground structure; confining pressure; coefficient of lateral pressure

1 INTRODUCTION

Recently, use of underground space is inevitable because pathways and roads on earth are saturated due to constantly increasing population and traffic. Demands of underground means of transportation and underground shopping centers are increasing for that reason. Construction of underground concrete structures began in 1950~1960 and deterioration in some structures has been severely progressed(Choi, 2013). Rapid and eco-friendly demolition by explosives is required for demolition of concrete structures(Jeong, 2010) these days. But cases of demolition for underground structure are few while there are many cases of demolition for on-earth structures.

In this study, blasting experiments on slab model which was essential for demolition of underground structure were carried out to consider failure aspect due to confining pressure from above. Concrete slab members were produced and blasted with different confining pressure varied with block type reinforced earth retaining wall and reinforced soil.

2 EXPERIMENTAL DESIGN

2.1 Materials for Experiments

2.1.1 Concrete Member

In this study, normal portland cement was used for concrete member according to concrete quality standards(KS L 5201), and to enhance durability and strength, AE water was used for reducing admixture according to chemical admixture standard(KS F 2560).

2.1.2 Explosive

In this experiment, TNT(tri-nitro-toluene) cartridge which has being used as standard reference explosive due to certain detonation force was used for experiment.

2.1.3 Charge Calculation

The formula of charge weight using TNT was applied to the concrete stripping charge formula from Headquarts Department of The ARMY

$$P=3.3(3.3h+0.5)^3 \quad (1)$$

where P——charge, lb;

h——breaching radius, m.

2.2 Designing Experimental Member

The dimension and structure of scaled and partial-modeled member of slab for the experiment was shown in Fig.1, and tensile portion of concrete was supplemented with steel reinforcement.

The diameter of main reinforcement considering real scale structure was more than 13 mm, center-to-center spacing was less than 20 cm, and each member was cured for 28 days to equalize to specified concrete strength.

Corresponding author E-mail: hsyang@chonnam.ac.kr

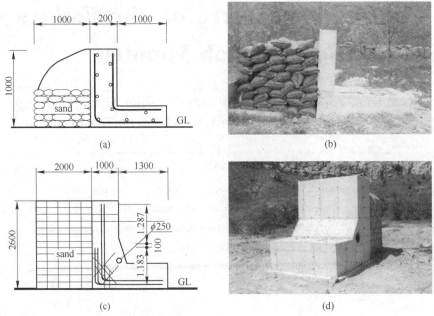

Fig.1 Dimension and structure of concrete member
(a) dimension of scaled slab; (b) structure of scaled slab; (c) dimension of slab member; (d) structure of slab member

3 SCALED SLAB MODEL EXPERIMENT

Before the partial-modeled member experiment, the scaled slab model experiment was conducted three times to determine the charge weight according to suitable breaching radius. The breaching radius was used to be 28cm (diagonal of scaled model), 20 cm (width of scaled model), respectively. Also, the maximum charge weight to charge in the blast hole of partial-modeled member was 27.6 kg per meter, it was equaled to charge weight with calculate the stripping charge formula by 65% for 1 m width of the partial-modeled member. Thus, the breaching radius was decided to be 0.13 m by 65% for 0.2 m width of the scaled model.

The result was shown at Fig.2 and Tab.1. In the case, the scaled model which adjust the charge weight 4.3 kg of 0.28 m breaching radius and the charge weight 2.3 kg of 0.2 m breaching radius was shown to destroy of the concrete and to cut of the steel reinforcement, also the scaled model to adjust the charge weight 1.2 kg of 0.13 m breaching radius was shown that concrete was destroyed and steel reinforcement was partial cut. Accordingly, the structure was unable to role by destroy of concrete and partial cut of steel reinforcement even in the breaching radius of 0.13 m.

(e)   (f)

Fig.2 Demolition of scaled models

(a) before experiment (case 1); (b) after experiment (case 1); (c) before experiment (case 2);

(d) after experiment (case 2); (e) before experiment (case 3); (f) after experiment (case 3)

Tab.1 The result of scaled model demolition

Dimension and result	Model	case 1	case 2	case 3
Breaching radius/m		0.28	0.20	0.13
Charge weight/kg		4.3	2.3	1.2
Result	Concrete	Destroy	Destroy	Destroy
	Steel reinforcement	Cut	Cut	Partial cut

4 EXPERIMENT FOR PARTIAL MODEL OF SLAB

This experiment was implemented to consider failure aspect of partial model of slab due to confining pressure, and block type reinforced earth retaining wall was installed behind the model and the gap was filled by reinforced soil. 2 times of experiment were carried out and each confining pressure was imposed by reinforced soil to 2 m and 6 m thickness, respectively. The figure 6 m was calculated by coefficient of lateral pressure(which mean normalized lateral stress by vertical stress), because if coefficient of lateral pressure is 0.33, secondary stress around roof and floor of structure becomes. The breaching radius 0.65 which is calculated from scaled slab model experiment, thereby charge weight was estimated to be 27.6 kg. And slab was wrapped by felt and wire mesh and covered by blasting mat. The result was shown in Fig.3 and Tab.2.

In case of 2 m thickness of block type reinforced earth retaining wall and reinforced soil, confinement was too small so the block wall was completely dispersed, the concrete slab was destroyed and the rebar was not cut but bended about 15°.

Tab.2 Blasting result of models for slab member

Dimension and result	Model	Case 1	Case 2
Shape of blast hole		Circle	Circle
Breaching radius/m		0.65	0.65
Charge weight/kg		27.6	27.6
Condition of confinement		Block type reinforced earth retaining wall and reinforced soil (2 m)	Block type reinforced earth retaining wall and reinforced soil (6 m)
Result	Concrete	Mostly destroyed	Partial destroyed
	Steel reinforcement	Uncut /15° bending of rebar	Uncut/ none

(a)   (b)

(c)   (d)

Fig.3 Demolition of the models for slab member

(a) before experiment (case 1); (b) after experiment (case 1); (c) before experiment (case 2); (d) after experiment (case 2)

5 CONCLUSIONS

In this study, experiments were carried out to examine the effect of confining pressure to the failure of the underground structure demolition. Before the experiment, the breaching radius and charge weight were decided through scaled model tests.

The result of the demolition of models for slab member, the failure mode of slab was largely influenced by confining pressure. Thus, factors for the confining pressure should be considered for underground demolition. They are location, structure, member section and strength of the target underground structures.Confining pressure due to concrete or asphalt should be considered as well as consolidated overburden in real structures.

REFERENCES

[1] Choi DY. 2013. Durability evaluation of underground concrete structures through the residual prediction. *Master's degree thesis.Daejon. University. Korea*:1~4.

[2] Headquarters Department of the ARMY. 2007. FM 3-34.214(FM 5-250) *Explosive and Demolitions*: 3-11~20.

[3] Jung M S, Song Y S, Park Y S, Heo E H. 2010. A Case Study of RC Rahemen Structure Explosives Demolition (Focusing Demolition at Chungang Department in Daejeon City). *Explosives&Blasting(J. KSEE)*28(2):99~107.

Successful Practice of Ultra 1 km Wells Increasing Water Blasting

GONG Wenxin, ZHANG Zhongyi, JIANG Guixiang, YU Haibin, LI Peng

(*Harbin Hengguan Blasting Engineering Co., Ltd., Harbin, Heilongjiang, China*)

ABSTRACT: Under water blasting main difficulty lies in the explosive brisance and reduce the waterproof difficult under water. This paper introduces the geothermal wells 1 km deep underwater, the successful implementation of blasting methods and measures of increasing water. Mainly elaborated and the use of blasting explosives underwater selection method, the shaped charge technology to improve blasting effect analysis and the design of priming network. At the same time also introduces the field making auxiliary treatment measures of glass steel casing is waterproof, decompression of charge, using thermoplastic pipe insulation of electric blasting network waterproofing measures. Through the auxiliary measures of blasting design, scientific and effective, ensure the successful implementation of the ultra deep water blasting, offering reference for similar blasting engineering.

KEYWORDS: geothermal wells; ultra km; increasing water blasting

1 INTRODUCTION

A hot spring wells of 2100m deep in Harbin fails to meet the requirements of the original design out of its water output. After argumentations from experts of water conservancy, exploration and so on, we finally decided to take the blasting method for well washing and surging. According to geological reports, it was a sandy strata at the 1700m deep with adequate amount of water, so we decided to implement the well washing and surging blasting at that position. We dived the gathering energy liner down to the designated position in the well with a rope, detonated the gathering energy liner to burst and shatter the well pipes, as well as wall rocks, so that the underground hot spring can flow into the well from the crevices with increasing flux of the underground hot spring.

2 SELECTION OF BLASTING SUPPLIES

2.1 Selection of Explosive[1]

Under the action of deep water and static pressure, both the detonation velocity and the brisance of explosive will be reduced. When the depth of water is 30m, the detonation velocity is reduced by an average of 26%, and the brisance is lowered by an average of 33%. In order to ensure an excellent blasting effect at the depth of 1700m (about 16.67MPa), we must select high-class and violent explosives with a high detonation velocity, a strong brisance and a sound water-resistance.

The chemical properties of TNT are stable that it is insoluble in water with little sensitivity, and its detonation heat is 4100~4580 kJ, its detonation temperature is 2737℃, its detonation velocity is 7000m/s (ρ=1.6 g/cm^3), and its detonation pressure is 19GPa. Because of TNT's moderate melting temperature, it can be mixed with other explosives ingredients for charging through various ways, such as press fitting, casting fitting and so on, whose power is better-than-average[1~5].

RDX is white powder that is insoluble in water with a melting point of about 205℃, its explosion heat is 5600 kJ, its detonation temperature is 3427℃, its detonation velocity of 8600m/s (ρ=1.77 g/cm^3), its detonation pressure is 33.8 GPa. As one of the best high explosives[1~5], it is more powerful.

The above two kinds of explosives are frequently-used in military, when they are fused together, which can not overcome the weakness of RDX of unable to be soaked for a long time, on the other hand, which enhances the detonation and brisance of TNT, forming a kind of composite explosive that is difficult to dissolve in water, of a high temperature-resistance, a violent detonation velocity, a strong brisance, as well as a high power, which is commonly used in the propellant of anti-tank mines and high explosive cannon (aerial bomb) [1~5].

According to the characteristics of the above two kinds of explosives, we mixed 40%TNT and 60% RDX for pro-

---

Corresponding author E-mail: 13804561597@139.com

pellant, and its density $\rho = 1.7$ g/cm^3, its detonation velocity $D = 8000$m/s, and the detonation pressure $P=\rho D^2/4=27.2$ GPa, which is rather suitable for this project. In order to ensure the reliability of detonation, we also added up to an accelerating detonation grain to guarantee an accurate detonation of explosives under a high pressure.

2.2 Selection of Detonators

Since the initiation point is located at the depth of 1700 m underground, we take the waterproof, compression-resistance, high temperature resistance of detonator and other important properties into consideration. Because the project time is urgent, we have to use instantaneously electric detonator, so we need to take strict measures on waterproof, compression-resistance, high temperature resistance, etc.

3 PROCESSING CRAFT OF GATHERING ENERGY GRAIN

3.1 Principle of Linear Gathering Energy Charge[1~9]

The diameter of well pipe at the initiation point is only 175mm. Due to the limited space, the charge amount is also limited, a scientific and reasonable employment of the characteristics of gathering energy charge structure can effectively deal with the lack of charge amount to make up the shortage of explosives' falling detonation velocity and brisance under the deep water and static pressure.

The products of high temperature and pressure after the blasting of gathering energy bomb quickly scatter along the direction of surface normal in the charging hole, and they are inevitable to get together to one point in the front of the hole under the influence of the hole (linear charge forms one line), the blasting product density of this point (line) can be increased to 4~5 times, and its speed is $(1.2\sim1.5)\times10^4$ m/s[4]. If the hole shell is made of metal liner, the detonation products will transfer the energy to the liner in the process of driving the grain wall to a movement of axis. Because there is litter compressibility of the grain, the internal energy increases too little, while a great part of the energy is showed as the form of dynamic energy, which can avoid the energy dispersion caused by high pressure expansion, so that the energy is more concentrated. At the same time, when the grain wall converges and collides in the axis, the energy density has been further increased, forming a metal jet and a low-velocity slug behind that. The slender metal jet flow has a high kinetic energy, and there is a speed grads along its length direction on each particle, namely, the end speed is as high as 7~8km/s or even tens of meters, while the tail speed decreases gradually to the carrot with only 0.5~1km/s. The end energy density of jet flow is 14.4 times of that of the explosive detonation wave surface, and the energy density of liner wall is 1.4 times of that of the explosive detonation wave surface, so the gathering energy of the liner is of great significance.

As we all know from the above analysis, the main characteristics of gathering energy effect is high energy density and strong direction with strong penetration, but they are just powerful on the taper direction, and have the same destructive effect to the ordinary charging on other directions. Therefore, gathering energy charge is generally applicable to the field of producing local destructive effect, and wells' surging blasting is using this characteristic to blast the well pipe wall and the whole external rock, or to add and enlarge their cracks so as to increase the output of ground water.

3.2 Parameters of Liner[6, 7]

(1) Shapes of liners: they are mainly axis symmetry and plane symmetry. According to the actual situation of the project, this design selects a linear type and three-sides symmetric gathering energy liner, as shown in Fig.1 and Fig.2. We made three-grooves gathering energy surface along the axial direction of explosive column, formed three vertical gathering energy cutting flows on the wall, and also cut through the inner wall to form three vertical cutting belts to appear more and longer fissures and water points. Considering there will be water pressure after water entry, we made a circular wall as an explosive column shell to protect.

Fig.1　The grain map

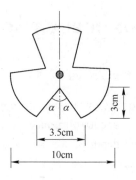

Fig.2　Medicine column cross-sectional schematic diagram

(2) The taper angle of liner: the size of the taper angle directly decides the penetration of gathering energy jet flow. According to the formula of hydromechanics theory:

$$v_j = v_0 \cot(\alpha/2) \quad (1)$$
$$m_j = m \sin^2(\alpha/2) \quad (2)$$

The velocity of jet flow v_j increases as the decreasing the taper angle of liner, while the quality of jet flow m_j decreases as the decreasing the taper angle. It has been demonstrated by the experiments that: when the taper angle is less than 30°, the perforation performance is rather not stable; when the taper angle is 30°~70°, the jet flow is equipped with sufficient quality and speed to guarantee the stable perforation effect and an excellent effect; when the taper angle is bigger than 70°, the perforation depth decreases rapidly, while the crushing effect is increased. Therefore, according to the needs of the project, it is reasonable to design the taper angle as 60°, as shown in Fig.2.

(3) Thickness of the liner: usually, we select as $\delta = (0.021 \sim 0.024)D_k$ that is the bottom diameter of the liner, and we finally select 0.5mm after calculations. According to material selection requirements of the liner, we determined to use copper plate that has small compressibility, high density, good plasticity and excellent ductility, and will not be vaporized during the process of jet flow as the liner.

(4) The blasting height. The metal liner converge and collide along the axis direction under the effect of detonation, then, it needs a proper space and distance during the process of gathering, extending and forming normal jet flow. When the distance is too long, the jet flow will radially disperse, swing, extend to a certain extent, and then creak, which makes the perforation effect reduced and even failed. According to practical experience, in general, a suitable bursting height is 1~2 times of the bottom diameter of the liner[6]. Integrating with the penetration test results of University of Science and Technology of China on wedge-type gathering energy liner at different heights of burst: when the opening width ratio of the height of burst and wedge-type shade is 1, the penetration depth is the biggest[6], and the bottom diameter of designed explosive column gathering energy hole is 35mm. Under 1700m in the well, the inner diameter of well tube is 175mm, the outer diameter of gathering energy blasting is 100mm width and 600mm length, so the height of burst in the designing is relatively reasonable.

(5) Geometric parameters' calculation of charging.

The upper explosive thickness of concavity $\lambda = 0.625D(1-0.25/\tan\alpha) = 12.4$ mm(Actually 15 mm);

Heights of the concavity $h = 0.47D/\tan\alpha = 28.5$ mm (Actually 30 mm); the bottom diameter D of explosive column concavity takes 35 mm;

The taper angle of explosive column concavity 2α takes 60°.

(6) Formula of grain's extended charging amount:
$$q = S\rho = 103 \text{ g/cm} \quad (3)$$

Among which: q represents the charging density of explosives, g/cm; S is the section area of explosive column, cm²; p is the explosive density, 1.7 g/cm³.

By calculating, grain is cylindrical. We made three-grooves gathering energy surface; the angle is 60°; the bottom diameter D of explosive column concavity takes 35mm.

4 CONNECTION MODE AND PROTECTION TREATMENT OF LINER

In order to avoid that the water-pressure, water-temperature and other factors effect the normal initiation of explosive column when the gathering energy column arrives in the blasting position, we took measures of waterproof, pressure-resistance, thermal-treatment, etc on the gathering energy column and detonation network, which is of the same importance to the production of gathering energy liner, and it is also the premise and guarantee of a successful blasting. We applied epoxy resin, curing agent, accelerant, gauze and handmade glass steel in the connections the explosive column and the network as an overall outer protective cover.

When the detonator is inserted into each ignition hole of each grain, we connected the grains well in the method of sequentially connection, we also directly made a layer of glass-steel protective cover with waterproof, heat-insulation and pressure-resistance. The glass steel is a kind of glass fiber reinforced material, it is a new type of composite material developed abroad in early twentieth century with advantages of light-weight, high-strength, pressure-resistance, anti-corrosion, waterproof, thermal-insulation, insulation and flexible operation and so on, which can ensure that when the grain soaks into the water, it detonates without any other effects.

Fig.3　Glass fiber reinforced plastic protection schemes

Fig.4　Insert schematic detonator

Because the glass steel has the same high-tension and strong pressure-resistance to the stainless steel and other metal materials, there is no doubt that, it will also stop and weaken the metal jet flow produced by gathering energy blasting in the blasting process. Through the surface blasting-test observation, it is the same to the explosive column without a glass-steel protection on blasting and penetrating the depth of the rock. Since there are particularities underwater, and our observation conditions are demanding, we did not test the blasting.

5 EFFECT OF THE INCREASING WATER BLASTING

After a successful vertical gathering energy cutting initiation at the 1700m deep with adequate amount of water, we compare the output amount of water between the pre-blasting and the pro-blasting. Before the blasting, there will occur flow interruption in the well water level when the pump has continuously dewatered for 20min in the well, which can be restored to the original level after 2 hours, and the pumping must be interrupted; and after blasting, when the pump has continuously pumped for 10 hours, flow interruption never occur, which has fully showed that the vertical gathering energy cutting blasting effect has achieved the expected goal of increasing water.

For the first time the technology adopted high energy explosives in deep well blasting to improve the detonation velocity, brisance and waterproof; the linear gathering energy charging makes the energy concentrates on the same point to play the biggest role when both the charging amount and space are limited; the glass steel has effectively guaranteed protective measures of the explosive column and detonation network, such as waterproof, pressure resistance, heat resistance and so on; series & parallel dual ignition independent initiation network provides a double insurance for the successful detonation when the explosives have been transported to the designated position; this method is cheap, effective, simple and flexible, which in indeed an economic and feasible way for the surging of some water wells that lack of water volume.

REFERENCES

[1] Xia Changqin, et al., BLASTING[M].The people's Liberation Army soldiers Chinese Changsha Engineering College,1988:255.

[2] Jin Shaohua,Song Quancai.Theory of explosive[M]. Xi'an: Northwestern Polytechnical University press,2010.

[3] Xu Gengguang.Properties and application[M].Beijing Institute of Technology,1991.

[4] Sun Yebin.Military explosives[M]. Beijing: Weapon Industry Press,1995.

[5] Chen Xirong.The performance of explosive and the charging process[M]. Beijing: National Defence Industry Press,1988.

[6] Wang Xuguang.Design and construction of blasting[M]. Beijing: Metallurgical Industry Press,2012:535~555.

[7] Yu Yalun.Theory and technology of Engineering Blasting[M]. Beijing: Metallurgical Industry Press,2007:381~389.

[8] Zhang Guowei.The end effect and its application technology[M]. Beijing: National Defence Industry Press, 2006.

[9] Qi Shifu.With the use of military blasting engineering design[M].PLA University of Science and Technology,2002.

Regularities of the Formation of Submicron Particles at the Explosive Destruction of Rocks

S.D. Victorov, A.N. Kochanov
(*Institute of Comprehensive Exploitation of Mineral Resources* (*IPKON*), *Russian Academy of Sciences, Russia*)

ABSTRACT: Results of experimental studies on assessment of dispersed and mineral composition of submicron particles after explosive rock breaking. While conducting experiments to assess dispersed composition of particles used optical and electronic microscopy, laser spectrometry and x-ray analysis was used to study the mineral composition. The regularities of distribution of the generated particles in the size range 0,1-10,0 microns. It is noted that the mineral particles for most rocks are maximum in the distribution by number in a range of sizes to a few microns. The resulting distribution can be explained from the standpoint of increasing the strength of the particle with the reduction of their sizes, and from the point of view of the mineral composition, one factor of which is the content of quartz. Considered the mechanisms of formation of submicron particles at the explosive destruction of rocks. It is noted that the mechanism of formation of submicron particles resulting fragmentation and the formation of new surfaces, is the most common and is implemented with the destruction of rocks regardless of the type of treatment. Developed the experimental method for the study of the patterns of formation of submicron particles directly at the time of destruction of rock samples.

KEYWORDS: rock; the explosive destruction; submicron particles; experiment; methods; mechanism; laser spectrometry

The destruction of rocks regardless of the form of action-always accompanied by the formation of separate particles, including submicron, less than 10.0 mm. The particle size distribution of particulate matter generated during destruction depends on the microstructure, the degree of heterogeneity, the stress-strain state of rocks. For the study of the formation of submicron particles experimental studies, the objectives of which was to study the possible mechanisms of their formation, establishment size distribution.

In experiments on explosive impact of the methodology, which is described in detail in papers [1,2].The explosive action was performed using the following procedure: Rock samples 14 mm × 14 mm × 10 mm in size were placed into special steel storage ampoules designed at the Institute of Problems of Chemical Physics, Russian Academy of Sciences. Using these ampoules, an investigated substance can be stored and comprehensively analyzed. A charge consisting of three or four trotyl blocks was put on the ampoule's cap. The amplitudes of shock waves passing through the sample were ∼100 kbar. After passage of the shock waves and the brief action of high pulse pressure, the samples became a monolithic compacted powder mixture. As a result of dynamic compression whose value was determined by the amplitude of the explosive wave, we observed the sample's disintegration into microblocks and then compaction. In this work, we used electron microscopy and laser spectrometry to analyze the microstructural changes in the rock samples before and after the explosive action and to obtain quantitative estimates of a number and sizes of microparticles formed by this action.

Most of the images were obtained using a LEO1450 VP scanning electronic microscope, while others were recorded on a JSM 5910LV electronic microscope. We estimated the number and sizes of the microparticles formed after disintegration of the samples as a result of the passage of shock waves using an original procedure developed at the Kurchatov Institute and used previously in studying the emission of microparticles upon the failure of metals [3]. The procedure was adapted to our experimental conditions: after the explosive action, a powdered mixture of the sample was taken from its storage ampoule and placed in a steel trough attached to an emitter ultrasonic dispersant. Under the action of an ultrasound wave, the mineral particles of the powder in question began to rise, and air delivered them to the entrance of a HAND HELD 3016 laser particle counter that allowed us to control the number of particles in the ranges of 0.3, 0.5, 1.0, 3.0, 5.0 and 10.0 μm.

Our electron microscopy studies allowed us to obtain images of fragments of rock sample surfaces before and after explosive action at different scales of magnification. To conduct an informative analysis of the fragments of sample surfaces, the maximum possible magnification did not ex-

---

Corresponding author E-mail: victorov_s@mail.ru

ceed(5000～10000)×under our experimental conditions. Fig.1 presents images of a granite sample, limestone and jasper after explosive action.

Fig.1 Electron microscopic image of (a) granite (b) limestone and (c) jasper after explosive action

Analysis of the images obtained by electron microscopy indicates that along with particles 1～10 μm and greater in size, there are smaller particles with sizes of less than 1 μm whose number and dimensions are difficult to estimate. The distribution of microparticlesaccording to size was qualitatively determined via laser spectrometry. Based on the data in the table showing the results from our studies, we obtained the distribution of mineral particles in the range of 0.3～10 μm, in the percent ratios of their total quantity. It was established that for the samples of marble, green marble, granite, and limestone, the percentage of mineral particles 0.3～0.5 in size was ～10%～25% of their total quantity in the range of sizes in question, and most particles were 1～5 μm in size. Jasper was characterized by the formation of particles with sizes of less than and greater than 1 μm in approximately equal quantitativeratios.

Tab.1 Microparticle size distribution for rocks after explosive action

Rock	Typical size of microparticles/μm					
	0.3	0.5	1.0	3.0	5.0	10.0
Marble	5.5	5.3	3.4	13.4	43.5	28.9
Green marble	6.0	4.9	3.8	12.1	41.8	31.4
Granite	10.0	15.4	44.6	25.4	4.6	0.0
Limestone	10.5	19.6	41.4	20.0	8.1	0.4
Jasper	18.6	26.5	35.2	13.6	5.7	0.4

CONCLUSIONS

The regularities for micro- and submicron particle distribution in the range of 0.3～10 μm after explosive action have been established. As was established experimentally a few mineral particles with minimum sizes of 0.1～0.3 μm formed during the disintegration of our rock samples, but the most of the rocks were composed of larger particles 1～5 μm in size, which were dominant in the investigated range of sizes. The obtained distribution of microparticles can be explained from the viewpoint of an increase in the strength of the particles when their size is reduced, or one of compositions that determine such structural features of rocks as the size and shape of mineral grains, the state of intergrain boundaries, and so on. Among the factors of mineral composition leading to the formation of submicron particles is the content of quartz prone to brittle failure. A procedure was developed to study the microstructure and disperse composition of rock samples atthe microlevel upon explosive action using special storage ampoules and the subsequent analysis by electron microscopy and laser spectrometry of particlesizes. The results from our investigations can be usedto develop concepts for the processes of rock deformation and failure, and to model a mechanism of microparticle formation under the dynamic action of high pulse pressures.

ACKNOWLEDGMENTS

This work was supported by the Russian Foundation for Basic Research, project no. 14-05-00446 a.

REFERENCES

[1] Viktorov S D, Kochanov A N, Odintsev V N. Sdudy of Conditions of Generation of Superfine Particles at Explosive Rock Destruction New Development on Engineering, China, 2007:124~126.

[2] Viktorov S D, Kochanov A N. Rock Microstructure disintegration in case of breakage by blast [C] // The 2nd Asian-Pacific Symposium on Blasting Techniques July 8-12, 2009, Dalian, China. MetallurgicalIndustryPress, China. 2009: 56~57.

[3] Aleksandrov P A, Kalechits V I, MNShakhov., inSb. nauchnykhtrudovnauchnaya sessiya MIFI (Proc.MIFI Scientific Session), Moscow, 2004, 9,: 224~225.(in Russian)

Research on Blasting Scheme Matching Semi-continuous Mining Process Used in East Open Pit

LIU Xiaoming, QI Maofu, CHENG Fei

(*Shanxi China Coal Pingshuo Explosion Equipment Co., Ltd., Shanxi, China*)

ABSTRACT: East open pit mine stripping semi-continuous process system has been put into use, but the existing blasting technology hardly meets the need of fully mobile crushing station's semi-continuous process system and complete construction equipment for the blasting rock lumpiness and the security. As a result, optimization of blasting scheme supporting semi-continuous mining process is proposed and different blasting schemes are invested for research. Results suggest that by optimizing blasting initiation network, adjusting the charging structure, using pre-split blasting and other comprehensive blasting technology, it can effectively reduce the blasting rock lumpiness as well as the pre-shoot and the backward-turning distance of the blasting rock, eliminating security risks during deep hole blasting and making MMD fully mobile crushing station's semi-continuous process system working efficiently in East open pit.

KEYWORDS: mining engineering; semi-continuous; mining; blasting scheme

1 FOREWORD

The fully mobile crushing station's semi-continuous process system of East open pit of Pingshuo group, china coal is designed as 9000 t/h, which is now the most powerful mining semi-continuous system around the world. The first MMD fully mobile crushing station semi-continuous process system was put into use at 2013.11.The system had large production capacity and high degree of automation, employed less workers and used electricity instead of oil, which contributed to energy conservation and environmental protection. And at the same time, it learned from the advantages of both continuous mining technology and intermittent mining technology .It was regarded as the most viable open pit mining system for the reason that it was applicable in hard rock and winter freezing soft rock mining, could operate continuously in rock transportation, expanded production scale and lowered production costs.

Many key technical issues in MMD fully mobile crushing station's semi-continuous process system are still badly in need of solving, such as the working face arrangement, definition of system operating parameters, mine rock lumpiness and its optimizing control. The face arrangement and the particularity of ancillary equipment require strict blasting control. During step level deep hole blasting, not only suitable rock lumpiness for the system is needed but also the minimum pre-shoot and backward-turning distance of the blasting rock in order not to destroy the equipment supporting the system. Optimizing control over rock lumpiness should be finished during open step loose blasting process for the purpose of security and efficiency of the semi-continuous system, which means optimizing blasting parameter, improving construction technology and elaborating construction to control the lumpiness and to eliminate step deep hole blasting's potential impact on MMD fully mobile crushing station's semi-continuous process system.

2 BLASTING REQUIREMENT FOR SEMI-CONTINUOUS MINING TECHNOLOGY

MMD fully mobile crushing station's semi-continuous process system now used in East open pit employs double-face tape system, consisted of single-bucket excavator, all mobile crushing plant, mining face moving conveyor (2 sets), end semi-fixed conveyor, fixed belt conveyor, dumping mobile face conveyor and dumping machine together with the level blasting pattern.

According to the parameter of fully mobile crushing station's semi-continuous process system used in East open pit and the mobile crusher models, quality of blasting scheme matching 15 m standard level was confirmed in order to improve blasting quality, control rock lumpiness and reduce pre-shoot and backward-turning distance. The confirmed blasting quality was as follows: (1)Rock blasting pre-shoot distance for no more than 40 m. (2)Back turning distance for no more than 7 m. (3) Radial direction for no

---

Corresponding author E-mail: jzzdbt@163.com

more than 2 m(smallest diameter of fully mobile crushing station inlet was 2.5 m).

3 STATUS AND BREAKTHROUGH DIRECTION OF BLASTING TECHNOLOGY

The step for rock loose blasting in East open pit is 15 m high .The drill hole diameter is 250 mm.The designing hole net parameter is 8 m(pitch)×7 m(row spacing).And the hole arrangement uses rectangle or triangle .High density hole is needed for particular rock condition. Continuous charging is the main charging way. Surface nonel detonator delay is 17 ms between the holes and 100 ms between the rows. Big fragment rate(the maximum radial length for more than 2 m)is about 0.04% and average muckpiles slope angle is 20.5 degree. The blasting pre-shot distance is 38~42 m and backlash(back-turning) distance is 4~8 m. The existing blasting technology can hardy satisfy the need for fully mobile crushing station's semi-continuous process system, as a result of which we experimented many pre-trials .Based on the confirmed parameter of 250 mm for hole diameter, 8 m×6 m for row distance and triangle for hole arrangement,we chose net initiating blasting pattern, surface time delay, and powder charging structure in the hole as the breakthrough.

4 THE INFLUENCE OF TIME DELAY ON BLASTING

4.1 Theoretical Foundation for Time Delay Choosing

4.1.1 Influence of Delay between Holes on Rock Blasting Lumpiness

For quite hard and brittle rock, because rock dynamic response time is short, the delay time between the holes must be shortened. And for the multi-pore large plastic soft rock, the delay between the holes must be increased. Because if the delay between the holes is too short, fissure will appear firstly in between the boreholes and the rock will be pushed to farther position; if the delay between the row is too long, the blasting hole work separately, leading to bad lumpiness. According to research results domestic and abroad along with practical experience, reasonable time delay between the holes for hard rock is 3 ms / m, while for common rock it goes to 3 ~ 8 ms / m.

4.1.2 Influence of Delay between Rows on Muckpiles Throw Distance

When blasting happens in difference rows ,it is needed to ensure sufficient time delay between rows to pursue perfect throw effect, under the condition of which the before-blasted rock can thoroughly get rid of its original position to make enough free space for the later to be blasted rock. If the time delay between rows is less than a critical value, the rock between rows will block each other after blasting which will make the back turning aggravate and the blasting pile higher, leading to high pressure and tightness on the bottom of the blasting pile not conducive to shovel operations. And different rocks have different dynamic response time, so the critical value varies a lot .When the value comes to less than 8ms/m, rock block after blasting will occur. Increase of time delay between rows will not affect the blasting quality however delay too long will make the pre-blasted rock sustain after throw, affecting the movement of the later row rock. And the ameliorative effect of wave overlay between rows during micro difference blasting and rock fragment collision can not be sufficiently expressed. According to the facto, the maximum time delay between row should be less than 15 ms/m.

To sum up, contradiction between the blasting pile throw distance and the blasting effect exists during the confirmation of time delay between rows. So we chose to change the throw direction to reduce the blasting pile distance in order not to block the rock movement and ensure the blasting quality.

4.2 Trial Results and the Confirmation of Time Delay between Rows

East open pit's surface detonator delay time commonly used in deep hole level blasting can be divided into four types: 17 ms, 42 ms, 65 ms, 100 ms .According to the research results domestic and abroad:(reasonable time delay is 3 ~ 8 ms/m between holes and 8 ~ 15 ms / m between rows), along with our experiences in open pit blasting ,we come up with the following conclusions:

(1) The net connection pattern of the middle row being the control row to initiate the blasting does harm to the blasting quality and the rock smashing degree. The rocks lack enough time needed for mutual collision force which results in uneven rock smashing .At the same time , due to the lack of free space, the rock between rows will block each other after blasting which will make the back turning aggravate and the blasting pile higher, leading to high pressure and tightness on the bottom of the blasting pile not conducive to shovel operations. And the improvement on pre-throw is not obvious. So, we denied the net connection blasting pattern.

(2) The front row being the control initiating, the distance of blasting pile is much farther when the between-hole delay is 17 ms than that when 42 ms. And the distance of blasting pile is much nearer when the between-row delay is 65 ms than that when 100 ms. So by increasing the time

delay between holes and decreasing the time delay between rows which can increase the angle between the throw direction and the free face, blasting pile throw distance can be efficiently reduced.

(3) By increasing the last row's time delay, it can make both the front blasting hole and the last blasting hole move to the free space more efficiently, reducing the pressure from the last blasting hole and shortening the last blasting hole's back-turning distance.

Comparing the results of each trial, we finally defined the surface micro difference time delay as: adjusting the controlled delay time from 17ms to 42 ms, middle delay time from 100 ms to 65 ms, and last delay time be 100 ms.

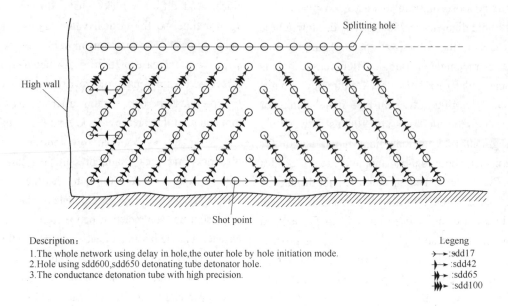

Fig.1   Blasting area network layout

The figured out angle between initiating blasting net and the control row is 20.556°, while that of East open pit was designed as 8.3°. Increasing the angle between initiating blasting net and the control row can make rock goes laterally, beneficial for reducing the front-throw distance. According to the trial statistics, the chosen initiating blasting net delay time can reduce the blasting pile distance by 3 to 4 meters and can basically eliminate back-turning.

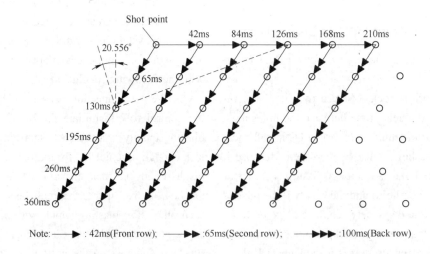

Fig.2   Isochron sketch map

5   THE INFLUENCE OF TIERED CHARGING STRUCTURE ON THE BACKWARD-TURNING DISTANCE AFTER BLASTING

Under the premise of ensuring the explosive energy can affect the part of hard rock, we can change the charging structure of the last blasthole, then change continuous coupling charge into stratified intervals charge. It can weaken the damaging effects of explosive stress wave and detonation gas on the next free surface by reducing the amount of charge in the final blasthole. For the soft rocks and joints, the effect of the rock before the final blasthole can be achieved by the backlash of front blasthole's blasting, and it will not affect the blasting effect due to reducing the dosage. Through the observation on long-term effects of blasting, Segmentation charge of the last row can significantly weaken the blasting destruction of the integrity of the free surface and get better rock blasting lumpiness.

6 THE APPLICATION OF AIR SPACE CHARGE STRUCTURE

6.1 The Principles of Air Space Charge

The principles of air space charge is as follows: (1) reducing the peak pressure of the blast and over crushing of rock around the hole. (2) In addition to affecting the rock by the blast, it can also affect the rock by the pressure wave formed by an explosion of gas and reflected waves from the borehole bottom of the hole. When the pressure of this secondary stress wave is more than rupture strength limit of rock(it represents the pressure required to expand the fracture), rock fissures will be further extended. (3)Extending the duration of the forcing action. Shock waves affect the bottom of the hole or blockage, then they return to the air interval. Due to multiple effect of shock waves, the field is enhanced and the action time of stress waves in the rocks is prolonged(about 2～5 times).

When the air gap is placed in the middle of the grain, enhanced stress field is gained by the effect of the peak of stress waves which is produced by explosion in both ends of the air gap interactions.Due to these three functions, the rock lumpiness is more even.

6.2 Results and the Definition of Powder Charging Structure

In the scheme matching MMD fully mobile crushing station's semi-continuous process system, two types of powder charging structure separated by air was designed. One is from the middle and another is from the orifice. As shown in Fig.3.

(1) The middle model structure's space height is basically 1.5～6 m. The upper powder counts for 33% to38%,and the bottom for 67%～62%.And for super high stage, employ three section of powder separated by two section of air. The bottom powder counts for 60% and the other two sections count for 20% respectively, which not only increase the powder height, avoid big fragment occurrence at the orifice but can efficiently avoid flying stones. During free face landslide and weakness, it can improve efficiency and reduce flying stones.

(2) Orifice gas separation can postpone the function time of the blasting gas, decrease the frequency of orifice big fragment, reduce the unit consumption and avoid the blasting gas bursting out of the orifice with stone, efficiently reducing the flying distance of the orifice stone.

7 APPLICATION OF LAST ROW PRE-SPLITTING BLASTING

Pre-splitting blasting can efficiently reduce blasting quake and control the back-turning distance in the main blasting region. To reduce the impact of the frontier blasting region on the back region's frontier air surface and to avoid the big fragment occurrence in the back row of the blasting pile during the late period of mining, semi-continuous mining technology matched blasting scheme uses pre –splitting holes laid out in the back row of the main blasting region, which blasts before the main blasting region .For the purpose of cost and unit consumption control, it is applicable to increase the pre-splitting hole distance as long as it can eliminate the backlash effect.

8 SUMMARY

Since MMD fully mobile crushing station's semi-continuous process system was put into use, East open pit blasting center has carried out 4 blasting trials in the 1305 working face of the west mine.The maximum pre-shoot distance was 34 m and the maximum backlash distance was 2 m. Blasting flyrock control within the safe range. Meanwhile, the diameter of the smashed rock were all below 2 m, effectively guaranteeing the safety of the system running and elevating the efficiency by 175%.The achievement of the trials provides the equipment matching the semi-continuous sys-

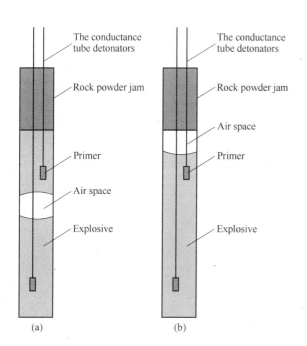

Fig.3 Charging structure
(a) middle position; (b) orifice location

tem now used in the East open pit with technical guarantee for safe operation.

REFERENCES

[1] Sun Gang, Li Aifeng. The development trend of open pit mine semi-continuous mining technology and equipment [J]. Journal of mechanical engineering and automation, 2011 (6): 201~202.

[2] Liu Zhongwei. The exact time of delay and hole by hole blasting in blasting practice in the application of [J]. Mining engineering, 2003,1 (5): 37~40.

[3] Shi Jianjun, Wang Xuguang, Wei Hua, et al. The hole by hole detonation technology. and its application [J]. Mining engineering, 2006,27 (4): 25~28.

[4] Yu Yalun etc. Engineering blasting theory and technology [M]. Beijing: Metallurgical industry publishers, 2004.

[5] Huang Bin, Wang Zhuping, Song Zhiwei. Air interval charge of engineering application [J]. Journal of coal blasting, 2009, (2): 26~28.

Relationship between the Fragmentation Effect and the Blast Wave on the Controlled Blasting for Breaching

Kana Nishino[1], Shiro Kubota[1], Yuji Wada[1], Yuji Ogata[1], Norio Ito[2], Masayuki Nagano[2], Atsuya Fukuda[3], Mieko Kumasaki[4]

(1. National Institute of Advanced Industrial Science and Technology, Tsukuba, Japan;
2. Sagami Kogyo Co., Ltd., Sahamihara, Japan; 3. Kayaku Japan Co., Ltd., Tokyo, Japan;
4. Yokohama National University, Yokohama, Japan)

ABSTRACT: The breaching is one of the supporting technologies of quick rescue at the time of a disaster. In order to rescue the survivors in need of help, a part of concrete wall is removed by the breaching. We have proposed the usage of the explosive to establish quick and safety technique for the breaching. A series of the blast experiments were conducted using model Reinforced Concrete (RC) wall. To clarify the influence to the blast wave caused by the difference of the fragmentation effects, the blast wave measurements and the high-speed photography were performed. There were three configuration of the fragmentation. First have the crater at the borehole side only, the second have that at the back side, and last cases have that at the both sides of the wall. The results of the first and last cases were compared to discuss the influence to the blast wave caused by the difference of the fragmentation effects.

KEYWORDS: breaching; reinforced concrete (RC) wall; blast wave; crater

1 INTRODUCTION

The breaching is one of the rescue techniques by firefighter. For example, first, the searching hole is bored on the reinforced concrete wall to confirm the situation of the back side of the wall and to measure the thickness of the wall. If the survivors are found at the back side, the fragments cannot be allowed during the breaching. If there is no person at the back side, the fragments are allowed. The former is called "clean breaching", and latter is called "dirty breaching". We are developing the new technology that efficiently crushes a Reinforced Concrete (RC) wall by using the explosive(Kubota, 2013, Nishino, 2014). In order to clarify the relationship between the loading conditions of explosives and the crater size, the model experiments were carried out. In this paper, to clarify influence to the blast wave caused by the difference of the fragmentation effects, the results of the blast wave measurements and the high-speed photography were discussed.

2 MODEL EXPERIMENTS

The specimen for the RC wall is shown in Fig.1. The width, height and thickness of the sample of concrete are 1000 mm, 1000 mm and 150 mm, and its density and compressive strength are 2300 kg/m^3 and 31.3 MPa, respectively. Rebar with 13 mm diameter was arranged 75 mm from the surface. A specimen was drilled for making a single borehole as shown in Fig.1 (b). The C4 explosive and the electric detonator were contained in the polycarbonate pipe to make the explosive device. The diameter of borehole was fixed as 16 mm. The clay was used as stemming material. The depth of the borehole and the weight of the C4 explosive were varied to investigate the relationship between the charge conditions of the explosive device and the fragmentation effect. The sizes of the craters after the shots were measured to estimate the fragmentation effect. The measurements by the blast pressure sensor (137A23, PCB Piezotronics Inc.) and a high-speed camera (FASTCAM-SA5, Photron Inc.) were performed to investigate the relationship between the fragmentation effect and the blast wave. The blast pressure sensor was installed on the front and the back side of the RC wall. The distance from the RC wall to the blast pressure sensor was 1m. Two types of the model experiments, i.e. the single shot and the pattern experiments, were conducted. The former employed single borehole per shot and the latter used seven boreholes per shot.

Corresponding author E-mail: kana118-nisihino@aist.go.jp

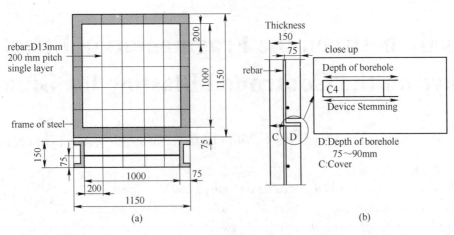

Fig.1 The specimen of Reinforced Concrete (RC) wall (1000 mm×1000 mm×150 mm)

(a) RC wall specimen; (b) cross-section

3 RESULT AND DISCUSSION

3.1 Relationship between the Conditions of Explosive Charge and the Crater Size(Kubota, 2013)

Tab.1 shows the crater depth under various loading conditions. The experimental results can be divided by three cases for the configuration of the crater. First cases have the crater at the borehole side only, the second cases have that at the back side, and last cases have that at the both sides of the wall. We call the first cases as the borehole side fragmentation and the last cases as the both sides fragmentation in this paper. The first cases were confirmed at the conditions from 75 to 87 mm borehole.

Tab.1 Result of single shot experiments (Loading conditions vs crater size)

Depth of borehole /mm	Weight of explosive /g	Depth of crater on borehole side /mm	Depth of crater on back side /mm
75	2.0	15	cracks
75	2.5	50	cracks
75	3.0	65	cracks
75	3.5	50	cracks
80	3.0	25	cracks
85	3.0	30	cracks
87	3.5	65	cracks
90	2.0	cracks	60
90	2.5	20	60
90	3.0	30	70
90	3.5	70	65

In the case of the 90 mm borehole, the crater was generated on both sides. It can be considered that 75 mm borehole with 3.0 g device is suitable for clean breaching and that 90 mm borehole with 3.0 g device is suitable for dirty breaching. In this discussion, the result of the condition of 75 mm borehole with 3.0 g device and the result of the condition of 90 mm borehole with 3.0 g device were compared.

3.2 Discussion on the Influence to the Blast Wave Caused by the Difference of the Fragmentation Effect

Figures 2(a) and 2(b) are the histories of the blast pressure measured on borehole side. It has to be emphasized that the RC wall did not penetrate, even though the condition was the both side fragmentation. Therefore, the detonation product gas does not leak to the back side for all conditions in this study. Nevertheless, the blast pressure of the condition of the borehole side fragmentation is remarkably large.

The depth of crater was 65 mm in the case of the borehole side fragmentation. On the other hand, the depth of crater was 30 mm on the borehole side and 70 mm on the back side in the condition of both sides fragmentation.

We consider the pressure profiles in Fig.2 based on the influence of the fragmentation at the borehole side only. Because the crater depth for the condition of the borehole side fragmentation was greater than that for the condition of the both sides fragmentation, the strength of the blast wave for the condition of the borehole side fragmentation must be relatively small. However, the pressure profiles show an opposite result. We considered the influence of total fragmentation. Because the total depth of the crater for the condition of the both sides fragmentation was greater than that for the condition of the borehole side fragmentation, there was no contradiction in the relation of the pressure profiles.

Fig.3 (a) shows the high-speed photography in the case of the condition of the borehole side fragmentation at 5 millisecond (ms) after the explosion of the device. Fig.3 (b) corresponds to the case of the condition of the both sides fragmentation at 20 ms after the explosion. In Fig.3 (a), it can be confirmed that the product gas blow out from the borehole side with small fragments. Therefore, it can be considered that the explosion energy does not contribute to generate the large size crater.

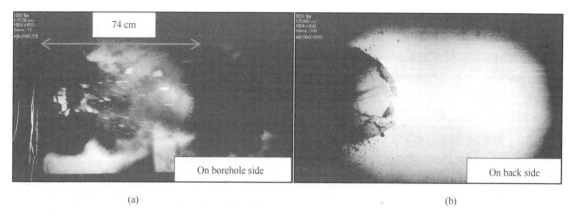

Fig.3 The result of the high-speed photography

(a) the borehole side fragmentation after 5 ms, the speed was set as 1000 fps;

(b) the both sides fragmentation after 20 ms, the speed was set as 5000 fps

Fig.4 shows the result of blast pressure measurement on the back side. The remarkable difference of both conditions is arrival time of the blast pressure. In the both sides fragmentation, the small peak with about 0.38 kPa can be confirmed at 2 ms in Fig.4. On the other hand, in the condition of the borehole side fragmentation, the first peak arrived at 4 ms. It agrees to the arrival time at the second peak of the condition of both sides fragmentation. Because the stress wave arrived at the back side of the RC wall interacts with the air, the weak shock wave propagates through the air. The weak shock wave corresponds to the fist peak for the condition of the both sides fragmentation. In contrast, in the condition of the borehole side fragmentation, such weak shock wave can not be confirmed. We consider that the first peak in the condition of the borehole side fragmentation.

Fig.5 shows the photography after it experiments on the condition of the both sides fragmentation of the pattern experiment. The amount of the explosive a hole is 3.0 g. The total crater depth was 100.8 mm on the average. Maximum overpressure of the blast wave was compared in the pattern experiment. As for the blast wave at the borehole side, it was 25.3 kPa in the condition of the borehole side fragmentation, and it was 14.4 kPa in the both sides fragmentation. The blast pressure in the borehole fragmentation was large as well as the result of a single-shot experiment. The blast pressure on the back side was 0.92 kPa, and it became a not different result from the result of a single-shot experiment.

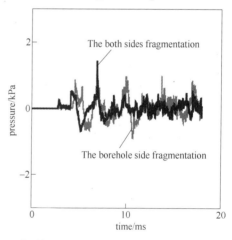

Fig.4 The blast pressure measurement result on back side

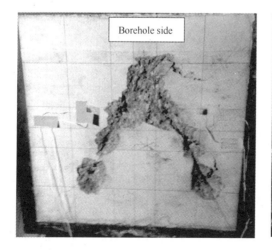
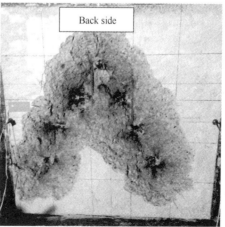

Fig.5 The crushing situation on the condition of the dirty breaching

4　CONCLUSIONS

A series of the blast experiments were conducted using model RC wall. To clarify the influence to the blast wave caused by the difference of the fragmentation effects, the blast wave measurements and the high-speed photography were performed. The results of the condition of the borehole side fragmentation and the condition of the both sides fragmentation were compared to discuss the influence to the blast wave caused by the difference of the fragmentation effects.

ACKNOWLEDGMENT

This study was supported by the Promotion Program for Scientific Fire and Disaster Prevention Technologies in 2012 and 2013.

REFERENCES

[1] Kubota S, Nishino K, Wada Y, Ogata. Y, Ito N, Nagano M, Nakamura S, Taguchi T, Fukuda A. Application of micro-blasting for rescue work[C]// *Proceeding of the 7th International Conference on Explosives and Blasting. Beijing*: Metallurgical Industry Press, 2013: 235～240.

[2] Nishino K, Kubota S, Wada Y, Ogata Y, Ito N, Nagano M, Nakamura S, Taguchi T, Fukuda A. Estimation of crater in reinforced concrete wall caused by mini-blasting[C]// *Explosion, Shock Wave and High-Energy Reaction Phenomena 2, Materials Science Forum* 767, 2014: 154～159.

[3] Nishino K, Kubota S, Wada Y, Ogata Y, Ito N, Nagano M, Nakamura S, Taguchi T, Fukuda A. Partial destruction of reinforced concrete wall by small blasting[C]//*Proceedings of the 8-th International Symposium on Impact Engineering, Applied Mechanics and Materials* 566, 2014: 268～273.

The Application of Pre-splitting Blasting near Slope Technique in Yuanjiacun Iron Mine

JIA Chuanpeng

(*Taigang Group Lan Country Mining Co.,Ltd., China*)

ABSTRACT: The Yuanjiacun iron mine is currently under the side-hill cutting, mining in the horizontal 1725 m of the eastern stope is about to reach the final mining Area. The pre-splitting blasting technique scheme should be used to sustain the slope stability. it is necessary to determine a pre-splitting blasting scheme suitable for the mine lithology. The suitable Hole network parameters, explosive payload, loaded constitution and network connection are determined by reference to the pre-splitting blasting parameter in homo-mining area in China and carrying out small scale targeted experiment near slope. By analyzing experimental effects and making specific improvement measures, the pre-splitting blasting reaches the expected effect.

KEYWORDS: slope; pre-splitting blasting; hole network parameters; explosive payload; loaded constitution

1 INTRODUCTION

Tisco group Yuanjiacun iron ore use top-to-down by mining by level slowly. The closed loop is 1455 m, the section is 15m height, and the final section is 30 m height, leaving space for security platform and clearing platforms. It is designed that there will be steps with 70° slope at the horizontal 1725 m of the eastern slope. The controlled blasting of slope is required and it is necessary to maintain good slope without excessive and decreased digging, electric shovel mining come into slope for one time after blasting. The error must be within the designed range.

2 THE ENGINEERING GEOLOGICAL CONDITIONS

The site of presplit blasting is located in the horizontal 1725 m of the eastern stope. Its geological conditions is complex and the lithology is mainly dolomite, interspersed with mixing metamorphic rock and diabase and fracture zone. The development of gap in blasting zone is weak and there is no groundwater.

3 THE DESIGN OF PRE-SPLITTING BLASTING PARAMETERS

3.1 The Hole Diameter

The existing DTH drilling rig in the mine is Xuanhua Jinke JK580 (D) - a crawler hydraulic driller - and the hole diameter is ϕ140 mm and ϕ115 mm by changing the aiguilles. So the pre-splitting hole diameter is ϕ115 mm, the buffer hole diameter ϕ115 m, and the main blasting hole diameter ϕ140 mm.

3.2 The Hole Network Parameters

3.2.1 The Pitch

Pre-splitting pitch [1] is not only affects the size of the explosive load, but also has a direct effects on the quality of pre-splitting wall. Practice shows that under the premise that the two holes splitting and forming gap are guaranteed, the quality of small spacing wall is better than that of large spacing wall. The choice of pitch is expressed by multiples of the hole diameter. It can calculated by the following equation:

$$a_{pre} = (7 \sim 10)d \qquad (1)$$

For the coefficient in above equation, its value is small for hard rock and broken rocks, while large value for soft rock. The rocks of this pre-splitting blasting area are mainly limestone and dolomite, the coefficient of hardness therefore is $f=10\sim12$, and since joint developed so the coefficient is 8.5, and $a_{pre}=1\,\mathrm{m}$.

According to the rock hardness coefficient and the previous blasting effects, main blasting hole spacing area: $a_{main}=5\,\mathrm{m}$.

Slow punching hole spacing: $a_{slow} = \dfrac{a_{main}}{2} = 2.5\,\mathrm{m}$.

3.2.2 Row Spacing

Row spacing [2] had a great influence on blasting effect and blasting vibration. The blasting of buffer cartridge blasting should have two free surfaces. If the rowing spacing between the buffer hole and the pre-splitting hole is unduly small, the blasting could make a gap between the pre-splitting hole and a buffer hole. The blasting direction of the buffer hole will go to the direction of pre-splitting direction, damaging the formed presplit face and increasing the pucker of surrounding rock and causing a strong vibration to the slope. If the row spacing between the buffer hole and pre-splitting hole is too large, the front rock of the pre-splitting hole is likely to leave blasting blind area or produce a large number of large pieces, making it has to increase the dosage in the buffer hole to improve the blasting effect, thus causing greater vibration to slope and different degrees of damage to surrounding rock. On the other hand, it will not leave enough compensation room for the rocks after blasting, resulting that the rocks cannot be threw fully, stuffy produced and the blasting vibration increased dramatically. In principle, the rowing spacing between the buffer hole and the pre-splitting hole is large for soft rocks and small for hard rocks, and a little bit less than the row spacing between the buffer hole and the main blast hole.

Going through the experiment of position of similar lithology, it can make sure that the row spacing between the buffer hole and the pre-splitting hole is 3 m, and the row spacing between the buffer hole and the main blasting hole is 3.5 m.

3.2.3 Borehole Inclination

Layout of the pre-splitting hole is on the end line of the steps, by 70°. Buffer hole applies in 70° and main blasting hole is 90°.

3.2.4 Hole Deep and Ultra Deep

The height of the stope stage is 15 m, the pre-splitting hole drilled by 70° without ultra depth, so the depth of pre-splitting hole is:

$$h_{pre} = \frac{15.0}{\sin 70°} = 16 \text{ m} \quad (2)$$

The buffer hole is designed according to the 70°, super 0.5 m deep, the buffer hole depth is:

$$h_{slow} = \frac{15+0.5}{\sin 70°} = 16.5 \text{ m} \quad (3)$$

The vertical drilling of the main blast hole, super 2 m, so the depth of main blasting hole is:

$$h_{main} = 15 + 2 = 17 \text{ m} \quad (4)$$

Hole layout and parameters as shown in Fig.1 and Tab.1.

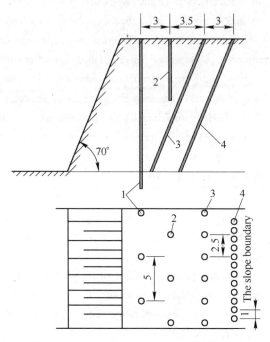

Fig.1 Layout of pre-splitting hole blasting
1—the first row of main blasting; 2—the second row of main blasting; 3—buffer hole; 4—pre-splitting hole

Tab.1 Pre-splitting blasting parameters (period height H=15 m)

	Hole diameter/mm	Pitch angle/(°)	Pitch depth/m	Ultra depth/m	Hole spacing/m	Row spacing/m
Pre-splitting hole	115	70	16	0	1	3
Buffer hole	115	70	16.5	0.5	2.5	3.5
The first row of main blasting hole	140	90	17	0	5.0	3
The second row of main blasting hole	140	90	7	2	5.0	3

3.3 The Main Blast Hole, Buffer Hole and Pre-splitting Hole

The main blasting hole calculation formula is:

$$Q_{main} = qaW_1 H \quad (5)$$

where q——unit explosive consumption, 0.65 kg/m³;
  a——pitch, take 5.0 m;
  H——step height, take 15;
  W_1——chassis resistance line, take 3.5.

To calculate the $Q_{main} = 170$ kg.

Buffer hole explosive load is 45%~55% of the main blast hole charge.

Take $Q_{slow} = 85$ kg.

The line density of pre-splitting hole is calculated and determined with 3 methods: the theoretical calculation method, the experience formula method and the experience numerical method. For the theoretical calculation method, many parameters are difficult to make sure or even inaccurate due to the immature presplit blasting theory. So the experience formula method and the numerical method is used to calculate, and mutual authentication.

$$q_{line} = 0.042[a_{pressure}]^{0.5}[a]^{0.6}$$

To calculate the $q_{line} = 0.48$ kg/m.

According to the charge density for $\phi 100 \sim 125$ mm recommended by Swedish U. Langefores, the charge density take $q_{line} = 0.90 \sim 1.40$ kg/m.

By taking reference from some successful engineering examples and the effects of small-scale experiments, comparing the geological conditions, the performance of explosive, the drilling aperture, blasting scale and construction etc. and taking all into consideration, the line density takes 0.9 kg/m.

Therefore $Q_{pre} = 0.9 \times 16 = 14.4$ kg, because of pre-splitting explosive column 750 g/a. In practice, taking the integral 14.25 kg.

3.4 Charge Structure

The main blasting holes are continuous charging structure, as shown in Fig.2.

are two specifications: diameter $\phi 45$ mm, quality 750 g and diameter $\phi 60$ mm, quality 750 g. Apart from that, $\phi 45$ mm explosive shells are also available.

Non-coupling coefficient is 2.87, with four pre-splitting explosive columns $\phi 60$ mm at the bottom for consecutive charging, and then using 5 groups of $\phi 45$ mm columns for 2-real 1-virtual loading, upper part with 5 groups of 1-real 1-virtual loading. Fix double strands of detonating cord in pre-splitting explosive columns on both sides, orifice backfill 1.5m, as shown in Fig.4.

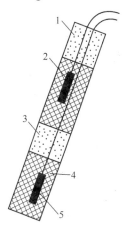

Fig.3 Slow punching charge structure diagram
1—upper orifice rock slag; 2—porous granular anfo explosives; 3—central lower interval of rock slag porous; 4—granular anfo explosives; 5—detonating tube detonator

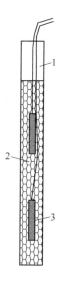

Fig.2 Main blast hole charge structure diagram
1—orifice slag; 2—porous granular anfo explosives; 3—detonating tube detonators

Buffer hole is piecewise charge structure. Installed at the bottom are the 45kg anfo explosives. The interval of the central rock slags is 2.0m. The top are 40kg anfo explosives, orifice backfill is 4.0m, as shown in Fig.3.

Non-coupling charging is used for pre-splitting hole. Pre-splitting explosive columns are customized by China Arm Shanxi Jiangyang Explosive Materials Co., Ltd. and there are turnbuckles on the explosive column so that the pre-splitting explosive columns can be screwed together. There

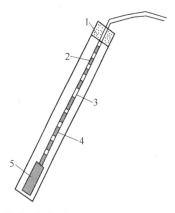

Fig.4 Pre-splitting hole charging structure schematic diagram
1—orifice slag; 2—reduced charge; 3—central normal charging; 4—detonating cord; 5—strengthen the charge at the bottom

Specific pre-splitting hole charging load are as shown in Tab.2.

Tab.2 Pre-splitting blast hole dosage parameter

Pitch depth/m	Blocked depth/m	Weaken top charging		Normal central charging		Strengthen charging at bottom		Charging full hole/kg	Full hole average line density /kg·m^{-3}	Varieties of explosives
		Length/m	Charging/kg	Length/m	Charging/kg	Length/m	Charging quantity/kg			
16	1.5	4.5	3.75	8	7.5	3	3.15	14.4	0.9	2 rock ammoniumnitrate

3.5 Detonating Network and Delay Time

Blasting equipments use common non-conductance tube detonators produced by Fenxi. The main blasting area and buffer hole apply in the hole use two segments, row use 5 segments, inner hole use two segments. Pre-splitting holes are connected by detonating cords.

Pre-splitting delay time goes 110 ms ahead of the first exposed hole of the main blasting area.

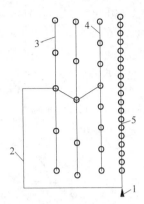

Fig.5 Connection diagram

1—initiation point; 2—5 periods of detonating tube detonator; 3—2 periods of detonating tube detonator; 4—5 periods of detonating tube detonator

3.6 Scope of Alert

The main risk factors of the blasting are flying objects by blasting and the blasting shock wave.

3.6.1 Blasting Flying Distance of Flying Objects

Sweden Tony Nick Research Foundation proposed the following empirical formula to estimate the deep the slungshot distance [4]:

$$R_{F\max} = K_\varphi D$$

The flying distance of slungshot is $R_{F\max}$ (m); K_φ is the safety factor, take 15~16; D is the hole diameter, 210 is calculated.

3.6.2 Safety Criterion of Blasting Shock Wave

The safety distance for the bunker blaster homework operators in the shelter to avoid the air shock wave[5] is determined by the following formula:

$$R_K = 25\sqrt[3]{Q}$$

where R_K —— the minimum allowed distance to avoid from the air shock wave ,m;

Q —— a blasting explosive quantity, kg ; Seconds delay blasting calculated by the max segmented charge load, millisecond delay calculated by the total dose of one blasting.

175 m is calculated.

Taking safe distance of blasting shock wave and blasting flying objects into consideration, and according to the requirement of the "Blasting Safety Regulations"(GB 6722—2003), the scope of the alert takes 300 m.

4 CONSTRUCTION WORK

4.1 Drilling

Slope measuring unreeling is an important guarantee to designed contour digging. The unreeling measurement should be done strictly before construction. Slope measurement should be twice, the first measurement mainly for the position of the rig operation platform, and the second slope position measurement occurs after the platform is constructed. Its measuring methods can be finished at one time by using total station and the pile points are set in 10m intervals.The connecting line of the pile points is the drilling contour line.

The drill platform is the site of drill shift and erection. In principle, the width of the drilling platform should be as wide as possible. It generally depends on the type of drilling machine, but the minimum width is no less than 1.5m. The platform should try to be the horizontal level, vertical flat.

Drilling must be erected by "counterpoint, direction, Angle is fine" to control the drilling accuracy. Drilling angle is controlled by angle ruler.

4.2 Charge

Since the pre-splitting explosive columns are customized by Jiangyang Explosive Materials Co., Ltd., it is fine to just screw the pre-splitting explosive columns together and then put them into the hole according to design requirements. The pre-splitting explosive columns are fixed in the middle of the holes by wooden supports.

4.3 Network Connection

When carrying the experiment, residues of detonating cord were found and the analysis of detonating cord shows that is caused by the nonstandard connections. Therefore, the detonating cord connection follows the requirements of detonating cord strictly to ensure that the blasting angle between branch line and main line is less than 90° and tape winding length >15 cm.

5 THE BLASTING QUALITY EVALUATION AND IMPROVEMENT

5.1 The First Pre-splitting Blasting

	Half-wall rate	Roughness of slope	Inclined slope rate
First time	60.5	58%	Inclined slope rate: 1.8° Vertical slope rate: 2°

Blasting effect: the pre-splitting blasting effect is of ordinary level and achieves the expected goals, but there also exist the following issues:

Pre-splitting hole is 117, half-wall holes are 71, half-wall hole at a rate of 60.5%. The half-wall hole are concentrated in the north blasting area, while there is no half-wall hole in the south area. The measurement of slope angle is 79°. The half-wall hole are rare to see at 2 ～ 3 m of the step roots.

5.1.1 Analysis

Not half-wall hole appears in the south blasting area: (1) borehole is of poor quality, (2) air-slake in south rock is serious, (3) south pitches and the density of the lines are unreasonable.

The reason for no half-wall hole at 2 ～ 3 m: the dose in the pre-splitting hole at the bottom is insufficient.

5.1.2 Improvement Measures

In north area, maintaining the original parameters without adjustment, while the south area is adjusted as follows:

(1) The line density of 0.9 kg/m is reduced to 0.7 kg/m.

(2) The charge of the bottom of the hole within 3 m is strengthened with 60 mm pre-splitting explosive columns.

5.2 Second Presplit Blasting after Improvement

	Half-wall rate	Slope roughness	Inclined slope rate
Second time	72.3%	65%	Inclined slope rate: 1.8° Vertical slope rate: 1.9°

From the blasting effect, it is obviously improved compared to the first time where problems appear, and half-wall holes are also appeared in the south area, without obvious bottom of foundation.

6 CONCLUSIONS

(1) From the practice of Yuanjiacun iron ore presplit blasting, it can be seen the perforation quality, especially pre-splitting hole quality, is critical. Looking at the form of half-wall hole, individual holes have deviation, and the bottom of foundation appeared in the corresponding placea. Its quality should be strengthened in future perforation.

(2) Carrying a small range of targeted experiments is of significance to choosing the suitable hole net parameter, charge, charge structure for the rock properties of the location.

(3) Analyzing the effect after blasting and correcting parameters of blasting scheme will help to constantly improve the pre-splitting blasting effect.

REFERENCES

[1] Wu Chengshuang, Sun Jikun, Qu jie. Research and practice of large pore size presplit blasting open-pit mine slope [J]. Blasting, 2011,1.

[2] Qin Pengyuan, Liu Jingguo, Cheng Suping. Application of large pore size presplit blasting in baiyun iron ore [J]. metal mine, 2007, 11.

[3] Wang Xuguang. The blasting design and construction [M]. Beijing: Metallurgical Industry Press, 2011: 252～255.

[4] Wang Xuguang, Yu yalun, Liu Dianzhong. Regulations of blasting safety implementation handbook [M]. 2004: 160～162.

[5] Gao Yushan, Zhuang Shiyong, Jiang Yufu, Liu Xinyu, Li Lanbin. Large aperture and large pore size presplit blasting in Nafen open-pit mine experiment and application [J]. China mining, 2007, 7.

Study of Energy Characteristics Gasifier Cartridges Onpreservation of Raw Materials Division Monoliths of Rock Mass

G.P. Paramonov

(*National Mineral Resources University "University of mines", Russia*)

ABSTRACT: This paper presents the results of experimental studies of the effect of the burning rate on the value of the pressure pulse in the charging chamber and breaking stress at Whence. The rates of combustion of the mixture composition density, dispersion components, dimensions of the cartridge housing. The interrelation of technological parameters of mining operations (the distance between the hole diameter of a hole depth of the cartridge) with the value of the failure stress of breaking through the block when triggered gasifiers.

KEYWORDS: pit; monolith; rocks; gas generator; sodium chlorate; hydrocarbons; the heat exchange gas-generating structure; the shot; burning speed; pressure

The main conditions for effective separation of the block of stone from the rock massif is the preservation of its monolithic blocks providing a flat surface with the least irregularities on the edges. In this case, it is necessary to ensure minimum disturbance developed an array of education in it induced fractures that can affect the quality of subsequent blocks separated.

There are quite a number of ways of separating blocks of rock mass, such as the use of explosives low blasting, non-explosive mixtures destructive - LDCs, gas generating compositions, etc. The choice of method depends on certain properties of the rock, fracturing and blocking array. Recently, the extraction of block stone quarries of building materials are increasingly used gas generating cartridges.

This is a continuation of studies of the whole complex of works dedicated to the development of gas generators (SG) based on sodium chlorate and hydrocarbons [1,2]. The principle of operation is based on producer gas cartridges during combustion of the gas generating mixture in deflagration (nedetonatsionnom) mode, with the release of a large volume of gases that create the necessary efforts to secede monolith along the lines of stress concentration (through holes). Components of the gas generators (oxidizer - sodium chlorate and fuel - hydrocarbons) by themselves are used for various industrial purposes and are not explosive. On the exposed surface, they do not work even when initiating explosive for their functioning is a necessary condition for a closed space. Structurally gasifiers may constitute two systems: the powder (powder mixture of oxidant and fuel) and layer (powder and oxidizer fuel honeycombs, tubes, strips, etc.).

In [3] presents the results of studies of the effect on the development of the separation fracture granite blocks from the array pressure gasifiers shpurovyh (BFS). BFS system constitute the powder. During the tests, analyzed the induced fracture near the charge plates GDSH samples obtained after calving. Analysis of the results showed that the maximum length of cracks - 14 hole radii. This leads to a reduction in the quality of finished products and the preservation of minerals.

To stabilize the properties and enhance security in the separation of minerals from the monolith of rock mass with the use of gas generators (different design systems) is necessary to investigate the influence of the combustion rate, properties and components of gas generating component of the dynamic characteristics of the HS.

Earlier, in [1,2] showed a good agreement of calculation results on equations autocatalytic reactions of the pressure in the bomb with the experimental data for these compounds (correlation coefficient of not less than 0.98). Assuming that the pressure in the bomb is determined by weight of the combustion gas charge (initial concentration - [A]), the concentration of "seed" - [B], we write the modified equation for the reaction in the form of

$$\frac{d\alpha}{dt} = k(1-\alpha)(\alpha+\beta) \qquad (1)$$

where α —— completeness of the combustion process (the ratio of the pressure by the complete combustion of the current);

k —— reaction rate constant;

Corresponding author E-mail: paramonov@spmi.ru

β —— autocatalyst share in the total mass ($\beta = [B]/[A]$).

The solution of this equation has the form

$$\alpha(t) = \frac{\beta e^{(\beta+1)kt} - 1}{1 + \beta e^{(\beta+1)kt}} \qquad (2)$$

Applying this equation to describe a process of combustion of polyethylene powder or diesel fuel shows that the correlation coefficient is not lower than 0.98. For the composition of sodium chlorate and polyethylene in the form of tubes (system layers), this equation requires taking into account the nonadiabatic process (burning period layered systems are more than an order of magnitude-based compositions of hydrocarbons in the form of powder or diesel). For layered systems (with sodium chlorate or not) use the equation that takes into account the heat losses (by Zel'dovich)

$$\alpha(t) = \frac{\beta e^{(\beta+1)k_1 t} - 1}{1 + \beta e^{(\beta+1)k_1 t}} \qquad (3)$$

$$k_1 = \frac{1}{\dfrac{a}{T_O} + bt} \qquad (4)$$

where T_O —— initial combustion temperature;

t —— time;

a, b —— coefficients determined experimentally.

Application of such a scheme allowing for losses enables description of the process with the correlation is not below 0.98. Using these equations when the parameters of the charge (diameter and height) and bombs (volume bomb and its present height) allowed to determine the process parameters such as the change in the linear velocity of the combustion bomb, determine the velocity of the outflow of combustion products from the burning surface and compare it with the speed of sound in combustion products. The magnitude of outflow rate characterizes the stability of the combustion process and can consider it as one of the criteria for deflagration to detonation transition: the lower the outflow rate, the smaller the possibility of moving into the explosion and vice versa. Tab.1 summarizes the process characteristics and parameters of the equation.

Test compounds were carried out in a constant volume bomb of approximately 300 cm^3 (volume varies depending on the number and size of the test rounds, in which the composition is loaded).

In the first two rows of the table shows the results of tests and calculations for the layered system (combustible material - polyethylene in the form of tubules) in the first row of data for a nominal reflected gasifier length 85 mm inner diameter polyethylene tube 23 mm rigidly fixed to the starting device PU-5 in a steel bowl $\phi 40 h 4$ mm and height $\sim$100 mm. Weight igniter composition in PU was 0.7 g.

Tab.1 Comparative parameters of compositions

system	cipher	a/β	σ/k_1	u/m·s^{-1}	g/kg·s^{-1}	W/m·s^{-1}	$\sum \tau$/s	τ/s	M_*/g	RT/kJ·kg^{-1}	note
	1	2	3	4	5	6	7	8	9	10	11
layered	40	5.85	0.087	2.79	1.74	129	0.17	0.016	0.7	599.5	modification
layered	38	3	0.039	0.67	12.79	0.29	0.29	0.08	1.5	599.5	modification
powder	303	10^{-6}	1900	16.62	28.02	442	0.01	0.0075	1.5	646.5	
powder	302	3.5×10^{-9}	2600	22.58	38.58	585	0.01	0.0075	1.5	646.5	
powder	304	10^{-10}	3100	27.05	46.03	721	0.01	0.0075	1.5	646.5	
powder	0.005	3×10^{-4}	3500	30.62	52.10	940	0.0045	0.003	3.0	686.5	body is split
powder	0.002	3×10^{-5}	7300	63.87	108.7	1960	0.002	0.0014	3.0	686.5	body is split
diesel fuel	21	1.6×10^{-9}	2500	18.0	30.55	555	0.01	0.008	3.0	537.6	
B.powder	36	—	—	7.88	8.76	319	0.014	0.007	3.0	387.6	

All other tests reported below were conducted in textolite cups with an inner diameter of 38 mm. In the second and third column for layered systems presents the empirical coefficients for calculating the equations (3), (4) (modified equations autocatalytic reaction, taking into account nonadiabatic processes layered systems - 12 column of the table). In 4, 5 and 6 columns represent estimates of the maximum values of linear (u), mass (g) the burning rate and magnitude of outflow rate of combustion products (w) from the burning surface. The total burning time bomb is presented in column number 7, number 8 in the column while maximum linear (mass) of the burning rate. Energy Values of the process parameters (RT) - specific combustion efficiency are shown in the column number 10. Next three rows (in the layer system) represent the results of calculations with three different pairs of the coefficients of equation (2) (without modification) by treatment of the experiment as seen by numerical values in columns No. 7,8,9,10 and 11. The presented data shows that reducing the share of "seed" (mathematical modeling) with a simultaneous increase in the share of the non-catalytic component of the reaction leads to the corresponding values of linear growth, mass-velocity combustion and outflow rate while keeping the overall duration of the process and the time to reach their maximum values. The next two lines reflect the results of the impact value of the mass of the igniter (3a against 1.5 g discussed above) and as a result, a sharp decrease in processing time with a corresponding rise, especially outflow velocities. Outflow rate of combustion products has reached a critical line at the top (940 m/s) and significantly higher in the subsequent line (1960 m / s). These results can be explained by the transition to a low-speed combustion detonation, which is confirmed by the destruction of the housing charge (column number 11). In the penultimate row shows the results for the composition based on sodium chlorate and hydrocarbon liquid (diesel fuel). Analysis of the results shows that the speed parameters are superior to those for the layered system, but inferior powder, especially when using the ignition module with a greater mass. The last line for comparison with experimental data gunpowder and calculations. As seen from the table, all dynamic parameters considered inferior gunpowder system based diesel fuel and sodium chlorate, based on the composition of the polyethylene powder and sodium chlorate. The analysis of the data from the point of view of the possibility of transition from combustion to explosion (main criterion - is the amount of outflow rate) should be noted that most of this is the furthest from the layers of the system. In contrast, a system based on polyethylene powder and sodium chlorate showed the possibility of transition into an explosion as the calculation and experiment. As for systems using as fuel diesel fuel, it is probably the increase in mass of fuel and power igniter is possible to create the conditions for transition of burning to explosion.

Presented the results showed that the velocity of outflow from the burning surface can serve as one of the criteria for the possibility of transition from combustion in the explosion.

On the other hand [4], the ratio between the rates of inflow of combustion products and their consumption is generally accepted criterion for assessing the stability of the combustion by KK AF Andreev and Belyaev. If coming gases (measured burning rate) is less than the rate of consumption (the expiration of the products of combustion gas dynamic critical conditions), the combustion process does not go into detonation and vice versa. We apply these criteria to the considered experiments and calculations. Based on the foregoing, we determine the rate of influx, knowing the linear burning rate, the charge density and the surface area of combustion. Believe that the process of burning mechanical, combustion products obey the equation of ideal gases. Then the mass burning rate is calculated by the formula:

$$m = u \times f \times \rho \tag{5}$$

where u —— inear burning rate, m/s;
f —— urning area, m^2;
ρ —— uel density, kg/m^3.

Considering the efflux critical in terms of gas dynamics flow rate of combustion products is defined as,

$$g = A \times f \times \rho(t) \tag{6}$$

where A —— gasdynamic coefficient taking into account the critical conditions for the expiration of the combustion products, $A = 7.723 \times 10^{-4}$ s/m.
$\rho(t) = p_{max} \times \alpha(t) + 101325$

where p_{max} —— maximum pressure in the bomb after combustion of the composition, Pa;
$\alpha(t)$ —— determined by equations (2) or (3).

It should be emphasized, first, consider the efflux of combustion products in a closed volume, and second, the combustion process of formulation based on sodium chlorate has pronounced autocatalytic character, and not a simple power-law dependence of the form:

$$U = A + B \times p^v$$

This means that a maximum burning speed is not reached when the maximum pressure in the bomb, and at the other, defined by equations (2) and (3), i.e. at the point of inflection of the parish. Fig.1 shows the dependence of the rate of income and expenditure for the combustion system layers depending on the pressure in the bomb tested in textolite cup inner diameter of 38 mm with a weight of 1.5 g of black powder used as an igniter.

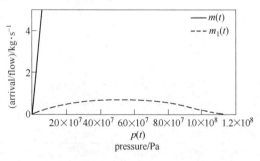

Fig.1 Depending on the mass burning rate of arrival of $m_1(t)$ and flow rate $m(t)$ from the combustion pressure for layered systems

The presented data suggest that the layered system burns steadily and it is not possible to go into the combustion explosion. Analysis of the powder system begin by considering the influence of the parameters of the equation by the amount of linear burning rate (Fig. 2).

A characteristic feature of the graphic dependency is their similarity δ-Dirac function: the same integrand area pronounced bell-shaped curves, the same time achieve maximum value (see Tab.1 column number 8. And curves). Experiment meets the first curve.

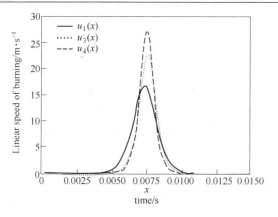

Fig.2 Change in the linear burning rate is shown depending on the parameters of the equation: $u_1(x)$ for a layered system with the code 303 in Tab.1, $u_3(x)$ c codes 302 and $u(x)$ c cipher 304 weight igniter (black powder) 1.5 g

Changing the parameters of the equation, and the effect of increasing non-catalytic reaction, thereby leading to the growth rate of the process - the growth of the linear burning rate and other parameters. For example, the supply of a powerful heat source can not lead to the predominance of an autocatalytic reaction at a higher speed. Let us compare the income and expenditure for the two extreme systems (shifr303 and 304 in Fig.3). Of Fig.3 shows that the flow curve is above the curve joining (code 303), ie system is stable combustion. Other relationship holds for the theoretical curve (code 304): first arrival curve is above the flow curve, and further intersects flow, which means the possibility of explosive development process. This confirms the above stated assumption of increasing the burning rate

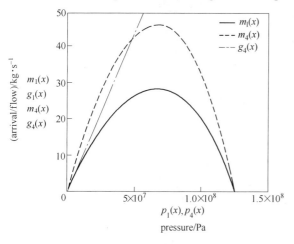

Fig.3 Depending on the mass burning rate of arrival of $m(x)$ and flow $g(x)$ from combustion pressure: $m_1(x)$ - code 303 and $m_4(x)$ - code 304 (Tab.1) $g_4(x)$ for the powder system (mass igniter (black powder) 1.5 g, charge diameter 38 mm)

for our case for increasing the usual reaction. As seen from Tab.1, the use of a more powerful ignition of gunpowder 3a 1.5 g instead can lead to a dramatic increase in the rate of outflow (higher than the calculated speed of sound), the destruction of the housing that features a low-speed detonation. Fig.4 shows a similar dependence for the case.

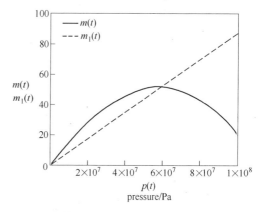

Fig.4 Dependence mass combustion rate (joining) $m(t)$ and flow $m_1(t)$ from the combustion pressure for the system based on sodium chlorate and polyethylene powder

In the illustrated chart (Fig.4) shows that the curve is much higher arrival rate curve than the case considered above (Fig.3), ie, with greater speed and weight of heat (see column 5 in Tab.1). Here passed burning to detonation. Obvious interest from the standpoint of the model combustion stability data for the system are based on sodium chlorate and diesel fuel as comfortable in equipment and relatively cheap.

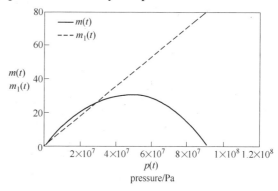

Fig.5 Dependence mass combustion rate (joining) $m(t)$ and flow $m_1(t)$ from the combustion pressure based composition chlorate sodium and diesel mass igniter 3a gunpowder charge diameter 38 mm

The analysis of the relations shows that both curves are in the initial part to the pressure in the bomb is about 3×10^7 Pa almost coinciding, and this despite the powerful igniter. But since the arrival dominates the flow, this can be attributed to the composition of the explosive systems.

Thus, by using both methods to determine the stability of the combustion process as compared to the speed of outflow from the burning surface with the velocity of sound in the combustion products, and by comparing the joining velocity (speed of combustion) with the magnitude of the flow can be more reliably estimate the stability of the combustion process and the possibility its transition to detonation (explosion).

To check the stability of the gas generator system layers were conducted industrial tests crushing boulders in the construction of an oil terminal separation granite blocks.

Fig.6 includes photographs of crushing boulders gasifiers for various periods of time, and Fig. separation unit before and after separation.

Fig.6 Application gasifiers for crushing boulders in the construction of the terminal in Primorsk

Fig.7 Application of gas generators for separation granite block at the quarry "Tervayarvi" unit volume of 72 m^3

On the basis of theoretical calculations and experimental-industrial tests of the parameters of location blast-hole inflator cartridges to separate granite blocks from a rock mass and specific mass flow gas generating composition, which are presented in Tab.2.

Tab.2 Calculated values of the charge mass gasifier for breaking granite depending on the distance between the hole of varying depth (The diameter of holes 32 mm)

		The distance between the hole/m									
		0.25	0.3	0.35	0.4	0.45	0.5	0.55	0.6	0.65	0.7
Hole depth /m	1	0.058	0.068	0.078	0.088	0.098	0.108	0.117	0.127	0.137	0.147
	1.25	0.072	0.085	0.097	0.11	0.122	0.134	0.147	0.159	0.172	0.184
	1.5	0.087	0.102	0.117	0.132	0.147	0.161	0.176	0.191	0.206	0.221
	1.75	0.101	0.119	0.136	0.154	0.171	0.188	0.206	0.223	0.24	0.257
	2	0.116	0.136	0.156	0.176	0.195	0.215	0.235	0.255	0.274	0.294

Continues Tab.2

		The distance between the hole/m									
Hole depth /m	2.25	0.13	0.153	0.175	0.198	0.22	0.242	0.264	0.287	0.309	0.331
	2.5	0.145	0.17	0.195	0.219	0.244	0.269	0.294	0.318	0.343	0.368
	2.75	0.159	0.187	0.214	0.241	0.269	0.296	0.323	0.35	0.377	0.405
	3	0.174	0.204	0.234	0.263	0.293	0.323	0.352	0.382	0.412	0.441
	3.25	0.188	0.221	0.253	0.285	0.318	0.35	0.382	0.414	0.446	0.478
	3.5	0.203	0.238	0.273	0.307	0.342	0.377	0.411	0.446	0.48	0.515

CONCLUSIONS

(1) Experimental results and calculations based on two criteria showed that combustion layered system is only in the deflagration mode. Transition from combustion to explosion is unlikely and practically eliminated.

(2) Powder system for both methods is a typical representative of the class of explosives and is explosive.

(3) Composition-based diesel fuel and sodium chlorate can also be attributed to the class of explosives, but much less powerful.

(4) A relationship of technological parameters of mining operations (the distance between the hole diameter and hole depth) along the block from breaking when triggered specific fuel gasifiers.

REFERENCES

[1] Paramonov G P, Kovalevsky V N, Kirsanov O N. Some features of autocatalytic reaction kinetics of combustion and ignition of sodium chlorate with polyethylene in volume bomb. Theory and practice of blasting. Sb.Blasting business. Issue number 102/59, Moscow: ZAO "IAC explosives in AGN", 2009: 40~46.

[2] Paramonov G P, Vinogradov Yu V, Kirsanov O N. Experience in the application of new non-explosive materials in the production of block stone and gentle destruction of natural objects. Modern problems of mining / St. Petersburg State Mining Institute (Technical University). St. Petersburg, 2006: 239~242. (Notes of the Mining Institute. T.168).

[3] Bychkov G V, Kokunin R V, Kazakov S V. Results of the study the destructive action of the separation of charges GDSH monoliths and blocks from the array fields grpnita. // Extraction, processing and utilization of natural stone: Sb.nauch.tr. - Magnitogorsk: HPE "Bauman im.G.I.Nosova", 2008: 67~76.

[4] Andreev K K, Belyaev A F. Theory of explosives. - M. Oborongiz, 1960, 596 s.

Application of Multi Luntai Millisecond Blasting Technique on Limestone Mine

OUYANG Hairong[1], SUN Guang[1], BAI Hongfeng[1], ZHANG Jinbang[2], XIAO Chun[2]

(1. Xinjiang Huituo Engineering Blasting Deveiopment Co., Ltd., Urumqi, Xinjiang, China;
2. Xinjiang Mountain and River Engineering Blasting Technique Deveiopment Co., Ltd., Urumqi, Xinjiang, China)

ABSTRACT: Multi row millisecond blasting technology can effectively reduce blasting cost, but also has a comparative advantage in terms of safety and efficiency. In this paper, based on the introduce of the design method and its good blasting results in limestone mine.

KEYWORDS: multi row millisecond blasting technology;deep hole blasting; blasting results

1 INTRODUCTION

Good blasting quality is always the ultimate goal of mining workers and the dynamite\rock and blasting technology are the major elements that affect the blasting quality. With the improvements of the blasting performance, multi-row millisecond blasting technology is becoming increasingly sophisticated. Though domestic big open pit mines have all been applying this technology in recent years, the blasting results are quite different because of different designing ideas. Under normal circumstances, the open pit mines have two standards to evaluate the blasting quality: one is the external form of blasting pile geometry which means the height should be suitable for loading and transport operations; the other is that the rock (ore) blasted off possesses uniform fragmentation, high mechanical efficiency, low energy consumption. In addition, complicated surrounding environment requires a higher guarantee of security of the blasting technology. Xinjiang Luntai limestone mine has used this technology for 4 years, resulting in good blasting effect and economic benefit.

2 MULTI ROW MILLISECOND BLASTING TECHNOLOGY ANALYSIS

Multi-row millisecond delay blasting is a blast caused an initial blast of more than 10 rows and a total blast of 20 to 40 rows, with the depth of the hole above 6m and the diameter of the hole between 76 and 150mm. In the inter row hole according to the set time delay in turn back row of detonation, detonation propagation time difference in 25 ～ 250ms, so, the blasting process in time according to a burst when the line forward, until the blasting process.

The multi-row millisecond delay blasting process, compared with smaller delay initiation, has the following characteristics.

2.1 Based on Multi Row Millisecond Delay Deep Hole Blasting is Squeezing Blasting

Multi row millisecond delay blasting and the general 3 ～ 5 rows of delay blasting are fundamentally different, millisecond delay blasting 3 ～ 5 rows can be brought into full play the role of airport reflected tension, throwing motion front can provide good for the rear free surface. But when blasting to the 8 row, free space limited rear force burst crushed fracture, upward movement, the formation of muckpile surface uplift, so is the basis of the technology of extrusion blasting.

2.2 Multi Row Millisecond Delay Blasting Reinforced Broken Rock

The explosion in the mine (rock) reflected tensile cracking body, in the blasting explosive gas can produce expansion pressure, forcing the expansion of cracks, so as to enhance the degree of fragmentation of ore (rock).

2.3 In Certain Conditions, the Lower Number, Bulk Rate is Low

The appearance of massive objects in multi row millisecond delay blasting in the position of the main part in the before and after, in the blasting process once detonating row number, the front and rear part ratio is low, then the probability of occurring is low.

Corresponding author E-mail: 969739142@qq.com

2.4 Infinite Partition in Large Scale, High Productivity

Multi row millisecond delay blasting, it once blasting hole number can reach hundreds of above, blasting volume can reach tens of thousands of square. Total charge from several tons to several dozen tons, so one blasting scale is There is nothing comparable to this general blasting technology, has obvious economic benefit.

2.5 Multi Row Millisecond Delay Blasting Vibration Generated by Relatively Small

Multi row millisecond delay blasting reduces the quantity of dynamite, vibration produced a smaller impact on the surrounding environment, especially for blasting in complicated environment, can be used as the preferred blasting technology.

3 APPLICATION OF MULTI MILLISECOND BLASTING TECHNIQUE ON LIMESTONE MINE

3.1 Mine Multi Row Millisecond Delay Deep Hole Blasting Scheme Design

Multi row millisecond delay deep hole blasting technology has matured, but in the concrete design and construction is due to various factors have differences, the blasting effect is not the same. In the mine blasting design and implementation, in addition to the blasting design program of the general, but also focus on the drilling, loading and packing, the delay time of detonator segments, choice of personalized design.

3.1.1 The Selection of Blasting Parameters

Luntai limestone mine is a typical semi hillside open-pit. The ore quality is good, protodyakonov coefficient of rock $f=6\sim 8$. Blasting in deep hole bench blasting method, step height is $H=15m$, vertical drilling, using triangular hole arrangement mode, aperture is $D=90mm$. The blasting parameters were: $a=4$ m hole spacing, row spacing of $b=3$ m, hole depth of $L=16.5$ m, the ultra deep $h_1=1.5$ m, chassis resistance line $W_1=3.5m$, stemming length is $3.5\sim 4m$, $q=0.35\sim 0.4 kg/m^3$ explosive.

The determination of blasting parameters, the need to choose and to pass tests according to different lithology and topography, make necessary adjustments. Each completed a large blasting cycle, to summarize, adjust, improve, improve, not limited to time and local.

3.1.2 Drill Hole

The borehole shall do a good job of work before the flatness, removal of surface debris, and used a bulldozer flattening, on a flat plane using pneumatic drill. According to the data of geological exploration of hole layout, relatively large changes in lithology and unstable rock surface, should adjust drilling parameters. Hole distribution principle, from the flat crest, ensure the rig safety, at the same time, in strict accordance with the design requirements, control the depth and the angle of the hole. Before loading to the pore wall and measure the depth of inspection, and make relevant records. In practice, the hole diameter not less than 90mm, depth should be controlled in 12m.

3.1.3 Charge and Packing

Charge to ensure each cartridge to design position, to prevent the card hole; when the water hole should choose to use waterproof explosives or take waterproof measures. In deep hole millisecond blasting delay in packing work requirements are relatively high, to ensure the packing quality and length, to ensure that the blasting effect.

3.2 Determination and Blasting Network Delay Time

Compared to the blasting vibration requirements high area, in the use of multi row millisecond delay blasting technology, in order to avoid the blasting vibration wave superposition, resulting in peak velocity increase. The former 10 detonator hop arranged in 1, 3, 5, 7, 9, ...Section by row ignition. This is mainly the vibration waveform amplitude attenuation when the need for more than 3 cycles (more than 20ms), if not for the sequence mode, easy to cause the vibration wave superposition effect, increase the vibration of surrounding regions, thereby causing accidents. But after the sequence processing, can make each row of the initiation time is more than 50ms, so you can stagger vibration peak, reduce the vibration intensity produced by the blasting.

Initiation network system using non electric detonation tube system, nonel tube detonator adopts four connecting, complex network, available electric detonator excitation.

4 THE BLASTING EFFECT AND THE MATTERS NEEDING ATTENTION

4.1 The Blasting Effect

The use of multi row millisecond delay blasting technology, won the unanimous approval of mine workers, and achieved good blasting effect and economic benefit, mainly for :

(1) The blasting block size, block rate low, second blast less.

(2) It can control the direction of muckpile movement , control and reduce the ore loss and dilution effectively.

(3) The muckpile profile rules, loose degree good, blasting off the block size uniform, easy to shovel, improve the economic benefits of the mine enterprises.

(4) The control of blasting vibration, noise, flying rock, mine safety has been strengthened.

4.2 Matters Needing Attention

(1) For sequence not only opened the time difference, but also conducive to the front after blasting were left shift, can create free face quite good for the rear, can effectively improve the effect of blasting.

In application, the 15 section after the relay section area front muckpile surface loose blasting pile, caused the uplift is relatively high. This was mainly due to the 12 hole to hole and delay of 1200ms, more than 15 segment 990ms is about 210ms, the time difference between the larger, more easily in the hard granite were broken and scattered.

In order to improve the blasting effect, should be in every 3~4 row arrangement between arranged a hop or using hole relay way increase the difference between the rows of blasting.

(2) To prevent the improper design, "crushed" phenomenon. Multi row millisecond delay blasting inside the extrusion pressure is relatively strong, if the technical design of the pre is not accurate or not appropriate is easy to produce "crushed" situation and influence blasting effect.

Application of that, when the depth of the hole between 8~12m, an initiation 10 row effect is better, but with the increase of the number of rows of initiation, when the row number is more than 15 rows of muckpile, produce upward uplift phenomenon. With the continuous increase in blasting engineering ranked number, blasting the generated compression is greater, the muckpile throwing distance ahead is more and more small, finally "crushed" phenomenon.

(3) Improve throwing displacement.

In order to change this adverse situation, improve blasting effect, not only can increase the time difference between the rows of initiation, but also can increase the hole density method. The multi row millisecond delay blasting can reach every 3 to 4 rows of improving a throwing displacement effect. Through the practice in strengthening the hole density in the interval, in the process of loose blasting blasting heap is good, and the surface of the block rate is relatively low, and achieved good blasting effect.

5 CONCLUSIONS

Method of multi row millisecond delay blasting technology for deep hole is relatively simple, effective, with a large amount of ore and rock blasting, fragmentation, bulk rate is low, reducing seismic hazards and other advantages, so it is widely used in engineering.

REFERENCES

[1] Chen Chen, Huang Wanqing. Superduper Multi row millisecond delay deep hole blasting technology in Xiangjiaba Hydropower station [J].Shang Qing, 2013(1):289~291.

[2] Liang Yuanhong. Multi row millisecond delay Subgrade deep hole blasting construction method [J]. Theoretical study on construction of city (electronic version) ,2013(9).

[3] Ji Jianhong, Yang Yiwei. Multi row millisecond delay deep hole blasting technology in Used in the mines a ditch blasting [J]. Science and technology innovation herald, 2010(16): 61.

[4] Yang Yiwei, Huang Bin. Squeeze blasting and millisecond blasting technology in chute descending [J]. Coal mine blasting,2010(1):33~35.

The Specification of Granulometric Composition of Natural Jointing in the Rock Massif by Their Average Size

B. R. Rakishev, Z. B. Rakisheva, A. M. Auezova

(*K.I.Satpayev Kazakh National Technical University, Almaty, Kazakhstan*)

ABSTRACT: The article according to the percentage of natural separately in the rock massif fitted to the experimental data. The average size of natural separately is defined and the inverse problem is solved by them. As in the technical documentation of mining enterprises there is only indicated distance between the natural cracks, the developed method is a reliable tool for determining the percentage of natural partings in the rock massif.

KEYWORDS: rock massif; a natural jointing; the distance between the cracks; an average size of a natural jointing; granulometric composition

There are three orders cracks in the rock massifs. The cracks of the first and the second order are determined resistance drilling in crushers. The third order cracks are the most significant effect on the efficiency of blasting rock massifs. The endogenous cracks refer to them arising from metamorphic rocks by reducing the amount of rocks. Also, they include tectonic cracks developing in the rocks under the influence of tectonic forces, artificial cracks are formed in the rocks during mining operations, and crack weathering. The third order cracks have a significant stretch, measured in centimeters, meters or even kilometers. Their size varies from opening 10-6 to 10-1 m. These cracks can be filled with other rocks or remain unfilled. The third order cracks are characterized that they divide the massif into the structural elements – jointing[1, 2].

So the massif consists of fractured natural jointing of different sizes and contents.

For the determination of the natural jointing in the rock massif, we carried out targeted research on quarries nonferrous metallurgy in Kazakhstan. Block sizing of massifs of Kounradsky, Akzhalsky and Sayaksky fields was measured on exposures immediately and photoplanimetric way[1]. The classification of rock massifs is offered on the basis of conducted researches by the block sizes is given in Tab.1[3]. The graphical representation of these features is shown on the Fig.1.

Tab.1 Classification of rock massifs by block-size

Classes of massifs by block sizes	Rock massifs block sizes (modularity) (the cracked degree)	The content of large jointing (%) in the massif/m							The average diameter of jointing/m
		<0.20	0.21~0.40	0.41~0.60	0.61~0.80	0.81~1.00	1.01~1.20	>1.21	
I	Small block size (high cracked)	82.0	10.3	7.0	0.5	0.2	—	—	0.15
II	Medium block size (heavy cracked)	48.0	27.0	10.5	6.0	4.2	3.3	1.0	0.31
III	Large block size (medium cracked)	29.5	20.2	14.0	11.8	10.6	8.7	5.2	0.50
IV	Very large block size (small cracked)	17.5	16.1	14.6	13.2	12.7	12.9	13.0	0.69
V	Highly large block size (almost monolithic)	—	3.0	8.0	13.0	18.0	26.0	32.0	1.00

We have shown[3] the content of natural jointing in different rock massifs block-sizes generally varies by the exponential law:

$$y = ae^{bx} \quad (1)$$

where a, b ——constants for each class of block sizes;
x ——natural jointing size.

Corresponding author E-mail: b.rakishev@mail.ru

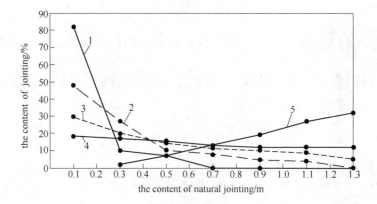

Fig.1 Specifications of the rock massifs by the block sizes

1—small block size (high cracked); 2—medium block size (heavy cracked); 3—large block size(medium cracked);
4—very large block size (small cracked); 5—highly large block size (monolithic)

These constants are set by experimental data. Values of the constants a, b, and the coefficients of determination (R^2)[4] for the massifs of rocks (Tab.1) are shown in Tab.2.

As can be seen, with the increasing resolution of the natural jointing these coefficients abruptly reduced. The only exception is the coefficient b for only large block size massifs.

Tab.2 Constants of massifs the block sizes and coefficients of the determination

Block sizes of rock massifs	Constants		Coefficients of the determination
	a	b	R^2
Small block size (d_e = 0.15 m)	154.6	−7.5	0.96
Medium block size (d_e = 0.31 m)	59.32	−3.0	0.96
Large block size (d_e = 0.50 m)	30.58	−1.3	0.96
Very large block size (d_e = 0.69 m)	17.05	−0.3	0.82
Highly large block (d_e = 1.00 m)	2.15	2.24	0.93

To determine particle size distribution of natural jointing in the rock massif in their average size is necessary to set the constants of equation (1) depending on the size of the natural jointing (0.1, 0.2, ..., 1.2 m). For this purpose there were built graphics of unknown dependencies, which are presented in Fig.2. The coefficients generally described by the equations[5]:

$$a = c\, e^{kx};\quad b = l\ln x + f \quad (2)$$

where c, k, l, f ——constants of the equation;

  x ——size of natural jointing.

For the considered conditions (d_e = 0.15, 0.31, 0.50, 0.69, 1.0 m), it is easy to see:

$$a = 264.9 e^{-4.52x},\quad b = 4.023\ln(x) + 1.694 \quad (3)$$

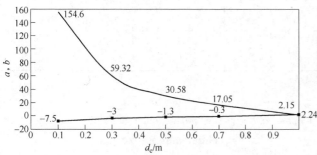

Fig.2 The dependence of the constants a and b of a medium-sized natural jointing

The percentage of natural jointing in massif, is calculated from formula (1) with the substitution of the obtained significance a and b of the expression (3) by blocking grade is shown in Tab.3. They are represented graphically in figures 3, 4 and 5 (the dotted line). The comparison of these data with the experimentally established natural granule composition jointing in massif (Tab.1) shows that they coincide quite closely.

Thus, the particle size distribution of natural rock massif jointing in the formula (1) can be calculated for any size fixed blocking. The significances of the coefficients a and b are determined by dependencies (3). For the following sizes for certain permanent natural a and b has the following meanings:

(1) for the small block size massifs with d_e=0.1m: a=167.1, b=−7.6; with d_e=0.2m: a=104.1, b=−5.0;

(2) for the medium block size massifs with d_e=0.3m: a=65.3, b=−3.2; with d_e=0.4m: a=42.2, b=−2.0;

(3) for the large block size massifs with d_e=0.5m: a=27.6, b=−1.1; with d_e=0.6m: a=23.4, b=−0.8;

(4) for the very large block size massifs with d_e=0.7m: a=17.5, b=−0.3; with d_e=0.8m: a=15.3, b=−0.1;

(5) for the highly large block massifs with d_e=0.9m: a=4.0, b=1.6; with d_e=1.0m: a=2.8, b=2.0.

Tab.3 Estimated particle size distribution of separately in natural rock massif

Classes of massifs by block sizes	The massifs and their block sizes (the average diameter of jointing) /m	The content of large jointing (%) in the massif/m						
		<0.20	0.21~0.40	0.41~0.60	0.61~0.80	0.81~1.00	1.01~1.20	>1.21
I	Small block size (d_e = 0.15 m)	76.77	17.83	4.14	0.96	0.22	0.05	0.02
II	Medium block size (d_e = 0.31 m)	45.63	25.04	13.74	7.54	4.14	2.27	1.68
III	Large block size (d_e = 0.50 m)	27.09	20.89	16.11	12.42	9.57	7.38	6.48
IV	Very large block size (d_e = 0.69 m)	16.93	15.95	15.02	14.14	13.32	12.55	12.17
V	Highly large block size (d_e = 1.00 m)	2.76	4.32	6.78	10.63	16.67	26.14	32.74

The percentage of natural jointing results by this technique, with dimensions d_e = 0.1, d_e = 0.2, ..., d_e = 1.0 m in the rock mass is given in Tab.4.

To visualize the data in Tab.4 graphically illustrated on figures 3, 4, 5. The Fig.3 shows the dependence of the percentage of natural building block jointing in massifs with average diameters: d_e = 0.1 m; d_e = 0.2 m (Fig.3(a)) and medium block size massifs with average diameters: d_e = 0.3 m; d_e = 0.4 m (Fig.3(b)).

According to the percentage of the natural specifications coarse-jointing are massifs with average diameters d_e = 0.5 m; d_e = 0.6 m (Fig.4(a)) and very large block size massifs with average diameters d_e = 0.7 m; d_e = 0.8 m (Fig.4(b)).

Fig.5 shows the percentage of natural jointing in large massifs only with an average diameter natural jointing d_e = 0.9 m; d_e = 1.0 m.

Tab.4 Content of natural jointing in rock massif and their block sizes

Classes of massifs by block sizes	The massifs and their block sizes. The average diameter of jointing/m	The content of large jointing (%) in the massif/m						
		<0.20	0.21~0.40	0.41~0.60	0.61~0.80	0.81~1.00	1.01~1.20	>1.21
I	Small block size (d_e=0.1 m; d_e=0.2 m)	78.1	17.1	3.7	0.8	0.2	0.0	0.0
		63.1	23.2	8.5	3.1	1.2	0.4	0.3
II	Medium block size (d_e=0.3 m; d_e=0.4 m)	47.4	25.0	13.2	7.0	3.7	1.9	1.4
		34.6	23.2	15.5	10.4	7.0	4.7	3.8
III	Large block size (d_e=0.5 m; d_e=0.6 m)	24.7	19.8	15.9	12.8	10.3	8.2	7.4
		21.6	18.4	15.7	13.4	11.4	9.7	9.0
IV	Very large block size (d_e=0.7 m; d_e=0.8 m)	17.0	16.0	15.1	14.2	13.4	12.6	12.2
		15.1	14.8	14.6	14.3	14.0	13.7	13.6
V	Highly large block (d_e=0.9 m; d_e=1.0 m)	4.7	6.5	8.9	12.3	16.9	23.3	27.3
		3.4	5.1	7.6	11.4	16.9	25.3	30.9

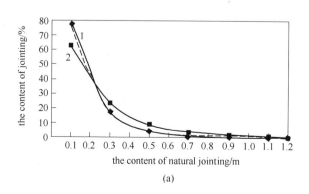
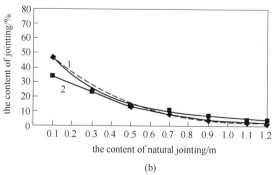

Fig.3 Percentage of natural jointing in small block size (d_e=0.1 m, d_e=0.2 m) (a) and medium block size (d_e=0.3 m; d_e=0.4 m) (b) massifs

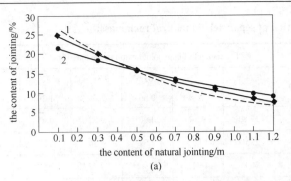
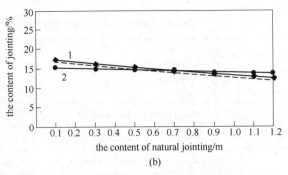

Fig.4 Percentage of natural jointing in large block size (d_e=0.5 m, d_e=0.6 m) (a) and very large block size (d_e=0.7 m, d_e=0.8 m) (b) massifs

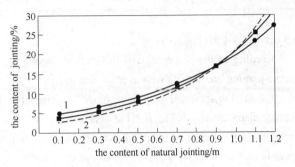

Fig.5 Percentage of natural jointing in highly large block size (d_e=0.9 m, d_e=1.0 m) massifs

The analysis of jointing graphs which is shown in figures 3~5 shows that the natural size distribution jointing in massifs of different blocking are rather different, and in massifs of the same name blocking varies in proportion to the natural basis. In general, established patterns are very useful for the disclosure of the natural properties of rock massifs, necessary for the design of mining systems.

CONCLUSIONS

(1) The conducted researches have shown that by the average extent of natural jointing it is possible to establish significance of coefficients a and b, and with the using the exponential law (1) to find percentage of natural jointing in the rock massif.

(2) Due to the fact that in geological reports and other technical documentation mining companies usually only indicated the average distance between the natural cracks of all orders (d_e), the developed method is a reliable tool for determining the particle size distribution of natural jointing in the rock massif.

(3) The percentage of natural jointing in the massif is, necessary to design parameters process mining the aim is to ensure their high performance.

REFERENCES

[1] Rakishev B R. The energy intensity of mechanical weathering of rocks. Almaty, 1998: 210.

[2] Victorov S D, Zakalinsky V M. Explosive destruction of mountains in Russia. Explosive affair. Edition #107/64. M.: 2012: 181~190.

[3] Rakishev B R, Rakisheva Z B, Auezova A M, Daurenbekova A N. The regression models of different block-sizes rock massifs. Herald KazNTU, #6 (100). Almaty, 2013: 104~110.

[4] Aivazyan S A, Eniukov I S, Meshalkin L D. Applied statistics. Study dependencies. – M.: 1985: 487.

[5] Rakishev B R, Auezova A M, Kalieva A P, Daurenbekova A N. The determination of particle size distribution of the rock massif at the average size of the natural jointings. The collection of works international scientific-practical conference "Innovative technologies and projects in the mining and metallurgical complex, their scientific and personnel support." Almaty, 2014: 186~190.

Examples of Blasting Using Emulsion Explosive in Mining in the 200 Meters Ultra Deep Water

ZHOU Ming'an[1], XIA Jun[1], OU Liming[2], REN Caiqing[1]

(1. College of Military Basic Education, National University of Defense Technology, Changsha, Hunan, China;
2. Hunan Lianshao Construction Engineering Group Co., Ltd., Loudi, Hunan, China)

ABSTRACT: Somewhere's manganese ore in Hunan Province is about 170 m below the surface. The roof is limestone and the thickness is 170 ~ 175 m. The groundwater connects with the underground river and the water level is about 3 m below the surface. The method of deep well pumping ore has been employed while the effect is poor for the manganese ore having the cementation phenomenon and the flowability being bad. The loosening blasting technology has been adoptted to enhance the effect by drilling blast holes around the deep well. The diameter of the blast holes is 130 mm and the depth is 220 ~ 250 m. The glass microsphere sensitized emulsion explosive has been applied and the detonation velocity is 4500 m/s. The pressureproof and waterproof PVC tube has been used as the shell of the emulsion explosive. The diameter of the PVC tube is 90 mm and the length is 40 cm. The TNT-RDX explosive and the general electric detonator are used in making the metal shell shaped charge initiation. The 50 ~70 m long charge has been delivered accurately to the orebody 200~250 m below the surface through self-made wire winch. After several experiments, the explosion of the emulsion explosive is reliable, the successful blasting is achieved.

KEYWORDS: deep well pumping ore; ultra deep water; emulsion explosive; blasting mining

1 INTRODUCTION

Somewhere's manganese ore in Hunan Province is in a relatively flat terraina. South and west to the ore there is the mountain as shown in Fig.1. The east side there are paddy fields and vegetable and the north side there are residents housing as shown in Fig.2. The distance between the operating point and the residents housing is about 60m. The ore is about 170 m below the surface. The roof is limestone and the thickness is 170 ~ 175 m. The core of the limestone is shown in Fig.3. The groundwater connects with the underground river and the water level is about 3 m below the surface. The method of deep well pumping ore has been employed while the effect is poor for the manganese ore having the cementation phenomenon and the flowability being bad. The mine had been proved long before. The hole mining and strip mining technology cannot be achieved. The method of deep well pumping ore has been employed while the effect is poor for the manganese ore having the cementation phenomenon and the flowability being bad. The manganese ore pumped out is shown in Fig. 4.

Fig. 1 The mountain

Fig. 2 The residents housing

Corresponding author E-mail: 1976009004@qq.com

Fig. 3 The core of the limestone

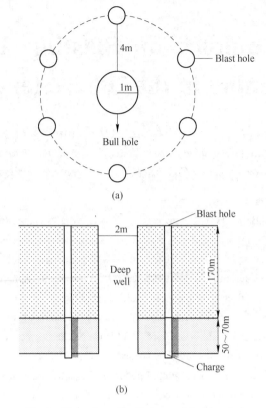

(a)

(b)

Fig.5 The schematic diagram of the blast hole

(a) top view; (b) side view

Fig. 4 The manganese ore pumped out

2 TECHNICAL SCHEME

A large-caliber vertical deep well is to be drilled to the bottom of the ore body to form the centric bull hole. The loosening blasting technology is to be adoptted to facilitate the ore extraction work by drilling blast holes around the deep well. The millisecond delay blasting technology is to be adopted to reduce the blasting vibration damage. The emulsion explosive is to be employed to reduce the mining cost.

3 BlASTING PARAMETERS

The diameterhe of the centric bull hole is 2 m. The 6 blast holes were drilled isometricly around the bull hole and the depth exceeds the bottom of the ore body as shown in Fig.5. The blasting parameters are shown in Tab.1.

Tab. 1 The blasting parameters

Property	Value	Unit
Hole diameter	130	mm
Hole deep	220～250	m
Hole spacing	4.2	m
Charge diameter	80	mm
Charging length	50～70	m
A single doset	251～351	kg
Unit explosive consumption	0.4	kg/m^3

4 SAFETY AND PROTECTION

The blasting point is in the deep underground and the water accumulation in the blast hole is rich.The damage from the air shock wave and the flying roch can be ignored. The max single detonation dosage is about 351 kg. The distance between the explosion source and the surrounding buildings needing protection is about 204 m. According to the international and domestic general formulas, $v = 1.28$ cm/s ($\alpha = 1.5$, $k = 200$). The corresponding vibration produced in the other sections is less than this value. All of the vibration velocity values are less than the national safety standard which is 2.0 cm/s and wouldn't cause harm to the surrounding dweller buildings.

5 OPERATING METHOD

The geological drilling rig was used to drill the well and holes as shown in Fig. 6. Affected by the thickness of the ore body and the roof, the holes' depth changes slightly basing on the design depth as shown in Fig.7. The glass microsphere sensitized emulsion explosive which the density is 1.0～1.3 g/cm^3 and the detonation velocity is 4500 m/s has been applied. The pressureproof and waterproof PVC tube has been used as the shell of the emulsion explosive. For the single section PVC pipe, the length is 40cm, the outside diameter is 90 mm and the inside diameter is 80 mm. The charge and shell are shown in Fig.8. The 50 ～ 70 m long charge composed of multiple single section charge has been delivered accurately to the orebody 200～250 m below the surface through self-made wire winch. The charge setting is shown in Fig.9. The TNT-RDX explosive and the general electric detonator are used in making the metal shell shaped charge initiation.

Fig. 6　The drilling machinery

Fig. 7　The blast hole marking

Fig. 8　The charge and shell

Fig. 9　The charge setting

6　THE BLASTING EFFECT

After several experiments, the explosion of the emulsion explosive is reliable, the problem of the explosion of the the emulsion explosive in the 200 m ultra deep water has been successfully solved. The surrounding buildings were unaffected. The successful blasting is achieved and the blasting effect is shown in Fig.10.

Fig. 10　The diagram of the blasting effect

7 EPILOGUE

The emulsion explosive having the most varieties and the fastest speed of development is one of the water resistant explosives of the industrial explosives in the current. The emulsion explosive will misfire in deep water owing to the effect of pressure. By changing the packages and adjusting the formula, a series of technical problems such as casing compressive strength and seal and charge form and initiation of the emulsion explosive have been solved. The emulsion explosive can explode reliably in the 200 m ultra deep wate and the scope of application has been expanded.

The engineering case study in this paper is preliminary and the hige drilling costs is to be solved. The further studies of the methods of continuous production of large quantities of emulsion explosive can be used in ultra deep water and the explosive loading easily in ultra deep hole.

REFERENCES

[1] ZHANG K Y, WANG X Y, ZHAN F M, XU M Q. Problems and countermeasures of blasting in deep water[J]. *Engineering blasting,* 2006, 12(4):57~59.

[2] QIN R P, GU W B, LI Y C, CHEN L. Comparative study on drilling blasting effect of deepwater rock with continuous and intermittent charge[J]. *Engineering blasting*, 2013, 12(4):57~59.

[3] WANG D S, GONG M, LI J S, ZHOU Q P, ZHANG G Z. Optimization technology of deep water borehole blasting[J]. *Metal mine*, 2008, 387(9):19~22.

The Town of Rock Blasting Engineering under Complex Environment

JIN Huishi, YUAN Maoyu, LIU Chongyao, TAO Yongsheng

(*Shenyang Xiaoying Blasting Engineering Co., Ltd., Shenyang, Liaoning, China*)

ABSTRACT: The East soup hot spring reservoir is located in Fengcheng city, East Tang town center, the surrounding environment is very complex, short construction period, a large blasting dosage. Through segment millisecond, of hole by hole initiation by row, the explosive charge controlled in allowable range, strictly control the blasting vibration and flying rocks in blasting is the difficulty of blasting engineering, based on the test data regression analysis, ratio with correct, accurate control of blasting vibration; through strict control of flying rocks, on the surrounding buildings and facilities did not cause any influence of blasting project successfully completed.

KEYWORDS: pre-splitting blasting; regression analysis; non coupling charging; blasting; blasting vibration

1 SURVEY

1.1 Project Overview

Earthwork blasting project is located in Fengcheng city, East Tang town centre, with the excavation of a length of 37 meters, width of 31 meters, depth of 8 meters of the reservoir, according to the geological drawings provided by the local quality, the structure is divided into three kinds of mixed soil weathering rock, weathered rock, in the excavation area, the upper 4 meters (for soil excavation has been completed), the lower layers of rock to blasting construction.

1.2 The Surrounding Environment

The project is located in the town center area, the town is a resort, personnel of many vehicles, blasting construction environment is complex, blasting region distance east east hot spring hotel is 20 meters, the air force sanitarium distance only 7 meters, West Mountain, north of the The Springs Hotel is 16 meters around the blasting area, as shown in Fig.1.

1.3 Project Difficulty

The project is located in the town center area, the blasting construction environment is complex, buildings around more, from the protection of only 7 meters, so the blasting vibration and fly control requires very high. And the time is short, such as pre used shallow hole blasting, the blasting vibration and flying rocks will can play high control, but because the project requirements, this plan will not be implemented. Such as pre application of deep hole bench blasting, a detonation quantity big, should strictly control the blasting vibration and flying rocks, strengthen communication and coordination with the surrounding units, residents, timely evacuation.

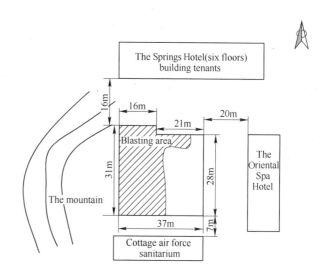

Fig.1 Blasting areas surrounding environment plan

2 THE BLASTING DESIGN

2.1 Scheme Selection

According to the previous experience of comprehensive blasting, the engineering characteristics and requirements, especially by the combination of deep hole bench blasting and presplitting blasting.

Corresponding author E-mail: syxyxzswb@163.com

2.2 The Blasting Parameters

Must the construction according to the "Regulations" the blasting safety regulations in construction, good quality, strict control of explosive consumption, reasonable packing, to ensure construction safety. Special preparation of the blasting parameters:

(1) The hole diameter: $D=90$ mm.
(2) The hole spacing: $a=2$ m.
(3) Hole row spacing: $b=1.5$ m.
(4) Light face: $a=1.5$ m cartridge diameter $d=32$ mm.

Tab.1 Hole blasting parameters and main gun

Hole depth /m	The lower charge /kg	The consumption of explosives /kg·m^{-3}	The middle packing /m	The upper charge /kg	The consumption of explosives /kg·m^{-3}	The upper packing /m	Total dosage /kg	The overall consumption of explosives /kg·m^{-3}
5.0	4.5	1.0	0.4	3.0	0.23	3.0	7.5	0.5
6.0	6.0	1.0	0.4	3.0	0.23	3.3	9.0	0.5
6.5	6.75	0.5	0.5	3.0	0.23	3.5	9.75	0.5
7.0	7.5	0.5	0.8	3.0	0.23	3.5	10.5	0.5
7.5	8.25	0.5	0.8	3.0	0.22	3.6	11.25	0.5
8.0	9.0	0.5	0.8	4.2	0.28	3.8	13.2	0.55

Tab.2 Pre splitting blasting parameters

Hole depth /m	The length of the charge /m	The lower charge /kg	The middle section of the air /m	The upper charge/kg	Filling length /m	Total dosage /kg	The overall consumption of explosives /kg·m^{-3}
5.0	2.3	1.9	—	—	2.7	1.9	0.25
5.5	2.6	2.1	—	—	2.9	2.1	0.25
6.0	2.8	2.3	—	—	3.2	2.3	0.25
6.5	2.9	1.5	0.3	0.9	3.3	2.4	0.25
7.0	3.2	1.6	0.5	1.0	3.3	2.6	0.25
7.5	3.4	1.7	0.8	1.1	3.3	2.8	0.25
8.0	3.6	1.8	1.0	1.2	3.4	3.0	0.25

2.3 Charge and Packing

According to the design of blasting charge and packing, the project main blasting zone take continuous charging and stratified charge combination, pre splitting hole to uncouple charge. The loess or rock powder packing, packing with bamboo pole gently tamped, without interruption, ensure the block length and the packing quality of bore hole. Charge structure is shown in Fig.2.

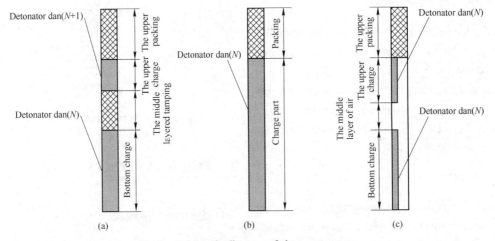

Fig.2 Schematic diagram of charge structure

(a) the main holes layered charge structure diagram; (b) the main hole continuous charging structure diagram; (c) schematic diagram of smooth blasting charge structure

2.4 Blasting Network

This project is a complex environment of blasting engineering, and an initiation dosage. Through segment millisecond, hole with 17 section, 18 section, 19 section, 3 section of micro hole external difference, inter row with 4, 5 millisecond blasting, hole by hole initiation by row, a hole or a hole how loud a sound. The explosive charge controlled in allowable range, the network connection diagram shown in Fig.3.

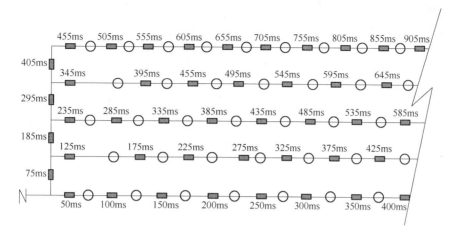

Fig.3　Schematic diagram of the network connection

3　REGRESSION ANALYSIS OF THE BLASTING VIBRATION

Because the protection object and the distance of blasting area center close, so the use of Sa Rodolfo J Ki formula for regression analysis, calculation formula of universal application of the engineering blasting formula and our country "blasting safety regulations" recommended:

$$v = K\left(\frac{\sqrt[3]{Q}}{R}\right)^{\alpha}$$

where　v ——ground peak particle vibration velocity, cm/s;
　　K and α ——for the terrain, geological conditions and the blasting point and the calculation among the related coefficient and attenuation coefficient;
　　Q ——explosive charges (Qi Baoshi as the total charge, delay blasting for maximum a charge), kg;
　　R ——observations (computing) to explosive distance, m.

According to the field test, data analysis, with the correct ratio, K value is 176, α value is 2.

4　THE STUDY OF PROTECTIVE MEASURES

4.1　The First Explosion Pre Splitting Hole, Play a Damping Effect

In the pre-splitting hole a hole Zuodi emptying two damping line.

4.2　Blasting Slungshot Calculation and Protection

Bench blasting flyrock distance according to the empirical formula as follows:

$$R_F = 70q^{0.58}$$

where　R_F ——fly dispersal distance, m;
　　q ——consumption, kg/m³.

The maximum 0.55kg/m³ consumption, the stone flying distance of 49 m. In order to effectively control the fly rocks scattered distance, according to the change of blasting conditions, rationally determine the unit consumption of explosive and blasting parameters, should be controlled strictly according to the following scheme, will fly control in the range of 5 m:

In order to prevent the damage caused by the blasting slungshot, flying rock protection measures we adopted by the company are mainly in the following 2 aspects:

(1) Direct overwrite protection. The specific measures are: first, covered with a rubber pad, and then soil (sand) bag cover, to ensure environmental safety.

(2) The blasting site clearance in place. The specific measures are: the blasting site of all movable shield all clear, to ensure the safe distance, and the surrounding units, residents timely communication, timely evacuation.

5　CONCLUSIONS

The blasting effect is good, to meet the design requirements. The blasting vibration was controlled in allowable range, no stone, no detrimental effects on the surrounding buildings. The successful experience is a complex environment of deep hole bench blasting. The blasting through the test data regression analysis, with the correct ratio, significant value on blasting a homogeneous explosion we charge, provided an important basis.

REFERENCES

[1] GB 6722—2003 Blasting safety regulations of the People's Republic of China State Administration of quality supervision, inspection and quarantine, 2004.

[2] Wang Xuguang. The blasting safety regulations implementation manual[M]. Beijing: China Communications Press, 2004.

[3] Wang Xuguang. The blasting design and construction [M]. Beijing: Metallurgical Industry Press, 2011.

[4] Wu Shengwen, Yang Xingguo. The Manwan hydropower station two engineering protection measures for excavation of rock anchor beam[J]. Blasting, 2011, 28(1): 53~57.

Explosives and Initiation Techniques

3

Sensitivity to Impact of Binary Mixes on a Basis Fluoropolymer F-2M

A.V. Dubovik, A.A. Matveev

(*Mendeleev Chemico-technological University, Moscow, Russia*)

ABSTRACT: The analysis of proceeding of explosion-like reactions in polymer F-2M and potassium perchlorate (PP) at the impact is made and critical parameters of its initiation are defined. Experimental indicators of explosion initiation in mixtures F-2M with aluminium and PP are received and thermodynamic characteristics of its explosion products are calculated. Depending on the nature of the second component polymer F-2M shows oxidising (in mixes with Al) properties or properties individual (in a mix with PP) substance.

KEYWORDS: fluoropolymers; explosion-like and explosion reactions; explosive mixes; sensitivity to impact

1 INTRODUCTION

The phenomenon of proceeding of explosion-like chemical reactions (ELCR) at impact on pressed (pressure of $0.3 \sim 0.5$ GPa) samples of some halocontaining polymeric materials (HCP) is investigated in works[1,2]. It is shown that distinctly observed explosion-like effect at impact was achieved the samples of polymer should possess a complex of physical and chemical and mechanical characteristics: strength on compression is not less 50 MPa, polymer is balanced on a hydrogen index (the relation of number of hydrogen atoms to number of haloid atoms in an elementary unit of polymer close to 1∶1 thanks to what in products of explosion reactions the maximum quantity gaseous halohydrides is formed), thickness of samples should be in rather narrow limits (the relation of a thickness h by the diameter D makes $0.01 \sim 0.1$).

Then at impact of moderate force (energy E is more 10 J, pressure P above 1 GPa) with use of the test device with the free radial flow of substance (No.2) in the sample of HCP occurs ELCR accompanied by sound effect, sometimes strong enough (on $5 \sim 10$ dB exceeds a sound of idle impact), occurrence of a caustic smoke, nigrescence of axial area of the sample and the colouring of its central zone quite often extending up to peripheral sites. According to State Standard RF 4545-88 on the tests of explosives (HE) on sensitivity to impact all listed effects fixed at impact are classified as explosion, and the tested substance admits explosive. And it has appeared, that critical parameters of impact initiation HCP are comparable to indicators of explosion initiation for typical regular HE - RDX, HMX and so forth.

However in our case HCP definition according to "standard" approach as HE is represented incorrect. The matter is that all HCP tested by us decay at heating with rather weak exothermal effect ($1 \sim 2$ MJ/kg)❶ which less or comparable with heat of explosion transformation ammoniac saltpeter (AN). Last is classified as insensitive and practically safe in handling the substance possessing extremely small detonation ability. At the same time well-known, that multiton weights of AN can detonate with strong destructive effect. As neither detonation ability, nor ability to self-propagation for ELCR in samples of HCP was not investigated by us, in[1,2] is specified, that observed at impact ELCR are local flashes of a thermal origin, and HCP are not considered as HE.

In the same place in [1,2] the mechanism of ELCR phenomenon at impact is considered. It consist that by the end of charge failure (during ≈10 mcs) in the central sites of HCP charge which yet have been not subjected to radial recession, collects enough considerable quantity of energy $\sim P^2/2k$ (k——rigidity of the striker), transferred by it from loading systems of drop-weight by means of a series of wave disturbances. Corresponding to the given density of energy lifting of temperature (≈500 K) initiates thermal decomposition in HCP with formation black carbonized products and gaseous halogen hydride. The last, expiring from area of compression and possessing high speed of movement (~ 0.1 km/s), make characteristic sound effect of explosion in surrounding atmosphere.

Because HCP possess though also small, but quite notable

Correspond author E-mail: a-dubovik@mail.ru

❶ To it precedes endothermal depolymerization with heat effect ≈ 100 kJ/mol.

power, and are capable to explosion-like thermo decomposition, the estimation of their contribution to sensitivity to impact of mixes of HCP with the typical oxidizers and fuels is of interest. For the permission of this question it is necessary to receive dependences of indicators of sensitivity, power intensity and the componental contents of thermo decomposition's products for the specified mixes on their structure.

2 EXPERIMENT

As a base component of tested mixes it is chosen fluoroplast F-2M - a copolymer vinyliden fluoride with tetrafluoroethylene [($-CH_2-CF_2-$)$_n$ ($-CF_2-CF_2-$)$_m$ at a parity components 95/5]. In a condition of delivery it represents a white powder with the size of particles ≈100 microns. According to DSC-analysis F-2M tests endothermal (ΔH=0.031 mJ/kg) phase transition in the range of temperatures 145~155 ℃, presumably connected with loss by it glassy state. The size estimation of enthalpy formations F-2M by data about enthalpy polymerization (–147 kJ/mol) and enthalpy formations of components (–344 and –659 kJ/mol accordingly) gives to value –360 kJ/mol. Before carrying out of experiments on impacts the charges from the powder under the pressure of 0.3~0.5 GPa were pressed.

On Fig.1 photos of face surfaces of rollers of the test device with samples F-2M after impact with energy E=49 J (weight of 10 kg) are presented. Initial thickness of charges h (mm) are shown in the top corners of frames. The turned black central sites of samples testifying to decomposition of polymer to carbonized products, and the turned yellow peripheral parts, characteristic for decomposition of polymer to polyene structures are well visible. Such picture remains for all charges with h=0.1~0.4 mm❶. By increase h the burning out zone is reduced, yellow coloration of residual layers disappears and at h > 0.5 mm appearance of charges after impact practically does not change. Also there are no characteristic for explosion the signals of the photodiode and sound effects.

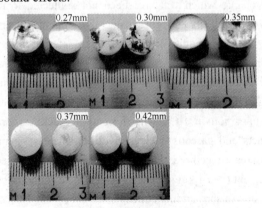

Fig.1 Rollers with a layer of substance after experiment carrying out

To definition of indicators of sensitivity to impact F-2M and mixes on its basis applied laboratory test methods - critical pressure (CP) and critical energies (CE)[3]. On Fig.2 the oscillogram of pressure record of impact (the bottom beam) with E=49 J on the charge F-2M with h=0.25 mm in the thickness is resulted. The top beams of an oscillograph start to fix signals of the photo diode and a tiny microphone at the moment of sharp recession of pressure (it is shown by an arrow). These records visually testify that flash and a sound signals of "explosion" arise during time of strength failure of charge. And the amplitude of the sound fluctuations which have been written down by a microphone, is more than amplitude of a sound of igle impact in 5 times (loudness excess on 10·lg5=7 dB).

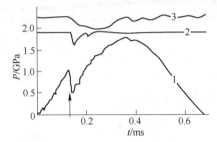

Fig.2 The oscillogram of impact pressure (1), luminescence of explosion-like reaction (2) and an audiosignal (3) in charge F-2M with h=0.25 mm at E=49 J

According to a data processing rule on method CP on Fig.3 a charge failure pressures for F-2M, accompanied by "explosion" (black points) and without its (light points), depending on a charge thicknesses h are presented. Under this schedule we define values of critical parameters of initiation F-2M: P_{cr}= 0.66 GPa, h_{cr}=0.37 mm. Dotted curve $P(h)$ is spent through all points of the schedule. On it we define size of charge strength on compression at impact σ=107 MPa.

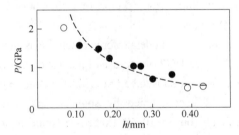

Fig.3 Dependence of failure pressures on a thickness of charges F-2M at impact with energy of 49 J

As a combustible component in a mix with F-2M aluminium in the form of powder PAP-2 (the size of plates 20×20×0.5microns) was used. Sensitivity of mixed (pressed) charges to impact was defined by method CE according to which for a mix of the set structure by a variation of a charge thicknesses and heights of dropping weight there

---

❶ At h <0.1 mm the area of superthin charges for which required failure energy of impact ≫49 J is begins[3].

was minimum for all h an energy of impact at which there is an explosion. It is called as critical energy E_{cr}, and a thickness of a charge corresponding to it – critical thickness h_{cr}. For set solid HE linear relationship between E_{cr} and P_{cr}^2 with proportionality factor $a=37.4$ J·cm^{-2}·GPa^{-2}[3] is found.

On Fig.4 dependence E_{cr} of a mix on the aluminium contents α is presented. In the field of $0.1<\alpha<0.5$ (stoichiometry at $\alpha_s=0.22$, equimolar point at $\alpha_e=0.29$) it has a flat minimum which specifies in presence of strong chemical interaction between F-2M and products of its thermo decomposition and Al thanks to which size E_{cr} becomes twice less than for F-2M. At $\alpha>0.5$ unlimited increase E_{cr} as Al is not explosive is observed. In given example F-2M has proved as the effective oxidizer capable to sensitize a fuel mixtures with its participation.

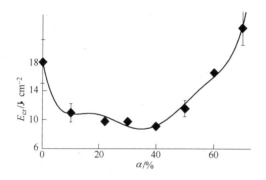

Fig.4 Critical energy of impact initiation for mix F-2M/Al depending on the aluminium content

Researches of sensitivity to impact for mixes F-2M with an oxidizer potassium perchlorate (PP) have yielded following results. First of all we will notice, that in the generally accepted opinion [4,5] pure PP does not explode at mechanical impacts. By us, however, it is established, that in the range of thickness from 0.15 to 0.5 mm the failure of PP charges at impacts with energy of 49 J (weight 10 kg) is accompanied by sound effect, comparable on the force with a sound of high-grade explosion. On the oscillogram of impact the powerful signal of the photo diode (Fig.5) though the explosive luminescence is not present is thus observed. Thus any traces of decomposition in scattering products of charge failure, and also a characteristic smell of explosion by us it is not revealed. On curve $P(h)$, plotted by analogy to a curve on Fig.3, we find for PP charges the value of strength on impact compression $\sigma=200$ MPa.

Obviously, that in this case we also deal with the phenomenon of explosion-like processes in charges PP similar to considered above ELCR in samples HCP, stimulated by impact. We will notice, that at $h>0.5$ mm on oscillograms of impact pressure on PP also are observed the pressure decays connected with failure of charges. However they proceed without characteristic sound effects, and photo di-

ode signals are not fixed any more. These data quite seriously allow to enter for PP charges the critical pressure of ELCR initiation $P_{cr}=1.0$ GPa and the corresponding size of a critical thickness of charges $h_{cr}=0.47$ mm.

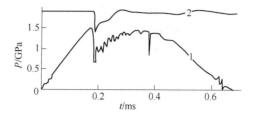

Fig.5 The oscillogram of impact pressure (1) and luminescences of power process (2) at impact with energy of 49 J on charge PP in the thickness of 0.25 mm

On Fig.6 the values of critical parameters of charge initiations for mixes F-2M with PP depending on size α — the relative contains PP, received by method CP are presented. From its consideration follows, that dependence $P_{cr}(\alpha)$ it is practically constant up to $\alpha=0.5$, has poorly expressed (at the level of an error of experiment) minimum at $\alpha=0.68$ (stoichiometric and simultaneously equimolar point of the mix) and further monotonously increases. On character of its growth it is visible, that for the given mix size $P_{cr}(\alpha \to 1) = P_{cr}(PP) = 1.0$ GPa. And it is obvious, that unlimited increase of dependence $P_{cr}(\alpha)$ in a small interval $0.95<\alpha<1$ which follows from representations that PP is substance nonexplosive and insensitive to impact, to expect does not follow.

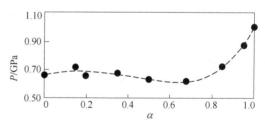

Fig.6 Dependence of critical initiation pressures for explosive reactions in mix F-2M/PP on PP contents

3 DISCUSSION OF RESULTS

As it was already specified in Introduction, check of polymer F-2M about display by its an oxidizing-reducing properties in mixed structures tested on sensitivity to impact was the purpose of the given work. The analysis of curves on Fig.4 we will spend together with discussion of the calculation results of parameters for thermo decomposition of investigated mixes received numerically under thermodynamic program Real, created for computer modelling complicated chemical equilibriums at high pressures and temperatures[6].

In calculations the values of maximum temperature for

explosion decomposition T_p (K) are received at different parities between F-2M and Al, and also the composition of products for their interaction is defined. We applied version 2.0 program at established parameters $P = 100$ MPa, $H = 0$ kJ/kg with use virial equation of state for decomposition products. In calculations preset value enthalpy formations of components: ΔH_f (F-2M) $=-7863$ kJ/kg, ΔH_f (PP) $= -3104$ kJ/kg, ΔH_f (PAP-2) $=0$. Calculation was spent with step on $\alpha=0.1$. As a result received: values of a thermal capacity (C_p, kJ/(mol·K)), composition of reaction products with their concentration (c_i, weights). The heat of thermo decomposition $Q_p=<C_p>(T_p-T_0)$ at $T_0=293$ K. On Fig.7 the results of calculations for mix F-2M with aluminium are resulted.

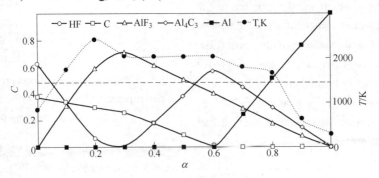

Fig.7 Calculated values of temperature and the basic products reactions between F-2M and Al depending on α

It is visible, that the temperature of reaction $T_p(\alpha)$ follows basically to the curve of formation aluminium threefluoride (TFA), characterised rather high enthalpy formation $\Delta H_f = -1510$ kJ/mol. This circumstance as a whole explains presence of a flat minimum on curve $E_{cr}(\alpha)$ in the range of size changes α from 0.1 to 0.5. Near to $\alpha=0.5$ the maximum of formation of aluminium carbide ($\Delta H_f = -209$ kJ/mol) takes place. Formation of these extremely enthalpy products supports high level of sensitivity of a considered mixes in the specified interval of changes α. With their decrease recession of temperature of explosive reaction in a mix and its sensitivity to impact sharply begins decreases to zero.

The drawn conclusion proves to be true also following results of calculations. Apparently from Fig.7, formation HF stops at $\alpha<0.3$ and on it, it might seem, explosions should end, as gas-formation is the basic form of explosion display (along with exothermicity and high speed of process). However the temperature in hot points of a mix at impact is so high, that exceeds temperature of sublimation AlF$_3$ (1550 K, it is shown by a dotted line on Fig.7). For this reason gaseous TFA, in exchange HF, becomes the manufacturer of a sound at explosion for $\alpha> 0.3$. The dotted straight line crosses temperature curve $T(\alpha)$ in points 0.08 and 0.82 i.e. approximately in points of the beginning and the end of an interval of a hypersensibility of a mix (Fig.4 see).

Using Fig.8 on which the temperature and structure of products of reactions between F-2M and PP are presented, at first sight it is inconvenient to explain a constancy of size P_{cr} characterising level of sensitivity of the considered mix at $0 <\alpha <0.8$. Here uniform recession of contains HF (with growth of α) is consistently replaced by formations and subsequent decreases CO, then CO$_2$ and at last O$_2$ so explosion gas formation is available at all α. From here follows, that at characteristic temperatures of explosive reactions the polymer F-2M possesses higher rate of decay than oxidizer PP. At $0 <\alpha <0.8$ polymer initiates explosion decomposition of a mix, and PP only actively supports it. However the situation changes in a range $\alpha> 0.8$ where the leading part in impact initiation of explosion in the mix begins plays decomposition PP. As a result sensitivity of a mix decreases to the level characterising mechanical sensitivity PP.

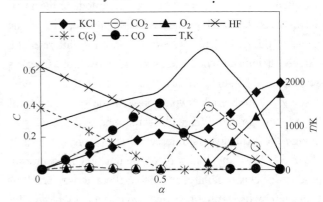

Fig.8 Temperature and product compositions of reaction between F-2M and PP depending on the oxidizer contents

In work [4] insensibility PP to mechanical influences accounted for endothermal character of low temperature decomposition PP. Later[7] it has been established, that at temperatures ≈500 ℃ PP undergoes two-steps decomposition: at first (slowly) decays to potassium chlorate and O$_2$ and then there is (fast) decomposition of chlorate to chloride and O_2. As a result it is possible to write down following equations of consecutive reactions $KClO_4 \rightarrow KClO_3 + 0.5O_2 - 39$ kJ and $KClO_3 \rightarrow KCl + 1.5O_2 + 45$ kJ. As a result possible exsothermal effect in both reactions should make 6 kJ/mol. However despite extremely low heat of thermodecomposition PP it is capable to show sensitivity to impact that follows from calculation of parameters of explosion initiation on the method[3]. Using the received value of explosion heat 43 kJ/kg, a thermal capacity of 812 J/(kg·K), $\sigma = 200$ MPa, temperature of fusion 610 ℃, pressure coefficient of fusion temperatures 0.26 K/MPa (as at ammonium perchlorate), thermo activation parameters for decomposition in solid phase - energy of activation $E = 412 \pm 44$ kJ/mol and pre-exponent $Z = 1.8 \times 10^{21}$ c^{-1}[8], we will find critical values $P_{cr} = 1.155 \pm 0.254$ GPa, $h_{cr} = 0.403 \pm 0.115$ mm and temperature in hot point $T_h = 1430 \pm 160$ K. The received values will not bad be co-ordinated with our experimental and settlement (on Real) data on critical parameters of explosion initiation PP by impact.

Let's address to Fig.5 and we will present the following sequence of the processes proceeding at mechanical initiation PP. Directly before charge destruction in the loading system of drop-weight machine collects considerable elastic energy (> 10 J) which owing to loss of bearing ability of a charge starts to be spent for disorder of formed fragments, and also on dissipative heating of substance which leads to its thermo decomposition. This decomposition is accompanied rather small own calorification and consequently comes to the end upon termination of action of a source of heating, that is by the end of charge failure. For the same reason smallness heat generation, and also small sizes of the initial centres of reaction, explosive process cannot extend for their limits and fades. Used in our work photo diode FD-10G having a pass-band of a signal 0.5~1.75 microns, apparently, reacts to the infra-red radiation arising in dying substance owing to course various recombination processes, and also at the specified above reactions proceeding at thermal decomposition PP

4 CONCLUSIONS

(1) Sensitivity to impact is investigated and critical parameters of mechanical initiation of "unexplosive" compounds F-2M are defined: $P_{cr} = 0.66$ GPa, $E_{cr} = 18$ J/cm^2, $h_{cr} = 0.37$ mm, $\sigma = 107$ MPa; the same for PP: $P_{cr} = 1.0$ MPa, $h_{cr} = 0.47$ mm, $\sigma = 200$ MPa;

(2) In mixes with fuel aluminium fluoropolymer F-2M shows properties of a strong oxidizer, entering with Al in chemical interaction in which result critical energy of explosion initiation in a range $0.1 < \alpha_{Al} < 0.5$ becomes twice less, than at initiation most F-2M;

(3) In mixes with oxidizer PP the polymer F-2M decomposes at impact with fair speed and in a wide range of changes $0 < \alpha_{PP} < 0.8$ plays a role of explosion initiator, supporting critical pressure of initiation on the level most F-2M.

The reported study was partially supported by RFBR, research project No 14-03-00333 a.

REFERENCES

[1] Dubovik A V, Matveev A A. Explosion-like Reactions in Poly(vinyl Chloride) on Impact // Doklady Physical Chemistry, 2012. Vol. 446, Part 2, 163~165 /M.: Pleiades Publishing, Ltd., 2012.

[2] Dubovik A V, Matveev A A. Explosion-like Reactions in Halovinyl Polymers at Impact // Russian J. Physical Chemistry B, 2014, v.8, N 2, 192~195 / M.: Pleiades Publishing, Ltd., 2014.

[3] Dubovik A V. The Sensitivity of Solid Explosive Systems to Impact / M.: Publishing house Mendeleev RCTU, 2011: 276 (in Russian).

[4] The power condensed systems. The short encyclopedic dictionary / Under the editorship of B.P. Zhukov / M.: JanusK, 2000: 596 (in Russian).

[5] Blinov I F. Chlorate and perchlorate explosives – M.: Oborongiz, 1941: 102 (in Russian).

[6] Belov G V. Termodynamical Modelling: Methods, Algorithms, Programs / M.: The Scientific world, 2002: 184 (in Russian).

[7] Schumacher J. Perchlorates. Properties, Manufactures and Uses. – NY: Reinhold Publ. Corp., 1960: 276.

Research on Non-primary Explosive Slapper Detonator System

SHEN Zhaowu[1], MA Honghao[1], CHEN Zhijun[2]

(1. Modern Mechanics Department, University of Science and Technology of China, Hefei, Anhui, China;
2. Anhui Lei Ming Hong Xing Chemical Co., Ltd., Hefei, Anhui, China)

ABSTRACT: After many year of hard work non-primary explosive slapper detonator has been manufactured inland. In this article many points of the slapper detonator have been systematically introduced, such as development, features, performance, advantage compared with primary explosive detonator. Basing on non-primary slapper detonator linear delay technique and green detonator idea have been brought in, which may be give some help for innovation in explosive materials.

KEYWORDS: slapper; non-primary explosive detonator; linear delay element; green detonator

1 INTRODUCTION

The non-primary explosive slapper detonator (NESD for short) is invented by Prof. Shen Zhaowu in USTC. With more than twenty years development this technique has been accepted and manufactured inland, which is to solve safety problem caused by primary explosive detonator. Referring traditional slapper detonator with electrical exploding foil in, the core part "exciting device" has been designed. This article describes the NESD's development process, technique feature, performance by test in detail. In addition the linear delay technique and green detonator idea have been introduced.

2 NESD TECHNIQUE

2.1 Theoretical Foundation

Traditional slapper detonator exploded by flying foil gave the inspiration, which was invented by J.R. Strond in the 1950s. The principles are as follows. Firstly when thousands of ampere current is passing through thin metallic conductor, the conductor would be gasified rapidly and produce plasma with high temperature and high pressure, which would drive the flyer. Secondly Walker and Wasley[1] proposed exploding criteria of heterogeneous explosives:

$$p^2 t = \text{Constant}$$

This criteria shows that whether or not the explosive is exploded is decided by two factors, pressure p and duration time t. Although this detonator is safe and pollution-free because of non-primary explosive, it needs very strong current with steep-sided pulse, whose initiation system is so huge to be used[2]. The traditional slapper detonator is not suitable for civil use, but it inspired researchers to make improvement on civil detonator.

Other studies[3] have shown that exploding product of PETN ($p^2 t$) varies with density, shown in Fig. 1. The slapper driven by explosive could be so fast to firing explosive by impact, implying that detonator could be exploded by fast flyer[4]. This guides the research[5~12].

Tab.1 Critical initiation condition of PETN

Density /kg·m^{-3}	Critical initiation pressure/Pa	Exploding product ($p^2 t$) /Pa^2·s	Critical exploding energy /J·m^{-2}	Critical exploding velocity of particle /m·s^{-1}
1.60×10^3	9.1×10^8	125×10^{10}	16.8×10^4	300
1.40×10^3	2.5×10^8	41×10^{10}	—	300
1.00×10^3	—	5×10^{10}	8.4×10^4	—

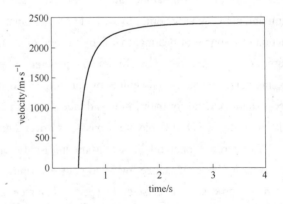

Fig.1 Velocity curve of flyer (explosive: RDX,1.6g/cm^3, ϕ20mm×40mm; flyer: steel, ϕ20mm×4mm)

2.2 Design Idea

The slapper detonator is exploded by impact of fast flyer on main explosive. According to exploding criteria of heterogeneous explosives we know that if the impact energy is larger

---

Foundation item: supported by the National Natural Science Foundation of China (51134012; 51174183).
Corresponding author E-mail: hhma@ustc.edu.cn

than the critical exploding energy, the detonator would be exploded. So it is obvious that flyer with high enough speed is the crux, which is realized by exciting device.

The exciting device is some small cylinder with exciting powder in. As the exciting powder is fired, gas with high temperature and pressure would cut the bottom part of cylinder, and the part would be accelerated, which is called the flyer. When the flyer impacts the explosive below the cylinder, hot spots have been made by compression and pressure, and explosive around the spots is ignited first. Obviously the exciting powder provides the energy, but there is no primary explosive because the main component is RDX. That's why NESD could avoid the problems of primary explosive.

At present kinds of NESD include instantaneous/delay electric detonator, instantaneous/delay nonel detonator. Structure of delay NESD is sketched in Fig. 2.

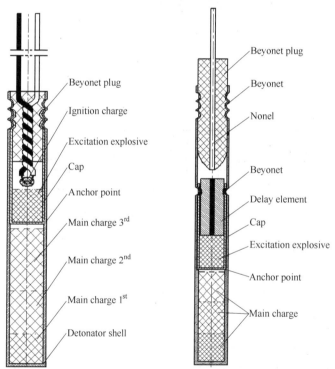

Fig.2 NESD: left one is electric kind and right one is nonel kind

2.3 Structure

There are three components of NESD: ignition part, exciting device, main explosive. Ignition part is used to transport ignition signal from outside to inside, generally ignition charge or nonel. Main explosive is the same as that in primary explosive detonator. These components in primary explosive detonator could be used in NESD directly. The difference is initiation part, one containing primary explosive, the other containing exciting powder without primary explosive.

Exciting device[9]:

The purpose of the exciting device is to produce flyer with high velocity. The components of the exciting device are cap and excitation explosive (in Fig.3). The cap is a thin shell cylinder with top open. The cap is fixed in the detonator shell by bayonet. The density of the excitation explosive in the cap is $0.5 \sim 2.5 \text{g/cm}^3$, and the height is $0.5 \sim 2$ times the external diameter of the cap.

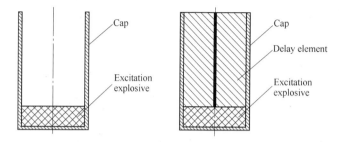

Fig.3 Sketch map of excitation setting: left one for instantaneous and right one for delay

2.3.1 Cap

The cap is made of Fe, Al or other materials. Caps in the experiments below are all made of Al. Its surface is smooth, without rip or pinhole porosity. The outside diameter is about 6.8mm and the shell is 0.5mm thick. Its height is often determined by the requirement of production, usually not less than 18mm. To make it easier to produce flyer, the cap's bottom can be finished with rift circle or thinning treatment.

2.3.2 Excitation Explosive

The excitation explosive can produce high pressure in the cap to separate its bottom and drive the flyer while it is burning rapidly. The excitation explosive is made from pure substance or mixture of RDX, HMX, TNT, PETN, etc. Comparatively the granulated RDX is better to be excitation explosive.

2.3.3 Excitation Explosive Design[13]

Excitation setting is the key component of NESD, and the excitation explosive is the most important element of the setting. So far RDX and PETN have been proved to be suitable for the excitation explosive. But the properties are different. The thermal stability of pure RDX is better than that of PETN[14,15]. The thermal explosion critical temperature of PETN in 1.74g/cm^3 is 197℃, which of RDX in 1.72g/cm^3 is 214℃. Cuneiform experiment[3] shows that in the same situation PETN is more sensitive to shock wave than RDX. Because the principle of firing excitation explosive in electric detonator is different from that in non-electric detonator, PETN and RDX used as excitation explosive act dissimilarly. It is proved that granulated RDX is more stable and reliable than PETN[16]. Except special explanation, granulated RDX is used as excitation explosive in this article.

2.4 Performance Study

2.4.1 Speed of Flyer Measuring

2.4.1.1 Experiment Design

The method "connection or not" is usually used to measure detonation wave speed. Signal panel which is made of Al foil is the element which is disconnect in circuit normally. Several panels in this experiment have been set in certain distance away from each other. When the flying object with some speed penetrates the panels, each one would be connected, and the electric signal would be produced and recorded. For distance is known and connecting moments have been recorded in sequence, the average speed between adjacent panel could be calculated by formula: $v=s/t$. The schematic diagram and device is shown in Fig. 4.

2.4.1.2 Data

The experiment data are shown in Tab. 2.

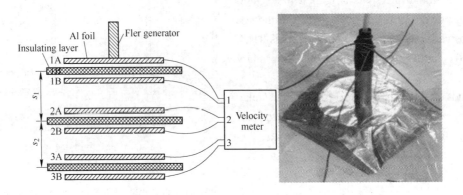

Fig.4　Experiment sketch map

Tab.2　Flyer speed data

No.	Cap/g	Excitation explosive/g	s_1/mm	s_2/mm	t_1/μs	t_2/μs	v_1/m·s^{-1}	v_2/m·s^{-1}
1	0.845	0.108	10.0	11.5	14.0	23.0	714	500
2	0.845	0.101	10.6	—	20.0	—	530	—
3	0.844	0.102	12.6	—	23.8	—	529	—
4	0.850	0.111	10.8	9.4	33.6	13.9	321	681
5	0.852	0.104	11.4	—	18.2	—	626	—
6	0.829	0.104	100.0	—	244	—	410	—
7	0.832	0.109	13.6	—	8.9	—	1528	—
8	0.843	0.112	11.2	—	13.7	—	817	—
9	0.729	0.126	40.0	—	33.7	—	1187	—
10	0.844	0.102	50.0	—	51.1	—	978	—

Remarks: there are no constraints in No. 5/6; diameter of constraints are a little larger, the detonator.

2.4.1.3 Discussion

Copper pipe has been used to constrain the detonator shell in order to reduce energy loss by rarefaction wave. The constraint strength in No.1~4 is obviously weak than No.7~10 because of larger copper pipe. General speaking the flyer moves faster with stronger constraint, so results in No.7~10 are larger than others. There is no constraint in No.5/6, of which the results agree well with No.1~4. Then results in No.1~6 could be called as actual speed of flyer. In contrast, results in No. 7~10 couldn't represent the real situation for the constraint, but could be named as extreme speed. Average speed in No.1 is 607m/s and in No.4 501m/s, so the actual speed of flyer is 534m/s, and the extreme speed is 1528m/s. Flyers and holes on Al foil penetrated by flyers in experiments are shown in Fig. 5, which show believability.

2.4.1.4 Summary

By experiments the flyer speed ranges from 500m/s to 1500m/s. Normally it is about 534m/s and up to 1528m/s.

2.4.2 Underwater Explosion Performance

2.4.2.1 Experiment Design

Explosion performance of NESD and primary explosive detonator are studied underwater separately. The samples contains instantaneous/delay NESD, instantaneous/delay primary explosive detonator, in which the primary explosive is NHN. The vessel is 5 meters in diameter, with water depth of 5 meters. Detonators and PCB sensors are set in position of 3 meters deep, shown in Fig. 6. Two PCB sensors (138A25) are set separately in distance of 0.9 meter and 0.7 meter from the detonator in the same horizontal plane, of which the sensitivity are 30mV/MPa.

2.4.2.2 Data

Data are shown in Tab. 3.

2.4.2.3 Analysis

Parameters could be calculated from P—t curves recorded by sensors, such as shock wave impulse, shock wave energy, bubble energy, total energy, shown in Tab. 3. The data could be grasped overall in Fig.7. Generally speaking NESD performs as well as primary explosive detonator.

If three factors A (delay or not), B (with primary or not) and C (measuring location) are classified, significance analysis could be done by variance analysis of pressure (in Tab.4 and Tab.5), shock wave impulse (in Tab.6 and Tab.7) and shock wave energy (in Tab.7 and Tab.8).The analysis tells us that there is no cross influence of the three factors on pressure and shock wave impulse, and the difference of shock wave energy caused by factor B is so small which can be ignored.

2.4.2.4 Summary

Though in underwater explosion tests data of NESD are smaller than primary explosive detonator slightly such as pressure, shock wave impulse, shock wave energy, bubble energy and total energy, the variance analysis show that the difference could be ignored. In other words NESD performs as well as primary explosive detonator.

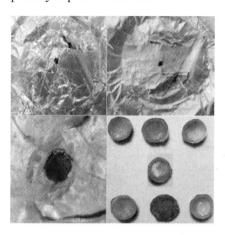

Fig.5 Flyers collected in experiments and penetrating hole on panel

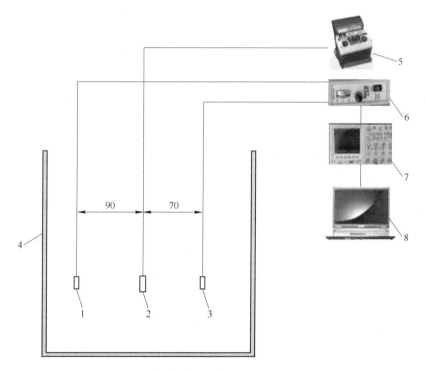

Fig.6 Experiment design

1—sensor; 2—detonator; 3—sensor; 4—vessel; 5—exploder; 6—charge amplifer; 7—oscilloscope; 8—computer

Tab.3 Data in underwater explosion experiments

No.		Channel	Pressure /MPa	Shockwave impulse /10^{-4}MPa·s	Shockwave energy /10^{-4}MJ	Bubble energy /10^{-3}MJ	Total energy /10^{-3}MJ	Pulsation period/ms
				Primary detonator				
Instantaneous ones		1	5.276	1.31	9.30	2.23	3.80	26.575
		2	5.300	1.40	9.45	2.23	3.82	26.568
		3	3.488	1.19	9.00	2.23	3.75	26.579
Delay ones		1	5.026	1.43	9.49	2.23	3.84	26.599
		3	4.038	1.24	10.60	2.23	4.02	26.586

Continues Tab.3

No.	Channel	Pressure /MPa	Shockwave impulse /10^{-4}MPa·s	Shockwave energy /10^{-4}MJ	Bubble energy /10^{-3}MJ	Total energy /10^{-3}MJ	Pulsation period/ms
NESD							
Delay ones	1	4.934	1.31	8.14	2.02	3.40	25.745
	2	4.918	1.42	8.32	2.02	3.42	25.740
	3	3.388	1.17	8.81	2.02	3.51	25.743
Delay ones	1	5.084	1.36	8.68	2.09	3.56	26.029
	3	3.369	1.18	8.78	2.09	3.57	26.031
Delay ones	1	5.000	1.35	8.44	2.12	3.55	26.164
	3	3.368	1.22	8.91	2.12	3.63	26.099
Instantaneous ones	1	4.665	1.43	8.70	2.11	3.58	26.100
	3	3.715	1.13	9.05	2.11	3.64	26.099
Instantaneous ones	1	4.545	1.42	8.64	2.11	3.57	26.112
	3	3.702	1.13	9.26	2.11	3.67	26.144
Instantaneous ones	1	4.913	1.47	9.61	2.12	3.74	26.159
	3	3.605	1.13	9.14	2.12	3.67	26.164

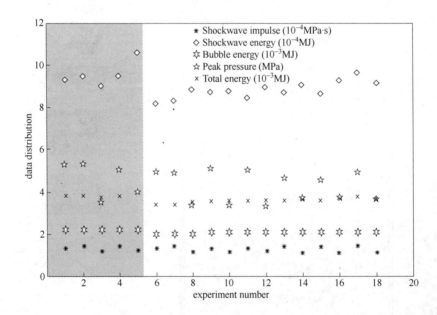

Fig.7 Scatter diagram of energy and pressure (dash area reprent priamry explosive ones)

Tab.4 Data of pressure peak according to factors A/B/C

	B1		B2	
	C1	C2	C1	C2
A1	5.272	3.488	4.665	3.715
			4.545	3.702
			4.913	3.605
A2	5.026	4.038	4.934	3.388
			5.084	3.369
			5.000	3.368

Tab.5 Variance analysis on pressure peak

Variance source	Sum of deviation from average	DOF	Mean square	F	Significance
A	0.0056	1	0.0056	0.0321	
B	0.2128	1	0.2128	1.2297	
C	7.2496	1	7.2496	41.8924	**
A×B	0.0170	3	0.0170	0.1185	
A×C	0.0615	3	0.0205	0.0979	
B×C	0.0023	3	0.00077	0.0045	
误差	0.5192	3	0.1731		
总计	8.0040	15			

Remark: $F_{0.05}(1,3)=10.13$, $F_{0.05}(3,3)=9.28$, $F_{0.01}(1,3)=34.11$, $F_{0.01}(3,3)=29.46$.

Tab.6 Data of shockwave impulse according to factors A/B/C ($\times 10^{-4}$ MPa·s)

	B1		B2	
	C1	C2	C1	C2
A1	1.40	1.19	1.43	1.13
			1.42	1.13
			1.47	1.13
A2	1.43	1.24	1.31	1.17
			1.36	1.18
			1.35	1.22

Tab.7 Variance analysis on shockwave impluse

Variance source	Sum of deviation from average	DOF	Mean square	F	Significance
A	1310^{-4}	1	1310^{-4}	0.0381	
B	0.0048	1	0.0048	1.8286	
C	0.1980	1	0.1980	75.438	**
A×B	0.0027	3	0.0027	1.0286	
A×C	0.0156	3	0.0052	1.9841	
B×C	0.00068	3	0.00023	0.0857	
误差	0.0079	3			
总计	0.2298	15			

Remark: $F_{0.05}(1,3)=10.13$, $F_{0.05}(3,3)=9.28$, $F_{0.01}(1,3)=34.11$, $F_{0.01}(3,3)=29.46$.

Tab.8 Data of shockwave energy according to factors A/B/C

	B1		B2	
	C1	C2	C1	C2
A1	9.45	9.00	8.70	9.05
			8.64	9.26
			9.61	9.14
A2	9.49	10.6	8.14	8.81
			8.68	8.78
			8.44	8.91

Tab.9 Variance analysis on shockwave energy

Variance source	Sum of deviation from average	DOF	Mean square	F	Significance
A	0.063	1	0.0625	0.172	
B	1.864	1	1.864	5.121	
C	0.360	1	0.360	0.989	
A×B	1.191	3	1.191	3.27	
A×C	0.331	3	0.110	0.30	
B×C	0.001	3	0.0004	0.001	
误差	1.092	3			
总计	4.902	15			

Remark: $F_{0.05}(1,3)=10.13$, $F_{0.05}(3,3)=9.28$, $F_{0.01}(1,3)=34.11$, $F_{0.01}(3,3)=29.46$.

2.4.3 Delay Performance

The delay time measuring results of various kinds of NESD are shown in Tab.10, which accord with requirement of GB 19417—2003.

Tab.10 Delay performance of NESD

Kinds	Sample	Min/ms	Max/ms	Average/ms	Range/ms	SD/ms
Electric instantaneous ones	48	6.9	10	8.1	3.1	0.63
Electric delay ones	90	237.58	457.01	345.46	219.43	45.04
Nonel instantaneous ones	60	3.9	5.3	4.66	1.4	0.29
Nonel delay ones	130	163.06	457.01	341.73	293.95	48.28

2.4.4 Test For Resisting Impact

The test device is one block of wood with 5 quin-cunx holes (Fig.8). The one in center is 1.3cm far from surrounding holes. NESDs are inserted in holes. After the center one is exploded, the others have been seriously distorted. But they haven't been detonated (Fig. 9). These 4 detonators could be exploded yet. This test shows that NESD can resist strong impact.

Fig.8 Device for impact resisting test

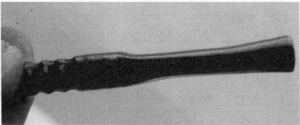

Fig.9 Distorted detonators

2.4.5 Test for Resisting Pressure

This test visually shows the safe performance of NESD. One NESD was placed on cement floor and a steel plate with 16mm thick lain on it. Some car with 2.3 ton weight slowly rolled over the steel plate and the NESD distorted seriously but didn't explode. Then increased NESD to nine and made it again. All the NESDs distorted seriously but none of them exploded. The distorted ones could be ignited normally. The process is shown in Fig. 10.

Fig.10 Test for resisting pressure

(a) one NESD under steel plate; (b) the detonator distorted seriously; (c) The distorted one exploded;

(d) same experiment with nine NESDs; (e) car used in the test; (f) nine distorted NESDs

2.4.6 Test for Resisting Sympathetic Detonation

This test is to simulate the situation that in factory detonator in mould explodes accidently.The center NESD would be ignited to observe what would happen to the 6 NESDs around. Millisecond delay NESDs and half-second delay NESDs are used separately.In fact the 6 ones were distorted seriously but not exploded, shown in Fig. 11 and Fig. 12 seperately.

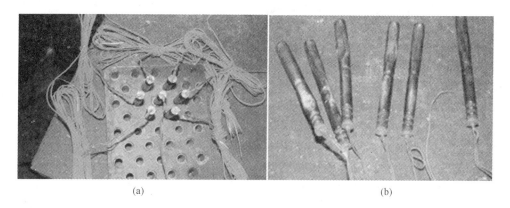

Fig.11 Test used millisecond NESDs

(a) before explosion; (b) after center one exploded

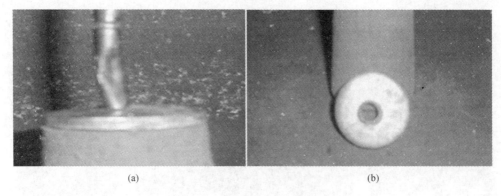

Fig.12 Test for half-second NESD

(a) distorted NESD; (b) distorted NESD exploded on Pb plate

3 DELAY TECHNIQUE

3.1 Traditional Delay Technique

At present the proven traditional delay element is drawn from lead tube which contains delay composition. First the tube with delay composition in is drawn several times until it can be put in detonator shell. Usually the internal diameter of the shell is about 7mm and the external diameter of the lead tube is 5.5~6.5mm. Then segments with proper length are made by cutting the lead tube. The segment is called delay element, which is finally fixed with the detonator shell by bayonet. Fig.13 shows the typical traditional delay element mentioned above. The diameter of the traditional delay element couldn't be too small. If not, it can't be bayoneted with the detonator firmly. So it's impossible to reduce the element's diameter.

Ordinarily, delay composition is designed as 0~60 time periods; delay time is distinguished to millisecond delay (0~3000ms) and half-second delay (0~10s). In order to meet all these delay requirements, time periods are classified to several team, and for each team one kind of delay composition formula is designed. Consequentially there would be so many kinds of delay composition formulas that it isn't convenient in production and it's difficult to ensure accuracy of each kind. And the long delay element must be needed for the long delay time case, which leads to long detonator shell and raised cost. In addition the slathered lead blocks the environment protection.

On the other hand, the delay composition has been researched. So much work on burning velocity of the composition has been done. Many kinds of composition have been developed, including ones for long time delay requirement. According to actual effect, the ignition ability and stability of the composition for long time is fallible. Sometimes fire cutoff and time period jumping happen.

In this article an innovation of linear delay technique is given, which can decrease the level of lead use, reduce cost and enhance the delay precision.

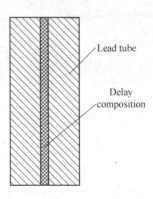

Fig.13 Sketch map of traditional delay element

3.2 Linear Delay Element

3.2.1 Design Idea

After times of drawing the traditional delay element to small diameter (usually 1.4mm), it becomes the delay line. Delay line is cut into proper length to be inserted into the vacuum of the detonation tube, and then we get the linear delay element (Fig. 14). The length of the linear delay element is determined by the delay time required.

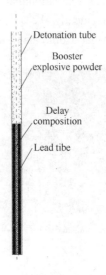

Fig.14 Linear delay element

3.2.2 Action Principle

The NESD including linear delay element is shown in Fig. 15. The production process is as follows. First the delay element is inserted into the vacuum of the bayonet plug, and the ends are on the same plane. The element and the plug are fixed with the detonator shell by bayonet. The open end of the plug touches the excitation explosive. When the detonation tube has been detonated, the delay composition would be fired by the shock wave arriving at the end of the delay element in the tube. After designed delay time, the composition at the other end of the linear delay element is fired and ignites the excitation explosive below[17]. Under the action of the high pressure in the cap produced by the excitation explosive, the detonator is finally exploded.

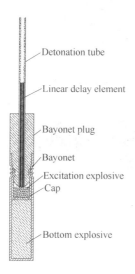

Fig.15 Sketch map of NESD

3.2.3 Advantages

The difference between traditional delay element and linear delay element isn't just the dimension but the delay idea. Obviously the traditional delay element is in the detonator, while the linear delay element is in the detonation tube which is outside the detonator. The advantages are as follows.

3.2.3.1 Lower Cost

Compared with traditional delay element, the diameter of delay line is so small, which means with equivalent lead the linear delay element can realize longer delay time than traditional delay element. For example, the diameter of delay line is usually 1.4mm (d_1); the diameter of traditional delay element is 6.2mm (d_2). According to the volume relationship, we can get:

$$\pi d_1^2 l_1 / 4 = \pi d_2^2 l_2 / 4 \tag{1}$$

That means l_1/l_2=19.612.

As the delay time is controlled by the length of the delay element. To realize the required delay time with same delay composition, the material amount of the linear delay element is approximately 1/20 of the amount of the traditional delay element. Of course the cost of material is just 1/20 than before. It's obvious that the linear delay element technique can lower cost significantly.

3.2.3.2 Precise

Experiments have proofed the linear delay element is more precise than the traditional delay element especially for long time delay required.

3.2.3.3 Independence

For there is no primary or sensitive explosive, the production, preservation and transportation are safe. So delay line could be produced separately, which is efficient with less cost.

3.2.3.4 More Safe, Less Process

Because the delay line is in the nonel, its length isn't limited by detonator shell length. It is possible to realize all the delay time requirements with only several kinds of delay compositions. It can increase security and reduce process during the detonator production.

4 GREEN DETONATOR TECHNIQUE

4.1 Background

Metallic lead is widely used in detonator industry, as pipe material in delay element, additive in delay powder, testing plate for brisance. For lead would cause heavy metal pollution to environment and human health, especially practitioner in detonator factory. That is why we bring out the green detonator idea, which is proposed to remove lead out from detonator industry.

The green detonator idea contains several techniques such as NESD, delay element, test method for brisance. It would be carried out as following: make NESD replace primary explosive detonator, linear delay element substitutes for lead delay element, use steel plate instead of lead plate in brisance test.

4.2 Green Detonator Technique

Linear delay element is the key shown in Fig. 15. When lead is used as pipe containing delay powder, lincar delay element structure could reduce the lead usage observably. If aluminium takes place of lead, of course there is no lead in the element.

4.3 New Testing Method for Brisance

4.3.1 Present Method

In detonator industry some convenient method testing brisance is used by measuring exploding hole on lead plate. If the diameter of the hold is larger than outside diameter of the detonator, it could be judged to be qualified.

Though this method is easy to carry on, every year abundant lead is consumed. Take standard lead plate for example, which is 40mm outside diameter, 4.0mm thick, weighs about 57g, values approximately 1.425yuan. If the detonator is qualified, there is about 3.56g lead loss by exploding on the plate. Totally three billion detonators are produced each year and suppose each in ten thousand would be sampled, there would be at least 1068kg lead loss and walk into natural circulation. The damage would be serious and irreversible.

4.3.2 New Method

According to present method of lead, Fe could be used. For Fe is non-pollution and widely distributed. The experiment proofs the feasibility, in which the steel plate with 1.5mm thick is used, and diameter of the hole is 8mm, larger than outside diameter of the detonator. The judgment rule of lead method could be also used in steel plate method.

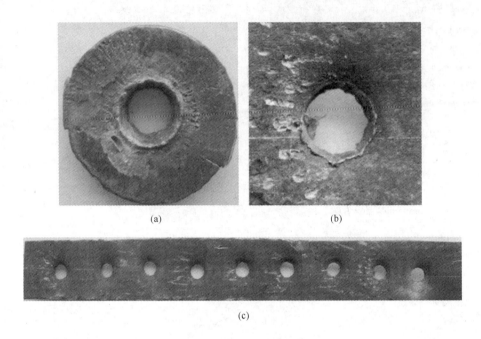

Fig.16 Brisance tests

(a) brisance test by lead; (b) brisance test by steel; (c) brisance tests by steel

5 CONCLUSIONS

Several techniques of detonator are introduced in this article, such as NESD, linear delay element, green detonator idea. NESD is the key technique, of which the most important feature is high safe. By this article the author hopes there would be more new techniques and proactive thoughts to improve the development of explosive materials.

REFERENCES

[1] Walker F E, Waslev R J H. Critical Energy for the shock Initiation of Heterogeneous Explosives [J]. Explosivestoffe, 1969, 17(1):9~10.

[2] YUAN Shiwei, ZENG Qingxuan, FENG Changgen. Design of a new kind of slapper detonator[J]. Initiators & Pyrotechnics, 2002(2):5~9.

[3] Zhang Guanren, Chen Danian. Detonation dynamics of condensed explosive[M]. Beijing: National Defense Industry Press, 1990.

[4] Jing Fuqian, Chen Junxiang. Principle and technique of dynamic high pressure[M]. Beijing: National Defence Industry Press, 2006:139~150.

[5] Shen Zhaowu etc. Hollow non-primary explosive detonator[P]. CN87101726.

[6] Shen Zhaowu etc.Simple slapper non-primary explosive detonator[P]. CN87106394.8.

[7] Shen Zhaowu etc. Slapper non-primary explosive detonator [P].CN87218986.X.

[8] Shen Zhaowu etc.Flyer non-primary explosive detonator[P]. CN93235810.1.

[9] Shen Zhaowu etc. Detonator with exciting device [P]. CN101029814.

[10] Shen Zhaowu, Zhou Tingqing, Ma Honghao. Exciting device and the detonator[P]. ZL200710019659.1.

[11] Shen Zhaowu, Ma Honghao. Detonator with multistage exciting device [P]. ZL201010151709.3.

[12] Ma Honghao, Du Jianguo, Shen Zhaowu. Non-primary explosive detonator ignited by laser[P]. CN201110321587.2.

[13] Ou Yuxiang. Explosive[M]. Beijing: Beijing Institute of Press, 2006.

[14] Zheng Mengju. The capability and test technique of

ex-plosive[M]. Beijing: Weapon Industry Press, 1990.
[15] Chen Rongyi. The detonation performance of granulated PETN and RDX. Changsha Mining and Metallurgical Research Institute, 2002.
[16] Ma Honghao, Shen Zhaowu, Sun Yuxin. Solution to drill jam with multiple underwater blasting in excavation of large diameter deep tunnel[J]. Engineering Blasting, 2008. (1):38~40.
[17] Li Ji. Research on improving the delay time precision[D]. Nanjing: Nanjing University of Science and Technology, 2006.

Methodological Aspects of Properties and Blast Energy Kinetics Control of Industrial of Explosives

N.N. Efremovtsev [1], S.I. Kvitko [2]

(*1. Institute of Comprehensive Exploitation of Mineral Resources (IPKON), Russian Academy of Sciences, Russia;*
2. Innovation Center Summit-XXI, Russia)

ABSTRACT: The paper addresses some aspects aimed at the perfection of test methods and procedures for commercial explosives and respective detonation systems. The authors present the results of the development and tests of new compositions of commercial explosives.

Methodical aspects control rheological properties of the emulsions used in the manufacture of explosives. Considered depending on the viscosity of emulsions from the main influencing factors. Control of the properties of industrial explosives.

Discusses and analyzes the factors that contribute to the relevance of the development of the methodology of optimization of explosive characteristics промышленные взрывчатые вещества and detonation systems on their basis, of the kinetics of an energy in the destruction of rocks by explosion. Presents diagrams, characterizing the dependence of the emission of harmful gases from the speed of detonation charges for the various ways to control the explosive characteristics of centuries.

Present models of multiple nano-micro disperse systems granulits and emulsion of industrial explosives and detonation systems on their basis. Is a classification of methods of management of the kinetics of allocation and transfer of power to an array with a view to the regulation of crushing of rocks.

KEYWORDS: detonation velocity; energy release kinetics control; rock breaking; rheological properties

Methods of rock blasting control, and analysis of operational and economic results of the variation of fragmentation intensity are analyzed in papers and monographs by K.N. Trubetskoy, V.V. Adushkin, S.D. Viktorov, L.I. Baron, G.P. Demidyuk. N.N. Kazakov, V.A. Belin, V.M. Zakalinsky, E.I. Efremov, N.S. Efremovtsev and some other researchers [1~7]. For the improvement of the quality of rock fragmented by blast and minimization of the yield of substandard size materials (fine-size materials in the construction sector) the trend today is toward smaller diameter boreholes and transition to medium diameter 80~120 mm boreholes (close to critical). Application of conventional composite granulated explosives in such non-ideal conditions of detonation results in significant reduction of their actual force, higher yield of oversize material, lower performance of excavation and haulage machinery.

It is a well-known fact that with specific tasks of blasting the proper choice of explosives and charge design options is very important for the efficient fragmentation of rock with certain physical and mechanical characteristics. The efficient mining of valuable crystalline materials and development of mineral deposits containing various ores and construction materials requires the application of explosive charges differing in energy release kinetics. It is a multiple-path task, as detonation velocity and force control of limited-size explosive charges can be solved by various methods [7,8].

The integrated effect of rock properties on their blastability (fragmentation by blast) is described in detail by N.S. Efremovtsev in Kachkanar deposit case studies. Rocks are classified as five categories by rock mass blastability with due account for natural fracturing and weathering of the rock mass, and as classes by structural features (grain size) and mineralogical composition /4/.

The field test analysis performed by the authors proves the feasibility of substantiation of such factors as dynamics, conditions, kinetics of energy release and transfer to the rock mass, and their practical application in the research activities. It is important to numerically evaluate in detail the effect of these factors on the efficiency of blasting of rocks with different physical and mechanical characteristics, as well as on the extent of blast harmfulness for the surrounding man-made entities, human beings and natural environment /9/. Moreover, it is necessary to take into account the effect of energy release kinetics and the parameters of a controlled detonation system on the efficiency and extent of comprehensive exploitation of mineral resources, as well as performance and energy intensity of the subsequent mining processes.

Corresponding author E-mail: noee7@mail.ru

For optimization of detonation systems it is important to perform a system multilevel analysis and control of the processes of blasting energy release:

(1) in a layer, where reactions go, and in the area of detonation high-temperature product emission;

(2) along a charge length in boreholes or blastholes;

(3) in a pattern of boreholes or blastholes within the boundaries of a block to be broken with due account for the optimal level of mineral losses and dilution.

An important part of the research is the formation of models of designed and tested commercial explosives as multiple micro- and nanodisperse energy-saturated systems and investigation of the effect of their parameters on the kinetics of blast energy release and transfer to the media to be broken and having different physical-mechanical characteristics.

Operational, economic, seismic and environmental factors of optimization of detonation system energy release kinetics control for efficient fragmentation of the rock mass and safety of blasting processes are described in detail by N.N. Efremovtsev and P.N. Efremovtsev in their papers mentioned in the Referenceы below /9,10 /. The authors point out the non-linear nature of the detonation velocity effect on the performance and economics of mining processes. The scientific rationale of the efficient energy release kinetic, velocity and patterns of charge detonation in rock blasting is a multi-factor optimization engineering and economic task requiring due account for the environmental effect of blasting and for the seismic effect of the blast action on specially protected areas.

Scientific approach to blast energy kinetics control for efficient mineral mining is based on the proposed categorization of methods of regulation of commercial explosives explosive characteristics and on the classification of methods of commercial blast control in an effort to regulate rock mass breaking /7,8/.

An important part of the research is the perfection of the test concept and technology for low-sensitivity explosives and the formation of methods and procedures aimed at the optimization of detonation systems for blast specific tasks. Consideration must be given to numerical estimates of the blast effective work redistribution due to minimization of the harmful effect (seismic action and shock air wave).

To the fullest extent possible the aspects of low-sensitivity explosives test are described in the Collected Works issued by the Institute of Chemical Physics of the USSR Academy of Sciences, in which the results of studies performed by K.K. Shvedov, V.V. Lavrov, B.N. Kukib, A.N. Afanasenkov, V.A. Sosnin, et al.

In general, it is important to once more emphasize the idea that earlier proposed methods and procedures for commercial explosives tests do not contain the estimation of the possible redistribution of blast energy spent on unproductive and dangerous blast action to useful forms of blast action.

An important component of the assessment of hazard of one or another explosive is the identification of conditions, which provide the propagation of the explosion process in a manner, which is most efficient for a specific task.

The methodological approach proposed by the authors for commercial explosives tests is peculiar for the following:

(1) The test objective is not only the assessment of safety and explosive characteristics of commercial explosives, but also the optimization of detonation system parameters for specific tasks in specified mining conditions.

(2) A multiple nano- and mictosystem model of a test commercial explosive composition is formed. Various methods of the regulation of explosive characteristics of commercial explosives are considered. Regulation methods (ECRM) are ranked by their importance for the task of explosives properties control. For the selected ECRM the possible and reasonable limits of energy-saturated material nano- and microdisperse system characteristics control are set. Variation of energy release kinetics of test charge designs is studied in detail for various methods of energy release kinetics control. Feasibility is also studied of the application of the wave method of explosives components modification.

(3) In the course of detonation system optimization for charges of a desired length to decrease labor intensity of the pretest procedures use is made of the proposed contributing factor analytical dependences of explosive characteristics.

(4) Detonation velocity in steel tubes is estimated by a fiber-optic method. In open-type charges, pilot boreholes and blastholes the variation of the detonation front propagation velocity is measured continuously by a SpeedVOD instrument along the whole length of charges.

(5) For the assessment of the comparative force in addition to blasting-cone volume measurement the velocity of ground seismic vibrations and characteristics of a shock air wave, height and radius of flyrock travel are measured in comparable conditions. Color and other characteristics of detonation products are recorded.

(6) On the basis of system analysis of the information on the variation of the force of test explosives and detonation systems on their basis the trends and numerical values of parameters characterizing the blast energy redistribution in useful and hazardous blast action forms are estimated.

(7) The proposed methods and procedures /12/ are used for the assessment of the energy release kinetics effect of a test explosive or its detonation system on the fragmentation

of rocks differing in brittleness and strength, and valuable crystalline materials.

The available instrumentation allows more precise and objective judgment on the relative variation of completeness and kinetics of explosive transformation. Moreover, the methods and procedures proposed by the authors for field tests require some changes in the pattern of charges and instrumentation location on the test site.

A test facility design has been developed, and now it is in the process of field trials, tests are also in progress of explosives with the assessment of their efficiency for specific tasks of blast application in the mining industry. The objective of the design effort: palpable enhancement of the performance and informative value of operations.

Test methods and procedures, and test facility design will provide an efficient solution with minimized time and labour inputs for the following tasks:

(1) assessment of the velocity of chemical reaction front propagation in an explosive charge and detonation system of a particular design and parameters;

(2) assessment of the blast breaking effect on crystalline materials and rocks having different properties, with simultaneous blast energy application to 90 rock samples;

(3) assessment of the variation of a blast seismic effect and shock air wave effect resulting from the variation of a detonation system energy release and transfer kinetics;

(4) assessment of the shattering action of an explosive and detonation product pressure in a blasting chamber;

(5) assessment of the content of harmful gases in detonation products, variation of their concentrations;

(6) comparative assessment of the variation of blast efficiency and redistribution of energy spent for blast dangerous effect in favor of its efficiency.

Application of the methodological approach proposed by the authors and the available hardware provides a significant additional scope of information of the hazard rate and efficiency of new explosives application with various detonation systems, as well as normalized test methods and procedures for commercial explosives.

The experimental efforts of the authors are aimed at the perfection (structural modification at the nano- and micro-level) of the available compositions of commercial explosives and development of new compositions, as well as the upgrading of charge design and improvement of the parameters of blasting processes to provide the most efficient solution for specific tasks of mining and construction:

(1) lower shattering effect of a blast to provide the undamaged condition of valuable crystalline materials and lower yield of fines in construction material mining;

(2) lower seismic effect of blast and shock air wave;

(3) higher intensity of fragmentation and radius of controlled rock fragmentation in ore deposit mining.

Fig.1 shows the proposed layout of a test site, which is peculiar for the fact that charges are placed in a ring pattern at an equal distance from a seismographer rather than in straight lines. For the assessment of the sensitivity of commercial explosives and completeness of their detonation the test open-type charges in paper and plastic envelopes are placed horizontally rather than vertically on the ground with vegetable soil removed.

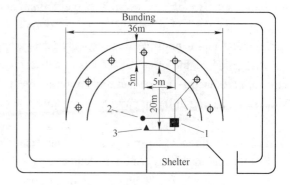

Fig.1 Field test site layout

1—armored case for a seismographer and detonation velocity measuring instruments; 2—microphone for measurement of shock air wave characteristics variation; 3—seismic pickup; 4—armored hollow shelter for optical fiber and cable; ○—charge location point

The proposed design of charges in cardboard envelopes for commercial explosives sensitivity tests and the assessment of transient detonation zone size variation is shown in Fig.2. Charges are placed horizontally on the ground and fired in turn. It makes possible the assessment of the comparative force judging by blasting-cone, seismic and acoustic components of test detonation system action in comparable conditions.

Application of the methodological approach and instrumentation proposed by the authors provides a significant volume of additional information on a hazard level and efficiency of new commercial explosives with various detonation systems and allows standardization of the commercial explosives test methods and procedures.

For the enhancement of blast efficiency, expansion of the range and capabilities of explosive characteristics controllability of cheap commercial explosives, as well as safety of their application the ANO National Organization of Explosives Engineers has performed laboratory research and field tests. The results of these tests have served a basis for the development of new compositions of Granulites (Granulite EF-P, Specs. TU 7276-002-94120064-2009, and Granulite EF-PN, Specs. TU 7276-003-94120064-2009), and elaboration of some aspects of the fabrication technology on use sites and borehole-charging areas.

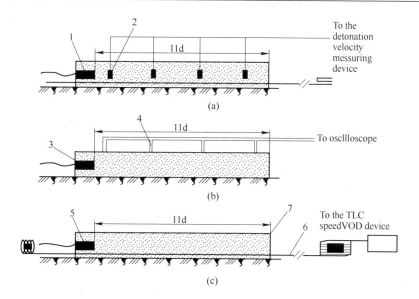

Fig.2 Charge design for horizontal placing to assess sensitivity to the initiating pulse, detonation velocity, variation of blast seismic action and shock air wave effect

(a), (b), (c)—design of charges in steel tubes for contact method of detonation velocity assessment (closing sensors for detonation front passage), fiber-optic method of detonation velocity measurement, and TLCSheedVOD instrument

1—primer consisting of electric detonator and detonation block; 2—steel tube envelope; 3—optic fiber;

4—detonation velocity measurement cable for TLCSheedVOD

Laboratory tests of new explosives compositions formed an important methodological part of the research. The laboratory tests pursued the following main objectives:

(1) changing the retention capacity of granules of porous and smooth ammonium nitrate with the use of a various emulsifier compositions and conditions of their exposure to physical fields;

(2) effect of the composition of various emulsifiers on the emulsion stability and retention capacity;

(3) effect of ultrasonic exposure and magnetic pulse treatment in fields of complicated spatiotemporal structure with the use of a GAN 5000 plant on the kinetics of ammonium nitrate decomposition and energy release in conditions of combustion and detonation, and on the absorption capacity of an oxidizer;

(4) fuel emulsion dispersion and stability;

(5) assessment of the Granulite EF sensitivity to impact according to the GOST 4545-88 standard and with a K-44-3 instrument;

(6) research into the kinetics of energy release of composite commercial explosives and their components.

The laboratory tests were performed in cooperation with experts from the Institute of Chemical Physics, RAS, the National Research Nuclear University MEPhI and the The Federal Center for Dual-Use Technologies Soyuz.

The research into the kinetics of thermal decomposition was carries out in MEPhIwithaNETZSCHSTA 409 instrument, as well as in the ICEMR RAS with the use of a Q-15000 thermogravimetric analyzer. The results were analyzed with Ecochrom software. The effect of Granulite exposure to physical fields and concentration of various energy additives on energy release kinetics was investigated in every detail.

Dynamics of the detonation front, kinetics of the processes of detonation transition to accelerated combustion, and characteristics of high-temperature gas detonation products (blasting medium), indices of blast seismic action and shock air wave were studied in the process of field tests in steel tubes with the use of fiber-optic method.

The effect of detonation velocity, charge design and layout, physical properties of rocks is investigated in the process of field tests with the used of methods and procedures proposed by the authors and envisaging the application of simulation models of blast media and rock samples differing in brittleness and uniaxial compression strength. The proposed methods and procedures include the assessment of energy release kinetics of special-design charges located at different distances from test samples of rocks having different physical properties in terms of crushability and residual strength.

Dispersion and through-time stability of fuel emulsion compositions used for the fabrication of Granulite EF compositions designed by the authors were studied with the use of a D-1200 electroacoustic spectrometer. Research into the microstructure of ammonium nitrate was carried out with the use of HITACHITM-1000 scanning electron microscope. The results of dispersion emulsion microstructure studies are shown in Fig.3. Photographs of the surface of smooth ammonium nitrate before and after treatment are shown in Fig.4 and Fig.5.

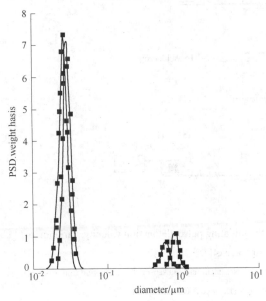

Fig.3 Results of Granulite EFs fuel emulsion nano- and microstructure studies with the use of D-1200 electroacoustic spectrometer

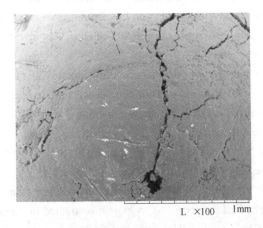

Fig.4 Photograph of the surface of ammonium nitrate, grade A, GOST 2-85 standard, taken with the use of a HITACHITM-1000 microscope

The performed laboratory research into viscosity of the emulsion and its mixtures with diesel fuel and mineral oil has proved an opportunity of controlling the rheological properties of the fuel component of EF-P Granulite that is particularly important for Granulite fabrication with the use of mix-pump trucks. The MCR-302 (AntonPaar) instrument was used for the assessment of the intensity and nature of the temperature effect in a -40℃ to +60℃ range, as well as the effect of a dynamic factor of stirring intensity, physical fields and content of the emulsion main components on viscosity. At +20℃ the viscosity of petroleum products used for EF-P Granulite fabrication is 100～152 MPa·s, while the viscosity of the emulsion with a petroleum product as its component is 9～15 MPa·s. If used at +20℃ the emulsion shows 10-fold lower viscosity. Application of the emulsion at subzero temperatures (e.g., at -10℃) shows 9-15-fold lower viscosity. Commercial-scale tests of emulsion at -25℃ were also successful.

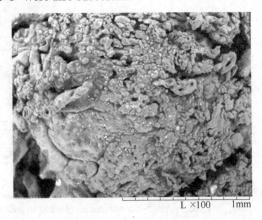

Fig.5 Photograph of the surface of ammonium nitrate, grade A, GOST 2-85 standard, taken after treatment with pore-forming emulsion in the process of Granulite EF-P and Granulite EF-PN fabrication. The image is taken with the use a HITACHITM-1000 microscope

Lower viscosity, higher penetrating power of the fuel component due to the formation of its nano-microdispersion structure, as well as porous framework structure in ammonium nitrate granules by many times expand the area of fuel and oxidizer contact and enhance the completeness and rate of primary and secondary chemical reactions at detonation.

Investigation of the surface of ammonium nitrate with a HITACHITM-1000 scanning electron microscope has shown that the contact of the proposed fuel mixture with the surface of ammonium nitrate granules gives rise to the formation of pores in the latter, and the fuel mixture evenly fills the framework structure of the oxidizer thus providing multiple expansion of the fuel and oxidizer contact area, with a granule shape being unchanged. The emulsion globule size is by several orders of magnitude less than that of ammonium nitrate pores, and therefore it easily penetrates deep into ammonium nitrate granules (see Fig.5). Fuel and catalysts of primary and secondary reactions, sensitizing agents and energy additives penetrate into ammonium nitrate granules, and they are evenly distributed in granule volume, thus by several orders of magnitude expanding the area of the fuel and oxidizer contact.

Investigation of Granulite EF fuel component dispersion with the use of D-1200 electroacoustic spectrometer has shown that the globule size of the micro-nano-dispersion emulsion proposed by the authors is 2.4 microns and 450 nanometers (see Fig.3). The emulsion is characterized by high stability and rather narrow range of globule size variation.

In experiment it was found that by way of in-process correction of the emulsion composition it was possible to decrease the flyrock travel by 50%～70% with high quality of the broken material being kept. The test results are validated by the joint test certificate of OOO Vzryvstroy and ANO National Organization of Explosives Engineers.

A program of mix-pump truck modernization is available for the expansion of the functional capabilities and in-process control of energy saturation and environmental safety of the application of detonation systems with an expanded range of their explosive characteristics control /14/.

Thus, the proposed commercial explosives and technologies of the formation of a framework nano- and microdisperse structures of explosive material components and detonation systems on their basis allow to significantly broaden the opportunities of in-process control of explosive characteristics, formation of charges with variable energy saturation and kinetics of blast energy release and transfer to the rock mass to enhance the efficiency of detonation system charges with no losses in safety of their application.

The effect of detonation velocity, charge design and its location, physical properties of rocks is studied in the process of field tests with the use of the proposed methods and procedures including the application of simulation models of blast media and rock samples differing in brittleness, viscosity and uniaxial strength. The proposed methods and procedures also include the assessment of the kinetics of energy release of charges of special design in terms of breakability and residual strength of rocks with different physical properties. Fig.6 provides a graphical illustration of the effect of energy release kinetics (charge detonation velocity) and a distance to a charge on the breakability index of rocks differing in ultimate strength. The nature and effect of a charge diameter on rock breakability index have been also investigated.

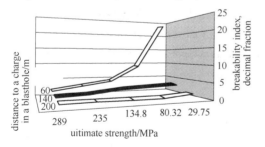

Fig.6 Dependence of a breakability index and fissuring on a distance to a charge and blast media material strength

In the course of commercial acceptance tests of EF-P Granulite the research was performed into opportunities of rock breaking intensity control.

The blast breaking action in a block has been assessed based on the results of the investigation of the particle-size distribution in different areas of the block:

(1) in the area of a hole mouth (within 1.25 m);
(2) between holes (at a 1.25～2.5 m distance);
(3) in a zone located at the maximum distance from a blasthole charge in different parts of the block, including bench edge between charges, as the most probable zone of grade yield of broken ROM material with high content of the maximum-size particles.

The particle-size distribution of the broken ROM material is assessed by photoplanimetric method.

The values of the average particle size and ROM size distribution were calculated. When approved Igdanite Granulite explosive was used the average particle size was estimated at 14.6～20.3 cm, with a breakability index of 6.8～5 accordingly. Application of EF-P Granulite showed the reduction of an average particle size to 6.8～10.5 cm, with the yield of fine size material of 0.5%～1.7% and breakability index of 10～14 accordingly.

Tests have revealed significant extra opportunities of in-process regulation of rock breakability indices with the application of the proposed technology of EF-P Granulite fabrication and use. Granulite EF-P test results have shown also an opportunity of a 50%～60% reduction of an average particle and a 43%～98% growth of a breakability index. The results of experimental studies and tests are supported by the Test Certificate and Test Record Sheets.

The dependence has been discovered of the breakability index and average particle size of broken ROM material on a distance to a charge (See Fig.7).

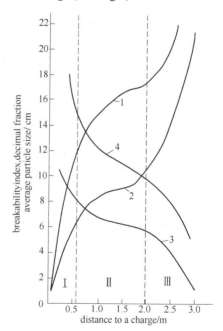

Fig.7 Nature of dependences of the breakability index and average particle size on a distance to a charge in a blast block by results of analysis of particle size distribution of blast ROM material in the process of Granulite EF-P commercial tests

1, 2—dependences of the breakability index and average particle size on a distance to a charge in a blasthole for the approved Granulite Igdanite;

3, 4—dependences of the breakability index and average particle size on a distance to a charge for EF-P Granulite; I, II, III—coverageeexplosion

REFERENCES

[1] Intensification of rock blasting for continuous operation machinery. 1967. *VzryvnoeDelo*, No 62/19: 316. (in Russian)

[2] Viktorov S D, Kazakov N N, Zakalinsky V A. 1995. Analysis of methods of rock blasting process control.*GornyZhurnal* No 7: 46~47. (in Russian)

[3] Trubetskoy K N, Melgunov V G. 1975. The effect of lump size of rock on the performance and economics of loaders. 1975. *GornyZhurnal*, No 11: 22~24. (in Russian)

[4] Efremovtsev N S. 1968. Investigation and on-the-job assessment of properties of rock as blast media in conditions of mineral deposit surface mining: Kachkanar GOK Case Study. 1968. *Author's abstract of a thesis.*Magnitogorsk: 43. (in Russian)

[5] Efremov E I, Nikiforova V A. et al. 2001. On the interrelation of fines yield in rock blasting and pulse parameters of various explosives.*Metallurgicheskaya I GornayaPromyshlennost*, No 1: 88~90. (in Russian)

[6] Trubetskoy K N, Viktorov S D, Kutuzov B N. 2009. Problems of blasting processes evolution on the ground surface. *VzryvnoeDelo*, No 101/58: 3~33. (in Russian)

[7] Efremovtsev N N, Kazakov N N. 2012. Revisited: Formation of the classification of control methods of energy release kinetics and transfer to the rock mass for its breakage with a desired degree of fragmentation. 2012. *Collected works: Voprosynauchnogoobosnovsaniyasovershenstvovaniyasredstvupravleniyaintensivnostyurazrusheniyagornykhporodvzryvom. No 12. Special Issue:* 27~32. (in Russian)

[8] Efremovtsev N N. Classification of methods of explosive characteristics control for commercial explosives. 2007. *GornyInformatsionno-AnaliticheskyBulletin.VzryvnoyeDelo Collected works. Vol. 2, MSMU Publishers. (in Russian)*

[9] Efremovtsev N N, Efremovtsev P N. 2012. Seismic and environmental factors of the optimization of detonation system energy release kinetics control for safety of blasting processes. *Collected works: Voprosynauchnogoobosnovsaniyasovershenstvovaniyasredstvupravleniyaintensivnostyurazrusheniyagornykhporodvzryvom.No 12. Special Issue:* 11~16. (in Russian)

[10] Efremovtsev N N. Engineering and economic factors of optimization of detonation system energy release kinetics control for reasonable fragmentation of rocks by blast. 2012. *Collected works: Voprosynauchnogoobosnovsaniyasovershenstvovaniyasredstvupravleniyaintensivnostyurazrusheniyagornykhporodvzryvom.No 12. Special Issue:* 3~10. (in Russian)

[11] Methods of low-sensitivity explosives tests. Best Practices. Under editorship of Shvedov K.K. 1991.Institute of ChemicalPhysics, USSRAcademy of Sciences, Chernogolovka: 149. (in Russian)

[12] Efremovtsev N N, Kvitko S I, Pozdnyakov A S. 2012. About the program of mix-pump truck modernization for the expansion of the functional capabilities of blast action control. *Collected works: Voprosynauchnogoobosnovsaniyasovershenstvovaniyasredstvupravleniyaintensivnostyurazrusheniyagornykhporodvzryvom.No 12. Special Issue:* 23~26. (in Russian)

[13] Efremovtsev N N, Sosnin A V, Belin V A, Efremovtsev A N. 2008. Development of simulation models for charge blast assessment in terms of valuable crystalline materials undamaged condition. *VzryvnoyeDelo Collected works. No 157.* (in Russian).

[14] Borovikov V A, Andreev A A, Efremovtsev N N. 2007. Characteristics of detonation of Granulites including low-density polystyrene-containing compositions. *GornyInformatsionno-AnaliticheskyBulletin.VzryvnoyeDelo. Collected works.MSMU Publishers.*(in Russian)

[15] Efremovtsev N N. 2012. Chatacteristics of the formation of multiple micro- and nanodisperse systems of Granulites EF and commercial water-resistant emulsion explosives on their basis.*Collected works: Voprosynauchnogoobosnovsaniyasovershenstvovaniyasredstvupravleniyaintensivnostyurazrusheniyagornykhporodvzryvom.No 12. SpecialIssue:* 17~24. (inRussian)

Research of Manufacturing Process Improvement of Expanded Ammonium Nitrate Explosive

ZHENG Bingxu, GAN Dehuai, SONG Jinquan, LI Zhanjun

(*Guangdong Handar Blasting Co., Ltd., Guangzhou, Guangdong, China*)

ABSTRACT: As Expanded AN Explosive is a new type of No Ladder Powder Explosive, it gets a wide range of application and development in china. Subject from powder explosive technology, it is based on the research of explosive theory, we put forward the thought that oil phase water phase to continuous mixed continuous puffed crystalline powder, we studied a new production process. The new process further enhance the technology of Expended AN explosive and simplify the production line greatly, reduce the storage quantity of production line, Improve the comprehensive performance of explosive and will have a wide range of significance to develop.

KEYWORDS: ammonium nitrate explosive; liquid mixed type; explosive producing

1 INTRODUTION

As Expanded AN Explosive is a new type of No Ladder Powder Explosive, its design idea are based on hot spot theory, application surface and surfactants theory. We have deep research on the balace of AN saturated solution, optimize chemical unit process and we estabilsh the process of forced crystallization bulking products under the actions of surfactants so that the Self-sensitization to AN can be succeed. Technology from 1995 through the appraisal of industrial in China since the industry get a wide range of application and development. At presents, the productions lines are more than 80, application of the technology is mainly divided into horizontal process vertical process and continuousbentroite mixed process. After twenty years of technology application practice, all kinds of process line technology is also exposed some problems and the insufficiency, such as low powder density, poor resistance to water performance, complicated production process etc.

Subject from powder explosive technology, it is based on the research of intermolecular mixed explosive, explosive theory and surfactants and disperse system etc, we put forward the thought that oil phase water phase to continuous mixed continuous puffed crystalline powder, we also study a new powder AN industrial explosive production process—Antioxidant Ammonium Nitrate and Combustible composite oil are mixed in the liquid molecular statefirst, thenvacuum drying, the products have less than 0.2 percentmoisture content. The pocess production of this industrial explosive have excellent explosion performance, water resistance performance, storage performance, it improve the perfomance of Expanded AN Explosive greatly, simplify the production process, so the new process can referred to liquid mixed type production process is a important breakthough in the process of Expanded AN Explosive and it open up a new way to product the explosive.

2 THE ANALYZE OF TECHNICAL CONDITIONS OF THE ORIGINAL EXPANDED AN EXPLOSIVE PRODUCTION LINE

The horizontal process and vertical process of the original Expanded AN Explosive and continuous bentonite mixed process production line and production process are all AN crystallization bulking, mixing of components, charging, packaging, the manufacturing process can be seen from Fig.1.

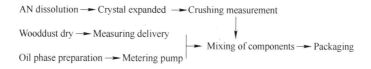

Fig.1 Expanded AN explosive production process flow diagram

The shortage analysis:

(1) Simplify the production process. At that time because of the technical way and technical condition the influence,
the production process consists of bulking various raw materials preparation and delivery ammonium nitrate crushing mixed medicine and subsequent charge package. Due to the

Corresponding author E-mail: 274991025@qq.com

various process, it caused many problems such as it demand more equipments and the safety of prodution process and the quality control point are increasing, so it incease the difficulty to the enterprise safety management, on the other hand, it was also disadvantageous to the energy saving and it do not conform to the new forms of saving energy and reducing production consumption.

(2) Low powder density and poor resistance to water performance. Due to the Expanded AN is based on the bubble sensitization theory that design successfully, in Expanded crystallization process water overflow and formate pore channel, the crystal surface present irregular edge crack and burr that comprise the hot source. Because of these hot source formation on one hand, it formed Expanded AN self-sensitization successful, but on the other hand it caused low powder density and poor resistance to water performance. It affect the effect of industrial explosive blasting and the acceptance of users. The author count, test and analyse according to the density of the Expanded AN Explosive and the water resistance of the process. The results can be seen from Tab.1.

Tab.1 The results of density and water resistance of several Expanded AN Explosive process

Process route	Product density range /g·cm^{-3}	Expanded AN hygroscopicity[①] /%	AN hydroscopicity[①] /%	Water resistance of Expanded AN explosive[②]
Horizontal process	0.85~0.90	17.6	15.5	Body of explosive soluted in water
Vertical process	0.85~0.90	18.8	15.5	Body of explosive soluted in water
Continuous bentonite mixed process	0.78~0.85	20.1	15.5	Body of explosive soluted in water seriously

① Hygroscopicity test amount of expanded AN and ordinary AN which use surface plate to hold and exposed in air to moisture absorption two hours, then through the analysis of balance weight and gain weight, we can calculate suction wet.

② Cartridge are put in the container which has 1m depth of water, then test after in the water two hours.

According to the results of Tab.1, we know the density of Expanded AN explosive is lower than original Ammonium ladder explosive, the density of Horizontal and Vertical process are higher than the Continuous bentonite mixed process, but it lack of unstable performance, strictly control in the process and so on. The hygroscopic of Expanded AN is higher than ordinary AN, the main reason are that these shortage are decided by the unevenness of porosity of AN and surface formed by bulking in the process.

(3) The online quantity of production line system explosive workshop are too large. As the production process include Expanded AN preparation grinding oil phase material preparation, wood powder preparation, raw material measurement conveying the raw material mix and so on. No matter horizonal process, vertical process, continuous bentonite mixed process which are existing prodution technology are still belong to half continuous process, its puffed discharge process are intermittent type, one-shot process, the output samples are large. Horizontal process, vertical process can meet with large capacity, while it can parallel many equipment and feed at the same time; although continuous bentonite mixed process is single set of equipment alternate in and out of stock, but one design charging is 750kg, obviously the quantity of these three process are too large. In order to solve the insufficiency, we use industry process isolation, control material thickness of the belt conveyor, reduce a charging rate and other techincal means, try to reduce and sympathetic detonation dose or reduce the drug with workshop, it obtained certain significance, but it can not ensure to reduce the dosage and online security risk control.

3 CHARACTERRISTICS OF LIQUID MIXED TYPE PRODUCTION PROCESS

The Liquid mixed type of AN explosive formula design and preparation principle: first it is according to the zero oxygen balance principle of industrial AN explosive, it can design formula of high performance Expanded AN explosive; second Antioxidant AN and combustible composite oil phase in liquid molecular state mixed together, thus the uniformity of solid state can improve, the performace of dynamite exolosionis obviously improved and has certain resistance to water performance; the third is water cover oil disperse system, it uses vacuum drying process in order to reduce drying the energy consumption and enhance the drying efficiency and made high performance powder anfo explosives. The Liquid mixed type AN explosive process can be seen in Fig.2.

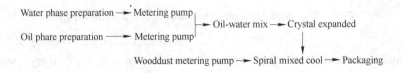

Fig.2 Liquid mixed type production process flow diagram

The innovation technology production line of the Liquid mixed type AN explosive has some features as follows:

(1) Through the fluctuation bin series, it can realize the single set of equipment of continuous puffed crystallization, discharge intervaly and the time is 90~100s, the calculation crystallization machine in the largest stock of material was only 98 kg in according with the theory that the annual output of prodution line is 12000 ton. Equipment in stock is greatly reduced when match with the original process and the one-shot sample is little, so ,it reduce the wave crest of production process material accumulation and provides a convenient for making subsequent eqipmentseletion and produciton organization smoothly.

(2) It first innovates the process route of continous mixing with oil and water first, then continuous puffed crystalline powder, after the crystallization is the powder products, it greatly simplify the subsequent process and equipment and cancel the original process of Expanded AN conveying screw, Expanded AN mill, after crushing AN conveying screw, three material mixer and other equipment. It reduces energy consumption and the online storage quantity of production line. The performance, water resistance and density of the explosive production are improved greatly. The measured performance of new process equipment explosive can be seen from Fig.2.

Fig.3　Photo of hybrid crystallization

Fig.4　Photo of hybird process

Tab.2　Actual performance of liquid mixed type of Expended AN explosive production line

Products＼Index	Rock Expanded AN explosive (no wooddust)	Rock Expanded AN explosive (wooddust)	Rock Expanded AN explosive (old process)
Formula/%	AN 94.5,Oil phase 5.50	AN 92,Oil phase4，wooddust 4	AN 92,Oil phase4，wooddust 4
v_d/m·s^{-1}	4000	3800	3500
Brisance/mm	16.5	16.2	14.5
Gap distance/cm	6	7	5
Density/g·cm^{-3}	0.92	0.88	0.88
Water resistance/cm	≥4	≥4	0
Friction sensitivity/%	0	0	0~4
Impact sensitivity/%	0	0	0~4

(3) The friction sensitivity, impact sensitivity of production explosive is zero, it quite same with the sensitivity of the original Expanded AN explosive; but as the result of the new process cancel the old one in the process of Expanded AN crushing device, the essence safety level of the production line is improved.

(4) It use the two stage water bath type atmospheric condenser, primary cooling, all the condensed water are recycling, secondary cooling, the condensed water is recycling, it eliminate the waste water treatment and discharge of the follow-up basically, at the same time it reduced cooling water circulation of subsequent pollution greatly and has siginificance meaning of environmental protection.

(5) It use new technology which simplify the production process and reduce the middle equipment, it improve the continutiy of production line and ensure the production line automatic control reliabiltiy.

(6) The develop resistance of system is strong and can provide Liquid AN can, Semi-finished suspension transportation and so on, it further reduce the operators in production line.

Fig.5　Photo of Liquid AN tank

Fig.6 Photo of Finished suspension conveying

4 CONCLUSIONS

Liquid mixed type of Expanded AN explosive production line technology further enhance the technology of Expended AN explosive and simplify the production line greatly, reduce the storge quantity of production line, Improve the comprehensive performance of explosive, Friction sensitivity and mechanical sensitivity of the production is zero, the new technology of Expended AN explosive is a important breakthrough for the production process. It open up a new way to the Expended AN explosive production process and will have a wide range of significance to develop.

REFERENCES

[1] LV Mingxu, LIU Zuliang, LU Ming and so on. Ammonium Nitrate Explosive[M]. Beijing: Weapon Industry Press, 2001: 6~92.

[2] LU Ming. Formula design of industrial explosive[M]. Beijing: Beijing University of Science and Technology Press, 2002, 76-103.

[3] Lu M, Lu C. A computer model for formulation of ANFO explosives[J]. Sci. Tech. Energetic Materials, 2007, 68(4): 117~119.

[4] LU Ming. Research of high performance powder explosive[J]. Demolitionequipments and material, 2007, 36(6):9~11.

Some Behavior Characteristics of Condensed Composite Materials in Explosive Processes

A.V. Starshinov[1], S.S.Kostylev[1], I. Y.Kupriyanov[1], N.Y.Yargina[1], JijigJamiyan[2]

(*1. Nitro-Technologies Sayany Co., Ltd. Russia;*
2. Mongolian-Russian Joint Venture MONMAG, Mongolia)

ABSTRACT: The results of determination of explosive characteristics in composite materials in various physical states. Much attention is given to experiments performed in real and quite specific conditions in Russia and Mongolia. Experimental data are compared with the results of other researchers. Reasons for obtained trends taking into account modern concepts of phenomenon mechanism in explosive processes with consideration of the chemical nature of mixture components and their structural features are introduced.

KEYWORDS: explosives; mixtures; liquid; coarsely dispersed system; emulsions; failure diameter and detonation velocity

1 INTRODUCTION

Almost all materials in condensed state used as explosives in production sector, in terms of component content are mixture systems and in terms of structure are heterogeneous systems. Component composition of mixtures is determined by chemical structure characteristics of basic materials (tautomeric and isomeric molecule forms) as well as additives which are incorporated in system deliberately in order to get desired properties or enter the system in the form of process or accidental impurities. Structure characteristics of explosives and charges from them are determined by complex factors ranging from short-range ordering in liquids, forms and sizes of primary formations – grains in crystal systems to complex physical-chemical heterogeneous systems, for instance, emulsions. Explosive composition and structure characteristics are interrelated.

The main factors for application of materials as explosives are their ability to explosive transition in predetermined conditions (explosion), technical and economic efficiency as well as hazard which can differ markedly from specified in initial conditions, in case of changes in composition and structure, including random effect.

2 ANALYSIS RESULTS

One of the examples of significant increase in detonability and hazard relatively due to changes in composition are liquid explosives, including C- and O- nitro compounds. The most significant results were obtained by different researches for systems based on nitromethane(NM)[1,2]. This material is widely used in different industrial sectors and before explosion of railway car in 1958 was considered as safe material.

A lot of research works on systems based on NM were carried out in D. Mendeleyev University of Chemical Technology of Moscow[1]. It was discovered that additives are put to amino class NM materials have sensibilization effect, concentration is about some parts per mille (10^{-6}). If diethylamine additive concentration is 0.001% detonation failure diameter (d) of NM in glass tube decrease by 2, if diethylamine concentration is about 2% failure diameter decrease up to minimum value which is about 2 mm. These characteristics with less critical effect were obtained for other liquid state nitro compounds, including, trinitrotoluene solution (TNT). It was discovered that organic compounds with olefinic functionality (dibond and triple bond) in carbon chain have "strong" sensibilization effect on NM, compound which has amino group and dibond (allylamine) was the most active.

The summarized results are presented on Fig.1. It should be mentioned that additives which lead to drastic change in detonability of the system can appear in it accidently from environment. For example, material contamination by amines can happen due to the contact with subjects coloured with aniline dyes.

Continuing research works dealing with systems based on NM[2], provide support for previous obtained results, which can be explained by structure changes of liquid explosives with formation of "active" centers where primary reactions start and spread along the whole explosive using chain type mechanism[1]. It appears probable that "active" centers can preliminary form in basic mixture, for instance, due to tautomeric transformation of nitroform NM into aci-

Corresponding author E-mail: kostylev62@yandex.ru

form, or occur due to explosion wave in the system, for example, due to breakdown of dibond (multiple bond), breakdown in terms of brittle facture mechanism due to limitation in rotation degree of freedom [4].

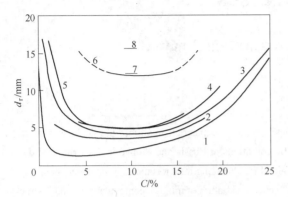

Fig.1 Amine effect on detonation failure diameter of nitro compounds and their solutions

1—nitromethane with pyridine;
2—trinitrotoluene solution in nitromethane (25/75) with pyridine;
3—trinitrotoluene solution in nitromethane (50/50) with pyridine;
4—trinitrotoluene solution in nitromethane (60/40) with pyridine(60/40);
5—solution 2,4 dinitrotoluene in nitromethane (47/53) with allylamine;
6—trinitrotoluene solution in pyridine, temperature is 83~850℃;
7—solution 2,4 dinitrotoluene in nitromethane(47/53) withpyridine;
8—solution 2-nitropropane in nitromethane (48/52) with allylamine

In contrast with liquid explosive in the form of regular solution in heterogeneous systems possible effect of material components chemical nature, generally, is surpassed by structure effects and component distribution equitability. In mixtures based on ammonium nitrate (AN), in solid state, system detonability is determined by ammonium nitrate crystal structure, presence of defects and impurities, density of primary crystal formations – grains, form and sizes of particles. Whereas, AN particles structure as well as detonability of mixtures depend on the method of producing these particles and mixture preparation. Advantages and efficiency of mixture explosives based on AN with particles in the form of porous granules (ANPP) using fuel materials in the form of liquid fuel among fuel oil (FO) or even expanded small particles in powdered systems are well known [5].

Solid particles structure analysis (grain, powder), with the help of electron scanning microscope shows that significant differences in body crystal structure and particles surface depending on the method of their occurrence and/or processing.

Fig.2 shows the examples of granule structure obtained by drop formation method – prilled with aftercooling using different modes: (1) monotonously obtaining high density product, (2) gasification due to evaporative additives producing porous ammonia nitrate (PAN) in Russia, step change in temperature and excess moisture elimination. AN granule structure influence on detonation behavior are approved by experiments where the critical diameter of ANFO mixture detonation is measured (dk): for AN "a" type (Fig.2) dk is more than 100 mm, "b" type is 80~100 mm, "c" type is less than 80 mm [6,7]. AN particles influence can be viewed when specify detonation behavior (dk) of ANFO mixtures based on powders. The result depends on material initial state, for example, granules according to Fig.2 as well as the powder production method: mechanical processing, thermal effect, "expansion" [5] and other.

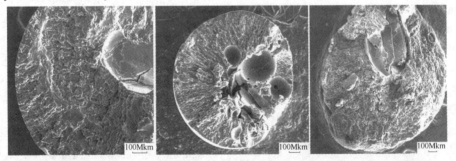

Fig.2 AN grades granule profiles (the first is "2a"- high density according to the GOST 2-85 Russian Federation, the second is "2b" –porous, the third is "2c" - PAN "Grande Paroisse" France)

Pictures of powders, produced by mechanical processing and cyclical thermal effect (5 cycles) using "2a" type granules according to Fig.2 are given in Fig.3.

Fig.3 AN particles produced by different methods using high density granules "2a" according to Fig.2: on the left – mechanical grinding, on the right – cyclical thermal processing within temperature interval 20~50℃ (5 cycles)

Dk experimental values of ANFO mixtures based on these powders with 0.5~1 mm particles sizes are 80 mm and 40 mm respectively. Specific conditions for raw materials supply in Russia and Mongolia requires finding special additives and techniques to provide quality. The most available in these conditions are methods based on initial heat treatment of AN granules with modification transformation of crystalline grid (theoretical temperature – 32.3℃) [7,8], and application of additives in the form of powder with high absorbing properties of FO: brown coal, turf, organic materials dust, including processing of foodstuffs wastes and others.

Modern mining industry uses emulsion explosives for blasting works. These explosives and production technologies are rapidly developing in China and now are introduced at several plants in Mongolia and in Lipovtsy plant, Primorski Krai, Russia.

These explosives are produced using technologies developed by BGRIMM and Shenzhen King Explorer Science and Technology Co. Ltd., effectively used in adverse climatic conditions and have excellent detonation behavior and potential for use:

(1) sensitive to primers in the form of prime – cap with weight quantity 1 g of explosive and detonating cord with weight quantity starting from 6 g/m of explosives;

(2) experiment value of dk, measured in different conditions (plants in China, in Mongolia, in Russia) is not less than 15 mm, but some experiments showed detonation in charges of about 10 mm.

(3) physical state (thickness) is the same as in initial state up to the – 50℃. detonation behavior and sensitivity remain;

(4) 32 mm diameter cartridges have steady state detonation after storing in climate conditions of Mongolia within 2.5 year;

(5) plastic consistency of explosives provide the possibility to form charges using different covers and use them for different purposes: blasting of oversized rocks and concrete, breaking down of metal metal constructions and others.

3 CONCLUSIONS

Present raw material supply conditions and technological capabilities for explosive production using non-explosive materials for industrial needs give opportunity to produce materials (explosives) with different properties, but possible changes in mixture properties should be considered under the influence of environment conditions and accidental effect, in particular, process impurities or impurities of uncertain composition.

REFERENCES

[1] Kondrikov B N, Kozak G D, Starshinov A V. *Proceedings of the USSR Academy of Sciences.* 1977, 233 (3).

[2] Mochalova V M, Utkin A V. *Blast waves in condensed matter.* 2012: 60~63 Kiev.

[3] Railroad Accident Investigation. MT PULASKI ILL. June 1958.

[4] Agafonov G D, Bilera I V, Vlasov P F. *Combustion and explosion* 6 (152~169), Moscow: Torus press.

[5] Lu Chunxu "The Aplication of Surfase Active Theory tu Energetic Materials – Research jf Expansions Ammonium Nitrate Explozives" NTREM, Ⅷ, 2005: 260.

[6] Qin Hu, Li Guozyjng, Hua Baolin, Song Jinquan. Study on Detonation Performance of Two Kind of ANFO Explosives. China. BGRIMM. 2004: 784~787.

[7] Starshinov A V. Neiman V R. JijigJamiyan et al. 2009 *Blasting work* 102\59 (145~155), Moscow.

[8] Pongovskyi V, Serafinovich S, Suboch B. Some characteristics of Polish porous ammonium nitrate: 1982. *Drilling and blasting VI conference materials* (4): Kiev.

The Comparison of the Commercial Explosives in United States to that in China and Discussion on Development of Chinese Commercial Explosives

YAN Shilong[1], GUO Ziru[1], SHEN Zhaowu[2], WU Hongbo[1], WANG Quan[1]
(1. Anhui University of Sci. & Tech., Huainan, Anhui, China;
2. University of Sci. & Tech. of China, Hefei, Anhui, China)

ABSTRACT: The annual consumption, species changes and the application of commercial explosives in United States are presented and analyzed. Based on these analyses, the development trend of commercial explosives in China is discussed and suggested.

KEYWORDS: explosives; development; comparison; trend

1 INTRODUCTION

Commercial explosive is a special energy used for mineral exploitation and rock excavation, which has been honored as "the energy for the energy industry". Commercial explosives not only relate to the construction of railway, road, water conservancy project and hydro-power but also to the exploitation of metal and none-metal minerals. So far, explosives are possibly the most versatile and hardest working of all tools in the mines exploitation and rock excavation. With the advance of science and technology, new materials and machinery equipments will be constantly appeared which maybe lead to the no application of explosives to mines exploitation and rock excavation. Nevertheless, according to the current technology level, blasting by explosives still is fundamental means for the obtaining resources from the Earth and sustained human civilization in the long time. Today, even in the United States, where the most advanced science & technology and the most powerful economy are presented, a great quantity of explosives have been used annually. Compared with other methods, blasting is the most effective and economical method in mineral mining and rock excavation, and perhaps the only method in certain situation.

Species and manufacturing methods of industrial explosives should be linked to the national geological conditions, mining methods, as well as its security requirements in application. Based on the commercial explosive consumption in United States, this paper explores the development direction of China's industrial explosive.

2 ANNUAL CONSUMPTION, SPECIES CHANGES AND THE APPLICATION OF COMMERCIAL EXPLOSIVE IN UNITED STATES

From 1987 to 2004, the total annual consumption of the United States industrial explosives is about 2000~2500 thousand tons. Annual consumption of 2005, 2006, 2007, 2011 and 2012 were more than 3 million tons. From 1993 to 2012, annual consumption, species changes and use fields of commercial explosive in United States are presented in Tab.1. Fig.1 and Fig.2 provide the annual consumption, species changes and application fields of commercial explosive between the year of 1986 and 1995. These data come from the official website of IME.

In Tab.1, the blasting agent means ANFO explosives, slurry and emulsified blasting agents, and their mixtures with ANFO in various densities, which are insensitive to one detonator. Ammonium nitrate means on-site mixed granular products and solutions. From Tab.1, it can be known that since 1993, over 98% (recent years, over 99%) commercial explosives in United States are blasting agents and on-site mixed explosives, which are insensitive to No.8 detonator. Coal-mine permissible explosives and other high explosive which are sensitive to detonator only account for a small percentage, not exceeding 2%. Underground coal-mine permissible explosive consumption has decreased for many years. For instance, coal-mine permissible explosive consumption was 4430 tons in 1993, but reduced to 1260 tons in 2006, and was only 860 tons in 2007. Coal-mine permissible explosive consumption was about 1000 tons during recent years in US.

Corresponding author E-mail: 13805543828@163.com

Tab.1 Annual consumption, species changes and use fields of commercial explosive in United States from 1993 to 2012

Year	Total consumption and main species percentage /%				Use fields percentage/%				
	Total consumption /kt	Blasting agents and ammonium nitrate	Permissible explosive	Other high explosive	Coal	Construction works	Metal ore	Quarrying and non-metal minerals	Others
1993	1880	98.05	0.23	1.72	66	7.23	11.12	12.93	2.76
1994	2320	98.39	0.16	1.45	66	7.24	10.52	13.02	2.84
1995	2280	98.25	0.15	1.60	66	7.23	10.96	13.51	2.51
1996	2240	98.47	0.11	1.42	65	7.14	10.67	13.84	3.17
1997	2670	98.81	0.09	1.10	66	7.23	10.34	13.82	3.07
1998	2910	98.87	0.09	1.04	67	7.15	9.45	14.16	2.57
1999	2120	98.45	0.08	1.47	67	7.41	9.58	13.73	2.77
2000	2570	98.62	0.06	1.32	67	7.55	9.18	13.50	2.84
2001	2380	98.49	0.07	1.44	68	7.52	7.77	13.36	2.86
2002	2510	98.43	0.05	1.52	69	7.53	7.77	13.35	2.83
2003	2290	98.40	0.05	1.55	68	7.86	7.16	14.10	2.76
2004	2520	98.31	0.04	1.65	66	10.32	7.10	14.33	2.38
2005	3200	98.96	0.04	1.00	65	11.00	7.41	14.00	2.28
2006	3160	98.78	0.04	1.18	66	11.58	7.50	12.75	2.03
2007	3150	98.69	0.03	1.28	66	11.59	7.52	11.74	2.44
2008	3420	98.83	0.04	1.05	70	10.94	8.04	10.08	2.25
2009	2270	98.69	0.07	1.04	70	10.35	7.75	9.21	2.51
2010	2680	98.88	0.04	0.84	71	9.18	8.43	9.40	2.51
2011	3000	99.33	0.03	0.73	71	8.83	8.60	9.30	2.50
2012	3380	99.11	0.04	0.93	68	10.38	8.70	10.27	2.60

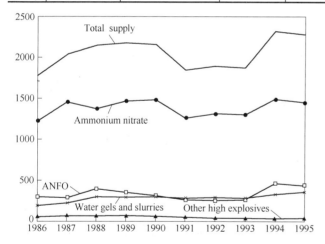

Fig.1 Annual consumption, species changes of commercial explosive in United States from 1986 to 1995(ordinate unit: thousands tons)

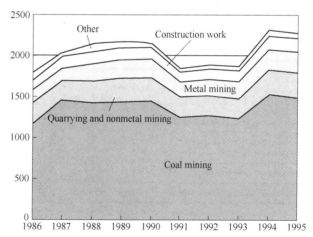

Fig.2 Annual consumption and use fields of commercial explosive in United States from 1986 to 1995(ordinate unit: thousands tons)

From the above description, coal mining is the most explosives consumed industry, which accounts for about 65%~71%. Additionally, its consumption is relatively stable and consumption change is not obvious annually. The second consumption fields are the quarrying and other non-metallic minerals, whose consumption accounted for 9%~14%. The percentage of explosive consumption in metallic mine was 7%~11% and decreased year by year. Meanwhile, the consumption in construction accounted for 7%~12%. Before 2007, the consumption in construction increased annually, but deceased by two percentage in last 5 years. It shows that no matter in what field in the United States, the main types of explosive consumption was detonator-insensitive explosive and on-site mixed blasting agents. The blast with large diameter and deep hole at open pit used detonator-insensitive explosives predominantly. Tab.2 listed the number of United States explosive manufacturers and the average annual production volume of each company in the last nearly 20 years. From that, it was shown that the explosive production volume of single company in United States is quite large.

In addition, in order to improve the fragmentation effect in the stiffness, hard-rock blasting, there is much kind of commercial explosives added aluminum powder in the United States. But in China, industrial explosive species with aluminum powder is very little.

Tab.2 Number of the last 14 years of the company's U.S. industrial explosives

Years	1994	1995	1996	1997	1998	1999	2000	2001	2002	2003	2004	2005	2006	2007	2008	2009	2010	2011	2012
Number of explosives	22	23	32	33	29	29	29	28	30	23	23	23	19	19	20	18	18	18	18
Average annual production capacity of single company/kt	110	100	70	80	100	70	90	90	80	100	110	140	170	170	170	130	150	170	190

3 THE SITUATION OF COMMERCIAL EXPLOSIVES IN CHINA

During the 10th five-year Plan in china, the total consumption of commercial explosive rapid increased from 1.28 million to 2.4 million tons. It reached over 2.6 million tons in 2006, and the annual output was about 3 million tons in 2007 and 2008. The annual consumption was increasing in last 5 years and was over 4.3 million tons in 2013. Because of incessant policy adjustments, the output of emulsified explosive increased very fast in 2006 and accounted for 41.88% in total explosive output, and the traditional TNT powdery explosive only accounts for 28.47%, the percentage proportion of ANFO category explosive was 27.87%. In 2008, the proportion of traditional TNT powdery explosive only was 0.2% and the one of emulsion explosive was reached 53.64%, while ANFO category explosive was 44.22%. The output of emulsion explosives and ANFO category explosive reached 2.33 and 0.63 million tons, and accounted for 53.63% and 15% in 2013, receptively. The output of bulk explosives was 1.93 million tons and the proportion was 44% of total explosive consumption. Generally, the consumption is increased annually for emulsions, ANFO, bulk explosives, and cap-insensitive explosives.

Tab.3 and Tab.4 listed the species change and total production output of Chinese commercial explosives in last over ten years. From Tab.3 and Tab.4, the commercial explosive species in china is similar to that in the United States and is cleaning and environmental friendly in manufacture.

Tab.3 The total output of commercial explosives and the species changes in china from 2002 to 2008

Years	Total output /10⁶t	The species distribution					
		Nitroglycerine explosives /%	TNT explosive /%	Water-gel explosives /%	Ammonium oil explosive /%	Emulsion explosives /%	Other explosives /%
2002	1.56	0.43	52.33	1.26	20.94	25.06	0.39
2004	2.16	0.31	41.85	1.13	24.35	32.28	0.38
2006	2.62	0.08	28.47	1.31	27.87	41.88	0.26
2007	2.86	0.07	10.01	1.45	38.05	46.45	3.99
2008	2.90	0.06	0.19	1.40	44.22	50.06	4.07

Tab.4 The total output of commercial explosives, the species changes and the application fields in china from 2009 to 2013

Year	Total output/10⁶t	The species distribution/%						The distribution of application fields/%				
		Bulk emulsions	Bulk ANFO	Water-gel explosives	Powder emulsion, expanded AN etc.	Emulsions	Other	Coal mining	Metal mining	Non-metal mining	Construction	Other
2009	2.96	6.1	10.4	1.3	39.0	40.2	1.0	29.3	22.3	23.5	18.4	6.5
2010	3.52	5.8	10.3	1.2	38.7	42.9	1.1					
2011	4.07	6.4	10.6	1.0	36.5	43.6	2.0	26.6	24.5	26	11.0	11.9
2012	4.19	7.2	12.6	1.0	32.1	43.8	3.3	29.4	23.7	24.9	8.8	13.2
2013	4.37	6.4	15	1.1	27.9	46.3	3.3	31.0	26.0	25.0	9.0	9.0

Tab.5 presented package specification of commercial explosives in china in recent years. The capacity of small diameter package explosives deceased and that of big diameter and bag package explosives increased annually, which is suitable for advanced blasting engineering.

Tab.5 The package specification of commercial explosives in china in recent years

Year or species	Package (diameter<40mm)/%			Package (diameter>40mm)/%			Bag package			On-mixed		
Year	2011	2012	2013	2011	2012	2013	2011	2012	2013	2011	2012	2013
Total explosives	42	37.5	33	18.1	20.1	23	22.1	22.6	24	17.8	19.8	21
Emulsions	58.3	54.4	—	37.0	40.7	—	4.7	4.8	—	0.0	0.0	—
Water gel	71.5	66.9	—	27.1	33.1	—	0.5	0.0	—	0.0	0.0	—
Expanded AN	47.4	44.5	—	2.5	1.7	—	50.1	53.8	—	0.0	0.0	—
Modified ANFO	30.8	24.9	—	6.3	9.4	—	63.0	65.7	—	0.0	0.0	—
Powder emulsion	42.9	38	—	3.6	3.5	—	53.5	58.5	—	0.0	0.0	—

Before 2008, the total output and main species distribution was only presented in the census of commercial explosives in China. According to the relevant information and the census of recent years, coal mining is main application fields of explosives in china, which is similar to that in US. But the explosive capacity used in Chinese coal mine is smaller than that in US. Most coal mines are open in US while coal mines in china are mainly in underground. There are relative much amount permissible explosives in china, which are in small diameter cartridge and sensitive to one blasting cap.

From the practice now, a large proportion of commercial explosives are cap-sensitive in China. Even for the big diameter packaged(more than 75 mm) or bagged explosives used for deep hole blasting, most of these species are sensitive to ordinary No.8 detonator(e.g. Powdery emulsion explosives, expanded AN and modified ANFO), which are manufactured by sophisticated production techniques and equipment in fixed factory from Tab.5.

Nowadays, the main species of permissible coal mine explosives are safe level II and safe level III emulsion explosives or water gel. Due to the intrinsic defects, it is hard to manufacture relative high safe level permissible coal mine explosives from dry, powdery AN explosives, which account for small percentage in all permissible explosives.

From the view of blasting technology, there is no need to employ cap-sensitive explosives in deep hole blasting. It will be good to use simple-produced or the onsite mixed ANFO, or the blends of ANFO and emulsions (Heavy ANFO), which have sufficient power to blasting and crushing rock or ores. The onsite mixed and cheap explosives, non-electric initiation system and booster technique means the advanced blasting technology, which was popular abroad early in the 1960 s of last century and now is just attached importance through arguments and efforts of many years in China.

4 DEVELOPMENT TREND OF CHINA'S COMMERCIAL EXPLOSIVES

(1) Be optimistic to China's demand for total consumption of commercial explosives in the coming decades. The United States is the most developed country in the world today. Its commercial explosives mainly uses in mining and the annual consumption reaches to 3 million tons today. China is still a developing country and will inevitably consume various resources for its industrialization and infrastructure construction in future. Thus it will still consume a large mount of explosives. It is expected that China will have great demand of commercial explosives in the coming decades or even a longer period.

(2) Vigorously promotion of the development of safer, cap-insensitive, simple-manufactured explosives. From the view point of cleaning and environmental protection, species and type of China's commercial explosives is similar to that of the United States, which is consistent with the overall trends of industrial technology development. However, as to the open-pit deep hole blasting, the explosive species does not match with blasting technology in China and not achieve the safer and more efficient in technology and economy.

As the improvement of productivity, the deep hole blasting technology has been widely and consciously accepted by society currently. Meanwhile, it has been widely used in basis engineering constructions of road, water hydropower etc. and quarrying, mining etc. Therefore, it should devote major efforts to developing safer, cap-insensitive, simple-manufactured explosives and promoting onsite mixed and simple-manufactured explosives, such ANFO in various density, cap-insensitive slurry and emulsion blasting agents, and their mixtures with ANFO (Heavy ANFO). Currently, it is not economic and safe, or even not necessary that emulsions, powdery emulsion, the modified ammonium nitrate fuel oil explosives with large diameter or bagged package and more over with cap-sensitive are used in deep-hole blasting produced by many factories.

(3) Developing coal-mining permissible explosives with the property of safer level against methane-air ignition and more reliable detonation at same times. Coal-mining in the United States is a major contributor of explosive consumption, but most of its coal mines almost is open-pit. Thus, the

amount of permissible explosives is few. This situation makes no one shows any interest in technology improvements, performance testing and government regulation for coal mine permissible explosives in the United States. For example, the test gallery and relative instruments for coal mine permissible explosives could not run normally. An explosion event in 2008 urged Institute of Manufacturers of Explosives to ask the United States Council pay attention and fund for the permissible explosives. In China, most coal mines are underground and under gassy condition while coal is primary energy consumption in China. In order to ensure the blasting safety of coal mining, it should not weaken the technologic improvements, performance testing and government regulation for permissible explosives. Blasting operation in coal mine requires that the permissible explosives should be cap-sensitive. Meanwhile, in order to prevent blast to cause the ignition of methane and coal dust, the energy content of permissible explosives must be controlled and chemical inhibitors are often needed to add to the permissible explosive compositions. Meanwhile, rock-breaking capacity of the permissibles also should be ensured. Based on the current situation of China's coal mine permissible explosives and the blasting practice of underground coal-mining, coal mine permissible explosives of water-gel type are the most safe against the ignition of methane and coal dust, and the most reliable in detonation, especially for the high-safe level permissible explosives. So, it is suggested that the high-safe level permissible explosives should be water-gel type explosives.

(4) The enhancement for production concentration of commercial explosives. The numbers of manufacturer of China's commercial explosives have fallen significantly for many years, and the production scale and concentration level have increased. However, Compared with the United States, there are still more manufacturers of China's commercial explosives. In order to improve safety and efficiency, it is the instinctive needs for enterprises and society to enhance the production concentration.

REFERENCES

[1] US interior department, US Geological Survey (USGS). Minerals Yearbook Explosives. http://ime.org.

[2] China Explosive Materials Trade Association, Bulletins, Series No. 149,157,185,224,249,262,273.

[3] Wang Xuguang. Current status and future prospect of engineering blasting and explosive materials in China[J]. Engineering Blasting, 2007, 13(4): 1~8.

Nonel Detonator Delay's Error and One by One Detonate Network

LIAO Xiaolin[1, 2], CAI Jianhua[2]

(1. Guiyang Xinxing Blasting Engineering Ltd., Guiyang, Guizhou, China; 2. The Institute of Judicatory Appraise for Explosive, Police Officer Vocational College of Guizhou Province, Guiyang, Guizhou, China)

ABSTRACT: This paper proceeds from the minimum standards of GB19417-2003 "NONEL DETONATOR" inspection of products, through the NONEL DETONATOR delay error analysis, explains the probability of simultaneously detonated at the same time exists in one by one detonate networks, points out the impact on the maximum segment dosed, and explain this calculation methods of probability in the table of normal distribution briefly.

KEYWORDS: delay error; probability of simultaneously detonated

1 INTRODUCTION

1.1 Issues Raised

There is always an deviation of detonator delay's accuracy, and deviation of qualified products always is limited in a range of provision. In time directional axis, milliseconds as the basic unit, Showing the length "section" segment (time range), That is "delay number" of the detonator; in accordance with existing safety distance calculation on blasting vibration requirements, blast hole's total dose within a certain period of time, is "a dose"(or" single-dose response ").

In this article, "Several holes Simultaneously detonated" refers to the number of blast holes initiated in a short period time; "Simultaneously detonated" refers that these boreholes are detonated within a certain some time range. According to a safe distance from the blasting vibration is calculated on the maximum period prescribed dose of anti-calculated, should be site-specific dose of a detonating cap control. In practical engineering blasting, when the charge of each blast hole is determined, the maximum amount of controlled charge is for the numbers of blast holes with the same delay time.

Now commonly used by hole initiation network, that is, "high segment in the holes, lower segment outside holes" which two segments combined detonator network (that called "tandem network"). It makes better blasting vibration member and easy to operate, increasingly extensive use of controlled blasting in towns and cities in the country, usually we use such a network at the state of the safety instance in the critical state, if all hole dose by single for the maximum period of review of the safety distance, suspect it is wrong, then write the article.

1.2 About Nonel Detonator Segment Number Partition and Error

The segment of detonator is divided with the extension of time and error, It should be broadly in line with the general rules of the shock wave propagation in rock. Currently, there has no in engineering blasting. Is actually a reference to the nonel detonator of the segment is divided; and thus to determine the maximum segment dosed.

Partition criterion of detonator segment is mostly in accordance with existing national standards, GB 19417—2003"NONEL DETONATOR" delay time" set forth in "the first series" implementation, Generally have 20 segments, upper and lower limits specified in paragraphs error is: "Any period of extension detonator's delay time limit specification (U) are the extension of time and the upper segments an extension of time value. The extension of the time limit specification (L) is an extension of the paragraph following paragraph extension of time and the time value. "

According to the regulations, visually see: 1segment from the lower limit of zero delay starting. Between all adjacent upper and lower segments are continuously milliseconds, Each paragraph maximum delay value between the minimum value of the delay period has no interval, and no overlap. That is to say in normal circumstances, no matter how much delay, always within the scope of a particular segment.

Corresponding author E-mail: liaoxiaolin526@qq.com

2 CURRENTLY CONTROLLING PRODUCTION DETONATOR DELAY ACCURACY

2.1 About Nonel Detonator Production Process

A combination of the detonator tube is connected to the delay detonators fire tube and, external Nonel way. The delay flameless drug into a special lead tube, after several tension, compression and extension become thinner, control the length of cut precisely, low detonation wave generated by the detonation tube ignited and spread to the extension of medicine primary detonator initiation, in order to achieve detonator delay control.

2.2 Many Factors Influence the Accuracy of the Delay

In the manufacturing process, delay finalized by the delay length of the tube, however, cursory analysis of the impact of these delays accuracy as well as the following: the types of extended drug type, drug impurities and moisture content, drug fineness, mixing time and the way of mixing drugs the way, composition and characteristics of lead pipe, number and the drawing stress, drug loading density, etc., and the initial temperature, atmospheric pressure, humidity, detonate excitation energy, etc. These factors are relatively affect the effects of delay variations. Each batch production detonator of delay, Performance on the delay length of the tube, Even there's differences between the same segments.

2.3 Delay Accuracy Errors of Normal Distribution and Current Production Acceptance Rules

Generally, If a variable is larger, unrelated factors misplaced priorities of the joint action of the results. So this variable is subject to Normal distribution, The delay error detonator is just such a situation. Theoretically, Nominal delay value of each segment and the lower limit is established, There is a standard differential of normal distribution. Under ideal conditions, If the accuracy of the various factors affecting the delay in production control well (Eliminates all systematic errors), Each segment of its products within the error condition is similar to time in milliseconds to a few basic unit driving shaft, Delay is the name of the midline (some segments do slightly different), A symmetrical normal distribution curve. When considering the range of 99% of the whole paragraph, Value corresponding to the ends of the curve can be regarded as the provisions of the upper and lower limits, For the rest of the range due to a "small probability event," It should not be considered in project.

In accordance with the normal curve on theoretical analysis shows that: the probability of either a segment Before the nominal delay value, or after each one standard deviation range of quality products in the third paragraph of the central region (ie "base") detonate, is 68%; and the probability of before the nominal delay value, or after each two standard deviation range of products qualified middle segment about two-thirds of the initiation, is 95%; the probability of before the nominal delay value, the standard deviation for each of the three segments within the range of quality products that detonate, is 99% (see related probabilistic analysis data), this is the ideal product delay nonel detonator situation.

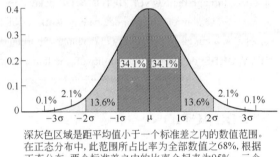

深灰色区域是距平均值小于一个标准差之内的数值范围。在正态分布中,此范围所占比率为全部数值之68%,根据正态分布,两个标准差之内的比率合起来为95%；三个标准差之内的比率合起来为99%。

Fig.1 Indicate the standard normal distribution

Nonel detonator delay precision reflected in two aspects, "accuracy" and "intensity". The former is the mean deviation of the sample delay production in test measured value and the nominal value of the delay. The latter is the sample standard deviation measured.

From the national standard GB 19417—2003 "NONEL DETONATOR" normative appendix A "delay time measured Supplementary Regulations" can be seen (see scan in "A.2"), its "Receive curve" taking the sample standard deviation and the mean deviation, when the sample mean deviation of the nominal delay value is too large. It is required to reduce the sample standard deviation which will pass, and vice versa. In quantities of 3201 to 10000 extract, 20 rounds as samples, from the "Receive curve" diagram can be seen, when the sample mean is equal to the nominal delay value (where $y = 0.5$). Sample standard deviation of the maximum control value: when normal inspection, sample standard deviation (S) and set the range ($U-L$) ratio X. Defined as no more than 0.230 (relaxing the test is not greater than 0.340, when tightened inspection is not greater than 0.205). Compared with this value is 0.167 standard normal distribution curve. Obviously, each segment maximum standard deviation (S) is: $s = 0.23 \times (U-L)$.

According to "normal 3δ principle": if the quality characteristics obey normal distribution, then within ± 3 standard deviations range contains 99.7% of the value of quality characteristics. Therefore, actual samples of qualified acceptance range (*R*) can also be range than the set (*U-L*) larger. The actual range (*R*) relax the maximum increases in multiples: 0.23÷0.167 = 1.38, this is the maximum increase factor of standard deviation (*S*).

附 录 A

(规范性附录)

延期时间测定的补充规定

A.1 延期时间的上、下规格限

任何一段延期导爆管雷管的延期时间的上规格限(*U*)均是该段延期时间与上段延期时间的中值，而延期时间的下规格限(*L*)则是该段延期时间与下段延期时间的中值。瞬发导爆管雷管在与延期导爆管雷管配段使用时，延期时间的下规格限为零。末段延期导爆管雷管的延期时间的上规格限，可用末段延期时间与延期时间的下规格限之差加上该段延期时间求得。

示例：

半秒 1 段：$L=0$；$U=(0+0.50)/2=0.25$。

半秒 10 段：$L=(4.00+4.50)/2=4.25$；$U=(4.50-4.25)+4.50=4.75$。

A.2 延期时间合格与否的判定

A.2.1 试验后按式(A.1)和式(A.2)计算图 A.1 所示接收曲线横坐标(*x*)和纵坐标(*y*)的值。

$$x=\frac{S}{U-L} \qquad \text{(A.1)}$$

$$y=\frac{\overline{X}-L}{U-L} \qquad \text{(A.2)}$$

式中：

S——试样延期时间的标准偏差值，单位为秒(s)或毫秒(ms)；

U——试样延期时间的上规格限，单位为秒(s)或毫秒(ms)；

L—— 试样延期时间的下规格限，单位为秒(s)或毫秒(ms)；

$\overline{X}$—— 试样延期时间的平均值，单位为秒(s)或毫秒(ms)。

延期时间及其平均值、标准偏差值在记录和运算中的有效数字应和表 1 中给出的有效数字相同。计算 *x*、*y* 时，应保留三位有效数字，下一位按 GB/T 8170 的规定进行修约。

A.2.2 把点(*x*、*y*)描在对应的接收曲线上。如该点落在接收曲线内(包括曲线上)，则为合格，否则为不合格。

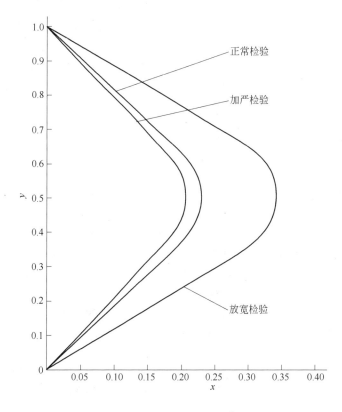

Fig.2 National standard GB 19417—2003 "NONEL DETONATOR" Appendix A, "whether or not an extension of time passing judgment"

Fig.3 National standard GB 19417—2003 “NONEL DETONATOR” Appendix A, "Extension of time to receive curve"

3 ABOUT "ONE BY ONE DETONATE NETWORK"

3.1 About “Simultaneously Detonate "Number of Holes Determine the Value of *Δ*

Tab.1 shows the name of the national standard GB19417-2003 "NONEL DETONATOR" to determine the nonel detonator of delay value *N*, the provisions of range *M*, and the calculated actual maximum range *R*.

Hypothesis: provisions hole detonators range as *M*, the hole outside the nominal delay detonator is *N*, $\Delta = M / N$, then: $0 \leqslant \Delta < \infty$.

discussed as below:

When the $0 \leqslant \Delta \leqslant 1$, if no consider the propagation velocity of nonel. Analyze the situation of time delay, this network really is "one by one" detonate. Such as $\Delta = 0$, hole is 1 segment (Excluding on deviation, delayed by zero), hole outside is any period ; Or the $\Delta = 1$ (That is $M = N$, such as the hole outside is two segments, a combination of hole 3 segments, or the outside the 3-hole section, 7- hole segment portfolio); or the hole, inside and outside the same paragraph. However, in rock due to explosion shock waves propagation speed greater than nonel propagation speed, simultaneously is easy flying rocks smashed near nonel, such a network is easy is algorithm of fault (smashed) off, network security is poor.

Tab.1 Nominal delay values *N*, predetermined poor *M*, and the actual maximum range *R* (ms)

	1	2	3	4	5	6	7	8	9	10	11	12	13	14	15	16
N	0	25	50	75	110	150	200	250	310	380	460	550	650	760	880	1020
M	0	25	25	30	37.5	45	50	55	65	75	85	95	105	115	130	160
R	0	35	35	41	52	62	69	76	90	104	117	131	145	159	179	221

When the $1 < \Delta < 2$, the probabilities of two holes "Simultaneously detonate "exist, and when the Δ value changes from 1 to 2(for example: hole of the outer segment unchanged, hole segment increases,) "Simultaneously detonate " probability changes from small to large.

When the $2 < \Delta < 3$, the probabilities of three holes "Simultaneously detonate" exist, (at this point two holes "Simultaneously detonate " probability of even greater), and the Δ value changes from 2 to 3, "Simultaneously detonate " probability changes from small to large, More than analogize.

Extreme cases are: hole's outside are 1 segment, hole's internal is arbitrary in the same paragraph that when Δ to ∞ (due to the presence of detonators "detonation growth period", to 1 segment deviation detonator there, Delay can not be zero), all bore the same probability of distribution curve segment blasting cap basically coincide, "Simultaneously detonate" probabilities can be 99% (overall detonate).

In theory, as long as M is larger than N, namely $\Delta > 1$, there is probability of several holes "Simultaneously detonate ", in other words, commonly used "high segment hole's internal, the hole outer lower segment" one by one detonate network, there are several holes "Simultaneously detonate" probabilities easily. In actual blasting, this probability of is small reasonable combinations (Or does not exist), otherwise larger.

Therefore, this reference to the NONEL DETONATOR's segment divided to determine the maximum doses practice, in one by one detonate network, obviously can not simply be regarded as the greatest single hole dose, in particular, when probability of a larger "Simultaneously detonate ", and a safe distance at the critical state, the holes should be combined together and the probability of size, blasting the number, considering the environmental safety requirements.

3.2 About Select the Hole the Outer Segment Number Combinations

In one by one detonate network, rational combination of hole's internal and external detonator segment, should be affiliated with rock hardness, structure and pore network parameters etc, we selected reasonable delay time according to these rules, that is a reasonable choice of hole external segment. In this case difference is too large flying rocks grave, too small probability of a large number of holes "Simultaneously detonate ". While also taking into account of the reliability of the booster, as possible to first hole detonate an instant booster farther; Engineering Practice in Blasting has proved, in general the deep hole blasting, for medium-hard rock, hole outside the detonator as in the above 6 paragraph is flying rocks more serious, should be 3 to 5 segments external the holes, corresponding holes 9 to 13 segments is appropriate. For the slightly soft rock, select appropriate paragraph 5 hole of external, if hole outside selected the 2 segments, hole inside is the high segment, while network security is better, however, the probability of "Simultaneously detonate " is a large number, to increase the maximum period of multiple dose (corresponding increase safety distance); such a network, if used in demolition blasting will, also be easy to cause detonation order reversed, blasting effect caused turnovers.

3.3 Accurately Calculate the Maximum Probability of Several Holes Simultaneously Detonated

Thinking rough calculation points are: in poor normal graph showing, pan right from the lower point (can be set to zero) the nominal delay value (equivalent to the delay), after the end of moving is still within the range, once for the presence of two holes "Simultaneously detonate "probability, quadratic is the presence of three holes "Simultaneously detonate "probability, by analogy until reached the reaches or exceeds the maximum point is reached, after partially accounted the percentage of the total area of the normal curve, that is the nominal number of holes "Simultaneously detonate " the maximum probability under appropriate conditions.

For a more accurate calculation of the nominal maximum probability, which can follow the "standard normal distribution table" to carry out, this time should standard deviation (in milliseconds) as the basic units in the translation. Fig.1 shows precisely the time delay of hole outer notional is delay range one sixth of hole inside. Nominal maximum probability of two holes "Simultaneously detonate "is the area from-3δ to 2δ beneath the curve, three holes "Simultaneously detonate "nominal maximum probability is the area from-3δ to 1δ beneath the curve, by analogy. "Nominal maximum probability" is the results "confidence interval" for the entire dwell range when obtained; when the "confidence interval" shrink, this will reduce the probability but confidence level is increased; if taking into account further delay error hole outside the detonator, when the delay value is not the same, the nominal maximum probability will be slightly changed.

With 3 segment (outside) and 9 segment (inside) combination as an example: 3 segment's nominal delay is 50 ms(280 ms to 345 ms), apparently 65 ÷ 50 = 1.3, 1 <Δ <2, "Simultaneously detonate " that the probability of two holes is exist, 9 section of a standard deviation up to 65 ÷ 6 = 10.83 ms, in 9 segment of the delay range, the 3 Section is equivalent to the nominal delay of 50 ms are: 50 ÷ 10.83 = 4.61 standard deviations, examine "standard normal distribution table." 4.61 standard deviation of the value should be 0.9463, that is the name of the greatest probability of two holes "Simultaneously detonate " are as follows: 1-0.9463 = 0.0537 = 5.37%; however, according to the preceding, acceptance standard deviation 1.38 times the maximum relaxation, actual 9 segment range（R）is R = 1.38 × 65 = 90 (ms), the actual maximum standard deviation was 90 ÷ 6 = 15 (ms), the 3 Section is equivalent to the nominal delay of 50 ms are: 50 ÷ 15 = 3.33 standard deviations, Examine "standard normal distribution table", 3.33 standard deviation of the value should be 0.6293, that is the name of the greatest probability of two holes "Simultaneously detonate " are as follows 37.07%, this is 9segment range 99% confidence region is calculated on the basis, its 95% confidence interval based on apparently more realistic, such result becomes 7.0%, this value is known as the "actual nominal maximum probability."

"Actual nominal maximum probability" calculation conditions are: (1) Minimum eligibility criteria for product acceptance (normal inspection "on the curve or curve"). (2) The actual confidence interval of range 95%. (3) The nominal delay value of holes outside.

Calculated according to the method described above, the actual nominal maximum probability of a combination of several typical number of "Several holes simultaneously detonated" shown in Table II （"—" means not exist, Blank space is not calculated）, for reference.

From the calculated results can be seen: (in the segment can be actually used) probability of "Several holes Simultaneously detonated" is increased with the increasing of hole inside segment number, and decreased with the increasing of hole outside segment number.

Tab.2 Several typical combination of actual nominal maximum probability

Inside \ Outside	2 segment			3 segment		4 segment		5 segment	
	II	III	IV	II	III	II	III	II	III
9 segment	63.5	7.0	—	7.0	—	—	—	—	—
11 segment	71.9	16.9	—	16.9	—	1.1	—	—	—
13 segment	84.9	47.1	11.7	47.1	—	11.7	—	—	—
15 segment				63.5	7.0	30.6	—	2.5	—
16 segment	92.6	75.1	48.4	75.1					

3.4 Set VAlue and Actual Value of Range

In order to adapt detonator manufacturing production levels, current GB is to relax the standard deviation, comparison of the standard normal distribution curve, normal curve of products become "chunky", resulting in range cross range of adjacent segment number increases. For example No. 16 segment, the actual poor R = 1.38 × 160 = 221 ms, with a range of 15 segment to 31 ms cross. For each segment number, there are actually expanding this range segment number of cases. This shows that even if the product is inspected, the upper and lower ends, delay values adjacent segment number is always overlap, that is contradictory with this article first "Between all adjacent upper and lower segments are continuously milliseconds, Each paragraph maximum delay value between the minimum value of the delay period has no interval, and no overlap" In this regard it is understood as the "actual state" NONEL DETONATOR errors, the GB is just "settings state", only a small probability of overlap at it.

Therefore, to avoid this small probability, requires "every segment" for use in blasting, Is necessary; when you can not avoid, and the segment number must be continuous used (For example, in paragraph No. be reluctant to use, and tunnel blasting cross section larger segment number more), for maximum security, when the reviewing a safe distance, It should be with two, three or even more than the total dose continuous segment number.

4 CONCLUDING COMMENTS

(1) This paper starts with the minimum standard of NONEL DETONATOR qualified products, in various combinations one by one detonate network, analysis and calculation the actual nominal maximum probability of "several holes Simultaneously detonated", demonstrated its influence on the maximum dose of segment number.

(2) In the design of one by one initiation network, it should be combined together of "simultaneously detonated" holes, the probability of size, the case of rock, blasting frequency, environmental safety requirements etc. Reasonable choice of segment number combination, and to determine the maximum period of dosing.

(3) Currently, the ways of using the NONEL DETONATOR segmented criteria determine the maximum dosing period practice, weather meets the propagation of blasting vibration or not ? How to determine the maximum period of reasonable dose? Worthy for blasting industry colleagues to discuss.

(4) The short term can not be changed NONEL DETONATOR production process, manufacturing principles are completely different, error Control accurate electronic detonators be used, in blasting engineering is the right direction.

(In this paper several times amendments, Get Shi Yayu researcher careful advice, In this very grateful)

REFERENCES

[1] GB 8031—2005 "industrial electric detonator".
[2] GB 19417—2003 "NONEL DETONATOR".
[3] "Blasting materials" in 2008 the second phase, Yan Honghao Etc. "Series and parallel network superposition blasting probability calculation".

Production of Nitroester-Based Explosives in Russia

A.S. Zharkov, E.A. Petrov, N.E. Dochilov, R.N. Piterkin

(JSC Federal Research and Production Center "ALTAI", Russia)

ABSTRACT: FR & PC "ALTAI" is one of the largest Russian scientific centers developing receipts and technologies of industrial explosives in Russia.

In 2013, it is 15 years since FR & PC "ALTAI" launched the production of high-safety nitroester explosives, which are applied at coal mines especially dangerous by the concentration of combustible gas and coal dust.

Production of high-safety explosives is distinguished by a high degree of explosion safety and survivability of technological operations. Closed water and acid-rotation, purification of rinsing water, catch of acid vapor and solid release by means of modern methods provide ecological purity at all stages of the process. The technological process is almost completely automated. Phase control of nitroesters production, the process of mixing and patronizing of explosives is performed remotely with the coordination at central control station.

Due to its flexibility and adjustability the technological complex is universal and able to produce all types of high-safety explosives.

KEYWORDS: explosion; explosives; nitroesters; uglenites; safety properties; technology; production process; efficiency of explosives

Federal Research and Production Center (FR & PC) ALTAI is one of the largest Russian scientific centers developing technologies and establishing productions of special chemistry, particularly, industrial explosives [1] including uglenites, carbatoles, detonites, hexoplasts [2]. Since 1978 FR & PC ALTAI has been playing a leading role in the sphere of industrial nitroester-based explosives in Russia.

Currently, powder explosives are mainly used: detonites and high-safety compositions with the content of nitroesters being not more than 20%. These explosives are used in cases when small-size charges (cartridges) with high sensitivity to detonation impulse are needed. Sensitization of explosives by liquid nitroesters is especially important in relation to high-safety explosives (uglenites, ionites), which are applied at coal mines dangerous by the concentration of combustible gas and coal dust. These are multicomponent selective-detonated explosives ion-exchange type, being the safest among the whole variety. Operating principle – very "clever" explosives, able to self-dose the release of energy, depending on process conditions of explosive process. In case of emergency, only a part of energy is released which is not able to ignite combustible mixture and in case of optimal operation – maximal amount, comparable to the energy of most industrial explosives. The necessary level of protecting properties is obtained by strict compliance of chemical and dispersive composition of the components.

After disintegration of the USSR, Russia was left without production of detonites and high-safety explosives. To eliminate the shortage of such productions, in 1998 FR & PC ALTAI launched one [3].

Nitroestersare made using injector method on the most up-to-date equipment in Russia, having capacity of 600 kg/h, which does not have analogues in the world. The main advantage of the unit is the low (not more than 10 kg) loading capacity and, hence, a high safety level. Not less than 200 kg [4] are processed on the most advanced foreign equipment.

The main advantages of the developed technology (Fig. 1) are high velocity of nitration process, separation of reaction mass into two phases of nitroestes control and purification as well as the use of efficient small-sized equipment. Nitration is performed inside the injector (3) cooled by nitrated mixtures (2). After cooling in the refrigerator (4), the reaction mass is separated in the centrifuge (5). Acid nitroesteris transported by means of injector (7) into the two-stage centrifugal extractor (14), where it is washed according to the counterflow scheme with 2%~3% of sodium solution. Purified from acids nitroesteris transported in the form of emulsion with warm water by means of injector (7) in mixing room(21) of industrial explosives. The quality of nitroester, taken out from washed extractor, is controlled uninterruptedly with the help of acidity indicator (20). Waste acid from

---

Corresponding author E-mail: frpc@secna.ru

the centrifuge is passed through the controlled separator (6), some part is back into the process (11), the other part is output onto the decomposition reactors of dissolved nitroesters (22) and transported to the denitration and concentration (10). Nitrogen oxides from reactors is neutralized on two absorption columns by means of urea solutions. Waste acids, scourage, containing nitroesters, are transported by means of sealless disk-shaped pumps (9). Control of equipment operation is remote. Duration of process cycle from the moment of alcohol loading into the nitrator up to the output of pure nitroester is 3~4 minutes, setup time and launching – no longer than 15 min. The unit is able to operate uninterruptedly for 30 minutes and up to 6 days.

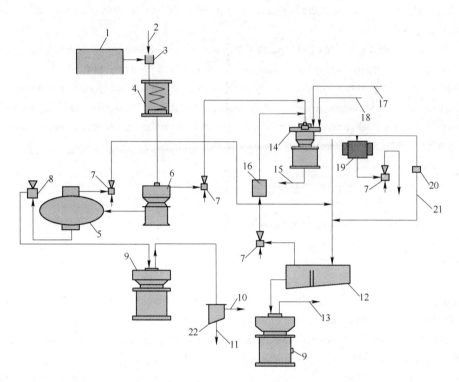

Fig.1 Technological scheme of nitroesters production

For replication in Russia and abroad, FR & PC ALTAI designed a unit with a capacity of up to 50 kg/h. Here, on the basis of sealless disk-shaped pumps for nitration of polyatomic alcohols, centrifugal nitrators with a loading of not more than 100 kg were developed. Due to the application of these nitrators, there has been created a two-stage process of alcohol nitration without returning to the waste acid.

On the whole, the production is characterized by a high explosion safety, reliability and viability of technological operations. Closed-loop water and acid circulation, treatment of scourage, collection of acid vapors and solid discharge using modern methods provide ecological cleanness at each stage of the process. Technological process is practically fully automated. Phase management of nitroester production, blending and cartridging of explosives is executed remotely with the help of the controlled computing complex coordinated on the central control station. Due to its flexibility and maneuverability of the equipment, technological complex is considered to be universal and is able to produce both standard (Tab.1) and novel (Tab.2) explosives.

Tab.1 Standard nitroester explosives

Parameters	Blasting gelatin	62%-dynamite	Detonite M	Uglenite E-6	Uglenite 12TsB	Ionite
Density/ g · cm^{-3}	1.55	1.4~1.5	0.95~1.2	1.1~1.2	1.0~1.3	1.0~1.2
Velocity of detonation/ km · s^{-1}	7.8	6.5	4.2~5.0	1.9~2.2	1.9~2.0	1.6~1.8
Exothermicity of reaction/kJ · kg^{-1}	6530	5333	5800	2680	2300	1930
Performance/cm^3	595	360~400	460~500	130~170	95~120	95~125
Brisance/mm	24	15~18	18~22	7~11	5~7	5~6

Tab.2 High-safety explosives (class 5)

Parameters	E-6 (mod)	S	P	M	13P
Exothermicity of reaction/kcal·kg^{-1}	640	700	780	707	630
Gas volume/L·kg^{-1}	560	600	600	685	665
Velocity of detonation/km·s^{-1}:					
of an open charge	2.3	1.8	2.3	1.8	2.0
of a cased charge	2.6	2.3	2.8	2.5	2.4
Sensitivity:					
to shock/%	68	30	70	62	60
to friction/kgf·cm^{-2}	2300	4500	2500	5400	2400
Mass of a charge limit/g	250	300	200	350	250
Efficiency/mm	6.8	7.8	8.3	7.6	6.8

Successful experience of production activity between 1998 and 2013 confirmed correctness of the chosen technological and engineering solutions. Over the whole period of time, application of these high-safety explosives by the consumers has not caused any accidents. Achieved output makes up 2 000 tons a year that makes it possible to satisfy the demand of coal industry establishments of Russia and the near abroad.

REFERENCES

[1] Zharkov A S, Dochilov N E, Litvinov A V, et al. Innovation Developments in the Sphere of Subsurface Management and Blasting Work;Expert-Tekhnika (Expert Technology), 2008(2): 40~49.

[2] Petrov E A. Industrial Explosives of FR & PC ALTAI;From the collection of scientific papers: Safety of Colliery Undertakings; Kemerovo, NC VostNII, 2002: 86~95.

[3] Zharkov A S, Dochilov N E, Petrov E A, et al.;Production of Nitroesters and Industrial Explosives on Their Basis;GornyZhurnal (Mining Journal), 2006(5): 37~41.

[4] Piterkin R N, Prosvirin R Sh, Petrov E A. Technology of Nitroesters and Nitroester-Based Industrial Explosives; Monograph, Biysk, Publisher: Altai State Technical University, 2012: 268.

Preliminary Discussion on Greening Design and Assesment of Industrial Explosives

ZHANG Daozhen, LI Yunxi, SONG Zhiwei, JIANG Tiansheng

(Zhejiang Gaoneng Blasting Engineering Co., Ltd., Hangzhou, Zhejiang, China)

ABSTRACT: As a special energy, industrial explosive occupies an important place among the special energy resources. This paper proceeds from general chemical process. Based on the twelve principles of greening chemistry proposed by Anastas P.T, it puts forward greening design ideas of industrial explosive on the basis of formulation design, product requirements, atom economy and energy consumption reduction. It also discusses greening quantitative assessment from atom economy, environment factor and environment coefficient, and production process of energy efficiency.

KEYWORDS: industrial explosive; greening chemistry; greening design; assessment

1 INTRODUCTION

In the 1990s, America Chemist Anastas P.T et al. proposed the concept of greening chemistry for the first time, and they also proposed the twelve principles of greening chemistry. Greening chemistry has become forefront research of international chemical study now. Due to its essential character, specialty of production and importance of safety, as basic chemical energy, the raw material, technology and equipment of industrial explosive also will develop in the direction of green with the development of military, engineering technology, mining industry and chemical industry and so on. In recent years, green industrial explosive has been proposed by scholars. However, they seldom do researches on how to asses and evaluate greenness on the whole life cycle of design, production, transportation, employment and abandonment.

2 THE CONNOTATION OF GREEN INDUSTRIAL EXPOSIVE

To design the route of the formation of chemicals and the corresponding molecule conversion, and increase the energy transmission efficiency and safety, and reduce energy consumption and unnecessary raw material consumption during the production of industrial explosive, and raise the continuity of productive process, the essential safety of production and the safety of productive process, green development of industrial explosive should be based on green chemistry, atom economy and the standard of friendly environment. The architecture of it should include green production design, green production process, and the development of clean energy and raw material.

3 THE MEASURES OF THE REALIZAITON OF GREEN INDUSTRIAL EXPLOSIVE

3.1 The Request of Going Green for Formulation Design and Production

The raw materials of industrial explosive and non-toxicity of reactants are important index in measuring greenness of industrial explosive. What greenness need is to apply the principle of reduction and elimination harm to the greatest extent in every aspect of chemical design. Green industrial explosive requests that the raw material should be renewable material instead of exhausted harmless material(reagent, solvent, catalyst). Besides, the production of industrial explosive is mainly H_2O, N_2, NO, NO_2, CO, CO_2 and NO, CO are poisonous gas. Green chemical requests the least poisonous gas after explosion, so the formulation of industrial explosive is always designed as zero oxygen balance. Oxygen balance not only has an influence on the product after explosion, but also has an influence on thermo chemistry parameters of explosive.

3.2 Atom Economy of Industrial Explosive

Atom economy of industrial explosive should be comprehensively considered from atom economy and the result of explosion. High atom economy requests high energy efficiency before and after explosion, more non toxic product, high detonation performance and good mechanical adaptability to blasting objective. The speed, strength, gap distance and the ability to work of industrial explosive is pro-

---

Corresponding author E-mail: zhangdz_007@163.com

portional to the explosion heat and specific volume of it. High performance asks for high explosion heat and specific volume. Atom economy of explosive requires that volume of unit mass of industrial explosive should produce the maximum heat during the process of the appearance of the explosion product. And it also require that the production of more non toxic gas and less unnecessary by-product.

3.3 The Reduction of Energy Consumption and the Use of Renewable Energy

The reduction of energy consumption and the use of renewable energy is a key step of green industrial explosive. The optimization of synthetic reaction, improvement of atom efficiency to adapt to the request of low energy, and fundamental change of the demand of inner energy by the design of the reaction system, they all make a contribution to reducing the need of energy and the use of renewable energy to the greatest extent. For example, the ingredients of combustible are mainly non-renewable fossil fuels. Using the grease and wax of flora and fauna to take the place of industrial oil, synthesizing them under normal temperature and pressure, the design of reactor, and strengthening the coupling of unit process are all the directions need to be studied.

3.4 The Production Process and Essential Safety of Explosive

The production process of explosive includes formulations, component ratio, production mode, equipment and the power to control the parameters online and so on. These factors not only have something to do with production quality, but also have something to do with safety. Proper design of production equipment of industrial explosive, safe operation, effective protection measures are necessary for safe production. The development of production equipment and technology of industrial explosive should aim at improving security of explosive production and the technology of the reduction of environment pollution. In addition, in the formulations of industrial explosive, the physical components and chemical compatibility should be good. In the process of chemical transformation, all the material and forms of material should try to reduce the possibilities of chemical accidents, such as fire and explosion.

3.5 The Chemical Combination of Computational Chemistry and Green Chemistry

The combination of numerical operation of simulation chemistry and logical operation of computational chemistry which is called rational design of molecules will be the characteristic of chemistry in the 21st. It refers to molecule, molecule cluster, interface, particle, system unit and process. In the process of green chemical craft and technology, as a basis for thermal chemical theory to establish the mathematical model and experimental methods of combining the theory, with the help of the results of quantum chemistry calculation from computer program operation, using multi-scale modeling ideas and methods, to choose more accurately substrate, catalyst, solvent and reaction ways. In this way, less experiment leads to more effect. Then it provides help for complicated and multi-aim non linear problems, such as the choice of raw material of chemical craft, the design of reactor, the development of process and safety.

4 ESTIMATION OF THE GREEN CHEMISTRY PROCESS OF INDUSTRIAL EXPLOSIVE

4.1 Atom Economy

Atom economy is a way to estimate the environmental influence of synthetic chemicals of different ways before and after reaction. The atom economy of industrial explosive should be decided by the characteristics of explosive reaction, the heat produced by detonation production, explosion heat of the whole industrial explosive and the contribution of specific volume. Tab. 1 is the physical and chemical parameters of detonation.

Tab. 1 Industrial explosive physical and chemical parameters of detonation products

Detonation products	Formula	Molecular weight	Formation heat (25℃) /kJ·mol^{-1}	Formation heat of unit mass /kJ·g^{-1}	specific volume of unit mass /L·g^{-1}
Carbon monoxide	CO	28	111.69	3.99	0.80
Carbon dioxide	CO_2	44	393.13	8.93	0.51
Water vapor	H_2O	18	240.35	13.35	1.24
Nitrogen	N_2	28	0	0	0.80
Nitric oxide	NO	30	-90.16	-3.01	0.75
Nitrogen dioxide	NO_2	46	-34.40	-0.75	0.49
Aluminium oxide	Al_2O_3	102	1664.48	16.32	—
Sodium oxide	Na_2O	62	414.57	6.69	—
Sulfur dioxide	SO_2	64	296.61	4.63	0.35

It treats the choice or the productivity of production as the standard of the estimation of chemical reaction process or the quality of some synthetic technology. However, it doesn't take the production yield, excess reactant, and the use of reagent, the loss of solvent, the consumption of energy and the effect to the environment into account.

4.2 Environmental Factor and CO-efficient

Environmental factor (E-factor) is a standard of measurement raised by R.A Sheldon, a professor of Netherland in 1992. The bigger the E-factor is, the more waste and the worse effect to the environment. Because of the complicated chemical reaction and operation process, E-factor must be got from the date derived from real productive process. E-factor is related to reaction and other unit operation. What's more, E-factor only takes the quantity of waste into account and it leaves the unfriendly factor Q behind. Thus the value of the effect to environment should be estimated by environment factor and environmental coefficient. Generally speaking, EQ is put forward to quantitative estimate the effect of the whole life cycle to the environment. Environmental factor (E-factor) and environmental coefficient are important parameters on condition of different test data for the comprehensive utilization, and meanwhile as also one of the basic approach and important means for environmental impact comprehensive evaluation. The process of the design, manufacture and using of industrial explosives should be based on engineering practice experience and testing data to establish more reasonable calculation theoretical calculation formulas of EQ for green chemistry requirements.

4.3 The Efficiency of Productive Process

The estimation of the efficiency of productive process mainly includes economic efficiency estimation, energy flow estimation, carbon footprint estimation and life cycle estimation. The economic efficiency of energy is always expressed by the strength of the energy. It contains unit production output value, energy consumption of unit GDP, energy consumption of unit production and energy service consumption. The way of energy flow estimation was raised by Rahimifard. He also analyzed it from unit process energy consumption, production energy consumption and manufacture energy production. Carbon footprint estimation calculates various GHG emission quantities by the activity level data and emission factor data of the whole life cycle. Life cycle estimation is a way to estimate the effect of production or service system (from raw material, production manufacture, sale, use, abandonment to the final disposal) to the environment. The productive process of industrial explosive hasn't formed a standard to reflect the energy utilization ratio. Thus we should learn more about the experience from the foreign scholars and combine with the characteristics of industrial explosive manufacturing technology to establish appropriate manufacturing process assessment indicators and methods of energy efficiency.

5 CONCLUSIONS

The greening design of industrial explosive is based on the greening production. But it also changes and studies the traditional technology. The greenness estimation of production and process should treat the raw material and substrate as core, and comprehensively thinking from thermal chemical raw materials, industrial explosives and explosive performance and craft process of industrial explosives. And the selection of parameters and measures should be based on atom economy, environmental factor and environmental coefficient and production manufacture and various aims. We should reduce and debate pollution from the source. We should also make use of every atom of molecule to the utmost and make it combine with the aimed molecule. Compared with the traditional industrial explosive chemistry and technology, the green design of industrial explosive will attach more importance to environment protection, energy saving, atom economy and resources utilization and so on.

REFERENCES

[1] Anastas P T, Farris C A. Benign by Design Chemistry. In Benign by Design: Alternative Synthetic Design for Pollution Prevention [J]. *American Chemical Society,* 1994.

[2] Anastas P T, Warner J C. Green Chemistry: Theory and Practice [M]. *United Kingdom: Oxford University Press,* 1998.

[3] Collins T J. Green chemistry in Macmillan encyclopedia of chemistry[M]. *New York: Macmillan Inc.,* 1997.

[4] Lv C X. Green manufacturing of explosive [M]. *Beijing: National Defense Industry Press,* 2010.

[5] Lu M, Lv C X. Green Chemistry and Industrial Explosives[J]. *Explosive Materials,* 2002, 31(4): 12~16.

[6] Trost. The Atom Economy: A Search for Synthetic Efficiency[J]. *Science,* 1991, 254 (5037): 1471~1477.

[7] Wang X G. Emulsion Explosive[M]. *Beijing: Metallurgical Industry Press,* 2008.

[8] Yun Z H. Thermochemical calculation of slurry explosive[J]. *Explosive Materials,* 1980, 9(2): 1~6.

[9] Wu L X, Xue H X, Li C M, He N. The Safety Research on Continue Production Process of the Emulsion Explosive[J]. *Explosive Materials,* 2003, 32(2): 3~8.

[10] Liu Z F. Research on Creating energy efficiency evaluation index system *energy and environment*, 2007, 5(2): 2~4.

[11] Rahimifard S, Seow Y, Childs T. Minimising Embodied Product Energy to Support Energy Efficient Manufacturing[J]. *CIRP Annals-Manufacturing Technology,* 2010, 59(1):25~28.

[12] Wang S B, Yang J X, Hu, D. Assessment Method and Progress of Life cycle[J]. *Shanghai Environmental Sciences,*1998, 8 (11): 7~10.

Properties of Emulsion Matrices of Explosives Based on the Best Russian, Kazakh and Chinese Emulsifiers

E.A. Petrov[1,2], P.G. Tambiev[3], P.I. Savin[4]

(1. "Polzunov's Altay State Technical University", Naukograd Biysk, Russia; 2. Federal Research and Production Center "ALTAI", Naukograd Biysk, Russia; 3. Research and Production Enterprise "INTERRIN" Almaty, Kazakhstan; 4. Federal Research and Production Center "ALTAI", Naukograd Biysk, Russia)

ABSTRACT: Strain-stress properties of emulsion matrix (EM) poremit-type, obtained on the best emulsifiers of Russian, Kazakh and Chinese productions were investigated in this study. Comparative results are given here concerning emulsifying efficiency of the following emulsifiers: polymeric grade REM (Russia), pigmental grade "P" (Kazakhstan), polymeric grade SPAN-80 (China). The following parameters were estimated: microstructure, electric capacity, viability, thermal stability. Microstructure was determined by the method of optical microscopy with zooming in 400 times; viability was estimated according to electric capacity changing; thermal stability was measured by the method of differential thermo-gravimetric analysis; measurement of electric capacity was room-temperature. Studies concerning emulsifying efficiency showed that under equal experiment conditions more higher stress-strain properties and EM quality were obtained with REM. To achieve the same results of EM with SPAN and P, it is necessary to increase the content of emulsifier or burning phase in composition.

KEYWORDS: emulsion matrix (EM); explosive; emulsifier; electric capacity; viability; microstructure; thermal stability; emulsifying efficiency

Emulsion industrial explosives are widely applied practically in blasting in all mining countries and successfully used for over 25 years [1~3]. Emulsion structure with nanosized film of fuel on the highly dispersed particles of saturated solution of oxidizer provides the explosive with high explosive and water-resistant characteristics, as well as low sensitivity to mechanical and thermal influences. Meanwhile, explosive characteristics and stress-strain properties of the emulsion matrices (EM) are largely determined by the composition of the emulsifier used in their production. Stress-strain properties of emulsion matrix (EM) poremit-type, obtained on the best emulsifiers of Russian, Kazakh and Chinese productions were investigated in this study. The following grades of emulsifiers were used: - polymeric grade REM-2, Spec. (ТУ) 7511903-631-93 (Russia);- pigmental grade "P" ST TOO 38441379-01-2006 (Kazakhstan);- polymeric grade SPAN-80 (China).The following parameters were estimated:- microstructure – by the method of optical microscopy with zooming in 400 times;- electric capacity – by the method of GosNII "Kristall", measured at room temperature [4]; - viability – according to changing of electric capacity at room temperature after circular loadings "hot-cold" [4];- thermal stability – by the method of differential thermo-gravimetric analysis on thermal analyze DTG-60 by "SHUMADZU".Emulsion matrix was prepared in the laboratory mixer at temperature of 80℃ with the use of double-level agitator turbine type. The rotation speed was 3000 rpm. EM compositions and the results of studies on electric capacity are given in Tab.1. Marking of compositions is given in accordance with the grade of applied emulsifier.

The best results on electric capacity were obtained with REM in case of equal compositions and under similar experiment conditions. The required quality of EM on SPAN and P under the same conditions are not obtained, because the given results of electric capacity exceed the limiting values 250 πF. When adding wax, increasing the emulsifying efficiency, the electric capacity of SPANV is almost equaled with REM one, and improved for P. The results on electric capacity are conform to optical studies of EM microstructure (Figs.1~4). The most large-dispersed structure is obtained on P. When adding wax into composition, the size of emulsion particles is reduced and so its electric capacity is reduced as well. When increasing in P composition the emulsifier and combustible phase, the electric capacity of $P_{10.5}$ is equaled with SPANV values and in case of further adding of wax ($PV_{10.5}$), it reaches the REM quality. The same situation is for EM viability (Tab.2). REM stands not less than 6 cycles, corresponding to about 6 warranty months of explosive storage.

Corresponding author E-mail: post@frpc.secna.ru

Tab.1 Compositions and electric capacity

Component content/%	SPAN	SPANV	REM	P	PV	$P_{10.5}$	$PV_{10.5}$
Ammonium nitrate	79	79	79	79	79	77	76.5
Water	14	13.5	14	14	13.5	12,5	12.5
Wax	—	0.5	—	—	0.5	—	0.5
Emulsifier	1.5	1.5	1.5	1.5	1.5	2.25	2.25
Dis. fuel	5.5	5.5	5.5	5.5	5.5	8.25	8.25
Electric capacity/πF	280	144	142	280	180	150	127

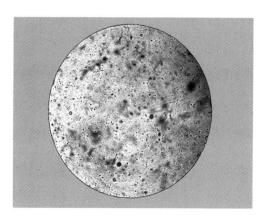

Fig.1 SPAN microstructure

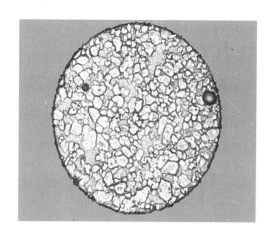

Fig.3 P microstructure

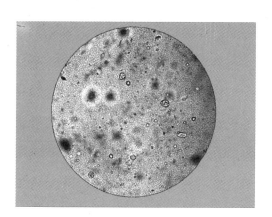

Fig.2 REM microstructure

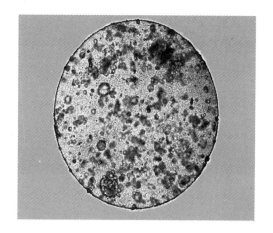

Fig.4 PV microstructure

Tab.2 EM viability

Number of cycles	Electric capacity/ πF				
	SPANV	REM	PV	$P_{10.5}$	$PV_{10.5}$
0	144	142	180	150	127
1	175	150	208	190	136
2	232	156	230	210	139
3	259	180	290	328	164
4	300	180	300	350	164
5	300	170	336	360	156
6	285	168	380	360	155

The vitality of SPANV and PV does not exceed two cycles and increases with the increasing of emulsifier content or burning phase in EM composition. The results on electric capacity showed the following. When heating up to 500 ℃ the main mass loss for all the species, connected with chemical interaction of EM components, is observed in the range of 200 ℃ with maximum heating rate at 240 ℃. There are some changes at the initial stage of heating up to 200 ℃, where the mass loss caused by expulsion of water or diesel fuel. More loss of mass is more in the species with low quality, i.e. in EM with higher electric capacity and coarse-grained structure. So, mas loss for REM is 10% and for P – 18%.

CONCLUSIONS

Studies on emulsifying ability of emulsifiers P, REM, SPAN-80 showed that in equal experiment conditions more higher physic-mechanical characteristics and quality of EM are obtained with REM. To achieve the same results of EM with SPAN and P, it is necessary to increase the content of emulsifier or combustible phase in composition.

REFERENCES

[1] Sosnin V A. World Tendencies of Development of Industrial Explosives. // Vzryvnoe Delo. –No 107/64. - M. – 2012: 107～121.

[2] Tambiev G I, Olshanskiy E N. Production Development of Industrial Explosives and Their Usage in RPE "Interrin". // Vzryvnoe Delo. - No 98/55. – M. – 2007: 192～203.

[3] Ilyin V P, Valeshniy S I. Emulsion Industrial Explosives in Russia // Vzryvnoe Delo. - No108/65. - M. – 2012: 174～190.

[4] Kolganov E V, Sosnin V A. Emulsion Industrial Explosives // Dzerzhinsk, Nizhegorodskaya obl. GosNII "Kristall". - 2009: 592.

Anti-dynamic Pressure Overloads Performance and Detonation Characteristics of Emulsion Explosives Sensitized by MgH₂

CHENG Yangfan, MA Honghao, SHEN Zhaowu

(*Department of Modern Mechanics, University of Science and Technology of China, Hefei, Anhui, China*)

ABSTRACT: Traditional emulsion explosives have pressure desensitization and low explosion power problems in utilization. In order to improve performance of emulsion explosives, MgH₂ sensitized emulsion explosive was invented. This type of emulsion explosives is sensitized by hydrogen storage material MgH₂, emulsion matrix will be sensitized in the process of detonation, and it is a dynamic sensitization. This special sensitization method makes emulsion explosives charged with a high density, which improves the explosion power density of blasting. What's more, dynamic sensitization can solve the safety problems of emulsion explosives in transportation and storage. Underwater explosion experiments and brisance testing experiments show that hydrogen sensitized of emulsion explosives has excellent anti-pressure ability, compared with Glass Microspheres sensitized of emulsion explosives, the shockwave total energy and peak pressure of hydrogen sensitized of emulsion explosives respectively increased over 30% and 20%, and the brisance of hydrogen sensitized of emulsion explosives even reaches 23mm in recent study.

KEYWORDS: emulsion explosives; pressure desensitization; explosion power; MgH₂

1 INTRODUCTION

Emulsion explosives were developed in the 1960s[1], and it was widely used in blasting engineering for its excellent water resisting ability, safety concerns and storage performance[2]. The matrix material of water-in-oil type (w/o) emulsion explosives is detonator insensitive until it is sensitized by a physical or chemical method converting it into emulsion explosive. However, traditional emulsion explosives have pressure desensitization phenomena and lower explosion power problems in utilization: detonation property of traditional emulsion explosives will be decreased if compressed by shockwaves or stress waves before detonation, this phenomenon is called "pressure desensitization"[3-5], and may lead to half explosion or misfire of emulsion explosives, which seriously affects blasting safety; researches on traditional high energy emulsion explosives mostly contain high explosives, perchlorates or high energy fuels (such as aluminum powders), which increased explosive sensitivity [6-8].

In order to solve the pressure desensitization problem of traditional emulsion explosives encountered in their utilization, a new type of emulsion explosives sensitized by hydrogen containing material MgH₂ was manufactured. Underwater explosion experiments and brisance testing experiments show that MgH₂ sensitized emulsion explosives have excellent anti-pressure ability and detonation characteristics.

2 EXPLOSIVES MATERIALS

In the experiments, the average particle size of Glass Microspheres (GM) is 55μm and its bulk density is 0.25g/cm³, purchased from Minnesota Mining and Manufacturing Company of USA; MgH₂ average particle size is 3μm, and its purity and bulk density are separately 98% and 1.45g/cm³, purchased from Alfa Aesar Company of USA; NaNO₂ dissolves in water easily, and its density is 2.17g/cm³, purchased from the Chinese Medicine Group Chemical Reagent Co., Ltd.; emulsion matrix density is 1.31g/cm³, purchased from Huainan Shun Tai Chemical Co., Ltd. The composition of emulsion matrix used in these experiments is described in Tab.1.

Tab.1 Composition of emulsion matrix (%)

Component	NH₄NO₃	NaNO₃	C₁₈H₃₈	C₁₂H₂₆	C₂₄H₄₄O₆	H₂O
Mass ratio	75	10	4	1	2	8

3 PRESSURE DESENSITIZATION EXPERIMENTS

Pressure desensitization phenomenon of emulsion explo

Foundation item: supported by the National Natural Science Foundation of China (51374189; 51174183).
Corresponding author E-mail: hhma@ustc.edu.cn

sives in millisecond delay blasting can be simulated by emulsion explosives compression equipment, and anti-pressure abilities of emulsion explosives sensitized by chemical sensitizers of $NaNO_2$, GM and MgH_2 are studied.

3.1 Preparation of Explosive Samples

3.1.1 Emulsion Explosives Samples

Formulation designs of emulsion explosives are listed in Tab.2, $NaNO_2$ and MgH_2 sensitized of emulsion explosives samples are put into thermostat at 322.15K for an hour, and then kept with GM sensitized of emulsion explosives at normal temperature for about 24 hours before experiments.

Tab.2 Different formulation designs of emulsion explosives

Emulsion explosives	Mass ratio/%			
	Emulsion matrix	GM	$NaNO_2$	MgH_2
GM sensitized	100	4	0	0
$NaNO_2$ sensitized	100	0	0.2	0
MgH_2 sensitized	100	0	0	1

3.1.2 Pressed RDX Booster Samples

Pressed RDX booster is composed of bulk RDX and paraffin with mass ratio of 100:5, and its weight and density are 10g and 1.65g/cm respectively.

3.1.3 Waterproof Treatment for Explosives Samples

Pressed RDX booster and emulsion explosives samples are cylindrical and spherical charges respectively, and are all encapsulated with polyethylene plastic bags, then entangled by rubberized fabric, and sealing is coated with Vaseline in order to achieve a better waterproof effect.

3.2 Experimental Methods

Pressed RDX booster is fixed in the center of the rectangular steel frame and emulsion explosives samples are placed at different distances from the pressed RDX booster by steel wires, and the emulsion explosives compression equipment is placed underwater, as shown in Fig.1(a). The pre-compression degree of emulsion explosive samples by an underwater shock wave induced by a RDX booster detonation depends on the spacing between the booster and the emulsion explosives. Finally, these compressed emulsion explosives samples will be detonated in the underwater explosion testing tower, and shockwave signals are recorded by oscilloscope; the structure of the underwater explosion testing tower is shown in Fig.1 (b). With regard to different types of emulsion explosives which are compressed at the same distance, we have tried to ensure the same delay between sample pre-compression and the subsequent detonation test.

The depth of water "H" is 5m and the charge position "h" is 2.5m underwater, the distance between charge and sensor "R" is 0.7m, as shown in Fig.2 (b). Testing instruments include Agilent5000A digital storage oscilloscope, 482A22 type of constant current source and ICP138A25 type of PCB pressure sensor (PCB Piezotronics Inc., USA.). Anti-pressure desensitization ability of emulsion explosives can be estimated by comparing output energy of non-compressed and pre-compressed emulsion explosive. Each type of emulsion explosive sample has been tested three times.

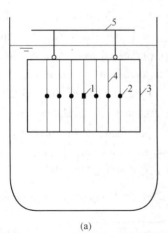
(a)

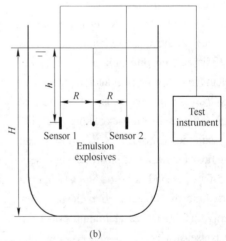
(b)

Fig.1 Assembly experimental system
(a) emulsion explosives compression device; (b) underwater explosion testing device
1—pressed RDX booster; 2—emulsion explosives; 3—steel frame; 4—steel wire; 5—support

3.3 Experimental Results

Fig.2 shows typical shockwave pressure-time curves of GM, $NaNO_2$ and MgH_2 sensitized emulsion explosives compressed by external shockwaves at different distances in underwater explosion experiments. The shorter the spacing between the booster and emulsion explosive, the stronger its pre-compression degree.

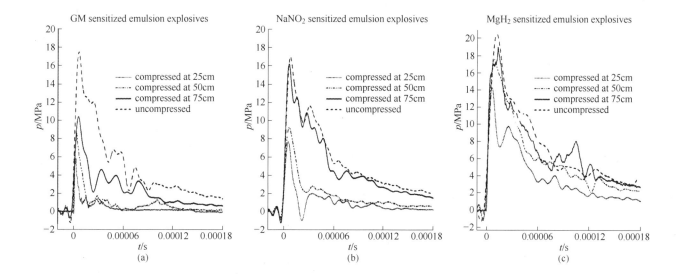

Fig.2 Pressure-time curves of compressed EMXs
(a) GM sensitized; (b) NaNO$_2$ sensitized; (c) MgH$_2$ sensitized

In underwater explosion experiments, shockwave peak pressures of two detonators were respectively 5.92MPa and 6.08MPa, and the average value is 6.0MPa.

3.4 Estimation of Desensitization Degree

As it is described in the ref.[9], use the parameter "desensitization ratio" to express the deduction degree of compressed emulsion explosives to the uncompressed. The smaller desensitization ratio is, the better anti-pressure ability that emulsion explosive has.

3.5 Discussion and Analysis

Tab.3 shows that when compressed by the same strength of shockwave, the desensitization ratio of MgH$_2$ sensitized emulsion explosives is the lowest of the three types of emulsion explosives, NaNO$_2$ sensitized emulsion explosives takes the second place, and GM sensitized emulsion explosives is the highest. At the compression distance of 25cm, desensitization ratio of GM sensitized is 100% which indicates that GM sensitized emulsion explosives is misfired, and desensitization ratio of NaNO$_2$ sensitized is also up to 88.11%, while MgH$_2$ sensitized is only 38.97%.

Tab.3 Desensitization ratios of EMXs at different compression distances

(%)

Desensitization ratio	Compression distance				
	25cm	40cm	50cm	60cm	75cm
GM sensitized	100	86.41	79.82	73.67	63.64
NaNO$_2$ sensitized	88.12	84.47	71.63	53.47	15.59
MgH$_2$ sensitized	38.97	18.89	12.11	10.45	11.76

Fig.3 shows that when the compression distance exceeds 50cm, the desensitization ratio of MgH$_2$ sensitized emulsion explosives is close to 10% and it approaches equilibrium. However, with regard to GM sensitized emulsion explosives, the desensitization ratio is 79.82% at compression distance of 50 cm and still greater than 60% when compressed at 75cm. As for NaNO$_2$ sensitized emulsion explosives, when compressed at 50cm, desensitization ratio comes to 71.63%, but influenced little by pressure desensitization effect at compression distance of 75cm.

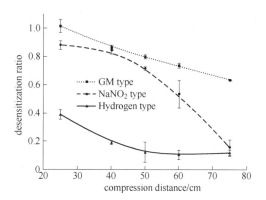

Fig.3 Relationship between desensitization ratios and compression distances

From the above analysis, we can see that MgH$_2$ sensitized emulsion explosives has the best ability of resistance to pressure desensitization.

4 EXPLOSION POWER EXPERIMENTS

In order to better reflect the detonation performance of emulsion explosive sensitized by MgH$_2$, we compared it with emulsion explosives sensitized by other materials.

4.1 Samples Fabrication

Emulsion explosives were made according to mass ratio in

the Tab.4, and the density of the pure emulsion matrix was 1.31g/cm³. Weights of emulsion matrix that emulsion explosives samples contain were all 50g, and each type of emulsion explosives tested three times.

Tab.4 Different formulation designs of emulsion explosives

Emulsion explosives	Mass ratio/%			
	Emulsion matrix	GM	Al powders	MgH$_2$
GM sensitized	100	4	0	0
GM and Al Sensitized	100	4	4	0
MgH$_2$ sensitized	100	0	0	2

4.2 Underwater Explosion Experiments

The emulsion explosives of underwater experiments were tested in the underwater explosion equipment, and testing requirements were the same as it was in ref. [10].

4.2.1 Experimental Results

Fig.4 shows pressure-time curves obtained from the underwater explosion experiments, and it is obviously shows that the shockwave peak pressure of MgH$_2$ sensitized emulsion explosives is the highest of the three types of emulsion explosives.

The shockwave peak pressure "Δp", shockwave attenuation time "θ", Shockwave specific impulse "I", shock wave specific energy "E_s", bubble wave specific energy "E_b" and total energy of shock wave "E" were obtained by calculation methods in literature, the results are listed in the Tab.5.

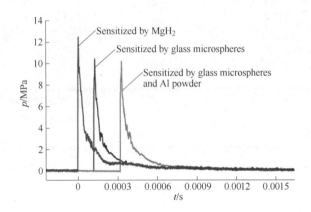

Fig.4 Pressure-time curves of three different types of emulsion explosives

Tab.5 Energy output parameters of emulsion explosives in underwater explosion

Emulsion explosives	Δp/MPa	θ/μs	I/Pa·s	E_s/kJ·kg^{-1}	E_b/kJ·kg^{-1}	E/kJ·kg^{-1}
GM sensitized	10.89	39.03	619.31	1085.62	940.83	2870.82
GM and Al sensitized	10.72	47.05	674.98	1129.65	1186.15	3187.21
MgH$_2$ sensitized	13.12	39.88	720.47	1302.71	1372.48	3761.52

4.2.2 Discussion and Analysis

Tab.5 shows that compared with GM sensitized emulsion explosives, the peak pressure, specific impulse and shockwave total energy of MgH$_2$ sensitized emulsion explosives were increased significantly, especially the shockwave total energy and peak pressure were respectively increased 30% and 20%. Adding aluminum to GM sensitized emulsion explosives improves detonation characteristics to a certain extent, but the effect is far less than MgH$_2$, and shockwave total energy increased only 10.0% while shockwave peak pressure decreased 1.6%.

4.3 Brisance Testing Experiments

Brisance is another important parameter of explosives that researchers concerned. In the brisance testing experiment, diameter and length of the lead column were especially 40mm and 60mm, and emulsion explosives used in the experiment were all 50g, which were made according to mass ratio in the Tab.6.

Tab.6 Brisance of different emulsion explosives

Explosive Performance	Emulsion Explosives		
	GM sensitized	GM and Al sensitized	MgH$_2$ sensitized
Density/g·cm^{-3}	1.14	1.24	1.26
Brisance/mm	16.12	16.24	19.08

From the Tab.6 and Fig.5, we can see that the brisance of MgH$_2$ sensitized emulsion explosives is obviously superior to the other two types of emulsion explosives.

Fig.5 Lead column compressed in sequence by GM sensitized emulsion explosives, GM and Al sensitized emulsion explosives, and MgH_2 sensitized emulsion explosives

5 RECENT RESEARCH RESULTS

5.1 "Dynamic Sensitization" Technology

The sensitivity of emulsion explosives sensitized by dynamic sensitization technology is very low, so its safety meets the requirement; hydrogen sensitized of emulsion explosive is sensitized along with the detonation process, its sensitization process needs high pressure environment, which solves the safety of transportation and storage of emulsion explosives and "pressure desensitization" problem of emulsion explosives. Meanwhile, "dynamic sensitization" makes the emulsion explosives charged with a high density, which increases the blasting energy density and improves the blasting quality.

5.2 Preparation of Explosive Samples

In our prophase research, MgH_2 powders were added into emulsion matrix without any treatment. When MgH_2 powders are added into emulsion matrix, they will be hydrolyzed to produce H_2 bubbles due to weak acidic and hydrous characteristics of emulsion matrix, and then emulsion matrix is sensitized. Although emulsion explosives sensitized by hydrolyzed MgH_2 has excellent detonation characteristics in the prophase experiments, but the hydrolysis process of MgH_2 is hard to control. In the recent study, we solved this problem by coating the MgH_2 with parafilm by a special technology (as shown in Fig.6), so the hydrolysis of MgH_2 is inhibited.

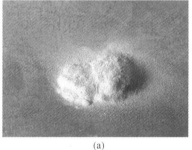

(a) (b)

Fig.6 MgH_2 powders

(a) without any treatment; (b) coating with parafilm

Add coated MgH_2 powders into emulsion matrix at mass ratio of 100 : 2, and add a little glass micro ball to guarantee the detonation of emulsion explosives, then emulsion explosives are made.

5.3 Experimental Results

Brisance of emulsion explosives sensitized by coated MgH_2 increased greatly, and its brisance even gets to 23mm, as it is shown in Fig.7.

Fig.7 Photos from left to the right are uncompressed lead column, lead column compressed by GM sensitized emulsion explosives and lead column compressed by hydrogen sensitized of emulsion explosives

6 CONCLUSIONS

Anti-pressure ability and explosion power of MgH_2 sensitized emulsion explosive have been introduced in this paper. MgH_2 plays double roles as sensitizer and energetic material in emulsion explosives. Desensitization ratio of MgH_2 sensitized emulsion explosives is far less than that of GM and $NaNO_2$ sensitized emulsion explosives when compressed by external shockwaves of the same degree, which suggests that MgH_2 sensitized emulsion explosives has excellent pressure desensitization resisting ability; what is more, compared with traditional GM sensitized emulsion explosives, shockwave peak pressure and shockwave energy of MgH_2 sensitized emulsion explosives are increased over 30% and 20% respectively, and its brisance even reaches 23mm.

REFERENCES

[1] Cudzilo S, Kohlicek P, Trzcinski V A, Zeman S. Performance of Emulsion Explosives[J]. Combustion, Explosion, and Shock Waves, 2002, 38:463～469.

[2] Takahashi K, Murata M, Kato Y. Non-ideal detonation of emulsion explosives, J. Mater. Proces. Technol, 1999, 85:52～55.

[3] Sumiya F, Hirosaki Y, Kato Y. Detonation velocity of precompressed emulsion explosives[C]// Proceedings of the 28th Annual Conference on Explosives and Blasting Technique. Cleveland: International Society of Explosives Engineers, 2002: 253～263.

[4] Wang Y J, Lü Q S, Wang X G. Experimental study on the desensitization of water-bearing explosives subjected to shock wave, Chinese Expl. Shock Waves, 2004, 24:558～561.

[5] Huang W Y, Yan S L, Wu H B. Experimental research on influence of emulsifier on crystallization quantity of emulsion explosives under dynamic pressure, Chinese J. COAL SCIENCE & ENGINEERING, 2011, 17:100～103.

[6] Jolanta Biegan'ska. Using Nitrocellulose Powder in Emulsion Explosives[J], Combustion, Explosion, and Shock Waves, 2011, 47:366～368.

[7] L Liqing. Use of aluminum in perforating and stimulating a subterranean formation and other engineering applications, U. A. Patent No.20030037692, 2003.

[8] Ustimenko E V, Shiman L N, Kholodenko T F. On environmental effects of emulsion explosives with products of processing of solid propellants in blasting works, Nauch. Vestn.Nats.Gorn.Univ.Ukrainy, 2010, 4:35～40.

[9] Cheng Y F, Ma H H, Shen Z W. Detonation Characteristics of Emulsion Explosives Sensitized by MgH_2[J]. Combustion, Explosion, and Shockwaves, 2013, 49(5):614～619.

[10] Yan S L, Wang Y J. Characterization of pressure desensitization of emulsion explosive subjected to shock wave, Chinese Expl. Shock Waves, 2006, 26: 441～447.

Explosive Energy Distribution in Explosive Column Case Study of Some Indian Mines

G.K. Pradhan

(*Deptt. of Mining Engineering & Dean of Faculty of Engineering & Technology, The IME Journal AKS University, Satna, India*)

ABSTRACT: Air-deck techniques had revolutionized blasting techniques in last few decades globally. This technique made its application in smooth wall blasting technique, and over the years had been adopted for main production blasting also. Additionally this has been an effective method for stabilizing pit walls, explosives charge distribution as a measure to ensure better fragmentation etc. Chemically activated gas bags, used in conventional air-deck had limited application and restricted usage in highly sensitive areas due to high cost component as well as limited flexibility. While researchers continued with its large number of applications in surface mine blasting, its use was site specific only. Authors in this paper have studied in depth the mechanism of air-gap technique vis-a-vis explosive energy distribution in an explosive column of surface mine blasting. With "energy factor" as a scientific tool to ensure proper level of explosive energy required for fragmentation, displacement etc, was adopted for placement of air-gap in the explosive column.

To make the concept user friendly and cost-effective empty plastic bottles were selected for placing air-gap. A number of field trials were undertaken before switching over to its application in production blasting. It is currently in use at some limestone, and one uranium mines in India. At the Tata Steel Iron Ore Mine in Joda, Odisha, India, since last April, this is being in use on regular basis in dry holes. The Joda East Iron Ore, is having very hard to soft strata in its mining excavation plan, and accordingly the use of plastic bottles have been made site specific. After detailed techno-economic evaluation, blast monitoring, the authors could implement this system on regular basis since June 2012. Till date over 35000 used plastic bottles of 70 mm dia. and 280 mm length have been used in most production blasts.

An attempt has been made in this paper to present the findings of the Joda East Iron Ore Mines (India) aided by photographs and blast details, in addition to the savings obtained vis-a-vis improvement in productivity, powder factor, reduction in generation of over size boulders, control of vibration level etc. explosive column and also in the sub-grade region of any blast hole.

KEYWORDS: airdeck technique; stemming column; explosive energy distribution; energy factor

1 INTRODUCTION

Blasting continues to be the cheapest and most popular method of rock breakage and loosening globally. Explosive energy which is primarily a chemical energy is the cheapest and easily available for any rock excavation. A number of studies undertaken had clearly indicated that the percentage of chemical energy available during the blasting process is around one fourth of the total explosive energy in a scientifically designed blasting round (Berta, 1990).

The percentage utilization of explosive energy in blasting change further when the blast is not proper, no free faces are available or the holes are over or under charged, with poor quality of explosive material. The wastage of energy not only is an area of concern but also contribute to serious environmental impact such as:

(1) Blast induced ground vibrations (in surface mines/excavations).

(2) Air blast of air - over pressure.

(3) Fly rocks.

(4) Noise.

(5) Blasting fumes.

(6) Blast induced ground vibrations, etc.

Apart from environmental problems the variation in explosive energy liberation, propagation and utilization cause the following serious operational problems thereby severely influencing the productivity and safety during mining. These are :

(1) Generation of oversize boulders.

(2) Extension of blast induced cracks into the sides and back of the blasting front.

(3) Poor blast profile leading to excessive or no throw.

Corresponding author E-mail: gkpradhan@aksuniversity.ac.in

(4) Difficulty in drilling.

(5) Excessive post-blast cleaning of the production face for smooth deployment of the excavator and movement of dumpers etc.

(6) Uneven floor conditions due to incidence of toes, penetrating into the floor surface etc.

In order to effectively utilize the explosive energy, the research has been now confined to effective utilization explosive energy inside a blast hole by improving quality of stemming, type of stemming material used, use of various inserts like water ampoules, stemming plugs, chemically activated blast hole plugs etc. An attempt has been made in this paper to highlight the various efforts made to effectively utilize explosive energy through use of cost-effective inserts within the blast hole and/or explosive column.

2 STEMMING MATERIAL

Stemming of any blast hole is essential so as to ensure proper use of explosive energy to break and move the rock the rock. Stemming material for blasting in underground coal mines is in the shape of a "clay plug". The clay plugs should be compact but not hard. It is a mixture containing 70% fine sand, 30 % clay and a small percentage of calcium chloride, to keep it in plastic condition. In most of our mines, clay plugs look like un-burnt bricks and are very hard, such clay plugs are not at all suitable. In coal mines, coal dust should not be used as a stemming material. In case of surface mines, drill cuttings are the prime source of stemming material. Several researchers have worked on use of alternate material such as crushed stone as stemming material and have reported considerable savings and improvements in blasting.

In order to effectively use the explosive energy, research were undertaken to use various plugs, locking material, inert material, air-pockets etc to improve blast quality. Melnikov (1970) conducted experiments with the use of air-gaps instead of inert material in explosive column, as 'deck material'. He had reported considerable increase in the efficiency of explosive energy utilization.

2.1 Stemming Plug

Paul N. Worsey, a researcher of University of Missori - Rolla had invented stemming plugs. The idea first came to Worsey when he was testing Resin grouted rock bolts in 1980. He observed that rock bolts similar in design to the plug would eventually lock up in the boreholes when attempting to with draw them. The bolts featured a cone-like configuration at the base, which wedged the cured resin tightly against the hole wall as the bolt was withdrawn frictional force against the borehole wall were so great that the bolt system failed, not the cone.

Worsey (1987) hypothesized that if high friction can be created by pulling a cone-shaped object through a bore hole in roc, it also can be duplicated by pushing it through the hole, Instead of the resin, however, muck/frill cuttings would be used.

Load blast holes with explosives. Fill muck/drill cuttings up to 2 borehole diameter, place the plug on the drill cuttings. This muck layer insulated the violent blast energy from the plug, which otherwise could have a damaging effect from the heat since the plug is the thermoplastic. The rest of the blast hole is then backfilled with the drill cuttings.

2.2 Borehole Plug

They are used for plugging long hole of a vertical crater retreat (*VCR*) method of mining, blast hole punctured through developed gallery (very much common in Indian Coal fields), plugging the stemming column so as to leave void between the charge and stemming length in AIRDECK method of pre-splitting as a method of providing decks etc. Bore hole plugs are either of wooden material, inflated tire or gasbag type(Poly-deck gasbag, manufactured by S.A. Pty. Ltd of Australia). The easy to type and inexpensive type of borehole plug is of sand filled bags having lowered up to the desired depths by a nylon cord line.

Other types of plugs:

(1) Rock Lock System – Parihar (2005) had reported the success achieved with this method of inserting an air-lock inside the explosive column in a captive limestone mine in Rajasthan.

(2) Use of funnels.

(3) Use of rubber balloons.

(4) Use of waste/fresh plastic bottles with lid etc.

3 AIRDECK TECHNIQUE

With the availability of plugs to have air inside, further research initiative was done to adopt air deck technique. At the bottom of the hole a specific quantity of explosive is loaded and a large part of it remains empty (expect for the stemming). When fired the rebounding shock waves produce a cleanly split rock. This technique of pre-splitting is now quite popular. In this method a single explosive charge (cap-sensitive - primed) and a bore-hole plug (either a power plug, air bag; bore plug - chemically activated or a jute bag filled with sand hung from the surface at the desired depth. Chironis(1989) and (1991) had first reported the application of "Air-deck" techniques in pre-splitting of high walls for better high wall stability in coal mines. Mead et el (1993) depicted the use of "air-decks" in production blasting.

In India, for the first time this technique of pre-splitting was utilized for blasting in a dragline bench at Black – II open cast coal mine of Jharia Coalfield (by the IBP and experts from Indian School of Mines).(Gupta et el, (1990) and Sandhu & Pradhan, (1991). Chakraborty & Jethwa (1995), reported the application of this technique at Dongri Buzurg Mn. mine of MOIL. Sarma & Saran (2005) dealt on the regular use of plastic spacers for air decking in production blasting in limestone mines. Pradhan (2011) had extensively studied the use of Sua make "chemically activated blast hole plugs", used at Rajashree Cement Limestone Mine, for decreasing the explosive column density along with high density bulk SME explosives so as to make SME economically viable without compromising with pow der factor and blasting cost.

(1) Fly rock control - Where due to over break or unfavourable geology the burden on the front row of holes has significantly reduced. In such situations AIRDECK loading begins below the reduced burden and extends upward to above the weakened zone. This results some degree of breaking and fragmentation of the burden.

(2) Tunnel over break protection: For perimeter holes in tunnels one third portion of the hole loaded with explosives and leaving an air deck followed by plugging and stemming. This leaves a smooth wall.

(3) Reduction of fines: Where fines generated are dumped as waste by introducing "air deck"the crushing around the blast holes are minimized leading to less fines generation. In these holes powder factor is 15%to 20 % less than normal; "air deck"gives the explosives reaction more time to work while contributing to increased gas volume. Fines have been reduced as much as 50 %.

(4) Blasting for Dimension Stone : By "air deck"load ing of specially prepared cartridges at the bottom of the holes and stemming with crushed stone, the dimension stones when blasted with electric or NONEL detonators result crack free surfaces.

(5) Ore-Waste Separation: Two different degree of fragmentation are obtained which helps in easily separating the ore and waste rock. Portion of the bore hole loaded with explosives and 'air deck' in zones of waste rock, coarser fragments help in segregating the ore and waste.

(6) No need for sub-grade drilling.

(7) Control of vibration : By substitution "AIRDECK" better high walls, faster loading, improved fragmentation and controllable levels of vibration could be achieved at Block - II (in Jharia Coalfields) for the first time in India. Since then this has been widely adopted in many mines.

4 AQUA DECK

Use of water ampoules in underground coal face blasting has been in use since 1950's. Yunmin, Wang, et el, 2013, had made detailed investigation *of Bore hole Aqua Stemming Blasting* at Sinosteel Ma'anshan Institute of Mining Research Co. Ltd, China. They have undertaken this investigation with a view to improve explosive energy utilization on the upper section (preferably in the stemming zone) of the bench and obtained 20%, improvement in fragmentation. The mathematical model developed by them had taken into account the propagation of blasting shock wave in borehole water, aqua-stemming height in the blast hole, extra available energy determination, etc. The Mathematical Model, can also be used for the use of empty plastic bottles, being used in the explosive column.

5 SCOPE FOR IMPROVEMENTSIN BLASTING

It was decided to adopt a system by which explosive energy is uniformly distributed. The standard practice adopted in other mines are :

(1) Lowering the explosive energy,

(2) Expanding the blast hole pattern,

(3) AIRDECK method by using gas bags etc (this was quite feasible but with chemically activated gas bags available in India, the final cost of blasting and mining gets increased significantly).

6 CASE STUDY OF AN IRON ORE MINE

Joda East Iron Mine is a captive open-pit mine of Tata Steel in Odisha state of India(Fig.1). The block of the mine lies between longitude 85025' & 85027' and 21059' & 22003' within the survey of India Topo sheet No. 73 F/8 and is covered by a ridge towards NNE-SSW direction. The mine is well connected by roads from Chaibasa (Jharkhand) and Barbil (Orissa). Iron ores of this region occur in the Iron ore series of Upper Dharwar age. The principal rocks found within the area are quartzitic sandstone, shales and cherty quartzites, both banded and unbanded. Sometimes colloidal quartzites contain bands of Haematite and are recognised as Banded Haematite Jaspars (BHJ). Shales in this area are often silicified iron ore deposits and these are mainly confined to the top portions of the series. The rocks show a general strike NNE-SSW direction with a westerly dip. Jointing, both along and across the bedding planes, is very prominent giving rises to rectangular to rhombic blocks of silicified shales and quartzites. Such joints have even affected the iron ore beds especially the compact and massive variety. The varieties of ore in the area are (1)Hard ore, (2)Soft ore, (3)Friable Flaky & Powdery ore and (4)Blue Dust.

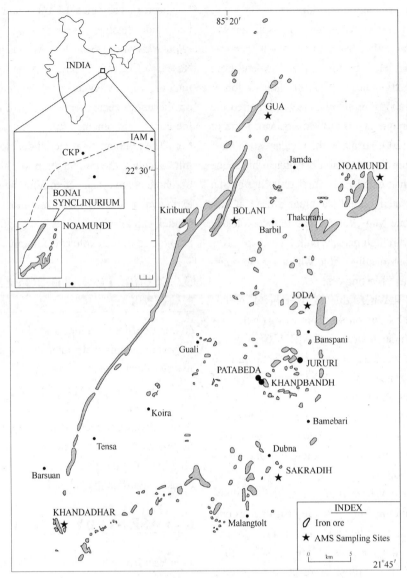

Fig.1 Shows location of the mine in the Singbhum-Bonai Synclinorium

This mine has about 40% of friable ore and powdery ore which has less than 2.5% of Al_2O_3. This category of the material is very much suitable for converting it into iron ore fines (-10mm) which will have average alumina of 1.8%. The balance 60 % material which primarily consisted Lateritic Ore, Hard Ore, high aluminous Flaky Ore (having >2.5% Al_2O_3) is proposed to be processed in existing Wet Beneficiation Plant.

Mechanized method of open cast mining has been adopted for mining iron ore in a series of 9 m high benches which will be reworked in future to 12 m high benches with the help of shovel-dumper combination. Drilling is done using 150 mm diameter drills with 10% sub grade drilling. Secondary drilling (if required) is done by Jackhammer.

Site Mixed Emulsion (SME) supplied by M/s Gulf Oil, down-the-hole is being used in this mine along with shock tubes to initiate the cast booster of 250gms weight. SME has a VOD of 4000±500 m/s and is charged into the blast holes at 1.12 g/cm³.

Some of the blasting related problems encountered are:
(1) Generation of oversize blasted material.
(2) High incidence of back break resulting in difficulty in maintaining uniformity in front row burden while drilling.
(3) Difficulty in drilling.
(4) Engagement of rock breakers for oversize material sizing during the production cycle, etc.

CIMFR, Blasting Research Group is regularly monitoring the explosive and blasting performance of this mine on long term basis. In some specific zones, where normal blasting has not been allowed by the regulatory agencies, Orica(India) had been entrusted with conducting blasts with the use of electronic delay detonators.

7 EXPLOSIVE ENERGY DISTRIBUTION

Explosive energy distribution study was initiated at this mine based on the "Energy Factor" basis (Pradhan, 2011). Energy factor for the benches where previously blasting has been undertaken using Site Mixed Emulsion explosives along with shock tubes. After ascertaining the "Energy Factor" for each bench, and studying the post blast observations from the record (in respect of extent of back break and side break; generation of oversize boulders per 000 T on blasted material, tightness of the muck pile etc), the following conclusions were drawn:

(1) There has been a considerable reduction in stemming column due to the use of high density emulsion explosives. On an average the stemming column recorded was 5.0 to 5.5 m.

(2) The mine adopted only one deck of 1.0 m to ensure column rise.

(3) Boulders were observed at the stemming column region.

(4) Difficult to maintain uniform front row burden owing to difficulty in drilling and drill placement. Resulted due to excessive/un-controlled back break from the previous blast.

(5) Uneven floor conditions reported despite 10% sub-grade drilling, resulted in more time spent on face preparation.

Pradhan (2011) recommended the use of AIRDECK in the explosive column for reducing the stemming column, for replacing the decking with drill cuttings, increasing the explosive occupancy with Site Mixed Emulsion explosive by decreasing the column density of the explosive inside the blast hole. The use of AIRDECK by inserting "used plastic bottles" (Figs.2(a)& 2(b)) was suggested, based on the success of the experiments undertaken at some limestone mines Pradhan et el (2012).

Fig.2(b) Photograph showing the blasting site and placing the bottle inside the blast hole

7.1 Experiments with Use of Plastic Bottles

Empty Plastic Bottles as shown in Fig.3, provided air deck:

(1) Inside the stemming column for stabilization of the stemming,

(2) Inside the explosive column replacing "decking", and

(3) Inside the explosive column for increasing the occupancy of explosive and reducing the stemming height.

These bottles also are low cost, easily available, user friendly, and do not contaminate the Ore quality with its total destruction in the process of blasting. With the use of plastic bottles inside the explosive column, there was no need to have additional booster since the gap allows continuity of the explosive column and each bottle replaces approximately 1.1 kg of explosives by weight. Site Mixed Emulsion explosives, when used in the blast with plastic bottles it was ruled out to reduce the fixed explosive density inside the blast hole. Fig.3, shows the placement of bottle(s) in explosive and stemming column, and Fig. 6 the photograph of a blast having used 12 ~14 bottles inside the explosive column in a medium-hard bench.

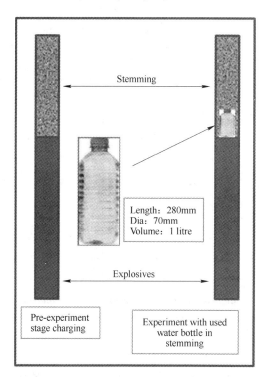

Fig.2(a) Schematic view of placing bottle in the stemming column and its dimension

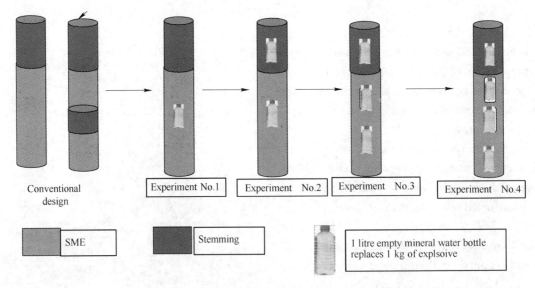

Fig.3 Showing placement of 1 litre empty plastic bottles in explosive & stemming column

7.2 Results

Each and every blast where bottles have been used were monitored and all pre-blast and post-blast data recorded and analyzed at regular interval so as to ascertain its impact on overall blasting performance. The summary of the results are presented at Tab.1, Tab.2, the savings generated with the use of plastics, and Table 3, Total savings on yearly excavation basis. Fig. 4, presents the various aspects of savings and benefits obtained during the use of bottles & Fig. 5, the photograph of a blast having used 12~14 bottles inside the explosive column in a medium-hard bench.

Tab.1 Presents the summary of (pre and post use of bottles) results

Description	Pre use of Bottles		Post use of Bottles	Use of Plastic Bottles during Jan-Dec 2013 period
	February 2012	March 2012 (up to 20th)	March 2012 (21st to 31st)	Jan-Dec 2013
No. of Bottles used	—	—	—	39696
No. of Blasts conducted	12	7	5	NA
Total Explosives consumed	134665	103100	49345	1852140
Total Meters blasted	10402	8102	4252	146486
Quantity of material blasted	—	—	—	25848533
Explosives consumption per m/kg·s·m^{-1}	12.95	12.73	12.17	12.64
Change in linear charging density /kg·m^{-1}				1.9
Quantity of explosive saved /kg·s	—	—	—	22880.8

Tab.2 Savings generated due to the use of plastic bottles

Average Explosives reduction /M. of blasting /kg·s·m^{-1}	0.66
Total Metres Blasted after 21 March	4252
Total Explosives reduction /kg·s[①]	2806
Total savings till date (INR)	126270 INR
Total savings during Jan to December 2013 has been on account of Explosives cost	1144040 INR

① INR/kg of Explosive is considered @ current rate of ▢ 45.

Tab.3 Total savings on yearly excavation basis

Cost Savings/M of blasting Rs·m^{-1}	29.9
Total meters to be blasted in a year	130450
Total Cost saving in a year (if bottles are used)	3900455
Total Cost of bottles(1 Lakh bottles @ Rs.3/bottle)	300000
Net Projected Direct Savings in a year (Rs)	3600455

On a conservative side an average of 5% savings has been considered. However, the savings can go to at least 10% with the continuous use of bottles on a sustained basis.

In addition to the direct savings, (1) The savings due to elimination of Sub-grade drilling in all benches except very hard rock, will be ascertained as the savings due to drilling will be quite high. (2) The savings due to improved fragmentation reflected as KL/Tonne saved for Diesel and KWH/T at Crushing plant etc will be additional.

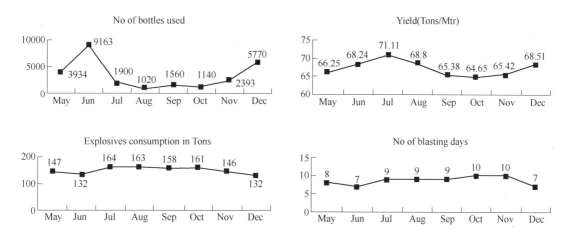

Fig.4 Presents the various aspects of savings and benefits obtained during the use of bottles

Fig.5 Photograph of a blast having used 12~14 bottles inside the explosive column in a medium-hard bench

8 OBSERVATIONS AND RECOMMENDATIONS

(1) In order to have uninterrupted availability of bottles, the mine management had tied up with one local vendor to supply empty bottles with lid in place.

(2) Placing one or more bottles inside the explosive column was recommended for blasts in 9 m height benches, in addition to the bottle in the stemming column. By having a fixed stemming column length, quantity of explosive saved with the bottle(s) space was monitored so as to quantify exact quantity of explosive saved per hole and per blast.

(3) Since the mine is facing toe problems, placing bottle in the toe region was recommended.

(4) Further it was suggested to take experiments in hard massive ore by use of bottles and also to establish the usage during rainy season.

(5) In a bench height of 9 m, a reduction of 10 kg of explosive in a hole is achievable without having any negative impact on any of blast output parameters.

(6) All the non-bio-degradable plastics being used in mineral water bottle is incinerated inside the blast hole which may have created problem to the whole of ecosystem.

(7) There is a considerable reduction in generation of Noxious fumes as the explosive consumption is reducing & better stemming stabilization.(Study by Maunsell Pty Limited- Australia shows that there is generation of 167.3 kg of CO_2, 16.3 kgs of CO, 3.5 kgs of NO_2 by exploding 1 Ton of SME).

(8) In blast holes having cracked zone, bottles when used helped in sealing the cracks by flowing towards the cracked zone during pumping of emulsion explosive.

(9) At Joda East Iron Mines, a series of experiments were conducted by using bottles in the stemming column just over the explosive column, inside the explosive column, at the toe region and the results were monitored for every blast. In-hole *VOD* was also recorded for the blasts using bottles in the blasts. The *VOD* recordings are presented at Fig.6.

(10) Majority of blasts having used bottles had shown uniform fragmentation. Fig. 8 : Photograph of a blast having used 12~14 bottles inside the explosive column in a medium-hard bench.

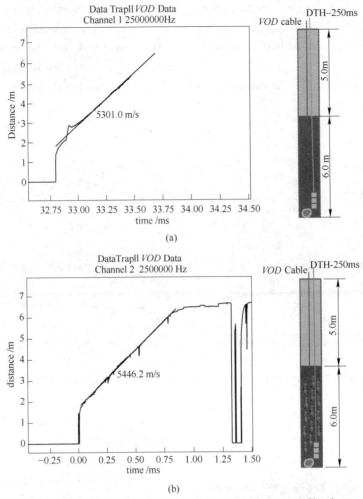

(a)

(b)

Fig.6 VOD recordings showing without and with the use of bottles

(a) without the use of Bottles; (b) with the use of 14 Bottles inside the explosive column

The results of the use of bottles were quite encouraging and Fig.7, presents the trend of its use in all these months where experiments were continued in normal production blasts. The figure also indicate the contribution of bottle use for reducing CO_2 emission. Fig.8, shows the amount CO_2 reduction by 2020 with its use at this mine.

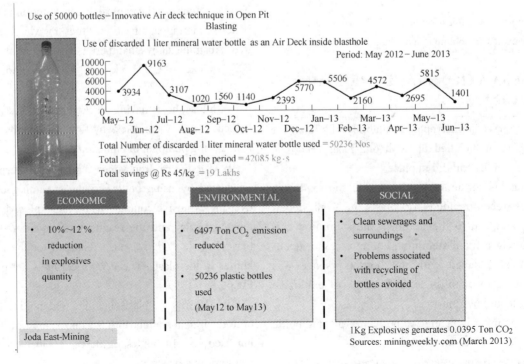

Fig.7 Presents the trend of its use in all these months and its impact

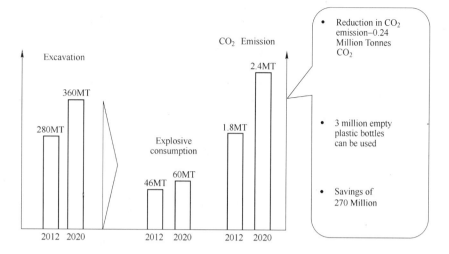

Fig.8　Presents the trend of CO_2 reduction with the use of bottles by 20120

9　CONCLUSIONS

The experiments which were carried out during last few months have been successful in achieving reduction in boulder generation, improving PF, improving fragmentation quality, excavator productivity etc. During the experiment stage, adequate care was taken to monitor and record the various parameters influencing blasting and post-blasting stages. Enbergfy Factor based explosive quantity reduction/optimization had greatly influenced the success of using bottles.

ACKNOWLEDGEMENT

The author record his sincere thanks to the mines for giving an opportunity to conduct experiments in the above study.

REFERENCES

[1] Berta G. Explosives : An Engineering Tool, *Italesplosivi*, Milano, Italy, 1986.

[2] Chakraborty AK, Jethwa JL. Air-deck blasting- a feasibility study, *The IM & E Journal*, 1995: 11~17.

[3] Cook, Melvin A. The Science of Industrial Explosives, *Utah Graphic Service Supply, Inc.* USA, 1974.

[4] Gupta RN, Ghose AK, Mozumdar BK. Design of blasting rounds with AIRDECK pre-splitting for dragline and shovel benches, near populated area as a case study, *Proc. of Intl. Symp, on Explosives and Blasting Techniques*, New Delhi, 17-18, Nov. 1990, 3: 11~27.

[5] Mead DJ. et al. The use of air-decks in production blasting, *Procc. of IVth Intl. Symposium on Rock Fragmentation by Blasting*, Austria, 1993.

[6] Melnikov NV. et al. Effective methods of application of explosion energy in mining and construction, *Dynamic Rock Mechanics (Ed: Clark, G.B.)*, 1970.

[7] Pal Roy P. et al. Air-deck blasting in opencast mining using low cost wooden spacers for efficient utilization of explosive energy, *Jl. of Mines, Metals and Fuels*, August, 1995: 240~243, 248.

[8] Pal Roy P. Rock Blasting Effects & Operations, *Oxford & IBH*, New Delhi, 2005: 345.

[9] Parihar CP. Eco-friendly blasting – A case study of Aditya Limestone Mines, *Procc. of 3rd National Seminar on Rock Excavation Techniques*, Organised by The IME Journal, Nagpur, 2005: 274~276.

[10] Pradhan G.K. Explosives & Blasting Techniques, *Mintech Publications*, Bhubaneswar, 2001: 386.

[11] Pradhan GK. et al. Effective utilisation of explosive energy – application of air-deck techniques in surface mine blasting, *Procc. of Golden Jubilee Seminar on Present Status of Mining and future Prospects*, April 6~8, 2007, Organised by MEAI, Hyderabad, 2007:169~176

[12] Pradhan G K. Energy Factor Based Explosive Selection and Performance Evaluation of Blasting in Surface Mines, *Unpublished Ph.D. Thesis, Visvesvaraya National Institute of Technology*, Nagpur, 2011.

[13] Pradhan GK, Pradhan Manoj. Explosive Energy Distribution in an Explosive Column through use of non-explosive material – Case Studies of some Indian Mines, Procc. of FRAGBLAST'10, New Delhi, 2012.

[14] Sandhu MS, Pradhan GK. Blasting Safety Manual, *IME Publications*, Calcutta, 1991: 280.

[15] Sharma RK, Saran P. Use of plastic spacers for air decking in production blasting in an opencast limestone mine, *Procc. of 3rd National seminar on Rock Excavation Techniques*, Nagpur, 2005: 125~130.

[16] Yunmin Wang et el. Investigation of Borehole Aqua Stemming Blasting, Rock Fragmentation by Blasting (Eds: Singh & Sinha), Fragblast 10 : Procc. Volume, Taylor & Francis Group, London, 2013: 823~826.

Application of Digital Electronic Detonators in Blasting Safety in Construction Process

CHENG Guangxiang, KANG Quanyu

(*Liaoning Chengyuan Blasting Engineering Co., Ltd., Liaoyang, Liaoning, China*)

ABSTRACT: In the trial of two areas show that, using the digital electronic detonator and field mixed explosive technology combined with the production mode of operation, has a significant effect in improving the blasting safety and economic benefit etc. This result is due to the digital electronic detonator santi illegal initiation and accurate delay function, charac-teristic of intrinsic safety has also benefited from the field mixed explosive technology good. Charge high efficiency. The model is worth in promoting the industry, from a long-term point of view, it is a developing trend.

KEYWORDS: digital electronic detonator; field mixed explosive; construction blasting; safety

1 INTRODUCTION

In the current process of blasting, demolition quality and safety issues, is the main problem facing production units blasting operations. These problems are with the detonator quality problems associated with certain non controllable, such as some nonel tube detonator are off drugs, charge is uneven, the pipe wall leakage, the quality problem of unequal diameter, it is prone to blind shot blasting safety accident, it cannot prevent and control blind shot production; in the cost of production, often due to uncouple charge, hole utilization rate is less than the ideal value and indirect costs. In the blasting safety, because no continuous or interval charge and easy to cause the explosive detonation, detonation phenomena such as half. So, in the blasting engineering construction, the quality is stable and reliable blasting equipment is the fundamental guarantee for the blasting effect and blasting safety.

Digital electronic detonator is developed in recent years, new technology and products, the field mixed explosive although technology development at home and abroad for decades, but in China only in recent years with the industry policy adjustment and get rapid development, highlight the characteristics of these two kinds of technology, is a high security, representing the two development direction in civil explosive industry.

2 THE ADVANTAGES OF DIGITAL ELECTRONIC DETONATOR AND FIELD MIXED EXPLOSIVETECHNOLOGY

2.1 Digital Electronic Detonator Technology Advantages

Digital electronic detonator is a according to the actual needs to set the delay time and can realize the accurate initiation of new electric energy blasting equipment, it has advantages of safety and reliability, high precision, flexible extension set etc.

Initiation system of the digital electronic detonator introduces the anti interference electronic isolation technology and the digital key initiation technique, ability has antistatic, anti RF, anti stray current foreign power, the blasting network, can be achieved online programming, accurate delay, network security detection and anti illegal initiation and other functions, to ensure the safety of the electronic detonator the production, transportation, safety in the process of using the reliability guarantee.

Digital electronic detonator has broken initiation function, namely: when the detonator initiation instructions issued after all the detonators, blasting network is in operation since the state, even if the blasting hole shock waves or flying cut blasting network, accurate initiation will not be affected after the shot hole, to ensure that the blasting effect and safety of construction.

2.2 Advantages of Field Mixed Explosive Technology

2.2.1 Safe and Reliable

Safety field mixed explosive technology mainly displays in: mixing loading truck transport

explosives but not the finished product of raw materials or semi-finished products; bin is filled and explosive of raw materials or semi-finished products, on-site mixing, no need to set the finished powder magazine.

2.2.2 Formula and Production Process

The formula is simple, wide raw material sources, low price.

Corresponding author E-mail: 61547cy@163.com

2.2.3 Environmental Protection

Little pollution to the environment, to ensure the health of workers.

2.2.4 Charge High Efficiency

From the nature of safety, environmental protection and low carbon perspective, it is of great significance; from the cost point of view, its advantage is obvious.

3 EXAMPLES OF DIGITAL DETONATORS AND ON-SITE MIXING TECHNOLOGY IN ENGINEERING APPLICATION

Our company since 2013, blasting construction operation mode using digital electronic detonator and field mixed explosive technique combining, in QianShan limestone mine, HanLing iron ore mine production test, results show that, the model can effectively prevent and reduce disaster and blind shot blasting accidents, improve the blasting safety.

HanLing company rock blasting test are as follows.

3.1 The Basic Situation of Blasting Area

Blasting test area for mine rock stripping project, the main component of rock mixed in fine grained granite, the fissure is developed along the crack surface, easy fragmentation, rock density of 2.80 t / m^3, consistency coefficient 5~7, stope for stripping phase, test area mining step height is 15 meters, the hole diameter is 120 mm, and the depth is 17 meters.

3.2 Blasting Equipment and Test Plan

3.2.1 Selection of Blasting Materials

By blasting test using adjacent hole area distance is 15 meters.

(1) Blasting area using high precision plastic nonel tube detonator and emulsion explosives were filled with 100 mm.

(2) Using digital electronic detonator blasting area and mixed explosive vehicle production system for filling.

3.2.2 Test Program to Verify the Contents

(1) The safety and reliability of digital electronic detonator.

(2) Verification of digital electronic detonator delay time.

(3) The blasting vibration, blasting toes, bulk rate quality of blasting effect.

(4) The analysis of economic comparison with the diameter of 100 mm emulsion explosive and mixed explosive vehicle production system fills the blasting cost.

3.3 Analysis of Burst Test Results

3.3.1 Analysis of Blasting Safety

After the test, compared with other detonator, digital electronic detonator with network error recognition, when the emergence of initiation network leakage even, wrong link, illegal access and other phenomena, initiation network cannot initiation, only up to and design of network match, can play a positive role in the initiation, Has a positive effect from the blasting safety aspects.

3.3.2 Analysis of Digital Electronic Detonator Delay Time

Compared with the traditional detonator, digital electronic detonator with online accurate setting delay time, the delay time error can be controlled within 1ms, the user can according to the geologic strata structure, geological structure, joint fissure, blasting environment determine the rational delay time.

In the open air rock blasting, digital electronic detonator delay time can be measured in meters, usually between hole extension of time to 5~15 ms/m, the delay time and 15~25 ms/m was the best, the best delay time according to the specific needs of blasting tests to determine the rock types, geological conditions, the blasting environment and other objective conditions.

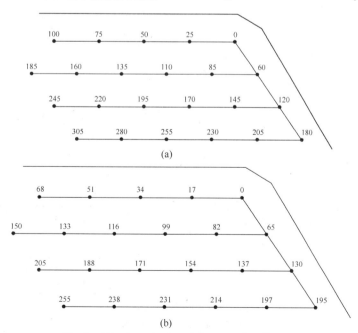

Fig.1 Blasting test network connection diagram

(a) schematic diagram of digital electronic detonator network; (b) schematic diagram of high precision detonator network

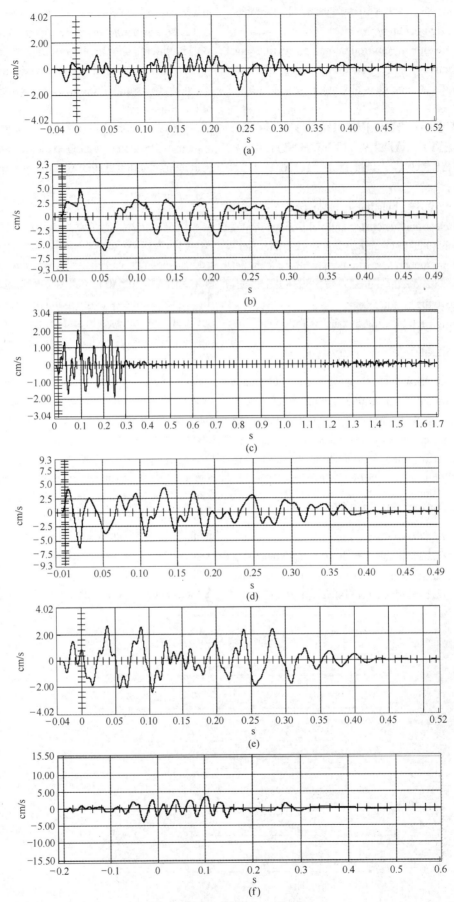

Fig.2 Curve of blasting vibration

(a) the first group: history curve of digital electronic detonator vibration;   (b) the first group: history curves of blasting vibration No.1 high precision tube;
(c) the second group: history curve of digital electronic detonator vibration;   (d) the second group: history curves of high precision tube blasting vibration ;
(e) the third group: history curve of digital electronic detonator vibration;   (f) the third group: history curves of high precision tube blasting vibration

3.3.3 Analysis of Blasting Vibration

In the "Blasting Safety Regulations" (GB 6722—2003), for the building (structure) to build the allowable safety vibration velocity for the vertical vibration velocity of particle, the vertical vibration velocity can be used to characterize the overall vibration velocity of blasting vibration.

The following three groups of curves for digital electronic detonator and high precision tube vertical vibration monitoring curve comparison chart.

Comparison of blasting monitoring data table.

Tab.1 Basic parameters of blasting and vibration monitoring results of comparative test

No.	Cap categories	The distance from the explosion center/m	Explosive charge/kg	The delay time between the hole /ms	The delay time /ms	The vertical vibration velocity /cm · s^{-1}
1	High-precision nonel detonator	50	110.5	17	65	5.090
	Digital electronic detonator	50	110.5	25	70	1.886
2	High-precision nonel detonator	60	115.0	17	65	4.670
	Digital electronic detonator	60	115.0	25	70	2.152
3	High-precision nonel detonator	80	102.5	17	65	4.342
	Digital electronic detonator	80	102.5	25	70	2.853

From the table can be obtained, in the same geological conditions and the hole net parameter, digital electronic detonator compared with other detonator, the use of digital electronic detonator blasting vibration reduction rate of 40% or more, for the slope protection of blasting is particularly useful.

3.3.4 Analysis of Blasting Effect

Fig.3 and Fig.4 are digital electronic detonator and the general high accuracy detonator blasting mining blasting pile, pile can be clearly seen from comparison, digital electronic detonator blasting pile with respect to the high accuracy detonator blasting pile is more uniform, loose degree is relatively large, relatively less. To facilitate production equipment dug, under normal conditions, digging machine production equipment efficiency can be increased by more than 15%.

Fig. 4 High precision detonator blasting pile effect

Fig.3 Digital detonator blasting pile effect

3.4 Analysis of Economic Benefits Compared with Differents Pecifications of Explosives

The ϕ100 mm emulsion explosive and on-site mixing technology filled with explosives, in the same density explosive premise, in the unit length of the hole, hole filled with explosives by on-site mixing technology utilization rate for 1.44 times with 100 mm explosives, when the rock blasting unit volume, can save drilling cost more than 30%.

4 CONCLUSIONS

Safety and production technology and comprehensive at present blasting production, using digital electronic detonator and on-site mixing technology combined with the mode of production, whether from the cost of production or from the blasting safety are the best blast- ing operation mode of production, be worth industry promotion.

Special Blasting and Demolition Blasting

Blasting Demolition of "L" Shape Frame Structure Building in Downtown District

XIE Xianqi[1,2], YAO Yingkang[1,2], JIA Yongsheng[1], LIU Changbang[1], WANG Honggang[1]

(1. Wuhan Blasting Engineering Co., Ltd., Wuhan, Hubei, China;
2. College of Civil and Transportation Engineering, Hohai University, Nanjing, Jiangsu, China)

ABSTRACT: Blasting demolition of 8-storey "L" shape frame structure building in downtown district was introduced in the paper. The building consist of North-South parts and East-West parts, was divided into 2 independent parts by a 120mm wide settlement joint in the corner. Based on characteristics of building structure and surrounding environment condition, the overall blasting demolition scheme that utilized settlement joint, reasonable delay and collapsed in the same direction was proposed firstly. According to the overall scheme, the relays initiating circuit was designed that the delay time between 2 parts of the "L" shape was 460 ms, the blasting demolition parameters and safeguard measures were also introduced detailed. Advises for similar building were given based on analyzing the building collapse process, the muck pile and vibration monitoring results, which has a certain reference meaning for similar projects.

KEYWORDS: blasting demolition; "L" shape frame structure building; settlement joint; collapse in the same direction; initiating circuit

1 ENGINEERING SITUATION

The 8-storey frame structure building was located in intersection of two arterial roads in downtown district, would be demolished because of urban construction and development.

1.1 Surrounding Environment

The surrounding environment of the blasting demolition building was complicated: a 2-storey building and a 7-storey residential building were located 4.0 m and 22.0 m east of the demolition building; in the south, the distance between electric cables, telecom cable, water supple pipe and the building were 4.0~10.0 m, besides, the major arterial and a 21-storey residential building was 15.0 m and 48.0 m form the building; the 10 kV electric cables and the minor arterial were situated 2.5 m and 3.0 m west of the building, a Flower & Pet market was 20.0 m from the building; the vacant space in the north was widest. The surrounding environment of the blasting demolition building was shown in Fig.1.

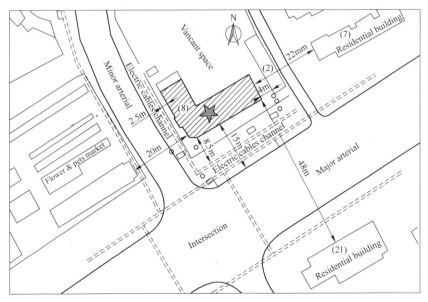

Fig.1 Surrounding environment graph

Corresponding author E-mail: 94394995@qq.com

1.2 Building Structure

The 8-storey frame structure building was built in 1980 s, which was 45.88 meters long, 28.9 meters wide and the maximum height was 32.7 m, the total architectural area was almost 7000 m². The building consist of North-South parts and East-West parts, looks like a "L", was divided into two independent parts by a 120 mm wide settlement joint in the corner. There were 3 staircases and 1 elevator well in the building. The section dimension of the building columns (LZ1, LZ2) was 400 mm×500 mm, reinforcement parameters of LZ1 was 10 Φ16 and Φ6@250, reinforcement parameters of LZ2 was 18 Φ25+2 Φ16 and Φ6@250. The planar graph of the building structure is shown in Fig.2.

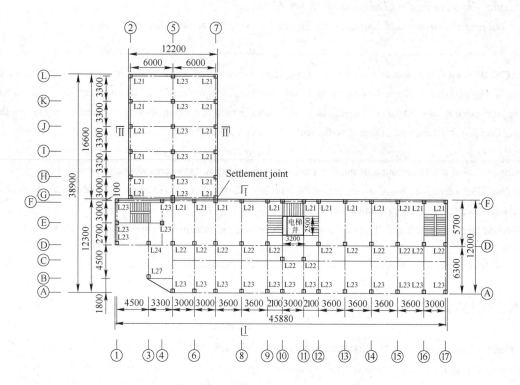

Fig.2 Structure planar graph

2 BLASTING SCHEME

2.1 Difficulty Analysis

(1) The "L" shape building consist of two independent frame structure parts, it's difficult to collapse thoroughly; shear walls distributed around the elevator well enhanced the integral rigidity, which resulted in fragmentation difficulty in processes of collapse.

(2) The building located in the intersection of two arterial roads in downtown districts, the adjacent 2-storey buildings, residential buildings, electric cables channel, and arterial roads may be destroyed in process of blasting demolition, thus, blasting demolition risks were distinct.

(3) Flowers & Pets market was 20m from the blasting demolition building, passengers, large fishbowls, and especial pets were sensitive to blasting vibration, collapse vibration, flying debris and blasting shockwave, it's difficult to control each blasting adverse effect.

2.2 Overall Blasting Scheme

According to the building structure and the surrounding environment condition, two blasting schemes could be compared: (1) directional collapse northward once; (2) separated two parts of the "L" in the settlement joint firstly, and then directional collapse twice.

High blasting cut is indispensable for Scheme (1), which increased the workload of drilling and safety and projection directly, besides, the back ward collapse distance and the blasting muck-pile range were larger than Scheme (2), which may be destroy protected objects. Transformed "L" shape building into two parts "1" building by preliminary demolition in settlement joint could decrease blasting vibration, collapse vibration and blasting muck-pile range, but preliminary demolition was long duration and unsafe, furthermore, twice safety evacuation and alert because of twice blasting demolition induced more adverse social influences.

Took safety, schedule, cost, social influence and other factors into account, overall blasting scheme of "utilize settlement joint, reasonable delay time, and directional collapse northward once" was adopted.

2.3 Blasting Cut

Based on the overall blasting demolition scheme, blasting cuts of two parts of the "L" shape building were $1^{st} \sim 4^{th}$ floor, blasting cut height of different floor and axis were shown in Tab.1

Tab.1 Blasting cut height (m)

Axis	L	K	J	I	H	G	F	E	D	C	B	A
1st floor	3.0	3.0	2.4	1.8	1.5	0.6	3.0	3.0	1.8	1.5	—	—
2nd floor	2.1	2.1	1.5	1.5	—	—	2.1	2.1	1.5	0.6	—	—
3rd floor	1.5	1.5	1.5	1.2	—	—	1.5	1.5	—	—	—	—
4th floor	1.5	1.5	1.2	—	—	—	1.5	1.5	—	—	—	—

2.4 Preliminary Demolition

2.4.1 Shear Walls

Transform shear walls into column by cutting method, the cutting height was $1^{st} \sim 6^{th}$ floor.

2.4.2 Staircases

Staircases located in support region should be destroyed partially, which means only the first step of different section staircase should be disposed that excavated the concrete and reserved the reinforcement, the dispose range was no less than 20cm. Staircases located in cutting region should be destroyed entirely.

2.4.3 Non Bearing Walls

Non bearing walls located in cutting region should be demolished, but the exterior walls and support region non bearing walls should be reserved properly.

2.5 Blasting Parameters

Blasting parameters were shown in Tab.2.

Tab.2 Blasting parameters

Column	Dimension /mm	Minimum burden w/mm	Hole distance a/mm	Row distance b/mm	Hole depth l/mm	Unit charge k/g·m^{-3}	Charge quantity per hole Q/g
LZ1, LZ2	400×500	150	300	—	350	1500	90
LZ3, LZ5, LZ6, LZ7	350×500	150	300	—	350	1500	75
LZ4	250×400	125	300	—	250	1300	40
Shear walls	200	70	300	300	130	1500	33

2.6 Initiating Circuit

Non-electric detonator with shock-conducting tube relay delay circuit of was used, high grade detonator of MS16(1020 ms) was set in each blasting hole, detonator of MS9(310 ms) was arranged outside for independent part of the "L", detonator of MS11 (460ms) was arranged in the settlement joint delay. Columns in the same axis of $1^{st} \sim 2^{nd}$ floor and $3^{rd} \sim 4^{th}$ floor initiating synchronously, detonator of MS9 (310ms) was used between the same axis of $1^{st} \sim 2^{nd}$ floor and $3^{rd} \sim 4^{th}$. Detailed delay time of the initiating circuit was shown in Tab.3.

Tab.3 Delay time (ms)

Axis		L	K	J	I	H	G	F	E	D	C	B	A
$1^{st} \sim 2^{nd}$ floor	Outside relay	0	0	310	310	620	620	1080	1080	1390	1390	—	—
	Inside initiate	1020	1020	1330	1330	1640	1640	2100	2100	2410	2410	—	—
$3^{rd} \sim 4^{th}$ floor	Outside relay	310	310	620	620	—	—	1390	1390	—	—	—	—
	Inside initiate	1330	1330	1640	1640	—	—	2410	2410	—	—	—	—

3 BLASTING SAFETY

The "L" shape 8-storey frame structure building was located in intersection of two arterial roads in downtown district, all kinds of protective targets should be ensured safety, and thus the problem of blasting safety was predominant. Blasting adverse effects should be calculated and evaluated before initiating; and safeguards measures were indispensable in case of high blasting risks.

3.1 Blasting Vibration

Equation for blasting vibration as follow[1]:

$$v = K \left(\frac{Q^{1/3}}{R} \right)^{\alpha} \tag{1}$$

where　v——the ground vibration velocity, cm/s;
　　　　Q——the maximum charge of one initiating, kg;
　　　　R——the distance between protective target and blasting center, m;
　　　　K——empirical coefficient concerned geology condition, value range 50~350;
　　　　α——attenuation coefficient of vibration wave, value range 1.3~2.0.

Based on the blasting demolition scheme and surrounding environment condition, the ground vibration velocity of eastern 7-stoery building(v_1), southern 21-stoery building(v_2), western 10 kV electric cables channel (v_3) were taken as protective targets to evaluate.

According to equation (1), Q =30.9kg, R_1=48m, R_2=53m, R_3=18m, k=32.1, α =1.54, calculated v_1=0.53 cm/s, v_2=0.41 cm/s, v_3=2.19 cm/s, all were lower than safety allowable value of relevant regulation.

3.2　Collapse Vibration

Equation for blasting vibration as follow[2, 3]:

$$v = K_t \left[\frac{(mgH/\sigma)^{1/3}}{R} \right]^{\beta} \quad (2)$$

where　v——the ground vibration velocity, cm/s;
　　　　m——the quality of the falling building component, kg;
　　　　g——the acceleration of gravity, m/s²;
　　　　H——the building component falling height, m;
　　　　σ——the failure strength of ground medium, MPa, generally 10 MPa;
　　　　R——the distance between protective target and ground center by impact, m;
　　　　K_t, β——attenuation coefficient, empirical value.

The same with blasting vibration evaluation, the ground vibration velocity of eastern 7-stoery building(v_1), southern 21-stoery building(v_2), western 10 kV electric cables channel (v_3) were taken as protective targets to evaluate.

The total quality of the 8-storey frame structure building was 7500000 kg, because of the relay delay initiating circuit, structural components collapsed and impacted the ground sequent, and thus the m = 2500000 kg. According to equation (2), H =13 m, R_1=45 m, R_2=53 m, R_3=26 m, K_t =3.37, β=1.66, calculated v_1=1.88cm/s, v_2=1.43 cm/s, v_3=4.68 cm/s, all were lower than safety allowable value of relevant regulation.

The 10 kV electric cables channel was located only 2.5 m west of the building, may be destroyed by collapsed structural components. Therefore, effective safeguard measures should be designed and implemented (Fig.3). The safeguard measures as follows:

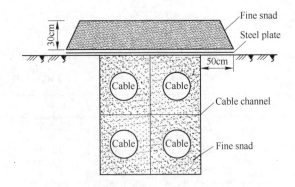

Fig.3　Safeguards measures sketch map

(1) Filled the electric cables channel with fine sand.

(2) Laid steel plate above the electric cables channel cover board, the thickness of the steel plate was 20mm, and the laid range exceeded the channel edge 50cm.

(3) Paved fine sand above the steel plate, the thickness of the fine sand was 30cm.

3.3　Flying Debris

The generate principle of flying debris was correlated with blasting parameters and unit charge. Equation for flying debris as follow[4]:

$$L_f = 70 k^{0.58} \quad (3)$$

where　L_f——debris flying distance without any safeguard measures, m;
　　　　k——unit chare, kg/m³.

According to blasting parameters and equation (3), k =1.5 kg/m³, calculated L_f =87 m, effective measures must be implemented.

To shield off flying debris, a comprehensive scheme adopting covering and proximal safeguard combined with protective safeguard was designed (Fig.4).

Fig.4　Flying debris safeguard

(1) Covering safeguard: specific columns of 1st~2nd that located in external of the building and all columns adjacent to Follower & Pets Market should be wrapped with three layers of quilts and one layer of bamboo raft from inside to outside.

(2) Proximal safeguard: put up a double layer bamboo

shelves 1.0m form outside of the building, than hanging two layers of dense-mesh protection networks from top to the ground, the height of the proximal safeguard was 6.0 m.

(3) Protective safeguard: for protecting large glass facilities in Follower & Pets Market, hanging 3.0m high two layers of dense-mesh protection networks along the wall of the market.

4 BLASTING VIBRATION MONITORING

According to regulations of the "Safety Regulations for Blasting" (GB 6722—2003), ground particle vibration velocity around key protective targets should be monitoring. Monitoring point arrangement was shown in Fig.5.

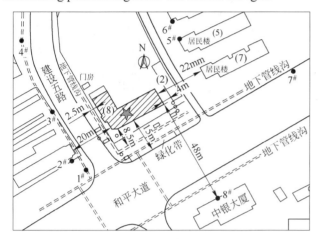

Fig.5   Planar graph of monitoring point

Mini-Mate Pro4 (made in Canada) was chosen as the blasting monitoring instrument. Monitoring results were shown in Tab.4; vibration waveform of $1^{\#}$ monitoring point was shown as Fig.6.

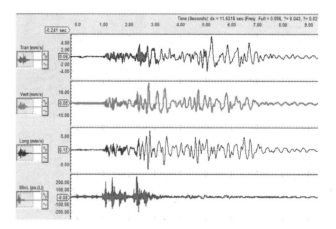

Fig.6   Vibration waveform of $1^{\#}$ monitoring point

As shown in Fig.6, the duration time of monitoring vibration was longer than blasting vibration; blasting vibration and collapse vibration lasted 7.5s; compared with blasting vibration, collapse vibration with characteristics of lower frequency and higher velocity.

As shown in Fig.5 and Tab.4, value of $1^{\#}$ monitoring point is the most among 8 points, but lower than allowable value of relative regulations.

Tab.4   Vibration monitoring data

Monitoring point	Horizontal tangential		Horizontal radial		Vertical	
	Velocity /cm·s^{-1}	Frequency /Hz	Velocity /cm·s^{-1}	Frequency /Hz	Velocity /cm·s^{-1}	Frequency /Hz
$1^{\#}$	0.58	3.0	1.8	6.1	0.72	7.3
$2^{\#}$	0.54	3.9	0.89	3.9	0.35	2.9
$3^{\#}$	0.21	3.75	0.21	3.75	0.21	3.75
$4^{\#}$	0.04	3.67	0.09	3.13	0.1	4.14
$5^{\#}$	0.13	3.59	0.06	3.98	0.31	3.48
$6^{\#}$	0.44	6.4	1.83	6.2	0.39	7.3
$7^{\#}$	0.44	3.3	1.05	3.5	0.33	2.1
$8^{\#}$	0.51	3.7	0.91	3.1	0.51	4.1

5 BLASTING EFFECTS AND EXPERIENCE

5.1 Blasting Effects

As the live video shown, the "L" shape 8-storey frame structure building collapsed in accordance with blasting scheme: after initiating 3.0s, two independent part of the building separated from the settlement joint, there was no collided in the air and deviation from the design direction, which indicated 460ms delay time was reasonable (Fig.7).

Fig.7 Collapse process
(a) $t=1.5$ s; (b) $t=2.5$ s; (c) $t=3.5$ s

Fig.8 Blasting muck-pile form
(a) whole form; (b) cross range form(east side)

The blasting muck-pile form was shown in Fig.8, all structural components collapsed and fragmentized; the blasting muck-pile exceed the horizontal projection of the building only 2.0m in the east, 3.0m in the south; under the influence of blasting muck-pile superposition of two independent parts of the building, the muck-pile exceed the horizontal projection of the building reached 4.0m; because of effective safeguard measures, the 10kV electric cables channel, adjacent residential building, large fishbowl of the Follower & Pets Market and other protective targets were intact; the blasting demolition proves successful.

5.2 Experience

(1) The collapse direction and space of "L" shape building were limited because of which located in intersection of two arterial roads in downtown district, the blasting demolition scheme should be determined by the structure of the building and surrounding environment conditions.

(2) The settlement joint of the "L" shape building should be utilized in blasting demolition, blasting cut height and delay time of the settlement joint are key parameters, should paid more attention.

(3) Cross-range of the blasting muck-pile could be controlled effectively via end wall preliminary demolition.

(4) Collapse vibration could be decreased effectively via relay delay circuit; the comprehensive safeguard system consisted of covering and proximal safeguard combined with protective safeguard can control flying debris in limited range.

REFERENCES

[1] GB 6722—2003 Safety Regulations for Blasting[S]. Beijing: Standards Press of China(*In Chinese*).

[2] ZHOU Jiahan, YANG Renguang, PANG Weitai, et al. Ground vibration caused by structural components collapse and impact in building blasting demolition [C]// Proceedings in Rock and Soil Blasting(2), Beijing: Metallurgical Industry Press, 1984 (*In Chinese*).

[3] ZHOU Jiahan, CHEN Shanliang, YANG Yezhi, et al. Safety distance ascertain influenced by blasting demolition vibration[J]. Blasting, 1993:165~170 (*In Chinese*).

[4] LI Shouju. Study on flying distance of flying debris in blasting demolition [J].1994,11(4): 10~12.

Design and Construction Technology of Large Dock Cofferdam Demolition Blasting

GUAN Zhiqiang[1,2], ZHANG Zhonglei[2], LI Qiang[1], WANG Lingui[2], YE Jihong[1]

(1.Zhejiang Ocean University, Zhoushan, Zhejiang, China; 2.Da Chang Construction Group, Zhoushan, Zhejiang, China)

ABSTRACT: This paper overviews the domestic demolition blasting technology of dock cofferdam in 10 years, including the introduction of dock cofferdam structure, the selection of blasting scheme and parameters, the identification of blasting harmful effects, especially the standards for blasting vibration control, blasting implementation technique and safety protection measures. The boatyard in this article weighted from tens of thousands to 0.5 million tons, and the cofferdam length was from dozens to hundreds of meters Blasting volume at a time was from thousands to 0.2 million cubic meters and the blasting fire dose was from several tons to 50 tons or even more. Having dozens of successful cases, we developed an entire set of design-construct and safety protection measures.

KEYWORDS: reserved rock ridge; compound cofferdam; blasting environment; harmful blasting effect; safety threshold; safety protection technique

1 INTRODUCE

Dock is a hydraulic structure artificially constructed at the shoreside for ship building, ship repairing and ship berthing and the quantity and size of it is one of the important indicators of a country's shipbuilding capacity. It is commonly divided into dry dock and floating dock. The dock in this paper refers to dry dock. Dock cofferdam means the temporary retaining structure for building of perpetual facilities in water conservancy and port construction projects. It is used to creative a dry environment for dock construction and employed immediately after building docks and removing cofferdam.

1.1 Structure of Cofferdam Dock

In order to save the cost of construction, coastal boatyards are usually built at the rocks foundation other than meeting requirement of water depth, typical dock is usually built around mountains and dock cofferdam make full use of the rock bar facing water to be a part of cofferdam or total cofferdam. When the size, shape and the rock organization of reserved rock bar can not exactly meet requirements, it's feasible to throw fill stones on the top or at the side of cofferdam, use steel casing cast-in-place piles embedded in rock and set buttress retaining wall upside. High pressure jet grouting waterproof curtain is set vertically in the middle of cofferdam for the sake of preventing cofferdam leakage, and high pressure grouting or cement mixing piles are used to reinforce the reserved rock bar with poor geological conditions so as to form the compound cofferdam with reserved rock bar as the main body.

1.2 Features about Demolition Blasting Technology of Dock Cofferdam

(1) Complex environment and strict controlment of blasting effect: the toe of dock cofferdam is close to dock floor, dock block, pump house and threshold, ships might have been assembling and building in dock when removing cofferdam after closing dock door, so did the complicated blasting environment. Blasting should make full disintegration of cofferdam rock and reinforced concrete retaining wall so as to clean and salvage easily, it also should fully consider the adverse effects of blasting.

(2) Dock cofferdam is usually adopted compound forms with complex cofferdam structure and difficult construction. Preprocess the support system is necessary before blasting, otherwise blasting effect will be influenced, but mishandling also can lead serious safety accidents.what's more, the design and excavation of outline face is complex and explosion of underwater shaped difficultly because some drilling depth of cofferdam removal exceeds 50 meters.

(3) Unable to exactly master the terrain and geological datas of cofferdam and reef explosion: It might difficult to drill, charge and connect network due to the poor geological conditions of dock cofferdam like low bedrock surface, broken rock layers. In addition, inaccurate distinguishment of original rock interface and complex underwater terrain, because dregs and sludge locate outside of the cofferdam with water nearby.

(4) High quality requirements for blasting equipments and complex blasting network. Cofferdam is blasted at a

---

Corresponding author E-mail: 13505803468@163.com

time but uneasily handle if misfired, what's more, part of steel supports are removed after connecting network, so detonators with high-intensity and high-precision, explosives with high-performance also well waterproof and reliable blasting network are necessary.

(5) Poor working environment and easily affected by climate: cofferdam construction sites are narrow, reef explosion, slag removal are marine engine operation underwater easily influenced by climate and tide. So we must fully predict the negative effects result from typhoon and take the relevant measures.

(6) High requirements of ecological protection because the creatures in water are sensitive to blasting shock.

2 DOCK COFFERDAM DEMOLITION BLASTING SCHEMES

2.1 Principles and Methods of Schematic Design

The overall plan is to be formulated reasonably adjusting measures to local conditions. Sufficient broken and reasonable blast heap shape are both to be fully considered. Effective safety protection measures should be adopted in advance for the possible extrusion on the dock gate induced by the blast slag produced by the cofferdam removal.

Blasting successfully for the first time should be guaranteed. The water-resisting property and anti-corrosion property of the blasting materials should be concerned in the design process, as well as the safety, reliability and simplicity in the working progress. Another factor that has to be taken into account is the reliability and safety of the priming circuit. Influence of water shall be considered when designing the blasting parameters, and the design principle is "high unit consumption and low single explosive".

Full demonstration for the influence induced by the blasting on the adjacent constructions is required. Criterion for permission and safety of blasting shall be formulated properly. Safety protection measures should be carried out to control the harmful effects of blasting in the allowed range.

Docks and cofferdams are constructed in the coastal region. The outside of dock cofferdams are close to the sea, and the scouring action of tides shall be considered sufficiently for the stability of the cofferdams in the working progress. Generally, low water time is chosen for the advantage of minimum waterhead inside and outside the dock opening unless the over-current flush slag, thus the threating caused by the blasting and the blasting slag flying into the dock can be decreased.

2.2 Design and Construction Process of Dock Cofferdam Demolition Blasting

The dock cofferdam demolition blasting is waterfront or underwater blasting, also belongs to demolition blasting. Compared with general blasting engineering, the program is more complex. The specific process as shown in Fig. 1.

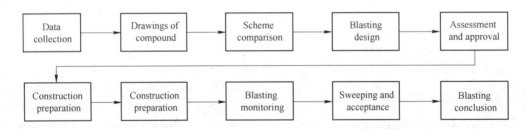

Fig. 1 The flowchart about design and construction process of dock cofferdam demolition blasting

2.3 Project Design of Demolition Blasting

There are varieties methods of cofferdam blasting. According to the different conditions like whether to close the gate or whether the cofferdam filled with water can form various schemes. everyone has its advantages and disadvantages and application conditions, thus should be confirmed by the comparison and selection of blasting surroundings, cofferdam structure and project duration.

Drilling inclined hole in the medial facade of cofferdam and drilling vertical hole in the top of cofferdam are it is general to adopt several versatile drilling forms. It is feasible to choose pre-splitting blasting in two different position for guaranteeing the well planeness of outline surface at the both ends and the bottom of cofferdam after blasting, as well as diminishing the blasting vibration.

2.4 Parameter Design of Blasting

Cofferdam blasting divide into main blasthole and auxiliary blasthole. Main blasthole is mainly used for blasting disintegration of cofferdam body, auxiliary blasthole including pre-split hole, cofferdam counterfort retaining wall and other building blasting holes within the scope of blasting areas, which playing a supporting role in cofferdam blasting.

(1) Drilling diameter and inclination: drilling on coffer-

dam often encounters rock joint fracture development, seawater sweeping into holes, deep depth of holes and other difficulties, placing PVC plastic tube into blast holes well drilled immediately, the diameter of holes, tubes and explosives are usually 140mm, 110mm, 90mm respectively. According to engineering practice experience, transverse-drilling inclination slight less than slope inclination outside the cofferdam is beneficial to control minimum resist line. Vertical drilling changes with the slope angle outside the cofferdam, near the inside the drilling inclination is usually 90°.

Fig. 2 Blast holes arrangement and drilling construction cofferdam

(a) transverse drilling blasting; (b) vertical drilling blasting

(2) Drilling derth: bing confirmed by the height of cofferdam and thickness of bottom, the domestic cofferdam inclined holes maximum over 50 m and vertical hole can deep to 15~20 m.

(3) Hole distance and row space: considering the convenience of construction and blasting network adjustment, rectangle or square arrangement are commonly used, with the hole distance from 2 to 2.5 m as well as the row space. Especially note that hole bottom distance should not exceed 3m when fan-shaped arranging the drilling holes.

(4) Dynamite unit consumption: if rocks by cofferdam blasting can't fully broken because of the inadequate unit consumption, shallow sport will arise and it's difficult to blast underwater again, which affect construction period and construction cost directly. Increasing dynamite unit consumption properly can ensure blasting fully broken.

According to the cofferdam demolition blasting statistics in recent 10 years, unit consumption is generally $0.8\sim1.8$ kg/m^3, taking large value for bottom of cofferdam and small value for upside, likewise, taking large value for hard rock and small value for soft rock. If the depth of hole exceeds 10m then select $1.5\sim2.0$ kg/m^3, however, special-site selecting more than 2.0 kg/m^3 is also advisable.

(5) Charge per hole: hole length decrease stemming length is charging length times line density.

(6) Stemming length: vertical drilling blasting, when the explosion is transverse low-dipping blasting, decreasing stemming length from top to bottom and row by row, with 4 m for the highest 2 m for the lowest. If exist Fan-shaped blast holes, part of holes should be widen appropriately on account of the concentration blast holes.

(7) Charge structure: generally holes are charged continuously, improving charge density in hole bottom and decreasing it upside. Typical security checking of blast holes is necessary before charging, if the blasting harmful effects over safety threshold then segment initiation, the bottom first with the top following.

2.5 Selection of Blasting Materials

Main blasthole is generally filled with emulsion explosives or seismic charges with plastic casing high precision detonating detonators or electronic detonators. In addition, with minority common emulsion explosives and detonating cord for support holes.

2.6 Design of Blasting Network

Dock important buildings are close to blasting point, we must strictly control the explosive charge and usually choose high precision plastic detonating tube millisecond delay hole-by-hole rally blasting network.

2.6.1 Principles of Network Design

The design principles of firing order and delay time are illustrate as follows:

(1) Detonator in holes are generally use high grade and hole between the detonator use low grade in hole by hole rally blasting network.

(2) Detonating time difference of front and back row corresponding blasthole should not differ too much, or it will Influence the blasting crushing effect which increasing the risk of the earlier blasting holes damaging the later blasting holes.

(3) The design of network should neat and clean which

can help check the network.

(4) Give the corresponding technical measures to improve the reliability of network detonation transmission.

2.6.2 Forms of Network Connection

Two detonators with feature of high-precision, time-lapse and non-electric are loaded into each blasthole, while their delay time is generally 400～600 ms. If the hole depth exceeds 20 m, every hole mounted three detonators. If it is segmented blasting in single blasthole, concatenating 9 ms or 17 ms detonators after nonel tube in the hole detonators. Time delay among holes adopts high-precision watch delay detonators, delaying 25 or 42 ms between the same row, whereas delaying 42 or 65 ms in the different rows, the shallow hole of buttress retaining wall of reinforced concrete is connected to the nearest main blast hole.

2.6.3 Position Choice of the Opening

There are different shapes of muck piles due to the different blasting open position. We should consider varieties factors like cofferdam structure, engineering geological conditions, distance that cofferdam away from dock port facilities, and dock doors whether to close, whether to allow over-current and flush slag after blasting when choosing initiation point.

(1) Blasting in the middle of cofferdam: detonation transmission appeared to be a V-shape from the middle blasting point to two both sides. When explosion is the vertical hole blasting, the V-shaped opening towards outside of dock entrance, and its initiation sequence is from the outside-in, as well as from the middle to two sides; when explosion is the transverse low-dipping blasting, the V-shaped opening towards up, and initiation sequence is from the top-down, as well as from the middle to the two sides; however, when blasting in the middle of cofferdam, the muck pile after explosion relatively concentrate to the center.

(2) Blasting in the opens of cofferdam: the end opening of cofferdam can be divided two different forms, blasting in the one opening side or two sides. When blasting is in the ends of cofferdam, the opening towards outside after vertical hole blasting, and its initiation sequence is from the outside-in, as well as slash transmission from one end to the other end, when explosion is the transverse low-dipping blasting, the opening towards up, and initiation sequence is from the top-down, as well as slash transmission from one end to the other. when blasting in the opens of cofferdam, the muck pile after explosion are relatively uniform.

2.6.4 Network Simulation Tests

It is necessary to have a network simulation test before formal blasting.During the process of test, test detonating network should exact explosion completely.

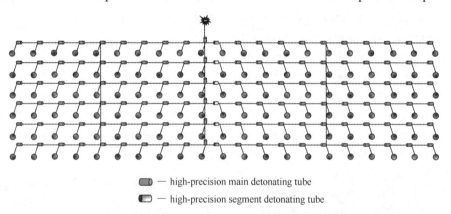

— high-precision main detonating tube
— high-precision segment detonating tube

Fig. 3 The blasting network of high-precision detonating tube

2.7 Design of Security Technology

2.7.1 Control of Blasting Harmful Effects

Cofferdam demolition blasting can result in harmful effects including blasting vibration, blasting flying objects, water shock and blasting surged waves, water stone debris, etc. So it's necessary to control the harmful effects in the allowed range.

(1) Blasting vibration: in view of the lack content of this part in "Blasting Safety Regulations" (GB 6722—2003), it is advised to adopt the following control standards by our repeatedly data collation, conclusion and analysis: water pump room, dock block and dock floor, wharf and dock threshold granite covers are blasted with the vibration velocity of 20 cm/s, 25～30 cm/s, 5～8cm/s respectively. Adjusting blasting parameters according to the vibration velocity test results during preliminary demolition process before cofferdam blasting.

(2) Blasting flying objects: cofferdam outside facing water, generally does not produce a lot of flying stones, mainly upper buttress retaining wall and so on which produce a small amount of flying objects influence the docking port facilities, the cofferdam blasting flying objects by adjusting the charging structure and safety protection measures.

(3) Water shock wave: summarizes many times through coffer demolition monitor, to dock gate, dock entrance of reinforced concrete, the granite cover, outfitting pier piles and so on, takes its static compressive strength as the secu-

rity threshold value which demolition instantaneous dynamic pressure surge wave overpressure, because the structure tendency compressive strength is far bigger than its static compressive strength, this standard used have already contained certain safety factor, might guarantee the security of the construction.

(4) Blasting swell and water-rock debris flow: the effects on the dock door, dock block, and dock threshold granite veneer from surge and water-rock debris flow should be noticed when dock cofferdam blasted with the door unopened. Its better to choose the bottom tide to blast which can minimize the difference between water level of inside and outside dock; cable should be used to locate before blasting, rubber fenders should be installed around the dock to prevent seawater swarm into dock so fast that leads to dock doors drift and hit dock wall mutually after cofferdam blasting; its also necessary to especially protect dock door with granite covers to prevent the damages from the shock of water-rock debris flow when dock cofferdam blasted without water.

2.7.2 Safety Protection Design of Hydraulic Structures and Facilities

In the process of dock cofferdam demolition blasting, the main safety protection objects are dock gate, water pump room, dock block and granite veneers, dock floor, quay, etc.

Blasting should be taken the measures about active protection based, supplemented by passive protection, while the different parts should be used of different security protection methods.

2.7.2.1 Active Protection

Active protection is achieved by adjusting the parameters of blast hole net, charging structure, selection of detonating blasting materials, initiation order, unit consumption of explosives, charge amount of single blasting and other ways. It is the primary means of blasting safety protection. it can not only make cofferdam rock bar fully loose that easy to be cleaned, but also minimize the scope of blasting accumulation body, which are decreasing the extrusion and strike to dock doors and dock entrance facilities, at the greatest extent, from blasting accumulation body and flying objects through effective active protection.

2.7.2.2 Passive Protection

Passive protection is divided into three categories: near body, coverage and protective protection, it should be taken the three different measures flexibly according to different parts of safety protection.

(1) Protection of dock door: most dock doors was built in docks during the process of dock construction, it's necessary to protect dock door which is an important facility when cofferdam blasting with door closed or within dock.

It may cause damages to dock door when blasting with door closed, they are mainly the strike of blasting flying objects, the extrusion to dock door from blasted-piles and blasting vibration, there also have the problem of overpressure from shock wave in water if blasting filled with water. The cliffside flower protective measures is generally taken as the protection of the facade of dock door, protection walls can be built between dock doors and cofferdam and buffer material can be filled between wall and dock door. It also can pile flexible buffer material at the main protective parts in front of dock doors, as to which protective method be chosen, it base on the field conditions, as long as can ensure the safety of dock door.

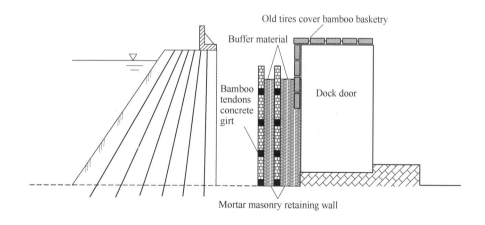

Fig. 4 The plot of typical protection retaining wall

When blasting with door opened, the location of dock door put in the dock is generally above 100m away from dock port. the possible damage to dock door when blasting are the strike of blasting flying objects, fast over current in cofferdam and surge produced by blasting with water filling or not, which possibly lead to dock doors drift and hit dock wall mutually. Protective measures are made to take fixed measure of dock door before cofferdam blasting, at the same time, rubber fender are fixed around dock and we initiate at low tide.

Fig. 5 Practical effect pictures of dock door protection

(2) Protection of dock threshold with granite covered: dock threshold granite veneer has the main role of sealing up and once damaged by blasting flying objects produced in the process of cofferdam blasting demolition, the veneer breakage will leak after closing dock doors, which is greatly difficult to be repaired, so it must be protected strictly. Generally, using sandbags, old tires, bamboo basketry and other materials to implement near body and cover protection.

(3) To Water pump room, dock block, outfitting wharf by the sea, important construction facilities which can not be transferred before blasting, we can establish protective shelving, cover or suspend old fishing nets, steel plates, sandbags, bamboo basketry, tires and other protective materials to implement passive protection.

2.7.3 Resources Protection of Aquatic Organism

The effect on aquatic organism is mainly the shock wave in water, in the complete revision and upcoming book "Blasting Safety Regulations" (GB 6722), demonstrating that the damage degree to fish from peak overpressure of water shock wave is controlled by Tab.1.

Tab. 1 Water peak overpressure impact on fish safety control standards (MPa)

	Fish species	Natural state	Cage culture
High sensitivity	Sciaenid family fish	0.10	0.05
Moderate sensitivity	Grouper, sea bass, barracuda	0.30~0.35	0.20~0.25
Low sensitivity	Winter cavefish, wild carp, sturgeon, flounder	0.35~0.50	0.25~0.40

Fish are sensitive to overpressure of water shock wave even if under a small overpressure. the different fish families have various capacity to bear water shock wave overpressure and the most sensitive fish is Sciaenidae. The resistance capacity is also related to living environment, this capacity of fish being free are superior to that being constrained. Fish in bombings have a certain ability to remember, we can adopt the method of small amount of explosive warning blast fish during blasting away from the blasting area, and thus to minimize the damage to fish. Fish have various spices and huge diversity of individual size. In engineering application, we should establish practical and reasonable security control standards according to comparative experiments in virtual case.

The main protection measures to the fish resources of the cofferdam blasting: generally, using small charge blasting or fish-catching equipments to drive fish outside the reach of shock wave, the time of quit driving should maximum close to cofferdam blasting time in order to prevent a few fish entrance or re-entry during the period from driving end to blasting start. It is also can be set bubble curtain outside of cofferdam to cut down water shock wave. Protective measures of water shock wave is shown in Fig. 6.

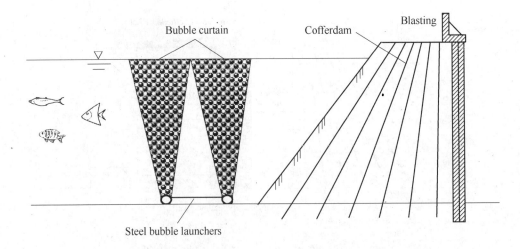

Fig. 6 Diagram of bubble curtain protection measures

3 INNOVATION AND PRACTICE OF ENGINEERING TECHNOLOGY

Through engineering practices of dock cofferdam demolition blasting in 10 years, dock cofferdam demolition blasting technology have matured gradually and formed a set of complete design construction and safety protection technology.

3.1 Engineering Technological Innovation

3.1.1 A Complete Set of Technologies Including Dock Cofferdam Demolition Blasting Throwing Accumulation and Slag Discharge Control

(1) Cofferdam demolition blasting with door closed: adopting the transverse slight inclined borehole to make blasting pile accumulate in the original cofferdam position reduce the threats of blasting body deposit and blasting shock wave to gate; in view of the characteristics of slag blasting relatively concentrated at opening position, we choose cofferdam at both head end position for initiating and preventing the gate deformation from the extrusion of blasting pile to central gate.

(2) For the dock without dock channel, cofferdam should be blasted at low tide after filling water into dock to high water level. Overflowing should be controlled strictly with the purpose of preventing blasting slags flowing into dock chamber, causing slags difficultly cleaned after gate closed.

(3) For dock with a dock channel, whose gate is not constructed in docks and lack of mat dolphin, cofferdam demolition blasting can use blasting overflowing flushing the slag at high water in order to reduce the quantities of underwater slag removal; dock implosion slag can implement dry slag after being closed the door and drying the inland water. The engineering progress is quick and the construction cost is only a third of the grab bucket ship slag removal.

3.1.2 Auxiliary Operation Technology of Dock Cofferdam Demolition Blasting with Rapid Construction

(1) Taking the pretreatment technology of dock cofferdam steel liners rock-socketed piles before blasting truncation and plain concrete filling replacement have solved the difficulty of the subsequent underwater manual cutting steel after applying the method of dealing steel liners rock-socketed piles by blasting, which ensure the safety of concrete structures in front of dock entrance, while the remains without treatment after being cut becoming a part of dock entrance structure, preventing the water flushing action to dock entrance, as shown in Fig. 7.

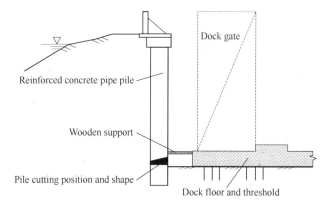

Fig. 7 The schematic diagram of pile cutting position and shape

(2) The preprocess technique of "deep cutting shaping control blasting" is used to solve the unstructured problem inside of cofferdam when excavating dock chamber. it not only benefits the precision control of drilling lofting, but also can effectively resolve the difficulties of insufficient space for subsequent drilling, improve the body blasting effects and operating conditions.

(3) Cofferdam coastal sidetrack drilling ship underwater blasting and mat slag drilling construction technology: for the challenges of drilling underwater rock near seawater when adopting vertical hole blasting scheme, the platform tide water reef blasting borehole is used and conducted along with the cofferdam body charging operation above water then blasting together with the main body of cofferdam after charging. For the scheme of transverse slightly inclined holes, compared with the traditional artificial structures of scaffolding drilling platform, the high wind pressure drilling operation platform formed by matting slags into cofferdam has advantages of low cost, fast speed, well safety and easy to be cleared.

(4) Preliminary excavation construction technology of dock cofferdam: seizing the moment flexibly based on the cofferdam stability before coming dock cofferdam blasting, the preliminary excavation of cofferdam near the water surface shall create a good free surface for throwing slags generated by cofferdam blasting into the near water side as far as possible, it will be good for the protection of the facilities in dock entrance and reduction of caisson on the base board of dock gate.

3.1.3 Comprehensive Demolition Technology of Dock Cofferdam Blasting Safety Protection

(1) The protection of cofferdam blasting without water filling. Buffer materials should be filled, between the cofferdam and the guarded body in the protective body, such as flexible thatch, straw and waste tires, for the intention of forming a "alternate kindness with severity" protection system, cutting the slag body momentum after the cofferdam blasting and delaying the acting time so that avoid blasting direct blow to the gate, and reducing the impact on the gate and ensuring the safety of the gate.

(2) The protection of cofferdam blasting with water filling. With the purpose of avoiding the threats to gate from shock wave in water, except for choosing reasonable drilling mode and initiating position, we can establish bubble launchers under water and closed to explosion source to from the bubble curtain, which can effectively reduce the intensity of shock waves in water.

3.1.4 Fine Construction Control Technology of Dock Cofferdam Blasting Demolition

(1) Technology of drill positioning and precise control. It is difficult to confirm the interface of mud and rock due to the complex geological conditions around dock cofferdam. To

ensure the accuracy of blasting minimal resistance line, it is necessary to use consciously wear part of drilling with drilling angle less than expected rock surface angle, and on the premise of efficient sealing plans with dry seaweed. Adjusting the actual drilling depth to ensure holes be drilled accurately and avoid the appearance of foundation after blasting.

(2) For the transverse slightly inclined holes blasting, we install PVC pipe and two threaded plastic housing for charging, which ensure the serviceability rate under the poor conditions like the poor integrity of rock, serious leaking and gushing of blast hole, and the length of blast holes more than 50m, as well as the charge continuity in harsh environments.

(3) The use of waterproof high precision millisecond delay detonating tube detonators or digital electronic detonators hole by hole relay system can make the extension of time accurate to a few milliseconds, initiating section number reaches hundreds of thousands which ensure the accuracy blasting of initiation system.

3.2 Brief Introduction of Dock Cofferdam Demolition Blasting Examples

Tab.2 list the projects of Dock Cofferdam Demolition Blasting the author participated in since 2005, including the time of blasting implement, the project designation, schemes and main parameter indicators.

Tab.2 Dock cofferdam demolition blasting scheme selection and the main parameter main parameter indicators

Blasting time	project designation	Description of the main parameters and indicators
2005.10	Zhoushan Shipyard IMC yongyue 300 +150000 ton dock cofferdam demolition	Vertical straight hole, horizontal inclined hole and shallow auxiliary hole filled with water before the blast With a total blasting demolition of about 25000 m^3. The amount of main explosive holes is 556, 23.46 tons of emulsion explosives is used, 2726 detonators, and detonating cord is 1500m. The initiation network carries forward from the middle to both sides, the hole blasting duration is 3510ms
2007.02	Golden Bay Shipbuilding Co., Ltd., No. 2 dock (300000 tons) cofferdam demolition	Vertical deep-hole one-time blasting demolition, filled with water before the blast dock, bubble curtain is arranged between the inner and the dock cofferdam gate. The blasting volume is 27000 m^3. Using local segmented high-precision millisecond delay detonating tube compound initiating network. High density seismic source column is 29600 kg and the high precision non-electric detonator is 1360 rounds
2007.01	Zhoushan Chengda ship West Xiezhi shipyard remove the 30000 ton dock cofferdam	Horizontal fan hole forming, flow cinder flushing, and artificial cofferdam shallow hole blasting. Blasting volume is 3000m^3. Emulsion explosives is 7227kg, non-electric detonators 575 milliseconds primer, electric detonators is 10 rounds, detonating cord is 250m, and the total duration is 1350ms
2007.10	Removal of Golden Bay shipyard 500000 ton Dock Cofferdam	Gently inclined hole scheme of Cofferdam Blasting at a rock ridge, buttressed retaining wall with shallow hole blasting auxiliary, dock is not filled with water. Composite cofferdam to be removed is 55000 m^3 with 56 rock-socketed pile and an abandoned pier is outside central cofferdam. The total dosage is about 52 tons (51 tonnes seismic charge), the use of high-precision detonators primer is 1700 rounds, divided into 265 sections, with a total delay 1640 ms
2007.11	Golden bay boat No.3, 4 dock (70000+150000) ton Cofferdam demolition	Gently inclined hole, vertical hole assisted one-time blasting at low water level, and the dock is filled without water when it is blasting. The full length is 165m, blasting amount is 31185m3 (including irrigation stone), concrete-filled steel tube pile is 197, and the reinforced concrete beam is 390m^3. The drilling is 1325, of which the deep hole is 327; the safety tire is 380, bamboo raft is 4525 and the bag is 5682. Total consumption of explosive detonator is 23544kg and 3439 detonators primer
2008.06	Japan Tsuneishi group zhoushan daishan xiushan shipbuilding base 100000 ton dock cofferdam demolition	Lateral tilt deep hole, close the gate, not filling water, and low water blasting. The total blasting is 15008m^3; using high precision millisecond delay detonator network, hole by relay blasting technology to reduce the harmful effect of blasting; blasting pile to avoid excessive concentration of dock gate. Total charge is 20.05T, using the high precision kay detonating detonator, by hole blasting . The use of high precision detonator primer is 875 rounds, it is divided into 130 sections, and the total delay 2142 ms

Continues Tab.2

2008.06	No.1 dock (30000 ton) cofferdam demolition in Dong Xiezhi Shipyard of Zhoushan Long-sheng Shipbuilding Co., Ltd.	Deep lateral tilt, over-current slag, and artificial weir shallow hole blasting. Cofferdam blasting engineering quantity is 9631m^3. Using ordinary plastic shell emulsion explosives, domestic non-electric detonator. Blasting using intermediate openings, V-shaped booster, from the middle to the ends of detonating non-electric blasting holes relayed by detonating pipe network
2009.01	Zhejiang Eastern shipyard Co., Ltd. 300000+150000 ton dock cofferdam demolition	Horizontal tilt pore, assisted partial vertical hole, filled with water weir design. The project is about 64736m^3 (including ± 0.00 riprap elevation above the amount 17046 m^3), reinforced concrete blast wall buttresses square is about 1028m^3. Assembly dose is 40.8T, by using high-precision detonators, detonating detonation hole (hole section above) technology. The use of high-precision detonators primer is 1582 rounds, divided into 240 sections, with a total delay 2150 ms
2009.05	Cofferdam demolition in Panzhi Island Shipyard of Zhoushan Big shenzhou shipbuilding-repairing Co., Ltd.	Close dock door, gently inclined deep transverse mainly shallow hole auxiliary, between the cofferdam and the dock door with masonry protection wall, filled with cushioning material "hardness with softness" of the dock door security system. The total project of cofferdam blasting and reef explosion is about 42000 m^3. Using seismic charge is 16.1 t, ordinary emulsion explosive is 550 kg, and the use of high-precision millisecond delay detonators (Orike) 1927 rounds, divided into 156 sections, blasting total delay 2075 ms
2009.09	Ship repairing base in ship industry Co., Ltd., Jiangnan mountain base dock 1 and 2 (400000+300000) ton cofferdam demolition	The bottom demolition elevation (-11.0m) above the pile rock ridge with gently inclined deep hole blasting, blasting pile bottom elevation above -11.0m without rock ridge of pre cut pile cofferdam; high pressure jet grouting curtain of vertical deep hole blasting, shallow hole blasting with buttressed retaining wall. Cofferdam filling water in the dock before blasting to reduce Pile cofferdam blasting after collapsing on the dock facilities impact damage. Blasting area of about 4200 m^2, engineering quantity is about 30000m^3, buttressed retaining wall blasting of about 505 m^3, and remove the pile is about 155 roots. Consumption of explosives is 23000 kg, high precision non electric millisecond delay detonator 3000 rounds, detonating cord is 3400m, detonating tube is 3000m, and blasting the total delay time of 2.6 seconds
2010.10	Zhoushan Pacific Ocean Engineering equipment and ship building dock cofferdam demolition Project	Using gently inclined and vertical bore holes combine dock doors are not closed, inadequate water in the cofferdam, and disposable one blasting program. Earth cofferdam dredging is 146191 m^3, blasting volume is 41172 m^3 (of which: concrete pile is 1010m^3, stone masonry is 961m^3, cofferdam rock ridge is 39201m^3). Explosives is 34086kg, Orike precision detonators is 2078 rounds, ordinary non-electric detonators is 1853 milliseconds hair, detonating cord is 4600m
2011.12	No. 1, No. 2 dock cofferdam demolition of Zhejiang Long-wen Shipbuilding Engineering Co.,Ltd.	Gently inclined and vertical bore holes combine dock doors are not closed, inadequate water in the cofferdam, and disposable overall blasting program. Cofferdam bedrock blasting volume about 29400m^3, square stone masonry approximately 2220m^3, blasting side wall buttresses amount 500m^3, slag and sludge throwing parties about 31600m^3. 929 millisecond delay detonators hole precision hair, Orike precision detonators 832 millisecond delay surface fat; ms13 section 1617 detonators rounds, detonating cord of 900m, and detonating tube of 980m. Main burst hole design loading dose (ϕ80 and ϕ75) is 39080kg, main explosive charge hole actual amount of 38390kg; and emulsion explosives (ϕ32) is 920.8kg. Cumulative assembly dose of 39310.8kg
2010.12 2012.01	The dock cofferdam demolition of Fujian province in east China shipyard Co., Ltd.	The vertical deep hole blast filled with water. The total amount of blasting about 5 to 60,000 m^3. The first phase of the use of blasting explosives is 3048kg, ordinary non-electric blasting detonators is 210 rounds, electric detonators is 4 rounds. The second phase of the use of blasting explosives is 14664kg, ordinary non-electric detonator is 120 rounds, Orike precision millisecond delay detonators 1471 rounds, detonating cord is 300m, and electric detonators is 6 rounds
Remark:		Main gun with 140 mm in diameter and 115 mm in diameter, hole built-in PVC pipe, and the vessel diameter smaller than the hole diameter of 2~3 cm

4 CONCLUSIONS

Engineering practices have proved blasting was the most economical and convenient way to remove boatyards. Construction enterprise developed throwing accumulation demolition blasting technology of boatyard cofferdam aided operational technique rapid construction safety protection integrated technique, a series of key technical research such as design, construction and safety control get a breakthrough from a set of demolition blasting technology of large boatyard cofferdam. The theory and key technology research fully reflects academic frontier in the field of engineering blasting and the progress of science and technology. Also the research has remarkable social and economic benefits and has important theoretical and practical significance for promoting the development of shipping industry.

REFERENCES

[1] Wang X G, Yu Y L, Liu D Z. Enforceable handbook of safety regulations for blasting[M]. Beijing: China Communication Press, 2004.

[2] Wang X G. *The blasting design and construction*[M]. Beijing: Metallurgical Industry Press, 2011.

[3] Zhao G. Study on technology of the cofferdam demolition blasting in deep water conditions [D]. Hefei: University of Science and Technology of China, 2008.

[4] Liu D S. *New blasting technology of China*[M]. Beijing: China Communication Press, 2008.

[5] Zhang Z L, Guan Z Q, Jiang Z B, Li C F. Demolition blasting technology of boatyard cofferdam of$(30+10)\times10^4$ tons in zhou shan mashi island[J]. *Engineering blasting,* 2006, 12(1): 29~34.

[6] Zhang Z L, Feng X H, Guan Z Q, Wang L G. Safe protection technology of explosive demolition of dashenzhou douck cofferdam[J]. *Blasting,* 2011, 28(3): 106~110.

[7] Zhao G, Wu C H, Wang E H. Research on blasting shock wave in water to damage of fish[J]. *Engineering blasting,* 2011, 17(4): 103~105.

Experimental Research on the New Technology of Explosive Cladding

MIAO Guanghong, MA Honghao, SHEN Zhaowu, YU Yong, LI Xuejiao, REN Lijie

(*Department of Modern Mechanics, University of Science and Technology of China, Hefei, Anhui, China*)

ABSTRACT: In order to ensure the quality of charge, explosives with structure of honeycomb was designed to resolve the current issue about the backward method of charge and low energy efficiency of explosives; explosives with structure of honeycomb and double sided explosive cladding, which significantly reduce the critical thickness of stable detonation of explosives, are used to increase of energy efficiency of explosives and save the amount of explosives. Emulsion explosives with the thickness of 5mm can stable detonation and reliable combination with two metal plates. In this paper(the high velocity explosives was selected in the experiment), the feasibility experiment of double sided explosive cladding for steel of 45 with thickness of 2mm to steel of Q235 with thickness of 16mm was successful investigated. The experiment of double sided explosive cladding for stainless steel with thickness of 3mm to steel of Q235 with thickness of 16mm was successful investigated too. Compared to the existing explosive cladding method, the consumption of explosives for steel of No. 45 to steel of Q235 and stainless steel to steel of Q235 are reduced by 83% and 77% in the case of cladding the same number of composite plates. The explosive cladding windows and collision speed of flyer plate are calculated before experiment. It has shown that the calculation prefigure exactly the explosive cladding for steel of 45/ steel of Q235 and stainless steel/steel of Q235.

KEYWORDS: double sided explosive cladding; honeycomb structure explosives; special structure explosives of explosive cladding; explosive cladding; energy efficiency

1 INTRODUCTION

The explosive cladding technique makes flyer plate produce plastic deformation, motion, collision with base plate, melting and interatomic combination with base plate through high energy which is released by the explosives[1]. Using the energy released by only one side of explosives, single sided explosive cladding has made many further calculations and researches at present[2~4]. Most of energy is released in the space in the form of shock wave due to only one side of explosive energy can be used. In such case, energy utilization rate is extremely low. The noise produced by explosion can still reach 80~90db at 5 kilometers away. The following problems, such as backward method of charge, heavy workload, serious dust pollution and harmful for workers' mental and physical health remain unresolved.

Double sided explosive cladding which combines with honeycomb structure explosives ensure the quality of charge is used to improve energy efficiency of explosives in the present study. Two groups of steel of No. 45 to steel of Q235 and two groups of stainless steel to steel of Q235 were prepared for this double sided explosive cladding.

2 EXPERIMENT MATERIALS

2.1 Materials of Base and Fly Plate

In the experiments, the same flyer plates and the same base plates are used. The flyer plates.

This paper is supported by the Natural Science Foundation of China (51374189, 51174183), and base plates materials used in this study were Steel of No. 45 to steel of Q235 and stainless steel to steel of Q235 respectively. Main mechanical properties of bonded materials are shown in Tab.1.

Tab.1 Main mechanical properties of bonded materials

Material	ρ/kg·m^{-3}	σ_b/MPa	v_{sf}/m·s^{-1}
Steel of 45	7800	355	5200
Steel of Q235	7800	235	5200
Stainless steel	7900	210	5790

2.2 Honeycomb Structure Explosives

Honeycomb structure explosives is made of aluminum honeycomb(see Fig. 1) filled with explosives in regular hexagon cell. Honeycomb structure explosives(see Fig. 2) can

---

This paper is supported by the Natural Science Foundation of China (51374189, 51174183).
Corresponding author E-mail: miaogh@mail.ustc.edu.cn

ensure the uniformity of explosives density, suit for mechanized mass production, long term storage and long distance transport. The charge thickness can be controlled by the height of honeycomb structure explosives. As Fig. 3 depicts the schematic representation of double sided explosive cladding setup, the setup of double sided explosive cladding is bilateral symmetry. Honeycomb structure explosives are sandwiched between two groups of flyer plates. Without constraints, the critical diameter of emulsion explosives is 14~16mm[5]. The critical thickness of explosives can be reduced by constraint of honeycomb material and two groups of flyer plates. Due to multi-directional constraint of honeycomb material and double sided flyer plates, the critical thickness of honeycomb structure explosives applied on double sided explosive cladding is 5mm. It can expand the application space of emulsion explosives and other explosives in this way.

Fig.1　Aluminum of honeycomb

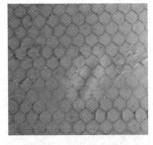

Fig.2　Explosive with structure of honeycomb

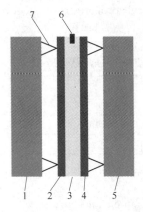

Fig.3　Schematic representation of double sided explosive cladding setup

1—base plate 1; 2—flyer plate 1; 3—honeycomb structure explosives; 4—flyer plate 2; 5—base plate 2; 6—detonator; 7—stand off distance

3　THEORETICAL CALCULATIONS OF DOUBLE SIDED EXPLOSIVE CLADDING WINDOW

Depending on the explosive cladding theory, there are three main parameters of explosive cladding: collision velocity v_p, collision angle β, collision point velocity v_c (in a parallel setup, collision point velocity v_c=detonation velocity of explosives v_d). There are certain geometric relationships among the three parameters, so two of them are independent variables. Any two variables can form a plane within which an area known as the "explosive cladding window" exists for each pair of explosive cladding materials. Collision velocity v_p and detonation velocity of explosives v_d as design parameters were calculated as follows.

3.1　Lower Limit of Collision Velocity of Flyer Plate

Lower limit of collision velocity should be calculated in terms of equivalent shock model[6].

$$v_{pmin}=2p_{min}/v_{sf}\rho_f \quad (1)$$

where, v_{sf} is sound velocity of flyer plate; ρ_f is density of flyer plate; p_{min} is minimum shock pressure required for explosive cladding.

For the cladding of steel of No. 45 to steel of Q235, the value of density of ordinary carbon steel, sound velocity of flyer plate and minimum shock pressure for explosive cladding of stainless steel and ordinary carbon steel were 7800 kg/m³, 5200m/s and 4.5GPa respectively. Due to the lack of the minimum shock pressure for ordinary carbon steel and ordinary carbon steel, we choose the value of stainless steel and ordinary carbon steel to approximate it. Choosing p_{min}=4.5GPa, v_{pmin}=222 m/s was calculated by Eq. (1).

For the cladding of stainless steel to steel of Q235, the value of density of stainless steel, sound velocity of flyer plate and minimum shock pressure for explosive cladding of stainless steel and ordinary carbon steel were 7900 kg/m³, 5790m/s and 4.5GPa respectively. v_{pmin}=197 m/s was calculated by Eq. (1).

3.2　Upper Limit of Collision Velocity of Flyer Plate

The Eq.(2) defines the upper limit of collision velocity of flyer plate as proposed by Wylie [7].

$$v_{pmax}=(2E/\rho_f t_f)^{0.5} \quad (2)$$

where, E is the maximum energy provided by unit area materials under weldability condition; ρ_f is the density of the flyer plate materials; t_f is the thickness of the flyer plate.

For the cladding of steel of No. 45 to steel of Q235, E takes the value of 7.54MJ/m² for ordinary carbon steel, the density of the flyer plate is 7800 kg/m³, the thickness of flyer plate is 2mm, v_{pmax}=983 m/s was calculated by Eq. (2).

For the cladding of stainless steel to steel of Q235, due to the lack of the E for stainless steel to ordinary carbon steel, we choose the value of ordinary carbon steel and ordinary carbon steel to approximate it. Choosing $E=7.54MJ/m^2$, the thickness of flyer plate is 3mm, $v_{pmax}=798m/s$ was calculated by Eq. (2).

3.3 Lower Limit of Detonation Velocity

Ezra[8] defined the lower limit of v_{cmin} according to the fluid hypothesis as follows:

$$v_{cmin} = (4.47 \sim 4.90)\sqrt{\sigma/\rho} \quad (3)$$

where, σ is the maximum yield strength of flyer plate and base plate; ρ is the minimum density of the flyer plate and base materials.

For the cladding of steel of No. 45 to steel of Q235, $v_{cmin}=1045m/s$ was calculated by Eq.(3) combined with Tab.1.

In a parallel setup, collision point velocity v_c=detonation velocity of explosives v_d, so $v_{dmin}=1045m/s$.

For the cladding of stainless steel to steel of Q235, $v_{cmin}=851m/s$ was calculated by Eq.(3) combined with Tab.1.

In a parallel setup, collision point velocity v_c=detonation velocity of explosives v_d, so $v_{dmin}=851m/s$.

3.4 Upper Limit of Detonation Velocity

Upper limit of v_{cmax} is predictable at 1.2 to 1.5 times the speed of sound. Generally speaking collision point velocity must be less than sound velocity of material[8], namely,

$$v_{cmax} = c_{min} \quad (4)$$

where, c_{min} is the minimum sound velocity of flyer plate and base plate.

In a parallel setup, collision point velocity v_c=detonation velocity of explosives v_d, so detonation velocity for the cladding of steel of No. 45 to steel of Q235 and stainless steel to steel of Q235 must be less than sound velocity of ordinary carbon steel. $v_{dmax}=5200m/s$ was calculated by Eq. (4) combined with $c_{min}=5200m/s$.

4 DOUBLE SIDED EXPLOSIVE CLADDING FOR STEEL OF NO. 45 TO STEEL OF Q235 AND STAINLESS STEEL TO STEEL OF Q235

Investigated in experiments, steel of No. 45 with the thickness of 2mm was cladded on steel of Q235 with the thickness of 16mm, stainless steel with the thickness of 3mm was cladded on steel of Q235 with the thickness of 16mm. Calculation results of the above show that detonation velocity for the cladding of Steel of No. 45 to steel of Q235 should stay between 1045m/s and 5200m/s, for the cladding of stainless steel to steel of Q235 should stay between 851m/s and 5200m/s. The detonation velocity of explosives applied on double sided explosive cladding is 4900m/s. Honeycomb structure explosives was detonated by a detonator placed in the middle of the short edge. Double sided explosive cladding setup which keep erecting was buried deeply under the ground. Two experiments were designed to investigate the double sided explosive cladding. Details of the selected cladding conditions are listed in Tab.2 and Tab.3.

Tab.2 Technological parameters of explosive cladding for steel of No. 45 to steel of Q235

Experimental group	Material size / mm	Stand off distance /mm	Explosive thickness /mm	Load radio (r)
No.1	Flyer plate×2: (300×150×2) Base plate×2: (300×150×16)	6	10	0.75
No.2	Flyer plate×2: (300×75×2) Base plate×2: (300×150×16)	6	5	0.45

Tab.3 Technological parameters of explosive cladding for stainless steel to steel of Q235

Experimental group	Material size / mm	Stand off distance /mm	Explosive thickness /mm	Load radio (r)
No.1	Flyer plate×2: (300×150×3) Base plate×2: (300×150×16)	9	10	0.49
No.2	Flyer plate×2: (300×75×3) Base plate×2: (300×150×16)	9	7	0.37

The collision velocity v_p is calculated from the Gurney equation[8], namely, Eq. (5). The outcome of double sided explosive cladding can be predicted whether the collision velocity falls in the area of explosive cladding window. Due to the absence of the Gurney energy of emulsion explosives, $E \sim 0.6Q_v$[9] is proposed to use. Components of the emulsion matrix are listed in Tab.4. The heat of explosion for emulsion explosives which equal to 2966.84kJ/kg is calculated by the calculation method in [10].

$$v_p = \sqrt{2E_0} \cdot \sqrt{\frac{3r}{6+r}} \quad (5)$$

where, v_p is the collision velocity of flyer plate, m/s; E_0 is the Gurney energy, J/kg); r is ratio of the mass of the explosive to the mass of flyer plate.

Tab.4 Components of the emulsion matrix

Component	Mass fraction/%	Component	Mass fraction/%
NH_4NO_3	75	$C_{12}H_{26}$	1
$NaNO_3$	10	$C_{24}H_{44}O_6$	2
$C_{18}H_{38}$	4	H_2O	8

Using the calculated heat of explosion Q_v and the load ratio r, the collision velocity of two groups of flyer plate were calculated as shown in Tab.5 and Tab.6.

Tab.5 Calculated result of flyer velocity for the cladding of steel of No. 45 to steel of Q235

Experimental group	No.1	No.2
Collision velocity /m·s^{-1}	1089	863

Tab.6 Calculated result of flyer velocity for the cladding of stainless steel to steel of Q235

Experimental group	No.1	No.2
Collision velocity /m·s^{-1}	898	787

For the cladding of steel of No. 45 to steel of Q235, as can be seen from the collision velocity of two groups of flyer plate, the collision velocity of the second group of flyer plate falls in the area of explosive cladding window. Double sided explosive cladding was carried out by using the parameters in Tab.2. Due to the explosive cladding setup was buried deeply under the ground, flyer and base plates on both sides of the explosives are subjected to the same constraint. The quality of the explosive cladding is relatively consistent, so the sample taken from either of two composite plates can be used for metallographic analysis. Fig.4 and Fig.5 are respectively the SEM micrograph of the sample obtained by wire cutting for the first group and second group.

Calculated by Gurney equation in the first group of experiment, the collision velocity of flyer plate equals to 1089 m/s. Due to the upper limit of collision velocity of flyer plate equal to 983m/s, it shows that the collision velocity is too high. The continuous melted zones with the thickness of 5～10μm appeared at the bond interface. It confirmed that the collision velocity is too high.

Calculated by Gurney equation in the second group of experiment, the collision velocity of flyer plate equals to 863 m/s. Due to the upper limit of collision velocity of flyer plate equal to 983m/s, it shows that good connection was created at the collision velocity of 863m/s. Via detecting, the bonding rate of cladding is 100%. The height and length of the waves at the bond interface were in the range of 5～8 μm and 15～18μm respectively (Fig.5). It is generally recognized that there are three bond interfaces[11]: big wavy, small wavy, micro wavy. The bond interface dimensions of the second group are less than the bond interface of micro wavy (the wavelength is generally around 100 μm, the height of wave is generally around 20 μm). Compared with bond interface of big and small wavy, almost no transition zone, no cracks and loose like "cavity" exist in bond interface of micro cavy. It shows that good connection was created in the second group of experiment.

Fig.4 Interface wave of sample No.1

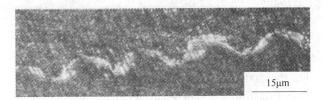

Fig.5 Interface wave of sample No.2

For the cladding of stainless steel to steel of Q235, as can be seen from the collision velocity of two groups of flyer plate, the collision velocity of the second group of flyer plate falls in the area of explosive cladding window. Double sided explosive cladding was carried out by using the parameters in Tab.3. Due to the explosive cladding setup was buried deeply under the ground, flyer and base plates on both sides of the explosives are subjected to the same constraint. The quality of the explosive cladding is relatively consistent, so the sample taken from either of two composite plates can be used for metallographic analysis. Fig.6 and Fig.7 are respectively the SEM micrograph of the sample obtained by wire cutting for No.1 and No.2.

Calculated by Gurney equation in the experiment of No.1, the collision velocity of flyer plate is 898m/s. Due to the upper limit of collision velocity is 798m/s, it shows that the collision velocity is too high. The result shows that two pieces of flyer plate are separated with two pieces of base plate. Analysis of reasons: due to the collision velocity is higher than the upper limit of collision velocity, the energy for cladding is too large. Excessively high energy exists in bond interface. After the explosion, bond interface is still in thermal softening state. Bond interface is separated by reflected rarefaction wave.

Calculated by Gurney equation in the experiment of No.2, the collision velocity of flyer plate is 787m/s. Due to the upper limit of collision velocity is 798m/s, it shows that good connection was created. Via detecting, the bonding rate of cladding is 100%. The height and length of the waves at the bond interface were in the range of 25~35 μm and 95~120 μm respectively (Fig. 6). Fig. 7 is the single interface wave of sample No.2, in which arrow A refers to the thin melted zone with different color from other regions. It shows that small melted regions exist in bond interface. It is generally recognized that there are three bond interfaces[11]: big wavy, small wavy, micro wavy. The bond interface dimensions of the No.2 are basically the same as the bond interface of micro wavy (the wavelength is generally around 100 μm, the height of wave is generally around 20 μm). Compared with bond interface of big and small wavy, almost no transition zone, no cracks and loose like "cavity" exist in bond interface of micro cavy. It shows that good connection was created in the experiment of No.2.

Fig.6　Interface Wave of sample No.2

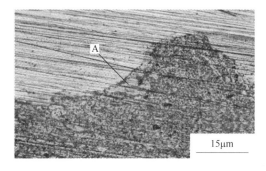

Fig.7　Single interface wave of sample No.2

As can be seen from the second group of experiments, the high velocity explosives can meet the requirements of explosive cladding. The bond interface of micro wavy with high bond strength was produced. Compared with the low velocity explosives, high detonation velocity leads to the higher detonation pressure and higher energy of detonation products. The acceleration of flyer plate provided by high velocity explosives is higher than the low velocity explosives. In order to generate the same collision velocity as the low velocity explosives, explosive consumption is relatively less.

Without constraints, the critical diameter of emulsion explosives is 14~16mm[5]. Applied on current single sided explosive cladding, explosives thickness should be at least 14~16mm, which take the value of 15mm. Using single sided explosive cladding, an explosion only gets one composite plate. The explosives with the thickness of 5mm and 7mm were used in double sided explosive welding which can get two composite plates simultaneously. Compared to the existing explosive cladding method, the consumption of explosives is reduced by 83% and 77% in the case of cladding the same number of composite plates.

5　CONCLUSIONS

(1) Due to multi-directional constraint of honeycomb material and double sided flyer plates, emulsion explosives with the thickness of 5mm can stable detonation and also can get two composite plates simultaneously. Generated by explosives, vast majority of energy is used for double sided explosive cladding. Compared to the existing explosive cladding method, the consumption of explosives for the cladding of steel of No. 45 to steel of Q235 and stainless steel to steel of Q235 is reduced by 83% and 77% in the case of cladding the same number of composite plates. By using honeycomb structure explosives and double sided explosive cladding, the energy produced by explosives has been fully utilized. It can also reduce public nuisance from blasting to keep the environment friendly in this way.

(2) High velocity explosives were selected in experiment. Compared with the low velocity explosives, high detonation velocity leads to the higher detonation pressure and higher energy of detonation products. In order to generate the same collision velocity as the low velocity explosives, explosive consumption is relatively less.

(3) The feasibility experiments of double sided explosive cladding for steel of No. 45 to steel of Q235 and stainless steel to steel of Q235 were successful investigated. Due to almost no transition zone, no cracks and loose like "cavity" exist in bond interface of micro cavy, good connection was created by double sided explosive cladding. It shows that double sided explosive cladding is feasible.

(4) The upper and lower limits of collision velocity of flyer plate were provided by calculation of explosive cladding window for steel of No. 45 to steel of Q235 and stainless steel to steel of Q235: 222m/s <v_p<983m/s, 197m/s <v_p<798m/s. Calculated by Gurney equation, the collision velocity of two groups of flyer plate for steel of No. 45 to steel of Q235 and stainless steel to steel of Q235 equal to 1089m/s , 863m/s and 898m/s, 787m/s respectively. It shows that the collision velocity of flyer plate in the first group experiment is faster than the upper limit of collision velocity and the collision velocity of flyer plate in the second experiment is in the range of upper and lower limits. So the bonding quality of the first group and the second group is poor and superior respectively. Two groups of

explosive cladding for steel of No. 45 to steel of Q235 and stainless steel to steel of Q235 show that experiment results can be better predicted by calculation.

REFERENCES

[1] Crossland B, Williams J D. Explosive Welding. Metals Review, 1970, 15:79~100.

[2] Wang X, Zheng Y Y, Liu H X, Shen Z B, Hu Y, Li W, Gao Y Y, Guo C. Numerical study of the mechanism of explosive/impact welding using Smoothed Particle Hydrodynamics method[J]. Materials and Design, 2012, 35: 210~219.

[3] Chen S Y, Wu Z W, Liu K X, Li X J, Luo N, et al. Atomic diffusion behavior in Cu-Al explosive welding process[J]. Journal of Applied Physics, 2013, 113: 044901.

[4] Sedighi M, Honarpisheh M. Experimental study of through-depth residual stress in explosive welded Al-Cu-Al multilayer[J]. Materials and Design, 2012, 37: 577~581.

[5] Song J Q. Research on detonation characteristics of emulsion explosives[D]. Beijing: University of Science and Technology Beijing, 2000: 45~47.

[6] Blazynski T Z. Explosive Welding Forming and Compaction[M]. London: Application Science Publishers Ltd., 1983.

[7] Wylie H K, Williams P E G. Further Experimental Investigation of Explosive Welding Parameters[C]// Proc of 3rd Int Conf of the Centerfor HEF, Denver: University of Denver, 1971: 1~43.

[8] Shao B H, Zhang K. Explosive Welding Principle and Its Application [M]. Dalian: Dalian University of Science and Technology Press, 1987: 8~287.

[9] Kennedy James E, Zukas Jonas A, Walters William P. In: Davison L, Hori Y, editors. The Gurney model of explosive output for driving metal explosive effects and applications. New York: Springer, 1998: 221~257.

[10] Lu Ming, Lu Chunxu. The mathematical model for the formulation design of emulsion explosive[J]. Explosion and Shock Waves, 2002, 22 (4): 338~342.

[11] Wang Yaohua. Research and practice of explosive welding of metal plates[M]. Beijing: National Edfence Industry Press, 2007: 31~38.

Study on the Blasting Technology of the Emergency Ice Breaking Used in the Ningxia-Inner Mongolia Reach of the Yellow River

YAN Junwei, YANG Xusheng, LI Xiukun, DONG Dekun

(*Shensi Design Institute of Engineering and Scientific Research, Shenyang, Liaoning, China*)

ABSTRACT: Aiming at the main ice-breaking problem faced by Ningxia-Inner Mongolia reach of the Yellow River, the characteristics and types of ice flood disaster of this reach have been studied firstly in this paper. Secondly, the advantages and disadvantages of the different ice-breaking technology have been analyzed comparatively. Based on this, through analyzing the characteristics and application conditions of the new research ice blasting equipment and comparing different position, different position ice breaking equipment, operation timing and implementation plan, a variety of the blasting technical schemes of the emergency ice breaking that were suitable for different ice, ice flood, ice dam formation and sectional characteristics have been put forward. Therefore, it provided technical support for enhancing the measures and emergency response capacity of the control and reduction of ice flood disaster for the Ningxia-Inner Mongolia reach of the Yellow River through the ice flood period safely.

KEYWORDS: the emergency ice-breaking; blasting technology; the Ningxia-Inner Mongolia reach of the Yellow River; ice flood; the control and reduction of ice flood disaster

1 INTRODUCTION

The Ningxia-Inner Mongolia reach of the Yellow River is located in the northernmost basin. Because of the terrain reasons, the river water flows from low latitude to high latitude. The temperature is warm up and cold down in the winter. The river has been frozen from bottom to above. In the spring of next year, due to the temperature is higher in the south than in the north, the river begins to melt from up to down[1]. In the ice-locked river season, the frozen ice under river causes an increased resistance to flow which makes the upper ice jam at the ice-locked place. At result, the plugged water causes the levee broken. When the upstream ice began to melt, the downstream was still frozen. At this time, the ices flowing from upstream to downstream were very easy to jam at the bending and narrow reach. Thus, the ice flood disaster has happened by the raised water level.

Although, the number and scale of the ice flood disaster can be reduce through the engineering of the dikes, the water and ice flood diversion and the ice flood reservoir protection[2]. Due to the ice flood of the Ningxia-Inner Mongolia reach of the Yellow River influenced comprehensively by many factors such as dynamic, thermodynamic, river boundary conditions and others which change continuously, interference mutually and interlace with one another, the ice flood disaster of this reach was very difficult to be forecasted, easy to burst, not easy to rescue. That has given a serious threat to people and property along the river. Therefore, it is necessary to study the emergency ice breaking technologies which are suitable for the different ice flood disaster.

2 THE ICE FLOOD DISASTER CHARACTERISTICS AND TYPES OF THE NINGXIA-INNER MONGOLIA REACH OF THE YELLOW RIVER

2.1 The Ice Flood Disaster Characteristics of the Ningxia-Inner Mongolia Reach of the Yellow River

The ice flood disaster of the Ningxia-Inner Mongolia reach is the most serious in the Yellow River., which has happened almost in different degree every year. Further, it often occurs in period of the river frozen up and began to melt[3, 4]. The disaster has the following characteristics:

Firstly, the sharp turn of the canyon river is easy to jam. Characteristics of the Ningxia-Inner Mongolia reach are small gradient, shallow broad river and bend road, sandwich beach and diversion ditch in the middle of river, extremely not smooth and frequent change. All these have made the ice flood drainage not free so as to lead easily to

---

Corresponding author E-mail: mf8507@163.com

the frequent occurrence of the flood disaster.

Secondly, Influenced by rapid temperature fluctuation, the formation time of ice is short. Due to lack of effective forecasting and monitoring technology, the happened places of the ice flood disaster are often difficult to be predicted. And the ice flood disaster evolution is very complex, change very quickly.

Thirdly, the ice flood disaster is easy to happen in the cold season. The ice refrozen after flooding increases the difficulty of rescue and relief.

2.2 The Ice Flood Disaster Types of the Ningxia-Inner Mongolia Reach of the Yellow River

The ice flood disaster has the following types.

2.2.1 Ice Jam, Ice Dam
2.2.1.1 Ice Jam

In the first ice-locked river season, influenced by the condition of upstream runoff and river boundary, a large number of ice squeeze into the ice blow to accumulate formation of ice jam under the action of water flow in the development process of the frozen ice front flowing to the upstream when the water surface slope become steeper from gentle. Thus, the ice jam have plugged water cross section and reduced section flow capacity so as to water level elevated to form flooding.

2.2.1.2 Ice Dam

In the period of ice-melted, upstream reach starts firstly, but downstream reach is not. When the flowing ice meet with the strong and no break ice sheet, or through the narrow river channel, bending, shoal, they have been blocked in the river. The blocked ices were accumulated to form ice dam so as to water level elevated to form flooding.

2.2.2 The Ice Flood of the Two-Ice-Locked or Two-Ice-Broken-Up River
2.2.2.1 The Ice Flood of the Two-Ice-Locked River

When the ice-locked river began to melt partly, because of the cold weather, part of the reach have been refrozen which made the cross section become small. If the ice-jam formed by the flowing ice from upstream under the ice, the cross section will be made smaller. This will cause the large number of cascading ice from upstream to be blocked so as to water level elevated to form flooding.

2.2.2.2 The Ice Flood of the Two-Ice-Broken-Up River

Influenced by different watershed latitude, there are only some reach began to two-ice-broken-up. The situation of ice broken up after locked in a short time is very prone to make the flowing ice accumulate to be blocked. At result, water level has been elevated to form flooding.

2.2.3 The Ice Flood of the Steady Ice-Locked River Season

After entering a stable ice-locked river season, affected by gradually narrowing cross reach of first ice-locked river part to the upstream water, the water has been refrozen. Because of the ice physical properties of brittle and easily broken, the refrozen ices are accumulated to form ice jams which make downward flowing poor. At result, it raises water level to cause the ice flood.

3 ANALYSIS OF ADVANTAGES AND DISADVANTAGES OF THE COMMON ICE BREAKING TECHNOLOGY

3.1 Analysis of the Foreign Common Ice Breaking Technology

At present, the foreign common methods of dealing with ice disaster are ice-melting, mechanical ice-breaking, icebreaker ice-breaking and blasting ice-breaking method[5]. The thermal ice-breaking method requires both heat resource and time that it deals with emergency condition poor. Due to the addition of chemical substances change the ecological environment of river, the chemical ice-melting method is not suitable for use in the Yellow River. Although it can be operated anytime and high work efficiency, the mechanical ice-breaking have the shortcomings of the device bulky, high energy consumption, high cost, more dangerous maneuver operations, and many department unable to expand and operation. The icebreaker can break the ice cover of river, but the channel depth of the Yellow River is not same and terrain complex which will make icebreaker easily run aground. Thus, the application of icebreaker is limited. The advantages of blasting breaking method are low technical difficulty, easy to control and organize, explosive amount adjusted according to actual demand, cost relatively low. And the disadvantages are large risk, sometimes limited by operating conditions that influences the effect of ice breaking.

Comparing the advantages and disadvantages of these, the blasting ice breaking method is only suitable in the Ningxia-Inner Mongolia reach of the Yellow River.

3.2 Analysis of the Common Blasting Ice Breaking Technology Used in the Ningxia-Inner Mongolia Reach of the Yellow River

The common blasting ice breaking technology used in the Ningxia-Inner Mongolia reach of the Yellow River are ice-surface blasting, ice-below blasting and artillery, aerial bomb ice-breaking method[6]. The ice-surface blasting and ice-below blasting method are all artificial blasting which have been restricted greatly above the ice and required of highly safe operation condition. Generally they are limited to break ice cover in the ice-locked period around the water engineering buildings nearby. Thus, their emergency ice breaking effects are not ideal. Although the artillery's method is operated from far range and not manual on the ice, it is not ideal from the blasting theory and ice breaking effect. Duo to the artillery method's killing main by shrapnel flying, it will cause secondary hazards. The explosion energy of aerial bombs ice-breaking is highly concentrated which will sometimes seriously damaged the riverbed, changed the channel and cause bank landslide. It is difficult for accurate implementation of the aerial bombing in a narrow river, bend, and the vicinity of hydraulic structures which formation of ice dam are extremely easy.

4 BLSATING TECHNOLOGY SCHEME OF EMERGENCY ICE-BREAKING

In order to reduce the ice flood danger of the Ningxia-Inner Mongolia reach of the Yellow River, it puts forward the blasting technology scheme of the loosening ice-breaking of the main channel full section, the ice-breaking of preventive eliminate hidden dangers and the ice breaking on short channel in ice-locked river period in the light of the danger phenomenon that often happens in the ice flood. There are using the new-research ice-breaking blasting equipment and creative blasting technique of emergency ice-breaking.

4.1 The Loosening Ice-Breaking of the Main Channel Full Section

4.1.1 The Danger Situation and Requirement

4.1.1.1 The Danger Situation

In the ice-melting period, the ice jam and ice dam formatted in the local river resulted in the upstream water leveled up gradually which put pressure to the dike defense. And the ice jam and ice dam are often located in the end position of river.

4.1.1.2 Requirement

It is required that the artificial active interventions on ice jam and ice dam should be carried out by loosening ice-breaking of the main channel full section. Thus, the ice jam and ice dam will be break down thoroughly. And the treatment of these to downstream and cross-strait dike can be removed.

4.1.2 The Scheme of Blasting Ice-Breaking

In view of the above situation and requirements, it uses the vehicle-mounted rocket launching large ice-breaking projectile system to carry out blasting ice-breaking. Its working principle is the large rocket ice bomb a group of ten are sequentially transmitted to a predetermined ice through a program controlled ignition, These rockets accelerate into the ice depending on the their body weight and the gravity. After, the trigger delay fuse bas been detonated to reached ice-breaking objective. This shows as Fig.1 and Fig.2.

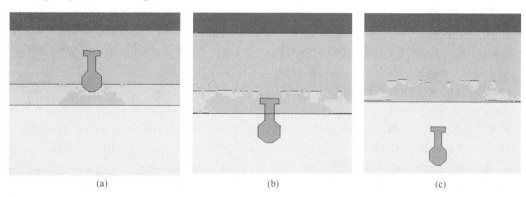

Fig.1 The numerical simulation of large ice-breaking projectile penetration ice cover

(a) $t=1113$ μs; (b) $t=6998$ μs; (c) $t=15679$ μs

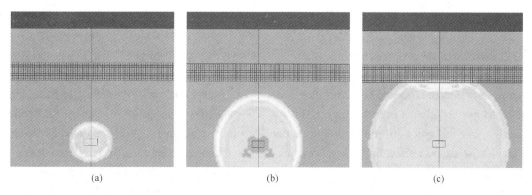

Fig.2 The numerical simulation of under-water ice-breaking by large ice-breaking projectile

(a) $t=137$ μs; (b) $t=716$ μs; (c) $t=1156$ μs

4.1.3 The Scope and Time of Ice-Breaking

This method can break ice within of 5m thickness, 500m width and within 5 km length. In general, the operation time is within 2 hours after the upstream water level elevated by the ice jam and ice dam be found. The safety influence of operation on the near embankment and bridge should be considered for. And the operation site should be away from the bridge and the village more than 500m.

4.1.4 The Expected Effect

When the ice dam is longer, more vehicles of this can be launched from downstream to upstream. When the two thirds of the dam are broken, the residual dam has been destroyed under the effect of hydrodynamic. At last, the river flowing is smooth.

4.2 The Ice-Breaking of Preventive Eliminate Hidden Danger

4.2.1 The Danger Situation and Requirement

4.2.1.1 The Danger Situation

In the ice-locked river period, influenced by river channel boundary condition, the reach in bridge pier or narrow, corners and the adjacent engineering river are easy to form ice jam and ice dam firstly.

4.2.1.2 Requirement

It is required that the ice-breaking of preventive eliminate hidden danger should be carried out on the position of the narrow corners, the near engineering of the river, or bridge piers. Thus, this will delay premature freeze-up or eliminate adverse freeze-up development situation.

4.2.2 The Scheme of Blasting Ice-Breaking

In view of the above situation and requirements, it uses the individual type rocket explosion belt to carry out blasting ice breaking. It depends on the rocket to carry explosion belt to the area above the ice, exhibition in a straight line and inline ice-breaking charge blasting when landing. Then, the horizontal line cutting blasting on the ice was carried out. It also can be used by combination. Under the control of blasting system, the synchronous array blasting has been carried out on the ice deposit arrangement of no rules. Thus, the task of ice drainage can be completed quickly and efficiently. The structure of the explosion belt shows as Fig.3.

Fig.3 The structure schematic of explosion belt

1—charge section; 2—detonating cord; 3—explosion skin; 4—flexible interval section

4.2.3 The Scope and Time of Ice-Breaking

This method can break ice within of 50 to 100 m width, 200 to 500 m length in the main bridge span. The single piece of ice area is within 15 sm. The local eliminate hidden dangers can be completed within 5 hours usually. It should consider the safety impact on the surrounding bridges, especially their piers.

4.2.4 The Expected Effect

Through the ice-breaking of preventive eliminate hidden danger, it can improve the adverse condition of ice jam and ice dam easily accumulated to formation on the narrow, steep bend, river shallow and bridge pier. Thus, it will improve the flood discharge capacity.

4.3 The Ice Breaking on Short Channel in Ice-Locked River Period

4.3.1 The Danger Situation and Requirement

4.3.1.1 The Danger Situation

In the ice-locked early period, influenced by short-time cold air, the local river appears shorter length to freeze for the first time. The first time and space distribution of ice are not ideal which may cause adverse effect on the ice flood prevention in later stage.

4.3.1.2 Requirement

It is required that forecast recent impossible continue to freeze, it can take the initiative to implement human intervention to break short river frozen ice, increase flowing capacity, delay the freeze-up, shorten the freeze-up period, reduce the tank water increment, strive initiative for the ice prevention work.

4.3.2 The Scheme of Blasting Ice-Breaking

In view of the above situation and requirements, it uses the vehicle mounted rocket dispenser small ice bomb to blast ice. It launch the rocket to the predetermined position, the rocket body opened, the small ice-breaking bomb by throwing into the air and parachute down using their own portable, accurate down to the predetermined ice surface. Under the control of initiation system, all the ammunition has been initiated. Thus, the purpose of ice breaking by synchronous multi-point explosive coupling damage is arrived. This shows as Fig.4 and Fig.5.

4.3.3 The Scope and Time of Ice-Breaking

This method can break ice within 5 to 15 cm thickness, 300 m width, and 3000 m length. Generally, the breaking should be carried out within 1 to 2 days after the freeze-up which is suitable for operation between 10am to 16pm. It should consider the safety influence of blasting on the embankment, bridge and near village. The operation site should be apart from the bridge, the village more than 500 m.

Fig.4 The distribution effect of small rocket ice breaking bomb dispenser

Fig.5 The effect of small rocket ice breaking bomb breaking ice cover

4.3.4 The Expected Effect

After the implementation of ice breaking blasting, it can quickly eliminate the ice, make the ice smoothly discharged. As a result, this postpones the frozen time of river.

5 CONCLUSIONS

Compared with the previous ice breaking technology scheme, the proposed blasting technology scheme of emergency ice blasting in this paper were with combination of waterproofing and drainage which were less affected by the weather, terrain and geological conditions. They can be common, emergency deployment, rapid setting out and work. And they have the advantages of ice breaking region accurate, blasting range controllable, less secondary hazards. Thus, these can be used to prevent flood disaster occurred by dredging river, lowering the water level when the Ningxia-Inner Mongolia reach of the Yellow River emergency ice flood danger happen.

REFERENCES

[1] ZHANG Baosen, JI Honglan, ZHANG Xinghong. Spatial and Temporal Distribution Characteristics of Ice in Inner Mongolia Reach[J]. Journal of Yellow River, 2012, 28(2): 36～38.

[2] CUI Jiajun, CAI Lin, XIN Guorong etc. The Research and Development of Yellow River Ice Flood Control Decision Support System[M]. Beijng: The Yellow River Water Conservancy Press, 1999, 1:176～179.

[3] FENG Guohua, CHAOLUN Bagen, YAN Xinguang. Analysis of Ice Slush Formation Mechanism and Ice Flood Causes of Yellow River in Inner Mongolia[J]. Journal of China Hydrology, 2008, 28(3):16～19.

[4] YAO Huiming, QIN Fuxing, SHEN Guochang. Ice Regime Characteristics in the Ningxia-Inner Mongolia reach of the Yellow River[J]. Journal of Advances in Water Science, 2007, 18 (6):893～898.

[5] XU Jianfeng. Ice Flod Disaster and Its Reduction Mearsures in the Inner Monogolia Reach of Yellow River[J]. Journal of Glaciology and Geocryology, 2005, 17(1): 1～7.

[6] TONG Zheng, MA Wanzhen, WANG Ning. Basic Methods of Explosive Icebreaking in Ice Run Period in Inner Mongolia Section of the Yellow River[J]. Journal of Yellow River, 2003, 28(12):9～13.

Research and Practice of New Technique for Blasting Demolition of Hyperbolic Cooling Tower

ZHU Jinhua[1], SUN Xiangyang[2], GUO Tiantian[1], WANG Qingtao[1]

(*1. School of Basic Education for Commanding Officers, NUDT, Changsha, Hunan, China;*
2. Zhongren Blasting Engineering Co., Ltd. of Hunan, Changsha, Hunan, China)

ABSTRACT: The hyperbolic cooling tower is tall and big with thin wall and small depth-width ratio. In the traditional blasting demolition, drilling and blasting are needed on herringbone column, ring beam and tower shell simultaneously which result in large quantity of drilling and charges, protection of high difficulty, wide range of flying rocks, recoiling and squatting without collapse occasionally. For the above problems, a new technique is proposed for the cooling tower blasting demolition. First, several longitudinal cuts are made on the shell of the cooling tower in the collapse direction which will accelerate the transverse deformation during the collapse process. Second, node blasting is conducted to both ends of the herringbone column. Then the upper structure will collapse to the giving direction by gravity.

The proposed technique is successfully applied in the blasting demolition of three 90m high hyperbolic cooling tower of Huarun Electric (Jinzhou) Co., Ltd. Practice results show, compare to the traditional technique the proposed technique has shorter construction time, smaller collapse vibration, less cost, better safety, reduce the diffusion range of flying rocks and has a guiding significance for the construction of similar projects.

KEYWORDS: hyperbolic cooling tower; blasting demolition; small cut; longitudinal cut; node blasting

In the explosive demolition of high aspect ratio asymmetric or atypical RC structures, the collapse could progress in an unintended direction, as instantaneous moment changes during collapse behavior due to the complexity and diversity of the members to which internal force is applied. The 119 safety center, which is located in Iui-dong, Yeongtong-gu, Suwon, Gyeonggi, has a width of 5.7m and a height of 32m. The ratio of width to height is 1 : 5.6, indicating that the aspect ratio is high. The safety center is an RC Rahmen structure with left-right asymmetry centered on the structure, and is a target structure for explosive demolition that has a limited overturning direction. In this study, for inducing the overturning of the target structure in a planned collapse direction, a simulation was performed to investigate the changes in the hinge point (where turning moment is generated) and twisting moment depending on the shape and position of pre-weakening and detonating delay. The extreme loading for structure (ELS), which is based on the applied element method, is a program that simulates the displacement, damage, and collapse behavior of a structure by applying an extreme load (ASI, 2006). The simulation was performed by assuming and modeling 4 different cases. Lastly, the reliability of the simulation was tested by directly comparing the overturning collapse behavior in the analysis and the overturning collapse behavior filmed with a high-speed camera at the millisecond (ms) level.

1 PROJECT OVERVIEW

1.1 Project Surrounding

The three cooling towers to be demolished are located in southwest of the Huarun Electric (Jinzhou) Co., Ltd. with a complicated surrounding as illustrated in Fig.1.

For cooling tower 1[#], there is a 80 m clear space in east, a using pump house 16m away in south, a bounding wall 25 m away in west and an office building 70 m away in north. For cooling tower 2[#], the nearest distance to the transformer substation in southeast is 70 m, 30 m away from the cooling tower 3[#] in southwest, 22 m away from the pump house in west, 90 m away to the office building in north. For cooling tower 3[#], the nearest distance to transformer substation in east is 100 m, 52 m away from the resident house in southwest, 25 m away from the pump house in northwest. The transformer substation, pump house, office building and resident house are all need to be undamaged. So many targets to be protected that increase the difficulty of blasting demolition.

Corresponding author E-mail: ttguo-mail@126.com

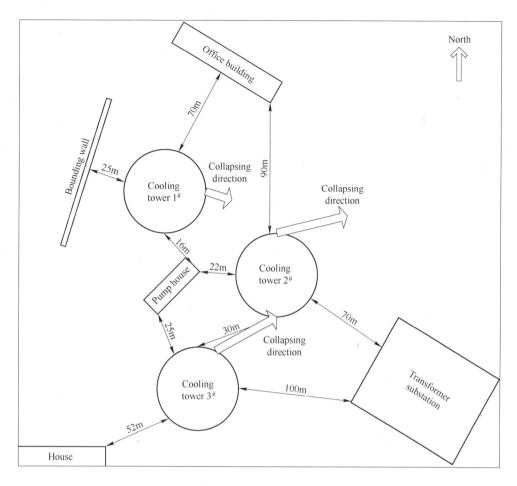

Fig. 1 Project surroundings

1.2 Structure of the Cooling Tower

The three cooling towers to be demolished have the same reinforced concrete hyperbolic structure which is a kind of tall and thin shell structure. The height is 90 m, the base surface diameter is 74.5 m, ventilator throat diameter is 38.8 m, the top diameter is 43.1 m, the water spray area is 3500 m². The reinforced concrete circular foundation is adopted, the design strength of concrete is 300, bottom of foundation depth is -3 m. The tower is supported by 40 pairs of reinforced concrete herringbone-columns on the bottom. The cross-section of the herringbone-column is an octagon, the length of each side of the octagon is 50 cm. The vertical height of the herringbone-column is 5.8 m. The thickness of the shell is 500 mm on the bottom, 140mm on the top, the design strength of concrete is 300. The total reinforced concrete volume is 5600 m³.

2 BLASTING SCHEME DESIGN

2.1 Design Principle

According to the structural characteristics and surroundings, the following principles are determined.

(1) In order to speed up the demolition progress and ensure the safety, the directional blasting is adopted to demolish parts of three cooling towers which above the ground.

(2) Open high trough on the shell of the cooling tower which is good for breaking and collapsing.

(3) In order to reduce blasting number, one directional blasting demolition for all the three cooling towers. priming sequences is 1#, 2# and 3#, the delay time of each cooling tower is 1000 ms.

(4) As illustrated in Fig.1, the collapsing direction of 1# is east, the collapsing direction of 2# is northeast 20°, the collapsing direction of 3# is northeast 30°.

2.2 Blasting Cut

The cut design is critical to the collapsing of the cooling tower. At present, big cut on the bottom is widely used in cooling tower demolition, the shape of the cut includes echelon, inverse echelon, triangle and compound shape. To choose a blasting cut, the following conditions must be considered.

(1) The thickness, structure, stress and strength of materials should be along the cut axis symmetry to prevent the eccentric instability after the pretreatment.

(2) The reserved cross-section on the cut should have high anti-press strength to ensure the directional collapsing.

(3) The height of the cut should meet the gravity failure angle for the collapsing.

(4) As far as possible to reduce the cutting height, to avoid high-altitude operations.

The common cut height is 12~25 m which is determined by the herringbone-column height, ring beam height and blasting shell. In other words, the traditional blasting demolition of the cooling tower must blast the herringbone-column, the ring beam and shell which means a lot of drilling, charge, stemming and protection on high-altitude. This method is very difficult and increase the cost.

In order to reduce the drilling, charging and stemming, improve the safety and reduce the cost, only the herringbone-column to be blasted. With the comprehensive consideration of the collapsing direction and anti-stress strength of the reserved shell, 21 pairs of the herringbone-column need to be blasted, and the cut length is 53% of the shell perimeter. The capsizing moment motivated by the herringbone-column blasting should be big enough to ensure the sufficient crushing, collapsing after the barrel touch the ground to reduce the ground vibration. So the trough should have enough height and area. The high trough scheme for this project is illustrated in Fig.2.

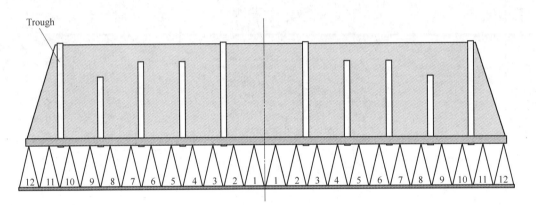

Fig.2　High trough scheme

Ten high troughs are going to be opened in the cut range and distributed on both sides of the collapsing center line. The height of the trough is 8~12 m, the width is 0.5~1 m. The two troughs in the center and two troughs on each side are the highest trough with the height of 12m, the height of the rest trough is 10m and 8m. Hydraulic hammer is used to open the high trough on the cooling tower shell.

2.3 Blasting Parameters

The blasting range of the herringbone-column are from the bottom up to 1.8 m and from the ring beam down to 1.8 m.

The blasting parameters are listed below.

(1) Minimal resistance line $W=(1/2)B=0.25$ m, B is the width of the herringbone-column.

(2) Hole diameter $D=38$ mm.

(3) Hole pitch $a=1.5W=1.5\times25=37.5$ cm, 40 cm is used.

(4) Hole depth $L=(2/3)H=33$ cm, H is the thickness of the herringbone-column.

(5) Hole number in one column $n=10$.

(6) Total number of the hole $21\times2\times10=420$.

(7) Explosive factor $q=1000$ g/m^3.

(8) single hole charge $Q=qBaH=1000\times0.5\times0.4\times0.5=100$ g.

Tab.1　Blasting parameters

Structure name	Short edge/cm	W/cm	a/cm	L/cm	Hole Number	q/g·m^{-3}	Single Hole Charge/g	Total Charge/kg
Herringbone-column	50	25	40	33	420	1000	100	42

Center of shear type hole is used in herringbone-column blasting which means the holes are staggered distributed on both sides of the vertical axis and 2~3cm away from the vertical axis of the column. The charge is $\phi32$ mm emulsion explosive.

2.4 Design of Detonating Network

2.4.1 Choose of the Blasting Material

The blasting is in a power plant, the environment is very complex. There are many electrostatic, stray current and inductive current. In order to ensure the safety, non electric detonator and non electric initiation system are used.

Inside the hole, the half second delay plastic nonel detonator is used(HS2, HS4, HS6). Outside the hole, the millisecond nonel detonator (MS1) is used for explosion propagation. Firing gun is used for ignition .The detonating network is grouped in clusters and connected by four-way connectors.

2.4.2 Blasting Division and Network Connection

In order to reduce the ground vibration, make a 1s interval on each cooling tower's ignition, so the three cooling towers will not touch the ground simultaneous. The ignition order is $1^\#$, $2^\#$ and $3^\#$, the blasting division and time delay is illustrated in Tab.2.

Tab.2 Blasting division and time delay

	Initiation site	Inside hole detonator	Outside hole detonator	Delay time /ms
1	$1^\#$ Cooling Tower	HS2	MS1	500
2	$2^\#$ Cooling Tower	HS4	MS1	1500
3	$3^\#$ Cooling Tower	HS6	MS1	2500

Nonel detonator, nonel and four-way connectors form a strengthening detonating network. The connecting methods are:

(1) Make every 20 nonel detonators as a cluster, each cluster is ignited by two non electricmillisecond nonel detonator.

(2) Use nonel and four-way connector to form a double strengthening detonating network.

(3) Use non electric firing gun as the ignite element.

3 SAFETY DESIGN AND PROTECTION

3.1 Blasting Vibration Control

According to the Safety Specifications of Blasting Operation (GB 6722—2003), the blasting vibration is calculated as equation 1.

$$v = K \left(\frac{\sqrt[3]{Q}}{R} \right)^\alpha \quad (1)$$

In Equation 1, v is the vibration velocity, cm/s; Q is the maximum charge in one sound, kg; R is the distance from blasting point to the target, m; K is a coefficient related to the blasting method, distance and geological conditions; α is an index related to the distance, geological conditions and propagation path. According to the empirical data, the K is 34, the α is 1.57.The allowed vibration velocity for ordinary house is 2.0cm/s, for power plant control room is 0.5 cm/s according to the GB 6722—2003.

Calculating results show that the vibration velocity in the distance 170 m away from the cooling tower is 0.07 cm/s. So all the targets are safe from the blasting vibration.

3.2 Collapse Vibration

The cooling tower is high-rising building, collapse vibration will be caused when it touches the ground. According to the empirical formula of Equation 2, the collapse vibration velocity is 0.27 cm/s at the location of transformer substation, lower than GB 6722—2003.

$$v_t = K_t [(MgH/\sigma)^{1/3}/R]^\beta \quad (2)$$

where M——mass, t;
  H——gravity height, m;
  R——distance, m;
  σ——breaking strength of the ground medium, MPa, take 10MPa;
  g——gravity acceleration, m/s², take 9.8 m/s²;
  β——attenuation coefficient, take 1.66;
  K_t——attenuation coefficient, take 3.37.

3.3 Shockwave Protection

The explosive charge is dispersed, and the multi millisecond blasting technology, therefore the air shockwave effect is very small, do not need special protection.

3.4 Throwing Rocks

According to the empirical formula, the maximum throwing distance of unprotected control blasting is illustrated in Equation 3.

$$S = k_f q d \quad (3)$$

where S——maximum throwing distance, m;
  k_f——safety coefficient, usually take 1.0~1.5;
  q——explosive factor, kg/m³;
  d——hole diameter, mm.

The result is

$$S = 1.5 \times 1.0 \times 42 = 63 \text{ m}$$

In the blasting demolition process, external wrapping protection using straw, barbed wire in the blasting site is implemented; individual throwing rocks will not exceed 30m and do no damage to the buildings.

3.5 Throwing Rock When Cooling Tower Touch Ground

When the cooling tower touches the ground, it will cause a certain range of throwing rocks. But at present there is no unified method to calculate the throwing rocks distance. According to the past experience, the throwing rocks will not do damage to the buildings at 50m.

4 BLASTING RESULTS

Fig.3 illustrates the collapsing process of the three cooling towers. It can be seen that the capsizing moment caused by the blasting made the cooling tower reverse and capsize, obvious crack appears on the location of 16m above the ring beam and expanding continuously until the cooling tower touch the ground and crushed. The collapsing direction of the three cooling tower is as same as the expectation without the rearward movement. The targets need to be protected are all untouched, the ground vibration is under control and no obvious throwing rock. The blasting results are perfect.

Fig.3 Blasting results

5 CONCLUSIONS

For the traditional blasting demolition method of the cooling tower, the blasting cut is a little bit high, herringbone-column, ring beam and shell need to be drilled an blasted simultaneous with a heavy drilling work, the large amount of blasting material, a lot of high altitude operation, high difficulty of protection, big range of throwing rocks and fail explosion sometimes. A new method for the cooling tower demolition which has a little and low height blasting cut is proposed which only requires node blasting on the top and bottom of the herringbone-column and open several troughs above the ring beam on both sides of the center line of the collapsing direction. The practice results of three cooling tower blasting demolition in HuaRun Electric (Jinzhou) Co., Ltd. show that the proposed method can reduce the drilling work greatly, shorten the construction time, reduce the collapse vibration, narrow the range of the throwing rocks, lower the construction cost and improves the safety and has a guiding significance for the construction of similar projects.

REFERENCES

[1] LIU Honggang, BAI Ligang,Li Jun. Discusssed the thinwall hyperbolic cooling tower directional blasting cut[J]. *Railway Construction*, 2005, B08 : 68~78(in Chinese).

[2] LIU Honggang, LIU Dianshu, DONG Taiming. The key technology of cut form in cooling tower demolition blasting[C]// *China Blasting New Technology II*. Beijing: Metallurgy Industry Press, 2008: 493~498(in Chinese).

[3] XIA Weiguo,WU Shuangzhang, TANG Yong. Explosive demolition of two 110m high hyperbolic cooling towers[J]. *Blasting*, 2011, 28 (2) :68~71(in Chinese).

[4] SUN Yueguang, ZHANG Chunyu. Blasting demolition of two cooling towers in a complex environment[J]. *Engineering Blasting*, 2010, 16(1) :55~58 (in Chinese).

[5] YANG Pu, BAI Ligang. Technique for demolishing high thin-wall hyperbolic reinforced concrete cool tower by directional blasting[J]. *Journal of Railway Engineering*, 2006(3):66~69 (in Chinese).

[6] HUANG Xiaoguang, ZHU Huaibao, GAI Sihai. Demolition of an 85 m high reinforced concrete cooling tower under complex conditions by controlled blasting[J]. *Engineering Blasting*, 2007,13 (3): 50~52 (in Chinese).

[7] WANG Hanjun, YANG Renshu. The technology of direction-alblasting demolition of thin-walled structure ofhyperbolic cooling towers[J]. *Coal Science and Technology*, 2006, 34(7): 36~40 (in Chinese).

[8] LIN Kai, PAN Qiang, LI Bo. Blasting demolition of large diameter hyperbolic reinforced concrete cooling tower[J]. *Engineering Blasting*, 17(1): 61~64 (in Chincse).

[9] LI Shouheng, SHANGGUAN Zichang. Study of collapsing condition for demolition of cooling tower[J]. *Journal of Liaoning Engineering Technology University*, 1999, 18(1): 9~14(in Chinese).

Removal of the Exhaust Tube Blasting in Complicated Environment

YU Hui, LIU Guixin, ZHANG Chao, FANG Limin

(Shenyang Xiaoying Blasting Engineering Co., Ltd., Shenyang, Liaoning, China)

ABSTRACT: The exhaust tube to be removed is high 55m, locate within Fushun Titanium Industry Co., Ltd. Risk is high and the environment is complex, a chlorine tank and pipeline is used around. Only the southwest side is open space. In this direction for dumping, to ensure the safety of blasting method and time, decided to adopt the South West 30° direction to remove.

KEYWORDS: accurate positioning; measures; chimney; directional blasting

1 PROJECT OVERVIEW

Because of the transformation of Fushun Titanium Industry Co., Ltd., a chlorine exhaust tube must be removed. In order to ensure the safety and schedule, the demolition blasting method was adopt. The exhaust tube was high 55m, in the elevation of +0.5m, diameter was 5.5m, perimeter was 17.3m, wall thickness was 0.54m. The outer layer was reinforced concrete structure, thickness was 30cm, lining (coating) was 24cm bricks. from top to bottom, there was no gaps between the two layers. There were two same size manholes in the northeast and southwest, the height and width respectively were 1.3m and 0.8m, in the exhaust tube wall there was an air stopping brick wall(back wall), thickness was 24cm and height was 2m. It was located in south side of the factory, and 28m from the south embankment, 32m from east vacant workshop, 4m from north railway fence, two factory railways were outside the wall, a chlorine tank and chlorine pipelines were on the north side of the railways. There were four chlorine storage tank at the southwest side 86m. The surrounding environment is shown in Fig.1.

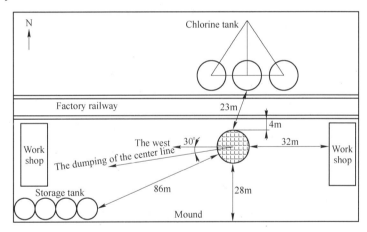

Fig.1 Schematic diagram of the surrounding environment

2 PROJECT DIFFICULTY

2.1 Explosion Protection, Prevent Damage

Because the environment was complicated, chlorine tank vents were on the north side only a distance of 23 meters, vibration and flying stones can cause chlorine leakage accident. Through effective protection (exhaust tube outer wall covering straw and wire binding measures), reducing vibration (millisecond blasting network, foreshadowing the soft protective), the maximum detonating charge controlling, the blasting vibration, touchdown vibration, air shock wave and fly rock can be controlled.

2.2 Blasting Construction Safety Measures

There was residual chlorine in the exhaust tube, chlorine is a toxic gas, it mainly invades the human body through res

---

Corresponding author E-mail: syxyxzswb@163.com

piratory tract mucosa and is dissolved in water, to generate hypochlorous acid and hydrochloric acid, which is harmful to the respiratory mucosa. 1L air allowed chlorine containing 0.001mg, more than this amount it would cause human body poisoning. So in the construction of blasting personnel must put on gas masks, and take turns to prevent poisoning of chlorine gas.

3 SELECTION AND DESIGN

3.1 Scheme Selection

According to the exhaust tube structure and the surrounding environment, the site is only suitable for directional collapse in the southwest side, south west 30° direction by precise measurement and localization to fulfill the requirements.

3.2 Gap and Pretreatment

3.2.1 Determining the Gap Position

The trapezoidal gap in the ground elevation 0.5m was set which was synunetrical with respect to dump center line.

3.2.2 To Determine the Length of Gap

The outer layer was a reinforced concrete wall (thickness 0.3m), the inside sleeve was a brick wall (thickness 0.24m). Therefore, the blasting center angle should not be too large, according to the practical experience, central angle 220° was selected. Incision length: $L=220°\pi D/360° \approx 10.6$m.

3.2.3 Determination of the Gap Height

In order to ensure the gap was formed, and exhaust tube can collapse, according to the similar engineering practice, gap height $h=5\delta=5\times0.3=1.5$ m, so the exhaust gap height was 1.6 m.

3.2.4 To Determine the Orientation Opening

In order to ensure the correct orientation of exhaust tube collapsed, generally the blasting orientation openings were set on both sides, the orientation opening angle $\alpha=40°$, the bottom of the orientation opening is 0.5m long. A groove was opened in the middle of the blasting gap, width was 0.8m, height was 1.6m. See Fig.2 and Fig.3.

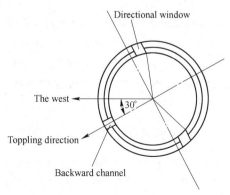

Fig.2 Sketch of blasting cut central angle, directional window, backward slot position

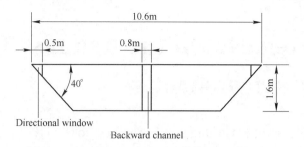

Fig.3 Schematic diagram of incision expansion

3.2.5 Pre-demolition

(1) The lightning rod, exhaust pipe, present ladder construction and all equipment connected with the ground were torn down before blasting.

(2) The directional windows were manually removed.

(3) Groove was cut manually.

(4) Lining was born down according to gap.

3.3 The Design of Blasting Parameters[2]

The minimum resistance line: $W=\delta/2$.

The outer layer (layer of reinforced concrete) $W=30/2=15$ cm.

Hole depth: usually $l=(0.65\sim0.68)\delta$, $l=270$ mm.

Spacing: $a=1.5W=300$ mm.

Row spacing $b=0.87$, $a=260$, and 250mm was selected.

Single hole charge: $Q=qab\delta$, $q=1.4$, $Q=50$ g.

3.4 Initiation Network[3]

To ensure the blasting network security, accurate, economical, reasonable, and avoid the influence of the stray current and other factors, millisecond delay detonator with shock-conduct tube was chose, and divided into three delay number. Each charge placed two same nonel tube detonator. Initiation network was double ring closed network.

4 SAFETY TECHNICAL MEASURES[4]

4.1 Blasting Vibration Calculation[4]

Peak particle vibration velocity is determined by the type:
$$v=KK'(Q^{1/3}/R)^\alpha$$

where R——allowable safety distance for blasting vibration, m;

Q——quantity of explosive, maximum charge weight per delay interval for delay blasting, 4 kg;

v——allowable vibration velocity, cm/s;

R——distance to the blasting center, m, $R=32$ m.

Coefficient of medium K, α according to engineering practice, relevant engineering analogy, $K=250$; $\alpha=1.8$; $K'=0.3$;

$v=0.34$ cm/s<2.8cm/s (general brick non seismic large brick building allowable safety vibration velocity for 2.8cm/s).

4.2 Collapse Touchdown Vibration Calculation[4]

The collapse touchdown vibration velocity formula:

$$v_t = K_t \left(\frac{\sqrt[3]{MgH/\sigma}}{R} \right)^\beta$$

where v_t——the surface vibration velocity of blasting collapse touchdown induced, cm/s;

R——the focus from building collapse touchdown distance, 32 m;

M——the quality of collapse, according to the formula of volume $(\pi R^2 - \pi r^2) \cdot H$, ignoring the exhaust cylinder diameter factors, exhaust tube outer diameter 5.5 m, wall thickness 0.54 m, high 55 m, $M = \rho V$, estimating exhaust the total mass of 1020t, the quality is the total mass of the 1/3, namely M=340 t;

g——the acceleration of gravity, 9.8 m/s²;

σ——failure strength ground off, the general 10 MPa;

H——collapse blasting building gravity drop, m, take 18.5 m;

K_t, β——attenuation parameter, which were K_t=3.37, β=1.66.

v_t=1.34 cm/s.

In order to reduce the influence of blasting vibration on the surrounding buildings, clay or other soft material were paved on the ground for the tilting direction angle on both sides of the line of 5° angle range.

4.3 Fly Rocks and Vibration Protection[4]

According to the mechanics of the Chinese Academy of Sciences example data analysis formula:

$$R = 70K^{0.58} = 104 \text{ m}$$

Results: proofreading needs strict protection of blasting off to ensure that fly-rocks were controlled within 5m. In order to prevent the damage caused by fly-rocks ,protection measures of this blasting the main protective method of multilayer were grass curtain and wire bundle. Specific methods are as follows:

(1) The protection range must be beyond the bounds of the hole more than 0.5m.

(2) The grass curtain lap length was not less than 0.3m.

(3) Each 3 layers of straw were tied , interval 0.5m, and must tied tightly .

(4) The operation was repeated 5 times, that is to say, the covering was 15 layers of grass curtain.

(5) In the north wall three layers straw were covered in order to protect the chlorine tank etc.

5 THE BLASTING EFFECT

Blasting was implemented according to the scheduled time. About 3s after initiation , the exhaust tube fell to the southwest predetermined direction, and accurately to the predetermined mound, without fly-rocks , and the ground vibration is very small.

6 CONCLUSIONS

Through the 55m high reinforced concrete exhaust tube successfully blasting demolition practice, the following conclusions were drawn:

(1) Pneumatic pick was used to get orientation openings, other openings, which improved the precision reduced the test effect on reservation supporting body, and effectively ensured the directional accuracy.

(2) Covering grass, sand, can control the spatter effectively.

REFERENCES

[1] Wang Xuguang, Yu yalun. Demolition blasting theory and engineering examples. Beijing.

[2] Wang Xuguang, Yu yalun. Blasting design and construction[M]. Beijing: Metallurgical Industry Press, 2011.

[3] Wang Xuguang. Yu yalun, Liu Dianzhong. Blasting safety procedures implemented manual[M]. Beijing: People's Communication Press, 2004.

[4] Wu Tengfang, Ding Wen, Li Yuchun, Yang Sha. Blasting materials and blasting technology[M]. Beijing: National Defence Industry Press, 2008.

Trial Blasting Technology in Demolition Blasting of Building and Its Analysis

FEI Honglu[1], ZHANG Longfei[1], HE Wenbin[2], YANG Zhiguang[1,2]

(1. Institute of Blasting Technique, Liaoning Technical University, Fuxin, Liaoning, China;
2. Fuxin Gongda Blasting Engineering Co., Ltd., Fuxin, Liaoning, China)

ABSTRACT: In demolition blasting of building, in order to verify the design of the blasting parameters and blast protective effect, select the best actual unit consumption of explosives, structural reinforcement and strength parameters and so on, usually selected the representative local wall body and column in building to blast before the formal loading explosives in order to adjust the blasting parameters and the strength of protection. In this paper, the basic methods and safety precautions in the trial blasting were introduced. Combined with engineering examples detailed analysis of the effect on the trial blasting and propose appropriate treatment measures. For the similar blasting engineering provides practical reference.

KEYWORDS: demolition blasting; trial blasting; blasting parameters; strength of protection

1 INTRODUCTION

With the development and progress of society, more and more problems in the modernization of city have appeared. For example, many building, which built in the 80's of last century building, need to be demolished and reconstructed, for their function and role have been difficult to meet the requirements of modern people or no longer meet the city modernization development direction because of the unreasonable intial-designment and city planning .But almost all of these buildings, which are the landmarks in those years, are in the downtown city lots, and the environment is complicated. As the traditional way of demolition has not only lots of noises but also long construction period, it will seriously affects the surrounding people's daily life and work, so the mostly used method of demolition is directional blasting. However, because different periods have different design code of building and different construction quality, and also, many construction drawings of archive file have lost, all of these factors bring great difficult for blasting engineer to demolish buildings. And as an important step in building demolition blasting technology, the trial blasting technology is getting more and more the blasting engineers attention, because it can directly influent the success or failure of blasting engineering and the result of demolition in the very great degree. This paper is intended to expatiate and analyze the related problems caused by trial blasting and put forward corresponding measures.

2 THE BASIC METHOD OF THE TRAIL EXPLOSION

2.1 Purpose of Trail Explosion

The purpose of trail explosion are mainly the following four aspects: first, testing the performance of detonators and explosives; second, the design of the trail explosion building's construction drawings is not completed, not providing the detailed reinforcement or the bearing column through the secondary reinforcement, providing the basis for the selection of special parts of the consumption of explosives; third, the trail explosion of buildings of different types of load-bearing cross-section of the column, verifying the selection of the value of the initial design of consumption of explosives, timely adjusting dose of single number to achieve optimal blasting effect by referring the effect of blasting; fourth, verifying that the protective effect and form of a single load-bearing columns of blasting, thereby adjusting the amount of protective material of the individual column and increase or decrease the intensity of the external overall protection in the appropriate position. In the case of ensuring safety that take into account the economy at the same time.

2.2 Selection of the Target of Trail Explosion

When we choose the target of trial blasting selection, two aspects are needed to pay special attention: first, choose the lower position of post to blast , if the post of lower position

Corresponding author E-mail: feihonglu@163.com

take the trial blasting in advance, it will lose weight capacity, which causes bad influence of the shape of the cut angle; second, the type of column section is greater or equal to the third, in order to reduce the adverse impact of the overall stability of the building, can choose the minimum and maximum section to take the blast, the other section with its test parameters can be the reference of difference calculation.

2.3 Select the Hole When the Network Parameters Explosions

With an example of 500 mm×500 mm square column: the blast hole drilling with the drill head $\phi 32 \sim 38$ mm generally used in blasting demolition of building on the column, blast hole diameter of $d=35\sim45$ mm; boreholes deep to $L=(0.6\sim0.8)\delta$, δ is the thickness of the column, $\delta=500$ mm, the calculated is $L=300$ mm with the take coefficient of 0.6; borehole spacing is $a=(1.0\sim1.4)W$, W is the minimum line, $W=250$ mm, $a=300$ mm with the coefficient is 1.2. Hole charge amount determined by the formula:

$$Q=qV \qquad (1)$$

where, q is the consumption of explosives, for reinforced concrete column, explosives consumption q is taken as 1.0 kg/m³; V is the volume of a single hole blasting should be affordable. Hole dynamite loading dose $Q=qV=1.0\times0.5\times0.5\times0.3=0.075$ kg, actually is 75 g.

Other section of the column of blasting choice hole does with this as a standard.

3 SAFETY TECHNOLOGY OF TRAIL EXPLOSION

3.1 Safe Distance Calculated by Blasting

Blasting individual flying objects allowed for personnel security distance calculation method is not explicitly stipulated in "Blasting Safety Regulations"(GB 6722—2003), usually determined by design. In this case, the distance of flying stone generated by the trial blasting is calculating in the empirical formula[1~3] which is used in the Chamber Blasting. This calculation is for reference only:

$$R_f = 20K_f n^2 W \qquad (2)$$

In this formula, K_f is the operating conditions coefficient with the construction(geological terrain, wind, clogging charge), generally is $K_f =1.5\sim 2.0$; n is the index of blasting, generally is $n=1$; W is the minimum resistance line, is $W=0.25$ m. The maximum flying distance of the stone can be calculated by the formula (2), $R_f = 20\times 2.0\times 1.0^2 \times 0.25 = 10$ m.

To be effective control the individual fragments splashing within 5m design requirement, must be strict for protection of all parts of charge protection with the design requirement.

3.2 Blasting Protection

3.2.1 Column Protection

In order to control the scattering stone in a certain range, to ensure the surrounding buildings, facilities and the personnel security, all the blasted parts usually will be blinding on several layers of straw and a layer of steel net. As shown in Fig.1, the blasted parts are blinding on three layers of straw and a layer of steel net, which formed the column protection, after the test of blasting, adjust the ultimate protection strength via observing the damage extent of the protective material.

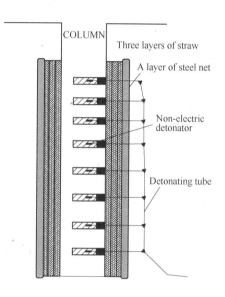

Fig.1 Schematic column protection

3.2.2 Whole Protection

In order to minimize the harm that the blasting scattering stone may cause on the surrounding residents, in addition to protect the loading position directly, but also protect and reinforce the whole building outside. Firstly, plug all the doors and windows within the scope of the blasting gap with a layer of straw and steel net, and hang a layer of straw and steel net not less than the height of the blasting cut on the building outside wall. Secondly, to the key protection direction of the residential area, a protection pipe should be provided with a height of not less than 15m, and hang a layer of straw and steel net to indirect blocking protection to ensure that control the scattering stone in a safe range.

3.3 Wholesale Construction Work Warning and Blast Warning Distance

The alert during the process of blasting construction is very important. Be sure to arrange the meeting in advance before the blast, let people realize the importance of vigilance.

According to the actual situation before blasting, create positions, personal, and responsibility. If the condition allows, alert to appropriate a larger range, not afraid of trouble, can't exist fluky psychology [4]. From the beginning to the end of blasting operation, the whole building site closure, by blasting the company personnel responsible for the alert, it is strictly prohibited to any other personnel to enter the site. Pillar blasting test position is lower, surrounded by a wall block, so the security range is 50m around the building. People in the alert area must be evacuated to safety alerts, 20 minutes before the trial blasting, personnel should stay away from windows and doors in the building, the local traffic police responsible for traffic control within the scope of the alert. After the test, until after blasting technical personnel confirm site safety can terminate the alert.

4　ANLYSIS OF ENGINEERING CASE

4.1　Engineering Situation

The Wuhai city of Inner Mongolia bank branch block being dismantled was founded in 1992, which is located in Wuhai city Haibowan heartland. It's 12 layer (including the underground layer) reinforced concrete frame shear wall structure. The building's main body is 54m in length, 12m in width and 48.35m in height. Its main facade columns have three rows of support and the bearing column is into four sections also with 24 square column s of 500 mm×500 mm. The 600 mm×600 mm, 700 mm×700 mm pillars and the column with a diameter of 1000mm each have four roots and they are symmetric distributed, a total of 36 roots. Then the building stairwells and staircase walls are for the shear wall bearing. Building construction cross-section is of V, whose east and west direction is to the long axis direction of the building, north and south direction is to the short axis of the building. The long axis symmetric is of distribution center, which makes the stability of the whole building is very good. So it brings certain difficulty to the demolition of the building. To determine the final blasting parameters, for building 500 mm×500 mm square columns and the column with a diameter of 1000mm for testing, building plan and the location test is as shown in Fig.2.

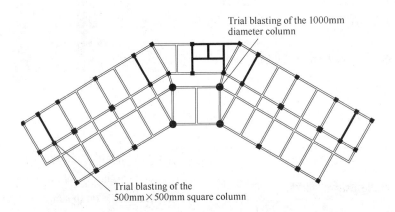

Fig.2　Locations of explosions pillars

4.2　Trial Blasting of the 500 mm×500 mm Square Column

There are 24 square columns of 500 mm×500 mm in the building. To make the testing explosion, we should select a column that connected to the stair shear wall. The shear wall has processed completely, but the connection that the column and the shear wall not be disconnected, which have a reinforcing effect for the column.

Blasting parameter is set, which is shown above, the depth of borehole is 300mm, the blast hole space is 300mm, the loading dose of a single hole is 75g, the charging length is 100mm ($2^{\#}$rock dynamite, the length of a single roll is 200mm, the weight is 150g), the length of filling is 200mm, and the protection uses three layers of straws and a layer of steel wire and uses wire to fix firmly on the outside. Taking into account the strength of the column higher than the ordinary square columns, decide to temporarily adjust the single hole to the dose 100g to ensure blasting effect. At this time the unit consumption is 1.33kg/m³. Abandon the bottom of the invalid hole when its trail blasting, charging up to four holes consecutively from the second beneath hole. The trail blasting photos of the 500 mm×500 mm square column are shown in Fig.3.

(a) (b)

Fig.3 Trail blasting photos of the 500mm×500mm square column

(a) before trail blasting; (b) after trail blasting

The result of testing explosion: the column stirrup is collapse and the longitudinal reinforcement is bended. The concrete in the column is crush uniformly and spill out from the steel reinforcement cage, the area of which is in the grass screen protection inside. The column loses its bearing capacity completely. The result of protection and explosion meet the intended purpose.

The analysis of problem: there are many residues concrete inside of the corner of the column; the reason is that the joint of the shear wall and column is incomplete. With the inner resistance line be longer, the medicine column eccentric distribution inside the column, not even work when explosives are used. As this column, we should increase the depth of the hole and make the medicine column in the middle of the column, which can avoid this kind of the phenomenon.

According to the result of the explosion, formal blasting with other section column charging parameter is adjusted for: five hole bottom adopt single consumption of $1.33kg/m^3$, namely 100g per hole charge, because of the difficulty of the protection of slingshots on the part of the column, the single consumption decrease $1.0kg/m^3$, namely 75g per hole charge. The protection of the single column is qualified, which of other section is three layer grass screen and one layer steel wire mesh.

4.3 Trial Blasting of the 1000mm Diameter Column

The center of the building is supported by four column with a diameter of 1000mm, shown on the construction drawings of the column diameter is only 750mm, do not tally with the actual size, the structure reinforcement are not sure, in order to ensure the building dumping, choose a pillar of the inside for the test. By figure pilot before blasting column figure can be seen the column drill hole is quite much, actually mostly residual hole drilling depth is not enough, because inside the column reinforced too dense, leading to a lower drilling success rate, so the test consumption relative is also enhanced.

Test parameters set as follows, blast hole depth increased to 800mm, Hole spacing is 300mm, single charge 375g, charging length is 500mm ($2^#$ rock explosive, single volume is 200mm , weighs 150g), fill length is 300mm, the unit consumption is 1.59 kg/m^3, careless protection adopts three layer and a layer of wire mesh with wire fixed tightly. When we test, giving up an invalid at the bottom of the hole, from the below second hole straight up charging five holes. Before and after the trial blasting, pictures column as shown is Fig.4.

Blasting effect: part of stirrup disconnect, longitudinal reinforcement deformation is not obvious, within the column concrete failed to throw, the column's failure to completely lose bearing capacity, failure to achieve expected blasting effect; Hole near the hulls are gravel smashing, slung shot fly out about 5m distance, careless by local position on both sides were destroyed.

Problem analysis: one of the reasons for poor blasting effect, many drill deep hole blasting test did not block processing, the minimum resistance line change, reducing blasting power. For another, column with double stirrups and longitudinal reinforcement, and have the trace of secondary reinforcement, column strength than ordinary column; Third, explosive consumption is low, lacking the ability to work.

(a)   (b)

Fig.4 Trail blasting photos of the 1000mm diameter column

(a) before trail blasting; (b) after trail blasting

With reference to the column of the blasting effect, the official with other section column charge blasting parameters is adjusted for: single-hole control adopt single consumption of $1.91 kg/m^3$, namely 450g per hole charge, at the same time, using cylindrical lateral stirrup reduce its binding capacity of concrete, and will not charge the gun mud filling; considering the local position protection strength is insufficient when the test , and puckering quantity increase, single column strengthening protection layer for six hulls and a layer of wire mesh.

5 CONCLUSIONS

Trial Blasting in demolition blasting of building is important to the quality and safety of construction in demolition blasting engineering, and the trial blasting should take attention to the following aspects:

(1) The column that we select in trail blasting should be at different positions in the building, to avoid multiple column at the same position losing carrying capacity, which will be disadvantageous to the successful collapse of building.

(2) Calculating charge length exactly to ensure that the explosive columns is always in the center of column, and to avoid the effect of blasting because of eccentric explosive columns.

(3) We can divide the charge area at the same column into two sections(the distance between them should be more than 500mm),and use different charge parameters for these two section, which allows us to make comparison by limited column to get the best blasting result.

REFERENCES

[1] LIU D Z. Engineering blasting practical handbook[M]. Beijing: Metallurgical Industry Press, 1999.

[2] ZONG Q. Effective technique of controlling fly-rock in demolition blasting in city[J]. Journal of Liaoning Technical University, 2002, 21(5): 598~601.

[3] XIONG Y F, DONG Z C, WANG X. Analysis of the formula about blasting flying rock's casting distance[J]. Engineering Blasting, 2009, 15(3): 31~34.

[4] LI Z J, ZHENG B X, WEI X L. Characteristics and corresponding measures of safety management in blasting demolition projects[J]. Blasting, 2007, 24(1): 97~100.

Directional Blasting Demolition of 60m Brick Chimney

YAO Jinjie, ZHAO Rundong, XIONG Yazhou, LUO Zhenhua

(*Department of Engineering Mechanics, College of Hydraulic & Environmental Engineering of China Three Gorges University, Yichang, Hubei, China*)

ABSTRACT: A 60m high brick chimney using directional blasting method was introduced. Blasting parameters and safety protection measures were determinated reasonably, aiming to achieve the desired result that the chimneys' collapsing to direction on the safety side.

KEYWORDS: bick chimney; directional blasting; blasting demolition

1 PROJECT DESCRIPTION

1.1 Project Features

There is a chimney at the Xiaoxi tower village. The chimney is brick structure. The chimney was builded twenty years ago, structural strength reduced, the surrounding factories and local residents have been threatened. ϕ20mm vertical steel bars distribute around chimney liner, the bottom outer diameter D is 4.6m, wall thickness is 0.80m, height is 60m. Top diameter is 3m.

1.2 Environmental Conditions

The brick chimney was located in the gas factory floor. The circumstance of blasting site is very complicated, as shown in Fig.1. The distance from the chimney to the eastern nearest living room is 1.7 m, the distance to the nearest north room is 12.5 m, the distance to the eastern side is only 20.3 m. If the folded collapsed demolition way is selected, it is easy to destroy living room, and difficult to control the collapsed direction, so the program is suitable for southwest directional blasting, the direction is S28.5° W.

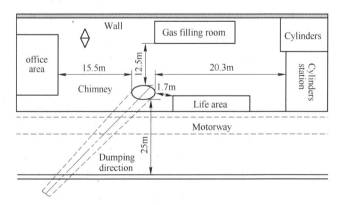

Fig.1 Schematic chimney environment

2 BLASTING PLAN SELECTION

The chimney collapsed in the south west direction, and was allowed a smaller collapse space to achieve a successful blasting, incision size should meet the following demands: first, not recoil phenomenon occurs in initiate of chimney blasting the chimney reserved sites must be able to support the total weight of the upper part of the chimney; second, when the incision closes after blasting, pouring torque should be greater than the limit anti overturning moment; third, when the incision closes the centre of chimne gravity must shift out the outer radius of the chimney bottom[1].

3 BLASTING PARAMETERS

3.1 Blasting Gap Parameter

(1) Cut Shape: to prevent the chimney backseat, trapezoid blasting gap was chosen.

(2) Cut Height: based on experience, blasting gap height $H=(1.5\sim3.0)B$ (B is the chimney wall thickness, here $B = 0.8$), because the chimney liner ϕ20 mm vertical steel bars distribute around, the gap must be higher, for the actual operation button, the gap is 0.8m from the ground, and cut height is 2.2m.

(3) Incision length: based on current experience. Often taking $L = (1/2\sim3/4)\pi D$ (D is the outer chimney diameter in the gap). Symmetrical oblique flanks type opening is used, the upper edge of the trapezoidal notch length $L_1 = 3.7$; lower edge of the trapezoidal notch length $L_2 = 6.5$ m.

3.2 Bore Hole Depth

Based on construction experience at home and abroad, reasonable hole depth δ is 0.65 to 0.70 times of wall thickness[2]. Drilling hole from the outside of the chimney, hole

---

Corresponding author E-mail: yaojinjie111@163.com

depth $h = (0.65\sim0.72)\ \delta$, here $h=0.68\delta$, $\delta=0.70$ m, $h = 0.68\times0.70$ m $= 0.48$ m.

3.3 Spacing and Row Spacing

Combined with the engineering example, there is $\phi 20$ mm vertical steel bars around in the lower liner part of the chimney, to ensure reinforced concrete blocks get out from the steel cage and have no supporting role, taking spacing $a = (0.85\sim1.0)\ h$, row spacing $b = (0.8\sim1.0)\ a$, here $a = 0.95h = 0.45$ m, $b = 0.88a = 0.4$ m, as shown in Fig.2.

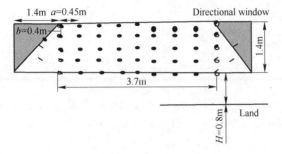

Fig.2　Blast hole schematic diagram layout

3.4 Dose and Structural Design

In order to ensure full incision formed to create favorable conditions for the chimney collapsed, borehole loading charge should enable incision completely broken and thrown out, as shown in Fig.3, the 2nd rock powdered ammonium nitrate explosives was chosen. Required loading charge was determined by the number of submissions, the amount of a hole charge $Q = qab\delta$[3], here charge consumption coefficient $q = 690$ g/m^3, a hole charge $Q = 86.9$ g, real taking $Q = 86$ g. Incision hole is 110, the assembly charge $\sum Q = 9.5$ kg.

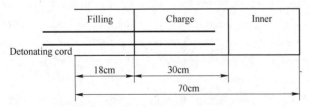

Fig.3　Schematic diagram of drug loading

3.5 Blasting Network

Electric detonators, duplex detonating cord hybrid blasting network was chosen. There were two detonating cord in the blast hole, and two detonating cords outside each row of holes. Horizontal opening paragraph blast hole detonating cord connected delay number 3 electric detonators, tilt opening paragraph blast hole connected delay number 5 electric detonators, as shown in Fig.4.

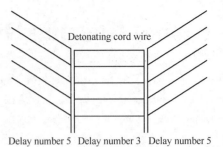

Fig.4　Schematic blasting delay network

4　SAFETY MEASURES

(1) No matter which direction dumping, open positions are 2.3 m above ground level and the drilling site height is 2.3 $\sim$3.9 m.

(2) The lower flue gaps in the east and west sides were blocked with brick, brick thickness is 48 cm.

(3) Part of the courtyardwall removed in advance, overhead cables and wires need to be cut before blasting, and re–connected after blasting.

(4) Three directional windows need to manually be opened before drilling, window parts must be measured accurately. 1$\sim$2 brick window width and a height of 1.6 m, windows must penetrate the cylinder.

(5) Drilling site should be accurately measured, to ensure accurate location, error should be less than 2 cm. Chimney boreholes are perpendicular to the surface and pointed to the chimney plane center.

(6) Blasting vibration velocity calculation.

$$v = k\left(\frac{\sqrt[3]{Q}}{R}\right)^{\alpha} = 160\times\left(\frac{\sqrt[3]{9.5}}{13.7}\right)^{1.8} = 5.5$$

Blasting vibrations will not damage adjacent buildings [4].

(7) Blasting safety distance calculation: flying rocks can be calculated based on non–covered blasting and the relationship between the amount of charge per unit $R_f = 70Q^{0.53}$ [5]. $R_f = 230$m, so this blasting safe distance is not less than 250m. The warning line should ensure to prevent onlookers injured by flying.

5　BLASTING EFFECT

After initiation, with a dull explosion, the main chimney collapsed in predetermined direction, the main chimney rupture at the 30m when the blast occurred about 2s, no folded ,no dumped back seat and no flying occured. The blasting achieved the desired effect, and was no damage on the surrounding buildings.

REFERENCES

[1] Jia H, Yin Y. Directional 40m-High Brick Chimney Blasting[J]. *Mine blast*, 2006, 2 (73): 34~35.

[2] Chen S G, Zhao J. A study of UDEC modeling for blast wave propagation in jointed rock masses[J]. *International Journal of Rock Mechanics and Mining Sciences*, 1998, 35(1): 93~99.

[3] M Y S, F G M. High 60m, 80m Two Brick Chimney Blasting[J]. Blasting, 2007, 9 (13): 56~58.

[4] Yang R U, Yu S J. Blasting Demolition of Buildings [M]. Beijing: China Architecture and Building Press, 1985.

[5] Li Z, L T Z. Blasting Brick Chimney complex environment[J]. Mine Blasting, 2008, 2 (81): 33~34.

The Irregular Structure of Meteorological Office Building Demolition Blasting

LI Wei, WANG Qun, ZHAO Jingsen, LI Dan

(*Shenyang Xiaoying Blasting Engineering, Co., Ltd., Shenyang, Liaoning, China*)

ABSTRACT: The Meteorological office building also known as the building of "thermometer" is located in the bustling downtown, in the southwest of it with main roads on the both sides. It is very close to the protected object too, and the building itself is a irregular structure. The centre of gravity is on the rear because of "thermometer" body on the southwest side. Use directional blasting delay control and accurately locate the position of the center of gravity to ensure the collapse with accurate direction. By using effective protection, such as damping ditch excavation, building flexible buffer layers for blasting stones, control the blasting effect for the ground shaking, to achieve the expected goal.

KEYWORDS: irregular and asymmetry; building quality uneven distribution; cylinder; pre-demolition

1 ENGINEERING SURVEY

According to the construction plan, Meteorological office building must be clean demolished, which is located in Shenyang. Due to the urgent schedule, blasting demolition was chosen.

1.1 The Surrounding Environment

The building is located in the center of the bustling downtown, in the west of Dongdian hospital, east of Qingnian street, north of Cultural road, south of land lines. The building is 140 m away from the Jianyuan street on the east side, buildings between No.8 and Jianyuan street will be removed. Qingnian street and the overpass is 28 m and 37 m away on the west side, the nearest underground gas pipeline is 25 m away on the west side, on the south side Cultural road is 32 m away, and a bridge is 49 m away, open space is on the north side. The gas pipeline goes along with Qingnian street from south to north, and on the west side of the building about 25 m to the roadside, and there is a water pumping station is 20 m on the west side of the building, there are underground cables 10m on the north side of building. All the pipelines both overground and underground would be completely cut off before the blasting, the surrounding environment is shown in Fig.1.

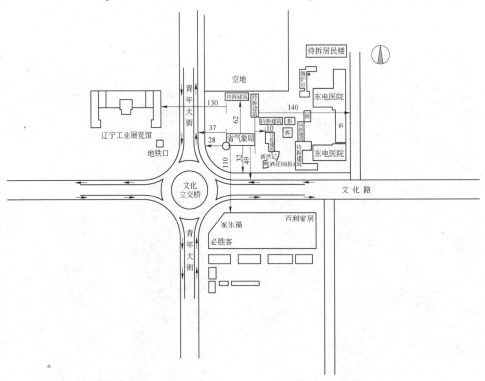

Fig.1　Sketch of blasting the surrounding environment (unit: m)

Corresponding author E-mail: syxyxzswb@163.com

1.2 Structure Overview

Meteorological office building is 40.2 m long, 18.6 m wide and is 17 floors, 69.3 m height. The building is the frame shear wall structure, there are seven row longitudinal column, and four row of transverse columns in another direction. In the southwest corner there are cylindrical shear wall with stair and elevator shaft, the circular diameter is 8.4 m, in the northeast corner there is semicircular shear wall with staircase, constitute the core tube, because the exist of shear wall, it would increase the rigidity and solidity.

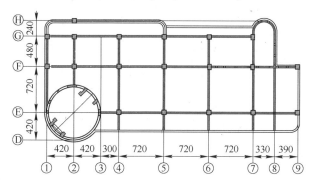

Fig.2　The first floor of Meteorological office building (+0.00 m) plan sketch(unit:cm)

1.3 The Center of Gravity Calculation

Use the northwest corner as the origin, build a three-dimensional right-angled coordinate system.

The whole building collapsed due to vertical solid only calculation of vertical shaft radial centroid position, does not need to compute the z axis.

Building the axial centroid position calculation:

$$x_G = \frac{\sum_{i=1}^{n} w_i x_i}{\sum_{i=1}^{n} w_i} = \frac{6.3+7.2+3.2+\ldots+14.5+3.3}{1.2+1.1+3.3+\ldots+4.3+2.1} = 10.25$$

$$y_G = \frac{\sum_{i=1}^{n} w_i y_i}{\sum_{i=1}^{n} w_i} = \frac{13.5+21.3+23.3+\ldots+12.2+21}{3.3+2.5+2.2+\ldots+3.6+1.8} = 16.4$$

After calculation, the center of gravity is on the north side of the building from the wall 10.25 meters, 16.4 meters away from the west side of the wall, as a basis for the setting of detonating sequence and delay time.

1.4 Blasting Requirements and Difficulties

(1) In the building southwest corner there is larger (outer diameter 8.4 meters) cylinder wall where stairs and elevator well is located, so southwest corner has great quality, with the opposite collapse direction, the cylinder and the internal staircase and elevator well must be correctly handled. The cylinder on the southwest must has enough supporting capacity before all the row column blow up.

(2) Meteorological office building is 25m above the underground pipeline, 37m away from the overpass, and near the subway, so blasting vibration must be strictly under control.

(3) The southwest corner weight is heavy, has an impact on the collapsing direction, so setting off sequence and the delay time must be carefully set.

2　BLASTING-DEMOLITION SCHEME

The meteorological office building is frame shear wall, the shear wall, increases the firm degree of the structure, and also increases the difficulty of demolition blasting. It is located in the downtown area, the surrounding environment is more complex, the collapse direction is limited.

There is open space in the north side the building, would not affect on the collapsing, fulfill the collapse distance. According to the building and the surrounding conditions, a variety of methods are compared, finally north directional collapse blasting scheme is chosen, the advantages of it is less work, better disintegration. The scheme is achieved by various vertical columns with different blasting height and the rational delay time, so scientific decide reasonable blasting height and the delay time becomes the key of blasting scheme set. According to the relevant information, previous experience in engineering practice, and choose the elevation angle with empirical formula to ensure the blasting height is 1～3 layer to get a triangle blasting gap, so that the building body is demolished in the direction, and has enough base reaction force. The objects to protect are the overpass, Cultural road, underground gas pipes. In order to prevent the back sit, delicate directional blasting technology for the last row column are in Fig.3, and at the same time, the time delay must be strictly controled. The pre-demolition of the cylindrical structure is shown in Fig.4.

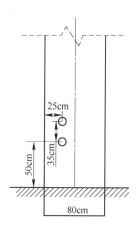

Fig.3　The rear column detail

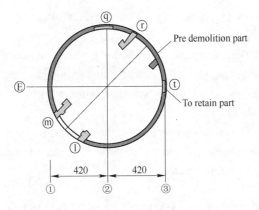

Fig.4　The cylindrical structure pre demolition

3　SET OF BLASTING PARAMETERS

Combined with practical engineering, blasting parameters of different structures are given in Tab.1, the charge should be adjusted according to the blasting test.

4　DETONATING SEQUENCE AND NETWORK CONNECTION

The initiation network consists of four number of delay. column G4, G5, G6 and G7 is Ms3 (50 ms), column G2,

Tab.1　The main pillar blasting parameter table

The blasting site	Size/m×m	Column number	Hole depth /m	Consumption /kg·m^{-3}	Line of resistance/m	Spacing /m	Row spacing/m	Single hole charge /kg
A layer of	0.8×0.8	G2~G6	0.7	1.4	0.28	0.56	0.24	0.2
	0.6×0.8	G7	0.7	1.4	0.3	0.45	—	0.3
	0.9×0.9	F2~F5F9	0.8	1.4	0.3	0.6	0.3	0.3
	0.6×0.9	F6~F7	0.8	1.4	0.3	0.45	—	0.3
	0.8×0.8	E4~E9	0.7	0.8	0.25	0.35	—	0.15
Two layer	0.8×0.8	G2~G6	0.7	1.4	0.28	0.56	0.24	0.2
	0.6×0.8	G7	0.7	1.4	0.3	0.45	—	0.3
	0.9×0.9	F2~F5F9	0.8	1.4	0.3	0.6	0.3	0.3
	0.6×0.9	F6~F7	0.8	1.4	0.3	0.45	—	0.3
Three layer	0.8×0.8	G5, G6	0.7	1.4	0.28	0.6	0.24	0.25

F4, F5, F6, F7 and F9 is Ms9 (310 ms), column F2, E4, E5, E6, E7 and E9 is Ms10 (380 ms). In order to protect the road on the west side and the underground gas pipes, pillar on the west side should play a supporting role in the building collapse process, so, column G2 is Ms9 (initiation at the same time with column F4~F9), column F2 is Ms10(initiation at the same time with column E4~E9), cylindrical column r is Ms9, column q and column t is Ms10, column m and column l is Ms10+ Ms3 (430 ms).

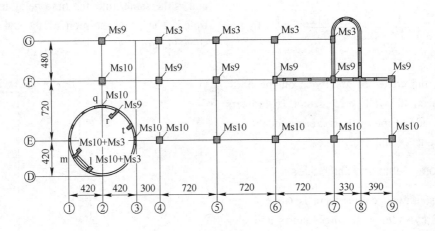

Fig.5　Number delay distribution

Double closed shock-conducting tube initiation network is used, and the net is initiated by the excitation pen and high pressure charging blaster.

5 MEASURES OF SAFETY PROTECTION

5.1 Blasting Flyrock Defence

According to the actual situation the following measures are taken:

(1) Direct protection of blast hole. The blasting parts of the column were surrounded by six layers of grass curtain, every three layers of grass curtain were banded with No.8 wire every 50 cm around the column, coverings must exceed hole boundary more than 50 cm and were tied tightly. Overlapping length cannot be less than 55 cm. After the first three layers of straw tied, a layer of barbed wire was covered. And tied, and then the other three layers of grass curtain were tied up, finally a layer of nylon wire bundling was tied up.

(2) Fencing. First doors and windows were covered with grass curtains in the gap blasting range and building on the south side, grass curtains must hang tight, and have no gap, and then covered with dense holes nylon net, which also must be tied tightly.

(3) Collapsing splash off stone protection. Two layers straw curtain were covered on the ground expected for the building collapse, pressed by the sandbags and water it before blasting, covering range should be out of expected range 5 m.

(4) Retaining wall was embanked 60 m away from the building in the collapse direction to prevent moving-up and flyrock, it was made of demolition waste and height was 2.5 m.

5.2 Dust Reduction Measures

(1) Residual slag on the ground and the floor was cleaned, and the floor was watered before blasting to reduce dust produced in the process of collapse.

(2) The mat on the collapse ground was watered before blasting to effectively reduce the collapse touchdown dust.

(3) Sprinkler sprayed water after initiation.

5.3 Collapse Vibration Prevention

We adopt the following measures to reduce blasting vibration:

(1) Two earth dykes (height 1 m, wide 1.5 m) were built 30 m and 40 m away from the building in the collapse direction and covered 2 layers straw curtain on it's surface.

(2) Excavation of the damping ditch, damping ditches were excavated in the west 10 m and south 10 m (height 2 m, wide 1 m), the blasting collapse vibration on the gas pipeline and the surrounding can reduced effectively.

6 THE EXPERIENCE

The building collapsed in the predetermined direction, had no backseat and, burst thoroughly pile height was about 3 m, the collapse length was 55 m, the fly stone was controlled within 5 m, underground pipelines and overpass were not damaged, achieve good blasting effect was achieved photos are given in Fig.6 and Fig.7.

Fig.6 Directional falling trend

Fig.7 After blasting the surrounding intact

Following experiences are achieved:

(1) After pretreatment, the elevator shaft shear wall had more free surface and was easy to blast.

(2) Laying sand, scarification, mats and other soft cushioning materials and excavation of the damping ditch, these measures effectly controlled the vibration.

(3) The cylindrical structure weight have a great impact on the collapsing direction, the reasonable pre-demolition, set of blasting sequence and the delay time these measures effectively realized the orientation effect.

The irregular building blasting direction is determined by the falling tendency early and rear overturning support.

REFERENCES

[1] Xiaorong Cui, Cansheng Zheng, Jianqiang Wen, et al. Irregular frame shear structure building blasting demolition of [J]. Blasting, 2012,29 (4): 95~98.

[2] GB 6722—2003 Blasting Safety Regulations [S]. Beijing: Chinese Standard Press, 2004.

[3] Tension. Engineering blasting technology [M]. Beijing: Metallurgical Industry Press, 2005.

Blasting Demolition of Urban Viaduct 3.5 km (2.175Mile) in Length

XIE Xianqi, JIA Yongsheng, YAO Yingkang, LIU Changbang, HAN Chuanwei
(*Wuhan Blasting Engineering Co., Ltd., Wuhan, Hubei, China*)

ABSTRACT: The viaduct, 3.5km in total length, consists of main bridge and approach road. The main bridge composed of prestressed concrete hollow slab with pretensioning method is 3.0 km (1.864mile) long. The piers made of reinforced concrete double-columns are in 180 rows, 360 piers in total. The viaduct is located in downtown area where the environment is extremely complicated. Alongside the viaduct, there are a mass of buildings and structures. Moreover, various municipal pipelines approximately 32 in quantity including a ϕ720 mm (28.346 in.) high pressure nature gas pipeline are buried under the viaduct. The overall scheme for demolition blasting of main bridge combined with mechanical demolition of approach road was proposed. One-time initiating blasting was used in main bridge. From center to southern and northern ends, the delay blasting implemented on piers row upon row lasts 24.77 s and 21.52 s in total respectively. In consequence, the viaduct collapsed in whole while the adverse effects of blasting get effective control and various pipelines underground are safe. During blasting process, ten blasting vibration monitoring lines were disposed within 200 m (656.2ft) alongside the viaduct, as a result, the law of vibration attenuation was obtained. Meanwhile, items such as dust concentration of blasting and ground stress were monitored in real time.

KEYWORDS: urban viaduct; blasting demolition; relay initiation; adverse effects of blasting control

1 PREFACE

The viaduct has the advantages of huge carrying capacity, rapid traffic diverging, safety and reliability, ect. It is available for alleviating urban traffic jam and improving moving efficiency. However, with the acceleration of the urbanization process, some viaducts built in the 1980's or 1990's last century, are being gradually revealed the ills of traffic function, load degree, running security and so on. A potential safety hazard may occur to the city. Therefore, the urban traffic layout and construction are in urgent need of safe, effective and economical demolition technology of viaduct. By comparison with mechanical and manual demolition, the blasting demolition has been the first choice of viaduct demolition due to the virtues of slight influence to traffic, high security, high efficiency and low engineering cost[1].

The Zhuanyang viaduct is located on Dongfeng Avenue in Wuhan Economy and Technology Development Zone (WETDZ), Hubei Province. It was built and opened to traffic in 1997. It was 4-lane two-way and 3.5km (2.175 mile) in total length, the design speed of which is 40km/h (24.856 mile/h). As social economy developing, the amount of cars increase rapidly. As a result, the existing viaduct cannot meet the using requirement, meanwhile there is no side of the verge and crash proof placed on the centre and both sides of viaduct, so the potential safety hazard may exist. Based on the comparison and research on technique and economy, the decision of demolishing the existing viaduct and expressway reconstruction was made.

2 ENGINEERING PROFILE

2.1 Surroundings

The Zhuanyang viaduct lies north and south across the Wuhan Economy and Technology Development Zone (WETDZ). It is the traffic throat within southwest where five urban arteries cross under it. The surroundings are extremely complicated: abundance of residential and office buildings and plants are distributed alongside the viaduct. The high voltage wires of 110 kV across the viaduct, and the high voltage tower stands only 24m (78.744 ft) from viaduct. There are 32 pipelines in total buried under the viaduct such as high pressure gas pipe of ϕ720 mm (28.346 in), water pipe of ϕ 800 mm (31.496 in) and high voltage wire of 110 kV. The surroundings of the Viaduct is shown in Fig.1.

Corresponding author E-mail: 94394995@qq.com

Fig.1 Surroundings of the Zhuanyang viaduct

2.2 Viaduct Structure

The Zhuanyang viaduct, 3500 m(2.175 mile) in total length, consists of main bridge and approach road. The main bridge 3000 m(1.864 mile) long is constructed as simply supported and continuous rigid frame structure. The north and south approach road with structure of U-shaped concrete gravity retaining wall is built 500m (1640.5ft) in length.

The main bride has 22 sections of which the length ranges 128 m (419.968 ft) to 144 m (472.464 ft). The group of 8~9 bridge openings is regarded as one section. There are 180 bridge openings in total including 26 bridge openings with the span of 18m (59.058 ft), 154 bridge openings of 16m(52.496 ft) and one bridge opening of 15.5 m (50.856 ft). The main bridge composed of prestressed concrete hollow slab with pretensioning method is made of C40 concrete. The piers with the structure of reinforced concrete double-columns are made of C30 concrete, of which the cross-section is 550 mm×1000 mm (21.654 in× 39.37 in). The foundation of main bridge is designed as spread foundation of C25 concrete, see Fig.2.

3 OVERALL DEMOLITION SCHEME

3.1 Difficulty Analysis

(1) The fundamental theory and design scheme of blasting demolition of urban viaduct is not yet mature; (2) There's no precedent case for one-time blasting demolition of the urban viaduct 3500m(2.175mile) in total length, of which the main bridge is 3000m(1.864mile) long; (3) the viaduct across five urban arteries is of heavy traffic so that it's difficult to ensure traffic flow during the periods of blasting demolition; (4) safeguards should be introduced to ensure the safety of pipelines underground, especially some of which are not deeply buried; (5) the demolition project has features such as long working plane, huge work amount, limit construction period and tough management.

3.2 Overall Scheme

In accordance with the engineering structure and surrounding features of the viaduct, the overall scheme for demolition blasting of main bridge combined with mechanical demolition of approach road was proposed. The demolition blasting scheme adopted drilling hole and initiating blasting in support components of the viaduct, which cause the superstructure instable and falling down to the ground. In sequence, the following disposal of blasting debris was carried out.

One-off initiating blasting was used in main bridge. From center to south and north ends, the delay blasting was implemented to piers row upon row. During the periods of construction, to determine various blasting parameters and safeguard and to ensure the whole stability of viaduct and safety of traffic, it is necessary to build a 1:1 physical model for test and numerical simulation.

4 PHYSICAL MODEL TEST

At present, the study on theories involved and key technique of blasting demolition of urban viaduct in complicated surroundings is not yet mature. The risk is relatively high provided that the blasting height of piers, unit volume consumption of explosive, delay time of initiating circuit and safeguard are determined just depending on the experience formula. The 1:1 physical model test is effective to be used for research on blasting instability mechanism of piers, determining blasting parameters and working out safeguard.

4.1 Model Design

According to the design drawings of viaduct and actual conditions on site, firstly we pick out the typical piers, and then built 12 detached piers as well as a 1:1 one-stride physical model with the same measurements, concrete intensity and reinforcement distribution in the test field. As shown in Fig.3, one precast concrete pipe and cast iron pipe are respectively buried under 1.5m (4.922 ft) of the ground of 1:1 one-stride physical model.

Through groups of detached pier blasting physical model test the holes parameters of blasting instability of piers, charge mode and optimal safeguard can be determined effectively. The optimal delay of each row of piers can be further determined in 1:1 one-stride physical model blasting test meanwhile the research into deformation effect generated by underground pipelines made of different materials under load of blasting and collapse vibration is conducted.

During the process of model built, pieces of apparatuses such as strain gauge, earth pressure gauge, accelerometer and blasting vibration meter(see Fig.4) were buried or disposed in piers, pipelines and soil under model, through which the further study on blasting destructive mechanism and adverse effects were carried out. When model test blasting occurs, the high-speed cameras from different angles capture the whole process of viaduct collapse. Combined with measurement data, the quantitative research on blasting destructive mechanism and collapse process can be implemented. Moreover, the method of dedusting by blasting water fog is employed to get the blasting dust under control.

Fig.2   Viaduct structure

Fig.3   1:1 one-stride physical test

Fig.4   Fix strain gauge

4.2   Test Results

4.2.1   Blasting Parameters and Safeguard Methods

By means of 12 groups of detached pier blasting physical model tests, the optimal blasting parameters and safeguard methods of C30 reinforced concrete rectangle piers with cross-section of 550 mm×1000 mm(21.654in ×39.37 in) were obtained: (1)Along the center line in long side of piers, holes of 40 mm(1.575 in) in diameter are arranged with an internal of 300 mm(11.811 in). The minimum blasting burden ranges from 250 ~ 300 mm(9.843~11.811 in) while the blast hole depth ranges from 700 mm(27.559in)to 750mm(29.528in). (2)The rock emulsion explosive with diameter of 32 mm(1.26 in) was chosen for the test. The optimal charge mode is: the five holes on the bottom were charged with explosive of 400g (0.882 pound) in quantity, the three holes in the middle of 300g (0.661pound) with an air internal of 20 cm (7.874in), and the holes on the top of 240g (0.529pound) with air internal of 26 cm (10.236 in). (3) The safeguard not only to blast the piers into fragments as required but also to control the flying debris is as follows: from inside to outside, the piers are wrapped with three layers of quilts, one layer of wire mesh and one layer of bamboo raft, around bottoms the sand bags piles up to 1.0 m (3.281ft) high. Proximal safeguard adopts hanging two layers of dense-mesh protection networks from the guardrails to ground alongside the viaduct.Fig.5 shows pier blasting physical model tests process and results.

Fig.5   Pier blasting physical model tests

4.2.2   Delay Time Between Rows

In physical model test of detached pier, it can be clearly observed by high speed camera that piers keep swaying when blasting occurs. The vibration velocity waveform of particles also illuminates that the swaying of piers will cause the particles vibration around. The peak value is many times bigger than that caused by blasting while smaller than that by deck impacting the ground. The vibration continues after blasting and the time of horizontal vibration lasts longer than that of vertical. Most of the peak values of horizontal vibration last in the order of 200~300 ms (Fig.6).

The analysis shows that if the initiating occurs at internal of 250ms between each row of piers, the particles vibration overlapping each other caused by piers blasting is able to be avoided. It is, therefore, suggested a choice of MS8 detonators（delay of 250ms） for initiating in relays between the rows of piers.

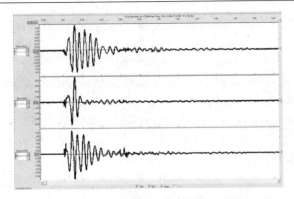

Fig.6 Velocity waveform of blasting vibration

4.2.3 Overall Instability of One-Stride Model

In 1:1 one-stride physical model test, high speed cameras are arranged respectively at front and on the side of the model (Fig.7, Fig.8) to observe the process of losing stability. Detonators of MS16 are set in holes while in reference of the test results of detached pier model, outside holes detonators of MS8 in relays are used.

Fig.7 High speed photograph at front

As shown in high speed photographs, the piers wrapped around with three lays of quilts and one layer of dense mesh wire net are almost intact. 120ms after initiating, piers begin to fall, while after 250ms, the distance in decline reaches as high as 10% of pier height. At this moment the center of gravity of deck slab tilts to the side of first initiating, but not too much to influence the stability of piers in back row. So the use of detonators of MS8 for initiating in relays proves reasonable.

Fig.8 High speed photograph on the side

In 1:1 one-stride physical model test, two piers in front row, in one of which the reinforcement stirrup is cut while in another reserved, are used to verified the stirrup function in blasting demolition of piers. Seen from Fig.8, it is obvious that the pier with broken thick stirrup is much easier to lose stability than that without. Provided that one side was cut while the other was not, it is possible to cause the deck slab incline towards the side of broken stirrup. In actual work, the property can be made use of to keep the important parts away from the inclined slab.

4.2.4 Monitoring Stress and Strain of Underground Pipelines

In light of the physical model test scheme, one precast concrete pipeline and one cast iron pipeline are buried respectively 1.5 m(4.922 ft)depth under the ground of 1:1 one-stride physical model, meanwhile the strain gauge is set as required to monitor the strain and stress state of underground pipes affected by vibration from blast and collapse. Based on the measurement data, the axial and hoop stress can be figured out. The monitoring and computing results are in reference to Tab.1.

Tab.1 Monitoring results of stress and strain of underground pipelines

Spot Number	S01	S02	S03	S04	S05	S06	S07	S08	S09	S10
Spot location	Cast iron pipeline (axial)	Cast iron pipeline (hoop)	Cast iron pipeline (axial)	Cast iron pipeline (hoop)	Cast iron pipeline (axial)	Cast iron pipeline (hoop)	Concrete pipeline (axial)	Concrete pipeline (hoop)	Concrete pipeline (axial)	Concrete pipeline (hoop)
Maximum strain /$\mu\varepsilon$	33.01	54.98	22.91	53.31	67.21	—	15.27	85.53	16.80	13.95
Maximum stress /MPa	4.85	8.08	3.37	7.83	9.87	—	0.59	3.34	0.66	0.54

As shown in Tab.1, the max compressive stress of concrete pipeline which appears at S08 reaches 3.34 MPa and the max tensile stress appears at S09 reaches 0.66 MPa. The max pull stress of iron pipeline which appears at S05 reaches 9.87 MPa. All the values are within the permitted range and a certain margin exists.

Based on the analysis of measurement data of strain, it is assumed that the security control criterion of blasting demolition is appropriate and the pipelines are safe and sound in that no destructive influence are made to the precast concrete pipelines and cast iron pipelines buried during the process of blasting demolition.

4.2.5 Test of Blast Water Fog Dedust

Blast water fog dedust is the method disposing explosive bags under water bags where abundance of water fog forms by means of blasting can be used for dedusting. The bags used in test are 5m (16.405ft) long, 0.9 m(2.953ft)wide and 0.15 m(5.906in)high with water filled. In test, rock emulsion explosive are selected. The 3 or 4 charges, each one of which is 50 g(0.11pound)in weight, are uniformly laid under water bags. The instantaneous detonators are used for initiating at the same time. To observe the range of water fog, one high speed camera is arranged respectively at front and on the side of water bags for record. As shown in Fig.9 and Fig.10, a comparison between blasting fog effect of three and four charges is illustrated.

Fig.9　Three explosive bags

Fig.10　Four explosive bags

Draw a comparison between blasting fog effect of 3 and 4 charges, it is found that: (1) the column-shaped water fog aroused by 4 charges initiating have wider spread and longer last in air; (2) after 4 charges blast, the water fog ranges in the order of 10m(32.81ft) along the bags axes, as well as along the perpendicular to the bags axes (5m(16.405ft) respectively on each end), so it is concluded that the fog forms by single water bag blasts in test can cover a range of 100 m^2. (1.196yd^2) or so; (3) observing the video captured by high speed camera, it is found that the fog in air lasts for 1.6s at the least and column-shape fog forms 50ms after initiating.

5　NUMERICAL SIMULATION

In order to not only directly find out the expected effect of blasting, but also continuously, dynamically and repeatedly see the detailed process of collapse in whole and breaks in part. Meanwhile the information obtained from the simulation feeds back into design, so the blasting design scheme gets rectified and improved[2, 3]. In this paper, software of SLM-DEM is employed to simulate the process of blasting demolition of viaduct.

5.1　Model Building

Geometry model mainly includes deck slab, piers and road surface as well as road foundation. All the measurements are derived from the original design drawings of viaduct. During the process of model building, the simplification of geometry properties with no effect on structure computing is made so it is convenient for time computing by means of mesh generation. To obtain the whole process of viaduct collapse and observe the load on the ground, select six-stride viaduct for calculating on basis of the computing times of model. The finite element model of six-stride viaduct is shown in Fig.11, of which there are 37321 units and 58577 nodes.

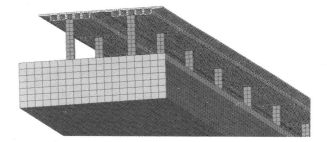

Fig.11　Mesh generation of six-stride viaduct

5.2　One Stride Collapse Simulation

The deck slabs of each stride are precast blocks. The deck slabs of adjacent stride are joint together with wet-joint of which the intensity is lower than the precast blocks. Therefore, when one end of piers is blasted, first of all, the section without support of deck slab falls down owing to gravity, which ruptures the wet-joint of the other end. The expansion joint used in adjacent deck has the same failure mechanism with the wet-joint. It can be seen from numerical simulation result that deformation of one end of deck occurs about 200~300 ms after initiating, then after 1.3 s one end drop to the ground, in sequence, 1.6s later the other falls down. Eventually, a hush follows 1.7 s later.

5.3　Collapse Process

According to the computing times of model, six strides of viaduct are selected for calculating so as to obtain the course of viaduct collapse. See Fig.12.

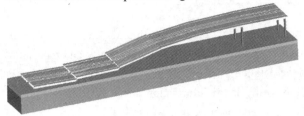

Fig.12 Continuous collapse of multi-stride viaduct

5.4 Simulation Result Analysis

(1) Collapse process analysis: The simulation result indicates that three rows of piers break and drop to the ground. Due to the box girders concreted into the piers with certain intensity, they collapse in whole rather than fall rapidly following piers break. However, once it gets bent in juncture, the corresponding box girders associated with piers which blasted in advance drop to the ground and break away from juncture immediately. It's possible that the adjacent box girders overlap each other after they fall down to the ground, but not too much. Seen from the pile shape, the muck pile is comparatively flat. The box girders don't overturn in air. Due to the low center of gravity, there is no flyrocks scattered as well. Seen from the collapse course and pile shape, the blasting scheme can be realized, that is, the demolition of the viaduct is able to go smoothly.

(2) Impact of drop to the ground: To estimate the impact generated the moment viaduct collapsed and impacted the ground, in simulation, the resultant per unit of road surface is recorded, when the first stride demolished by blasting drops to the ground. When deck slabs hit the ground, the maximum impact of each stride measures in the order of 22500kN(5062.51bs).

(3) Delay time: Conclusion drawn from physical model test shows that the blasting interval between each adjacent pier measures 250ms. Because the process of piers blasting has not been taken into account in simulation, the broken piers will not be involved in following computing during the model building. In simulation it is found that by mean of 250ms delay in initiating and blasting row upon row, the purpose of blasting demolition of the whole viaduct can be achieved. The interval of delay blasting proves rational.

6 BLASTING PARAMETER

6.1 Blasting Parameters

Blasting Parameters are shown in Tab.2. ϕ32 mm (1.26 in) PVC tube is used to control the charge interval. The charge structure is shown in Fig.13.

Tab.2 Viaduct pier blasting parameter

Hole diameter /(mm/in)	Section Area /(cm/in)	burden /(cm/in)	Distance between Holes /(cm/in)	Hole depth /(cm/in)	Quantity of explosive per hole /(g/pound)	Charge form	Remark
40/1.575	55×100/ 21.65×39.37	30/11.811	30/11.811	70/27.56	400/0.882	continuous charge	5 holes at the bottom
		25/9.843	30/11.811	75/29.53	150+150/ 0.331+0.331	Interval charge (interval of 20cm(7.874in))	3 holes in the middle
		25/9.843	30/11.811	7575/29.53	120+120/ 0.265+0.265	Interval charge (interval of 26cm(10.236in))	Remaining holes At the top

6.2 Initiating Circuit Design

The initiating in relays circuit of delay outside detonators is used. To realize delay blasting between rows in relays, for each hole high grade detonator of MS16 (1020 ms) is set in and outside low grade detonator of MS8 (250 ms) is arranged. Besides, in the very front of the circuit, three instantaneous detonators are used for initiating. See Fig.14.

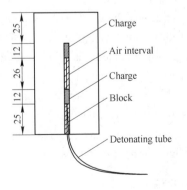

Fig.13 Interval charge structure chart

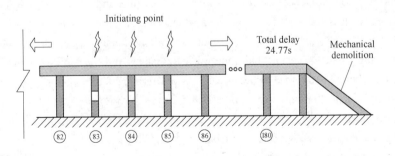

Fig.14 Initiating circuit schematic diagram

7 SAFEGUARD METHODS

The Zhuanyang viaduct is located on artery where problem related to security is extremely obvious owing to the heavy traffic, crowd people and complicated underground pipelines. So it is necessary to take blasting safety into delicate account and make relevant steps in order to reduce adverse effects on surroundings to the maximum extent.

7.1 Underground Pipeline Safeguard

Based on the results of physical model test and numerical analysis, the vibration from blast and collapse of viaduct did no harm to various underground pipelines. However, in order to ensure absolute safety of the underground pipelines, initiative and passive measures are taken as required by proprietors concerned.

(1) Initiative vibration reduction: Through reducing blasting height of piers nearby the underground pipelines, the pier top, cap beam and deck first directly impact the allowance of the bottom of the piers rather than the ground where they rest against the piers. As a result, the impact on underground pipelines caused by viaduct superstructure weakens initiatively.

(2) Passive safeguard: 1) To protect gas pipelines, they are laid from bottom to top in sequence with a layer of sand bags about 2 m(6.562ft) wide and 20 cm(6.562ft) thick, a 2 cm(0.787in) thick steel plate of 2 m(6.562ft) wide and four layers of abandoned tires. In addition, stack up sand bags forming a protection wall of 0.6 m (1.969ft) high and 1.5 m (4.922ft) wide respectively on each side of the plate, See Fig.15. 2) To protect drain pipelines, stack up sand bags forming a protection wall of 0.6 m(1.969ft) high and 1.5 m(4.922ft) wide respectively, along each side of the pipelines. 3) To protect the pipelines of electricity power and telecom, first arrange a layer of sand and steel plates on the pipelines and then along each side of the pipelines, stack up sand bags forming a protection wall of 0.6m(1.969ft) high and 1.5m(4.922ft) wide, aimed at vibration reduction which extend 16m(52.496ft, the horizontal width of the viaduct) at the least.

7.2 Safeguard against Flying Debris

To shield off flying debris, a comprehensive scheme adopting covering and proximal safeguard combined with protective safeguard is introduced. It is verified by physical model test that the safeguard not only to blast the piers into fragments as required but also to control the blasting flying debris is as follows: from inside to outside, the piers are wrapped with three layers of quilts, one layer of wire mesh and one layer of bamboo raft, around bottom the sand bags piles up to 1m(3.281ft) high. Proximal safeguard adopts hanging two layers of dense-mesh protection networks from the guardrails to ground alongside the viaduct (Fig.16).

7.3 Safeguard against Dust

By experience and test, the comprehensive safeguards against dust in demolition blasting of viaduct can be determined: (1)To clean and wash the deck before charging and forming circuit. (2) On the sensitive regions where transformer substation is sited, water fogs and drops brought about to absorb dust when water bags have been cracked hit by flying debris in blasting. (3) The large water bags, each one of which is 6 m(19.686ft) long, 0.9 m(2.953ft) wide and 0.15 m(5.91in) high with water filled, are laid within 100 m^2(1.196yd^2) on the viaduct. Four exclusive bags are used for initiating blasting. The initiating blasting of water bags are 250 ms prior to the front row of piers (Fig.16, Fig.17).

Fig.15 Pipeline safeguards

Fig.16 Fly debris safeguard

Fig.17 Dust safeguards

8 BLASTING VIBRATION MONITORING

8.1 Monitoring Scheme

The monitoring scheme includes such items as: (1) vibration effects during blast and collapse; (2) earth vibration and dynamic earth pressure(induced) caused by viaduct collapse and impact; (3) air shock wave and noise; (4) blasting dust.

8.2 Monitoring Results Analysis

(1) Analysis of ground vibration velocity: To monitor the

velocity of ground vibration, ten inspector trunks and 49 measuring points were set. The monitoring result shows that: 1) The max peak value of vibration velocity of measuring points associated with the residential and commercial buildings and transformer substation of 110 kV distributed alongside the viaduct are all within the permitted ranges. 2) Each peak value which appears along the vertical is much higher than that along the horizontal. The max monitoring value in constructions reaches 1.15 cm/s (0.453in/s). 3) The peak value of vibration velocity around 15 m (49.215ft) away from the viaduct approaches or exceeds 2 cm/s(0.7874in/s). The max reaches 2.39 cm/s (0.941in/s) in the first stride at initiating, however, the further the less. The vibration velocities beyond a distance of 60 m (196.86ft) are all less than 1 cm/s (0.3937in/s).4) The vibration dominant frequency ranges 3.8~100 Hz. The statistics suggest that vibration dominant frequency of 90% measuring points is within 10 Hz. Only one point dominant frequency is less than 4Hz (the value is frequency corresponding to the minimum spectrum characteristic period of Wuhan Aseismatic Design Code). 5) Along the same inspector trunk, the dominant frequency declines as the distance grows.

The max vibration effect is caused the moment viaduct block drops to the ground. The vibration caused by viaduct collapse has a correlation with gravitational potential energy, namely the quality and the height of viaduct block. It declines as the spread distance extends.

By means of regression calculation of monitoring data, the empirical formula[4] for vibration velocity generated by viaduct impacting the unprotected concrete ground is as follow:

$$v = 3.55 \left(\frac{R}{(MgH/\sigma)^{1/3}} \right)^{-1.13} \quad (1)$$

where v——the ground vibration velocity, cm/s;
M——the quality of falling viaduct block, kg for this project, select the quality of single stride;
g——the acceleration of gravity, m/s^2;
H——the viaduct block height, m;
σ——the failure strength of ground medium, MPa, generally 10 MPa;
R——the distance between measuring point and ground center by impact, m.

(2) Monitoring result analysis of dynamic earth pressure: Because of the protective cushion, the courses of viaduct block dropping to the ground are divided into two stages, one is called "impacting the cushion" and the other "finally dropping to the ground". Therefore, the max peak value splits into two peak values which drops the max single peak value. The peak value of earth pressure reaches 53.66 kPa (7.781psi), the moment viaduct block impacts productive cushion while that comes out at 41.40 kPa (6.003psi) when the block finally drops to the ground. However, in 1:1 model test, the measuring value of earth pressure appears 111.6 kPa (16.182 psi) in the same depth, namely, the max peak value approximates 48% of model test and the impact drops by about 52%. So it's obvious that the protective cushion has a great effect.

(3) Monitoring result analysis of air shockwave and noise: Monitoring result analysis of air shockwave and noise proves that 1) Beyond 29m (95.149ft.)away from viaduct, the peak value of air shockwave reaches 496 Pa(0.07192psi), much less than the permitted value for safety of human body and constructions, 2000 Pa (0.29psi). 2) The max noise in constructions reaches142.4dB, although which exceeds the permitted value of 120dB, the persons has withdrew from the constructions to security zone already. Therefore, blasting does no harm to persons and constructions.

(4) Monitoring results analysis of dust: The monitoring result analysis of dust indicates that:1）In form of puff, dust keeps diffusing and diluting. Within 0 to 30minutes after blasting, the dust content reaches the max and the nearer they close to the spot, the more dust content grows, which indicates that at this time, the puff center approximates the blast spot. 2）Within 30 to 90minutes after blasting, the dust content dramatically drops. Meanwhile, within the range of 120 m(458.52ft) around the blast spot, the nearer they close to the spot at a distance of 120 m(458.52ft), the more dust content grows, which illuminates that during this time, the puff center transfers approaching to measuring point of 120 m(458.52ft). 3）Within 90 to 150minutes after blasting, the dust content keeps almost the same within the range of 120m around the blast spot, but by comparison to the background value which is considered high. At this moment, the whole measuring zones are covered by dust and puff. Basically, the fluctuation scales of dust is as the same as that of puff. Thus, the puff keeps diluting rather than transfers and expands.

9 BLASTING EFFECT

The viaduct initiated blasting punctually at 22'o clock on 18 May, 2013. The viaduct tumbled longitudinally rows and rows like failing dominoes within 24.77 s. In less than 30min after blasting, the safety of constructions around and running status of various municipal pipelines were checked carefully by technical personnel and the proprietors, no destruction was caused at all. The blasting proves successful and the effect is well. The blasting process and effect of viaduct are illustrated in Fig.18 and Fig.19.

Fig.18 Blasting course of viaduct

Fig.19 Blasting effect of viaduct

10 CONCLUSIONS

(1) It is proved that demolition blasting technique used for one-time demolishing viaduct over 3.0km (1.864 mile) long in city is feasible, credible and economical.

(2) The relay shock-conducting tube initiation system was used in this demolition successfully. The total delay time of south part reaches 24.77 s. It is, therefore, shown that this initiating method can satisfy the demand for safety and accurate by demolition blasting using ultra-long-delay initiating circuit in city.

(3) The measures taken to protect underground pipelines, shield off flying debris and control dust are reasonable, which is able to meet safety and environment protection requirements proposed by urban blasting.

(4) Physical model test, numerical simulation and blasting monitoring, all of which are regarded as systematic methods to solve the difficulties faced with blasting demolition of urban viaduct in complicated environment, are effective in ensuring safety. All the same, the fundamental theories involved need to be studied even further and deep yet.

REFERENCES

[1] XIE Xianqi. Precision Demolition Blasting of Urban Viaduct [M]. Beijing: Science Press Ltd., 2013(*In Chinese*).

[2] XIE Xianqi. Demolition Blasting Numerical Simulation and Application [M]. Wuhan: Hubei Science & Technology Press, 2008(*In Chinese*).

[3] JIA Yongsheng, XIE Xianqi, LI Xinyu. Numerical Simulation of Structure Control Blasting Demolition[J]. Rock and Soil Mechanics, 2008, 29(1): 285~288 (*In Chinese*).

[4] ZHOU Jiahan, YANG Renguang, PANG Weitai, et al. Ground vibration caused by structural components collapse and impact in building blasting demolition [C]// Proceedings in Rock and Soil Blasting(2), Beijing: Metallurgical Industry Press, 1984 (*In Chinese*).

Water Pressure Blasting in Demolition of Domestic Architecture Analysis and Practice

YAO Jinjie, XIONG Yazhou, ZHAO Rundong, LUO Zhenhua

(*College of Hydraulic & Environmental Engineering of China Three Gorges University, Yichang, Hubei, China*)

ABSTRACT: The article introduces the basic principle of water pressure blasting, the characteristics and main advantages, analyses the method of water pressure blasting in demolition of houses building. Through a example of three layers houses' demolished engineering, this paper introduces the hydraulic pressure in the early period of the demolition blasting arrangement, blasting network, the formula of explosive package quantity and safety protection.

KEYWORDS: waterpressure blasting; domestic architecture; cartridgearrangement; dose calculation; safety protection

1 INTRODUTION

Filled with water in the container shape structure, detonating charge of hanging a certain position in the water, and water is used as the intermediate medium, burst pressure and to destroy structures, and the blasting vibration and flying rock and noise generated in the harmful effect such as effective control of the construction method, known as the water pressure blasting.The water shock wave on the surface reflection and detonation gas rushed out of the water splash form a blast of water spray.Spray water when it dropped the water gravity waves to spread around, causing wave effect on the surface of the water meter.Water pressure blasting is a new blasting technology, is not widely used at present.But the practice has proved, whether from the blasting effect, utilization rate of explosives, or from a security perspective, the water pressure blasting are operable.

2 THE BASIC PRINCIPLE AND MAIN CHARACTERISTICS OF WATER PRESSURE BLASTING

Decorate charge in the water, water is used as the intermediate medium, bubble pulsation, burst pressure, reach the purpose of damaged structures, at the same time and the blasting vibration and flying rock and noise generated in the harmful effect such as to be able to get effective control.Its main characteristics are as follows:

(1) water is used as medium, because of its small compressibility,high transfer efficiency, surface pressure, the use of explosives with high efficiency;

(2) compared with air, water as its medium compressibility, explosive in underwater explosion, the deformation energy consumed by a small water itself, so the water's transfer effect is good. Through the results of numerical calculation compared with the results of acoustic analysis, it can be seen that using water pressure blasting in deep hole charge radius is not less than 80% hole radius (coupling coefficient is less than 1.25), the charge is less than 64% of the coupling charge, charge blast initial shock pressure of not less than the initial shock pressure of hydraulic coupling charge blast.So, the use of deep-hole water pressure blasting, which can be a common coupling charging deep hole blasting save about 30% of the dose;

(3) cushion, shock wave in water evenly located on the blasting medium, the medium makes rupture, but do not produce contact blasting in rock of plastic flow and grinding, not only improve the utilization rate of energy, but also play a buffer action;

(4) water wedge effect: for vessels that have been cracked on high pressure water wedging into the crack, the expansion and extension.And water wedge is much larger than air wedge effect, because water carrying energy is greater than the energy carried by gas.

3 THE CHARACTERISTICS OF DOMESTIC ARCHITECTURE AND THE FEASIBILITY OF ADOPTING WATER PRESSURE BLASTING DEMOLITION

The basic characteristics of domestic architecture is low, remote, and building materials is brick hybrid structure in general.Houses generally for 2～3 layers, is located in a remote rural areas, suitable for blasting demolition, water

---

Corresponding author E-mail: 450856732@qq.com

injection and water will not be too serious, will not have big impact on the surrounding.As building materials is a brick hybrid structure, so only need to destroy the wall on the ground floor, can achieve the purpose of demolished buildings, and save the dosage.

4 A PRACTICAL LAYER 3 HOUSES DEMOLISHED

4.1 Project Summary

Located inYaoZhiHe village of baokang county, which is in Shennongjia of Hubei province, a half two-story houses is to be destroyed which blasting method is adopted.The wall of houses is the brick structure, points first and around two wing, covers an area of 291 m².

4.2 Facilities around the Main Building

Oblique left houses in recent 20 m, on the right side of overhead cable 8.8 m.

4.3 Hydraulic Demolition Method

According to underwater explosion shock wave theory, the structure is the water after the impact, the structure of the outer wall there must be a reflection of shock waves, if the outer wall is air, will reflect a strong tensile wave, if it is other media (e.g., geotechnical), shock waves can be absorbed by the medium in different degrees, the intensity of the reflected wave will be greatly weakened, so any airport in the face of the container structure damage.According to the research proves that no international airport airport surface and has a surface dose difference between 2～2.3 times.So choose interval type decorate water storage charge.

4.4 The Water Depth, Loading Height and the Number of Charge

Indoor water depth of 1.0 m, the total volume is 580 m³.

Charge generally below the surface two-thirds of deep water, medicine bag into the water depth, water depth by 0.6 to 0.7 times the door build by laying bricks or stones after sealing the actual water injection depth of 0.9～1.0 m, so the medicine bag into the water depth of 0.6m.Each cartridge from the corners of wall distance of 0.7m.

Using single cartridge arrangement more, the room layout according to actual condition 4 or 2 medicine package.

4.5 Dose Calculation and Cartridge Layout Diagram

According to the empirical formula, the room charge for medication:

$$W_0 = K_0 \sigma_1^{2.65} R^{1.5} \delta^{1.5} (2+f)^{-2.65}$$

Among them, W_0 is the explosive package quantity required for room, kg; K_0 is the constant, $K_0 = 17.8$; σ_1 is for the tensile strength, MPa; R is the half length of the structures, m; δ is for the wall thickness of the structures,m; f is the international airport for the plane shock wave reflection coefficient.

Room 9 cases, σ_1 =2.1 MPa, δ=0.27, f=0.8, R=1.9.

By calculating W_0 =3.04 kg，take 3 kg.

Tab.1 Room charge medicine inventory

Room	The room of total dose /kg	The room of single dose /kg	Room	The room of total dose /kg	The room of single dose /kg
1	2.1	0.5	8	3	0.75
2	2.8	0.7	9	3	0.75
3	3.8	0.95	10	1.2	0.6
4	1.2	0.6	11	3	0.75
5	3	0.5	12	1.2	0.6
6	3	0.75	13	3	0.75
7	8.3	2.1	14	3	0.75

Cartridge arrangement is as follows:

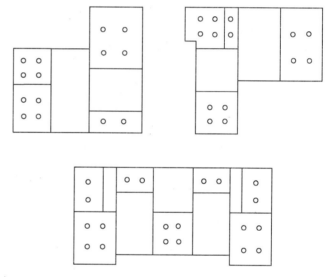

Fig.1 Cartridge arrangement

4.6 Detonating Network

Blasting use howeseconds delay blasting from the inside out, with a period of non-electric millisecond detonators networking, medicine bags with high period of non-electric millisecond delay detonator.The blasting network connections are shown in Fig.2 below.

4.7 Analysis of Blasting Effect

The blasting demolition of houses, the effect is very ideal.After blasting, homes were fully broken, explosive residue in the range of forecast control, nophenomenon of huge fly block.Houses, northwest and southeast direction of overhead cable also do not have any damage, the demolition process safety, high efficiency, no dust phenomenon, to achieve the ideal effect(Fig.3).

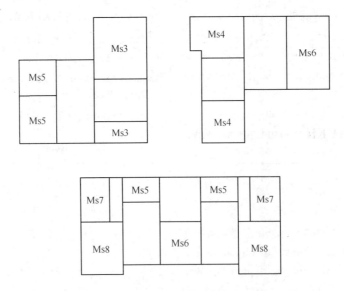

Fig.2 Detonating network diagram

Fig.3

4.8 Construction Problems that Should be Paid Attention to

Because the non-electric detonating network should strictly in accordance with the design of detonating tube detonating network connection, construction should be paid attention to the following questions:

(1) the detonating tube shall not knot, and strive to straight.

(2) the detonating tube shall not be folded in half and prevent misfire.

(3) wall damage do not use.Longer detonating tube shall not be cut off, in order to prevent foreign body into the tube.

(4) shall not pull detonating tube, lead to pipe diameter is fine.

(5) net jobs well qualified through inspection, care, will alert.

(6) connected detonators placed direction: explosive detonator shaped hole should be back the explosion transfer direction of the guide tube, detonating tube should be evenly distributed around the primer detonator, the distance must have about 10 cm.

5 CONCLUSIONS

Import the water pressure blasting demolition technology to houses demolished, simple, efficient, maximum limit to control the production of dust, save manpower and material resources, at the same time, the remote environment more suitable for water and drainage, provides mass of water pressure blasting was introduced successful experience.

REFERENCES

[1] Yao J J. Considering the airport surface water pressure blasting blasting shock wave reflection of the composite structure, 2001: 35~38.

[2] Song P. Explosive air and underwater explosion shock wave overpressure conversion relation, 2008, 31(4).

[3] Yang Y T, Jiang T, Zhan F M. Deep hole hydraulic blasting hole initial pressure numerical calculation, 2006, 23(2).

[4] Lin D N, Liu X C. The development and status quo of water pressure blasting, 1999: 19.

[5] Gu W B, Ye S X, Wang Z L. Underwater explosion shock wave under the action of concrete pier preliminary analysis of dynamic response, 1999, 19(1).

[6] Xiao C, Song P, Liang A D. The research progress of underwater explosion of explosive rule, 2006, 29(6).

Blasting Vibration and Safety

Blast Vibration Impact Analyses for Waste Dump Stability

P.K. Singh, A. Sinha

(*CSIR–Central Institute of Mining & Fuel Research, Dhanbad, India*)

ABSTRACT: The stability of waste dump slope depends on a number of geological and mining factors. In this paper, the data collected to study the impact of production blasts in terms of blast vibrations on waste dump stability has been analyzed. The ground water and geo-technical properties also influence the stability of waste dumps which have also been investigated.

Blast induced vibrations were recorded at various locations at waste dump with the existing blast designs practiced in the mine as well as with modified blast designs. Altogether, 34 blasts were conducted and 123 blast vibration data were recorded at different locations of the waste dump. The total explosive weight detonated in the blast round varied between 561 and 136559 kg whereas explosives weight per delay was 120 to 7800 kg. Maximum vibration recorded at waste dump was 74.4 mm/s with dominant peak frequency of 13.6 Hz. Maximum change of 23 cm was recorded in the dump in continuous slope monitoring of 2 years. The stability analysis shows that with the adopted parameters, the 90 m high internal dump is stable in drained condition. The lower portion of the dump, including the left out coal rib, may have slope angle of 70° to the height of the coal seam. The middle portion of the dump with a height of 38 m can have 40° slope angle. The top portion of the dump (35 m) can have a slope angle of 37° which was considered as angle of repose for the dump material.

KEYWORDS: blast vibration; dump stability; blast design; dump failure; vibration impact

1 INTRODUCTION

Mineral share in Gross Domestic Product (GDP) of India is around 2.1%. The contribution of minerals in the GDP of other mineral producing countries is 4%~6%. The government of India and private sectors are infusing huge capital in this sector. It is expected that there will be three-fold increase in share of contribution from mineral sector in GDP by 2017-18. Globally, coal accounts for 26% of the primary energy consumption whereas it accounts for 56% of India's total energy supply. The Integrated Energy Policy (IEP) of India had reaffirmed dominance of coal in the coming decades [1]. India will need 905 Mt of coal by the financial year 2016-17 to meet the energy demand. Open-cast mining plays a vital role in meeting of this huge demand. Coal production from opencast mines have increased phenomenally and presently about 92% coal is being produced from opencast mining to meet the energy demand of the country. The integrated energy policy of India has also indicated that in years to come the coal will be major contributor in energy supply. This indicates that large volume waste material will be generated from mining operations in opencast coal mines.

One of the major problems in opencast mining is disposal of a large volume of overburden waste material or spoil. The storage of waste materials from mining operations should be done keeping in view the danger to men and machines due to failure of waste dumps[2]. Failure of waste dumps is a serious and complex problem. In addition to environmental considerations, it directly affects the resource recovery, mine safety and mining cost. Thus, the stability of dump is now an important aspect of designing large open-pit mines [3]. The overburden material around the world has been traditionally disposed off in most economic way with little consideration given to technical and environmental aspects [4]. The random disposal of spoil material has resulted into disruption of water courses, stream pollution, slope stability etc. [5]. Rock slopes consist of several types of instabilities, depending on loading conditions, rock mass properties, excavation geometry and topography [6]. After finalizing the location for dump accumulation, it is mainly the design parameters that can be modified as per the existing foundation conditions. Dumps with low height and flatter slopes are ideal from the stability point of view but these not only occupy lot of ground space but also prove to be expensive due to transportation and other handling costs involved. Hence, a benchmark for an allowable dump height, slope angle and number of benches for a long term stable dump accumulation of the material type is necessary to obtain and should also be most economical. This not only provides with safe and stable conditions of mining but also minimizes the ill effects that a slope failure may have on the surrounding ecosystem [7, 8]. Different types of the toppling failure are still a major instability for slopes all

Corresponding author E-mail: pradeep.cimfr@yahoo.com

over the world. Circular failure is one of the most probable instabilities on slopes with severely crushed rock or soil slopes [6].

It has been estimated that over 70% of the larger mining operations had waste dump failures of some kind or others [9, 10]. Despite the fact that the parameters like resources recovery, mining cast, safety, environment, etc. affects stability of waste dumps. Very little work has been done in this area. Very limited published information is available on spoil dumps stability. The physical and strength properties of spoil, appropriate field and laboratory tests procedure or appropriate methods of stability analysis are not published regularly [11]. Economically, stable outside and backfilled dumps will have to be designed to address on-going issue of dump failure in big opencast mines. The benefits of designing waste will manifest themselves in the stability of the waste under any external or internal unfavourable conditions [12]. Assessment of the primary factors and techniques for evaluating slope stability is a necessary step in the development of suitable procedures for predicting and enhancing spoil stability [13].

Ground vibrations are an inevitable, but undesirable by-product of opencast blasting operations. The vibration energy that travels beyond the zone of rock breakage is wasted and can cause damage to structures (including rock mass and waste dump) and annoyance to the residents in the vicinity of the mines [14~17]. The stability of a rock mass under seismic loading has not received much attention despite the fact that natural rock slopes have failed during earthquakes and heavy blasting [18~20]. The undesirable known side effects of detonation of explosives are vibration, noise/air over-pressure, flyrock, dust and fumes [21,22]. The paper deals with a systematic study conducted to evaluate the impact of blast vibration generated due to big blasting operations on the safety and long term stability of waste dump.

2 EXPERIMENTAL SITE DETAILS

Field study was carried out at Jayant opencast coal mine of Northern Coalfields Limited in India. The area geographically lies between latitude of 24°0' to 24°12' and longitudes 82°30' to 82°45' and comprises Gondwana rocks covering about 312 km^2 of which coal bearing Barakars occupied 225 km^2. There are 3 coal seams viz Purewa top, purewa bottom and Turra coal seams. The total mineable coal reserves are 348.93 Mt and volume of the overburden is 907.20 million cubic metres. The average stripping ratio is 2.6 m^3 of overburden per tonne of coal. The coal production from Jayant opencast mine in financial year 2013-14 was 11.7 Mt and overburden removed was 36 Mm^3. The targeted production of coal is 12 Mt per annum. The direction of strike is towards E-W with broad swings. The main working seam is Turra seam. The coal is non-coking type. The dip of the coal seam is 1°~3° in northerly direction. The mining is being done with 10 m^3 shovels and 85 t/120 t dumpers. Three 24/88 R and one 15/83 R draglines operates for removal of over burden. The dumpers are being used to dump the overburden over dragline heaps at least two dragline cuts away from the existing ones. The total length and width of the one cut, by dragline, are about 1800 m in length and 80 m in width respectively. It takes about 16 months to complete one cut. Typical overview of the operating benches of the mine and waste dump is shown in Figures 1 and 2 respectively.

The soil water regime is generally influenced by heterogeneity of waste, soil properties, climatic and management factors. The impact of these factors cannot be studied independently of each other. Other properties, such as bulk density, porosity, content of organic matter and hydraulic conductivity, vary in the course of the year, and they have an interdependent effect on the development of water storage and flow [23]. The overburden rocks in the area are mostly medium to coarse grained sandstone, carbonaceous shale and shaley sandstone. The type of material distribution and constituent size are not uniform in waste dump. As a result the assessment of the internal strength of the waste dumps is difficult and often challenging due to inadequate engineering design [24]. Geo-technical study was carried out at the mine to determine the optimum internal dump and its overall slope angle. It was also aimed to know the interplay and effect of the input parameters of slope design on the factor of safety, which tells the importance of the parameter in the slope.

The mine area stands out as a plateau overlooking the northern flange. The southern flange at the foot of the plateau area is 280 m above MSL. The surface RL varies from 300 m in the south-east to 500 m above MSL in north-west portion of the mine. Major part of the area is 400 m above MSL. The southern part of the area-forming escarpment is very steep and traverses by a number of nalas and the tributaries forming deep ravines and gorges thus presenting a rugged topography. The average annual rainfall is 900 mm out of which 95% is during monsoon season from June to September only. There have been failures of dumps in the mine. Study regarding dump management gather momentum after three major dump failures in 1994 (2 failures) and 2008 (1 failure) at Jayant opencast mine.

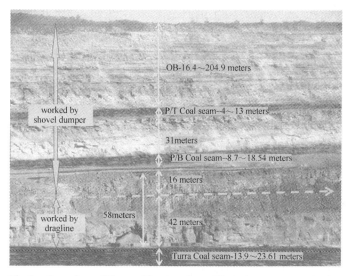

Fig.1 Overview of the working benches of Jayant opencast coal mine

Fig.2 Overview of the waste/spoil dump of Jayant opencast coal mine

3 EXPERIMENTATION AND MONITORING DETAILS

Blast induced vibrations were recorded at the various locations at waste dump of the mine with the existing as well as with modified blast designs. Altogether, 34 blasts were conducted and 123 blast vibration data were recorded at waste dump.

The field trials were conducted at coal, shovel and dragline benches and its drill diameters were of 159, 259/ 269 and 311 mm. The hole depths varied from 4 to 40 m. The burden was in the range of 4 to 10 m. Similarly, spacing varied between 4 and 12.5 m. The number of holes detonated in a blast round varied from 3 to 78 holes. The site mixed emulsion (SME) explosives detonated in a blast round was from 560 to 136559 kg whereas the explosive detonated in a delay was 120 to 7800 kg. The blasts were initiated with detonating cord as well as Nonel initiating system. The delay interval between the holes in a row (17 ms or 25 ms) and between the rows (65 ms to 142 ms) was provided by MS connectors, cord relays and trunk line delays. The distance of the monitoring points at dump from the blasting face varied between 150 and 910 m. The drilled blasting patch and nearest waste dump is shown in Fig.3. The summarised blasting details are given in Tab.1.

Fig.3 Typical drilling pattern at Jayant opencast coal mine

Tab.1 Summarized blast details

Blasting details	Details of data
Number of blasts	34
Number of peak particle velocity data recorded	123
Range of total explosive weight detonated/kg	560~136559
Range of explosive weight per delay detonated/kg	120~7800
Range of blast vibration monitoring distance/m	150~910
Range of recorded peak particle velocity/mm·s^{-1}	1.03~74.4
Range of dominant peak frequency/Hz	2.69~13.6

Slope monitoring observation stations along with base stations were constructed along the coal rib roof and at dragline sitting level of the mine. The monitoring was done with the help of electronic distance meter and laser profiler in overburden dump.

The samples of dump material were collected from different parts and depth of the existing dumps. The specimens were tested. The cohesion, angle of internal friction and bulk density of dump material, interface material, coal rib and floor material of coal rib have been tested and are presented in Tab.2.

Wipjoint and Wipfrag software were also utilised to document the blast face condition before and after blasts. Fig.4 depicts the monitoring of dump movement with Electronic Distance Measuring device (EDM).

Tab.2 Geotechnical parameters for coal mine dump design

Strength Parameters	Dump material	Interface material	Coal rib	Floor of the mine (sandstone)
Cohesion/kPa	74	40	265	230
Angle of internal friction/(°)	25	21	26	29
Bulk density/kg·m^{-3}	1830	2240	1410	2020

3.1 Blast Vibration Monitoring

A dragline blast with 44 holes were detonated with explo-explosives weight of 136559 kg. The maximum explosive weight per delay was 3300 kg. There were 13 rows and the designed duration of the blast was 1383 ms. The blast layout is depicted in Figure 5. The blast resulted with excellent fragmentation (Fig.6) and the recorded blast vibration at 675 m on the dump was only 20.2 mm/s. The recorded blast wave signature is presented in Fig.7. There was minor failure in the coal rib portion (Fig.8). The fragmentation resulted from this blast were analysed. The process involved in analyses of fragmentation is shown in Figures 9 and 10 respectively. The average mean size of the blocks is 0.893 m (diameter of an equivalent sphere) and the most common size of the block is 0.857 m. The maximum size of the boulder is of 1.154 m which was suitable of dragline mucking. Some localized failure were recorded in the waste dump after big blasts. One of the recorded localized failure in dump due to dragline blasting is shown in Fig.11. The maximum vibration recorded in this study was 74.4 mm/s with dominant peak frequency of 13.6 Hz. The vibration recording station was at 350 m on the waste dump from the blasting face. The total explosive weight and explosive weight per delay were 23870 and 4340 kg respectively. The deformation/failure in the dump was recorded (Fig.12) after the above mentioned blast.

The stability analyses show that the factor of safety decreases due to the presence of wet interface material and rainwater at the toe of the dump. In the absence of coal rib, the dump material stands at an angle of repose, which increases the stability of the dump. The small-scale dump failures are common just after the blasting in the coal rib. The percolation of rainwater to lower levels also causes unavoidable small-scale failures.

Fig.4 Monitoring of changes after blasting with the help of EDM at Jayant opencast coal mine

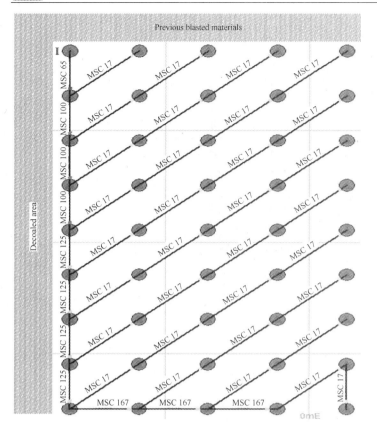

Fig.5 Lay-out of blast conducted at dragline bench at Jayant opencast coal mine

Fig.6 View of the blasted material resulted due to the dragline bench blasting

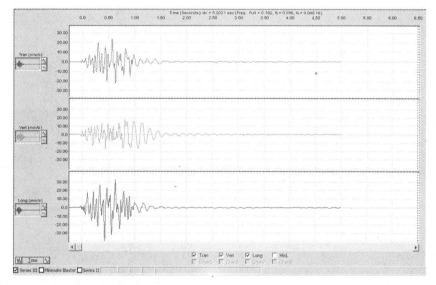

Fig.7 Blast wave signature recorded at 675 m on the dump

Fig.8 Small scale dumps failure near coal rib

Fig.9 Netting and contouring of fragmented rock blocks

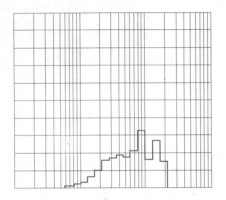 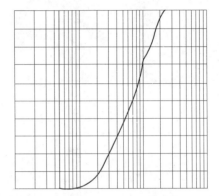

Fig.10 Histogram and cumulative size view of blocks

Fig.11 Deformation occurred in the dump due to dragline blasting

Fig.12 Deformation occurred in the dump due to 3rd & 4th bench blasting

3.2 Waste Dump Slope Monitoring

Disposal of mining waste and waste dump stability becomes challenging when the area available for placement is limited or restricted [25]. Slope monitoring method allows failures to be predicted and safe working conditions to be standardized. Slope monitoring can be used to confirm failure mechanisms. The review of monitoring results, visual inspection and regular briefing of field people help to detect the onset of failure.

The monitoring of waste dump slope to record its movement was performed with the help of total station. The observations were taken on monthly basis. Initially the monitoring stations were installed along the coal rib roof and dragline sitting level. A few slope monitoring observations are presented in the Tab.3. Fig.13 depicts the observed changes in the readings taken at an interval of one month at respective monitoring stations.

The monitoring data confirmed a maximum change of 23 cm due to blasting which was subsequently followed by failure (Fig.12). The observed changes up to 1 cm are related with personal and instrumental errors usually involved with the slope-monitoring job. The accuracy of the instrument was ±5mm.

Tab.3 A few slope monitoring observations recorded at the mine

Instrument locations	Station number	Horizontal distance/m	Changes in horizontal distance/cm	Remarks
CM 4	CM4	52.052	+0.7	Reading were taken at an interval of one month
	6	177.680	−19.5	
	7	182.177	−17.9	
	8	185.930	−22.5	
	9	192.957	−22.8	
CMW 1	CMW2	44.257	−0.4	Reading were taken at an interval of one month
	1	187.144	−1.8	
	2	175.175	−2.6	
	3	169.462	−4.1	
	4	162.384	−1.2	

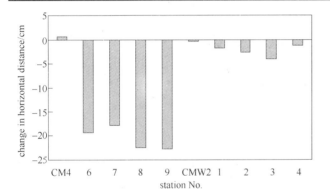

Fig.13 Observed changes for the readings taken at an interval of one month at respective stations

4 RESULTS AND ANALYSES

4.1 Blast Vibration Analyses

The maximum change in waste dump was 23 cm after the 31st blast. This blast was conducted on combined bench of 3rd & 4th. The view of the said bench to be blasted is presented in Fig.14. The joint orientation and in-situ block sizes of the blasting face are shown in Fig.15. The apparent block sizes (diameter of an equivalent sphere) were recorded. The minimum in-situ block size is of 0.215 m whereas the maximum in-situ block size is of 5.995 m. The mean in-situ block size is of 4.099 m. The maximum spacing of the joints is 1.347 m. Fig.16 represents the 3D scanning view of the blasting face just before the blast.

The maximum vibration recorded was 74.4 mm/s with dominant peak frequency of 13.6 Hz. The vibration recording station was at 350 m on the waste dump from the blasting face. In the same blast, vibration recorded at 450 m at the other part of the dump was 31.7 mm/s. The total explosive weight and explosive weight per delay were 23870 and 4340 kg respectively. The blast wave signature recorded at 350 m is depicted in Fig.17. The deformation/failure in the dump was recorded (as shown in Fig.12).

The continued study of two years indicated that maximum changes of only 23 cm were recorded. Although there were local failure due to deep hole blasting in a number of times. The recorded blast vibration data at the different locations in the waste dump were grouped together for statistical analysis. Propagation plot of recorded vibration data on the waste dump with their respective scaled distance is presented in Fig.18. Plot of recorded dominant peak frequency at respective distances are presented in Fig.19.

Fig.14 The view of the Purewa bottom coal and 3rd & 4th combined bench

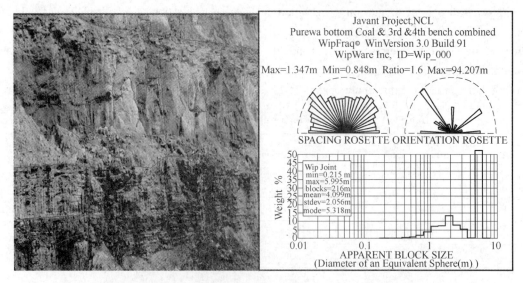

Fig.15 Rock joint analyses output at purewa bottom coal and 3rd & 4th combined bench

Fig.16 View of the 3D-laser scanning of the Dump face just before the blast

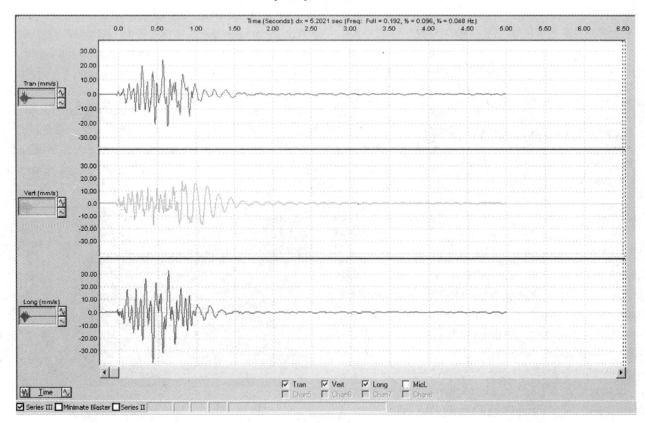

Fig.17 Blast waveform recorded at waste dump due to the blast conducted on 3rd and 4th combined bench

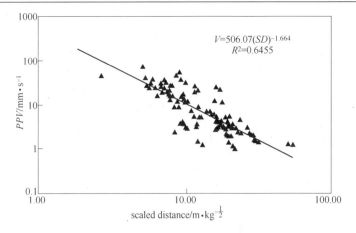

Fig.18　Propagation plots of ground vibration data recorded at various locations on waste dump with their respective scaled distances

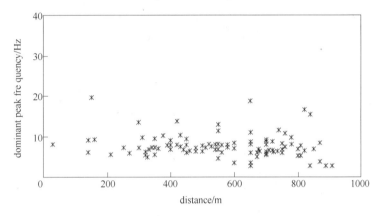

Fig.19　Plot of dominant peak frequencies recorded at various locations on waste dump

4.2　Waste Dump Stability Analyses

The stability analysis was done with the help of GALENA software based on limit equilibrium method with the following considerations:

The overburden rocks in this area are mostly medium to coarse-grained sandstone, carbonaceous shale and shaly sandstone. Though, the spoil dump by dragline stands on shale and shaly sandstone, which provides a competent foundation, the floor of the dragline dump is covered with a thin layer of wet mixture of coal dust, carbonaceous shale and sandstone. The failure will be in dump material only. Hence, the mode of failure in the slopes of the internal dumps was categorized as circular type of failure.

For the dragline dump, 90 m height is considered for the stability analysis based on the following details.

(1) From coal floor to coal roof (coal thickness) =18 m;
(2) Coal roof to dragline sitting level = 38 m;
(3) Dragline sitting level to dump top = 35 m.

The region falls under the earthquake zone-II. So a horizontal seismic coefficient value of 0.02 was taken during slope stability analysis (IS:1893–1975). The ground acceleration generated within the slope mass due to ongoing blast vibration was measured in the field and it is envisaged that the horizontal co-efficient of 0.04 will take account of ground acceleration generated within the slope mass both due to seismicity and blast vibration on slope mass. There is a variation in the seam floor inclination within the mine property from 2° to 4°. Dump floor inclination of 3° is considered here for stability calculation.

The most likely geo-mining condition of the dump was adjudged to be "a 3 m high phreatic surface at the toe of the dump", based on the field observations. It is because of the fact that the rainwater will percolate into the dump in rainy season and all the water may not seep out from dump to its toe drain. However, analysis with 6 m high water table was also done to know its adverse effect of un-drained condition on the stability of the dump slopes. This situation is possible during the extraordinary rainfall when the collection of water within the dump exceeds the seepage of water from the dump.

The cut-off value of safety factor was selected to be 1.2 for the internal dump slope design. The results of the analyses of the stability of dump in different condition are presented in Tab.4. The analyses of dump slope with coal rib in dry condition (factor of safety 1.27) and without coal rib and wet interface material in untrained condition (factor of safety 1.14) are presented in Figures 20 and 21 respectively.

The stability analyses clearly show that the factor of safety decreases due to the presence of wet interface material and rainwater at the toe of the dump [26]. So, dump should be kept in drained condition. In rainy season, the factor of safety decreases in un-drained condition. In the absence of coal rib, the dump material stands at an angle of repose, which increases the stability of the dump.

The small-scale dump failures are common just after the blasting in the coal rib. The percolation of rainwater to lower levels also causes unavoidable small-scale failures. The compaction of the top surface, the climate and the age of the dump characterize rainfall infiltration. The dump will gradually wet up over time, as a function of the climatic conditions, the rate of construction of the dump and the height it achieves, and the nature of the waste rock dump and how it breaks down with time. The wetter the climate, the more rapidly will the dump wet up. The faster the dump is constructed and the greater its' final height, the longer it will take for seepage to break through [27].

The continuous mining operation, blasting and changes in groundwater conditions continuously disturb the existing stress condition in the field. The whole system tries to come into equilibrium by stress redistribution and adjustment, which results into movement of the slope. The impact of blasting in the dump analysis was taken into consideration and accordingly safety factor was calculated.

Tab.4 Stability analyses in different dump slope conditions

Dump slope condition	Factor of safety
Dump slope with coal rib and wet interface material in drained condition	1.20
Dump slope with coal rib in drained condition	1.24
Dump slope with coal rib in dry condition	1.27
Dump slope with coal rib and wet interface material in un-drained condition	1.14
Dump slope without coal rib and wet interface material in dry condition	1.27
Dump slope without coal rib and wet interface material in drained condition	1.22
Dump slope without coal rib and wet interface material in un-drained condition	1.14

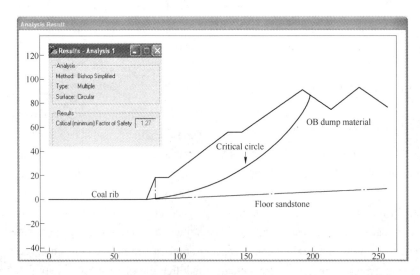

Fig.20   Slope stability analysis of dump slope with coal rib in dry condition

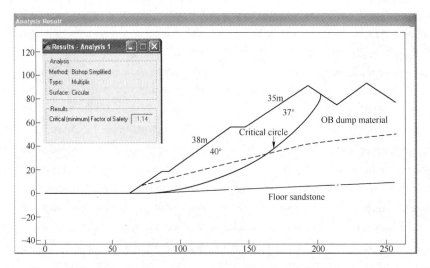

Fig.21   Slope stability analysis of dump slope without coal rib and sludge in un-drained condition

5 CONCLUSIONS AND RECOMMENDATIONS

The dragline blasting at the mines yielded excellent fragmentation and the operational efficiency of the dragline was exceptionally good which indicates that the blast designs finally implemented were optimum for the mine. The continued blast vibration and slope monitoring on the waste dump indicated that the maximum changes of 23 cm were recorded in the dump when the peak particle velocity reached at a level of 74.4 mm/s. The localized failures were recorded due to onset of such vibration levels in the dump. The other dragline bench blast had generated vibration level of 46.3 mm/s which caused minor failure in the waste dump. Although, the explosives detonated in the blasting rounds were more than 50000 kg in most of the blasting rounds. The study concluded that only local dislodgements in the loose materials of the dump were recorded. The impact of dragline blasts were documented on the nearby operating pit-walls too. Fractured rock mass due to repeated blasting got detached from the pit-wall and topping of boulders were documented.

The stability analysis had shown that with the adopted parameters, the 90 m high internal dump is stable in drained condition. The lower portion of the dump, including the left out coal rib, may have slope angle of 70° to the height of the coal seam. The middle portion of the dump with a height of 38 m can have 40° slope angle. The top portion of the dump (35 m) can have a slope angle of 37° which was considered as angle of repose for the dump material. Two corridors of 10 m each should be formed, one at the level of coal rib top and another at the dragline sitting level. The lower and middle portion of the dump will get consolidated under its own weight and have more strength.

It is recommended that during dragline blasts the heavy machinery should be placed at least 15 m away from the operating pit-walls in the close proximity to the dragline blast face. The presence of sump water at the toe of dump may cause high water pressure at the toe of the dump. The sump and drain should be in hard and in-situ natural ground mass. The sump/drain should be at least 5 m away from the toe of the dump to avoid liquefaction of the dump material near the toe due to its continuous contact with water. The water accumulation in the de-coaled floor of the mine should be minimized by ensuring natural gravitational drainage of water towards the main sump. It will prevent the dumping in water to increase the dump slope stability condition. The interface layer i.e. debris of coal dust, fragmented rock, soil mixed with water should be cleared as far as possible from the de-coaled floor before dumping of overburden by draglines.

ACKNOWLEDGEMENTS

The authors express their gratitude to Director, CSIR-Central Institute of Mining & Fuel Research, Dhanbad, India for his encouragement and support during the field study. The authors are also thankful to the officials of Jayant opencast mine for providing necessary facilities during the course of this study. The financial support of Ministry of Coal, Government of India for this study is also thankfully acknowledged.

REFERENCES

[1] Indian Budget, 2009 & 2014.

[2] Turer D. Turer A. A simplified approach for slope stability analysis of uncontrolled waste dumps. *Waste Management and Research*, 2011, 29: 146~156.

[3] Bishop A W. The use of the slip circle in the stability analysis of slopes. Geotechnique 1955, 5:7~17.

[4] Hoek E, Londe P. The design of rock slopes and foundations. Gen. Report for 3rd Cong. ISRM, Denver, 1974:1~40.

[5] Hoek E, Brown ET. Practical estimates of rock mass strength, Int. J. Rock Mech. Min. Sci. 1998, 34(8):1165~1186.

[6] Mohtarami E, Jafari A, Amini M. Stability analysis of slopes against combined circular-toppling failure. Int. J. Rock Mech. Min. Sci.2014, 67:43~56.

[7] Pradhan SP, Vishal V, Singh TN, Singh V K. Optimisation of dump slope geometry vis-à-vis flyash utilisation using numerical simulation. American J. Mining & Metallurgy, 2014, 2(1):1~7.

[8] Kainthola A, Verma D, Gupte SS, Singh TN. A Coal Mine Dump Stability analysis-a case study. Geomaterials, 2011, 1:1~13.

[9] CMPDIL project report for Jayant opencast mine, Singrauli (M.P.), 2007.

[10] Upadhayay OP, Singh DP. Different factors affecting stability of slope in opencast mines. Nat. Symp. Emerging Mining and Ground control Tech., BHU. Varanasi.

[11] Tiwari SN. Waste control in Mines. J. of Indian Min. and Engg, 1990:16~19.

[12] Javier M. Waste design in mining. Prepr. Soc. Min. Metall. Explor, 2011(11-041):1~10.

[13] Bradfield L, Simmons J, Fityus S. Issues related to stability design of very high spoil dumps, 13th Coal Operators' Conference, University of Wollongong, The Australasian Institute of Mining and Metallurgy & Mine Managers Association of Australia, 2013:376~386.

[14] Siskind D E, Stagg M S, Kopp J W, Dowding C H. Structure Response and Damage Produced by airblast from Surface Mine Blasting. U. S. Bureau of Mines, 1980, RI 8485: 111.

[15] Valdivia C, Vega M, Scherpenisse C R, Adamson W R. Vibration simulation method to control stability in the Northeast corner of Escondida Mine. Int. J. of Rock Fragmentation by Blasting, FRAGBLAST, 2003, 7(2):63~78.

[16] Singh PK. Blast vibration damage to underground coal mines from adjacent open-pit blasting. Int. J. of Rock Mech. and Mining Sci. 2002, 39(8):959~973.

[17] Singh P K, Roy MP. Damage to surface structures due to blast vibration. Int. J. of Rock Mech. and Mining Sci. 2010, 47(6):949~961.

[18] Ling HI, Cheng AHD. Rock sliding induced by seismic force. *Int. J. Rock Mech. Min. Sci.*, 1997, 34:1021~1029.

[19] Zhang De, Yan LG, Yuan Xu. Impact analysis of blasting vibration on the slope and dump. *Information tech. J*, 2014 13(4):730~737.

[20] Singh PK, Roy MP, Paswan R K. Controlled blasting for long term stability of pit-walls. Int. J. of Rock Mech. and Mining Sci. 2014, 70:388~399.

[21] Singh PK, Vogt W, Singh RB, Singh D P. Blasting side effects - investigations in an opencast coal mine in India. Int. J. of Surface Mining Reclamation and Environment, The Netherlands, 1996, 10:155~159.

[22] Singh PK, Mohanty B, Roy MP. Low frequency vibrations produced by coal mine blasting and their impact on structures. Int. J. of Blasting and Fragmentation, USA, 2008, 2(1):71~89.

[23] Kuraz V. Soil properties and water regime of reclaimed surface dumps in the North Bohemian brown-coal region - a field study. Waste Management, 2001, 21: 147~151.

[24] Watters R J. Influence of internal water and material properties on mine dump stability. Geological Society of America Abstracts with Programs,2005, 37(7): 394.

[25] Galla V, Zacaria P. Designing, Scheduling and managing waste rock facilities: a case study on the waste rock facilities at the Phoenix mine. NV 2011: 4.

[26] Galena. Slope Stability Analysis Software, Clover Technology, Australia, Version-3.1, copyright, 1982~2006.

[27] William D J. In closing a surface waste rock dump it is not simply a matter of constructing a cover retrospectively. Proc. of the seventh int. conf. on min. closure, 2012: 379~392.

Measurement and Analysis of Environmental Vibration Caused by Blasting Excavation of Foundation Pit in Town

TIAN Yunsheng, ZUO Jinku, LIU Weihua, WANG Ning

(*Shijiazhuang Tiedao University, Shijiazhuang, Hebei, China*)

ABSTRACT: According to two national standard that "Standard of Vibration in Urban Area Environment" (GB 10070—1988) and "Measurement Method of Environmental Vibration of Urban Area" (GB 10071—1988), the environmental vibration testing on blasting excavation of foundation pit in town was carried out. And the change characteristics and the propagation laws of environmental vibration caused by blasting excavation of foundation pit were discussed, through processing and analyzing the testing data and calculating the level of vibration acceleration. Proposed the point that the environmental vibration should be lead into the research on blasting vibration, in order to meet higher demands of the people for environmental safety. This approach can also help to reduce blast vibration disputes, and to ensure the safety and smooth implementation of the blasting excavation of foundation pit in town.

KEYWORDS: foundation pit blasting ; environmental vibration ; vibration level ; vibration testing

1 INTRODUCTION

Blasting vibration is one of the adverse effects that the engineering blasting implementation can't eliminate, it will cause environmental vibration pollution. With the rapid development of urbanization in our country, the circumstances in blasting engineering become more and more complex, combined with the improvement of the citizen's consciousness of security and safeguard rights, higher requirements to the blasting vibration control and safety standards are putted forward[1]. Therefore, the influence of environmental vibration caused by blasting on the living and working environment has attracted general concern.

The current "Rules of Explosion Safety" (GB 6722—2003) just specifies safety control standard about the damages to the buildings(structures) caused by blasting vibration, it doesn't cover for the environmental vibration control standards about people living and working environment. As a result, in the course of many blasting engineering's implementation, nearby residents reaction is very strong, although the blasting vibration velocity is not up to security allowed standard, thus it can cause civil disputes [2]. Therefore, in order to avoid law dispute, it is necessary to add the blasting environmental vibration to the blasting vibration study. In determining the blasting safety control standard, it not only considers the impacts of damages to building (structure), but also more attention to human feelings and psychological endurance and the like humanization indicators.

In the international, vibration has been listed on one of the seven environmental hazards, and studies were carried out on the law of vibration pollution, the cause of emergence, route of transmission, control method, and the harm to human body, etc[3, 4]. "Standard of Vibration in Urban Area Environment" (GB 10070—1988) was issued in 1988 in China, and the "Measurement Method of Environment Vibration of Urban Area" (GB 10071—1988) is the companion standards. The standard stipulates environment vibration limitation according to the different areas.

2 THE DANGER OF ENVIRONMENTAL VIBRATION AND SAFETY EVALUATION METHODS

2.1 Dangers of Environmental Vibration

The influence of environment vibration on man is very complicated, the feeling about vibration is related to someone's age, gender, health status, physical quality and environment conditions, etc. The main influence factors include vibration intensity, vibration frequency, vibration exposure time and vibration direction and so on. Vibration to the person's influence is mainly manifested in three aspects: damaging the human body, disturbing person's life, and reducing work efficiency.

Fund Project: Natural Science Foundation of Hebei Province (E2011210013).
Corresponding author E-mail: tianyunsh@126.com

When environmental vibration reaches a certain intensity, it may cause destruction and damage to the building (structure). Slight vibration may cause against peel, structure crack and seriously, it may cause foundation deformation and even sink and so on. Especially in some areas that earthquake has happened, after withstanding to the earthquake damage, building structures are more likely to be affected by environment vibration.

Environmental vibration can affect the normal using of precision instruments. For example, when you operate laser equipment, electronic microscope, electronic balance and the like precision and ultra-precision instruments, the vibration can cause these instruments and equipment in signal-to-noise ratio decrease, inaccurate data, lower repetitive, diminution of accuracy and even can't work normally, etc[5].

2.2 Environmental Vibration Safety Evaluation Methods

The influence of environmental vibration to the surrounding buildings, residents or precision instruments can be evaluated by acceleration or velocity, but generally not by displacement. And acceleration index is commonly used in inland.

According to Standard of Vibration in Urban Area Environment[6], the formula of acceleration vibration level is defined like this:

$$VAL = 20\lg\left(\frac{a'_{rms}}{a_{ref}}\right) \quad (1)$$

where a'_{rms} —— effective value of frequency weighted acceleration, m/s^2;

a_{ref} —— reference acceleration, $a_{ref} = 10^{-6}$ m/s^2.

The effective value of frequency weighted acceleration a'_{rms}, calculated as follows:

$$a'_{rms} = \sqrt{\sum a_{rms}^2 \times 10^{0.1C_f}} \quad (2)$$

where a_{rms} —— the effective value of acceleration in the ith center frequency, m/s^2;

C_f —— the correction value of different frequency-weighted in whole body vibration regulated by ISO2631, dB.

The effective value of acceleration in the i th center frequency a_{rms} can be calculated as follows:

$$a_{rms} = \sqrt{\frac{1}{T}\int_0^T a^2(t)dt} \quad (3)$$

where T —— integration time, s;

$a(t)$ —— the acceleration value in a certain time, m/s^2.

In order to simplify the calculation of $a(t)$, we can simplify the blasting vibration as pulse excitation damping vibration. The wave form of pulse excitation damping vibration acceleration in time domain shown as Fig.1.

In the figure, A_0 is the maximum peak of the whole vibration process, $A_1, A_2, ..., A_i$ is the 1th, 2th, ..., ith peaks adjacent to A_0, T is the vibration period:

$$T = 1/f \quad (4)$$

In the formula, f represents the inherent frequency of the vibration system, Hz.

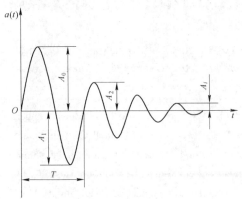

Fig.1 The wave form of blasting vibration acceleration

Supposing the relative damping coefficient in vibration system is ψ, then:

$$\frac{A_0}{A_i} = e^{\frac{\pi\psi}{\sqrt{1-\psi^2}}\cdot i} \quad (5)$$

when $\frac{A_0}{A_i} \geqslant 5$, the peak value of the i th peak decreased 14 dB or less to the maximum peak, it can be considered that, including the i th and the subsequent peaks they do relatively smaller effects on human health, and the effects can be ignored. In this time, the time of vibration which is not negligible defines as the effective exposure time:

$$T_2 = \frac{T}{2}\cdot i \quad (6)$$

The vibration before the i th peak can be processes as the plus sinusoidal vibration which have the same fundamental frequency, the same maximum peak and the same exposure time, it expresses as follows:

$$a(t) = A_0 \sin 2\pi f t \quad (7)$$

where $0 < t \leqslant T_2$.

By this way, the evaluation processes of the influence caused by blasting environmental vibration can be simplified[7].

3 ANALYSIS ON THE TEST RESULTS OF EXCAVATION BLASTIOG IN TOWN

3.1 Blasting Engineering of Foundation Pit Profile

Xichenghuafu Project in Pingshan county is located on the south side of West Yehe road, In this project, the No.2 and No.3 building foundation pit should be blasted, and the pit in the flat land. After excavating by excavator, spun off about 2 m topsoil and soil rock were stripped off. And primary foundation pit were formed. In the part to be excavated, the elevation is −2∼−7 m, deepness is 4∼5 m, foundation pit is 90 m long, 25 m wide, totally blasting volume is about 22000 m³. The rock to be blasted is limestone, developed joint fissure, rock consistence coefficient $f = 8∼12$.

According to the engineering condition and the surrounding environment, determined the overall blasting design scheme is horizontal layer-step loose controlled blasting. The parameters of bench blasting identified as: aperture $\phi=42$ mm; hole pitch $a=1.2$ m; row pitch $b=1.0$ m; explosive consumption $q=0.32$ kg/m³; the first bench height $H_1=2.3$ m, hole depth $L_1=2.5$ m, the charge in single hole $Q_1=0.88$ kg; the second bench height $H_2=2.7$ m, hole depth $L_2=3.0$ m, the charge in single hole $Q_2=1.04$ kg. Adopted duplicate non-electricity nonel tube network, delayed in the hole.

The surrounding environment is depicted on Fig.2.

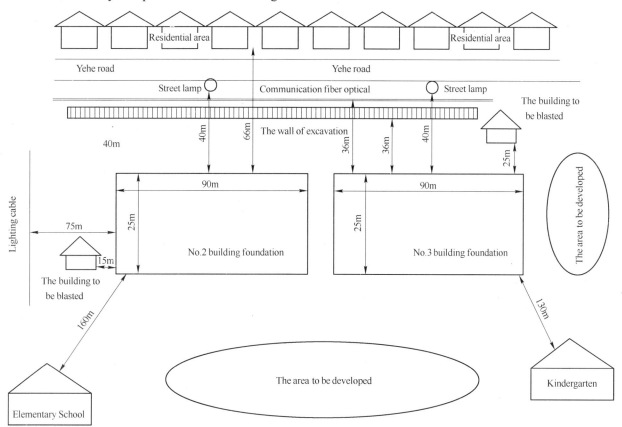

Fig.2 Sketch map of the surrounding environment

3.2 Monitoring Instruments and Measuring Point Layout

The blasting vibration monitoring instrument is TC - 4850 type of blasting vibration measurer which provided by Chengdu Zhongke Monitoring and Control Ltd. In order to accurately determine the influence and the propagation laws of environmental vibration caused by blasting excavation, placed the measuring points in 10 m, 20 m, 30 m, 50 m, 80 m, 100 m and 120 m point; tested the change rule of acceleration vibration level when the weight of dose is 24 kg, 36 kg, 48 kg.

3.3 Analysis of Test Result

3.3.1 The Acceleration Vibration Level Attenuates with Distance

The Fig.3 shows that: in the different dosage, if the section dose are more, the acceleration vibration level is higher; in the same dosage, the acceleration vibration level will reduce gradually as the distance from the monitoring points to the explosion center increases, and the vibration level within 50 meters from the monitoring points attenuates faster than the reverse.

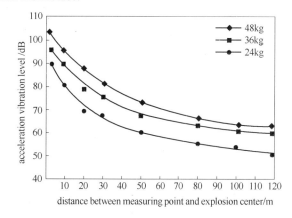

Fig.3 The attenuation curve of vibration level velocity with distance

3.3.2 The Change of Vibration Level with the Distribution of Dosage

Tested the rule of vertical vibration velocity and the vibration acceleration changed with the section dosa in three different distance : 30 m, 50 m and 80 m. By the formula of the vibration acceleration level, the relevant value of the vibration level can be got. The distribution curve of vibration level changed with dosage as shown in Fig.4.

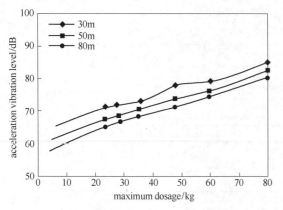

Fig.4 Distribution curve of vibration level changed with dosage

3.3.3 The Change of Vibration Level with the Distribution of Frequency

The Fig.5 shows the distribution of the vibration level in the 1/3 octave center from different measuring point. As shown in the Fig.5, the vibration component which frequency is less than 20 Hz do less attenuation with distance change, and the times distance attenuation is about 3 dB; the vibration component whose frequency is within 20~40 Hz attenuates slightly bigger and the times distance attenuation is about 7~9 dB; the vibration component whose frequency is within 40~60 Hz attenuates bigger and the times distance attenuation is about 12~14 dB; the vibration component whose frequency is within 60~80 Hz decreases even bigger and the attenuation is about 25~28 dB[8].

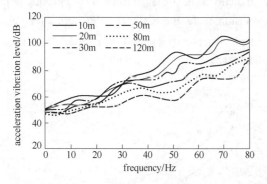

Fig.5 Distribution curves of vibration level in different points changed with dosage

4 CONCLUSIONS

(1) The environmental vibration problems caused by urban foundation pit blasting have become the key obstacle whether the blasting can be successfully implemented. Putting the blasting vibration into the environmental analysis, designing and controlling the blasting according the environmental vibration standard are promising methods to resolve disputes over vibration blasting and blast successfully.

(2) The tests and the results show that vibration level will gradually decrease with the increase of distance between measuring point and explosion center at the same doses; the vibration level will increase as the dose increases for the same point; the vibration component of high frequency whose attenuation of vibration level is faster, the vibration component of low frequency whose attenuation of vibration level is slower.

(3) Based on processing and analyzing the testing data, the conclusion acquired by the calculation rules of the vibration level is consistent with the result of blasting vibration research.

REFERENCES

[1] Zhang Z Y, Yang N H, Lu W B, etc. Progress of blasting vibration control technology in China [J]. *Blasting,* 2013, 30(2):25~32.

[2] Yan Y F, Chen S H, Zhang Q H, etc. Evaluation method and application of blasting vibration comfortableness[J]. *Engineering Blasting,* 2012, 18(1): 78~81.

[3] Xia H, Wu X, Yu D M. Environmental vibration induced by urban rail transit system [J]. Journal of Northern Jiaotong University, 1999, 23(4): 1~7.

[4] He H X, Yan W M, Zhang A L, etc. Human-structure dynamic interaction and comfort evaluation in vertical ambient vibration [J]. *Journal of Vibration Engineering,* 2008, 21(5): 447~451.

[5] Du Y Y. Vibration characteristic of human body and comfort evaluation underambient vibration[D]. Beijing: Beijing University of Technology, 2012.

[6] PRC National Standard GB10070. Standard of Vibration in Urban Area Environment[S]. 1989.

[7] Gao S X, Yu Y S, Jiang Y L, etc. Using ISO 2631 to evaluate human health in vibration environmeut excited by dulses[J]. Journal of Vibration and Shock, 1997, 16(2): 72~76.

[8] Yan W M, Nie H, Ren M, etc. In situ experiment and analysis of environmental vibration induced by urban subway transit [J]. Earthquake Engineering and Engineering Vibration, 2006, 26(4): 187~191.

State of the Art Review of R&D on the Blast Vibration at KIGAM

Chang-Ha Ryu, Byung-Hee Choi, Hyung-Su Jang, Myoung-Soo Kang

(*Geologic Environment Division, Korea Institute of Geoscience & Mineral Resources* (*KIGAM*)
124 Gwahang-no, Yuseong-gu, Daejeon 305-350, Korea)

ABSTRACT: Explosive blasting is very useful tool used for breaking rock in mining, quarrying and civil engineering constructions. However, the use of explosives always produces undesirable environmental impact due to the ground vibration, air-overpressure and flyrocks. Blast vibration may cause damage to adjacent structures, and has been a serious environmental issue. A series of R&D works on the blasting area have been done at Korea Institute of Geoscience & Mineral Resources(KIGAM) for past 20 years. A new method was developed to predict the ground vibration for the evaluation of safety of blast design in its relation to the generation of ground motion. It is based on the non-parametric source identification method, and provides the information on the history of ground motion as well as peak level of vibration. Parameter study was performed on time-domain using some commercial software like LS-DYNA, MIDAS, etc. Sensitivity analysis was carried out to set up the guidelines for the preparation of the basic input parameters in numerical modeling of blasting problems. In order to establish the National Standard for safe blasting work, allowable ground vibration level was suggested. Extensive field measurements have been done at various mining and construction sites to investigate the characteristics of blast-induced ground motion and structural response. In order to control the ground vibration more effectively, a new fracturing device was developed, which is applicable to the site where ground vibration has come to the front. Software for blast design was developed too. This paper summarizes the results of what have been done at KIGAM, including some current R&D work.

KEYWORDS: blast vibration; prediction; safe criterion; structural response; R&D work

1 INTRODUCTION

Blast-induced ground vibration, the so-called blast vibration, may cause an environmental impact such as neighbor's complaints or damage on adjacent structures and facilities. Complaints associated with blasting have often become a target of public grievances. Such problems have come to the fore as a social issue in Korea since 1980's in the midst of subway construction at the capital city, Seoul. The key to solving the problems involves finding a safety level of the target structure against the blast vibration, predicting the ground vibration induced from the specific blasting work and how to control the blasting so as not to exceed the allowable level. A series of R&D works on explosive blasting techniques has been performed at Korea Institute of Geoscience & Mineral Resources(KIGAM) for about 20 years. Some of the results are summarized in this paper.

2 SAFETY LIMIT OF VIBRATION LEVEL

2.1 Allowable Level

A safety limit of ground vibration in Korea is provided by the Environmental law only for the pleasant life of residents as shown in Tab.1. A vibration level which is compensated for human response is used as an indicator. The law is applied to the ground vibration generated from the general life works including the civil and construction work, plant operation, mining work, etc. The vibration levels in Tab.1 is those for blasting work, which are 10 dB bigger than those for other general works, by considering the transient characteristics of blasting and working time.

Tab.1 A safety limit of ground vibration by Korea Environmental law

Area	Time zone (06:00~22:00)
Residential area, green belt area, recreational area, environmental	Less than 75 dB(V)[①]
Preservation area, school zone, area of hospital and public library other area	Less than 80 dB(V)

① dB(V): Vibration level in vertical direction.

dB(V)= 20 log (A/A_o), dB.

$A_o = 2 \times 10^{-5} \times f^{-1/2}$ for $1\,Hz \leqslant f \leqslant 4\,Hz$; $A_o = 10^{-5}$ for $4\,Hz \leqslant f \leqslant 8\,Hz$; $A_o = 0.125 \times 10^{-5} f$ for $8\,Hz \leqslant f \leqslant 90\,Hz$.

where A——root mean square value of acceleration, m/s²;

A_o——reference acceleration;

f——frequency.

Corresponding author E-mail: cryu@kigam.re.kr

Tab.2 shows a safety limit for structures suggested in early 1980's for urbane subway construction works in densely populated capital city. It had been used as a guideline of allowable ground vibration level for blasting works for quite a long time in Korea.

Tab.2 Allowable vibration level suggested for urbane subway construction works

Structure type	Allowable limit/mm · s^{-1}
Antique, computer facility	2
Residential structure, apartment building	5
Commercial building	10
Factory, reinforced concrete building	10~40

The former U.S. Bureau of Mines recommended a safe particle velocity maximum of 5 cm/s for residential structures for frequencies above 40 Hz (Siskind et al., 1980). The regulation of the Office of Surface Mine specifies that the maximum ground vibration shall not exceed 0.75~1.25 in/s (1.9~3.2 cm/s) at the location of any dwelling, depending on the distance from blasting site (30 CFR part 175). Most US States adopt the blast vibration level of 1 ~1.2 cm/s as an allowable level for residential structure. The German vibration standard, which is known to be very conservative, gives varying levels of 0.5~2 cm/s for residential structures depending on the frequencies (DIN 4150, 1986). Some other countries have their own National Standard such as Swiss Norm 640 312a, British Standard BS7385, Australian Standard AS 2187.2, China GB6722—2003, etc. A variety of National vibration limits are listed elsewhere (Skipp, 1977).

2.2 Frequency Consideration

Peak particle velocity criteria have often been widely used in blast design. There is a growing tendency to adopt vibration criteria varying with respect to the frequency of the vibrations. We have tried to prepare new criteria considering the frequency of vibration, but have difficulties in establishing the lower and upper limit of frequency and vibration level. The frequency boundaries in each National Standards are different, and technical data or rational grounds for establishing the limits are hardly provided. In order to understand the frequency effects, simple structure models were analyzed under blasting load by using a step-by-step procedure. The first model is four-story frame without in-filled panels. The second one is that with in-filled panels. Fig.1 shows the numerical examples. The summary of the results are shown in Tab.3. Calculations of structural response show that the response depends largely on the frequency and type of structures and that principal frequency might be the critical factor in safety evaluation under impact loading. In this example, the responses are amplified to a considerable level of underground motions when the principal frequencies are 31.3 Hz and 6.25 Hz. The longer duration time may also give harmful effects. A new allowable level of ground vibration is suggested for mining work as shown in Tab.4, of which level is less conservative than the former guideline as shown in Tab.2 (Choi et al, 2011). It is still a question if the suggested threshold limit of vibration level and associated frequencies are optimum. Efforts have been made to accumulate the field data for structural response subject to blasting load, expecting to establish more rational standard for domestic structures in the near future.

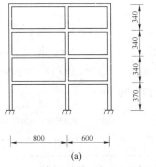
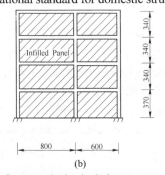

Fig.1 Simple structure models for numerical calculation

(a) void frame model; (b) in-filled panel frame model

Tab.3 Results of numerical calculation of structure response

Principal frequency/Hz	Structure type	Maximum floor velocity(amplification ratio)/cm · s^{-1}			
		2nd floor	3rd floor	4th floor	roof
62.5	In-fill	1.52 (0.65)	1.19 (0.51)	1.7 (0.72)	1.54 (0.66)
62.5	void	1.7 (0.72)	1.7 (0.72)	1.7 (0.72)	1.78 (0.76)
31.3	In-fill	3.1 (1.32)	7.76 (3.3)	13.7(5.83)	20.3 (8.64)
31.3	void	11.9 (5.06)	11.2 (4.77)	9.17 (3.9)	9.17 (3.9)
6.25	In-fill	2.65 (1.13)	3.23 (1.37)	3.86 (1.64)	4.76 (2.03)
6.25	void	1.89 (0.9)	1.99 (0.85)	1.39 (0.59)	1.56 (1.34)

Tab.4 Suggested limit of ground vibration

Structure type	Peak particle velocity/mm·s⁻¹		
	< 20 Hz	20~50 Hz	> 50 Hz
Industrial building, reinforced concrete structures	20	25	40
Residential building, masonry structures	10	15	20

3 PREDICTION OF BLAST-INDUCED GROUND VIBRATION

Ground vibrations associated with blasting in rock are attenuated with distance from source. Peak amplitudes are usually expressed as particle velocities, and distances can be scaled. In practical use, peak particle velocity can be plotted as a function of scaled distance of which concept is dividing the distances by a scaling factor of the charge weight. The most general form used for the prediction of ground vibrations is given by:

$$PPV = K (D/W^b)^n \quad (1)$$

where, PPV is the peak particle velocity in cm/s or mm/s; W is the charge weight per delay in kg; D is the distance from a blast source in m; The constants K, n and b are empirical and site specific.

Although peak particle velocity has been widely used to quantify the damage potential of a vibration, velocity itself is not sufficient to evaluate structural damage. There is a tendency that at close in distances from a blast, high frequencies predominate the vibration record and that low frequencies do far from a blast. The problem is that plug-in type prediction model such as equation (1) cannot define this decrease in predominant frequency with increasing distance. A new technique was developed to predict the information on the frequency characteristics of ground motion as well as peak particle velocity (Ryu, 2002). It is based on Source Identification Technique. Briefly introducing the algorithm, relationship between input source and response in a linear system where principles of superposition is applied can be expressed as follows:

$$U(i\omega) = H(i\omega) P(i\omega) \quad (2)$$

where, $U(i\omega)$ and $P(i\omega)$ are complex Fourier spectra of response; $U(t)$, at a point and input motion $P(t)$, respectively; $H(i\omega)$ is transfer function defining the relationship between input and response; ω is frequency; and i is $\sqrt{-1}$. Because the equation is composed of frequency dependent three complex functions, one of the functions can be easily determined if the other two functions are given. When $U(i\omega)$ and $H(i\omega)$ are given, source function, $P(i\omega)$, is calculated as follows:

$$H(i\omega) = U(i\omega)/P(i\omega) \rightarrow P(i\omega) = U(i\omega)/H(i\omega) \quad (3)$$

Numerical analyses coupled with non-parametric source identification method were performed to predict the ground vibration (Ryu, 2002). Fig.2 shows the selected results of calculated ground responses and field measurements. The frequency spectrum of the vertical ground motion at 60 m from the blast source shows only about 5 Hz difference in peak frequency. Fig.3 shows the velocity history at 60 m. One of the problems involved in this method is that the blast source estimated from equation (3) has no physical meaning when compared to the real blast source.

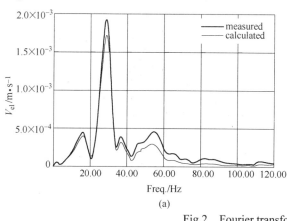
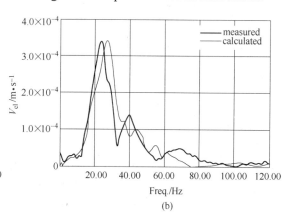

Fig.2 Fourier transform of velocity history

(a) horizontal ground motion at 20 m; (b) vertical ground motion at 60 m

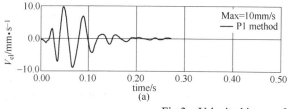
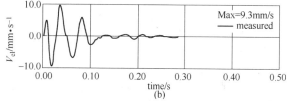

Fig.3 Velocity history of vertical ground motion at 60 m

(a) velocity history of vertical ground motion at 60 m, calculated;

(b) velocity history of horizontal ground motion at 60 m, measured

Another approach is to simulate the ground motion under blast loading numerically using an appropriate software based on FEM, FEM or DEM. Numerical analysis has been performed to model the blast-induced ground vibration and to figure out the significance of the input parameters and control variables of the software (Ryu et al, 2013). The maximum blasting load on the borehole wall is calculated by using the equations (4) and (5) suggested by NHI (National Highway Institute, 1991). The loading history is assumed to be equation (6) suggested by Starfield et al. (Starfield & Pugliese, 1968). It is assumed to excavate a tunnel with 10 m in diameter and 200 m in length. Fig.4 shows the dynamic pressure history of the blasting load applied to the boundary wall. Fig.5 shows a calculated particle velocity history of ground surface, located at 50m(h) and 100m(v) from underground blast source. Fig.6(a) and Fig.6(b) show the displacement distribution 20 ms and 30 ms after detonation, respectively.

$$P_{\text{det}} = 4.18 \times 10^{-7} S_{\text{ge}} V_e^2/(1+0.8 S_{\text{ge}}) \quad (4)$$
$$P_B = P_{\text{det}}(d_c/d_h)^3 \quad (5)$$
$$P_D(t) = 4 P_B[\exp(-Bt/\sqrt{-2})-\exp(-\sqrt{-2}\, Bt)] \quad (6)$$

where P_{det}——detonation pressure, kbar;
P_B——borehole wall pressure (Decoupled detonation pressure), kbar;
V_e——detonation velocity, ft/s;
d_c——charge diameter, mm;
d_h——borehole diameter, mm;
S_{ge}——specific gravity of the explosive, g/cm³.

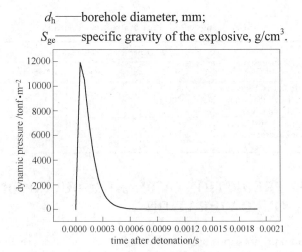

Fig.4 Pressure history of blasting load

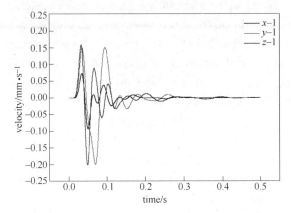

Fig.5 Particle velocity history at ground surface

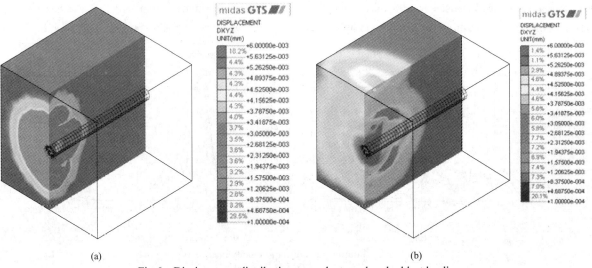

(a)   (b)

Fig.6 Displacement distribution around a tunnel under blast loading

(a) 20 ms after detonation; (b) 30 ms after detonation

This kind of approach is very useful to predict the blast vibration without test blasting in design stage, and to confirm the safety of the proposed blast design. There are, however, some difficulties in the numerical analysis. The dynamic ground motion is very sensitive to the input parameters such as modeling of blast source, the borehole pressure, dynamic properties of rock and constitutive model for dynamic stress-strain relationship, etc. It is not easy to prepare the input data due to lacks of information on the parameters. More quantitative phenomena on the blasting source and dynamic response of rock mass still remain to be solved. Some R&D work is going on at KIGAM. Fig.7(a) and (b) show the numerical calculations of the FEM based LS-DYNA and DEM based PFC model, which is performed to simulate the results of 3 hole blasting experiments (Choi et al, 2014). The main purposes of the experiments and numerical analysis are to study the propriety of the blast source modeling and to improve the modeling technique for predicting blast-induced damage and fracturing phenomena.

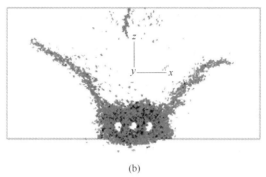

(a)　　　　　　　　　　　(b)

Fig.7　Numerical modeling of 3 hole blasting experiment

(a) FEM based LS-DYNA; (b) DEM based PFC3D

4　CONTROL OF BLAST VIBRATION

When the allowable level of ground vibration is given and the propagation characteristics are predicted, blast engineer should provide an optimum blasting method to control the ground vibration under the allowable level. Most general method is to control the maximum charge weight per delay which can be calculated from the prediction model. It is also very important to provide an optimum specific charge for effective breaking rock mass. A new algorithm was developed for design of tunnel blasting (Choi et al, 2010). The proposed algorithm yields a drilling and charging pattern with a proper specific charge and required maximum charge limit. A plasma blasting technique was developed as an alternative to the explosive blasting so as to be applied to a special case where the use of explosives are restricted due to its adverse effects (Lee et al, 1997; Synn et al, 1999). It utilizes the plasma energy transformed from the electrolyte solution. The peak level of ground vibration induced by plasma blasting decreases more rapidly and associates higher frequency which is more favorable to the structural damage potential.

5　SUMMARY AND CONCLUSIONS

Blast induced ground vibration may cause damage to adjacent structures, and has been a serious environmental issue. A series of R&D works have been carried out at Korea Institute of Geoscience & Mineral Resources(KIGAM) for past 20 years. This paper summarizes some R & D results at KIGAM and efforts in the area of blast vibration. Efforts are in progress to establish a more rational National Standard for safe blasting work; to improve numerical modeling technique; and to provide the useful software and hardware tools for blast design and performance control.

ACKNOWLEDGEMENT

This research was supported by the Basic Research Project of the Korea Institute of Geoscience and Mineral Resources (KIGAM) funded by the Ministry of Trade, Industry and Energy (MOTIE) of Korean Government.

REFERENCES

[1] Choi B H, Kang M S, Ryu C H, Lee I J, Kim J W. 2014. Rock plate blasting experiments and PFC modeling, *Proceedings of the 2014 Spring Symposium, Korean Society of Explosives and Blasting Engineering*: 7～12 (in Korean) .

[2] Choi B H, Ryu C H, Jeong J H. 2010. Development of a Designing Program for Underground Blasting, *Proceedings of The 5th International Conference on Explosives and Blasting*, The Japan- China- Korea Technical Committee of Explosives and Blasting, Sapporo, Japan: 22～27.

[3] Choi B H, Ryu C H, Jeong J H 2011. A Study on the Blast Vibration Produced by Mining Activity and Safety Criteria, *3rd Asian Pacific Sympo. on Blasting Techniques*, Xiamen, China: 413～416.

[4] DIN 4150 Teil 3: 1986. *Ersch Åtterungen im Bauwesen - Einwirkungen auf Bauliche Anlagen.*

[5] Lee K W, Ryu C H, Synn J H, Park C H. 1997. Rock Fragmentation with plasma blasting method, *Proc. of 1st Asian Rock Mechanics Symposium*, A Regional Conference of ISRM, (Environmental and Safety Concerns in Underground Construction) ed. H. Lee, et al. v1, Seoul, Korea: 147～152.

[6] Ryu C H. 2002. Computer modeling of dynamic ground motion due to explosive blasting and review of some modelling problems, Science and Technology of Energetic Materials, *J. of the Japan Explosives Society* 63(5): 217～222.

[7] National Highway Institute. 1991. *Rock Blasting and Overbreak Control*, US DOT, Publication No. FHWA-HI-92-

001: 430.

[8] Ryu C H, Jang H S, Choi B H. 2013. Sensitivity analysis of input parameters in numerical modelling of dynamic ground motion under blasting impact, *Proceedings of The 7th International Conference on Explosives and Blasting*, The Japan-China-Korea Technical Committee of Explosives and Blasting (ICEB2013): 12~21.

[9] Siskind D E, Stagg M S, Kopp J W, Dowding C H. 1980. *Structure Response and Damage Produced by Ground Vibration from Surface Mine Blasting*. USBM RI 8507.

[10] Skipp B O. 1977. Ground Vibration Instrumentation - A General Review. Instrumentation for Ground Vibration and Earthquakes, *Proc. Conf. for Earthquake and Civil Eng. Dynamics*, Keel, UK.

[11] Starfield A M, Pugliese J M. 1968. Compression waves generated in rock by cylindrical explosive charges: A comparison between a computer model and field measurements, *Int. J. Rock Mech. and Min. Sci. & Geomech. Abstr.* v5: 65~77.

[12] Swiss Association of Standardization. 1992., *Effects of vibration on construction*, Seefeldstrasse9, CH8008, Zurich.

[13] Synn J H, Ryu C H, Park C H, Choi S O, Lee K W. 1999, Numerical Simulation of Shock- and Non Gas-driven Rock Fragmentation, *99 Japan-Korea Rock Mechanics Symposium*, Fukuoka, Japan: 413~418.

Engineering Blasting Safety Monitoring System Based on Internet and Cloud Computing

YANG Min, YANG Xun, XU Qiang

(*Sichuan Top Measurement and Control Technology Co., Ltd., Chengdu, Sichuan, China*)

ABSTRACT: This paper describes the blasting safety monitoring system of Sichuan Top Measurement and Control Technology Co., Ltd., based on Internet and Cloud Computing technology in vibration, shock and noise monitoring of Engineering Blasting. It discusses the features of the system, the composition of the structure and working principle in detail. The system uses a distributed measurement, short-range wireless communications and remote telemetry, etc. It can achieve features such as on-site monitoring data automatically transmitted, query data at any time through remote terminal, measured by remote control, and online monitoring of secure of blasting construction through the client(PC, tablet computers, smart phones, etc.), Internet, cloud server and database system software, etc. You can complete the remote control, telemetry signals of blasting safety monitoring, and upload data, read data, and analyzing data at multiple test points of blasting construction in the intercity, inter-provincial or even international. Fully demonstrated the role of engineering blasting safety and environmental impact and control by engineering blasting safety monitoring products and Internet communications, Cloud Computing technology of Sichuan Top Measurement and Control Technology Co., Ltd., provides the quantized data protection of research on the safety criterion for engineering blasting.

KEYWORDS: safety monitoring of engineering blasting; cloud computing; remote telemetry; short-range wireless communications; on-line monitoring

1 INTRODUCTION

Currently, during blasting construction for large and medium engineering, measuring points for safety monitoring on blasting vibration, shock wave and noise are constantly increased in number and. Also, they are more widely distributed and some of them are distributed in inter-city scope and even inter-province scope. Owners, supervision or blasting company and other relevant units also expect to duly know safety monitoring results in each construction site at any time and adjust engineering blasting parameters and monitoring instrument parameters to ensure construction safety and monitoring sequence. Meanwhile, more convenient and faster data reading is needed on the testing site. Hence, a blasting safety monitoring system is needed which can realize not only wireless telemetering on the testing site but also remote monitoring, control and data reading in inter-city or inter-province scope.

In addition, at the present time, most of blasting safety monitoring equipment is installed on the construction site. Setting is needed prior to blasting and it is needed to recycle read data after completion of blasting. There are many problems existing such as "Inconvenient settings for multiple equipment, Cumbersome read data recycling, Difficult data management and sharing".

With the development of network technology which is represented by "Internet, wireless data transmission, cloud computing, etc.", networked measurement technology and new type testing instruments with network function rise in response to the proper time and conditions. In more applications, the local testing system can not meet requirements of blasting safety monitoring users already. A distributed remote monitoring system based on cloud computing technology is needed. With its central themes "Distributed Acquisition, Wireless Data Transmission, Data Resource for Cloud Sharing", it can realize organic combination of computer technology, sensor technology, network technology, wireless data transmission technology, database technology and measurement and control technology and can establish networked, wireless, distributed engineering blasting safety monitoring system to achieve data sharing and safety monitoring cloud computing.

Based on many years of development on engineering blasting safety monitoring instruments and products, by using the a generation of NUBOX series intelligent blasting monitor and wireless communication, Internet and other modern technologies, Sichuan Top Measurement and Control

---

Corresponding author E-mail: sales@tdec.sina.net

Technology Co., Ltd., forms the "engineering blasting safety monitoring system based on Internet and cloud computing technology" and provides an all-round inter-city or inter-province multipoint engineering blasting safety monitoring solution as well as basic service for cloud data management. Thus, users can duly view remote blasting site monitoring data and "point-to-point" remote monitor working process at any time.

Meanwhile, for the demand of distributed testing system, TDEC also develops "TopCloud Cloud Comprehensive Monitoring Platform". It can be able to organically and uniformly combine engineering blasting site testing system and network communication, cloud service and other links to realize "data acquisition, uploading, downloading, processing, analysis" and "remote access (online monitoring, querying data, getting data, controlling equipment), virtual instruments, database technology" and other system functions, and further form a special database to realize cloud data sharing and online expert system and to provide basic data service for cloud computing in further blasting monitoring industry.

2 SYSTEM WORKING PRINCIPLE

2.1 System Composition

(1) NUBOX series intelligent blasting monitor (onsite testing equipment).

(2) Data communication network.

(3) TopCloud cloud data center.

(4) Remote user terminals (PC, tablet computer, smart phone, etc.).

2.2 System Architecture

(1) The system mainly uses distributed data acquisition, networked data transmission mode. It mainly consists of four parts "onsite testing equipment", "data communication network", "TopCloud cloud data center" and "remote user terminals", as shown in Fig. 1.

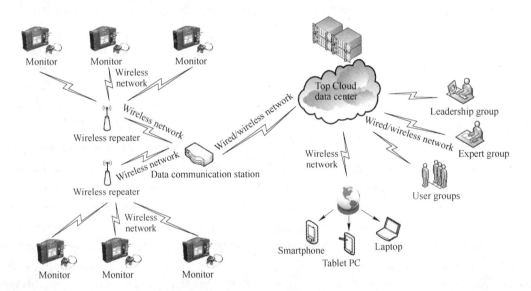

Fig. 1

(2) Onsite testing equipment is mainly composed of NUBOX series intelligent blasting monitor. The user not only can use the NUBOX series intelligent blasting monitor to get local test data on the engineering blasting site, but also transmit test data to the cloud serer through wired or wireless network. Through the cloud database management system, it can realize data storage, query, downloading, processing, etc.

(3) "Date communication network" refer to wired and wireless data communication links that are provided by site short-distance wireless communication equipment, repeaters, communication stations and various large network communication operators. It is mainly used to transmit test data.

(4) "TopCloud cloud data center" consists of cloud server and database management system, provides "user management", "equipment management", "data management" and "configuration management" and other functions, as well as ensure safety of user account and test data.

(5) The user can install the special client application software (APP) on terminal equipment like desktop PCs, laptops, tablet computers or smart phones. The user account and password with a specific authority can be used for login. By using wired or wireless network, it can access the cloud server at any time and at any place to realize blasting monitoring data query, downloading, processing and remote control for site equipment.

2.3 System Functions

2.3.1 NUBOX Series Intelligent Blasting Monitor

(1) Having all functions needed for blasting safety monitor-

ing, e.g.: color LCD touch screen; built-in chargeable lithium battery; high-strength engineering material shell, waterproof, dustproof, light, small and portable; one-key start for measurement, no need to set parameters; one-key auto setting of trigger conditions; 16384-segment extra large capacity for data storage, no data loss between two records; measured data and waveform can be displayed on the monitor in real time and recording and monitoring can be realized at the same time; a printer can be directly connected on the site to print the testing report; one-key USB flash disk data import/export.

(2) The monitor provides 2/3/4-way parallel acquisition channel. Multiple types of analog signal conditioning modules can be built in. It can be directly connected to vibration velocity, shock wave and noise sensors.

(3) For this monitor, WIFI module or short-distance wireless communication model (TWL) can be built in and 3/4 G wireless communication stations, Internet nodes and corresponding software system can be provided to complete inter-province or inter-city remote telemetering; also, GPS module can be built in to realize synchronous recording of multiple equipment and the synchronization precision is superior to 100 μs.

(4) Through wireless kit, it can realize unattended operation as well as remote instrument control, parameter settings, waveform display, data reading and other functions.

(5) Supporting monitor management by computers, tablet computers, mobile phones and other peripheral devices; built-in system of the monitor supports multiple network communication protocols. So, it is convenient to access TopCloud. Also, it supports clouds storage and realizes cloud computing. It is shown in Fig. 2.

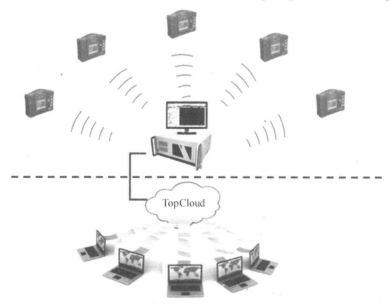

Fig. 2

2.3.2 TopCloud Cloud Data Center

(1) TopCloud cloud database management system has perfect user authority management mechanism as well as equipment system and data system backup safety mechanism to ensure cloud data safety. Also, it has improved user account management mechanism to effectively manage equipment, configuration, data and other resources.

(2) Supporting auto uploading of data at each measuring point.

(3) Providing multiple data query mechanisms, e.g.: query by time, query by device, etc.

(4) Supporting SMS notification, WeChat ID binding, import, export and management.

(5) Supporting "point-to-point" remote control function. The user can directly make remote control on instruments at engineering blasting site.

2.3.3 User Terminal

(1) The system provides client application software under Windows XP/7/8, WinCE, IOS (Apple) and Android platform to realize various functions including user login, user management, equipment query, equipment management, remote control, data query and data management.

(2) There is a special analysis module for engineering blasting monitoring built in the PC software to realize remote control, hardware settings, acquisition control, parameter calibration, engineering calibration, waveform display, data reading, data processing and analysis, saving, printing and communication functions. It duly displays monitoring data in the form of graphs and charts, fast generates the testing report with user-specified format, as well as provides data storage, management, query, output and dissemination.

(3) Through PC Web service, the user can log onto the

system by using the web browser, query equipment state, control equipment sleep or wakeup, set equipment working mode and parameters (you can set whether the data is automatically uploaded or not), as well as control equipment to start data acquisition.

(4) Through Wechat Web service, the user can connect Web service link with user ID in the browser built in the Wechat to realize all equipment, data and user management function under that user's name, including: Querying equipment status, controlling equipment work, completing data query and plotting, user management, etc.

2.4 System Working Process

2.4.1 Diagram on System Working Process

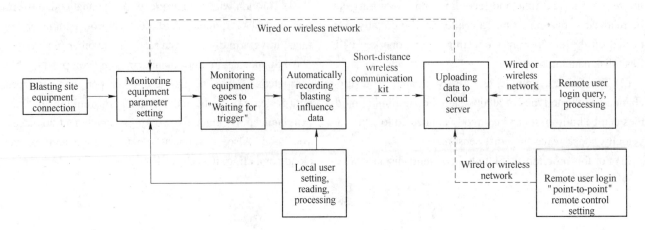

Fig. 3

2.4.2 Description on System Working Process

As is shown in Fig.3, the user completes monitoring equipment connection and parameter setting on the blasting site; then the equipment goes into "Waiting for trigger" status. After initiation, the monitoring equipment automatically records vibration, shock wave, noise and other monitoring data and saves them to internal memory. The user can immediately view monitoring results locally through the color LCD. At the same time, the monitoring equipment automatically uploads monitoring data to cloud server by using data communication device like short-distance wireless communication kit.

The remote user can make a remote access to the cloud server at any time by means of terminal equipment like office PC or tablet completer, smart phone, etc. through the wired or wireless network. With authorized user account and password, you can log in to query all monitoring data at any time and to download it locally for subsequent analysis and processing. Also, "point-to-point" connection to online equipment at the blasting site is available for remote control on parameters setting, data reading, etc.

3 CONCLUSIONS

We draw the following conclusions upon the above exposition:

(1) For demand of distributed cloud computing monitoring in engineering blasting industry, Sichuan Top Measurement and Control Technology Co., Ltd. has developed "engineering blasting safety monitoring system based on Internet and cloud computing technology". On the testing site, the NUBOX series intelligent blasting monitor is adopted and "TopCloud cloud comprehensive monitoring platform" is deployed. All functions completely meet engineering blasting safety monitoring requirements. It can conduct signal conditioning, acquisition and data transmission and storage on multiple blasting influence parameters including blasting vibration, shock wave and noise at the same time. Also, it can allow the user to query monitoring data at any time and any place and provides an all-round solution for forming the cloud computing system in the engineering blasting industry.

(2) This system is flexible and convenient in assembly, easy and fast in using, stable and reliable in working, strong in anti-interference capacity, powerful in software function, as well as rapid in data processing. It has been deployed and applied in projects. Practice has proved that with data reading through wireless network telemetering and cloud computing functions, it greatly reduces working strength on the site, quickens data transmission speed, improves working efficiency, ensures reliability of blasting monitoring data as well as further promotes intelligent, informationalized and humanized development of blasting monitoring instrument.

REFERENCES

[1] Wang Xuguang, Yu Yalun, Liu Dianzhong. Implementa-

tion Manual for Blasting Safety Regulations (GB6722-2003)[M]. Beijing: People's Communications Press, 2004.
[2] GB 6722—2011 Blasting Safety Regulation[S]. China Society of Engineering Blasting, State Administration of Work Safety, 2012.
[3] White Paper on Cloud Computing Architecture Introduction, SunMicrosystems, 2009.
[4] Product Manual/ Catalogue of Sichuan Top Measurement and Control Technology Co., Ltd., TDEC, 2013.

Blasting Harmonics

Adrian J. Moore, Alan B. Richards

(*Terrock Consulting Engineers, Australia*)

ABSTRACT: Blast holes fired in a pattern with a constant initiation delay create a frequency in the ground vibration similar to that from a percussion instrument, such as a drum, in the air. A constant firing delay generates a prime or forcing frequency into the ground motion. The forcing frequency may then be modified by ground transmission characteristics into sub harmonics or super harmonics of the forcing frequency. The moving vibration source between blast holes creates a frequency shift because of a Doppler Effect. Other wave interactions may form beats. The effects are exacerbated by the accurate firing times of electronic detonators.

This paper uses observations of wavetraces and frequency spectral analysis to demonstrate:

(1) Doppler Effect and frequency ellipsoids;

(2) Sub (and sub-sub) harmonics of the forcing frequency;

(3) Super harmonics of the forcing frequency;

(4) Beat formation that may double or triple PPV levels;

(5) The effect that some geological conditions have on wave shapes.

An understanding of the science involved has a practical application that has been successfully applied to initiation timings to reduce Peak Particle Velocity and adverse human and structure response from blasting operations.

KEYWORDS: beats; doppler effect; frequency ellipsoid; ground vibration; sub harmonics; super harmonics

1 INTRODUCTION

Ground vibration resulting from a blast is a function of many factors including:

(1) Charge mass per hole or per delay;

(2) Distance from the blast;

(3) Drilling pattern, burden, spacing and blast shape;

(4) Initiation sequence;

(5) Ground transmission characteristics;

(6) Degree and depth of surface weathering.

The main model used for analysis and prediction of ground vibration is the Scaled Distance Site Law model, of the general form:

$$PPV = k\left(\frac{\sqrt[n]{m}}{D}\right)^e$$

where PPV——Peak Particle Velocity, mm/s;

m——charge mass per hole or per delay, kg;

D——distance from blast, m;

k——site constant;

e——site exponent;

n——fractional indice.

At many sites, $n = 2$ and $e = 1.6$ are usually appropriate.

At these sites, all factors other than charge mass and distance are represented in k, i.e.

$$PPV = k\left(\frac{\sqrt[2]{m}}{D}\right)^{1.6}$$

When conducting detailed analyses of blasting events, to come to an understanding of the factors involved, the wavetraces are subjected to a forensic analysis to determine:

(1) Where in the wavetrace the peak occurs (times of peaks);

(2) Where this time occurs in the blast sequence compared with the blast duration;

(3) What type of wave the peak is associated with (P, S, R, A analysis);

(4) If beats or other signs of harmonic interactions are present.

An FFT frequency analysis is often conducted as part of the examination to determine:

(1) Dominant frequencies as a function of initiation timing and amplification factors;

(2) Frequency band width;

(3) The significance of Doppler Effect, sub harmonic splits and super harmonics;

Corresponding author E-mail: terrock@terrock.com.au

(4) Rogue or orphan frequencies not associated with the initiation timing that may indicate poor geophone coupling or ground transmission properties.

Mines are often obliged to monitor vibration to demonstrate compliance with regulation limits and providing the PPV's are within their limits, the Environmental Officer reports the result and that is the end of the process.

Yet there is a wealth of information available to the Blasting Engineer, Mine Manager and the Shotfirer on the performance of a blast, if the trouble is taken to conduct detailed examination of the wave traces; a process we call forensic examination.

During the forensic examination process, observations have been made that show harmonics have a significant impact on the shape of the wavetrace, the dominant frequency and frequency bandwidth, and ultimately the Peak Particle Velocity. In order to explain some of the observed phenomenon, it has been necessary to compare the vibration from a blast with vibration from a musical percussion event such as drumming. An understanding of the science involved has been used to enable recognition of the significance of the other factors and to develop techniques to lower PPV levels and move dominant frequencies away from the natural frequencies of houses to reduce response annoyance.

2 WAVE TYPES AND SEPARATION

Ground vibration consists of three (or four) different wave types each with its own characteristic motion, seismic velocity and attenuation. The principal waves are Compressional (or P) waves; Shear (or S) waves and Rayleigh (or R) waves. Some sources reference Love waves as a fourth wave type, but because they have the same velocity as R waves but different motions, for our purposes the two motions are not separable by simple observations and are considered as the one wave.

Close to a blast the three waves are interwoven, but with increasing distance the wave types separate because of the different transmission velocities. For example, consider the ground vibration from a single blasthole measured on the surface. The blast was a single hole containing 1000 kg of explosives and the wavetrace was recorded at 1800 m. Fig.1 shows the separation of the wave types; the comparatively faster attenuation of the P waves and the dominance of the S and R wave with increasing distance.

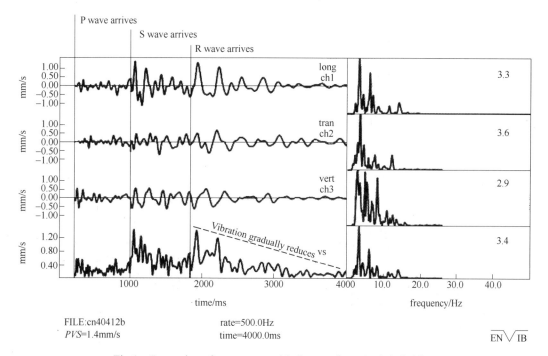

Fig.1 Separation of wave types with distance for a single hole blast

Similar observations are made with ground vibration from complex blasts of many holes. The arrival of the slower waves is often recognisable on the wavetraces by a change of motion (Fig.2). The velocity of the different waves can be determined if the distance is known, by using the velocity of sound in the air as the calibrator (PSRA analysis), e.g. typical speed of sound in air is 340 m/s.

The R wave dominance at a distance has been observed at sites where the geology consists of sub horizontal structures such as bedding in coal overburden or layering in basalt flows.

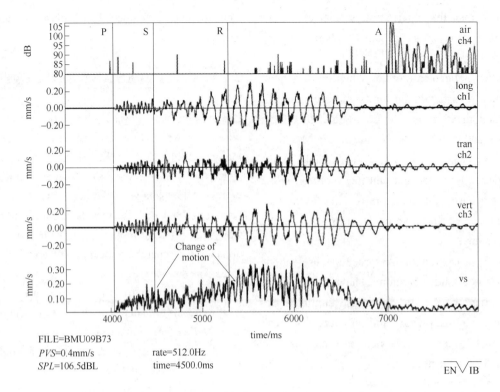

Fig.2　Separation of wave types with distance for a large blast, R waves dominate

At other sites where the geology includes sub vertical structures such as tilted bedding, faults and shears, Rayleigh waves, being surface waves, do not become dominant and may not be recognisable in the wavetrace. An example of this is shown in Fig.3.

At many sites recognition of the wave types is simple and reflected and/or refracted waves do not appear to have a significant influence of the wave shape. However, at other sites the geology has a significant influence.

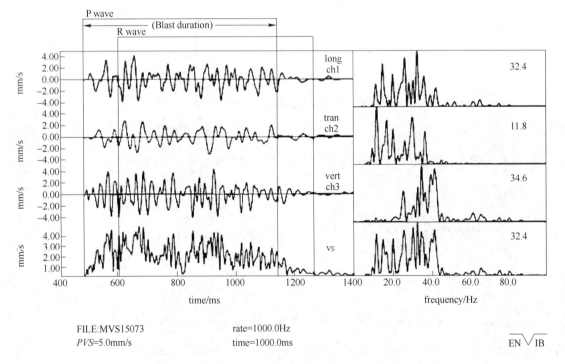

Fig.3　Wavetrace with no recognisable R wave development

3　FORCING FREQUENCIES

Constant timing delays create forcing frequencies. The forcing frequencies associated with the commercial signal tube detonators are shown in Tab.1. The sub harmonics are also listed.

Tab.1 Delay timing, forcing frequency and sub harmonics

Timing delay /ms	Forcing frequency/Hz	Sub harmonics/Hz			
		1	2	3	4
9	111	55.5	27.8	13.9	6.9
17	58.8	29.4	14.7	7.4	3.7
25	40	20	10	5	
42	24.4	12.2	6.1		
67	14.9	7.5			
100	10	5			
109	9.2	4.6			

The range is expanded by the large possible range of electronic detonators. With signal tube initiation "det scatter" may introduce a blurring of the frequencies. The accurate timing of electronic initiation may produce a tighter forcing frequency band if the same timing is used.

4 DOPPLER EFFECT

The Doppler Effect is a change of frequency that results when the source of the vibration moves. If we consider a row of blastholes fired with 25 ms delay (40 Hz frequency), the frequency in the initiation direction is higher than in the opposite direction. Say the holes are 10 m apart and the P wave velocity is 2000 m/s or 2 m/ms. It takes 5 ms for the P wave to travel between holes. In the direction of initiation the waves arrive 25 − 5 = 20 ms apart (50 Hz), and in the opposite direction 25 + 5 = 30 ms apart (33.3 Hz). Perpendicular to the row, the frequency is 25 ms or 40 Hz. The resulting frequencies can be considered as an ellipsoid with the following frequency dimensions (Fig.4).

The frequency in any direction from the blast can be determined from scaling the frequency ellipsoid, e.g. 45 Hz and 48 Hz frequencies are transmitted in the directions shown in Fig.4.

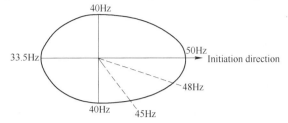

Fig.4 P wave frequency ellipsoid

Frequency ellipsoids for the other wave types are different because of the different velocities.

For Rayleigh waves, velocity 700 m/s, 0.7 m/ms, the frequency ellipsoid is as shown in Fig.5.

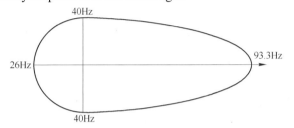

Fig.5 R wave frequency ellipsoid

5 SUB HARMONICS

An observation made from coal overburden blasts (with sub horizontal bedding) is that a high forcing frequency may not persist for a great distance from a blast and undergoes what we call the sub harmonic split, i.e. the forcing frequency splits into sub harmonics and then sub-sub harmonics with increasing distance.

For example, control row delay 42 ms; dominant frequency 23.8 Hz→11.96 Hz→5.9 Hz.

The distance from the blast where this may occur can be seen in this graph (Fig.6).

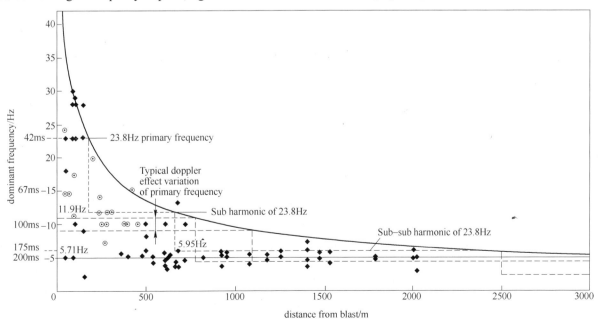

Fig.6 Sub harmonic split observations

The net effect on the above example is that the Doppler frequencies listed above may also split into sub and sub sub harmonics with increasing distance from a blast.

6 SUPER HARMONICS

A recent observation of the wavetraces of underground blasts on the surface is that super harmonics may also result from the initiation sequence. The delays used in firing the rings were 20, 40, 60 and 70 ms. The forcing frequencies and super harmonics are listed in Tab.2.

Frequencies Present in FFTs in Fig.7.

The geology of the site is tightly folded Ordovician sediments with major folds about 150 m apart with blasting at depths of about 600 m. The development of super harmonics was recognised by comparing the initiation sequence of the rings in the stope blasts with the dominant frequencies

Tab.2 Delay timing, forcing frequency and superhamonics

Delay timing/ms	Forcing frequency/Hz	1st super harmonic /Hz	2nd super harmonic /Hz
20	50	100	
40	25	50	
60	16.7	33.3	66.6
70	14.3	28.6	57.2

in the FFT. Often there was little or no energy at the nominal forcing frequency, but the dominant frequency was double what was expected. An example can be seen from the wavetrace and FFT shown in Fig.7(a) where "beats" were observed in the wavetrace. Beat formation is described in Section 7.

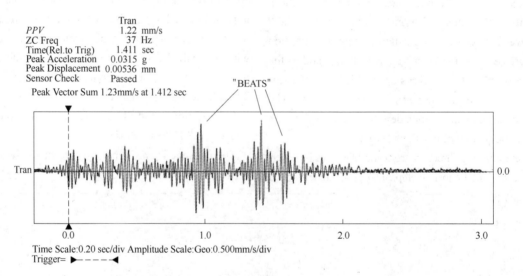

Fig.7(a) Surface wavetrace from underground blast showing beats

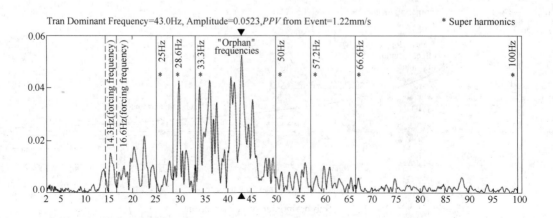

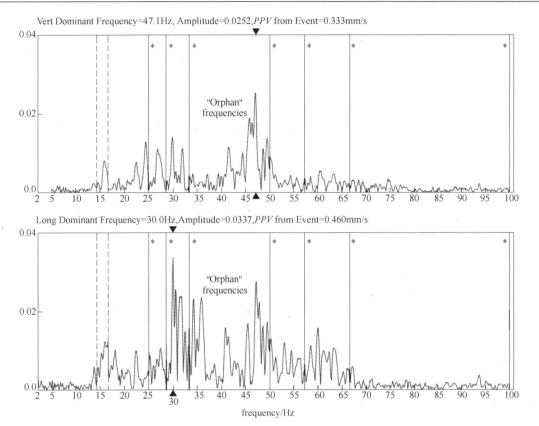

Fig.7(b) FFT analysis from Fig.7 (a) showing forcing frequencies, "orphan" frequencies and super harmonic frequencies

7 FORMATION OF BEATS

Beats form when the vibration consists of two or more closely aligned frequencies. The frequency of the beat is the difference between the two contributing frequencies. For example, if the vibration consists mainly of 30 Hz and 25 Hz, the beat frequency is 5 Hz and the wavelength of the beats is 1000 ÷ 5 = 200 m. This is demonstrated in Fig.8.

Another recorded wavetrace showing beats is shown in Fig.9. The dominant frequencies are shown in Fig.10. The source of the closely aligned frequencies is that the same delays were used on either side of an opening hole in the ring of a stope blast which resulted in a frequency shift. The Doppler Shift produced slightly different frequencies in the direction of the surface monitor which is further compounded by super harmonic development. The two frequencies resulting in the beats are shown as dominant frequencies in the FFT analysis (Fig.10).

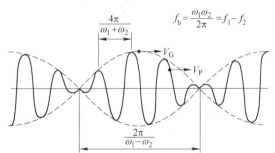

Fig.8 Explanation of beat formation

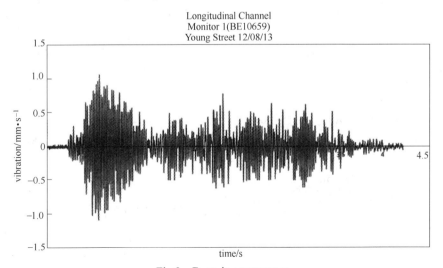

Fig.9 Beats in a wavetrace

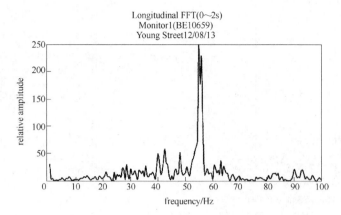

Fig.10 FFT analysis of wavetrace with beats from Fig.9

In the first two seconds, the main contributing frequencies are 55 Hz and 56 Hz. The beat frequency is 1 Hz and the first beat is about one second long.

If a beat consists of two contributing frequencies, the amplitude of the beat will be double that of the individual waves and if three frequencies the amplitude will treble. If beats occur in the wave motion, there is the potential for the *PPV* to be more than double. Recognition of and elimination of beats can dramatically lower resulting *PPV* levels. Beats are eliminated if the two contributing frequencies are separated by an appropriate initiation sequence. Two cases are presented where the initiation was altered to reduce the *PPV* at important locations.

7.1 Coal Overburden Blasts

Beats were identified in the motion resulting from coal overburden blasts fired with a centre stitch where it was essential to control *PPV* levels at a section of roadway made unstable by movement on a fault. A site sketch is shown in Fig.11.

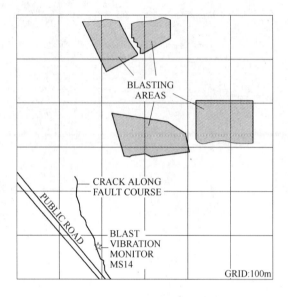

Fig.11 Site sketch – blast and monitoring locations to control vibration at the cracks

A typical wavetrace recorded at MS 14 shows the beats is shown in Fig.12.

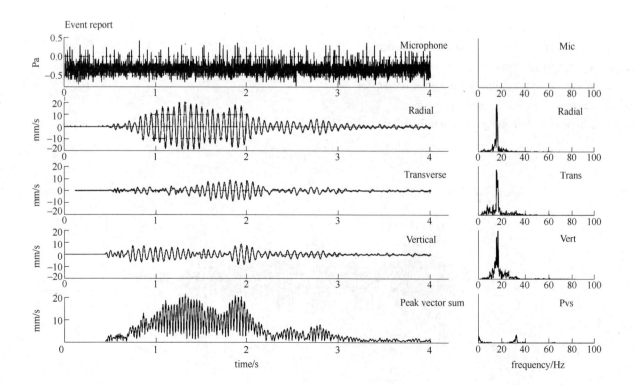

Results

Peak Vector Sum	21.19	mm/s	at 1.33	s
Peak Air Overpressure:	0.79	Pa		
	91.97	dBL	at 2.54	s

Max Peak Particle Velocity

Radial	20.87	mm/s	at 1.32	s
Transverse:	9.04	mm/s	at 1.82	s
Vertical:	8.68	mm/s	at 1.94	s

File:	Sile15_c320120918T111349E-MP02-L21.03
Date:	2012-09-18T11:13:53.039+10.00
Rate(Hz):	1000
Length(s):	4.01

Site=3
Location Name=Crack 3
Location Code=c3
Location Coordinates =S3230.2039E 151 6.9205
Serial Number=C 008
CF Card Label=CF068
Calibrabon Until=2013-03-29

Dominant Frequency

Air Overpressure:	0.00	Hz
Radial:	15.87	Hz
Transverse:	15.87	Hz
Verticat:	16.85	Hz
Peak Vector Sum:	0.00	Hz

Fig.12 Wavetrace showing beats and FFT analysis

The Road Authority placed a vibration limit of 15 mm/s at MS 14 and the resulting PPV measured was 21.19 mm/s, clearly an exceedence.

The radial channel shows a strong "beat" that has effectively doubled the *PPV* compared to the other two channels, where beats are present.

The frequency spectra of the wave motion is also shown in Fig.12. The dominant frequencies of the motion are 16.85 and 15.87 Hz.

The drill pattern and initiation sequence resulting in beats is shown in Fig.13.

The beat analysis details are summarised as follows:

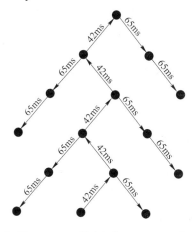

Fig.13 Drill pattern and initiation sequence from Fig.12

Frequency ellipsoids of the 65 ms delays in the echelon rows are shown in Figs.14(a) and (b).

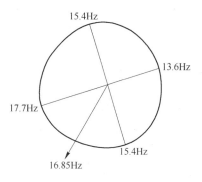

Fig.14(a) Frequency ellipsoid 65 ms delay down the left echelons

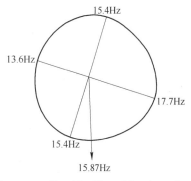

Fig.14 (b) Frequency ellipsoid – 65 ms delay down the right echelon

Combining two waves with dominant frequencies 16.85 Hz and 15.87 Hz produces the theoretical beats shown in Fig.16. This is compared to the Radial Channel detail reproduced and enlarged from Fig.12 in Fig.15, which shows intermeshed beats. The beat lengths are about one second long.

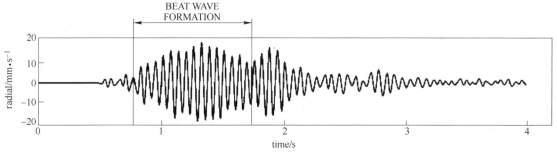

Fig.15 Radial Channel detail from Fig.12

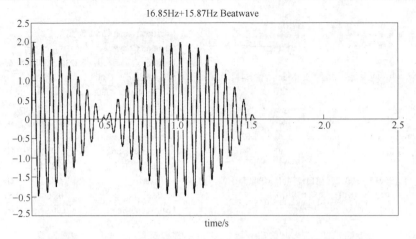

Fig.16 Theoretical beats from combined 15.87 and 16.85 Hz motion

These predicted waves have remarkable similarity to the measured wave. The formation of the beats in the Radial channel wave has doubled the *PPV* in this channel. The beats in the T and V channels are not as significant. The beats were eliminated by firing a single control row at the side of the blast with the echelons extending on one side only which resulted in only one dominant frequency being generated.

7.2 Underground Stope Blasts

In another case, beats were observed in the surface wavetraces from underground stope blasts that were particularly annoying to some house owners. A typical wavetrace with beats is shown in Fig.9. A wavetrace from another blast is shown in Fig.17. The transverse channel motion shows as a series of beats about 200 ms long.

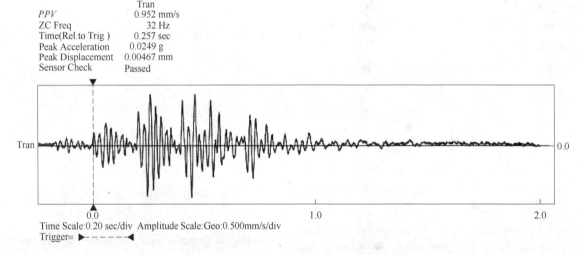

Fig.17 Surface wavetrace showing beats from an underground blast

The FFT of this wave is shown in Fig.18. The dominant frequencies are 39.6 and 44.3 Hz. The beat frequency is 4.7 Hz with a wavelength of 213 ms.

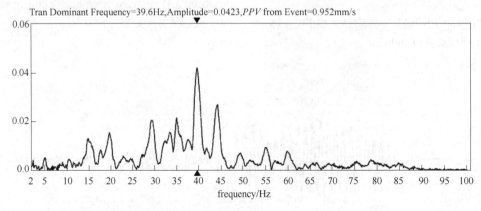

Fig.18 FFT frequency spectra from underground blast in Fig.16

This blast consisted of rings fired in a stope. The dominant frequencies resulted when holes were fired on either side of the opening hole with the same 25 ms delay in between. The Doppler Effect accounted for the slight frequency shift.

The solution was to design an initiation sequence with a range of totally different delays either side of the centre opening to broaden the frequency spectra and prevent the formation of beats and by spreading the energy in the motion to reduce the amplitude at any frequency. An FFT resulting from this frequency manipulation approach shown in Fig.19. Note the broad frequency spectrum and a single dominant frequency of 16.6 Hz.

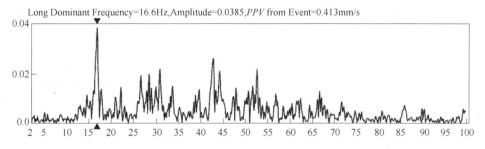

Fig.19 FFT spectra from blast with frequency manipulation

8 WAVEFRONT REINFORCEMENT AND SYNCHRONICITY

Wavefront reinforcement occurs when the combination of drill pattern and initiation timing results in the wavefronts from two or more blast holes coincidentally merging together in a direction from the blast. Consider blast holes with a burden and spacing of 2.2 m fired with a control row delay of 17 ms and an echelon row of 25 ms.

The wavefront reinforcement diagram is shown in Fig.20. This shows the relative positions of the wavefronts from all blast holes at a nominal time.

Three wavefronts reinforce the direction shown which, if synchronous, could treble the *PPV*. A monitor set up in the reinforcement zone recorded the wavetrace 800 m from the blast is shown in Fig.21.

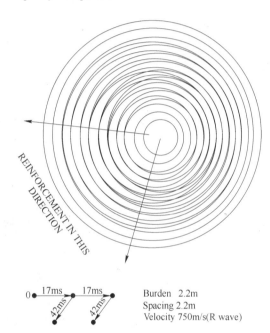

Fig.20 Reinforcement diagram for rayleigh waves

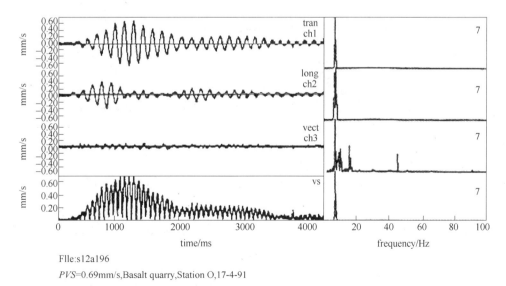

Fig.21 Dominant rayleigh waves produced by reinforcement and geological layering

The arrival of a number of wavefronts at a constant interval seems to generate its own harmonics on the surface of a basalt flow about 10 m thick isolated from lower basalt flows by 1.2~2 m of clay. The FFT shows an almost pure 7 Hz is the dominant frequency. The dominant frequency appears to be a sub sub harmonic of the 17 ms control row delay, i.e. 17 ms = 58.8 Hz→29.4→14.7→7.35 Hz. The ground motion was annoying because the home owner would arrive home at the end of the day to find all the mirrors and pictures hanging on the wall were tilted.

Not only is the drill pattern and the initiation timing contributing to the wave trace, but the thickness and isolation of the basalt flow seems to be important. In this case, the *PPV* at closer monitors was controlled to regulatory limits by using a non-reinforcing initiation sequence, e.g. alternate 17 and 25 ms control rows and 42 ms echelon rows, which broke up the wavefront reinforcement and produced a more complex FFT spectrum to which the house did not respond.

Another example of the site geology having a harmonious influence on ground vibration was at a basalt quarry where the ancient basalt was overlain by 20~30 m of transported clay, as shown schematically in Fig.22. A typical wavetrace from a blast at this quarry is shown in Fig.23. The vibration was recorded at a house 570 m from the blast. The blast consisted of 14 holes in two rows with a blast duration of 263 ms. The first wave arriving was the P, S and R waves travelling by the shortest seismological path (through the basalt) with a *PPV* of about 0.3 mm/s. The peak *PPV* (1.3 mm/s) results from the low frequency secondary waves which travels the long path through the clay and has a seismic velocity slower than the speed of sound in air. The vibration was very annoying to the home owners because their house was subject to these distinctly separate vibrations lasting for over 6 seconds. The 3 Hz vibration resulted in comparatively large displacements of the house in response to the ground vibration.

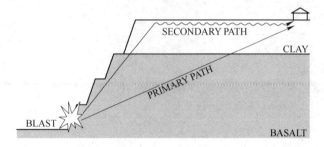

Fig.22　Schematic cross section quarry to houses

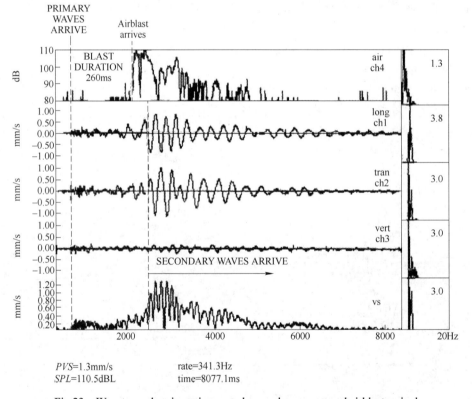

Fig.23　Wavetrace showing primary and secondary waves and airblast arrival

9　CONCLUSIONS

(1) The ground vibration resulting from a blast is often influenced by factors other than charge mass and distance.

(2) Controlling ground vibration by charge mass reduction alone has its limitations – a 50% charge mass reduction reduces the *PPV* by about 40% without affecting the other contributing factors.

(3) The initiation sequence produces forcing frequencies which may then be modified by
- Doppler Effect;
- Directional Frequency variations;
- Sub harmonic splits;
- Super harmonic development.

(4) Beats are formed when two or more generated frequencies are closely aligned, e.g. 16.8 Hz and 15.8 Hz.

(5) The beat frequency is the difference between the two

frequencies.

E.g. 16.8 Hz and 15.8 Hz – the beat frequency is 1 Hz.

(6) The beat wavelength is 1000 ÷ beat frequency = 1 second long.

(7) Beats consisting of 2 frequencies may double the *PPV* in the channel affected and three frequencies may treble the *PPV*.

(8) To reduce the *PPV* by preventing beat formation, first they have to be recognised and then the initiation sequence and direction of firing modified.

(9) Forensic wave trace examination to identify the causes of the peak *PPV* measured is an essential part of the blasting review process, especially if the *PPV* levels are approaching regulatory limits.

(10) The geology between the blast and a surface receptor can influence to wave shape, the *PPV*, the frequency spectrum and the vibration duration. This can also be evidenced by detailed forensic examination of the wavetrace.

(11) In situations where the geology is resulting in harmonious wavetraces, experimenting with variations in control timing to break up the forcing frequencies may be beneficial.

Application and Research on Slight Vibration Controlled Blasting Technology for Metro Tunnel Adjacent Intermediate High Pressure Gas Pipe

ZHANG Junbing[1], ZHENG Baocai[1], LI Zihua[2], HU Yunfeng[2]

(1. The Third Engineering Group Co., Ltd. of China Railway, Taiyuan, Shanxi, China; 2. Guangdong Construction Engineering Co., Ltd., The Third Engineering Group of China Railway, Guangzhou, Guangdong, China)

ABSTRACT: According to the blasting construction for Shenzhen City Rail Transit the 11th line of Fuyong station~Qiaotou Metro tunneling interval tunnel adjacent intermediate high pressure gas pipe(1.6 MPa), Optimization of blasting parameters can control blasting vibration velocity, which ensure the safety of high pressure gas pipe and reduce the cost of construction, at the same time the preference can be provided for the similar projects. Through the practice of construction, the controlled blasting technology of digital electronic detonator and millisecond delay electric detonator that are assembled in use of cutting and smooth blasting are used in the construction of Metro tunneling interval tunnel adjacent high pressure gas pipe, it provides good effect; the result of analysis about blasting monitoring shows that the largest blasting vibration velocity and main vibration frequency are in the permission scope. Using the initiation mode, blasting parameters are reasonable, the maximum charge per interval is suitable and it can reduce the blasting vibration velocity and comply with the control requirements of high pressure gas pipe, make good economic benefits.

KEYWORDS: intermediate high pressure gas pipe; metro tunnel; digital electronic detonator ; controlled blasting technology

1 BACKGROUND

In the process of the construction of urban subway, especially in the drilling and blasting method is adopted to improve the urban subway tunnel in hard rock tunnel excavation process, the surrounding environment is complex, most facing urban controlled blasting under complicated environment construction, blasting vibration control demand is high, is difficult. Reasonable blasting way in the process of selecting is to make sure that one of the important means of rapid construction, whether a measurement of blasting technology is feasible, safety guarantee, one of the important measures of economic and reasonable, i.e., blasting vibration monitoring and analysis become the inevitable requirement of construction process control and security guarantees.

Research achievements of controlled blasting and monitoring is more, some scholars study found blasting seismic wave on the surface, the influence of the structure is the largest, and put forward to adopt millisecond controlled blasting technology, to reduce the adverse impacts on the buildings [1]. Blasting vibration monitoring for the optimization of blasting parameters design, the maximum advantages of blasting technology plays an important role in guiding [2]; Blasting vibration velocity and vibration frequency of the surface take an effect to the safety control of buildings [3]. Advanced pilot tunnel segment in shallow buried large span tunnel blasting excavation way through the city blocks play an obvious result in the process of controlling blasting [4], etc. Especially in recent years, following with the advance of blasting material, the development and application of high precision electronic detonator provides technical support to the security of detailed control blasting, the electronic detonator can realize accurate control of the time difference, so it can achieve both before and after the cartridge blasting vibration wave troughs overlay, reach the purpose of interference decreasing vibration [5~7], Electronic detonator in the tunnel blasting process achieved good vibration damping effect [8].

Monitoring in the process of tunnel blasting excavation, especially of blasting vibration data acquisition, data processing, analysis, and the result feedback in time, play a very important role in the process of blasting parameters construction adjustment, real-time data analysis and feed-

---

Corresponding author E-mail: zhengbaocai@126.com

back of blasting construction are needed in the process of constrcution along building (structure) in adjacent, information feedback can be used to judge that whether the blasting method and parameters meet the control requirements or not, and then adjust and improve the blasting parameters.

2 PROJECT SUMMARY

An urban rail transit line 11 engineering underground excavation period of interval is mainly through the stratrum of sand clayey soil, full ～ strong weathering hemp mixed granite, strong breeze ～ grain of rock. Tunnel cross section of the groundwater is rich, extremely complex geological conditions. Tunnel buried deep 10.133 ～ 6.895 m, horseshoe cross section, double line tunnel minimum spacing is 12.78 ～ 12.78 m. In the construction site on the west side there are 1.6 MPa high-pressure gas pipeline gas group, buried depth is about 1.86 ～ 2.39 m, and it is gas transmission trunk line, the minimum distance of tunnel excavation boundary distance is 34 m, gas pipeline don't move in the process of construction.

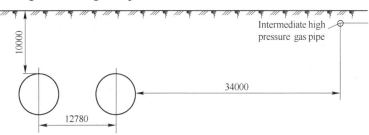

Fig.1 Sketch map of tunnel and gas pipeline

3 CONTROLLED BLASTING INUNDERGROUND EXCAVATION OF TUNNELS

3.1 Blasting Scheme of Key Principles

According to the actual characteristics of the project, in the process of blasting scheme and blasting, excavation surrounding building security is needed to focus on and ensure, especially the intermediate high pressure gas pipeline safety.

(1) the blasting scheme need to meet the urban complex environment harmful effect of blasting, blasting vibration control requirements, in particular, ensure the safety of the gas pipeline construction.

(2) the construction process of blasting excavation construction cost control requirements.

General principles: feasible, safety control technology, reasonable economy.

3.2 The Blasting Excavation Scheme

(1) The main characteristics of digital electronic detonator, safe and reliable, high precision of time delay, flexible extension set, many research results show that the vibration velocity of blasting excavation control core is a priming dose control, for the tunnel blasting excavation, the most main is to control the maximum initiating explosive cutting process.

(2) With the underground tunnel of the geological characteristics of the surrounding rock and tunnel section design, the method of upper and lower method blasting excavation is used, digital electronic detonator v-cut + smooth control blasting scheme is used in the construction.

(3) According to the peak interference decreasing vibration control thought, in the process of cutting, especially in the near time blasting cut high pressure gas pipeline construction process, through the preliminary test set reasonable cut hole blasting reasonable digital electronic detonator delay was set, to ensure that blasting vibration velocity meet the control requirements in the process of blasting cut.

3.3 Blasting Vibration Control Indicators

According to the "Blasting Safety Regulations" [9] GB 6722 —2011and the municipal department of high pressure gas pipeline blasting vibration control requirements, combining with the characteristics of shallow buried tunnel, nearly, blasting vibration velocity control standards was determined within 2.0 cm/s.

3.4 The Main Blasting Parameters

The blasting tunnel cross section area of 32 m^2, surrounding rock for III stage. Using upper and lower blasting excavation, in the upper steps, in the cut eye v-cut, a total of 4 rows(8), the initiation of the cut hole were detonated with digital electronic detonator initiation, using the 2 holes as a group to initiation, firing interval respectively set to 2 ms, 7 ms, 7 ms. Peripheral eye and easer eye were detonated by using millisecond delay electric detonator, peripheral eye charge structure for coupling interval charge form. According to the blasting source distance protected buildings and cut a detonation maximum dosage, the eye reference Sa-DaoFu formula, to carry on the back calculation theory of blasting vibration velocity, the largest compare to allow maximum blasting vibration velocity, judge whether reasonable blasting parameters design.

$$v = K\frac{Q^{\frac{a}{3}}}{R^a} \quad (1)$$

where R ——the distance between explosive source and the need to protect the buildings R, according to the current the buried depth of the tunnel (10 m), 35 m;

Q ——the amount of explosive, 6.6 kg, blasting a biggest quantity;

K ——geology, terrain influence coefficient, 160;

a ——attenuation coefficient, through the field test take 1.6.

Through theoretical calculation, the maximum value of blasting vibration velocity of the gas pipeline directly above the tunnel is 1.48 cm/s < 2 cm/s, which meet the control requirements.

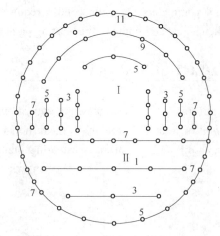

Fig.2 Blasthole plan of up and down the steps of excavation

Tab.1 Bench blasting parameter table

The blasting area	Hole name	Dan	Hole number	Deep hole /m	A single charge /kg	Subtotal controll /kg
The upper	Cutting hole (digital electronic detonator)	—	2	2.3	0.8	1.6
		2 ms	2	2.3	0.8	1.6
		7 ms	2	2.3	0.8	1.6
		7 ms	2	2.3	0.8	1.6
	Expansion slot hole	3 ms	6	1.5	0.6	3.6
		5 ms	6	1.5	0.6	3.6
	Supplementary eye	7 ms	4	1.5	0.6	2.4
	Bottom eye	7 ms	9	1.5	0.6	5.4
	Two (2) full eye	9 ms	7	1.5	0.6	4.2
	Perimeter hole	11 ms	22	1.5	0.3	6.6
The lower level	Drifting hole	1 ms	5	1.5	0.6	3
		3 ms	3	1.5	0.6	1.8
	Bottom eye	5 ms	5	1.5	0.6	3
	Perimeter hole	7 ms	8	1.5	0.3	2.4
	Total		83			39.4

4 BLASTING VIBRATION MONITORING

4.1 Monitoring Programm

Monitoring points distribute across directly above, tunnel excavation axis and above the intermediate high pressure gas pipeline, the east side of foundation pit, the blasting vibration velocity are tested.

4.2 Monitoring Results and Analysis

Monitoring results mainly aimed at the blasting vibration velocity in the process of blasting, which is at above the construction of tunnels, gas pipeline, the east of foundation pit monitoring. Monitoring content includes three direction maximum vibration velocity, frequency, duration, etc., through the analysis of the influence of blasting construction on the surrounding environment, the result measure whether the blasting parameters meet the requirements of safety control or not. Blasting vibration monitoring data are shown in Tab.2.

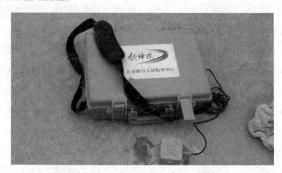

Fig.3 Blasting vibration recorder

Tab.2 Blasting vibration testing data statistics

No.	Distance between buliding and blasting center /m	Measuring point location	The largest vibrating velocity /cm·s^{-1}	Dominant frequency /Hz	Duration/s
1	10	Above the tunnel	1.16	25.6	0.0655
2	20	Above the gas pipeline	0.6821	23.5	0.0385
3	25	Above the gas pipeline	0.5897	41.7	0.0805
4	25	Above the gas pipeline	0.4389	32.3	0.2370
5	25	Above the gas pipeline	0.5907	83.3	0.4605
6	30	Above the gas pipeline	0.2568	29.4	0.1525
7	30	East of foundation pit	0.2092	250	0.1222
8	25	Above the gas pipeline	0.3439	37.9	0.3146
9	10	Above the tunnel	12.3517	22.7	0.0105
10	20	East of foundation pit	0.9249	37.0	0.0665
11	30	Above the gas pipeline	0.4262	54.5	0.1465
12	20	Above the gas pipeline	0.5591	22.7	0.0870

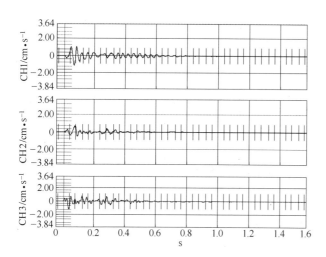

Fig.4 Blasting vibration velocity waveform 1

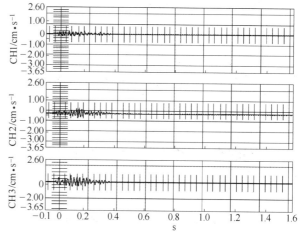

Fig.6 Blasting vibration velocity waveform 3

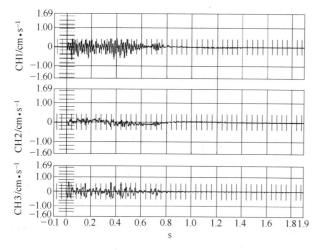

Fig.5 Blasting vibration velocity waveform 2

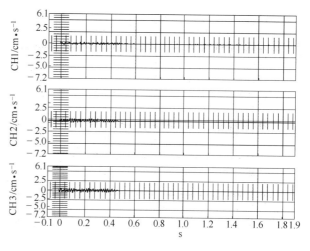

Fig.7 Blasting vibration velocity waveform 4

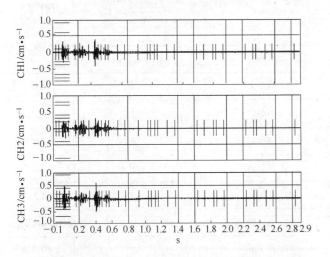

Fig.8 Blasting vibration velocity waveform 5

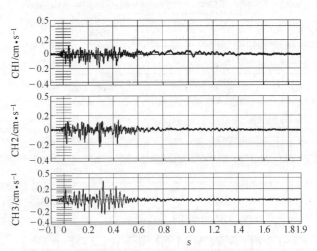

Fig.11 Blasting vibration velocity waveform 8

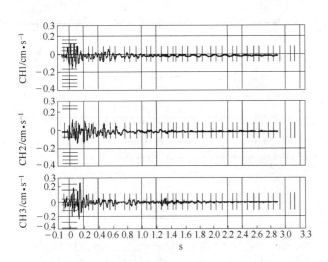

Fig.9 Blasting vibration velocity waveform 6

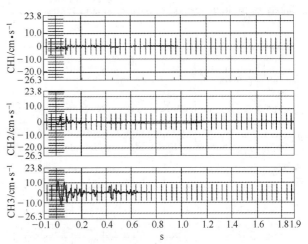

Fig.12 Blasting vibration velocity waveform 9

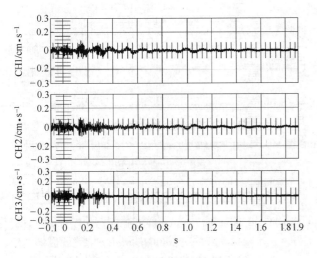

Fig.10 Blasting vibration velocity waveform 7

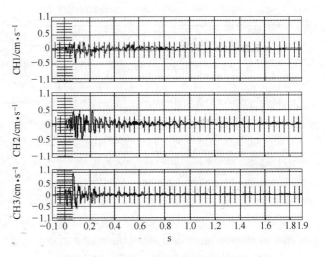

Fig.13 Blasting vibration velocity waveform 10

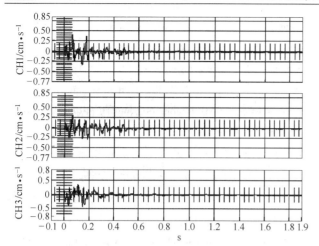

Fig.14　Blasting vibration velocity waveform 11

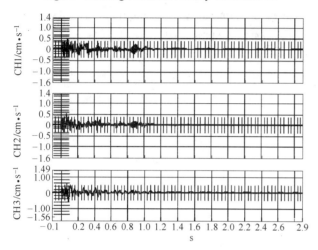

Fig.15　Blasting vibration velocity waveform 12

4.3　The Monitoring Analysis of Blasting Vibration Velocity

Known from the blasting vibration monitoring data statistics, the maximum blasting vibration speed in be protected at the top of the high pressure gas pipeline did not exceed the permitted maximum blasting vibration speed of 2 cm/s. Some monitoring data above the tunnel exceeded 2 cm/s, but the location of the blasting vibration velocity with the same time in gas pipeline detection value does not exceed allowed maximum blasting vibration velocity (2 cm/s).

It can be seen in the Fig.16, the value of blasting vibration velocity at the top of the intermediate high pressure gas pipeline numerical is less than or close to 80% in 0.6 cm/s, the blasting vibration value is only 30% of the allowable maximum blasting vibration velocity, blasting vibration velocity meet the requirements of intermediate high pressure gas pipeline safety control.

It can be seen in the Fig.17, the main vibration frequency of blasting monitoring sites above intermediate high pressure gas pipeline are more than 20 Hz, monitoring data of more than 80% of the measuring point frequency are between 20～60 Hz, the existing research and monitoring results show that when the blasting frequency less than 10 Hz is not conducive to the structure stability, monitoring results in the main vibration frequency are not less than 10 Hz, blasting vibration main frequency satisfies the requirement of safety control.

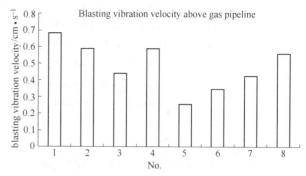

Fig.16　The blasting vibration velocity distribution diagram above gas pipeline

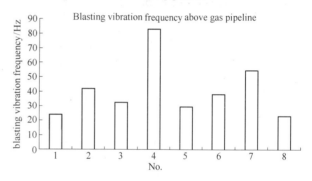

Fig.17　The blasting vibration frequency distribution diagram above gas pipeline

It can be seen from the single vibration waveform diagram, after using digital electronic detonator, vibration waveform shows more gentle wave (valley), and show a good effect in blasting vibration peak vibration reduction, it is advantageous to the control of blasting vibration velocity and the safety of the pipeline protection.

5　CONCLUSIONS

(1) With the monitoring results and site construction of gas pipeline, it can be known that, maximum vibration velocity is less than the allowable blasting vibration speed, blasting vibration frequency can satisfy the demands of pipeline safety control, the reasonable parameter of blasting scheme can ensure the safety of the gas pipeline and satisfies the requirement of on-site safety construction.

(2) The methods used in the process of cutting the digital electronic detonator blasting method and determined the adjacent hole delay time, can realize the superposition of blasting vibration peak valley, it shows obvious effects of vibration reduction, it solved the bottleneck problem of blasting construction technology, ensure the intermediate high pressure gas pipeline tunnel blasting safety construction.

REFERENCES

[1] Yu Y Q, Wen G C. Effect of blasting digging on surface buildings in metro tunnels[J]. *Journal of Liaoning Technical University*, 2007 (12):871~873.

[2] Shen Y S, Gao B, Wang Z J, Meng F J. Efect of Blasting in Double Line Tunnel on Existing Tunnel[J]. *Chinese Journal of Underground Space and Engineering*, 2009 (10):980~984.

[3] Li X S, Fu Y C, Sun X L.A Finite Element Analysis of Blasting Vibration in Qingdao—Huangdao Cross Harbor Tunnel Guide Line Project[J]. *Journal of Shijiazhuang Railway Institute(Natural Science)*, 2007 (12):71~74.

[4] Zhang J B, Zheng B C, Yu H T, Kang K. Blasting Control Technology for Shallow Large Span Tunnel under Passing Buildings and Monitoring Analysis of Blasting[J]. *Journal of Railway Enginnering Society*, 2011 (7):78~82,98.

[5] Yang N H. Blasting Experimental Study on Vibration Reduction Through Interference Waveform Caused by Electronic Detonator[J]. *Engineering Blasting*, 2013 (6) :41~45.

[6] Zhang G X, Yang J, Lu H W. Resaarch on Seismic Wave Interference Effect of Millisecond Blasting[J]. *Engineering Blasting*, 2009, 15(3) :17~21.

[7] Wei X L, Zheng B X. Analysis of Control Blast-induced Vibrationg & Practice[J]. *Engineering Blasting*, 2009,15(2) :1~6.

[8] Fu H X, Shen Z, Zhao Y, Wang S Q, Tang Y, Yan C X, Cheng L X. Experimental Study of Decreasing vibrion technonlogy of tunnel blasting with digital detonator[J]. *Chinese Journal of Rock Mechanics and Engineering*, 2012, 31(3) : 597~603.

[9] GB 6722—2011 Safety Regulations for Blasting[S]. Beijing: Standards Press of China, 2011.

Gas-Dynamic Hazard in Presentday Highly Productive Mines—Prediction Problems and Solutions

Zakharov V.N., Malinnikova O.N., Feit G.N.
(*Institute of Comprehensive Exploitation of*
Mineral Resources (*IPKON*), *Russian Academy of Sciences, Russia*)

ABSTRACT: Currently, the high-productive mining, with rapid advancing of the extended and high capacity stoping drastically increases geodynamic activity of rock masses, and results in elevated risk of hazardous and even disastrous geodynamic phenomena. In addition, the hazard of gas-dynamic phenomena is greatly contributed to by higher natural gal content and increased strata pressure in deep level mining, as well as by peculial physic-mechanical and gas-kinetic properties of coal-bering rock mass with tectonic faults. This being the situation, neither Russia nor rest of the world possesses sufficient knowledge and practice of deep and high-productive mining in highly gassy and faulted coal fields.

In order to elicit the mechanism of origination of the most information al predictors of gas-dynamic phenomena, the integrated research is now in progress to find out specific and regular characteristics of geomechanical, seismic, acoustic, physicochemical and gas-kinetic processes that take place in high stress, gas-bearing faulted rock mass exposed to in tensive mining and, for this reason, industrially impactad.

The greatest promise in this context is offered by the designed seismo-geomechanical automated forecasting and mine safety control system that is now passing the full-scale tests in mines in the Kuznetsk Coal Field(Kuzbass) and Vorkuta Coal Field.

KEYWORDS: cas-dynamic events; avalanche self-sustained failure; coal seam, face zone; limit stress state; outburst; technological impact

High-productive and high-ratemining nowadays evokesgeo dynamic activity inrock massesand causesh azardous gas-dynamic events in mines. The gas-dynamic hazardisalso contributed to by higher natural gas content and increased strata pressure in deep mines, physico-mechani cal characteristics and gas-kinetic properties of faulted coal and rock mass. Neither Russia nor other countries possess sufficient work experience in deep-level high-production mining of high gas-bearing faulted coal.

In this connection, it is necessary to develop the theory of initiation of gas-dynamic events in the altered natural and industrial environment and create reliable and appropriate monitoring and prediction methods for gas-dynamic safety in mines.

Aiming at revealing prognostic and informative indicants of gas-dynamichazard, complexresearch (physicalandnumericalmodeling, insitu observation) has been carried out into the features and patterns of geomechanical, seismic, acoustic, physicochemical and gas-kinetic processes in highly stressed, gas-bearing faulted rock masses under mining impact [1, 2]. According to the research findings, there are three kinds of hazardous gas-dynamicevents subject to their mining-geological and energy nature as well as determinal factors of their origination:

(1) Rock bursts and tectonic earthquakes;

(2) Out bursts of coal (rock) and gas in faces of excavations and development headings;

(3) Sudden collapse of roof and rupture of floor and gas outbursts in excavations and development headings.

Geoenergy and determin alfactorstaking partininitiation of the listed gas-dynamic events, to be taken into account in the gas-dynamic hazard prediction, are described in the table below.

Tab.1 Gas-dynamicevents, their energy nature and determinal factors

Gas-dynamic events	Energy and determinal factors of initiation
1	2
Rock bursts and tectonic earthquakes	Elastic energy of coal and rock mass. High strength, elasticity and brittleness of rocks (coal). Displacement of rock blocks, seismic vibration energy. Increased strata pressure zones
Coal (gas, rock) outbursts	Elastic energy of coal and rock mass. Energyoffree, liberating and beinggenerated methane in coal (rocks). High gas content and low strength of coal. Low gas per meability and high gas recovery of faulted coal. Zonesofplicativeanddisjunctive low-amplitude geological dislocations. Highandlowstratapressurezones
Sudden roof collapse and floor rupture and gas out bursts in excavations	Elastic energy of coal and rock mass. Energy offree, liberating and being generated methane in coal (rocks). Lamination of rock sinundermined and overcut areas, formation of gas traps. Displacement of rock blocks. Dislocations

Based on the performed theoretical and experimental studies, we have accepted two geomechanical models of initiation of gas-dynamic events in mining-impacted rock masses.

Coal and Gas Out burstsand Rock Bursts in Faces of Excavations and Development Headings. The main determinal fact or of out burstand rock bursthazardin rocks is the accumulated potential energy and gas content of compressed rocks. Elastic energy density in rocks should exceed acertain thresholdvalue for the given geo-material, certain limit stress–strain state and failure energy intensity.

Out burst sand rock bursts are caused by liberation of the excess elastic energy and gas release in altered rock mass under mining. The instability and initiation of avalanche self-sustained failure are conditioned by rapidorigination of new free surfaces during heading advance associated with unloading wave where failure wave develops under condition that the elastic energy density is always higher than the failure energy intensity. Thereforethe energy intensity and behavior of failure incoal (rocks) are defined by the ratio of the effective minimal and maximal principal stresses: σ_3/σ_1.

Some statements of the described model maybe illustrated by physical modeling of failure of coal in triaxial stress state [3, 4]. Fig.1 shows the experimental curve offailure of gas-saturated coal specimens, medium-dislocated, in triaxial stress state, with methane pressure $P = 8\sim10$ MPa ($\sigma_1' > \sigma_2' = \sigma_3'$)under varied lateral compression σ_3'. Here, σ_1', $\sigma_2'$and $\sigma_3'$are the effective stresses determined from the formula: $\sigma' = \sigma - mP$, where, m is the void factor of coal subject to coal jointing and porosity.

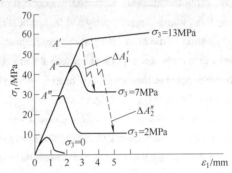

Fig.1 Failure curve for coal intriaxial stress state ($\sigma_1 > \sigma_2 = \sigma_3$) under varied lateral compression σ_3

A', A'', A'''—energy in tensities of failure at the ultimate strength and varied lateral compressive stress σ_3;

$\Delta A'$, $\Delta A''$—excess elastic energy releases under sharp drop of σ_3, MPa

It follows from the figure that when the lateral compression σ_3 is high enough, up to$10\sim13$ MPa, medium-faulted coal starts failing plastically, with the high failure intensity A';and upon decrease in σ_3', the failure energy intensity decreases abruptly from$A''$to$A'''$and coal failure becomes brittle. The similar process runs in the face zone of gas-bearing coal bed during heading advance.

By the physical modeling data and insituobservations inmines, usinghydraulic pressure cells[3~6], the criterion condition for the avalanche failure of face-adjacent rocks is:

$$\Delta W = W_\sigma + W_x \geq \Pi \qquad (1)$$

where ΔW——excess energy release under change in the ratio σ_3/σ_1;

W_σ——elastic recovery energy of unit weight (volume) of coal;

W_x——gas expansion energy inporesandfissures of unit weight (volume) of coal;

Π——surface energy of fractured material per unit weight (volume) of coal.

The experimentally found criteria for the avalancheself-sustained failure in coal under outbursts are:

(1) Elastic energy margin:$W_\sigma + W_x \geq 0.3\sim0.5$ MJ/m^3.

(2) Lateral compression decrease velocity: $V \geq 1\sim3$ MPa/s.

Sudden Roof Collapse sand Floor Ruptures with Gas Outbursts in Excavations and Development Headings. Themain cause of the seevents to initiate are mining and geologicalconditions and geomechanical processes of roof and floor rock lamination in the vicinity of contiguous gas-bearing beds associated with formation of gas traps, accumulation of gas in the traps and, later on, methane rush in mine workings.

Based on the generalized known and published insitudata on Kuzbass mines, mining and geological conditions for sudden roof collapse and floor rupture with gas outbursts in minc workings arc:

(1) All hazardous gas-dynamiceventsoccurinextended(150 m and over) overall-mechanized longwalls with rapid heading advance;

(2) Methanerushesfrom enclosing rocks in the face-adjacent- mined-outareatake place in roof and floor rocks enclosing gas-bearing coal layers or contiguous coal beds, in undermined or overcut zones;

(3) According to the behavior of methane liberation recorded using automated monitoring equipment for methane content in mine air, gas outburst in enclosing rocks is akin to coal and gas outbursts;

(4) Ultimate rockburst- and outburst-hazardous mining depth as well as appreciable elastic potential energy of rocks and energy of gas, $W_\sigma + W_x \geqslant W_{limit}$.

(5) Gas content of coal beds and layers is $X \geqslant 10$ m³/t and gas pressure is $P \geqslant 0.8$ MPa.

Interaction of Geomechanical and Geophysical Fields. Origination and development of gas-dynamic hazardous situation during mining are analyzed using geomechanical and geophysical research data. By the experimental data obtained by hydraulic pressure cells [5, 6], the first geomechanical indicants of initiation of a gas-dynamic hazard source in face-adjacent rock mass in the course of mining can be assumed the increase of the stress concentration coefficient k in the increased strata pressure zone and the diminution of the "buffer zone" (stress decrease zone) l_k.

The geophysical observations showed that rockburst- and outburst-hazardous areas in rocks are characterized by the increased acoustic emission and the change in the coefficient or decrement of elastic wave absorption.

The decrement of elastic wave absorption is:

$$\beta = \alpha_s C_s / \omega \quad (2)$$

where α_s——absorption coefficient;
C_s——elastic wave velocity;
ω——vibration frequency.

It follows from the in situ observations and numerical modeling that the elastic wave absorption coefficient α_s is 1.5~2 times higher in the jointing zone and 1.5~2 times lower in the increased strata pressure (ISP) zone than the average absorption in quiet zones.

The experimental data on stress state and geophysical field in face zone of coal beds during mining, in outburst- and rockburst hazardous and nonhazardous zones are generalized in Fig.2.

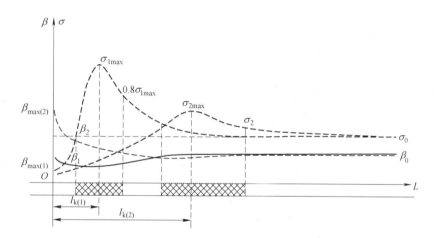

Fig.2　Change in the stress σ and elastic wave absorption decrement β as functions of stress state in faces in hazardous (1) and nonhazardous (2) zones: ISP—increased strata pressure zone

It seems worth noting that the curves in Fig. 2 are plotted based on overall averages. The variation ranges of the data in time and space in a real rock mass are wide but it is highly probable that they will periodically verge toward the first or second type in the course of heading advance, depending on outburst or rockburst hazard at the mine site.

The influence exerted by coal face zone properties and parameters on mining-induced vibrations in enclosing rocks has been modeled numerically with the purpose of forecasting hazard of gas-dynamic events. Heading entails acoustic vibrations in coal and rock mass in the mining area. Kinematic and dynamic characteristics of the vibrations are governed by geological structure, physico-mechanical and seismic-acoustic properties coal and enclosing rocks, as well as by geometrical parameters of excavations, stress-strain state of the coal face zone and amplitude-frequency distribution of the vibrations [7].

Modeling of the effect exerted by the increased strata pressure zones during drivage on amplitude-frequency characteristics (AFC) of the vibrations in rocks has shown that AFC depend on geological structure of coal and rock mass as well as the location of the increased strata pressure zones relative to mining site.

As mining advances into an increased strata pressure zone, high-frequency component of AFC grows faster than its low-frequency component, overruns the latter at a certain time, and the spectrum splits into two groups of vibrations in ranges 550~850 and 850~1100 Hz (see Fig. 3). In fact, a mine working acts as a divider of the vibration amplitude spectrum into high-frequency and low-frequency components, which has been first experimentally confirmed by Mirer, S.V.

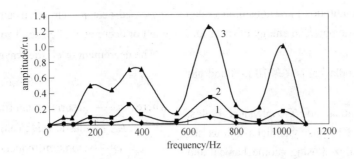

Fig. 3 Amplitude-frequency characteristic of vibrations for different locations of the increased strata pressure zone (ISP)

1—5 m distance to ISP; 2—0 m distance to ISP; 3—5 m inside ISP

Effect of the positive abnormality (increased strata pressure zone) on the amplitude spectrum of vibrations in coal and rocks in the actual mining area shows itself as follows:

(1) maximum amplitude of vibrations grows on approach of the mining site to the abnormality zone and upon advance into it;

(2) on approach of mining to the abnormality zone and upon advance into it, frequencies transfer from low-frequency band to high-frequency band, which describes the effect of the positive abnormality with increased velocities and slow attenuation.

Informative parameters for remote monitoring and outburst hazard prediction on are al-time basis may be amplitude and frequency or energy characteristics of vibrations in rocks, as well as their ratios.

Gas-Dynamic Hazard Prediction by Geoenergy of Coal an- d Rock Mass. According to the geoenergy model, the gas-dyn- a- michazardcriterion, experimentally proved in Kuzbass min- es [8] and recommended as the estimate standard, is:

$$B = (W_x + 0.33W_\sigma)\frac{l_0}{l_k} - 0.18 \quad (3)$$

where B——outbursthazardindex (hazardwhen $B \geqslant 0$; no- nhazard when $B < 0$);

W_x——activeenergyofgasincoalbed;

W_σ——elastic recovery energy of coal bed;

l_0——entry waylength (headingadvanceratepershift);

l_k——size of de-stressing and degassing area in the face zone.

The active energy of coal methane is:

$$W_x = (CX - n)Xb_{30}d\eta \quad (4)$$

where X——gas content of coal bed, m/t;

b_{30}——gasrecoveryfactor;

d——coaldensity, kg/m³;

η——methanedensityinnormalconditions, g/m³;

C, n——empiricalcoefficients.

The elastic recovery energy of coal bed is:

$$W_\sigma = \frac{(k_1 k_2 \sigma_{av})^2}{2E} \quad (5)$$

where σ_{av}——averagestressofintactrockmass: $\sigma_{av} = g\gamma H$, MPa;

k_1——stress concentration coefficient;

k_2——elasticefficiencycoefficient;

E——elasticmodulus of coal, MPa;

γ——average bulk weight of overlying strata;

$g = 9.8$ m/s²;

H——miningdepth.

The quantitative tools for finding the geoenergy (W_x, W_σ) and natural-technical (l_0/l_k) indicants of outburst hazard are feasible and based on geomechanical and geophysical data obtained in the course of geological exploration and regular routine monitoring in mines.

Thepotential (natural) gas-dynamichazard of coal is defined as the sum of the active energy of coal gas, W_x (the work of fast liberated gas from fractured coal) and the energy of strata pressure, W_σ (elastic recovery of coal). The dimension less technological parameter l_0/l_k characterizes stability of the edge area of coal seam at the most dangerous time—heading advance (formation of a new free surface in rocks).

The mining safety prediction interms of rock burst and outburst hazard should be taken as the problem on forecasting safety of a complex nature-and-engineering system of rock mass and mining-impacted environment. The successivehandling of the problem is only possible based on the integrated approach comprising geomechanical, geophysical and geodynamic control techniques and the state-of-the-art IT systems. This approach has been implemented in the automated system of mine safety forecasting in terms of rockburst and outburst hazard, developed by the Institute of Comprehensive Exploitation of Mineral Resources, Moscow, and introduced in mines of Kuzbass. This automated systemismeant for monitoring of any dynamichazard, from coal (rock) and gas outburst storockbursts in development the adingsandexcavations (current and local forecast) as well as for stress–strain state and gas-dynamic control of coal and rock mass in time and space, within the mine field (regional forecast).

REFERENCES

[1] Zakharov V N, Feit G N. Seism-GeomechanicalResearch, PredictionandMonitoringofGas-Dynamicand Geodynamic Hazards in Mining, Georisk-2009 Conference Proceedings, Moscow: RUDN, 2009: 131~135.

[2] Ruban A D, Zakharov V N, Malinnikova O N, Feit G N. Development of the Theory and Methodology for Integrated Seism-Geomechanical Monitoring of Gas-Dynamic Safety in Presentday High-Productive Mines, Conference Proceedings on Geodynamics and Stress State of the Earth's Interior, 2011, Novosibirsk: IGD SO RAN, 2011: 125~130.

[3] Feit G N. Strength and Avalanche Self-Sustained Failure of Stressed Gas-Bearing Coal Bed, Rock Mechanics, 1999, Issue 3113/ 99, Moscow: Skochinsky Institute of Mining, 1999: 63~69.

[4] Feit G N, Malinnikova O N. Characteristics and Regular Patterns of Geomechanical and Physicochemical Processes during Formation of Gas-Dynamic Hazard Sources in Mines, Gorn. Inform.-Analit. Byull., Topic Enclosure: Methane, 2007, Moscow: MGGU, 2007: 192~205.

[5] Dokukin A V, Kusov N F, Feit G N etal. Decrease of Outburst Hazard in Coalunder Dynamic Impact, Moscow: Nauka: 184.

[6] Gaiko E I, Feit G N. Influence of Backfilling on Coal Outburst Hazard, Bezop. Truda Prom., 1983 (1): 54–57.

[7] Zakharov V N, Feit G N, Malinnikova O N, Averin A P. Vibration Energy of Rocks in Zones of Underground Coal Mining, 20th Session of Russian Acoustic Society, Physical Acoustics. Nonlinear Acoustics, Wave Propagation and Diffraction, Geological Acoustics, Moscow: GEOS, 2008, 1: 309~312.

[8] Feit G N, Malinnikova O N, Zykov V S, Rudakov V A. Prediction of Rockburst and Sudden Outburst Hazard on the Basis of Estimate of Rock Mass Energy, Journal of Mining Science, 2002, 38(1): 61~63.

The Implementation and Application of Demolition Data's Informationalized Managerial System

OU Liming[1], XU Cheng[2], ZHOU Ming'an[3]

(*1. Hunan Lianshao Construction Engineering Ltd., Loudi, Hunan, China; 2. Engineer Academy of People's Liberation Army, Xuzhou, Jiangsu, China; 3. National University of Defense Technology, Changsha, Hunan, China*)

ABSTRACT: The management of blasting operation unit contains engineering management, personnel management, financial management and so on. Demolition data's management currently still relies on paper intermediary, so the informationalized management becomes an necessary requirement during the enterprise's development. According to the informationalized managerial requirement of demolition data in the process of blasting operation unit's management, database application system with the SQL language is adopted and the demolition data's managerial system based on the database technology is proposed and implemented. This dissertation researches on analyzing system structure, designing the progress of the system's logic implementation, testing the system's implementation, implementing the storage and the search for varied demolition cases, financial management and fresh data of personnel management. All of those contribute to provide accurate theoretic justification in engineering designing group's implementation and provide timely analysis materials about enterprise operations as well as information about enterprise's talent reserve. The test on system's implementation states that the excellence of system lies in high-capacity storage, stable operating state and keeping pace with the requirement of demolition data's informationalized management from blasting operation unit.

KEYWORDS: demolition data; informationalized management; system establishment

1 INTRODUCTION

At present, civil blasting operation units have established a registration system of civil explosive materials that inputs accurately the explosive materials' data about type, quantity and flowing direction in the purchasing, distributing, storing and applying process into a computer, which realizes the informationalized management of explosive materials. Demolition data's management currently still relies on paper intermediary, and data management contains engineering management, personnel management, financial management and so on. Demolition data's informationalized management is an inevitably demand in the corporate development and the informationalized management of blasting operation unit determines the unit's potential development and scientific management. Hence, it is a urgent need to solve it by exploiting demolition data's informationalized managerial system.

According to the informationalized managerial requirement of demolition data in the process of blasting operation unit's management, database application system with the SQL language is adopted and the demolition data's managerial system based on the database technology is proposed and implemented. The managerial system stores the data of various blasting works implemented by blasting operation units, the units' personnel files and financial data, which facilitates the invocation and consultation at any time.

2 DESIGN REQUIREMENTS

On account of the actual demand of the managerial system's establishment, operation and maintenance, three functions below should be realized in the informationalized managerial system.

2.1 The Management of Employees' Information

Despite being in a small scale, the blasting operation unit is professional with complex staff composition and large personnel mobility. It is easy to lose the personnel data in paper forms. How to solve the overall process of managing dynamically employees' data and monitor duly staff's studying situation and traveling business situation becomes the core point of achieving informationalized managerial system.

2.2 The Management of Blasting Engineering Data

The system classifies and stores the data created in each bl-

---

Corresponding author E-mail: xucheng91@foxmail.com

asting operation, including design proposal, approval data, supervision and assessment data, the summary data of blasting and so on. It is required for the data base to consider fully application needs, design scientifically conceptual pattern and logical structural pattern in designing process, which is convenient to classify and store various blasting engineering data in a database.

2.3 The Inquiry of Blasting Unit's Operation

It directly relates to the unit's benefits and the right operation whether the enterprise's manager promptly and accurately masters the operation, especially of financial status, which is convenient for manager to adjust the operating-scale according the actual occasion and scientifically and accurately manage. The managerial system is required to provide the timely financial status of each department and compare with the operating information in the same period of former years, which functions as browsing data, analyzing statistics, printing report forms and etc.

3 THE OVERALL PROJECT

3.1 The System Structure

The various components of the system should be connected and cooperated internally as a whole. Demolition data's informationalized managerial system collects the information including various blasting designs, excepts' information and managerial and operating data in order to provide a basis for the unit's management and blasting project. System structure is presented as following. It consists of three parts: the software and hardware equipment, application system and the process of collecting data.

The software and hardware equipment includes systematic software, developed software and operation environment and data acquisition equipment. Database application system is the core of the whole system, which applies the database of technique to realize the processing of collected information, user data management, engineering data management, database management and personnel and financial management. Data acquisition process is the source of data stored within the system, including collecting engineering data and managerial information.

3.2 The Structural System of Hardware and Software

The whole system is made up of administrator terminals, user terminals at all levels, sever and network switch. Administrator terminal is mainly to complete the maintenance of system and the system's data addition included users' information at all levels, unit's operation data and engineering design data; user terminals at all levels as the end of the whole system is used for helping all users to complete tasks by this system; sever provides the storage for storing informations, as the source of report forms, recorded withdrawal and designing proposal. The system is exploited by the operating system of Windows 7 at the platform of NetBeans 6.9.1, whose the environment of hardware and software is presented as Tab.1.

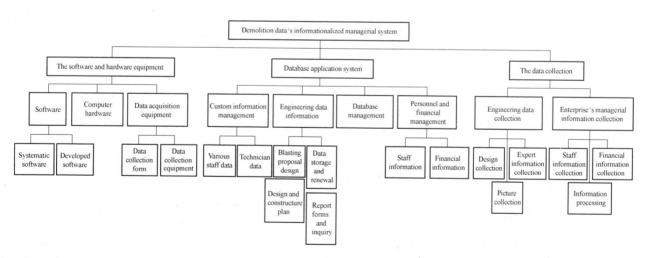

Fig.1　System structure

Tab.1　The environment of hardware and software

System's exploiting environment		System's practicable standards	
Exploiting tool	NetBeans6.9.1	Industrial standard	SW-CMM 1.1
Database	MySQL5.5	Communicating standard	TCP/IP
Operating system	Windows 7	Security standard	System meets C2 at least, information security standard

Continues Tab.1

System's exploiting environment		System's practicable standards	
CPU	P-III 450 256 M RAM	Database	Sybase
Operating system	WINDOWS(at least XP), WINDOWSMOBILE		
Server	HP DL380, Intel Xeon 5410/2.33 MHZ/Quad-core/4 GB/250 GB		

4 THE UTILIZATION OF THE SYSTEM

Demolition data's informationalized managerial system consists of three subsystems, engineering information system, personnel and financial information system and data maintenance system.

4.1 Engineering Information System

The process of engineering information system is at large showed by system flow chart as Fig.2. The utilization of engineering information system starts with the blasting demand and the application delivered to a blasting unit and works by blasting designers allocated by the blasting unit, which are to do a survey of construction site and register the operating situation forms and then to measure and record data about various engineering in the operating site in the collection form, at the same time, to achieve the uses-permission of engineering information system and log into the system. Finally, according to the information in the collection forms and blasting design data stored in the system, the system provides the suggestions to the blasting design and the advices from relevant professors and engineers in the way of comparing circularly informations and successive approximation.

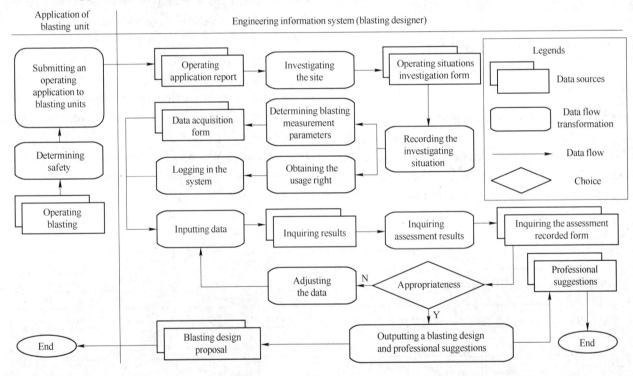

Fig.2 The process of engineering information system

4.2 The Personnel and Financial Information System

The personnel and financial information system is a system of collecting, storing, managing, describing and analyzing staff mobility, working condition and the data of operating situation. The system combines the analysis of human resources and the analysis of finance and rolls collection, management, renewal, comprehensive analysis and operating situation into one. On the one hand, it stores staff information according to the categories and the positions in order to support the data basis for other systems' establishment; on the other hand, the managerial department is easy to retrieval, update and record the operating data, enhance the knowledge of the staff's status in the unit and timely find talents, so that each member of staff in the unit can get the opportunity of studying and improving. Besides, it facilitates the mastery of the current operating situation by the managerial department, helps to work out a development plan and timely adjusts business scale.

4.3 The Data Maintenance System

The data maintenance system will realize higher requirements since the whole system is related to a variety of information, has the susceptibility and significance in an extent and updates in a short cycle. It requires not only to properly protect the system data, but also to strengthen the backup and recovery in the database, which effectively manages the database's security. The data maintenance system needs to implement better the control of the user session, allowing users to access the granted data rather than all the data. Therefore, the system needs to establish the register and entry module. The logging process is showed as Fig.3.

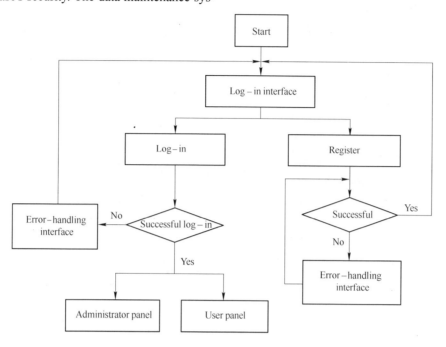

Fig.3   The logging process

The data maintenance system does automatically the back-up on a regular basis throughout the system to secondary storage devices such as the disk and the CD-ROM discs. When data changes in the database, it should be promptly backed up. A database's back-up is stored in a safe place, usually protected in different places. When the database breaks down, the system will return to the recent consistent state as what is before the system crash. In another word, it should be performed a recovery operation on a database in a timely manner to ensure the safe and smooth operation.

5   THE SYSTEM BENEFITS ANALYSIS

The establishment of the system efficiency analysis effectively resolves the difficulties of managing design data and unit data, and realizes the blasting unit's informationalized management as well as the scientific utilization of the blasting information. This establishment covers the whole scale of the informationalized managerial system.

5.1   Social Benefits

The establishment of the informationalized managerial system is the implementation of the specific requirements in The Safety Management of Civilian Explosives Ordinance as well as the favorable attempt and the method and idea innovation of the informationalized management in a blasting unit. What's more, it contributes to implement the informationalized management of engineering blasting and provide a new way of blasting unit management.

5.2   Economic Benefits

By database technology the system has implemented on the blasting design and the scientific store of blasting unit operation and the large save of the human and material resource used for sorting and filing the paper materials; financial supervision ensures that the unit's various appropriation is more precise to reduce the financial loss with the data errors; detailed personnel information management enhances the staff's enthusiasm that motivates staff to work more actively; the system provides a large amount of predicted data for blasting operation unit in the managerial decisions and blasting designs to improve the scientificity, reasonability and accuracy of those.

6   CONCLUSIONS

With the establishment and utilization of the system, many problems in the management and the design have been effectively resolved but at the same time many new problems have appeared. For example, the stored blasting design can not effectively correspond to the investigation site that loses the guiding significance and abnormal problems in the system's working process can not receive the timely solution, which reflects the lag in the awareness of systematic functions and in the consciousness and so on. With the development of social construction, blasting design is increasingly complex, which demands perfecting the original system, supplementing new subsystem and collecting the new blasting design. At the managerial level, with the development of an unit, staff components are more and more complex, so that the system needs to adjust its models to provide accurate data basis for the unit's decision. Besides, the system it self needs to continuously improve its functions and delete its redundant functions.

REFERENCES

[1] Jiang Xiaoyuan. The Integral Designing Study on the Technical System of Military Electronic Information System [J]. Journal of China Academy of Electronics and Information Technology, 2009, 4(1): 7~12.

[2] Zhou Ming'an, Xia Jun, Xiao Zhiwu, etc. The Discussion on the Inspection and Destruction of Explosives Remnants in Wars[J]. Explosive, 2007(2): 82~86.

[3] Zhou Ming'an, Li Shiliang, Xie Linxi, etc. The Design and Realization of Recognition System of Explosives Remnants in Wars [J]. Mining Technology, 2011(5): 100~102.

[4] Miranda D Stobbe, Gerbert A Jansen, Perry D Moerland, Antoine H C van Kampen. Knowledge representation in metabolic pathway databases[J].Briefing in Bioinformatics, 2012 (10):110~114.

[5] Ma Jun. The Function Design in GIS of Urban Drain-pipe Network[D]. Shanghai: Tongji University, 2006: 67.

Effect of Blast Induced Ground Vibrations on Green Concrete

M. Gurharikar[1], M. Ramulu[2]

(1. Rocktech Systems & Projects, Nagpur, India;
2. Central Institute of Mining & Fuel Research, Regional Centre, Nagpur, India)

ABSTRACT: In civil construction projects, the strive for a more time-efficient construction process naturally focuses on the possibilities of reducing the time periods of waiting between stages of construction. As an example, the rock excavation in open cut excavations in metro rail projects requires coordination between the bed concreting and adjacent rock blasting so that blast induced vibrations would not affect the green concrete which is hardly at a distance of 5~10 m. There also arise similar problems in mining, where concrete supporting is required in drivages of weak rock. The grid of drifts in recent mines is very dense for excavating as much ore volume as possible. This means that supporting systems in one drift are likely to be affected by vibrations in a neighbouring drift. A criterion for how close, in time and distance, to the young concrete, blasting can take place would be an important tool in planning for safe and economical mining projects. Another example is the foundation excavation of thermal power projects, where the concreting and rock excavation go simultaneously, adjacent to each other, to crash the time.

In actual practice, numerous occasions arise when it is necessary to build concrete structures at the same time when excavations by blasting are being carried out. For example, lining during tunnel driving, foundation for the primary crushing buildings near open pits, etc. Tab.1 gives prevention criteria given by Oriard (1980) depending upon general curing or hardening time of the concretes. This criterion is based on some assumptions of curing and cannot be made extensive to all types of concrete.

The basis for the existing criterion, during the hardening period of 0 to 4 hours, the concrete is still not hard and the admissible levels are relatively high. From 4 to 24 hours, it begins to harden slowly, and after 7 days it reaches the strength that is approximately 2/3 of the final product (28 days), allowing a progressive intensification of the vibrations.

The guidelines given by the previous research findings are broad and general in nature. Therefore, the authors that it requires further experimentation to correlate the strength acquired by the concrete with the threshold maximum vibration level. With the advent of insitu concrete strength measuring instruments, now a days, it is convenient to monitor strength and corresponding effect of vibrations. In this way the exact effect of vibrations on green concrete can be clearly known to avoid possible damage and deterioration to the structure in long run.

Some experiments were carried out during base slab concrete works of a Metro construction works in India to find out the vibration criteria for freshly poured concrete of age 0 to 4 hours. From the temperature history graphs it was found that the temperature of concrete starts increasing after 2 hours, which is the indication of curing process. At the experimental site the structural concrete was exposed to a blast induced vibration with peak particle velocity (*PPV*) level of 46 mm/s, which was still less that the threshold limit suggested by Oriard (1980). The vibration level was exposed at 3 hour 15 minutes from concreting. The vibration exposure took place in twice with a gap of 5 minute. There was another site with same structural concrete kept without any exposure of vibrations. After 15 days of curing samples of both the sites were tested for strength. Very interestingly the strength of vibration free concrete sample measured was 40.2 MPa and that of vibration exposed samples was 31.5 MPa. This indicates the vibration exposure resulted in strength reduction of 22%, which is very substantial one.

Similar experiments were conducted on samples of 3 hours age and with exposure of *PPV* 48.5 mm/s and observed 23.5% reduction in strength. This indicates that no chance of exposure of vibrations should be taken for the freshly poured concrete of age 2 to 4 hour, although the existing global vibration criteria is liberal for concrete up to 4 hour age. Therefore, the paper recommends that the green concrete up to 4 hour age is also vulnerable for vibrations above 40 mm/s and stresses for need of more research and experiments in this direction.

KEYWORDS: green concrete; Young shotcrete; blasting; vibration; peak particle velocity (*PPV*)

Corresponding author E-mail: more.ramulu@gmail.com

1 INTRODUCTION

Explosive are being used as primary source of energy for rock and concrete breakage in most of the mining and civil engineering applications. The effective utilisation of explosive energy has been reported to be around 15 per cent (Hagan and Just, 1974; Berta, 1990). Major part of explosive energy, which is not used for rock fragmentation, results in generating undesirable side effects like ground vibrations, air overpressure and flyrock. The vibration energy that travels beyond the zone of rock breakage is wastage and results in structural damage and human annoyance. In a properly designed and executed production blast, in general, only 20%~25% of the total explosive energy is utilised in useful work, and remaining 75%~80% of energy is being wasted in generation of ill effects (Rollins, 1980; Berta, 1990).

Ground vibration is an elastic deformation of ground due to blast-induced forces and air overpressure is a transient increase in air pressure (above or below the atmospheric pressure) arising from the detonation of an explosive. The unproductive explosive energy used for generation of vibrations and air overpressure can travel many kilometres under favourable geological conditions and cause disturbance to the structures and habitats in the vicinity. Surface blasting in general, results in both positive and negative effects with regards to the safety and productivity (Fig. 1).

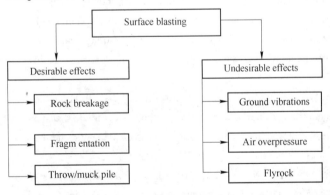

Fig.1 Desirable and undesirable effects of surface blasting

Since blasting is the cheapest method of rock breakage, mining and civil construction projects, in general, adopt the method for rock breakage and excavation. Even small scale blasts near structures and vital installations may generate dangerous vibration levels. Now-a-days prediction of blast induced ground vibrations is becoming increasingly important because of the fact that mining activity is spreading towards populated areas as well as the urban sprawl near mines and quarries. Therefore fine-tuning of the causative and control parameters of the vibrations and air overpressure has become a thrust area of investigation in the present day context.

2 REVIEW ON EFFECT OF VIBRATIONS ON GREEN CONCRETE

Hulshizer and Desai (1984) use the term "green concrete" to describe conventional concrete having an age of less than 24 h after completion of placement. Byfors (1980) defines "fresh concrete" as concrete before setting, "early age concrete" as concrete between setting and approximately 1~3 days of age, "almost hardened concrete" as concrete older than 1~3 days, but younger than 28 days which defines a "hardened concrete". In the following, "young shotcrete" will be used for shotcrete up to one day old.

The effect of vibrations on green concrete has not been fully investigated by researchers in the past. Except few researchers (Oriard, L. L. and Coulson, J. H., 1980 and Issac & Bubb, 1981), no published material on the effect of vibrations on young shotcrete has been found. There are, however, a number of interesting reports on *in-situ* tests conducted in tunnels and mines where fully cured shotcrete on rock have been subjected to blast induced vibrations. Some research attempts were made in the area of influence of vibrations on young shotcreting, which is similar to the effect of green concrete (Ramulu, 2010, Ramulu, 2010).

Kendorski et al. (1973) carried out *in-situ* tests to determine how a concrete lining was affected by standard drift blasts at various distances from the lining. A standard blast that consisted of 409 kg premixed ammonium and fuel oil (ANFO) showed that there was no bond failure at the shotcrete–rock interface. The tests revealed that cracks started to appear in the shotcrete when the detonations occurred at a distance of 16.5 m and that the function of the lining was considerably reduced when detonating from 12.2 m.

In civil construction projects, the strive for a more time-efficient construction process naturally focuses on the possibilities of reducing the time periods of waiting between stages of construction. As an example, the rock excavation in open cut excavations in metro rail projects requires coordination between the bed concreting and adjacent rock blasting so that blast induced vibrations would not affect the green concrete which is hardly at a distance of 5~10 m (Ramulu, 2013). There also arise similar problems in mining. To be able to excavate as much ore volume as possible, the grid of drifts in a modern mine is dense. This means that supporting systems in one drift are likely to be affected by vibrations in a neighbouring drift. A criterion for how close, in time and distance, to the young concrete, blasting can take place would be an important tool in planning for safe and economical mining projects. Another example is the foundation excavation of thermal power projects, where the concreting and rock excavation go simultaneously, adjacent to each other, to crash the time.

In actual practice, numerous occasions arise when it is necessary to build concrete structures at the same time when excavations by blasting are being carried out. For example, lining during tunnel driving, foundation for the primary crushing buildings near open pits, etc. Tab.1 gives prevention criteria given by Oriard (1980) depending upon curing or hardening time of the concretes, although such recommendations cannot be made extensive to all types of concrete.

As can be observed, during the hardening period of 0 to 4 hours, the concrete is still not hard and the admissible levels are relatively high. From 4 to 24 hours, it begins to harden slowly, and after 7 days it reaches the strength that is approximately 2/3 of the final product (28 days), allowing a progressive intensification of the vibrations. Other factors to take into account are the characteristic frequencies of the vibrations, external hardening conditions, area of rock-concrete contact, etc. On the other hand, Issac & Bubb (1981), summed up all their experiences and those of Scandinavian investigators in a graph where, according to strength acquired by the concrete, the maximum vibration level is determined.

Tab.1 Setting time of concrete versus admissible levels of particle velocities

Setting Time/h	Admissible level of PPV/mm·s^{-1}	
	Structiral concrete	Fill and mass concrete
0~4	50	100
4~24	5	25
24~72 (1~3 d)	25	30
72~168 (3~7 d)	50	75
168~240 (7~10 d)	125	200
>240 (>10 d)	250	375

The guidelines given by the previous research findings are broad and general in nature. Therefore, it requires further experimentation to correlate the strength acquired by the concrete with the threshold maximum vibration level. Therefore, it is very essential to know the exact effect of vibrations on green concrete to avoid possible damage and deterioration to the structure in long run.

3 FIELD EXPERIMENTATION

Blasting of hard rock was carried out as part of the excavation works in connection with proposed under ground stations for Bangalore Metro Rail Corporation Limited (BMRCL), at different locations at the majestic station in Bangalore city, India. Controlled blasting operations were carried out keeping in view of the adverse effects like ground vibration and air overpressure and flyrock. Blasting and concrete works used to go hand in hand to meet the time schedules and targets of the project. Therefore, freshly poured concrete was exposed to blast induced ground vibrations at the distance range of 10 to 25 m from the blasting sites as shown in Fig. 2. In such situations it was imperative to know the exact levels of ground vibrations, which cannot cause any internal and external damage to the green concrete.

The following sets of experiments carried out for assessing the effect of blast induced ground vibrations on green concrete:

(1) Close field ground vibration monitoring at freshly poured concrete by triaxial geophones.

(2) Measurement of curing characteristics like strength and temperature release of freshly poured concrete/green concrete by a device called Zone-cure at vibration free zone.

(3) Measurement of curing characteristics like strength and temperature release of freshly poured concrete/green concrete at vibration exposed zone.

Fig.2 Freshly poured concrete near blasting sites at Bangalore Metro Rail Project

Some experiments were carried out during base slab concrete works of a Metro construction works in India to find out the vibration criteria for freshly poured concrete of age 0 to 4 hours. From the temperature history graphs it was found that the temperature of concrete starts increasing after 2 hours, which is the indication of curing process. At the experimental site the structural concrete was exposed to a blast induced vibration with peak particle velocity (PPV) level of 46 mm/s, which was still less that the threshold limit suggested by Oriard (1980). The vibration monitoring was carried out by engineering seismograaph of Instantel make (Canada), which contains triaxial geophones as shown in Fig. 3. The vibration level was exposed at 3 hour 15 minutes from concreting. The vibration exposure took place in twice with a gap of 5 minute. There was another site with same structural concrete kept without any exposure of vibrations. After 15 days of curing samples of both the sites were tested for strength. The in-situ strength was tested by a Concrete Maturity Testing Systems called ZoneCure. Very interestingly the strength of vibration free concrete sample measured was 40.2 MPa and that of vibration exposed samples was 31.5 MPa. This indicates the vibration exposure resulted in strength reduction of 22%, which is very substantial one.

Fig. 3 Vibration monitoring by engineering seismograph near the green concrete

Similar experiments were conducted on samples of 3 hours age and with exposure of *PPV* 48.5 mm/s and observed 23.5% reduction in strength. This indicates that no chance of exposure of vibrations should be taken for the freshly poured concrete of age 2 to 4 hour, although the existing global vibration criteria is liberal for concrete up to 4 hour age. Therefore, the paper recommends that the green concrete up to 4 hour age is also vulnerable for vibrations above 40 mm/s and stresses for need of more research and experiments in this direction.

4 RESULTS AND DISCUSSIONS

Both strength and temperatures were monitored by fixing concrete curing measurement system Zone-cure at the site. The strength was measured for every 12 hours for a period of 15 days and temperature was measured for every 2 hours for a period of 24 hour. The experiment of strength and temperatures was carried out for a green concrete, which was exposed to a vibration level of 46~48.5 mm/s. The same experiment was carried out for a green concrete, which was not exposed to any vibration level. The experimental results of strength and temperatures are given in Tab. 2 and 3 respectively. The temperature monitoring indicate that curing process i.e. cement bonding starts between 2 to 4 hours as the raise in temperature is evident. The fall in strength after 15 days to the vibration exposed concrete indicate that there is substantial influence of vibration even at the levels at of 46~48.5 mm/s.

Tab. 2 Experimental results of strength versus time in green concrete

Sl No.	No. of hours	Strength of green concrete without exposing vibrations	Strength of green concrete with exposing vibrations (46~48.5 mm/s)
1	12	7.5	4.5
2	24	15.5	11.5
3	36	20	14.75
4	48	25	19.5
5	60	27.5	21.75
6	72	30.3	24.3
7	84	33.5	27.25
8	96	35	28.5
9	108	36	29.25
10	120	37	30
11	132	38	30.75
12	144	38.5	31.25
13	156	38.75	31.25
14	168	39	31
15	360	40.2	31.5

Tab. 3 Experimental results of temperature versus time in green concrete

Sl No.	No. of hours	Strength of green concrete without exposing vibrations	Strength of green concrete with exposing vibrations (46~48.5 mm/s)
1	0	28	28
2	2	25	25
3	4	27	26
4	6	32	29
5	8	40	37
6	10	51	45
7	12	58	50
8	14	55	48
9	16	50	45
10	18	47	42
11	20	45	40
12	22	40	39
13	24	36	35

5 CONCLUSIONS

The effect of vibrations on green concrete is an important issue in most of the civil construction projects where both blasting and concreting go simultaneously for crashing the duration of projects. Some experiments were carried out during base slab concrete works of a Metro construction works in India to find out the vibration criteria for freshly poured concrete of age 0 to 4 hours. From the temperature history graphs it was found that the temperature of concrete starts increasing after 2 hours, which is the indication of curing process. At the experimental site the structural concrete was exposed to a blast induced vibration with peak particle velocity (*PPV*) level of 46 mm/s, which was still less that the threshold limit suggested by Oriard (1980). The vibration level was exposed at 3 hour 15 minutes from concreting. The vibration exposure took place in twice with a gap of 5 minute. There was another site with same structural concrete kept without any exposure of vibrations. After 15 days of curing samples of both the sites were tested for strength. Very interestingly the strength of vibration free concrete sample measured was 40.2 MPa and that of vibration exposed samples was 31.5 MPa. This indicates the vibration exposure resulted in strength reduction of 22%, which is very substantial one.

Similar experiments were conducted on samples of 3 hours age and with exposure of *PPV* 48.5 mm/s and observed 23.5% reduction in strength. This indicates that no chance of exposure of vibrations should be taken for the freshly poured concrete of age 2 to 4 hour, although the existing global vibration criteria is liberal for concrete up to 4 hour age. Therefore, the paper recommends that the green concrete up to 4 hour age is also vulnerable for vibrations above 40 mm/s and stresses for need of more research and experiments in this direction. The lower time limit of 0~4 hrs should be kept of 0~3 hrs hours as the cement starts curing after three hours and the vibration threshold of 50 mm/s should be changed as 40 for structural concrete.

ACKNOWLEDGEMENTS

The authors thankfully acknowledge the cooperation extended by Coastal Projects Ltd & Bangalore Mtro Rail Corporation during the field studies. The authors also express their thanks to Director, CSIR-CIMFR for his cooperation and encouragement for this studies and for permitting to publish this paper. The views expressed in the paper are those of the authors and not necessarily of the organizations they represent.

REFERENCES

[1] Byfors J. Plain concrete at early ages, Research Fo 3:80. Swedish Cement and Concrete Research Institute, Stockholm. 1980: 345.

[2] Hagan T N, Just J D. Rock breakage by explosives-theory optimisation, Proc. 3rd Cong. Rock Mech., 2, 1974: 1349~1358.

[3] Hulshizer A J, Desai, A J. Shock vibration effects on freshly placed concrete. J. Constr. Eng. Manage. 110, 1984: 266~285.

[4] Issac I D, Bubb C. "Geology at Dinorwic", Tunnels and Tunnelling, British Tunnelling Sot., 1981, 11(3): 20~25.

[5] Kendorski F S, Jude C V, Duncan W M. Effect of blasting on shotcrete drift linings. Min. Eng. 25, 1973: 38~41.

[6] Oriard L L, Coulson J H. "TVA Blast Vibration Criteria For Mass Concrete", Proc. Conf. of ASCE, Portland, OR, Preprint 1980: 80~175.

[7] Oriard L L. "Near Source Attenuation of Seismic Waves From Spatially Distributed Sources", Proc. 8th Annual Symp. on Explosives and Blasting Research, Soc. of Explosives Engineers, Orlando, FL, 1992: 83~96.

[8] Ramulu M, Raina A K, Choudhury P B, Sinha A. Special tunnel blasting techniques- CIMFR contributions, November, 2010, Workshop on Tunnelling Quo Vadis? Kolkata,2010: 1~18. Books & Journals Private Ltd.

[9] Ramulu M. Controlled blast design, review and monitoring of ground vibrations, air overpressure and flyrock at BMRC project, Phase1, UG1, Bangalore, CIMFR Report, 2013.

[10] Ramulu M, Sitharam T G. Blast induced damage due to repeated vibrations in jointed gneiss rock formation, Int. Journal of Geotechnical Earthquake Engineering, IGI publications, USA, 2010(5): 31~39.

[11] Rollins. Energy partitioning, Rock mechanics and explosive research center, University of Missori-Rolla, 1980.

Achievement of Safe Distance of Air Shock Wave in Tunnel Blasting

YU Haihua, YANG Haitao

(*Maanshan Institute of Mining Research Blasting Engineering Co., Ltd., Maanshan, Anhui, China*)

ABSTRACT: Safe distance is computed by using overpressure function of underground mining blast air shock wave, which is employed inconveniently in the actual site, Nomograph method can solve effectively the problem, overpressure calculating equation of blast air shock is converted into line chart or graph in order to solve related problems. In accordance to the calculation formulas of the air overpressure underground mining, the human body can bear the overpressure, that is as a fixed value, and we can obtain graphic relationship of explosive amount, safety distance, tunnel structure parameters by employing Nomogram method, soon after, in the engineering background of tunneling, we assure that safe distance of air shock wave of man is 195m, and according to the characteristics of the tunneling blasting air shock wave, the measures of prevention and control of blasting air shock wave is put forward.

KEYWORDS: tunnel blasting; air shock wave; safety distance; nomograph

1 INTRODUCTION

In the tunnel blasting excavation process, with the explosion of explosive, high pressure gas is rushed up from blast hole and surrounding air is condensed by throwing rocks, the air shock wave is formed. Tunnel blasting air shock wave has the characteristics of one dimensional linear transmission, transmission distance, intensity attenuation slower and so on along tunnel, it leads to scaffold collapse down equipment damage cracking collapsed buildings/structures, etc., and endangers the safety of the workers. The destruction of the air shock wave is mainly related to peak pressure of shock wave front, shockwave effect time positive pressure area, shockwave impulse, shock wave effect to the natural vibration period of buildings, shape, strength and so forth[1, 2]. Hence, determination of the safe distance of explosive blasting air shock wave is very important.

The strength of the air shock wave is comprehensive result of many factors, that include initiating charge, hole filling, tunnel area and the surface roughness, explosive source center distance and so on. From the viewpoint of mathematics, overpressure wave intensity equation belongs to the multi-dimensional function of many variables, it is difficult to calculate, the calculation of formula is also much multifarious, Nomograph method could effectively solve the problem. Based on a certain geometry, Nomograph method depict function relation between the several variables of mathematical equations as a straight line or curve corresponding calculation chart[3].

This text serves the safety distance of blasting air shock wave to human body as constant value, according to a certain tunnel excavation engineering practice, the blasting air shock wave overpressure formula is combined with Nomograph method, we obtain the relationship between safe distance and the quantity of explosive, then people's security allowed distance is calculated. According to the harm of blasting air shock wave, the protection measures of tunneling blasting air shock wave are further put forward.

2 BASED ON THE SAFE DISTANCE OF AIR SHOCK WAVE WITH THE NOMOGRAPH METHOD

2.1 Blasting Air Shock Wave Formula and Simplified

Air shock wave overpressure calculation formula[4]:

$$\begin{cases} \Delta P = \left(3270\dfrac{qm_y}{r\sum S} + 780\sqrt{\dfrac{qm_y}{R\sum S}}\right) e^{\frac{\beta R}{d_n}} \\ d_n = \sqrt{4S_n/\pi} \end{cases} \quad (1)$$

where, ΔP is tunnel blasting air shock wave overpressure, kPa; d_n is tunnel diameter, m; S_n is the sectional area of the tunnel, m²; q is the equivalent of explosive TNT, kg; R is away from center distance of explosive source, m; $\sum S$ is adjacent to the total area of the tunnel of cartridge, m²; m_y is explosive energy conversion coefficient, see Tab.1; β is tunnel road surface roughness coefficient, see Tab.2.

Type (1) shows that, the relation between air shock wave overpressure ΔP and distance of explosive source R is a complex relation of nonlinear function. When overpressure

---

Corresponding author E-mail: yht 03121 @ 163. com

ΔP is known, the distance from blasting source R can't be solved out through simple mathematical operation, therefore, the analytical calculation method is difficult to use at the scene.

Tab.1 Explosive energy conversion coefficient

Blast hole stemming	Explosive energy conversion coefficient m_y
Stemming length of shallow hole is 1m	0.013~0.016
Uncharged shallow hole's length is 2.5m	0.015~0.019
Uncharged shallow hole's length is 1m	0.020~0.024
Stemming length of shallow hole is 2.5m	0.024~0.027
Shallow hole is all charged	0.025~0.030
Tunnel boring hole blasting	0.050~0.100
Clamp system under the condition of shallow hole blasting	0.300~0.350

Tab.2 Tunnel road surface roughness coefficient

Support type	Tunnel road surface roughness coefficient β
Concrete support	0.010~0.015
Wooden frame of incomplete stent	0.025~0.034
Arch support	0.025~0.060

To simplify the formula, we suppose that $K = q\, m_y / \sum S$, $i = \beta / d_n$, K is the amount of explosive per unit area into the air shock wave; i is the coefficient of air shock wave attenuation speed, i value is grater, the attenuation of blasting air shock wave is quicker. Simplified air shock wave overpressure formula:

$$\Delta P = \left(3270 \frac{K}{R} + 780 \sqrt{\frac{K}{R}} \right) e^{-iR} \qquad (2)$$

Generally, the safety design of the tunnel blasting is based on ΔP, K, i to calculate air shock wave security allowed distance. Because i value depends on the geometry size and surface roughness of the tunnel, specific blasting environment is as a fixed value, ΔP has also its corresponding safe distance for the specific protection object[5]. According to above analysis, value of ΔP, i is as constant value, the relationship of number or graphics between K and R is established, nomograph of air blast safety distance is traced out. By the mathematical transformation, the formula (2) is conversed as fellow:

$$K = \left(\sqrt{\frac{\Delta P e^{iR}}{3270} + 0.014} - 0.12 \right)^{2R} \qquad (3)$$

2.2 Drawing of Nomograph

According to the shock wave overpressure to the damage degree of human body, when the human body bears shock wave overpressure, that is 0.002×10^6 Pa, it is harmful to human body. The quantitative relation between explosive charge K and distance from blasting source R can be obtained by selecting different fixed value of i (see Tab.3), and workers' security allowed distance nomograph Fig. 1 is drawn.

Tab.3 The quantitative relation between i, R, K

i \ R \ K	50	100	150	200	215	230	250
0.001	0.030	0.056	0.093	0.194	0.223	0.246	0.289
0.002	0.035	0.083	0.149	0.272	0.309	0.367	0.475
0.003	0.038	0.104	0.187	0.328	0.417	0.502	0.613
0.004	0.041	0.121	0.250	0.462	0.573	0.668	0.812
0.005	0.044	0.135	0.311	0.638	0.778	0.987	1.214
0.007	0.049	0.192	0.520	1.229	1.613	2.169	2.663
0.008	0.058	0.231	0.671	1.687	2.215	2.962	3.881
0.009	0.067	0.276	0.853	2.304	3.589	4.734	5.611
0.010	0.073	0.325	1.121	3.138	4.327	5.958	7.987
0.012	0.085	0.451	1.733	5.562	7.139	10.229	15.786
0.014	0.105	0.621	2.703	9.648	12.758	18.673	29.993
0.015	0.112	0.726	3.362	12.603	18.997	29.548	40.905

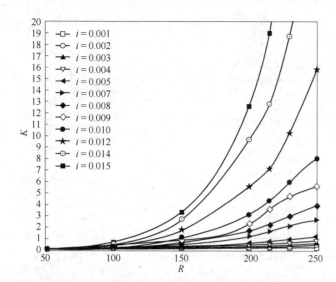

Fig.1 Blasting air shock wave overpressure is action to the safety distance nomogram of human body

2.3 Application of Nomograph

A tunnel uses shallow hole blasting of wedge cut way, tunnel sectional area is 32.0m², the concrete scaffold is used, a priming dose converts to TNT that is equivalent to 350kg, the length of hole filling is 1m. According to tables 1 and 2 and air shock wave attenuation adverse situation, we choose $\beta=0.015$, $m_y=0.025$, so $d_n=6.4$, $i=0.002$, $K=0.27$.

According to Fig.1, the ordinate $K = 0.27$ corresponds to value $R=195$m, that is the safe allowed distance in curve $i = 0.002$. K, i, R value are substituted into the formula (2), we calculate and obtain $\Delta P=0.018\times 10^5$Pa, the value is close to the human body supporting standard 0.02×10^5Pa of overpressure, it explains that application of nomogram calculations reliably safe distance, that meets engineering design.

3 TUNNEL BLASTING AIR SHOCK WAVE PROTECTION MEASURES

It is based on the characteristics of tunnel excavation blasting, in order to ensure the safety of personnel, buildings, facilities, etc, blasting air shock wave must be controlled and protected, the overpressure value is lower than the safe allowable value. When operating conditions can't meet the requirements of explosion and the safe distance, we need to build near obstacles near by being protected objects, the strength of the blasting shock wave is weakened.

There are four kinds of methods for being response to blasting air shock wave: preventing a strong shock wave, decreasing the strength of the shock wave, expanding shock wave channel, setting the discharge wave channel. Resistance wave wall is set in the blasting area or near protected object, air shock wave is generating or transmitting to be weakened. The specific measures are as follows [5].

3.1 Sandbags Wave Walls

Sandbags resistance wave wall has the characteristics of the low construction price, short construction time, according to the protected object size, importance and intensity of shock waves, its long, high and thick size are determined. It is the use of sand soil to be piled, that can be applied to the ground and underground blasting, tunneling blasting. Generally, when the intensity of shock wave is increased, the length of the sandbag wave walls will also be lengthened to prevent sandbag be blew away. Therefore, in order to ensure reliability, on its outer surface cover or tunnel wall we firmly fix with the ground of barbed wire, cable, etc, a hybrid wave wall is built to improve its wave ability.

3.2 Prevention Wave Row Column

Prevention wave row column is made up of lumber, whose diameter and distance is between 200～250mm, the lumber is positioned into checkerboard along tunnel's length direction. In order to improve the columns' stability, the columns should advance needling from the shock wave direction, the lumbers' length is longer than tunnel's height to 200 mm or so. Prevention wave row column's length is general 10~20m, the one can be up to 50m. Because it can bring shock wave's energy consumed in the column circulation or the increase of tunnel resistance, air shock wave could be weakened.

3.3 Pigsty Resistance Wave Wall

The pigsty resistance wave wall is made up of round log or sleeper, their diameter is 100~300mm. In order to improve the intensity of prevention wave wall, component or end face is fixed by nail with the wedge tight on the both sides of roadway. When shock wave is much intense, two or three layers prevention wave wall are constructed along the roadway. When air shock wave is upon pigsty resistance wave wall, shock wave may be great weakened, attainability 2/3, because the resistance wave wall components' partial reflection, diffraction, and the longitudinal wave and transverse wave. In order to protect particularly important object in the pit and prevent explosive magazine to blast, adjacent tunnels and facilities are destroyed, coagulation soil resistance wave wall, wall and other large resistance wave wall rock resistance are constructed. In order to protect distant independent structure away from explosive source, we may select activities or flexible wave walls light wave walls.

3.4 Protective Bent

The stake or bamboo pole can be used as bracket, coverings straw is set up into the protective shelf in the control blasting, it has an effect on shock wave in reflection on, guide and cushion, that can better play the role of weakening the air shock wave, single row can general lower the shock wave's intensity, that is about 30% ~ 50%. hence, its application is very broad.

According to being protected object, protective bent frame's size is confirmed, its strength is for the intensity of shock wave, the ability of the resistance shock wave and its importance of the protected object to decide. The protective bent frame's shape is general herringbone, the angle of blasting surface and the ground should be bigger 60° ~75° than the angle of back and ground, and blasting surface should be strengthened. In order to delay protection bent displacement time under the effect of shock wave and enhance its stiffness, protection bents' brace should be imbed. In order to focus on protected objects, double rows or more support should be set up, the row spacing is 4~6m.

3.5 Other Measures to Prevent and Control

Besides the air shock wave control measures, mantle covers in the blasting source, for example, covering sand, straw bag, hose shade, waste tires shade or rubber curtain and so on. For construction, the windows are opened and managed to fix, or it is picked off. If indoor equipment is protected, we may use plank or sand packet to seal door and window.

4 CONCLUSIONS

(1) According to combining blasting air shock wave overpressure calculation formula of underground mining with Nomograph method, we can effectively solve the blasting air shock wave overpressure formula calculation. Based on the blasting air shock wave overpressure ΔP and K, i, R function relation, ΔP and i will be set constant value, curve graph of variable K and R is plot, we plot tunneling blasting air shock wave acts on multiple curve nomograph of person body safe distance. In view of tunneling blasting engineering background, according to plotting nomograph, we obtain that person body safe distance of tunneling blasting air shock wave is 195m.

(2) According to tunneling blasting characteristic, the security arrangement of prevention blasting air shock wave is come up with, we may select sandbags wave walls, prevention wave Row column, pigsty resistance wave wall, protective bent and other measures to prevent and control.

REFERENCES

[1] Li Z. Air shock wave function person's safe distance[J]. *Explosion and Shock*, 1990(2): 136~144.

[2] Li Y M, Hu F. Down hole blasting air shock wave propagation law of dimensional analysis and fitting[J]. *Mining Metallurgical Engineering*, 1993(02): 13~17.

[3] Li K S. Nomograph analytical method[J]. *Journal of dongguan institute of technology*, 2001(1): 26~33.

[4] Lan C G, Wang, C G, Zhang Chengchao. Downhole blasting air shock wave safe distance Nomograph achievement[J]. *Blasting*, 2013(3): 39~42.

[5] Wang X J. Blasting air shock wave and its protection[J]. *Journal of the Chinese people's public security university(Natural Sciences)*, 2003(4): 41~43.

[6] Wang X G, Yu Y L, Liu D Z. Implementation of the manual blasting safety procedures[M]. Beijing: *China Communications Press*, 2004: 157~158.

[7] Gu Y C, Shi Y Y, Jin J J. Blasting safet[M]. Beijing: *China Science and Technology Press*, 2009: 480~481.

Technology for Safety and Environmental Control in Blasting Operations

Sushil Bhandari[1], Reema Bhandari[2]

(1. Earth Resource Technology, New Pali Road, Jodhpur; 2. Continuous Excellence, Melbourne, Australia)

ABSTRACT: Blasting operations cause several adverse environmental effects and may result in safety problems. With the development of new explosives systems and initiation devices, blast design and execution techniques, the blasting process has now become more efficient and safer than before. Use of tools for blast design, support in execution, blast monitoring and analysis make it possible that damages and dangers from blasting can be predicted before blasting. Several software are available for prediction of vibrations, airblast, wave front reinforcement, flyrock, dust plume movement resulting from blasting. This paper also gives examples of prediction tools can be used before carrying out blasts to control environmental impacts and follow safety norms.

KEYWORDS: blasting software; blast information; environment impact; blast predictors; blasting safety

1 INTRODUCTION

Global market for explosives is projected to reach 15.8 million tonnes by 2018 driven by growing demand from emerging economies and technology advancements focussed on development of safe blasting technologies. More than 75% of the production is consumed in mining operations alone across the globe. Despite the immense volume of explosives used, serious incidents involving the legal use of explosive materials are rare. Unfortunately, when incidents involving explosives arise, they are well publicized, whereas the great majority of blasting projects—even the most challenging ones—quietly occur without incident or public recognition. However, there are many examples, throughout the mining industry where, through an apparent lack of understanding of the basic principles governing efficient drilling and effective blasting, mines are not operating at their optimum—in terms of overall production costs, energy utilization, resource management and rock stability whilst, at times, jeopardizing safety or creating adverse environmental impacts such as ground and air vibration hazards, dangerous fly rock incidents, dust and fume hazards.

Although unfair, media coverage focuses on a few incidents and/or accidents, and as a result skews the public's perceptions of blasting. Many projects have been halted or delayed by community groups concerned about the potential damage or environmental impacts of blasting. As regulations and specifications become increasingly more restrictive, the net effect is an increase in the cost of rock excavation. Managing risks associated with blasting is becoming an ever increasing challenge as many mines are located close to populated areas. Not only is the work closer to people, structures and utilities, but environmental concerns about blasting effects on animals are also increasing. Inadequate control of ground and air vibrations and flyrock leads to complaints and agitations. Adequate attention has not been paid to reduce the generation of fines. Many fines get airborne causing environmental dust problems.

In order to improve performance, the drilling and blasting industry the world over is rapidly adopting new technology in all forms. This has been possible because the basic steps in blast engineering are followed — design the blast according to rock face conditions and results required with environmental considerations, execute the blast while controlling the actual drilled pattern and then appropriately load explosive products, monitor several parameters during the blast event and evaluate the outcome of the blast. Results are stored online for retrieval and for data analytics. Recently developed tools and technologies can enhance and streamline the process for the optimization of blast results with safety and environmental controls. The objective of this paper is to enumerate innovative blasting practices which could be adopted specially information technology related for safety and controlling adverse environmental controls in blasting operations.

Corresponding author E-mail: sushil_bhandari@hotmail.com

2 POTENTIAL BLASTING IMPACTS

Before blasting begins in new areas, it is important to define how the blasting might impact neighbors, animals, structures and the environment in general. Obviously, the degree of risk and impacts will vary depending on the nature of the blasting work. For instance, a downtown building implosion project will have very different risks from a proposed mining project in a national forest. Blasting programs in urban areas must control flyrock, vibration and air overpressure, whereas, the project in the national forest will likely have strict environmental controls on water quality and animal impacts. Due to the growing involvement of governmental agencies in the permitting process for new projects, especially large ones, potential environmental impacts are usually identified beforehand and mitigation measures are explained in environmental impact studies. The risks associated with blasting are wide ranging and some are quite unique. Tab.1, although not complete, can be used as a general aid for identifying and developing measures to manage blasting risks.

Tab.1 Impacts of blasting

Concern	Primary Impacts	Controls
IMPACTS DUE TO VIBRATION		
Startled or scared people	Complaints	Inform neighbors before each blast. Carry out blasting at fixed time. Take appropriate safety measures
Damage to buildings	Damage claims, work delays or suspension	Blasting controls, blast plan reviews, careful inspection of work, monitoring and pre-blast condition surveys
Damage to rock slopes and final excavation walls	Rock falls, remedial slope repairs and work disruption	Evaluate in situ condition of slopes and install additional support if needed. Develop blasting controls and carefully monitor the work
Damage to buried pipes and utilities	Unrealistic restrictions on blasting or total ban on blasting	Pre-qualification requirements, blasting controls, blast plan submittals and reviews, careful inspection of work, monitoring and blasting effects evaluation study by expert
Damaged water wells or aquifers	Blasting prohibition or project delays	Blasting controls, pre-blast and/or post-blast inspections, monitoring and blasting effects evaluation study by expert
Ground Vibration effects on historical buildings, structures of importance	Damage to historical monuments, dams, high rise buildings	Great care must be taken to avoid damage to historical buildings as they may be sites of special scientific or tourist interest
ENVIRONMENTAL IMPACTS		
Impacts on Animals	Blasting prohibition or project delays	Blasting impacts and mitigation study, pre-blast and/or post-blast inspections
Fines	Both environmental issue and an economical loss for the producers	Regularly conduct Process Optimization Audits to ensure that they produce the lowest achievable proportion of fines
Noise (Airblast)	Potential of disturbance to local communities	Measure airblast noise and design blasts and appropriate stemming
Agriculture	Reduction of agricultural land may reduce crop yield due to soil pollution & air pollution	Strict control measures to prevent air & water pollution
Water pollution	Leaching of chemicals in blast holes	Use water resistant explosive and provide proactive sheath to explosive such as liner
IMPACTS ON SAFETY		
Flyrock	Damage and Injury to personnel and neighbours, damage to machinery and buildings	Blasting controls (blast mats—burden requirements—stemming requirements), blast plan submittals and reviews, and careful inspection of work
IMPACTS ON HEALTH		
Fumes and gases	Harmful oxides (NO$_2$) may be released as fumes. The fumes dissipate readily, but if there is a temperature inversion and little wind, then they can linger	While using ANFO appropriate prill and the right amount of diesel be used ensuring that it is thoroughly mixed. Incomplete detonation of the explosive or incorrect mixing of the ANFO and presence of excess water may occur
Dust	Dust released is a health concern for on-site personnel and surrounding communities	Watering of the blast area. Delaying blasting under unfavorable wind and atmospheric conditions. Use appropriate blast design and explosives
OTHER IMPACTS		
Work or business disruption	Financial damage claims and/or organized opposition	Public education, blasting controls, monitoring and schedule blasting during non-working period

3 TOOLS AND TECHNIQUES

Several tools can be used in assessing blast face conditions, assist in designing blast, executing blasts, during and after blast monitoring, recording all results and analysing records to predict outcome and suggest remedial techniques for controlling resulting risks. Many tools and techniques have taken the guesswork out of explosives loading and blasting operations (Rodgers, 1999). Some of them are:

Face Profiler systems using laser technology profiles the rock face by pointing profiler to the floor, toe and crest—then take extremely accurate measurements and calculates bench heights, minimum and optimum burdens, computes drill hole angles and offsets, and hole depths. Drills may have a tendency to follow the faults, weakness planes, weak rock, cavity or similar other geological weakness creating borehole deviation. When operators start using electronic

Hole Deviation measurement tools, operators discover light burden areas and are able to thwart safety issues. The cable deviation measurement tools uses sensors that measure borehole deviation at fixed intervals from the collar position and transmit the findings instantly to a field computer.

The second tool is the 3-D laser profiling system, which includes equipment and software that allows the user to make precise burden measurements without venturing near the crest of the bench. This system, too, transmits its findings instantly to a field computer. From these field computers, the data from both tools are merged in blast design software to adjust the blasting plan by compensating for any initial "guesstimation" of the burden and for drilling inaccuracy.

Another technology that has great importance for drilling accuracy, and the integration of drilling and blasting operations is GPS as applied to drill positioning on individual blastholes. GPS based systems allow the blast plan with hole locations to be downloaded to the drill. Some form of moving display is used to guide the drill onto the designed hole location. The drill can be positioned to within about one third meter of the designed location. The system records exactly where the hole is drilled. The final design is downloaded to the drill equipped with GPS capability. The drill machine can then drill the blastholes accurately on the designed locations without the need for extensive field surveying.

If the mine is new, selection will have to be made between cartridge and bulk explosives depending on the proposed volume of excavation per month. Once this is made, the choice can be exercised amongst ANFO, HANFO, slurry and emulsion depending on the characteristics of rock vis-à-vis its response under dynamic loading during blasting. Selection of initiation system and timing need to be made keeping in view the environmental constraints.

The most significant changes in blast technology have taken place in product-delivery systems. One factor is the continuing trend away from the use of cartridged products in favor of bulk products for both surface and underground operations: new surface and underground delivery-vehicle technologies that boost blast accuracy and safety: high-precision pumps and blending and measurement devices, robotic arms that place the product in the hole, and remote controls. When considering blasting technologies, operating companies tend to be highly cost conscious, which mitigates opportunities to develop value-added or innovative products.

Use of Shock Tube/Signal Tube initiation has indeed made down hole initiation possible allowing much safer blasting operations. Electronic detonators are available. Advantages include more-precise delay timing (resulting in increased blast efficiency and control) and greater compatibility with remote-controlled loading of explosives and wireless detonation. However, these initiating systems have higher costs. Whilst most of the effort by the manufactures is being directed towards enhanced fragmentation, there are undoubtedly areas relating to environmental control that can bring benefits to operators, regulators and local residents. The trend towards use of electronic detonators has not been to replace other systems such as signal tube or shock tube which has its own advantages.

4 BLAST EXECUTION, MONITORING AND CONTROL

Optimum blasting just does not happen. It requires suitable planning, good blast design, accurate drilling, the correct choice of explosives and initiation system and methods, adequate supervision and considerable attention to detail. The rock type and structure; size, length and inclination of blast holes, drilling pattern and accuracy, type, quantity and distribution of explosives; charging and initiating techniques all play a significant role in the overall efficiency of a mining operation. During the design stage environmental constraints such as vibration limits or flyrock restriction with respect to any structure can be prescribed. Blast design software can be used which considers all the above aspects.

Face profile can be drawn with the help of face profiler survey instrument. Front row burden variation is the important cause of poor fragmentation, release of explosive energy and undesired flyrock. The charge in the holes of first row should be based on profile of the face.

The stemming length should be according to requirement of blasting. Presently drill cuttings are used as stemming material. The use of coarse aggregate material as stemming material rather than drill cuttings improves blast results and reduces dust being raised in atmosphere.

Correct delay sequence and timing of delays between the rows and between the holes are required. Timing of delays between the rows and between the holes often can result in successful or failed blast. Ideally, the blast need to be designed with computer assistance using information from laser surveys, geotechnical and other data and hole locations are placed on blast plan maps. Also several software are available to keeps records of blasts. In mines blast records are maintained but are generally inadequate for analysis and study. In general only blast location and average parameters are maintained. Post blast performance details, site information such as geology, geometrical information of blast site and detailed parameters are not recorded. It is important to record blast performance zone wise, explosive and accessories used and also unusual happenings. Routine mine blast operations experience must be gathered and stored in a systematic way so that it can guide present and future design changes. Regular feedback is required to ensure that the blasting objectives are routinely achieved, and if not, to determine what aspect of the process requires attention. Blasting audit should be carried out to monitor blast execution and performance and then quantify blast results.

5 INFORMATION TECHNOLOGY APPLICATIONS

Information technology can be used in every step of drilling and blasting operations. Based on customized blast design tools for any operation blast design, charging and execution can be planned. The design can incorporate environmental restrictions and result goals. After holes are drilled then measurements regarding burden, spacing, hole depth need to be made either by using GPS, deviation tools or be measured manually. There would always be difference between designed hole location and inclination and actual holes drilled. After actual drilling and blasting parameters are available then predictive tools may be used for fragmentation, vibrations, flyrock and dust. If drilled blast is likely to exceed respective limits then charging, initiation timing and sequence can be changed to keep adverse environmental impacts within defined limits.

5.1 Blast Design Software

A mine can be set up in one of two ways; either the plant is set up to accept whatever fragmentation the blasting group produces, or the plant dictates to the mining teams the fragmentation distribution they will accept. Several blast design software are commercially available, many include those from explosive suppliers. Several operations use their own software. One such blast design software gives design of blasts according to requirements and is able to predict resulting fragment size distribution. Blast fragmentation size distribution prediction software, based on the Kuz–Ram fragmentation model (Cunningham, 1987) has been developed to predict the entire fragmentation size distribution, taking into account intact and joints rock properties, the type and properties of explosives and the drilling pattern. Using these predictive tools blasting parameters and explosive usage can be optimised.

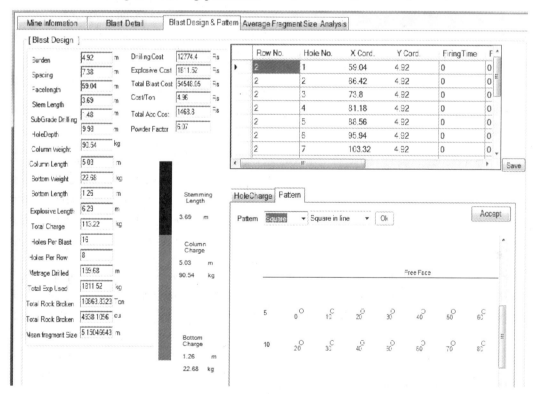

Fig.1 Blast Design Software based on experience gained from previous blasts

5.2 Blast Data

BIMS is software which helps to store, access and manage the information needed to take critical decisions for their mine/quarry operations (Bhandari, and Bhandari, 2006). The system stores blast details, blast parameters, blast pattern, face profile, explosive consumption, charging details, costs, weather information, pre-blast survey, post-blast evaluation data, fragmentation information, photograph(s), videos, accidents, misfires, vibration record and information for vibration analysis (Fig. 2). The stored blast information data can be retrieved quickly for analysis.

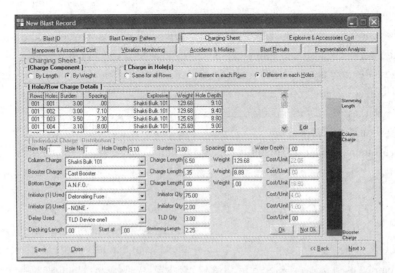
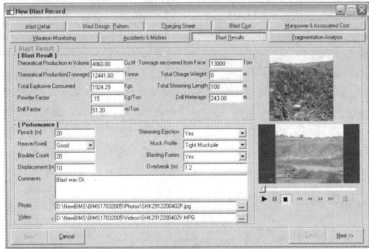

Fig.2 Blast Information Management System (BIMS) provides storage of all blast parameters and photographs, videos and performance of blast

The system generates reports for individually identified blast, monthly explosive consumption report, cost report, vibration monitoring report, and monthly blast performance report. The storage of this information in database format allows querying to retrieve scenarios, which meet certain criteria, and to use this information to further optimize the outcome from a blast.

If data is continuously recorded then large number of data becomes available, the system can up-date scaled distance relationship, based on location variations and ultimately provide the blasting engineer with an interactive means to assist with planning of future blasts.

5.3 Environmental Considerations

Increasing numbers of mining operations are coming under pressure to monitor and reduce blasting related safety and environmental hazards. Ground vibrations, air overpressure, fly-rock, dust, blasting fumes and in some cases leaching of chemicals in the blast holes and polluting ground water are some of the undesired events associated with blasting which collectively affect the surrounding environment adversely.

Much work has been carried out on the environmental aspects such as ground vibration and airblast control (Richards and Moore, 2002). Norms and standards regarding ground vibration and air blast as specified by regulating agencies must be complied with. Self regulatory limits need to be set by the organizations while taking neighbors into confidence. There is a need for severe penalty and stoppage of work if the limits are exceeded. Current norms were initially intended to prevent structural damage to adjacent properties, however nowadays they are being employed in an attempt to minimize human nuisance. Thus these values are now set at much lower levels than those based on damage criteria but still above human perception level and as a consequence complaints still arise. A statutory vibration limit be included on a sites operational license which must be adhered to at a specified confidence level at the nearest occupied property. It is therefore, vital for the industry to do all that it can to reduce the vibration levels experienced at these adjacent properties without imperiling the financial viability of the enterprise.

A comprehensive blasting analysis and reporting software VIBRATION PREDICTOR meets the needs of both operators and regulators. It supports and improves compliance with blasting related planning conditions, and contributes to improved blast performance and blast design. Regular updating of predictions using ongoing site data, providing minimum instantaneous charge (MICs) to the operator that ensure compliance with vibration level restrictions by design rather than by accident.

Scaling methods have been used for many years to determine relations between charge mass distance and blast vibration levels. The vibration is predicted by either square root or cube root scaling formulae relating vibration to charge mass and distance for a particular site. Excessive air and ground vibration are then controlled by a reduction in the explosives charge mass being fired at the one instant of time, or within a small time period of up to 8 ms. These scaling methods do not allow for the time taken for vibration wave fronts to travel from each blasthole, and cause reinforcement of vibration wave front. Wave front reinforcement has been found to cause substantial increases in both air and ground vibrations (Richards and Moore, 1995). This graphical software program can be used to alter design and initiation patterns to avoid reinforcement and thus lowering the vibration levels. This tool allows blast to be designed to reduce exceedance of vibrations both for airblast and for ground vibrations (Richards, 1995). This software also uses time window to see how many holes are firing at one time so that one can keep charge per delay under prescribed limits (Fig. 3).

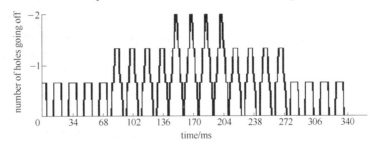

Fig.3　Number of holes detonating in 8 ms time window for a given pattern

5.4　Blast Safety

Explosives handling and blasting operations are high consequence risk activities. There are many safety hazards associated with use of explosive, transport and storage. One of the safety hazards during blasting is flyrock. Damage due to flyrock from blasting is one of the main causes of strained relations between mining operation and neighbors. Flyrock distances can range from zero for a well controlled mine blast to nearly 1.5 km for a poorly confined large, hard rock mine blast and many fatalities have occurred.

Thus, where large diameter blasting is carried in hard rock mining, extra precautions are required to control the flyrock damages in the surroundings. There is a "safe" blasting area in blasting is dependent on the knowledge of distance to which flyrock will propel.

Software for predicting distance to which a flyrock will travel has been developed (Richards and Moore, 2004). Inputs to the software are charge mass, burden or stemming height, and a site constant that lies within a general range that can be tightened by site calibration. The output is the distance that rock will be thrown, and this quantification can be used to establish both safe clearance distances, and the critical range of burdens and stemming heights where the situation changes rapidly from safe to hazardous. Using safety factors danger zones for machinery and persons respectively. If it is not possible remove any structure or person then one can change charging of holes. Use blast clearance zone predictor before charging may help in controlling flyrock by altering charging in a hole with reduced burden (Fig.4).

The output is the distance that rock will be thrown, and this quantification can be used to establish both safe clearance distances, and the critical range of burdens and stemming heights where the situation changes rapidly from safe to hazardous. Using safety factors danger zones for machinery and persons respectively. If it is not possible remove any structure or person then one can change charging of holes.

Determination of Throw in front face (m)		Determination of Throw in behind face (m)	
Burden(m)	Throw in front of face(m)	Stemming(m)	Throw Behind face(m)
1	1680.01	1	1680.01
2	277.1	2	94.77
3	96.56	3	33.03
4	45.7	4	15.63
5	25.59	5	8.75
6	15.93	6	5.45
7	10.67	7	3.65
8	7.54	8	2.58
9	5.55	9	1.9
10	4.22	10	1.44
11	3.29	11	1.13
12	2.63	12	0.9

Fig.4　Flyrock Prediction in front and behind the face based on blast parameters, explosive and rock

Blasting operations can generate large quantities of dust. However, dust is as yet not assumed to be causing problems. Whereas this dust when released in an uncontrolled manner, can cause widespread nuisance and potential health concerns for on-site personnel and surrounding communities. Though the blasting dust plume is raised for few minutes but most of the dust settles in and around mining area and some of it is dispersed before settling down. Depending on meteorological conditions the dust dispersal can travel to substantial distances endangering health of communities. Generation of fines and dust is influenced by several blasting and rock parameters.

Meteorological conditions such as wind speed and direction, temperature, cloud cover and humidity will affect the dispersion of airborne dust. Atmospheric stability affects dispersion of the emitted plume, determining the extent of the vertical and horizontal, transverse and axial spread of the emitted particulates. Thus, dispersal of dust plume resulting from blasting is an important area which needs attention. A computer model has been developed to simulate the dispersal of dust (Kumar and Bhandari, 2002).

However, production of fines and dust need to be reduced and controlled for better environmental conditions (Hagan, 1979). Dust generation and dispersion from blasting operations depends on factors such as meteorology, bench height, blast design information, and rock (Bhandari, et al., 2004). Concern is expressed about nitrogen-oxide (NO_x) releases from blast sites and their potential health impacts on workers, as well as their aesthetic and environmental impacts on nearby communities.

5.5　Data Mobility

Data Mobility is an end-to-end technology framework packaged as a simple data service to monitor remote data points. In the mine there are large numbers of faces spread over a large geographic area where blast is carried. This requires the application of a new creed of technologies that are capable of reaching further to acquire more data and sending it directly to those in the organisation that require it, all in real time. Some of the challenges are

(1) Blast data management to ensure blast meets environmental standards and reporting.

(2) Blast environmental monitoring such as ground vibration, airblast etc. to meet safety and regulatory requirements.

(3) Fragmentation monitoring across the site in real time to pin-point wastage and allocate costs.

(4) Acquisition of data from single points is not cost effective.

(5) Very wide area coverage is required.

(6) Limited or no power available at the measurement point.

All of these challenges mandate a new way of doing things. Fortunately we are living in a connected world. Communications is becoming ubiquitous. Power is a bigger challenge today than communications in many of these applications. Data2Desktop is an end-to-end technology framework for collecting data at bench face and to monitor remote data points and transfer data to desktop software for prediction and control of impacts. Data2Desktop can also use most communications means including mobile phone and satellite networks to send data from remote locations to the Data2Desktop cloud based servers. Thus an integrated Blasting ICT solution becomes available which integrates design, blasting parameter and measurement records, capture actual blast site data using mobile applications, use these input for altering explosive charging of holes and delay pattern thereafter use various predictors and calculations. Proposed solution can be used for controlling environmental impact, safety and optimised operations and increase mining productivity.

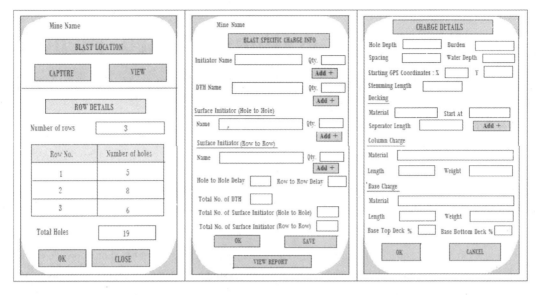

Fig.5 Collection of data at blast site for transfer to desk top based blasting software

6 CONCLUSIONS

Information and communications technology can be applied to the entire value chain and lifecycle of mining and mineral extraction. The paper has identified those key factors which impact upon optimum blasting performance, whilst highlighting, in particular, use of several predictive tools for blasts can be fired with the minimum amount of noise and airblast overpressure by effective distances associated with surface mine operations, is very unlikely to create ground vibrations which could cause structural damage and flyrock damage. An integrated Blasting ICT solution which will integrate design, blasting parameter and measurement records, capture actual blast site data using mobile applications, use these input for altering explosive charging of holes and delay pattern thereafter use various predictors and calculations. Proposed solution can be used for controlling environmental impact, safety and optimised operations and increase mining productivity.

REFERENCES

[1] Adhikari G K. Studies on Flyrock at Limestone Quarries[J]. Rock Mechanics and Rock Engineering, 1999, 32(4): 291~301.

[2] Bhandari S. Engineering Rock Blasting Operations, A.A. Balkema Publishers, Rotterdam, Netherlands / Brookfield, U.S.A., 1997: 370.

[3] Bhandari S, Bhandari A, Arya S. 2004 Dust Resulting from Blasting in Surface Mines and its Control, EXPLO 2004 Conference, Perth, August.

[4] Bhandari S, Bhandari A. Blast Operations Information Management System, Journal of Mines, Metals and Fuels, 2006, 54(12).

[5] Bhandari S. Information Management for Improved Blasting Operations and Environmental Control, 3rd Asia- Pacific Symposium on Blasting Techniques, August 10~13, Xiamen, China, 2011.

[6] Cunningham C V B. Fragmentation estimations and the Kuz-Ram Model – four years on, 2nd International Symposium of Rock Fragmentation by Blasting, Keystone, Colorado, 1987:475~487.

[7] Hagan T N. The control of fines through improved blast design, Proc. Aust. Inst. Min. & Metal, 1979: 9.

[8] Kumar P, Bhandari S. Modelling dust dispersal near source after opencast mine blast in weak wind conditions over flat terrain in tropical conditions, Explo 2001 Conference, Hunters Valley, October 28~31, 2001.

[9] Kumar P, Bhandari S. Modelling dust dispersal near source after surface mine blast over undulated terrain in weak wind conditions, APCOM –2002, Phoenix, February 25~27, Proceedings of the 29th International Symposium on Computer Applications in the Minerals Industries, 2002.

[10] Richards A B, Moore A J. Blast Vibration Control by Wave front Reinforcement Techniques in Explo 1995, 1995: 323~327 (The Australasian Institute of Mining and Metallurgy in Association with The International Society of Explosives Engineers: Brisbane).

[11] Richards A B, Moore A J. Flyrock Control – By Chance or Design? Proc. 30th Ann. Conf. on Explosives and Blasting Technique, International Society of Explosive Engineers, 2004.

[12] Rodgers J A. Measurement Technology in Mining, Proceeding MINEBLAST 99, Duluth, Minnesota, 1999: 17~23.

Study on Blasting Operations Hazard Identification and Risk Assessment Based on Improved Evaluation Method LEC in Xiaocishan Mine

HUANG Kaihe [1,2], ZHANG Xiliang [1,2], XIE Liangbo [1,2]

(1. Maanshan Kuangyuan Blasting Engineering Co., Ltd., Maanshan, Anhui, China;
2. Sinosteel Maanshan Institute of Mining Research, Maanshan, Anhui, China)

ABSTRACT: The study on blasting operations hazard identification and risk assessment based on improved evaluation method LEC in Xiaocishan mine was conducted to ensure the safety of blasting operations. According to the basic scoring criteria of the improved LEC method, the hazard risk assessment form was obtained. By assessing of risk size for the identified hazard during blasting operations, the risk evaluation results of hazard sources could be acquired. Then the blasting hazard assessment on the basis of improved LEC method was achieved in Xiaocishan mine, providing guidance for safety management in the subsequent blasting.

KEYWORDS: blasting operations; hazard identification; LEC evaluation method

1 INTRODUCTION

As a simple semi-quantitative method of safety evaluation, *LEC* evaluation method could beused to measure work risk of a potentially dangerous environment. *LEC* evaluation focuses on the people thinking. It evaluates the staff casualties risk by product value *D* of three independent variables (*L* — accidents likelihood; *E* — frequency of human exposure to hazardous environments; *C* — accident severity of consequences). Then when compared with the risk classification criteria, it could determine the risk degree of operating conditions. The formula is as followed, *D* = *LEC*. The greater the *D* value, the greater the risk. When the probability of hazards occurrence is introduced, and hazard expands to other property except environmental hazards casualties, then *LEC* evaluation method is improved.

2 PROJECT OVERVIEW

Xiaocishan mine is a pen pit mine, designed production 500000t/a. Industrial site locates in the north of the stope.

3 THE IMPROVING PROCESS AND APPLICATION IDEAS OF LEC

There are two main aspects to be improved of *LEC* method.

(1) *L* and *E* are relevant. *E* is one of the determinants of *L*.

(2) *LEC* method to casualties as a consequence of *C* grading factor is still not comprehensive enough.

Therefore, it try to introduce the probability of hazards occurrence to LEC method, while the range of possible consequences *C* that accident occurs is extended to form the safety management standards and criteria. Then it has a stronger applicability and more in line with the actual situation.

3.1 Level Improvement of Hazards Occurrence Probability

LE represents the hazard factors leading to the possibility of accidents in a certain time, basically as same as the probability of risk occurrence probability *P*. Specifically a corresponding relationship is shown in Tab.1.

Tab.1 Corresponding standard of *LE* and *P*

Level	Very low probability	Low probability	Medium probability	Higher probability	High probability
Accident description	Impossible	Rarely occur	Occasional	possible	Frequent occurrence
Interval probability	$P<0.01\%$	[0.01%, 0.1%)	[0.1%, 1%)	[1%, 10%)	$P\geqslant 10\%$
LE	<6	[6, 18)	[18, 36)	[36, 60)	$\geqslant 60$

Corresponding author E-mail: yht 03121 @ 163. com

3.2 Improvement of the Accident Consequences Severity C

The traditional *LEC* method only takes casualties as the standard when mentioned *C*. It does not meet the actual scope caused by security incidents, which is necessary to join the economic losses, production delays and environmental impact, etc. to better reflect the actual consequences.

Tab.2 C value standard of improved LEC method

Level	Large catastrophic	Catastrophic	Very serious	Severe	Larger	To be considered
Fractional value	100	40	15	7	3	1
Casualties	Deaths≥9	Deaths≥3	Deaths =1	Severe disability	Severely injured	Minor injuries
Economic losses (Ten thousand yuan)	>10000	5000~10000	1000~5000	500~1000	100~500	<100
Production delays (Days)	>60	30~60	20~30	10~20	5~10	<5
Environmental impact	Involving a very large range, severe function loss of surrounding environment, emergency transfer more than 1000 people, normal economic and social activities being seriously affected	Involving a large range, part function loss of surrounding environment, emergency transfer 500~1000 people	Involving a large range, regional normal economic and social activities being affected, emergency transfer 100~5000 people	Involving a small range, general population impact, emergency transfer 50~100 people	Involving a small range, no group effects, emergency transfer less than 50 people	Involving a minimal range, no group effects, basic non-emergency transfer population

4 RISK ASSESSMENT MODEL CONSTRUCTING BASED ON IMPROVED LEC METHOD

According to the improved LEC method, the accident risk of an identifying safety risk *S* in Xiaocishan mine was given a comprehensive evaluation by the 10 invited experts.

4.1 L Score Assessment of Accidents Possibility

Based on Xiaocishan mine general overview and the basic properties of safety hazards *S*, combining their expertise and knowledge, 10 experts give the l_i (i=1,2,…,10) scores of an accident possibility respectively. Then the calculating equation of *S* is shown as followed.

$$L = \sum_{i=1}^{10} l_i / 10 \tag{1}$$

4.2 E Score Assessment of Frequency of Human Exposure to Hazardous Environments

The 10 experts give the e_i (i=1,2,…,10) scores of frequency of human exposure to hazardous environments respectively. Then the calculating equation of *E* is shown as followed.

$$E = \sum_{i=1}^{10} e_i / 10 \tag{2}$$

4.3 C Score Assessment of Accident Consequences Severity

The 10 experts give the *C* score from four dimensions such as casualties, economic losses, production delays and environmental impact. They are c_i^1, c_i^2, c_i^3 and c_i^4 (i=1,2,…,10). The results calculating equation are shown as followed.

$$C^1 = \sum_{i=1}^{10} c_i^1 / 10 \tag{3}$$

$$C^2 = \sum_{i=1}^{10} c_i^2 / 10 \tag{4}$$

$$C^3 = \sum_{i=1}^{10} c_i^3 / 10 \tag{5}$$

$$C^4 = \sum_{i=1}^{10} c_i^4 / 10 \tag{6}$$

4.4 Solving of Final Risk Score D for Risk S

On the basis of safety risk *S* in Xiaocishan mine, *L* value and *E* value are obtained by judgment of 10 invited experts, as well as c_i^1, c_i^2, c_i^3 and c_i^4. Then the risk scores *D* can be calculated.

$$D = LEC = LE \times \max(C^1, C^2, C^3, C^4) \tag{7}$$

According to the accident hazard classification criteria of LEC evaluation method, the hazard level of dangerous source *S* could be determined, which provides a fundamental basis for the production construction and safety management of Xiaocishan mine.

5 EXAMPLE ANALYSIS OF RISK ASSESSMENT BASED ON IMPROVED LEC METHOD

Taking into account the risk source "unreasonable blasting parameters, triggering the rolling stones, flying rocks and vibration overrun" existed in Xiaocishan mine, the risk assessment was carried out by 10 invited experts based on the improved LEC method. And the risk evaluation form was obtained as shown in Tab.3.

Tab. 3 Risk evaluation form of "unreasonable blasting parameters, triggering the rolling stones, flying rocks and vibration overrun"

Rating category		Expert group										Final weighted score
		Expert 1	Expert 2	Expert 3	Expert 4	Expert 5	Expert 6	Expert 7	Expert 8	Expert 9	Expert 10	
L		3	3	6	6	3	3	1	1	3	3	$L = \sum_{i=1}^{10} l_i / 10 = 3.2$
E		3	3	3	6	6	3	3	3	3	3	$E = \sum_{i=1}^{10} e_i / 10 = 3.6$
C	C^1	15	15	7	15	15	15	15	40	15	15	$C^1 = \sum_{i=1}^{10} c_i^1 / 10 = 16.7$
	C^2	15	15	7	15	40	7	7	15	15	15	$C^2 = \sum_{i=1}^{10} c_i^2 / 10 = 15.1$
	C^3	15	15	15	15	15	7	15	40	15	15	$C^3 = \sum_{i=1}^{10} c_i^3 / 10 = 16.7$
	C^4	15	15	7	15	15	7	15	15	7	15	$C^4 = \sum_{i=1}^{10} c_i^4 / 10 = 12.6$
D		\multicolumn{10}{c}{$D = LEC = LE \times \max(C^1, C^2, C^3, C^4) = 192.38$}	

The risk score was 192.38 by Tab.3, belonging to highly dangerous. It need to be revised and improved immediately. And it should cause sustained attention in the late process of safety management.

Simultaneously the LE value was 11.52, corresponding to an interval probability $0.01\% \leqslant P < 0.1\%$, being low occurrence probability. But the score of accident consequences severity was relatively high, especially in casualties and production delays. It showed that the hazard probability and damage was not high based on the description of the invited experts, but the event was highly likely to have a huge impact on the surrounding personnel or production. Therefore it should maintain adequate attention and focus on coping measures in late safety management.

For the other safety hazards identified in mine blasting operations, they were assessed by the same method. The evaluation results were shown in Tab.4.

Tab.4 Evaluation results of identified safety hazards in Xiaocishan mine

Number	Risk source	L	E	C	D	Hazard level
1	Not standardized of blasting equipment, transportation, storage and blasting operations	3.1	2.7	13.5	113.0	Significant risk
2	Not blasting alert or too small	2.9	2.8	15.3	123.2	Significant risk
3	Blasting parameters unreasonable, triggering rolling stones and flying rocks	3.2	3.6	16.7	192.4	High risk
4	Not in accordance with the design requirements to control the scale and blasting charge	2.4	3.2	14.7	112.9	Significant risk
5	No hide blasting facilities	2.6	2.8	15.8	115.0	Significant risk
6	No blasting operations permit	1.7	3.2	14.3	77.8	Significant risk

6 CONCLUSIONS

(1) The probability of hazards occurrence was introduced to LEC method. It contributed to the risk assessment results in risk management and control work.

(2) For the insignificance of LEC method, it was improved taking into account L and E being relevant and risk probability level. It made the accidents consequences caused by risk sources expand from injuries to casualties, economic losses, production delays and environmental impact.

(3) Based on the improved LEC method, the safety assessment model of Xiaocishan mine was established, providing a fundamental basis for the mine production and construction and safety management as well.

(4) The mine safety risks assessment were conducted, achieving the mine safety hazards evaluation based on the improved LEC method, which laid the foundation for the late mine safety management.

REFERENCES

[1] Shi Y G, Fu Z Q, Zheng M. Application of LEC evaluation method on non-coal mine safety evaluation[J]. Gold, 2009, 30(9): 33~36.

[2] Zhu Y Y, Fu X H, Li K R, et al. Application of improved LEC method to hazard evaluation in hydroelectric project construc-

tion[J]. *Journal of Safety Science and Technology*, 2009, 5(4): 51~54.

[3] Mao Y P, Guo J F. Technique and Practice of Safety Evaluation for Non-coal Mines[J]. *Metel Mine*, 2003(4): 7~10.

[4] Shi Y F, Wu J. Research of modified LEC method for high-danger industries based on base-period standard[J]. *Journal of China University of Mining & Technology(Social Sciences)*, 2012(3): 69~74.

[5] Yuan C Y, Mi Y Q. The application of LEC method on dangerous source identification and assessment in the metal mine construction[J]. *Journal of Safety Science and Technology*, 2011, 7(8): 175~180.

Monitoring of Coal and Rock Mass Conditions, Coal Mine Air and Extraction Equipment State

S.S. Kubrin

(*Institute of Comprehensive Exploitation of Mineral Resources Russian Academy of Sciences* (*IPKON RAN*), *4 Kryukovsky tupik, Moscow Russia*)

ABSTRACT: The article discusses development and application of monitoring of rock mass, mine air and mining machinery in the course of mining. The author considers methods of detecting hazardous events based on various physical parameters taken by sensors in rock mass (stresses, seismic emission, acoustic emission, thermal radiation) and in mine air (concentrations of gases, air velocities, pressure, drawdown). The proposed mathematical model takes into account the cause-and-effect relationship in the mining machine—rock mass—mine air—mining machine circuit.

KEYWORDS: monitoring; rock mass; mine air; mining equipment; mathematical model

1 INTRODUCTION

Monitoring of mine air and geodynamic monitoring of enclosing rocks exist separately from the production methods approved in a mine and, what is more, conflict and restrain basic productive activities. That is to say, mine safety monitoring surely weakens accident risk but, the mine safety can never be granted when the technology and safety control disagree. The way out of this conflict is integration of a monitoring system in the production control within a coal mine.

The true and sustained development of mines requires to boost and promote practical application of state-of-the-art safety control systems in order to abate accident risk in hazardous mine sites down to actually admissible level. Underground mine safety greatly depends on prompt identification and adequately supported prediction of hazardous natural and mining-induced processes in local geomechanical systems.

2 THE MONITORING OF COAL AND ROCK MASS CONDITIONS. EQUIPMENT

Now available is a variety of procedures founded on different physical phenomena that show under natural or mining impact on geo-systems, which allows to register the related alterations and obtain information on strength and duration of the dynamic events. However, it is difficult to reach even 75% reliability of the forecasting in complicated geological and mining conditions due to diversified physico-mechanical properties and states of geo-systems. In addition, inception, growth and happening of a hazardous undesired event takes place as a part of change in the state and characteristics of the enclosing rocks, mine air and hydro-dynamic regime. Thus, a hazardous dynamic event evolves concurrently with alteration of physical properties and characteristics of rocks, namely, frequency spectra, acoustic emission energy, methane emanation rate, rock mass temperature, etc. Monitoring of one or a few parameters, for instance, methane concentration in mine air, allows, though partly, handling the mine safety issues. Nevertheless, the time history of one or two characteristics is an insufficient source for determining the nature and history of origination, propagation and relaxation of dynamic events in mines.

The seismic emission sensor (SES) is meant for locating seismic sources of ultimate strains in rocks and their distressing zones (rockbursts, outbursts, roof collapses in roadways, etc). An SES is designed as a hollow metal cylinder (see Fig.1(a)). Vibrator inverter is an electric circuit with an accelerometer based on micro electromechanical systems. Accelerometer is oriented along three mutually perpendicular axes (z-x-y) for three-component recording of seismic signals. SES enables measurement of seismic characteristic (vibration) in rocks within frequency range from 1 to 200 Hz.

The acoustic emission sensor (AES) is intended to locate acoustic sources of ultimate strains in rocks and their distressing zones (subsidence, additional loading and relief of rocks, rockbursts, outbursts, roof rock falls in mine roadways, etc). An AES is designed as a hollow metal cylinder with two end caps and a polyamide cylinder with three mutually perpendicular pockets to accommodate three electro-dynamic geophones oriented along the standard axes V-H1-

---

Corresponding author E-mail: s_kubrin@mail.ru

*H*2 (*z-x-y*) for three-channel recording of acoustic signals (see Fig.1(b)). The GS-20DX geophone is in line with the axis *V*, and GS-20DX-2B geophones are in line with the axes *H*1 and *H*2. The AES measurement range is from 28 to 2000 Hz.

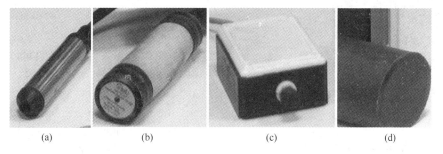

Fig.1 Sensor probes

(a) seismic emission sensor; (b) acoustic emission sensor; (c) heat emission sensor; (d) stress sensor

The heat emission sensor (HES) locates shows of ultimate strains in rocks and their relief zones by measurement of radiating heat. An HES includes housing and a radiating heat-sensitive element RTN-30G (see Fig.1(c)). The sensor probe HES measures continuous optical radiation in constant spectrum, in spectral wavelength range from 2 to 14 μm and in irradiance range from 2 to 200 W/m^2.

The sensors SES, AES and HES are installed in holes 46 mm in diameter and 10 to 20 m in depth, drilled using portable drill. Holes with SES and AES are plugged to create a uniform vibro-seismic (acoustic) environment for the sensor and rock mass (coal bed, host rocks). The HES placement hole is drilled in front of production face, 10～20 m apart of it. The HES hole is hermetically sealed to prevent from dust generation in the measurement zone.

The stress sensor (SS) detects shows of ultimate strains in rocks and their relief zones (see Fig.1(d)). An SS consists of 3 units each of which houses cells with strain gage converters (see Fig.2). Two cells 70 mm×70 mm×70 mm are installed sequentially, one after the other, to take measurement along *z* and *x* at an angle of 45° relative to these axes. The third cell made as 35 mm×35 mm×70 mm prism, is placed at an angle of 45° to the axis *y*. The SS sensor measures changes in the stress–strain state of rocks and generates analog output toward the recorder controller. This sensor is installed in holes 100 mm in diameter and to 5～15 m in depth, drilled with portable drill. The hole with the sensor is plugged using a special mixture having similar strength characteristics as the enclosing environment (coal bed, host rocks), in order to create the uniform medium in terms of strength. The plugged hole with the sensor inside it are aged for 5～15 days so that the mixture consolidates, develops the required strength and takes up load (static ground pressure).

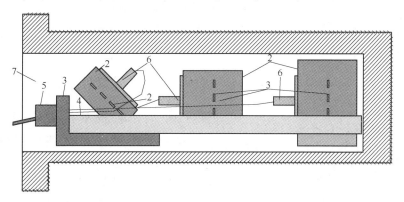

Fig.2 Stress sensor design

All sensors SES, AES, HES and SS generate analog signal and transmit it to the recorder controller. The recorder controller (CCTTP) synchronously records the analog signals (discretization interval to 4000 Hz) taken from the lower level sensors, amplifies them with keeping dynamic range, converts them in digital format (24 bit capacity AD converter) and transmits them in bursts in format TCP/IP to communication links. Spacing between the CCTTP installation and the locations of SES, AES, HES and SS must be not more than 10 m. If there is no mine-wide information network, the recorder controllers are equipped with high speed modem communication device to transmit the data from the underground sensors and recorders to a surface data processing center. The high speed modem enables two way information interchange with the other high speed modems via copper coiled pairs of wires with SHDSL sup

Fig.3 Recorder controller designed for processing analog signals obtained from lower level sensors

port. Rate of the data exchange reaches 300 000 bauds. Data traffic in SHDSL is translated to a TDM interface using a field-programmable gate array (FGPA). Outwards UART data streams are connected via Device-Master Serial HUB with one interface. Power supply is intrinsically safe. Total power consumed by a CCTTP and a sensor is under 2.5 W.

The SES sensors are arranged so that to supervise the entire mining site. The AES sensors are installed by threes in the headings adjoining the extraction stope, in front of the production face, at a distance enough for the day-long stoping front advance. Distance between AES sensors installed in the same heading fits with the face advance per day. When the powered mining machine complex approaches the AES installation place, the sensor is removed and installed in front of the AES sensor located farthest from the face. This repositioning of the AES sensors allows maintaining control over coal and rocks at a distance from 60 to 90 m in front of the production face (given the face advance for 20~30 m per day). The sensors HES and SS are arranged lengthwise the extraction site, 50 to 100 m away from one another, depending on geological and mining conditions (see Fig.4).

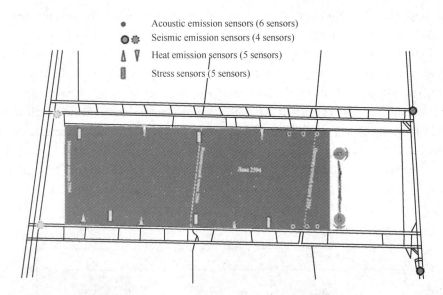

Fig.4 Arrangement of sensors on an extraction site

3 THE MONITORING OF COAL AND ROCK MASS CONDITIONS. MATHEMATICAL TREATMENT

The integrated synthesis monitoring analyzes simultaneously seismic and acoustic signals, thermal radiation and deformation data collected in rock mass, the mining equipment operation data and mine air control data. The analysis includes detection of hazardous geomechanical, geodynamic and gas-dynamic events based on records of acoustic emission in rocks, studies of amplitude–frequency characteristics of probing signals, geophysical tomography, gas release dynamics, geoenergy of rocks mass, taking into account operating regimes of mining machinery, assessment of stresses in rocks etc. Based on the analysis, it is possible to define: dynamic sources of critical strains in rocks and the areas of the ultimate strain relaxation (in the form of rock bursts, outbursts, collapse of excavation roof etc.); quasi-static movement of rock mass (subsidence, trough formation etc.); gassy fluid flows in mine air (dust–gas mixtures). As a result, the state of a coal bed, enclosing rocks and overall rock mass can be determined. The analytical results are stored in data bases and entered on reports. Besides, the obtained and processed data allow forecasting the state of the coal bed, enclosing rocks and entire rock mass.

The applied standard mathematical treatment involves:

Data packages on seismic emission along three axes of all seismic sensors;

Data packages on acoustic emission along three axes of

all acoustic sensors;

Strain measurements along six axes of all strain metes;

Thermal emission data from all heat sensors;

Methane concentration using sensors M1-32, M2-32, M3-32, M4-32, M6-32 installed in stoping face;

Air velocity using sensors AF1-32, AF2-32;

Carbon dioxide concentration using sensors OU1-32, OU2-32, OU4-32, OU5-32, OU6-32, OU7-32;

Data on operating mode of mine equipment;

Data on indentified seismic events;

Data on identified acoustic events;

Data on seismic emission rate;

Data on acoustic emission rate;

Data on rate of change of thermal radiation in rocks;

Amplitudes of high-frequency components and low-frequency components in the acoustic signal spectrum;

Data on gas volumes released in the first ten minutes and last ten minutes of dead time of a mining machine;

Gas-dynamic index;

Data on concentrations of methane after blasting;

Hazard criterion of gas-dynamic events;

Hazard criterion of gas-dynamic events, considering geotechnology;

Data on nonequicomponent stress state in face-adjacent rock mass;

Data on highest stress (strain) area in face-adjacent rock mass.

The measured and preprocessed data are used to calculate:

Seismic emission rate;

Acoustic emission rate;

Rate of change of thermal radiation in rocks;

High-frequency component amplitude to low-frequency component amplitude ratios in the acoustic signal spectrum;

Ratios of gas volumes released in the first ten minutes and last ten minutes of dead time of a mining machine;

Gas-dynamic index;

Maximum concentration of methane after blasting;

Hazard criterion of gas-dynamic events;

Hazard criterion of gas-dynamic events, considering geotechnology;

Nonequicomponent stress state in face-adjacent rock mass;

Distance between the highest stress (strain) area in face-adjacent rock mass and the face.

Next thing, the calculated values are compared with the threshold values. In the event of a number of indications of a geodynamic or gas-dynamic hazard, the integral assessment of the data is carried out. The actually observed and calculated data are introduced in a model, and time series of each parameter is analyzed. This study is aimed at finding regular component and random noise (chance variations due to heterogeneity of rocks, coal, mining machine operation etc.). The regular components and the trend of the time series of a parameter are determined (which is a general change in data during monitoring), or a "seasonal" component is defined (which is a cyclic component associated with work or repair shifts). The trend is a general regular linear or nonlinear component that varies with time. The systematic trend and the regular component are then correlated.

The systematic trend is defined using data-smoothing methods selected subject to characteristics of rock mass, coal bed, enclosing rocks, occurrence conditions as well as a geotechnology and mining rate as described below:

(1) moving average where each term of series is replaced by a simple average or weighted average of n adjacent terms, where n is a width of "window";

(2) median smoothing, using medians of values in a running window, which allows more stable and runout-resistant results (within the window); if data contain deviations (runouts) due to heterogeneity of rock mass, coal bed and enclosing rocks, the median smoothing will provide smoother curves for the measured or calculated data as compared to the moving average in the same window, and the more reliable systematic linear or nonlinear component (trend) as a result;

(3) least squares, to be weighted relative to distance if a measurement or calculation appears insufficiently reliable due to complicated mining and geological processes caused by mining-induced de-stressing of rocks;

(4) negative exponential weighted smoothing if a measurement or calculation appears insufficiently reliable due to complicated mining and geological caused by mining-induced de-stressing of rocks;

(5) bicubic splines, at the early stage of data processing.

The smoothing methods enable noise filtering-off and relatively smooth curve plotting. The general systematic component of the analyzed data is then determined, and the approximate functional connection of the time series is found using linear and polynomial functions and their combinations. Based on the obtained relationship of the measured or calculated data, the forecast over a pre-set period is made.

The periodic components of the time series are analyzed based on autocorrelogram of an autocorrelation function. Autocorrelation coefficients (and their parameters) are determined in a certain time range for a succession of time series and by studying the partial autocorrelation function to avoid influence of mediate observations. The periodic component is found in each k period by diminishing each i by the $(i - k)$th value. Removal of the found periodic component from a time series makes the time series homogeneous in time.

The steady time series are then entered in autoregression of integrated moving average. Later on, it is assumed that time series contain elements that successively depend one on the other:

$$xt = \xi + \varphi_1 x(t-1) + \varphi_2 x(t-2) + \varphi_3 x(t-3) + \ldots + \varepsilon$$

where ξ ——constant;

$\varphi_1, \varphi_2, \varphi_3$ ——parameters of autoregression.

The resultant estimates are used in prognostic modeling, calculation of new values of time series and definition of confidence interval for prediction. After new actual data on measurements or calculated parameters have been accumulated, they are compared with the prediction results, deviations are determined and the model parameters are assessed based on the actual data. The results are entered in the further modeling stages.

The integrated modeling uses the group method of data handling (GMDH) and the neural networks method. First, significant factors for modeling are defined from the correlation analysis. For each couple of time series, coefficients of correlation, covariance, Kendall's and Spirman's correlations of grades, multiple concordance of grades are determined. Accordingly, the integrated mathematical model is obtained for estimating geodynamic or gas-dynamic hazard per time series.

This approach bases on the successive choice of the best model out of generated set. Structure of the model, unlike in regression analysis, is not fixed but selected out of set of alternatives. The optimum model rests upon self-organization principles, subject to characteristics of coal and rock mass geology and the accepted mining technology. The significant parameters are chosen in a few stages. Each of time series is divided into two intervals: learning sequence and representative (check) sequence of data. The GMDH takes learning sequences of time series: $x = x_1, x_2, \ldots, x_n$. The first stage is formation of units composed of two variables from the input time series: $y_1 = f_1(x_1, x_2)$, $y_2 = f_2(x_1, x_3)$, …, $y_L = f_L(x_n-1, x_n)$. Out of the set of the generated units, the best variant in accord with a chosen criterion is selected and entered in the second stage, where the second-order units are generated: $Z_1 = f_2(y_1, y_2)$, $Z_2 = f_2(y_1, y_3)$, …, $Z_S = f(y_p-1, y_p)$. The best representatives of the latter are fed to the next stage and so on, until the criterion estimation is successively diminished. The obtained model is checked using the representative sequences of time series. The resultant integrated model, including the cause&effect of the mining machine—rock mass—mine air—mining machine loop, is applicable to forecasting of the geodynamic and gas-dynamic hazard.

4 THE MONITORING OF COAL AND ROCK MASS CONDITIONS. OPERATOR'S STATION

The surface data processing center includes a server unit (Stratus ftServer, warm standby, internet) and an operation unit (automation-equipped working places). The Automated Integration&Synthesis Monitoring (AISM) collects raw data by SCADA-iFix and iHistirian archiving system. To ensure standard data communication OPC DA between CCTTP, SCADA, MES and other subsystems, the CCTTP OPC-server has been designed for real-time processing of data fed simultaneously from several CCTTP recorder controllers (OPC DA version 2.0 and/or 3.0). The CCTTP OPC-server generates data text files, varies data stream modes and synchronizes them in time. The OPC-server translated information goes to iHistorian file collector for filing and storing.

Visualization of data in graphic form and animation is made using standard tools of SCADA iFix (Fig.5 and 6). Mine air is monitored by a mine air gas control system (AGC). The data from each air gas sensor installed on the extraction site are taken by the AGC server. The mining machine complex operation parameters are measured using standard hardware–software packages.

The actual data fed from sensors of seismic, acoustic and heat emission, stresses, methane concentration, carbon oxide content air flow rate in face zone, mining equipment operation, cutter-loader position in face zone are pre-processed on a real-time basis. In the course of the pre-processing, seismic and acoustic events are detected, seismic and acoustic emission intensities are calculated together with heat emission growth rate, high-frequency / low-frequency ratio in the acoustic emission signal spectrum as well as methane content released during cutter-loader operation to methane content liberated during the cutter-loader dead time. Later on, current geomechanical situation and mine air condition are estimated based on various criteria (see Tab.1). In case that several indications of a geo- or gas dynamic hazard are revealed, the integrated assessment of geomechanical situation in the mine becomes a must (Fig.7).

The data for processing are the time samplings (time series) meant for standard mathematical treatment. Based on the recorded (observed) and calculated data, the analysis of time series of each parameter is performed, aimed at nailing a regular component and a random noise due to heterogeneity of rocks, coal, machine operation, etc. (Fig.6). The regular component is found using a smoothing method, chosen in accordance with the coal and host rock characteristics, occurrence conditions as well as mining technology and its rate.

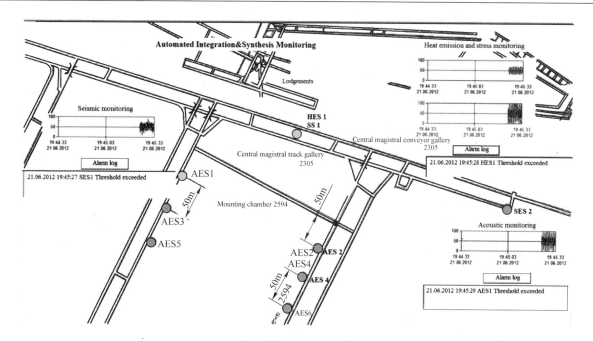

Fig.5　Symbolic circuit of the Automated Integration & Synthesis Monitoring

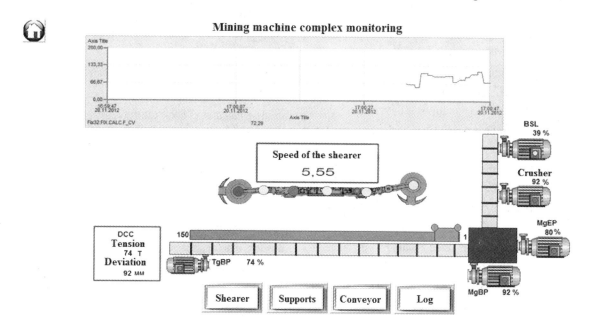

Fig.6　Symbolic circuit of the Mining Machine Complex Monitoring

Tab.1　Evaluation criteria for geomechanical condition of rock mass and coal mine air

Seismic / acoustic criterion	Strain criterion	Heat emission criterion	Mining machinery criterion	Mine air criterion
Seismic emission activity Amplitude–frequency response	Maximum stress control Nonequicomponent stress state hazard Location of concentrated maximum stresses	Rock mass temperature variation rate	Gas release (ration of methane release at machine operation to methane release at machine downtime) Gas dynamics index	
Determination of seismic and acoustic wave velocities	Location of concentrated maximum stresses in the face zone, considering heat emission		Determination of seismic and acoustic wave velocities	
	Gas release dynamics estimation based on mine air monitoring during drilling-and-blasting			Gas release dynamics estimation based on mine air monitoring during drilling-and-blasting
	Rock mass geo-energy assessment			Rock mass geo-energy assessment
Rock mass geo-energy assessment, taking account of mining technology				

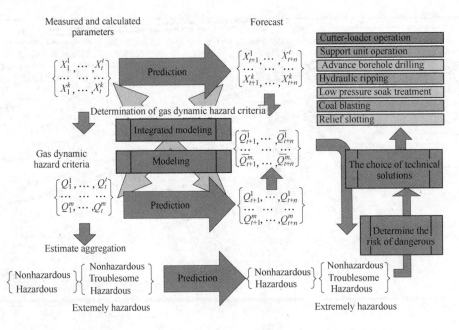

Fig.7 Sequence of data collection and analysis and rock mass state prediction

(1) moving average where each term of series is replaced by a simple average or weighted average of n adjacent terms, where n is a width of "window";

(2) median smoothing, using medians of values in a running window, which allows more stable and runout-resistant results (within the window); if data contain deviations (runouts) due to heterogeneity of rock mass, coal bed and enclosing rocks, the median smoothing will provide smoother curves for the measured or calculated data as compared to the moving average in the same window, and the more reliable systematic linear or nonlinear component (trend) as a result;

(3) least squares, to be weighted relative to distance if a measurement or calculation appears insufficiently reliable due to complicated mining and geological processes caused by mining-induced de-stressing of rocks;

(4) negative exponential weighted smoothing if a measurement or calculation appears insufficiently reliable due to complicated mining and geological caused by mining-induced de-stressing of rocks;

(5) bicubic splines, at the early stage of data processing.

The smoothing methods enable noise filtering-off and relatively smooth curve plotting. The general systematic component of the analyzed data is then determined, and the approximate functional connection of the time series is found using linear and polynomial functions and their combinations. Based on the obtained relationship of the measured or calculated data, the forecast over a pre-set period is made.

5 CONCLUSIONS

The proposed model allows early prediction of adverse and hazardous dynamic events in the course of hard mineral mining. The described geodynamic and gas-dynamic hazard prediction model improves reliability of the forecast and allows checking prognostic data. The timely identification of a potential accident enables its early prevention and precaution, or mitigation of the accident aftereffects. It is important to carry out the integrated rock mass, mine air and mine equipment monitoring in order to reliably and objectively estimate underground mine situation, taking into account the analysis and forecast of the mine air and rock mass parameters.

Effect Mechanism of Excavation Blasting in an UnderCross Tunnel on Airport Runway

LU Shiwei, ZHOU Chuanbo, JIANG Nan, XU Xing

(*Faulty of Engineering, China University of Geosciences, Wuhan, Hubei, China*)

ABSTRACT: Blasting excavation in a tunnel passing beneath an airport is quite rare at home and abroad. In this paper, the blasting excavation for the construction of a four-arc double-track tunnel passing beneath Xujiaping Airport in Enshi is introduced and in-site tests are carried out to monitor the vibration in airport runway. Field monitoring data indicated that, throughout the tunneling operation, frequencies between runway's self-vibration and blasting vibration differed appreciably, and no resonance would be caused. In terms of the influence mechanism, the dynamic finite element program LS-DYNA was adopted, whose reliability was verified by field monitoring data. The numerical results showed that the airport runway was affected little and an existed cave could amplify the amplitude of blasting vibration to an extent. Meanwhile, based on the stress wave theory and the failure law of maximum tensile stress, the safety threshold for the runway induced by blasting vibration was obtained. Then the safety threshold of *PPV* was also decided through the statistic relationship between the peak value of tensile stresses and the *PPV* on the runway. As the safety thresholds derived from both fore mentioned methods was compared, with Safety Regulations for Blasting in China (GB 6722—2011) taken into account, a final safety threshold was decided.

KEYWORDS: airport runway; under-cross tunnel; blasting vibration; resonance; safety threshold

1 INTRODUCTION

In recent years, due to the rapid development of society and economy, a lot of public facilities such as traffic tunnels and underground sewerage systems have been constructed. Because of the limited space, more and more tunnels must be constructed under existing transport lines. In these constructions, interaction effects between existing transport lines and tunnel constructions exist objectively. How to ensure the safety of both of them is a critical subject in engineering constructions.

It has been extensively studied at home and abroad to excavate tunnels under existing transport lines. Fang et al stated that the advantages for a DOT shield tunnel under the river are resulted by short tunneling duration and low risk (Fang et al., 2012). Gui and Chen estimated the DOT shield induced ground surface settlement in combination with the Taoyuan International Airport Access Mass Rapid Transit system in Taipei (Gui and Chen, 2012). Wang et al analyzed the settlement of the express way induced by an undercrossed tunnel by employing the 3D finite element method (Wang et al., 2009). Wu et al studied the effect of blasting excavation on existing structures by using the software UDEC (Wu et al., 2010). Fang and He analyzed the displacement, deformation and internal forces of the existing tunnel affected by an undercrossing tunnel by means of 3D-FEM (Fang and He, 2007).

As described above, many tunnels are constructed under existing structures, roads and railways with non-explosive excavation methods. However, blasting excavation of tunnels under an airport is still rare at home and abroad up to now. And correlated researches are insufficient. In this paper, in combination with the construction of Xujiaping Tunnel crossing under Xujiaping Airport, the effect of tunneling under the airport runway with explosive method was analyzed based on field monitoring data and numerical simulations, and the safety criterion was also given.

2 GEOLOGICAL CONDITIONS

Xujiaping Tunnel is located in Enshi Tujia and Miao Autonomous Prefecture, Hubei province, PR China. As illustrated in Fig.1, the tunneling was between Shizhou Avenue and Jingui Avenue. This project started from Hongmiao, which is near the Shizhou Avenue, passed beneath Xujiaping Airport, and terminated at Jingui Avenue. The total route length is 1490 m and the tunnel gradient is 3%.

Corresponding author E-mail: lushiwei 364 @ 163. com

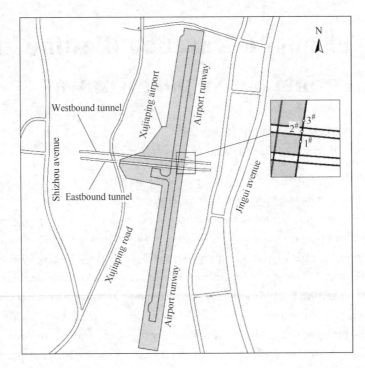

Fig.1 Plan of the double-track tunnel under Xujiaping Airport and arrange of monitoring points

The tunneling under Xujiaping Airport was carried out in Enshi basin, which is an area of Cretaceous clastic rocks. As reported in the reconnaissance report, in descending order, the clastic rocks consist of a highly weathered layer (0.8~12m) and a weakly weathered layer, which both belong to the Zhengyang Formation. The tunnel was constructed larger than 30.0m below ground level; hence the construction was carried out in the weakly weathered layer.

Fig.2 shows the cross sectional area of the double-track tunnel with a separation of 28m. The Permanent support is 40cm thick.

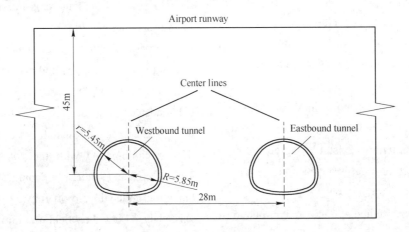

Fig.2 Section of the double-track tunnel

3 ANALYSIS ON FILED MONITORING DATA

Blasting vibration is monitored by using TC-4850 self-recording instrument (developed and manufactured by Zhongke (Chendu) Instruments Company Limited) during the tunneling in this project. And No.2 emulsion explosive was used for the excavation. The field monitoring data are listed in Tab.1.

Tab.1 Blasting vibration velocities

ID	Blast center distance/m	Peak velocity/cm·s^{-1}			Main frequency/Hz		
		x	y	z	x	y	z
1	80	0.048	0.19	0.23	34.188	35.398	43.478
2	80.78	0.06	0.205	0.176	51.282	40.816	52.632
3	80.72	0.07	0.146	0.137	38.462	45.455	52.632

The maximum velocities in Horizontal and vertical directions are less than 0.25cm/s, and the main frequencies are larger than 40Hz. The natural frequency of airport runways is within the range of 10~20Hz (Li and Xu, 2005). The small velocities indicate that the airport runway is affected little. Meanwhile, the main frequency of blasting vibration is much larger than that of airport runways, and no resonance would be caused. As a result, the airport runway is safe.

4 NUMERICAL ANALYSIS AND PARAMETERS SELECTION

For deeper understanding of the effect mechanism, 2 numerical models for exploring the attenuation in N-S and W-E directions are established respectively, as shown in Fig.3. The side surfaces except tunnel face and the bottom surface are applied for non-reflection boundary. The top surface, tunnel face and surfaces around tunnels are free. For simplification and symmetry, half of the model for axial analysis is calculated.

The JWL state equation can simulate the relationship between pressure and specific volume in the explosion process (Yang et al, 1996). The equation is as follows:

$$p_{eso} = A\left(1-\frac{\omega}{R_1 V}\right)e^{-R_1 V} + B\left(1-\frac{\omega}{R_2 V}\right)e^{-R_2 V} + \frac{\omega E_{0s}}{V} \quad (1)$$

where, A, B, R_1, R_2, and W are material constants, p_{eso} is pressure, V is relative volume and E_{0s} is specific internal energy. The physical and mechanical parameters of the dynamite are the same with that of the field test and are listed in Tab.2.

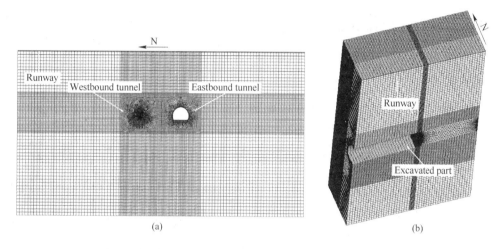

Fig.3 Dynamic finite element model for the calculations

(a) model for lateral analysis; (b) model for axial analysis

Tab.2 Mechanical parameters of explosive

Density /g·cm^{-3}	D /cm·μs^{-1}	A /GPa	B /GPa	R_1	R_2	ω	E_{0s} /GPa
1.09	0.4	214.4	18.2	4.2	0.9	0.15	4.192

The physical and mechanical parameters of surrounding rock, airport runway, primary support and permanent support are listed in Tab.3. The liner and airport runway are regarded as elastomer and fitted with the MAT_ELASTIC material model. MAT_MOHR_COULOMB is chosen to simulate the constitutive relationship of surrounding rock.

Tab.3 Mechanical parameters of different part

Type	Density/kg·m^{-3}	Young's modulus /GPa	Poisson's ratio	Cohesion/MPa	Frictional angle/(°)	Tensile strength /MPa
Surrounding rock	2300	10	0.25	0.8696	43.77	0.55
Runway	2500	35.5	0.15	—	—	1.89
Primary support	2500	32.25	0.20	—	—	1.27
Permanent support	2500	23	0.20	—	—	1.57

4.1 Attenuation in N-S Direction

With different designed cycle footages of 1.0m, 1.5m, 2.0m, 2.5m, 3.0m taken into consideration, the peak resultant velocities on runway with different horizontal distance from the axis of the westbound tunnel are shown in Fig.4. The maximum peak resultant velocities with different designed cycle footages are shown in Fig.5.

Fig.4 shows that the resultant velocity with an increasing distance in N-S direction distributes nearly symmetrically, and the largest velocity is just above the blast source. This indicates that the adjacent tunnel influence little on the distribution of peak resultant velocity. Fig.5 shows that the velocity increases with an increasing footage, but the rate of increase decreases, and the turning point is 2m cycle footage. The maximum peak resultant velocities are 2.03cm/s at 1.0m cycle footage and 3.57cm/s at 3.0m cycle footage. The maximum peak resultant velocity means even the blast source is just below the runway, the blast excavation under airport affects the runway little, and the airport runway would not be damaged by blasting vibration.

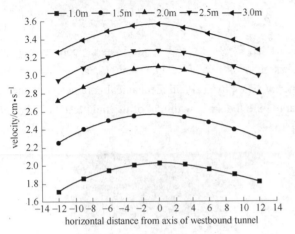

Fig.4 Peak resultant velocities with different horizontal distances at each cycle footage

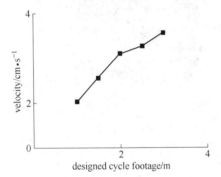

Fig.5 Maximum peak resultant velocities with cycle footage

4.2 Attenuation in E-W Direction

Based on numerical calculation results, the attenuation of resultant velocity on airport runway in east-west (axial) direction is obtained and as shown in Fig.6.

Fig.6 shows the resultant velocity decreases nonlinearly with an increasing horizontal distance from the tunnel face, and at a same distance, the resultant velocity on ground surface behind the tunnel face is larger than that in front of the tunnel face. This indicates that an existing cave can relatively amplify the surface vibration, which maybe because stress wave reflects at the surface of existing caves, and multiple superposition of stress wave takes place at the ground surface. With an increasing distance, the difference of resultant velocities behind and in front of tunnel face degenerates to zero. This means that the amplification to surface vibration wakens with an increasing distance, and only takes place within a limited region of a bit more than 20m away from the tunnel face.

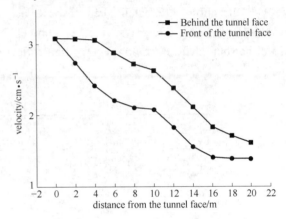

Fig.6 Peak resultant velocities with an increasing horizontal distance from the tunnel face

4.3 Verification

Numerical simulations must be verified by field monitoring data to ensure its reliability. Because the in-site cycle footage is 1.5m, simulation results with 1.5m cycle footage were selected to be verified by field monitoring data.

Field monitoring data and simulation results are listed in Tab.4. Both of their velocities approximately share the same magnitude. Field monitoring velocities are in the interval between 0.048cm/s to 0.23cm/s, and simulating velocities are in the interval between 0.054cm/s to 0.254cm/s, which are slightly greater than the field monitoring data. Meanwhile, both of their main frequencies approximately share the same magnitude.

Tab.4 Field monitoring data and simulation results

Means of monitoring	ID	Peak resultant velocities /cm·s⁻¹			Main frequency/Hz		
		x	y	z	x	y	z
Field monitoring data	1	0.048	0.19	0.23	34.188	35.398	43.478
	2	0.06	0.205	0.176	51.282	40.816	52.632
	3	0.07	0.146	0.137	38.462	45.455	52.632
Simulation results	H7966	0.054	0.203	0.254	36.74	37.17	45.05
	H7246	0.085	0.211	0.178	52.52	45.03	53.69
	H7066	0.109	0.149	0.136	41.30	52.53	56.94

5 SAFETY CRITERION OF RUNWAY

5.1 Safety Criterion of Stress Wave Theory

Consider the airport runway as a free surface, the reflection of P-wave is illustrated by Jiang (Jiang et al., 2012). The maximum velocity due to the tensile strength $[\sigma_t]$ is given as follows:

$$V_P(\sigma) = \frac{[\sigma_t](2\cos\alpha_1 \sin^2\beta_2 \sin 2\beta_2 - \cos^4\beta_2 \sin\alpha_1)}{2\cos^2\alpha_1 \sin^2\beta_2 \sin 2\beta_2 - \cos^4\beta_2 \sin\alpha_1 \cos\alpha_1} \sqrt{\frac{(1+\mu)(1-2\mu)}{E_d(1-\mu)\rho}} \quad (2)$$

where $[\sigma_t]$ —— dynamic tensile strength, $[\sigma_t] = 1.2\sigma_{t0}$;
α_1 —— incident angle for incident P-wave;
β_2 —— reflected angle for reflected SV-wave;
E_d —— dynamic young's modulus, $E_d = 1.2E$.

In case of unexpected damage, the maximum velocity muse be modified according its importance. 2.0 was chosen as the modifying coefficient in this paper, and the safety criterion is 8.83cm/s.

5.2 Safety Criterion of *PPV*

The statistical relationship model of the peak effective tensile stresses and the PPVs in airport runway is established and shown in Fig.7. The statistical relationship models established from Fig.7 are as follows:

$$\sigma_t = 0.0751PPV - 0.047 \quad (3)$$

where σ_t —— peak effective tensile stress, MPa;
PPV —— resultant velocity, cm/s.

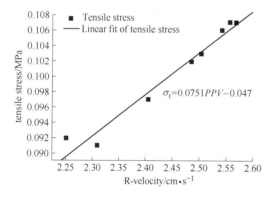

Fig.7 The statistical relationship of the peak effective tensile stresses and resultant velocities in airport runway

Eq. (3) indicates that a linear relationship exists between the PPVs and the peak effective tensile stresses (Chen et al., 2007). Based on the maximum tensile strength theory, when the *PPV* reaches 25.66cm/s, the peak effective tensile stress will approach the maximum tensile strength. 2.0 was also chosen as the modifying coefficient in this paper, and the safety criterion should be 12.83cm/s.

5.3 Comparison Analysis

According to the Safety Regulations for Blasting in China (GB 6722—2011), the safety criterion of blasting vibration velocity for 28-day fresh concrete is 10~12 cm/s. 12.83cm/s and 8.83cm/s are derived as safety criterions for airport runway based on numerical simulation and stress wave theory respectively. The results are consistent with that in the Safety Regulations for Blasting in China (GB 6722—2011). However, numerical result is more accurate and reliable. In order to be more practical, the safety criterion for the runway is determined to be 8 cm/s.

6 CONCLUSIONS

(1) Field monitoring data shows that, throughout the tunneling operation, peak velocities in every direction are quite small, frequencies between runway's self-vibration and blasting vibration differed appreciably. So no resonance would be caused and the airport runway would not be damaged.

(2) The existence of the adjacent tunnel does not influence much on the traveling of stress wave, but in the same tunnel, the excavated part can amplify the amplitude of blasting vibration to an extent in a limited region of a bit more than 20m away from the tunnel face.

(3) Based on the maximum tensile strength theory and stress wave theory, the safety threshold is calculated in the elastic range.

(4) Combined with numerical simulations, the relationship between dynamic stresses and PPVs is established for airport runway. Based on the maximum tensile strength theory, the safety thresholds of stress wave theory and *PPV* in the project are determined.

(5) With Safety Regulations for Blasting in China (GB 6722—2011) taken into account, a final safety threshold of 8.0cm/s was decided.

ACKNOWLEDGMENTS

The study was sponsored by the National Natural Science Foundation of China (Grant No. 41372312 and No. 51379194), the Fundamental Research Funds for the Central Universities, China University of Geosciences (Wuhan) (Grant No. CUGL140817) and the China Postdoctoral Science Foundation (Grant No. 2014M552113). We are also grateful to the China Scholarship Council (CSC) for supporting.

REFERENCES

[1] Chen M, Lu W B, Wu L, Xu H T. Safety threshold of blasting

vibration velocity to high rock slope of Xiaowan hydropower station[J]. *Chinese Journal of Rock Mechanics and Engineering*, 2007, 26 (1), 51~56.

[2] Fang Y S, Kao C C, Shiu Y F. Double-O-tube shield tunneling for Taoyuan international airport access MRT[J]. *Tunnelling and Underground Space Technology*, 2012, 30: 233~245.

[3] Fang Y, He C. Analysis of Influence of Undercrossing Subway Shield Tunneling Construction on the Overlying Tunnel[J]. *Journal of the China Railway Society*, 2007, 29(2): 83~88.

[4] Gui M W, Chen S L. Estimation of transverse ground surface settlement induced by DOT shield tunneling[J]. *Tunnelling and Underground Space Technology*, 2013, 33: 119~130.

[5] Hallquist J O. LS-DYNA keyword user's manual. *Livermore Software Technology Corporation*, 2007.

[6] Jiang N, Zhou C B, Luo G, Miao G J. Study on blasting dynamic characteristics of open pit to underground mining slope[J]. *Journal of Central South University (Science and Technology)* 2012, 43(7): 2746~2750.

[7] Li W M, Xu J Y. Analysis of Airfield Pavement Random Vibration and Dynamic Response. *Proceedings of the 14th National Conference on Structural Engineering*, Shandong, Yantai, 2005: 257~260.

[8] Wang Z, Du S J, Zhang W B, Li Y J, Wu X P. Analysis of Construction Settlement of Shallow Railway Tunnel under Crossing the Highway[J]. *Chinese Journal of Underground Space and Engineering*, 2009, 5(3): 530~535.

[9] Wu X Y, Jiang Z Q, Gao L, Zhang D L, Zhu H B. Study on blasting vibration of Highway caused by the Under-cross Jinniushan Tunnel[J]. *Railway Engineering*, 2010: 60~62.

[10] Yang R, Bawden W F, Katsabanis P D. A new constitutive model for blast damage[J]. *International Journal of Rock Mechanics and Mining Sciences & Geomechanics Abstracts*. Pergamon, 1996, 33(3): 245~254.

Spreading out Dust and Gases, Formed by Blasting Operations at Erdenetiin-Ovoo Ore Open Pit Mine and Application of Remote Control (Firing) System for Blasting

B.Laikhansuren[1], G.Tulga[2], D.Nyamdorj[3], N.Davaakhuu[3]
(1. Mongolian University of Science and Technology; 2. Mongolian Mining Corporation LLC;
3. Erdenet Mining Corporation Compan)

ABSTRACT: Blasting operations, using large amount of explosives (100~200 tons and over) and specific climate condition around the mines can create problems, associated with toxic gas and dust spreading and safety conditions of the employees.

The paper presents methodic, estimates and calculations for spreading out toxic gases and dust, formed by blasting at Erdenetiin-Ovoo ore mine. Determined are safety zone radius for toxic gases, weight of dust, falling down from dust-gas cloud etc. Also, some results of measurement and analyses for chemicals and substances at the mine site by the laboratory of Erdenet Mining Corporation Company are given in the paper.

The paper describes basic indices and specifications and advantages of remote control firing system for blasting Rottenbuhler enjinering®, selected and used at the mine. Estimates of economic effectiveness from introducing remote control firing system are provided. Conclusions and suggestions, based on test and studies are summarized.

KEYWORDS: safety zone radius for toxic gases, formed by blasting; contents, possible actions and consequences of some toxic gases; weight of dust, falling down from dust-gas cloud; measurement and analyses for chemicals and substances, remote control firing systems for blasting; "Dyno Rem" "BlastPET ST" "Rottenbuhler enjinering 1670"

1 SPREADING OUT DUST AND GASES, FORMED BY BLASTING

1.1 Determining Safety Distance for ToxicGases, Formed by Blasting

Present practices of blasting operations, using large amount of bulk explosives (100~200 tons and over) dictate requirements for careful determining safety distance for toxic gases and dust, formed by blasting.Decision for entering people to mine area, where carried out blasting should be done in accordance with projected blasting procedures. According to"Unified regulation for blasting operations", only leader of blasting operation shall give permission to people for entering to blasted area at least in 30 minutes after blasting and appearing clear visibility in an area due to dust falling down,in case of decreasing toxic gas concentration to allowable level and after the completion of technical inspection of blasting site.

We would like to introduce some results of trial blasts at several blocks of the mine since 2012 and estimates and calculations for spreading out toxic gases and dust, carried out by methodic of Russian scientists.

There were drilled 100~120 blast holes in 3 rows at one selected blasting block, depending on mine depth and water level in blast holes. Volume of blasted rock mass was 80~120 thousand.m³ and amount of used explosives–60~75 thousand tons. Powder factor of emulsion explosive is 0.62~0.75 kg/m³, explosive charge for 1 blast hole–500~650 kg, blast round–7.0×7.0 m; 8.0×8.0 m; 9.0×9.0meters, depending on rock properties.

It's recommended to determine safety zone radius for toxic gases, formed by blasting as follows in case of non-windy period:

$$R_{xx} = 1.5\sqrt[3]{CQ} = 1.5\sqrt[3]{200 \times 75000} = 367.5 \text{ m}$$

where R_{xx}——radius of toxic gaseszone, m;
C——amount of toxic gases in CO, L/kg(C=200 L/kg);
Q——mass of explosive charge ,kg(Q=75000 kg).

Safety zone radius for toxic gasesincreases along the wind direction during windy days.If wind velocity is V_c= 4 m /s for a given day,safety zone radius for toxic gases can be determined as follows:

Corresponding author E-mail: nyamdorj@erdenetmc.mn

$$R_{xx} = 1.5\sqrt[3]{CQ(1+0.5V_c)}$$
$$= 1.5\sqrt[3]{200 \times 75000(1+0.5\times 4)} = 1102 \text{ m}$$

where V_c——wind velocity before blasting, m/s.

The estimates show that safety zone radius for toxic gases, formed by blasting is equal to 367.5 meters for non-windy period, but it has increased 3 times (to 1102 meters) along with wind direction and velocity. Therefore, it's recommended to strongly abide safety regulations of blasting operations, specially not to enter in blasting area in 30 minutes period after blasting, stand along the wind direction etc.

Generally, 1 kg of industrial explosives can generate about 100 liters of toxic gases, that accounts for 10% of total blast gas products. There are formed various components of toxic gases after the blast, depending on explosives composition, which include carbon mono oxide (CO), nitric oxide (NO), nitrogen dioxide (NO_2), dinitrogenpentoxide (N_2O_5), sulfur dioxide (SO_2), hydrogen sulfide (H_2S) etc.

Tab.1 show contents, possible actions and consequences of some toxic gases, that may be formed after the blast, Tab.2 – possible amounts of carbon mono oxide and nitric oxide, formed by blasting.

Tab.1 Contents, possible actions and consequences of some toxic gases, formed by blasting

Toxic gas components	Toxic gas content and poisoning symptoms					
	Maximum content/mg·L^{-1}		Minimum content/mg·L^{-1}		Allowable level	
	Appearing poisoning symptoms for a long period	Appearing symptoms after breathing 1 hour	Appearing dangerous symptoms during 0.5~1 hour period	Loss of life for a short period	by content /mg·L^{-1}	by volume /%
CO	0.1,...,0.2	0.6,...,1.5	1.6,...,2.3	5.0	0.02	0.0016
NO,NO_2,N_2O_5	0.007,...,0.2	0.2,...,0.4	0.2,...,1.0	0.5	0.005	0.00025
SO_2	0.015,...,0.025	0.06,...,0.26	1.0,...,1.25	—	0.02	0.00035
H_2S	0.01,...,0.2	0.25,...,0.4	0.5,...,1.0	1.2	0.016	0.00066

Tab.2 Possible amount of carbon mono oxide and nitric oxide, formed after the blast

Category of rock drilling classification (by SNiP)	Soil condition	Amount of carbon mono oxide /L·kg^{-1}	Amount of nitric oxide/ L·kg^{-1}
5	Wet	10.5	5.0
5	Dry	7.0	5.5
6	Dry	19.0	1.5
7	Dry	30.0	1.5
8,...,9	dry, with intensive jointing	20.5	4.5
8,...,9	dry, with jointing	34.0	1.2
9	Wet	33.5	2.0

1.2 Estimating Amount and Weight of Dust, Falling Down from Dust-Gas Cloud, Formed by Blasting and Their Spreading Boundaries

Blasting area becomes dusty because of dust-gas cloud, formed by blasting, depending on wind direction and velocity. Dust particles size is very small (less than 200 μ) in most case, but coarse sized dust is spreading out in an area, 2~3 times greater, than initial falling area.

Weight of dust-gas cloud, formed by blasting can be determined as follows:
$$M = 2.39\rho Q V_c K^{-0.27} r^{-2} = 2.39 \times 2.55 \times 0.6 \times 1 \times 10^{-0.27} \times 0.5^{-2}$$
$$= 7.8 \text{kg}$$

where M——weight of small sized dust, falling down in 1 m^2 area, kg;
ρ——density of blasted rock, t/m^3 (ρ=2.55 t/m^3);
Q——mass of explosive charge, t (Q=75000 kg);
V_c——wind velocity, m/s (V=1 m/s);
K——coefficient, taking into account turbulent flow;
r——distance, allotted to M weight, m.

Amount of M weight decreases, as R distance gets away from the charge center.

Size of dust falling zone can be obtained, using following formula:

$$\int_L^r 2\pi r 2.39 pQ L^{-0.27} r^{-2} dr = \int_L^r 2\pi r^{-1} 2.39 pQK^{-0.27} dr$$
$$= 2\pi 2.39 pQK^{-0.27} l_T(r|L) = 30Q$$

where r——size of dust falling zone, m;
L——distance of dust generating source, m.

To determine dust weight in 1 square meter area, following indices and values can be taken into calculation practices: wind velocity at initial period V_c=1 m/s, coefficient, taking into account turbulent flow K=10 m^2/s, not depending on rock strength factor. Also, it's recommended that blasting of 1 kg explosives can generate 30 kg of fine (sized less than 200 μ) dust.

Size of dust falling zone can be determined as follows:

$$R = L\exp(3.72/p)$$

where L——distance of dust generating source, m;
$L\exp(3.72/p)$——calculation values, dependent on rock density(4.43).

Said zone and distance are determined :

$$L = B + 0.6H = 24 + 0.6 \times 15 = 33 \text{ m}$$

where B——block width, m(B=24 m);
H——bench height, m(H=15m).

$$R = L\exp(3.72/p) = 33 \times 4.43 = 146.2 \text{ m}$$

According to estimates, size of dust falling zone equals to 146.2 m, that largely depends on rock density and block width.Calculation values, dependent on rock density are given in Tab.3.

Tab.3 Calculation values, dependent on rock density

Rock density /t·m^{-3}	exp(3.72/p) values	Rock density /t·m^{-3}	exp(3.72/p) values	Rock density /t·m^{-3}	exp(3.72/p) values	Rock density /t·m^{-3}	exp(3.72/p) values
1.8	7.89	2.4	4.71	2.8	3.78	3.1	3.32
1.9	7.08	2.5	4.43	2.9	3.61	3.2	3.2
2.0	6.42	2.3	5.04	2.7	3.97	3.3	3.09
2.1	5.88	2.6	4.18	2.8	3.97	3.4	2.99
2.2	5.42	2.4	4.71	2.9	3.61	3.5	2.89
2.3	5.04	2.7	3.97	3.0	3.46		

1.3 "Mobile"laboratory of Occupationalhealthand Safety Department of Erdenet Mining Corporation Company

The laboratory was certified according to reqtirements of ISO/IEC 17025:2005 (MNS ISO/IEC1702552007) standard.

AVA-1 type dust aspirator is used for measurement of dust, formed by blasting at different horizons and places of the mine. For air dust sampling to determine heavy metals and toxic elements, there are used AFA-HA-20, AFA-HA-10,AFA-HP-20,AFA-VP-20 type paper filters.

Voltammeter type measuring apparatus (ModelAKV-07) is capable to determine 15 kind of chemicals, contained in water, soil and air. We made measurement and analysis for following chemicals, including non-metal elements (As, Se), transition metals(Cd,Cu,Zn,Ag,Fe,Co,Ni,Hg,Mn), and base/parent metals(Pb,Bi,TI,Sb).Results of measurement and analysis were processed with use of Polar 4.0 software.Some results of measurement and analyses by the laboratory of Erdenet Mining Corporation Company is given in Tab.4.

Tab.4 Results of measurement and analyses by the laboratory of Erdenet Mining Corporation Company

Date of sampling and measurement	Place of sampling and measurement /m	Wind velocity /m·s^{-1}	Air pressure /mm	Dust/μg·m^{-3} MNS 4585-2007		Pb/μg·m^{-3} MNS45852007		C/μg·m^{-3}	Cd/μg·m^{-3}	Zn/μg·m^{-3}	As/μg·m^{-3}	Tew/°C
				Maximum allowable limit	Measurement	Maximum allowable limit	Measurement	Measurement	Measurement	Measurement	Measurement	Measurement
2012.01.20	1370	3.7	653.3	500	2.91	1.0	—	6.04	0.032	1.83	1.52	−18.3
2012.02.17	1445	3.8	659.5	500	2.32	1.0	—	—	0.022	—	—	−6.8
2012.04.06	1310	15	651.4	500	1.21	1.0	0.088	0.08	barely enough to appear	0.0	—	15.6

Above mentioned measurement and studies show, average dust content during 30 minutes – 2.14 μg/m^3, lead content −0.088μg/m^3, cupper content – 3.06 μg/m^3, cadmium content – 0.02 μg/m^3, zinc content – 1.83 μg/m^3. But comparison was complicated because of absence of standards, except MNS 4585-2007.

2 REMOTE CONTROL (FIRING) SYSTEM FOR BLASTING

Conventional method of blasting at the mine requires blast shelter, located at least 100 m from blast site to protect people from blast effects. However, toxic gas and dust may influence on health and safety of workers. Therefore, the study was aimed on selection and introducing of advanced methods and accessories for remote control of blasting operation to create more safe and sound working condition at the mine. This kind of methods and systems are widely used in world mining practices.

As a result of initial study, there were selected and compared 3 different remote control firing systems for blasting, namely "Dyno rem" from Sweden, "BlastPET ST" of Australia and "Rottenbuhler enjinering 1670" from U.S.A. General view of selected remote control (firing) systems are shown on Figs.1,2,3.

Fig.1 View of DynoRem™(NONEL®) system

Fig.2 View of BlastPET® ST system

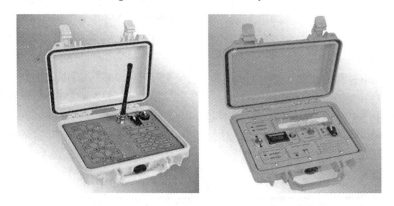

Fig.3 View of Rottenbuhler engineering® system

Basic indices and specifications of above illustrated remote control (firing) systems are presented in Tab.5.

Tab.5 Basic indices and specifications of remote control (firing) system for blasting

No.	Indices	Name of remote firing systems and their specification					
		DynoRem™		BlastPET® ST		Rottenbuhler enjinering®	
		Master control unit	Remote receiver initiator	Master control unit	Remote receiver initiator	Master control unit	Remote receiver initiator
1	Case material	Durable plastic with hard finish	Durable plastic with hard finish	Durable plastic with hard finish	Durable plastic with hard finish	Durable plastic with hard finish	Durable plastic with hard finish
2	Dimensions/mm	75×230×40	355×265×155	410×336×178	141×497 180×500	260×310×120	260×310×120
3	Initiation	Key and PIN code	Key and controller	Key and PIN code	Key and controller	Key and PIN code	Key and controller
4	Operating temperature range	−25～55℃	−25～55℃	0～70℃	0～70℃	−30～60℃	−30～60℃
5	Weight/kg	0.71	5.8	4.2	4.5	3.0	3.3
6	Battery	4.8 V, 1300 mAh	4.8V, 1800 mAh	12 V, 2700 mAh	12 V, 2700 mAh	12VDC	2VDC
7	Transmission range/km	1.0	1.0	4.0	4.0	4.0	4.0

Based on comparison, considering wide transmission range, operating temperature etc., Rottenbuhler enjinering® system was selected as more suitable to our mine condition.

Advantage of the selected system:

(1) Increased safety for blasters and workers, possibility to stay outside of toxic gas and dust zone.

(2) Economic benefit, no need to move blast shelter every time.

(3) Enhanced reliability.

(4) Possibility to determine blasting sequences in a optimal way.

(5) Possibility to blast 8 blocks simultaneously.

(6) Possibility to control blasting operation from the long distance (up to 4 кilometers).

2.1 Safety Considerations from Recent Practices

The Rothenbuhler remote control firing system should be operated by specially trained person in accordance with safety and operating manuals of the system, other occupational safety rules, instructions and regulations, issued by the company, central and local government organizations.High frequency radio wave may initiate electric detonators, so control units of the system should be placed at distance 25 meters or above from electric detonators.Shock tube sparkler of the system will generate up to 3000 v voltage.It's not allowed to be fiddling with initiation sparkers in case of equipment connection and conducting blasts.The system is guaranteed for 2 years (batteries -1 year, spark tips need to be replaced) against defects in workmanship and materials.Repair during warranty period will be done at manufacturer's factory or at the nearest authorized facility by authorized person.The remote firing system is designed to initiate electric and non-electric detonators. The system will be located at designated place.

One of advantage of the system, comparing with others is that there is two-way signaling: Master control to Remote receiver and Remote receiver to Master control. For this advantage, all equipment of the system can come to neutral position automatically, that ensures increased safety and check possibility to users.For underground blasting, the remote control firing system will transmit information via antenna system.

2.2 Working and Storage Condition

All system cases are hard anodized, extremely tough, corrosion and impact resistant, water tight and air tight. During site operation, case cover needs to be closed. Due attention and careful action are required to electrical connections inside of the case.

Case can resist rather substantial external load or impact. Case durability was checked by falling down from 6feet (1.8 meters) height on concrete floor.

Master control unit and remote receivers are connected to each other, but signaling and access to initiation will be held just before blasting.

The system should be stored in a cool and dry place. Storage temperature is 4°F-aac 86°For (−20°C to 30°C). Violation of storage temperature will have adverse impact on battery life.

Overall operating temperature of the system ranges from−22°F to 140°For (from−30°C to 60°C).However, temperature range, ensuring long battery life is recommended to be from 23°F to 122°For (from−5°C to 50°C).

3 ECONOMIC BENEFIT AND EFFECTIVENESS

Remote control firing system allows the users to stay at long distance from blasting site, that ensures much increased safety. Observation of blasting process also is possible from remote distance. There is no need to prolong wave transmission cord, that decreases cord consumption by 90%.

For underground mining, after ensuring relevant safety requirements and transportation of all people to the surface,blasting at each underground working can be controlled from the surface area.

Moreover, remote firing system offers more economic efficiency, comparing to conventional blasting system.

Economic estimates based on actual mine data and provided below:

(1) In most case of blasting, blast shelter is needed to be moved to a suitable place. Crawler and wheel dozers are used for this purpose. According to mine practice, moving blast shelters takes 4 hours of CAT -824 wheel dozer operation,depending on actual mine situation, road and traffic condition etc.

(2) Yearly operation time of wheel dozer can be estimated as follows:

$$T = 52T_{d.o} = 52 \times 4 = 208 \text{ motor hours}$$

where   52——number of weeks in a year;

$T_{d.o}$——averagedozeroperation time per week.

(3) Expenses for remote control firing system:

1) Depreciation cost of the firing system (as of 2013)

$$Z = 67023000 \times 10\% = 6702.3 \text{ thousand tugrigs}$$

where   67023000——price of the remote firing system, tugrigs;

10%——depreciation rate.

2) Other relevant shop expenses:

$$Z = 6702.3 \times 0.365 = 2446.3 \text{ thousand tugrigs}$$

where   0.365——rate of shop expenses.

3) Total expenses of remote control firing system:

$$Z = 6702.3 + 2446.3 = 9148.6 \text{ thousand tugrigs}$$

(4) Economic effectiveness from introducing remote control firing system at the mine(as of 2013):

$$E = TC - Z = 208 \times 104.265 - 9148.6 = 12538 \text{ thousand tugrigs}$$

where C — cost of 1 motor hour operation for CAT824 wheel dozer.

Cost estimation of 1 motor hour operation for CAT824 wheel dozerCAT 824 includes cost of diesel fuel, lubricant materials, transportation cost, salary and insurance expenses, depreciation cost, equipment operating and other shop expenses. Results of estimate are given in Tab.6 .

Tab.6 Cost of 1 motor hour operation for CAT824 wheel dozer

No.	Cost elements	2010 tugrigs	2011 tugrigs	2012 tugrigs	2013 tugrigs
1	Diesel fuel cost	40947	54123	53292	65287
2	Lubricant materials cost	4621	4872	8292	10158
3	Transportation cost	234	243	308	377
4	Basic and additional salary	5597	7592	9110	10021
5	Social insurance fees	700	1025	1184	1303
6	Depreciation cost	17297	17297	17297	17297
7	Shop expenses	7634	9367	9843	9400
8	Operating cost	34004	41724	43847	41777
9	Equipment utilization factor	0.8	0.8	0.67	0.67
10	Cost of 1 motor hour operation	88826.8	108994.5	95926	104265

Economic effectiveness of introducing remote control firing system during recent 4 years is summarized and presented in Tab.7.

Tab.7 Economic effectiveness from introducing remote control firing system during 2010~2013

No.	Indices	2010 thousand tugrigs	2011 thousand tugrigs	2012 thousand tugrigs	2013 thousand tugrigs
1	Economic effectiveness by each year	9130.1	13521.3	10804	12538
2	Total economic effectiveness	45993.4 thousand tugrigs			

Based on investigations, test and studies, carried out for Erdenetiin- Ovoo ore mine, following conclusions and suggestions can be summarized:

(1) Presently, the mine has normal aerologic condition. However, further mine development and deepening will require intensification of all mine operations, including drilling and blasting. Accordingly, number of blasting and amount of used explosive materials should be increased, that will influence on conventional blasting practices and overall work condition of the mine.In the meantime, present blasting practices need improvements and introduction of more advanced, efficient and safe methods and facilities for blasting. Blast shelter was one of problem. At the same time, safety zone radius for toxic gases, formed by blasting presently is estimated to 367.5meters, but it can be increased to 1102 meters along the wind direction at wind velocity 4 m/s. Therefore, creation of more safe and sound working condition at the mine is important.

(2) Use of remote control firing systemRottenbuhler enjinering® at Erdenet ore mine for recent years showed its technical advantage and economical effectiveness. The system offers features and advantages, such as increased safety and reliability, suitable operating range and temperature, optimal sequence of blasting, economic benefit.

(3) Economic effectiveness of introducing remote control firing system during 2010~2013 was 45.9 million tugrigs. Economic efficiency mostly gained because of eliminating works, associated with frequently moving of blast shelters.

REFERENCES

[1] Ganopoliskii M I, Baron V L, Belin V A, Pupkov V V, Sivenkov V I. "*Methods of conducting blasting works. Special blasting works*" Publishing house of Moscow Mining University, Russian Federation, 2007.

[2] Laikhansuren B, Tuvkhuu L, Jargalsaikhan Kh, Nyamdorj D "*Studies on ore massive and blasting operations*", Ulaanbaatar, Mongolia, 2009.

[3] Nyamdorj D. "*Studies on optimization of blast design parameters and blasting technology at Erdenet ore open pit mine*", Dissertation, Ulaanbaatar,Mongolia, 2004.

Mechanism of the Short-delay Controlled Blasting in the Air and Its Application in Anti-terrorist

YAN Honghao, ZHAO Tiejun, LI Xiaojie, WANG Xiaohong
(*State Key Laboratory of Structural Analysis for Industrial Equipment, Dalian University of Technology, Dalian, Liaoning, China*)

ABSTRACT: The single point and two points short-delay controlled blasting in the infinite air were simulated with the nonlinear dynamic analysis software LS-DYNA. Blasting on the simulation results of the short-delay blasting mechanism, a method of the short-delay controlled blasting in the air was designed, and the six points short-delay controlled blasting was made in the same method also. The experimental results were consistent with the results of numerical simulation, so LS-DYNA finite element software can be used to analyze the short-delay controlled blasting. In this article, the method of the short-delay controlled blasting was used to break into a door in anti-terrorist. In the case of the same distance and explosive charges, this method could reduce the air shock wave peak and the injury of personnel. It was found that the expansion bolt in a security door was the vulnerability of a door. In the experiment of breaking into a door, the reasonable dose of breaking into a door was obtained, according to the overpressure peak under different doses, the safe distance for anti-terrorist personnel was calculated.

KEYWORDS: short-delay controlled blasting; blasting; shock wave; security; anti-terrorism

1 INTRODUCTION

In recent years, with the increasingly rampant international terrorist activities, the anti-terrorist situation is increasingly grim. Terrorists often implement violence in densely populated places and take hostages as a threatening method. When anti-terrorist personnel rescue hostages, or are in the hunt for terrorists, they are often blocked by the limited space or solid buildings, so they need to break into doors or walls to open up a fast track to gain favorable opportunity.

With the development of science and technology, breaking tools have been widely developed. People have invented many tools, such as: breaking wimble, fusing style cutter, hydraulic cutting device, breaking gun, hydraulic expanding door device, breaking shells, shaped break device and so on [1]. With the merits of light weight, small size, convenience, high energy, high efficiency, blasting tools can effectively avoid the defects of other type tools. Not only can blasting tools instantly explode a door, the huge noise of the explosion also can effectively threaten criminals in the room [2].

In this article, a security door was set as the blasting target. There are two weaknesses for common doors: the door hinges and the door locks. Basing on the two weaknesses, most of the blasting tools are researched and developed. With the technological innovations of safety doors, it is diff- icult to break into the door by the shaped break device or other tools. On the premise of ensuring hostages' and anti-terrorist personnel' safety, it is not easy to dismember the hinges and locks with small-charges, hence the expansion bolt which links the door frame and wall can be taken as the weak spot of the safety doors to study.

Although it is effective to break into a door with the blasting method, it also has a dangerous. The explosive charges to break into a door needs calculation, otherwise it would cause unnecessary harm to anti-terrorist personnel due to excessive dosage. The harm that explosions cause to anti-terrorism personnel is mainly the direct effect of shock wave, followed by explosive noise, blasting debris and the adverse effects of explosive dust. People have done lots of researches on the effects that the shock wave causes to biology, and it is generally considered that the shock wave wouldn't cause organic damage to human when the overpressure peak is less than 20 kPa. So the method of multipoint short-delay blasting is used to break into a door.

Many scholars have been doing the theoretical and experimental researches of the short-delay controlled blasting for a long time [3~5]. The mechanical study of the short-delay controlled blasting in the rock engineering is more exhaustive. However, with the expansion of the application range of the blasting field, the theory of short-delay controlled blasting has been gradually applied to blasting de-

---

This project was financially supported by the National Science Foundation of China (No.10872044, No.10602013 & No. 10972051& No. 10902023) and the Natural Science Foundation of Liaoning Province of China (No. 20082161).
Corresponding author E-mail: zhaotiejunfe@126.com

molition, and so on. For certain special blasting situations, explosive packages need to be exposed to the air and the distance between explosive packages is very small. So it is necessary to study and analyze the mechanism of the short-delay controlled blasting in the air, and then to work out a short-delay controlled blasting method that can be applied in the air.

From the current study situation of the air shock wave, it was found that people have already got hold of the spread rule of the air shock wave, and various kinds of procedures they have adopted to effectively simulate the various complex problems of the shock wave in the air. However, most of these researches are about the single point blasting or multipoint simultaneous blasting [6~10]. In this article, the dynamics analysis software LS-DYNA was used to simulate the numerical value of the short-delay controlled blasting of single point and two points in the infinity-air under different working conditions. According to the simulation results, the mechanism of the short-delay controlled blasting was obtained, and then the method of the multipoint short-delay controlled blasting was also worked out. According to this method, the six points short-delay controlled blasting in the air was experimented. The results of the simulation were consistent with that of the experiment, which validated the mechanism and the method of the short-delay controlled blasting in the air.

In this paper, the method of the short-delay controlled blasting in the air was devised, compared with group charge, this method could effectively reduce the overpressure peak of air shock wave under the condition of the same distance and the same explosive charges.

2　NUMERICAL SIMULATIONS

The numerical value of single point blasting and two points short-delay controlled blasting in the infinity-air were simulated with ALE algorithm and the grid partition mapping method. The explosive charges (TNT, with 1.6 g/cm^3 density and 0.8193 cm/μs detonation velocity) of single point was 20 g, the material model was defined by *MAT_HIGH _EXPLOSIVE_BURN keyword, and the equation of state was defined by *EOS_JWL keyword. Air models was defined by *MAT_NULL keyword to define, the equation of state was defined by *EOS_LINEAR_ POLYNOMIAL keyword. In order to reduce the computing time, it was simplified with two-dimensional problem to simulate, but its algorithm required the establishment of three-dimensional solid element, therefore, one-eighth of three-dimensional sheet structure model with size 180 cm × 180 cm × 20 cm was established. The three surfaces near the coordinate origin imposed symmetry constraints. Transmitting boundary was imposed by the other three surfaces.

2.1　Numerical Simulation of the Single Point Blasting

Fig.1 was the x direction overpressure history of element which was 30 cm distance from explosion center. Fig.1 could be seen when the shock wave through this element, the pressure increased instantly. Pressure rapidly decreased after the shock wave peak. The over expansion of the explosion products generated a negative pressure zone, the internal pressure of explosion products was less than the initial atmosphere pressure $p_0(1.0 \times 10^5$ Pa), the reverse compression appeared and generated a sparse wave. Explosion pressure was greater than atmospheric pressure p_0, as the second time expansion of the explosion product began, the second shock wave appeared. The simulation results were consistent with the shock wave propagation theory, and the air shock wave velocity is about 30/221=0.1357 cm/μs, which proved the reliability of numerical simulation.

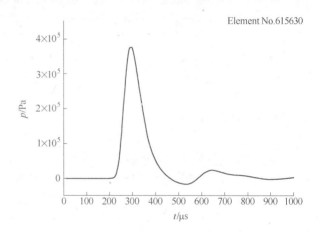

Fig.1　Overpressure history of element which is 30 cm distance from explosion center

2.2　Numerical Simulation of Two Points Short-Delay Controlled Blasting

In the process of simulation, explosive package was laid down along the x axis direction horizontally with 80 cm space. According to the numerical simulation results of a single point blasting, it could be obtained that the shock wave front produced by the first explosive package, and it just reaches 80 cm along x direction (just to reach the second explosive package point) in 1370 μs. Supposed the initiation time interval between the two explosive packages was t_0, because the initiation time delay of the two explosive packages was different, two points time–lapse blasting problem was divided into two kinds of working conditions to be simulated. Condition 1: after the detonation of the first explosive package, the shock wave front did not reach the explosive package, but the second package started initiation. Namely: $t_0<1370$ μs. Condition 2: after the detonation of the first explosive package, shock wave front just reached or exceeded a second package, and then the second package started initiation. Namely: $t_0 \geq 1370$ μs.

2.2.1 Condition 1

Fig.2 was the detonation overpressure nephogram when $t_0 = 100$ μs. The shock waves generated by the two explosive packages encountered when $t = 480$ μs, where the overpressure increased instantly. When $t = 1950$ μs, the shock wave produced by the second initiation point spread about 100 cm along the x axis, and two shock waves were integrated into one shock wave. The overpressure peak value of the first shock wave was 32.7 kPa, the overpressure peak value of the second shock wave was 33.3 kPa, and the overpressure peak value of the middle region of pressure superposition was 55.5 kPa. According to the simulation results of single-point air detonation, it was known that the overpressure peak value of the area that was 100 cm away from the blasting center was 33.5 kPa. Comparison showed that the pressure peak value of the two shock waves in the 100 cm non-overlapping area was almost the same as the value of shock wave of single point detonation; in the overlay region the pressure was significantly increased, and the pressure increased by 65.7%.

Through the simulation with different t_0, it was found that when t_0 was infinitely close to zero, namely the two points almost simultaneously detonate, two shock waves encountered in the middle of the connection of the two explosion centers, and in this area the pressure increase reached the maximum. As t_0 increased, the two shock waves encountered in the center-right position of the connection of the two explosion centers, in the meeting region the pressure increase gradually decreased.

When t_0 changed between 0 and 1370 μs, the pressure increase of overlay zone varied between 26.0 and 3.0 kPa. Through the above analysis, it showed when t_0 changed between 0 and 1370 μs, the two shock waves would be eventually converged into one shock wave; there was the pressure area enhanced in the middle of the two explosive packages. The closer t_0 was to 1370 μs, the smaller the amount of pressure increase was.

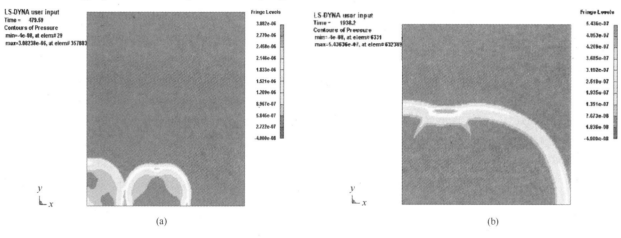

Fig.2 Overpressure nephogram when $t_0 = 100$ μs

(a) overpressure nephogram when $t = 480$ μs; (b) overpressure nephogram when $t = 1950$ μs

2.2.2 Condition 2

In condition 2, the delay time t_0 is supposed to 1370 μs, 2000 μs, and 2500 μs respectively. When $t_0 = 1370$ μs, the shock wave front showed like Fig.3. Fig.4 was the overpressure history of element with coordinate point (180, 0, 0). As the second shock wave front was beyond the first shock wave front, in the figure the first overpressure peak value was generated by the second shock wave and the other was generated by the first shock wave. At that moment the second shock wave overpressure peak was 13.5 kPa, the first shock wave overpressure peak was 8.4 kPa.

Tab.1 recorded the overpressure peak value of shock wave with different values of t_0, when two shock wave fronts reached the coordinate point (180, 0, 0). By comparing the super pressure value in the table, it could be seen that with the increase of t_0, the overpressure peak value of

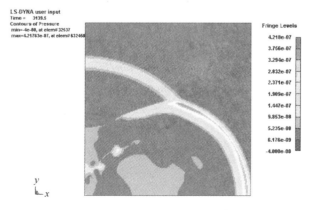

Fig.3 Overpressure nephogram when $t_0 = 1370$ μs

the second shock wave front decreased, and the overpressure peak value of the first shock wave front increased. Based on the above trend, it could be seen that the second shock wave couldn't catch up with the first shock wave front if t_0 was further increased. But influenced by the impact of the end of the first shock wave and the negative pressure zone, the energy of the second shock wave would decrease in certain degrees. When t_0 was very large, the spread of the second shock wave was almost free from the impact of the first shock wave, so the two shock waves spread separately, which equaled to twice initiations of explosion of single point, and the two initiations were large interval.

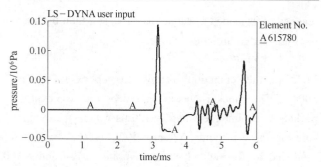

Fig.4 Overpressure history of element with coordinate point (180, 0, 0) when t_0=1370 μs

Tab.1 Overpressure value of shock wave with different value of t_0

The delay time between explosives /μs	The overpressure peak value of the second shock wave/kPa	The overpressure peak value of the first shock wave/ kPa
1370	14.5	8.3
2000	13.5	8.4
2500	12.5	9.6

3 SHORT-DELAY CONTROLLED BLASTING EXPERIMENTS

3.1 Short-Delay Controlled Blasting Methods and Experimental Design

Through the above mechanism study of short-delay controlled blasting in the air, the method of the short-delay detonation was designed, namely the effect of the short-delay controlled blasting was achieved by using the different length of nonel tube. The detonation velocity of nonel tube was about 0.2 cm/μs, the limiting deviation was ±0.0050 cm/μs, the detonation velocity was relatively stable, and the stability of delay interval could be guaranteed [12]. The experiment was done with single point with 100 g emulsion explosive (with 1.0 g/cm³ density and 0.33 cm/μs detonation velocity), the length of nonel tube of each detonator was increased by 150 cm, and the Fig.5 showed the position of explosive and the length of nonel tube. The length difference of the nonels between two horizontal adjacent explosive was 150 cm, namely the delay time was about 750 μs, the length difference of nonel between two vertical adjacent explosives was 300 cm, the delay time was about 1500 μs. In the simulation of the single point blasting of 100 g emulsion explosives in the infinite air, it was drawn that the time which it took for a shock wave to spread 80 cm was about 970 μs. This design could ensure that the initiation of explosion of two horizontal adjacent points were in line with that of condition 1 in the simulation of two points initiation of explosion, namely: before the first shock wave spread to the second explosive package, the second explosive package began blowing up, and then two shock waves were converged on one shock wave along the x direction. The two adjacent vertical initiation points were in line with that of condition 2 in the simulation of two points initiation, namely the first shock wave had reached the second blasting points for some time before the second initiation point started blowing up, the shock wave generated by the second initiation point was rapidly over the first shock wave, and then the first shock wave weakened the partial energy of the second shock wave.

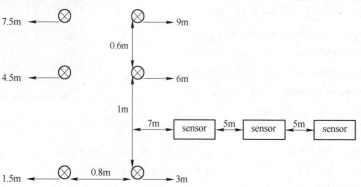

Fig.5 The position of explosive and the length of nonel tube

3.2 Experimental Results

3.2.1 The Simultaneous Initiation Experiments of Two Points

The purpose of simultaneous initiation experiments of two points was to measure the delay error among the detonators of the same batch and the same section. Single dose was 100 g and the detonators of the delay-time blasting were three section detonators of the same batch and the same box. The distance of explosion of sensors from the center of the connection of two blasting hearts was 7 m. The overpressure peak pattern that tester collected was with only one shock wave, indicating that the shock wave produced by the two explosive packages almost spread in the same speed, so the delay error between the detonators of the same section and the same batch could be ignored. Fig.6 (a) was the ultra-pressure curve obtained by the sensor, and its peak value was 23.373 kPa.

3.2.2 Six Points Short-Delay Detonation Experiments

The experiment was done with the designed delay detonation methods, and explosive package was fixed on the surface and exposed to the air to do the short-delay controlled blasting experiment. Single dose was 100 g, the specific length of the tube and the specific location of sensors were shown in Fig.5. Through on-site survey, after the explosion six tidy pits could be clearly seen on the ground, the location of explosion was almost not changed, so it could be determined that the explosives were all detonated. By Figs.6(b)~6(d) it could be seen that the shock wave measured in the experiment was with only three distinct peaks, and because two transverse waves met in the middle and were converged on one shock wave, so the three peaks should be produced by the three explosive packages which were close to the sensor side. In Fig.6(b) three wave peaks were 17.697 kPa, 9.848 kPa, 6.812 kPa, the last two peaks were smaller than the first peak, at the same distance, at the same time and the same position the first peak was much smaller than the overpressure peak 23.373 kPa, compared with the group charge and multipoint simultaneous detonation, this method could reduce shock wave overpressure peak. Compared with the delay blasting in the rock, this method could also reduce six wave peaks to three, so the overall action duration of positive pressure area could be effectively reduced.

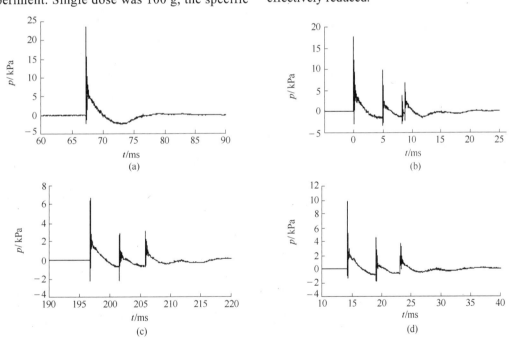

Fig.6 Overpressure history

(a) two points simultaneously detonated, overpressure history curve at 7 m;

(b) six points controlled blasting, overpressure history curve at 7 m;

(c) six points controlled blasting, overpressure history curve at 7 m;

(d) six points controlled blasting, overpressure history curve at 17 m

4 THE EXPERIMENT ON BOLTS RUPTURE AND EXPLOSIVE IMPULSION

In reality, the security door was fixed on the wall by the six expansion bolts. The specification of expansion bolts in security doors is M8×100, the outside diameter of the expanding tube is 14 mm, and the diameter of the countersunk head bolt is 8 mm. The wall in this experiment was simplified to a steel frame structure. During the experiment the level 4.8 M8 hex bolts were used to connect the door and the steel frame. The intensity of two kinds of bolts were showed in Tab.2 [13, 14]. Both the allowable tensile stress and the allowable shear stress were larger than those of the expansion bolts of security doors. In this study, the high-intensity hex bolts were used instead of expansion bolts, so that the experiment was more credible.

Tab.2　The Table of intensity of bolt

Expansion bolt specification	Allowable tensile stress/kN	Allowable shear stress/kN
M8	4.44	3.29

In the experiment of single bolt rupture, the steel short side that was small and groove-like was connected to a steel frame with level 4.8 M8 hexagon bolt, and then 30 g emulsion explosive was used to the experiment of single bolt rupture, experiment results were shown in Fig.7. The results showed that bolts were broken up, and the failure mode was that screws were cut off. Therefore it could be drawn that expansion bolts could be used as the weakness of breaking into a door.

Fig.7　The partial magnified drawing of the experiment on the bolt fracture

5　BREAKING INTO DOOR WITH SHORT-DELAY CONTROLLED BLASTING

The bolt rupture experiment has proved that 30 grams of emulsion explosive was able to break up a single expansion bolt, but in reality to blow up a door, the door distortion and wall damage will consume partial energy, therefore, the single point dose for the experiment of breaking into a door will be much larger than that for the bolt rupture experiment. The methods were used to carry out the experiment of breaking into door with six-point short-delay controlled blasting, and the single point dose was 100 g and 50 g respectively. The emulsion explosive cartridge was placed to the section of six expansion bolts and was perpendicular to the frame. The specific locations among explosive cartridges and the specific length of the tube were shown in Fig.5. In the experiment, only one sensor used to collect shock wave, and it was 5 m far away from the door.

5.1　Condition 1: Single Point Charge with 100 Grams

The results of the experiment of breaking into door showed in Fig.8. Through the damage caused to the security door, it could be seen that the door was completely broken and deformed seriously, the middle and both sides of the door missed large parts, the door frames and the door body were completely separated, the door flied out of 7 meters and the steel frame was also deformed on certain degrees. Based on the above results, it could be seen that single point of 100 grams of explosives was too much to break into a door, so hostages inside the door were easily hurt.

Fig.8　The picture of experimental effect in working condition 1

Fig.9 was the overpressure curve of the air shock-wave, a total of 4 peak was recorded. The first two peaks were very small, they should be the disruptions of the initiation end and the latter two peaks were generated by the emulsion explosive. Although the experimental results achieved the expected results, the door was severely damaged. Single point charge should be reduced.

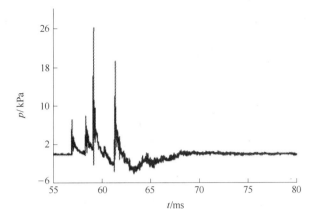

Fig.9 Overpressure curve for condition 1

5.2 Condition 2: Single Point Charge with 50 Grams

Because of the overlarge dosage as shown in condition 1, the door was severely damaged; therefore the single point dose 50 g was used to do the experiment. The results of the experiment of breaking into a door showed like that in Fig.10. Through the damage caused to the security door it could be seen that the security door was completely broken, six-point explosives were all detonated, the four corners of the door bent inwards, the door body was not penetrated and no missing parts appeared on the door, the security door flew out of 4 meters, so the dose was moderate. In Fig.11 there were three distinct peaks, the overpressure peaks gradually became smaller, no obvious superposition between peaks, the experimental results achieved the desired results.

Fig.10 The picture of experimental effect in working condition 2

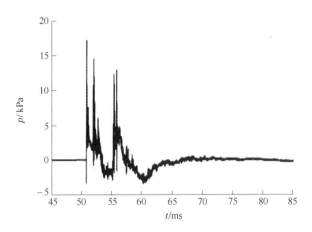

Fig.11 Overpressure curve for condition 2

5.3 The Safe Distance

When TNT explosive blows up in rigid surfaces like rock, concrete, etc., it can be seen that double explosive blows up in the infinite space. So the air overpressure peak can be calculated [15]:

$$\Delta p = 106\frac{\sqrt[3]{m}}{r} + 430\left(\frac{\sqrt[3]{m}}{r}\right)^2 + 1400\left(\frac{\sqrt[3]{m}}{r}\right)^3$$

$$(1 \leqslant \frac{r}{\sqrt[3]{m}} \leqslant 10:15) \quad (1)$$

In Tab.3, the average peak was the average of the three shock wave peaks in Fig.11. By comparison, the difference between the overpressure peaks was smaller, so the dosage of TNT could be obtained by air overpressure empirical formula (1), and overpressure average, it was about 94 g. Generally, when the overpressure is less than 20 kPa, people will not be affected by the organic damage. Therefore, if 94 g TNT and 20 kPa overpressure are substituted into the air overpressure empirical formula (1), it can be drawn that 3.6 m is the safe distance of breaking into a door by the short-delay controlled blasting under the condition of the single point 50 g emulsion explosive.

Tab.3 The contrast Table of the overpressure

Charge /g	Air shock wave overpressure value/kPa					The average /kPa	Formula/kPa
50	6.235	6.982	6.137	4.384	4.238	5.595	6.657
100	17.697	17.203	14.532	12.252	12.988	14.934	14.666

6 CONCLUSIONS

(1) The nonlinear dynamic analysis software LS-DYNA can be used to simulate the short delay controlled explosion in the air, and the mechanism of short delay controlled blasting in the air was gotten with the simulation of two points blasting under different time delay.

(2) According to the mechanism of the time delay controlled blasting in the air, an air delay initiation method was designed. The different length of tubes of the same section detonators can be used to achieve the time delay, and the delay time was below 1.5 ms.

(3) Compared with the group charge, the method of short-delay controlled blasting could effectively reduce the peak overpressure of air shock wave under the condition of the same distance and the same dosage, so it could effectively decrease the anti-terrorist personnel harm.

(4) Through the experiment of the bolt rupture, it could be concluded that the expansion bolts of security doors could be taken as the weakness of breaking into a door to study.

(5) Through the experiment results and the calculated results by the empirical formula, it was obtained that the single point with dose 50 g emulsion explosive was well to break into a door, and 3.6 m was the safest distance.

(6) As the restrictions of experimental conditions, the experiment only collected and analyzed the overpressure data of one side, the overpressure of other sides needs to be further analyzed and researched in subsequent experiments. Under the current condition biological damage experiments have not been started yet, and about the best dosage to break into different doors, the in-depth analysis has to be done.

ACKNOWLEDGMENT

This project was financially supported by the National Science Foundation of China (No.10872044, No.10602013 & No. 10972051& No. 10902023) and the Natural Science Foundation of Liaoning Province of China (No. 20082161).

REFERENCES

[1] YAO J Y. An Elementary Discussion on Counterterrorism Publicity and Education to the Public[J]. JOURNAL OF BEIJING PEOPLE'S POLICE COLLEGE, 2009, 4:50~53.

[2] Fan J T. U.S. fast break tool. Small Arms, 2003, 11:20~21.

[3] Zhou X, Huang P, Yao D. Broke into the house on break equipment and its use of tactics[J]. Global Military, 2007, 6:46~49.

[4] Zhang Q, Yang Y Q, Yu B. The time of rock blasting and crushing and the time delay optimization of the short-delay detonation[J]. Explosion and Shock, 1998, 3:77~81.

[5] Zhang K, Liu S X. The theoretical analysis of blasting interval of rock tunnel digging short-delay blasting[J]. Mine explosion, 2001,1:17~19.

[6] Zhao X T, Huang H L. The application of deep standing delay controlled blasting in rock excavation engineering[J]. Mine explosion, 2005,3:37~39.

[7] FOTIS R, SPYROS S. Experimentally validated 3-D simulation of shock waves generated by dense explosives in confined complex geometries[J]. Journal of Hazardous Materials, 2005, A121: 23~30.

[8] SPYROS S, FOTIS R. Computer-aided modeling of the protective effect of explosion relief vents in tunnel structures[J]. Journal of Loss Prevention in the Process Industries, 2006, 19:621~629.

[9] Ning J G, Wang Z Q, Zhao H Y, et al. The numerical simulation of blast wave flow around. Beijing University of Technology, 1999, 5:543~547.

[10] Li X D, Zheng Y R, Li L S, et al. The numerical simulation of the propagation rule of explosive shock wave in long tunnel[J]. Blasting Equipment, 2005, 5:4~7.

[11] Wang F, Wang W, Wang Y H, et al. The research on the clipping effect of block wave wall on air shock wave[J]. Blasting Equipment, 2004, 1:1~5.

[12] Wu H M, Song J P. The proposal on revising nonel tube standard[J]. Blasting equipment, 2003, 3:4~6.

[13] Sun Z L, Leng X J, Wei Y G, et al. Mechanical Design[M]. Shenyang: Northeast University Press, 2000.

[14] Zhao Y G, Gao C S. DISCUSSING ABOUT THE BEARING CAPACITY OF THE METAL EXPANDED BOLT[J]. Industrial Construction, 1993,12:31~34.

[15] The Eighth Department of Beijing Institute of Technology. Explosion and Its Action[M]. Beijing: National Defense Industry Press, 1979.

Development of Damage Criteria for Underground Mine Structures Subjected to Blast Vibration from Neighbouring Surface Mine

A. K. Jha[1], D. Deb[2]

(1. Blasting Cell, CMPDI, Ranchi, India; 2. Department of Mining Engineering, IIT Kharagpur, Kharagpur, India)

ABSTRACT: Blasting in a surface mine may cause ground vibration in neighbouring underground mine workings affecting their structural stability and resulting in loss of coal production. The effects of surface blasting on adjacent underground workings have been studied using peak particle velocity (*PPV*) measured at various locations in an underground Bord and Pillar mine. Predictor equations of *PPV* have been developed based on distance *D* from the source of blasting and charge (explosive quantity) per delay *Q*. This study finds that depending on the number and composition of rock strata through which stress waves are transmitted from the surface to underground, $Q^{0.35}$ factors may be appropriate for estimating the scaled distance, which is considered as a measure of *PPV*. Further a new approach is proposed to classify damage of underground structures into three groups using the concept of "Blast Damage Factor" and linear discriminant functions. Site-specific charts between charge weight and distance have been developed so that blasting at a surface mine can be conducted without causing damage to the adjacent underground mine.

KEYWORDS: surface blasting; peak particle velocity; predictor equation; blast damage factor

1 INTRODUCTION

Coal accounts for about 70% of total electricity generation in India and is likely to remain a key source for at least the next 30~40 years. The manifold increase in demand for coal puts a huge pressure on augmenting production from opencast mines. In general, near surface (upper) coal seams are mined by opencast methods, while deeper (lower) coal seams are excavated using the Bord and Pillar method. The increase in production within a short period of time demands heavy blasting in overburden and coal benches of opencast mines causing technical as well as socio-political problems due to ground vibration. In this regard, there is a danger to the safety and stability of underground (UG) mine openings, coal pillars, water dams, ventilation and isolation stoppings located in close proximity to operating opencast mines. Prediction of the peak vibration level caused by neighbouring surface mine blasting is important for the safety of underground structures in terms of pillar spalling, roof collapse and junction failure and is normally measured by the peak particle velocity (*PPV*) (Deb and Jha, 2010). Vibration prediction also helps the surface mine operators to optimise controlled surface blasting with regard to the safety of the underground mine structures.

The root of the problem lies in the nature of vibration that is experienced in underground structures such as pillars, roofs and floors due to blasting conducted in an adjacent surface mine. The problem can be addressed by understanding the characteristics of wave propagation and its attenuation characteristics which are reflected in the wave form received and monitored at observation sites. The attenuation of vibrations chiefly depends on the charge weight, frequency content of wave motion and geomechanical properties of the transmitting medium. The interrelationship among charge weight, distance and amplitude of the motion forms the basis of an attenuation law. Several predictor equations (attenuation laws) of *PPV* have been developed based on quantity of charge per delay and distance from the source of blasting (Langefors et al., 1958, Duvall and Petcoff, 1959, Davies et al., 1964, Birch and Chaffer, 1983, Roy, 1993). These equations are mainly used for forecasting *PPV* at a surface point resulting from blasting at a surface mine bench.

Corresponding author E-mail: ajayk_jha_in@yahoo.com

Peak particle velocity has also been used to evaluate blast damage index at an underground location caused by surface blasting (Singh, 2002). In most predictor equations, the square root of charge per delay $Q^{1/2}$ is assumed to be related to the scaled distance (SD). However, a study by Fourie and Green, 1993 demonstrates that peak vibration (acceleration and velocity) caused by surface blasting is lower at an underground location compared to a surface point at the same SD. It was proposed that PPV relates to one-third power of charge per delay $Q^{1/3}$ for underground locations.

In this study, vibration has been monitored at different locations in roofs, pillars and floors in an underground coal mine, while blasting was conducted at nearby surface coal mine benches. The blast induced vibration data were generated under a Science & Technology project sponsored by Ministry of Coal, Govt. of India with Central Mine Planning and Design Institute (CMPDI) as the nodal agency and Central Institute of Mining and Fuel Research (CIMFR) and CMPDI as implementing agencies. Peak particle velocities in the underground coal mine roofs, pillars and floors were monitored using geophones. Surface blasting and underground monitoring were synchronized so that the measured vibrations were only due to surface blasting. The monitored data were analysed using statistical techniques, and a new predictor equation of PPV based on distance R and explosive quantity per delay Q was developed. It was found that the power of charge per delay varies with local geological conditions in the best fit model. However, on average, 0.33 power of charge per delay provides a reasonably good estimate of PPVs measured in underground locations if the parting (transmitting) medium is composed of one or two rock strata.

On the other hand, due to repeated surface blasting, underground structures may also experience loading and unloading phenomena which may be detrimental to the stability of UG structures. As a result, surface mine management may force to restrict the maximum explosive charge per delay leading to planning and carrying out smaller size surface blasts in adjacent surface mines to control the blast vibration in underground within a certain threshold limits. This sub-optimal blasting operation has led to various downstream problems affecting the productivity and economics of the mining activity. Hence, there is an urgent need to understand and determine the threshold PPV up to which underground structures would be safe and can tolerate the blast induced vibration without any significant damage. This paper also elaborates on the development of "Blast Damage Factor (BDF)", based on classification of damage using estimated PPVs, rock mass parameters, pillar and room dimensions, for underground workings arising out of surface blasts carried out in adjacent surface mines (Jha, 2010). These threshold limits are determined for both the underground mines so that safe and economic surface blasts can be planned without any significant damage potential to underground workings.

2 MINE SITE DESCRIPTION

Blast vibration measurements were collected from Lajkura opencast mine (OCM) and Orient Mine No.2 underground mine operated by Mahanadi Coalfields Ltd., a subsidiary of Coal India Ltd. Fig.1 depicts a schematic view of typical vertical geological sections of the mine site. The case study mine is located in the IB valley coalfield area, which is a part of the large synclinal Gondwana basin of Raigarh-Hingir and Chattisgarh coalfields (Mahanadi valley). The Barakar and Karharbari formations are the major coal bearing formations. The area is generally free from major faults. In case study mine site, the Lajkura seam was excavated on the surface, and the HR seam IV was mined underground using Bord and Pillar mining. The average vertical distance between these seams was about 86 m. The inter seam rock layers are mainly composed of sandstone and shaly sandstone. Tab.1 lists a brief description of the underground mines and related geotechnical parameters.

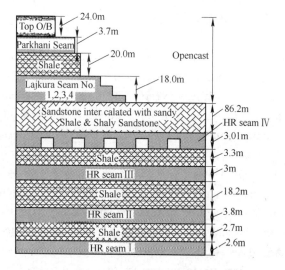

Fig.1 Typical layout of rock strata at case study mine site

3 METHOD OF MINING

The conventional Bord and Pillar mining with depillaring by slicing method was practiced. The average dip direction of the HR seam IV was S80°W with a gentle gradient of 1 in 13. The gallery size in the Orient Mine No. 2 mine was 2.6 m×4.5 m with pillar size of 25 m. The P-5 permitted explosive having 32 mm diameter was used for blasting coal face and blasted coal was loaded onto tubs manually. The average pull per round of blasting was about 1.2 m. Roof bolting, wooden props and cross bars were used as roof support system.

Tab.1 Description of underground mine and related geotechnical parameters

Particulars	Case study mine site
Name of the seam	HR – IV seam
Pillar size (corner to corner)/m×m	25×25
Gallery width/m	4.5
Gallery height/m	2.6
RQD (roof)	54.5
GSI (roof)	50.8
Support system	Roof bolting, props and cross bars
Average rock density/t · m^{-3}	2.13
Weighted P wave velocity/m · s^{-1}	2602
Rock layers in between surface and underground mines	Lajkura coal seam sandstone intercalated with sandy shale and shaly sandstone
Weighted UCS/MPa	23.0
Dynamic tensile strength/MPa	6.39
Immediate roof layer up to 2.0 m	0.4 m – coal, 1.6 m - shale

In the Lajkura OCM, 10/70 Dragline was deployed for excavating 18~20 m thick shale bench in order to expose the Lajkura seam number 1, 2, 3 and 4. The drilling in overburden bench is done by 250 mm electric drill. The blasted material is removed by dragline operation for shale bench and by conventional shovel and dumper combination for other benches. The exposed coal bench is mined with 7 m bench height. Blast holes are 160 mm diameter and made by diesel and electric drills. The blasted coal is loaded by shovel and transported out from the mine by dumpers. The average coal production from the Lajkura OCM was 1.0 Mt per annum with a stripping ratio of 2.98 m^3/t. This mine was commissioned after the Orient no. 2 mine came in production. As the Lajkura seam was excavated, the distance between the blasting face and the underground monitoring stations gradually decreased. The influence of surface blasting could be felt in underground pillars and roofs when the blasting bench was closest to the monitoring panel. As the mine progressed further, the distance between the source of blasting and monitoring stations increased. Fig.2 shows the Orient no. 2 mine overlaid by the Lajkura OCM along with underground vibration monitoring stations. Tab.2 lists explosive and drill parameters used in the Lajkura OCM. The burden and spacing was 6.0 m×7.0 m. During monitoring at the Orient no. 2 mine, 20 m shale strata above the Lajkura seam was blasted. The underground monitoring panels were located at least 800~1000 m away from the active underground workings. Hence, the influence of blasting conducted in other locations of the underground mine on the monitoring panels would be negligible.

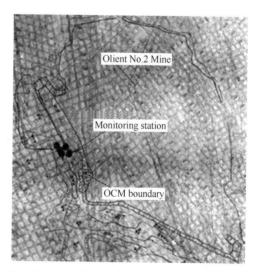

Fig.2 Mine plan showing the opencast and UG workings at case study mine site

Tab.2 Blasting details in case study mine site for overburden bench

Particulars	Case study mine site
Strata blasted	Shale
Hole diameter/mm	250
Hole depth/m	19.0
Subgrade length/m	NIL
Burden×spacing/m×m	6.0×7.0
Top stemming/m	5.0
Initiation system	Non-electric
Explosive type	Emulsion
Explosive density/g · cm^{-3}	1.18
Explosive quantity per hole/kg	575
Charge factor/kg · m^{-3}	0.72

4 MONITORING OF VIBRATION AT UNDERGROUND ORIGINATING FROM SURFACE BLAST

Three directional transducer/ standard geophones were mounted in the roof, pillar and floor by proper mounting arrangement. Firm contact between rock/coal strata and geophone surface was ensured by placing plaster of Paris as grout material. The mounting stations in the roof were at the junction as well as in between the two junctions of galleries. Geophones were mounted in pillars about 1.0~1.3 m below the roof surface and by cutting recess/duggy of 0.5 m inside the pillar from the roof line. Geophones were mounted at least 0.5 to 1 m inside the roofs and floors. These sensors were connected to the seismographs which were located at a safe location in underground.

The composite mine plan depicting the opencast and underground workings is shown in the Fig.2. The figure shows the benches of OCM and experimented Bord and Pillar panel. It may be noted that Bord and Pillar panel was only in the development stage when surface mine was in operation. Hence the radial distance from the blasted bench to the measurement stations varied continuously.

Vibrations in terms particle velocity and acceleration were recorded by geophones and stored in the base unit. Seismographs, namely Blastmate III and Minimate Plus, were used in both case study mines to record vibration data, and underground monitoring stations are marked in Fig.2. Surface blasting and underground monitoring timing was planned by observing proper coordination between surface blasting team and underground monitoring team so that vibration in underground structures occurred only due to surface blasting. In case study mine site, a total of 73 observations were recorded at different locations in the roof, pillar and floor. Apart from vibration monitoring, fall of roof, damage in permanent ventilation stoppings and spalling of pillars were also recorded underground right after surface blasting.

5 PEAK PARTICLE VELOCITY (PPV) AT UNDERGROUND MONITORING STATIONS

The average, maximum, minimum and standard deviation of *PPVs* measured during the field experimentation are listed in Tab.3~Tab.5 respectively. It can be noted that geophones were mounted in the roof and floor, roof and pillar, and pillar and floor simultaneously during field measurements. From the measurements, it is found that on an average *PPV* of roof is twice that of floor and one and half times that of pillar.

Tab.3 Average, max., min. and std. dev. of vibration records with geophones mounted at roof and floor

Particulars	Top priming		Bottom priming	
	Roof/mm·s^{-1}	Floor/mm·s^{-1}	Roof/mm·s^{-1}	Floor/mm·s^{-1}
Average	83.20	33.0	89.10	36.30
Maximum	214.70	57.1	200.0	57.79
Minimum	20.63	11.4	31.50	21.20
Standard deviation	56.61	12.81	50.53	13.24

Tab.4 Average, max., min. and std. dev. of vibration records with geophones mounted at roof and pillar

Particulars	Top priming		Bottom priming	
	Roof/mm·s^{-1}	Pillar/mm·s^{-1}	Roof/mm·s^{-1}	Pillar/mm·s^{-1}
Average	24.43	16.33	29.50	19.44
Maximum	57.75	38.50	81.80	50.80
Minimum	4.04	2.99	5.0	3.17
Standard deviation	17.90	12.44	20.88	13.75

Tab.5 Average, max., min. and std. dev. of vibration records with geophones mounted at pillar and floor

Particulars	Top priming		Bottom priming	
	Pillar/mm·s^{-1}	Floor/mm·s^{-1}	Pillar/mm·s^{-1}	Floor/mm·s^{-1}
Average	18.51	10.79	12.21	7.38
Maximum	44.06	26.70	26.45	17.40
Minimum	4.91	2.92	4.13	2.50
Standard deviation	12.35	7.02	7.17	4.74

6 CLASSIFICATION OF OBSERVED DAMAGE

The severity of the vibration in pillars, roof and floor can be gauged from the fact that the utilization of 10/70 dragline deployed in the Lajkura OCM has also reduced to as low as 35% due to blasting technique for ensuring adequate safety measures.

To assess the blast damage accurately, the study area was properly whitewashed so that the fresh fall from roof or pillar, development of new crack or extension of new crack can be visually noticed. Coal blocks detaching from roof having maximum dimension measuring up to 0.25~0.30 m^3 is assumed as "Severe damage" type and is shown in Fig.3. The average size of coal blocks in severe damage type ranged between 0.10~0.15 m^3. Some noticeable crack extension and fresh crack development was prominently witnessed in ventilation stoppings. There were number of instances when few loosened chips detached from roof or pillar and coal dust was generated after surface blast in UG workings. This type of damage is termed as "Moderate damage". The instance of no spalling from roof or pillar as well as no new visible crack formation in ventilation stopping and other structures is categorized as "No damage".

Fig.3 Photograph showing damaged roof due to blasting

7 DEVELOPMENT OF PREDICATOR EQUATION

Authors have developed the attenuation equations of *PPV* based on flexible scaling law for both the Minesites and details are mentioned in earlier references (Jha, 2010, Deb and Jha, 2010). In this paper, the concepts and results are mentioned below. The attenuation law can be written in general form as

$$PPV = KQ^m D^{-n} \quad (1)$$

where Q——charge weight/delay, kg;
D——distance of the measuring transducer from blasting face, m;
K, m, n——site constants to be determined from the measured data. Equation 1 can be rewritten in terms of scaled distance as

$$PPV = K\left(\frac{D}{Q^s}\right)^{-n} \quad (2)$$

where, $s = m/n$. Taking natural log in both sides, equation 2 becomes

$$\ln(PPV) = \ln(K) - n\ln\left(\frac{D}{Q^s}\right) \quad (3)$$

In equation 3, K, s and n are unknown. By applying least square method, K and n can be estimated if s is known. In the following, equation 3 has been used to determine value of K and n for both the mine sites by varying values of s from 0.1 to 0.75. The s value which provides the highest F statistic and $R^2_{adjusted}$ is considered to form the best predictor equation.

A total of 73 sets of vibration data were recorded at mine roofs of Orient No.2 mine. Out of these data, 6 data sets were recorded at the time of roof damage. The *PPV* of these 6 data sets are in the range between 169 mm/s and 243 mm/s. The range and average *PPV* of rest of the data sets are 7.16~156.6 and 73.8 mm/s respectively. In addition, roof fall, pillar spalling and cracks in ventilation stoppings had occurred due to high *PPV*. Hence, these 6 data, being outliers, are omitted in the subsequent statistical analysis. Using duplex method, 67 *PPV* data are divided into 50 training data sets and 17 validation data sets. A predictor equation is developed based on the training data set and then it is validated using the remaining 17 data. Fig.4 shows the plot of $R^2_{adjusted}$ and F statistic for various values of s. It can be clearly seen that the best fit model is achieved when $s = 0.35$. The site constant K and n have been estimated from the training data set. The best fit predictor equation of case study mine site is found to be

$$PPV = 12397\left(\frac{D}{Q^{0.35}}\right)^{-1.573} = 12397(SD)^{-1.573} \text{ (mm/s)} \quad (4)$$

From Fig.4(a), it can be inferred that $s = 0.5$ or square root of charge per delay does not provide the best fit equation, while vibration is measured at an underground location. Fig.4(b) depicts the relationship between *PPV* and scaled distance (*SD*) of the best fit equation. It shows that *PPVs* estimated by equation 4 matches fairly well with the measured data. The scatter plot between measured (both training and validation data) and estimated *PPVs* also confirm the fact that the predicator equation 4 can be applied to forecast *PPVs* at underground locations due to surface blasting under similar geological conditions (Fig.4(c)).

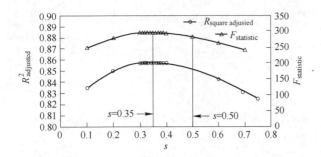

Fig.4(a) Relationship of R_{adjusted}^2 and F statistic with parameter s for case study mine site

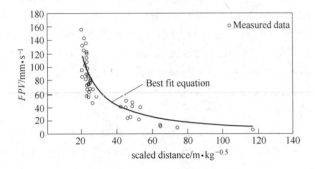

Fig.4(b) Relationship between PPV and scaled distance of case study mine site

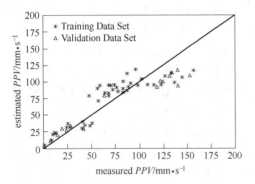

Fig.4 (c) Relationship between measured and estimated PPV for case study mine site

8 DEVELOPMENT OF BLAST DAMAGE FACTOR (DBF)

In general, several blasting factors (explosive type, explosive charge per delay), rock mass factors (dynamic tensile strength, P-wave velocity and Geological Strength Index of rock strata lying between surface mine to underground mine) and mining factors (size of rooms, distance from blasting site to monitoring station, pillar dimensions and others) can influence damage to UG structures due to surface blasting. A new concept of Blast Damage Factor (BDF) has been developed to assess the damage of underground structures using linear discriminant functions. A chart showing the relationship between Q and D are prepared for different values of PPV and BDF. The relationship can be used as a handy tool for determining safe blasting practices by estimating the explosive charge/delay at any given distance for no damage to the UG structures.

8.1 Definition of Blast Damage Factor (BDF)

Blast Damage Factor (BDF) is defined to assess the damage of underground mine workings caused by surface blasting. Yu and Vongpaisal (1996) suggested the concept of Blast Damage Index (BDI) for the same purpose. In this study, BDF is defined in terms of induced stress, damage resistance, together termed as Strength Factor, and Mining Factor and given as a dimensionless indicator of damage as:

$$BDF = \underbrace{\left(\frac{Induced\_Stress}{Damage\_Resistance}\right)}_{\text{Strength factor}} \underbrace{\left(\frac{Pillar\_Height}{Pillar\_Width}\right)}_{\text{Mining factor}} \quad (5)$$

$$BDF = \left(\frac{PVS \times \rho \times C_p}{GSI \times \sigma_{dts}}\right)\left(\frac{h}{W_p}\right) \quad (6)$$

where, blasting factor as PVS = Vector sum of peak particle velocity (PPV) in mm/s (blasting factor), Rock mass factors as ρ = Density of rock mass in kg/m³, C_p = Compressional P-wave velocity of rock mass in m/s, σ_{dts} = Dynamic tensile strength of rock mass in N/m², and GSI = Geological strength index of rock mass between blasting source and underground mine.

As the name suggests BDF must be inverse of factor of safety. It has two components. The Strength Factor component is a measure of inverse of factor of safety of the underground structures when subjected to blast induced dynamic loading. The numerator, the induced stress is a product of PVS, density of rock mass and compressional P wave velocity of the medium (rock mass). The denominator consists of dynamic tensile strength of intact rock multiplied by the GSI of rock mass. Dynamic tensile strength of rock mass can be approximated by $\sigma_{ci}/3.6$ where σ_{ci} is the uniaxial compressive strength of the intact rock (Mohanty, 1987, Yu and Vongpaisal, 1996). The Mining factor is inverse measure of the strength of coal pillars. The mine working factor is incorporated in BDF to evaluate the contribution of pillar geometry in the stability. In general W_p/h denotes the slenderness ratio of coal pillar and has been used in pillar strength equation proposed by Bieniwaski and others (Herget, 1988). Hence, the composite factor will give an indicatory measure of blast induced impact assessment of surface blasts on adjacent underground structures.

For any given mining condition, the variables ρ, C_p, GSI, σ_{dts} may be assumed as nearly constant if the roof rock remains the same. The above parameters define the geotechnical properties of rock mass. Under such assumption, it may be inferred that BDF is directly related to PVS. It may be approximated, mathematically, that $BDF = f(PVS)$ and $PVS = h(D, Q)$ where $f(PVS)$ and $h(D,Q)$ denote the arbitrary functions to be determined from datasets.

8.2 Concept of Linear Discriminant Function or Minimum Distance Classification for Generating BDF

Discriminant analysis builds a predictive model for group membership. The model is composed of discriminant functions for more than two groups based on the linear combinations of the predictive variables that provide the best discrimination between the groups. The functions are generated from a sample of cases for which group membership is known. For example, if x_i denetoes the centroid or prototype impact pattern of ith class of data sets then minimum distance linear discriminant function of ith class becomes (Zurada, 1992),

$$g_i(x) = x_i^T x - \frac{1}{2} x_i^T x_i \quad \text{for} \quad i = 1, 2, \ldots, k \quad (7)$$

where, k denotes the number of class. The function can then be applied to new cases that have measurements for the predictor variables but have unknown group membership. Thus discriminant analysis is used to investigate variables for group separation. A minimum distance classifier computes the distance from pattern x of unknown classification to each known prototype, x_i. Then the category number of that closest or smaller distance, prototype is assigned to the unknown pattern, x. This concept is also called correlation classification because a closest match is sought between the known prototype pattern and the unknown input pattern.

8.3 Damage Prediction by Linear Discriminant Functions

As mentioned earlier, damage has been classified into "Severe, Moderate or No damage" categories. Linear discriminant functions are estimated for these categories or damage classes using predicted PVS. A class Severe or Moderate or No damage is assigned to an unknown observation (BDF) if the estimated value of discriminant function of a particular is the maximum.

Before carrying out the discriminant analysis, the total data has been divided into three parts i.e. training data, validation data and test data in ratio of 75%, 15% and 10% respectively. The total 73 vibration data recorded at roof of the Orient Mine 2 due to surface blasting carried out at the Lajkura OCP, MCL has been divided into 55 training data,

11 validation data and 7 testing data.

Linear discriminant functions are evaluated using estimated PPV data. Based on the criteria mentioned above, training data sets were assigned with a damage class based on the observed phenomena in underground. Then using equation 6(b), BDF is estimated for each data set. The geotechnical parameters such as density, P-wave velocity, dynamic tensile strength and GSI for computing the BDF is taken from Tab.2. Equations 8a-c denote the linear discriminant functions for case study mine site.

$$g_{severe}(BDF) = 13.43\, BDF - 90.22 \quad (8a)$$
$$g_{moderate}(BDF) = 11.11\, BDF - 61.69 \quad (8b)$$
$$g_{unaffected}(BDF) = 4.97\, BDF - 12.36 \quad (8c)$$

Linear discriminant functions of Severe, Moderate and No damage classes of a given dataset with known D and Q have been determined based on the estimated PPVs. The results mentioned above are further analyzed to provide working chart for case study mine site based on Q versus D plots for different BDF as shown in Fig.6. Threshold PPV and BDF values for three damage classes are listed in Tab.4 for case study mine site. The results given above can be used as a guideline to determine the damage class and BDF can be obtained if D and Q are known using Fig.5.

Tab.4 Threshold BDF and corresponding PPV for mine site
(mm/s)

Mine location	Threshold BDF and PPV		
	Severe	Moderate	No
Case study mine site	6.72, 54.04	5.55, 44.68	2.49, 20.00

The severe roof damage in case study will take place when PPV will exceed 54.04 mm/s and moderate damage is expected if PPV ranges between 44.68 mm/s and 54.04 mm/s. Any PPV less than 20.00 mm/s will produce no damage to underground structures.

9 CONCLUSIONS

In India, there are several locations where coal seams are excavated simultaneously in surface (opencast) and underground (Bord and Pillar) mines. There are considerable stability and safety concerns where surface mine blasting occur in vicinity of underground mines, as this can result in pillar spalling, roof collapse and junction failure as well as an associated loss of coal production. In this study, roof, pillar and floor vibrations were monitored in the Orient no.2 mine while blasting was carried out at adjacent surface mines. New predictor equations of the PPV have been developed based on the flexible scaled distance law. The roof vibration data of case study mine site was analysed to develop a new predictor equation of PPV at an underground location resulting from surface blasting. Scaled distance based on one-half power (square root) of charge per delay is generally applicable if vibrations are measured at surface points but may not be suitable for predicting PPV at underground installations. The study concludes that 0.35 power of charge per delay can be used to calculate the SD, suggesting use of variable scaling law for vibration predictor equation.

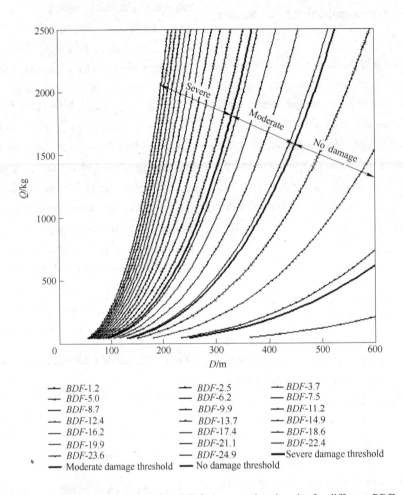

Fig.5 Relationship between Q and D for case study mine site for different BDF

A new dimensionless blast damage factor has been developed for damage prediction of underground roof so that safe blasting can be planned at surface mines with due regard to the safety of underground workings. The threshold BDF and corresponding PPV values of Severe damage, Moderate damage and No damage have been estimated using the linear discriminant function for case study mine site. From the study, it may be concluded that PPV less than 20 mm/s is unlikely to cause damage to underground structures. Relations between Q and D have been developed for different values BDFs for calculating safe explosive charge per delay at any given distance in underground mine workings. These charts can be a handy tool for practicing blasting engineers to ascertain safe charge for any known distance.

REFERENCES

[1] Birch W J, Chaffer R. Predictions of ground vibrations from blasting on opencast sites. *Trans. Inst. Min. Metall.* 1983, A 92A: A103~A107.

[2] Deb D, Jha A K. Estimation of blast induced peak particle velocity at underground mine structures originating from neighbouring surface mine. *Mining Technology*, 2010, 119(1): 14~21.

[3] Davies B, Farmer I W, Attewell P B. Ground vibrations from shallow sub-surface blasts. *Engineer*, 1964, 217: 553~559.

[4] Duvall W I, Petcoff B. Spherical propagation of explosion generated strain pulses in rock. *RI 5483, US Bureau of Mines, Pittsburgh, PA, USA,* 1959,21.

[5] Fourie A B, Green R W. Damages to underground coal mines

caused by surface blasting. *Int. J. Surf. Min. Reclam.* 1993, 7(1): 11~16.

[6] Herget G. Stresses in Rock, Rotterdam, Balkema. 1988.

[7] Jha A K. Evaluation of the Effects of Surface Blasting on Adjacent Underground Mine Workings. *Ph.D. Thesis, IIT Kharagpur, India.* 2010.

[8] Langefors U, Kihlstrom B, Westerberg H. Ground vibrations in blasting. *Water Power., 1958:* 335~338, 390~395, 421~424.

[9] Mohanty B. Strength of rock under high strain rate loading conditions applicable to blasting. *Proceedings of the 2nd symposium on rock fragmentation by blasting, Keystone, USA, 1987:* 72~78.

[10] Roy P P. Putting ground vibration prediction into practice. *Colliery Guard,* 1993, 241(2),:63~67.

[11] Singh P K. Blast vibration damage to underground coal mines from adjacent open-pit blasting. *Int. J. Rock Mech. Min. Sci.,* 2002, 39: 959~973.

[12] Yu T R, Vongpaisal S. New blast damage criteria for underground blasting. *The Canadian Institute of Mining Bulletin, 1996:* 139~145.

[13] Zurada J M. Introduction to Artificial Neural systems[M]. *New York: West Publishing Company, 1992.*

Blasting Open Pit Mining Impact on the Surrounding Forest Ecological Benefits

ZHOU Xiaoguang[1], JIANG Zhou[2]

(*1. Hunan Iron Troops Engineering Company, China; 2. Hunan Provincial Administration of Work Safety, China*)

ABSTRACT: In recent years, the state has increased the intensity of the ecological environment, a number of serious ecological damage to the mining areas effectively restore and improve. But overall, the prevalence of extensive economic growth mode and some even predatory exploitation of resources a fundamental change in the way yet, at the expense of the environment and local interests in exchange for the immediate phenomenon in mining activities still exist. Mineral resources development has become an important means of economic growth, but mining has triggered a series of ecological and environmental problems, ecological benefits severely restricted mining and sustainable socio-economic development. Vegetation Survey by open pit surrounding forest ecosystems and soil physical and chemical properties, analysis of the ecological environmental effects of mining in the open and damaged forest community structure based on open-pit mining caused by the loss of biodiversity, geological disasters, water pollution, soil degradation, pollution and erosion and other effects of the proposed open pit blasting process corresponding countermeasures and proposed mining must be considered in the loss of valuable forest ecosystems, to measure the scale mining on the ecosystem arising ecological and economic benefits.

KEYWORDS: strip mine; blasting exploit; forest ecosystems

1 INTRODUCTION

Opencast mining blasting a certain impact on the environment, mainly in blasting vibration generated by flying rocks, dust, harmful gases and their impact on forest structure, geology and geomorphology, groundwater resources and soil conservation in these areas. Dust which comes mainly from open pit blasting, drilling and scraping process, the dust generated can spread to a larger area; harmful gases, including carbon monoxide, nitrogen oxides and sulfides. Generally open-air blasting vibration generated by flying rocks is obvious, but also focus on the content and control people, but very easy to dust and harmful gas diffusion and dilution, and therefore the impact is not obvious, often not being seriously in particular the study of mining the surrounding forest is little ecological impact. This study is based on such factors surrounding the mining area of the ecological benefits of forest experiment sought to find factors blasting exploitation of forest ecosystems.

2 TEST THE PROFILE

Make testing ground limestone mine is located in Xiangxi Tujia and Miao Autonomous Prefecture in Hunan Guzhang county, under Guzhang County Redstone Town. With an annual output of about 1.2 million ton per year. According rock properties and engineering requirements, in order to obtain a good blasting effect, determined the way the cloth hole, the hole network parameters, charging structure, filling length, initiation methods, detonating explosives order and unit consumption and other parameters.

Tab.1 The hole blasting parameters

Steps Height (H)/m	Blast hole Depth (L)/m	Burden Line (W)/m	Hole spacing (a)/m	Row spacing (b)/m	Dip $(\theta)/(°)$	Diameter (D)/mm	Deep (h)/m	q/g·m^{-3}	Q/kg
15	16.5	4.5	6	4	75	150	1	0.35	128

Notes: Optimization of the actual blasting parameters in the implementation process.

2.1 Meteorological and Hydrological Conditions

No major rivers and ponds within the mine reservoir. And Qiu River is a tributary of eggplant from Datong mine closer and larger flow of the river, located in the north eastern side of the mine at 3 km nothing to do with the mine hydrogeological conditions. Other surface water bodies are main-

---

Corresponding author E-mail: 10276272@qq.com

ly reservoirs, all located in the mining area, but had no effect on the deposit mining conditions. Region subtropical monsoon climate, four distinct seasons, abundant rainfall, leading to southwesterly winds.

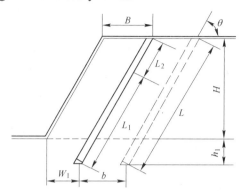

Fig.1 Blasthole layout

According to meteorological data Guzhang County (1980 to 2009), the mine meteorological parameters are as follows: the annual average temperature 15.9 ℃; extreme minimum temperature of –9.1 ℃ (February 2, 2008); extreme maximum temperature of 40.2 ℃ (1992 August 09 day); average annual evaporation is 1015.2 mm; an average annual rainfall of 1405.3 mm; maximum daily rainfall of 128.5 mm (21 August 2007); rainy period from March to September; dry time of the rainy season ends to December; winter time months to February next year.

2.2 Topographical and Geological Conditions

The area is eroded hilly terrain, the overall trend of the northeast ridge to the northwest and southeast high terrain slope is generally 25°～30°. Northwest elevation of 500 m or more generally, southeast Mountain Peak elevation 511 m, usually in the middle of the mine elevation 370～390 m.There are a small number of venues superficial type red clay eluvial layer distribution, more exposed bedrock. Mine vegetation coverage of 40%.

3 METHODS

3.1 Sample Set

Around the study area in the immediate vicinity of the stand, set the standard plot 3, based on the distance and the open-pit mine, mine A (200 m), mine B (100 m), mine C (50 m), 500 m away from the mine outside the set two blank control samples, the plot area are 400 m² (20 m× 20 m). Each survey carried wooden stand in the plot, the average measured tree stand calculating factor.

3.2 Measure of the Litter Layer

Be provided in a random sample of four quadrats 1 m × 1 m, not by decomposition, the decomposition half-split three levels, and the whole harvest fresh weight was measured, and then extract a small sample 1.0 kg, placed in drying oven at 80℃ to constant weight.

3.3 Soil Samples Collected

Each sample set three soil sampling point, each sample point by 0～15 cm, 15～30 cm, 30～45 cm levels, soil samples were 1 kg, 45 soil samples were collected. Removal of gravel and debris, air-dried over No. 20 sieve and No. 100 sieve. Sampling to use the ring blade, take the points layers of soil samples for the determination of soil bulk density, measured with a small aluminum box dug soil moisture.

3.4 Methods for Chemical Analysis

pH value was determined by potentiometric method; organic matter was determined by potassium dichromate oxidation temperature hot outside - Capacity Act; total N was determined by semi-micro Kjeldahl method; total P was measured using chromogenic method of molybdenum, antimony, total K was measured using NaOH melt - Flame spectrophotometry; available P double acid leaching - Mo-Sb colorimetry; quick K using ammonium acetate extraction - atomic absorption spectrophotometry.

4 BLASTING IMPACT ON THE SURROUNDING FOREST ECOLOGICAL BENEFITS OF MINING

4.1 The Impact of Water Conservation Benefits of the Value of Forests

Soils and vegetation as a forest ecosystem important two factors, both of which are closely related and influence each other.

Depends primarily on the amount of water holding capacity of forest woodland culvert water and litter layer. Increased forest plant roots, change the soil structure, water holding capacity increases capillary and non-capillary, the woodland litter and humus layer, increasing the water retention capacity.

Tab.2 The condition of soil water in stands

Sample type	Soil layer /cm	Volume /g·cm^{-3}	The total capacity /t·hm^{-2}	Capillary moisture capacity /t·hm^{-2}	Non-capillary water /t·hm^{-2}
Contrast	0～15	0.6235	76.47	65.00	11.47
	15～30	0.9855	62.81	52.12	10.69
	30～45	0.9845	62.85	52.64	10.21

Continues Tab.2

Sample type	Soil layer /cm	Volume /g·cm^{-3}	The total capacity /t·hm^{-2}	Capillary moisture capacity /t·hm^{-2}	Non-capillary water /t·hm^{-2}
Mine A	0~15	1.0172	61.62	52.35	9.27
	15~30	1.0394	60.78	52.20	8.58
	30~45	1.2334	53.46	46.79	6.67
Mine B	0~15	1.3899	47.55	39.41	8.14
	15~30	1.3819	47.85	40.32	7.53
	30~45	1.4186	46.47	41.61	4.86
Mine C	0~15	1.4265	46.17	38.61	7.56
	15~30	1.3999	47.17	42.15	5.02
	30~45	1.5566	41.26	38.92	2.34

Limestone mining area surrounding plots compared with the control samples, the ecological environment of woodland affected, changes in soil physical properties, particularly in surface soil changes significantly, the nearer the soil bulk density with distance from the mine increases.

The findings suggest that increased soil bulk density, reduced porosity and soil aeration. Especially in 0 ~ 15 cm soil changed greatly, soil bulk density increased significantly decrease 19.42 percent of total water holding capacity. Since the benefits of water conservation function of forest available "shadow engineering method" to calculate, according to the "forest ecosystem services assessment norms" reservoir unit capacity cost of 6.111 yuan/t, can be derived 0~30 cm depth soil water conservation benefits worth about for the still to 411.76 yuan/hm^2, mine A plot 358.23 yuan/hm^2, mine B plots 288.99 yuan/hm^2, mine C plots 274.20 yuan/hm^2.

4.2 The Impact of Water Efficiency Value of Holding on Forest Litter

Litter is a major source of forest litter layer of the soil is an important factor affecting forest water balance and soil development. The number of litters and water holding capacity, water holding timber depends primarily on the biological characteristics and forest growth environment.

The number of litters and water holding capacity, water holding timber depends primarily on the biological characteristics and forest growth environment. Use "shadow engineering" calculated: water conservation benefits litter control plots worth about 225.07 yuan/hm^2, mine A plot 321.93 yuan/hm^2, mine B plots 210.16 yuan/hm^2, mine C plots 145.93 yuan/hm^2.

Tab.3 Litter reserves, water holding capacity and water holdup

Sample type	canopy density	The litter average thickness /cm	The litter average quality /kg·m^{-2}	Natural moisture rate/%	Maximum water holding capacity/t·hm^{-2}
Contrast	0.90	1.50	7.45	61.38	34.97
	0.85	1.66	8.66	76.44	36.83
Mine A	0.75	2.10	7.71	81.54	49.36
	0.80	1.38	9.53	89.56	52.68
Mine B	0.70	1.10	5.40	54.03	26.50
	0.75	1.00	9.38	70.82	34.39
Mine C	0.5	1.70	7.34	54.85	19.74
	0.5	2.00	9.13	78.32	23.88

4.3 The Impact on the Effectiveness of Soil and Water Conservation Value Forests

Soil and water conservation features include reducing the amount of sediment runoff and prevent soil erosion in two ways.

4.3.1 Reduce Soil Erosion Affect the Value of Benefits

A close relationship between vegetation and soil erosion, vegetation cover and soil erosion was significantly negatively correlated.

Tab.4 Soil conservation status

Sample type	Canopy density	The litter average thickness /cm	Reduce soil erosion/t·hm^{-2}	Reduce sediment loss/10^5t·hm^{-2}
Contrast	0.90	1.50	35.80	6.19
	0.85	1.66	34.13	6.22
Mine A	0.75	2.10	33.36	5.24
	0.80	1.38	32.20	5.86
Mine B	0.70	1.10	33.58	5.35
	0.75	1.00	34.91	5.43
Mine C	0.5	1.70	32.10	4.69
	0.5	2.00	32.78	4.57

According to the "Forest ecosystem services assessment norms" artificial excavation and soil type Ⅰ Ⅱ required 42 hours per 100 m^3, each 30 yuan artificially daily basis, get digging earthworks cost per unit area of 12.60 yuan/m^3. In calculating the average value of each soil bulk. Can be drawn each year to reduce soil erosion control plots worth about 433.11 yuan/hm^2, mine A plot 419.11 yuan/hm^2, Mine B plots 308.91 yuan/hm^2, mine C plots 279.77 yuan/hm^2.

4.3.2 Reduce the Impact of Sediment Loss Benefit Value

According to "China's national conditions biodiversity research report," China 1 m^3 capacity of sediment project costs 0.75 yuan.Can be estimated to reduce annual sediment loss benefits worth about 110100 yuan to the control plots/hm^2, mine A plot 95700 yuan/hm^2, Mine B plots 69900 yuan/hm^2, mine C plots 57800 yuan/hm^2.

4.4 Effect of Soil Conservation Value Forests Benefit

4.4.1 The Impact on Soil Fertility Benefit Value

Growth process and the continuous forest vegetation metabolites lend physical and chemical effects on soil and soil involved in the internal circulation of materials and energy conversion; huge forest biomass, cultivated soil fertility; forest soil profile unique litter its good ecological effects, but also on forest soil formation processes have a significant impact.

Tab.5 Soil nutrients of woodland

Sample type	Soil layer	Nutrient element/g·kg^{-1}			Total/t·hm^{-2}			Converted into fertilizer/t	
		N	P	K	N	P	K	(NH$_4$)$_2$HPO$_4$	KCl
Contrast	0~15	1.77	0.91	14.01	6.951	3.593	74.968	49.6500	149.936
	15~30	1.65	0.87	18.42					
	30~45	1.61	0.82	21.82					
Mine A	0~15	1.59	1.11	14.63	6.286	4.270	76.729	44.8968	153.458
	15~30	1.51	0.98	19.41					
	30~45	1.39	0.96	20.77					
Mine B	0~15	1.56	0.74	15.29	5.501	2.442	69.810	39.2929	139.620
	15~30	1.47	0.63	18.87					
	30~45	1.43	0.61	22.44					
Mine C	0~15	1.39	0.59	13.86	3.686	1.626	52.911	26.3286	105.822
	15~30	1.23	0.55	18.06					
	30~45	1.12	0.51	21.77					

Diammonium phosphate price was 3410 yuan/t, potassium chloride average price of 2980 yuan/t; ammonium dihydrogen phosphate, nitrogen content of 14.0%, 15.01% phosphorus; potassium chloride potassium content of 50.0%; you can still get on the ground soil fertility worth about 616,120 yuan/hm^2, mine A plot 61.0403 ten thousand yuan/hm^2, mine B plot 550,060 yuan/hm^2, mine C plots 405,130 yuan/hm^2.

4.4.2 Elements of Litter Affect the Return Value of Benefits

Forest litter produced not only an important source of soil organic matter, soil and is most active, the surface layer of the work plays an important in maintaining soil stability, soil structure, improve soil physical and chemical properties as well as improved soil fertility, etc. effect. Litter from the elements making restitution to evaluate aspects of the soil to improve soil fertility benefits of improved value of forests.

Tab.6 Litter nutrients return

Sample type	Thickness /cm	Canopy Density	Nutrient element/kg·hm^{-2}			Converted into fertilizer/t	
			N	P	K	$(NH_4)_2HPO_4$	KCl
Contrast	2.10	0.80	85.2	3.1	13.9	0.6086	0.0278
Mine A	1.70	0.50	84.8	2.8	11.3	0.6057	0.0226
Mine B	1.10	0.70	76.1	2.2	11.5	0.5436	0.0230
Mine C	1.66	0.90	61.3	2.1	11.8	0.4379	0.0236

Litter nutrients available to return for benefits still be worth about 2158.07 yuan/hm^2, mine A plot 2132.83 yuan/hm^2, mine B plots 1922.12 yuan/hm^2, mine C plots 1563.42 yuan/hm^2.

4.5 Pairs of Forests Purify the Atmosphere Affect the Value of Benefits

4.5.1 Impact on the Value of Fixed CO_2 Benefits

Forest through photosynthesis, absorbs and fixed the main greenhouse gas CO_2, into organic matter. Meanwhile, the release respiration O_2, thus plays a vital role in the balance of the atmosphere. 1 g of dry matter production per plant needs to absorb 1.84 g CO_2; each producing about 1 m^3 timber to absorb 850 kg of CO_2 or equivalent to 230 g of carbon. Forests can be calculated according to the amount of photosynthesis CO_2 fixation, calculate the amount of CO_2 per dry matter production 1 g can absorb, and then calculate the year in the fixed amount of CO_2 per hectare. And the amount of O_2 release by the same calculation. Lang Kui Jian et al calculated that the number of indicators to convert dry matter production is related to the accumulation of photosynthesis-based data indicators, namely the production of 1 m^3 per net accumulation of forest to absorb CO_2 0.95355 t, releasing O_2 0.702 t.

Tab.7 Carbon fixation and oxygen benefit

Sample type	Accumulation/m^3·hm^{-2}	Net absorption CO_2/t	Release O_2/t	Content C/t	Carbon sequestration value/yuan	Release oxygen value/yuan
Contrast	133.3	127.1082	93.5766	34.6624	32314.38	93576.6
Mine A	129.25	123.2463	90.7335	33.6093	31332.58	90733.5
Mine B	101.55	96.8330	71.2881	26.4064	24617.59	71288.1
Mine C	99.95	95.3073	70.1649	25.9903	24229.72	70164.9

4.5.2 The Impact of Air Filtration and Purification Efficiency of Forest Values

Forest air purification features include: forest to absorb carbon dioxide, produce oxygen function; absorption of sulfur dioxide, hydrogen fluoride, chlorine, ammonia, mercury vapor, and a certain amount of lead vapor copper, iron, zinc, cadmium and other harmful gases and heavy metals.

Meanwhile forest photosynthesis in plants absorb CO_2, as well as to absorb a lot of air, American plant physiologist Saltz and Ross projections, plants need the air to absorb 8050 kg fixed 1 kg of carbon, approximately 6222 m^3. Dry wood CCR 450~500 kg/t, photosynthesis each producing 1 kg of dry organic matter can absorb 3622~4025 kg air, ie air filter 2780~3111 m^3. Evaluation of the value of forest filtration and purification of the atmosphere here using the cost method. 10000 m^3 air filter cost is $ 10, which can calculate the environmental benefits of purifying the atmosphere is changing.

Tab.8 Air purification function

Sample type	Type dry weight /kg·hm^{-2}	Absorb the amount of air /kg	Filtered air volume/m^3	Air purification efficiency/yuan
Contrast	53599.8233	204938924.6	157878279.7	157878.2797
Mine A	51971.3216	198712348.1	153081527.8	153081.5278
Mine B	40833.1738	156125639.9	120274113.3	120274.1133
Mine C	40189.8150	153665757.8	118379100.2	118379.1002

4.6 Open Pit Mining Combined Effects of Blasting on the Ecological Benefits of Forest Values

The value of ecological benefits, including water conservation value forests, water and soil conservation, soil conservation value, the value of air purification. The ecological benefits of consolidated total magnitude of value, compare the surrounding limestone mine in the overall scale of the difference between forest plots still far from the mine to the ground, the result is: to still be 1013200 yuan/hm^2, mine A plot 984500 yuan/hm², mine B plot 838900 yuan/hm², mine C plots 678000 yuan/hm². So we can see that the area surrounding the mining function changes in the value of forest ecological benefits are obvious. Therefore must be considered in mining the value of the loss of forest ecological benefits, to measure the ecological and economic benefits generated by mining in scale ecosystem. Not just the pursuit of high yield and high efficiency, but also give full consideration to the protection of the ecological environment in mining areas.

5 MINE ECOLOGICAL ENVIRONMENT PROTECTION MEASURES

5.1 To Develop a Reasonable Construction Scheme

Design departments and construction units to meet on the basis of ecological protection, the development of rational construction program. In designing the program, in addition to blasting blasting parameters, initiation network and security technologies, such as design, environmental design should be coupled with measures to protect the ecological environment. Blast hazard control not only the protection of buildings, personnel should pay attention to the mechanism and method of vibration, flying rocks, harmful gases, dust and other hazards from the protection of the ecological aspects of study control.

5.2 Dust Control

Dust generated during mining is mainly limestone particles, the main component is calcium carbonate, blasting, crushing, dust is generated during transport are. Dust on the human body, plants and soil influential: the main part of the human body is harmful particulates 10 μm particle size, the main harm people's respiratory system; major impact on the plant surface for dust to land plants, the accumulated effects of direct sunlight, if dust when wet, the surface layer of the crop, "shell", will affect the use of plants to light, reducing the efficiency of photosynthesis, its elements also have some impact on plant growth; impact on the performance of soil chemical elements in the soil, pH, pore and impact the surrounding environment.

Mine production in each step, the larger the amount of dust produced perforation. Under the conditions of dust control measures are not set, the amount of dust in the air to work long hours in workplaces rig up 600~800 mg/m³. Dust blasting is about 25 g/m³, after blasting dust particle size larger settlement in a short time, particle size <10 μm airborne dust is not easy settlement, but only 1% of the amount of dust produced. Mining dump truck during transport will produce some dust, its strength and road dust production types, wet and dry climate, and driving speed and other factors. Because of different mining location, climatic conditions, the differences are large amounts of dust production. When shipped ore cars running at 16 ~ 20 km/h speed in the dry mud knot gravel road, the amount of dust in the air between 10 ~ 15 mg/m³. So in the process of drilling and blasting and shipping them to pay attention to dust measures, which was previously neglected mining operations. To use the drilling rig has a dust collector, blasting work surface installed sprinkler watering facilities, road maintenance and attention to watering.

5.3 Engineering Measures

Spoil, stone should focus on waste rock to minimize pressure area table vegetation. In the field of waste rock bottom first with chunks of waste rock bottom, thickness greater than 1.5 m, in order to facilitate the penetration of water, waste rock venue divert rainwater. Taken from the top down segmentation level accumulation of waste rock timely formation. Step by step set up dams, to ensure dam safety and stability. Ministry set up the field of waste rock cut avoid waste rock field by flood erosion. New mining and transformation of both sides of the road, take the slope and road protecting group measures to prevent soil erosion and landslides, landslides. Caused by temporary construction on steep slopes, dams, take simple protective measures, and set the erosion fence to divert drainage, reduce soil erosion.

5.4 Biological Measures

Set isolated green belt boundaries between mines and mining blasting safety realm, a width of about 50 ~ 150 m, grown locally adapted plants; greenery on both sides of mine haul roads, planting trees and grass on the slope and roadbed; the final step slope compaction, planting trees, shrubs, grass, vegetation recovery. End of mining, it is timely casing, restore vegetation, the entire mine water conservation forest construction.

5.5 Environmental Management and Monitoring

Construction units should establish specialized environmental organizations, environmental construction personnel training and education, and to take the lives of minimal environmental impact, production as much as possible; supervise the implementation of environmental management plans and construction units, the implementation of environmental management regulations, standards, coordination good work among the various departments of environmental protection, ecological protection is responsible for project construction inspection facilities, inspection and operation, and supervision and management.

6 CONCLUSIONS

Limestone mining ecological impact is multifaceted, water conservation through research and calculation values, soil conservation value, the value of soil conservation, air purification value of these benefits worth indicators for forest ecology, come to open pit mining would be blasting the surrounding forest ecosystems have been affected, close to the mining area of forest ecological benefits valued at 678000 yuan/hm^2, lower than the value of forest ecological benefits away from the mining area of 335200 yuan/hm^2. Proposed to consider the issue of environmental protection in the grip safety, while promoting production. To combine environmental impact assessment guidelines and state of the environment in your area, and the development of engineering characteristics, environmental issues catch, good analysis of the environmental impact of the work, in order to fully understand the environmental impact of the characteristics of your area, and propose appropriate ecological environmental protection measures, so that the human resources, the environment in harmony in order to achieve sustainable development of the mining area.

This study is only through research to illustrate the ecological value of forests bursting open pit mining area will have some impact, but the blast damage to the mechanism of the impact of forest ecological benefit is not enough, since only a valuable role. Due to factors that affect the value of forest ecological benefits is complex, not just a single interfering with blasting, and will be affected by altitude, climate, geology and other aspects of the situation, so that the blasting exploitation of forest ecosystems influence the mechanism needs further in-depth study.

REFERENCES

[1] Pan Z, Z J. Different forest types Litters function. Northeast Forestry University, 2002 (5).

[2] Cheng Y J. Bournei leaf litter decomposition and nutrient release dynamics law of. Fujian Agriculture and Forestry University, 2003.

[3] Zhu J, Y J. Masson pine superba Mixed results unevenaged study. Forestry Science and Technology, 2010 (5).

[4] Wang H J. Subtropical area of several typical plantation ecosystem services research. Hunan Agricultural University, 2007.

[5] Kang W, X J. Hunan Forest charity Economic Evaluation - II solid soil of the forest fertilizer, soil and air purification efficiency. Central South Forestry University, 2001 (4).

[6] Wu E, H J. Casuarina northern coast of Hainan Island Shelter litter and soil nutrients, study the relationship between Casuarina fine root biomass. Hainan Normal University, Hainan Normal University, 2010.

[7] Wu X J. Solid Xiaolongshan forest area of forest carbon benefits of research. Northwest Forestry University, 2008 (5).

[8] Shi S, Y J. Assessment of the Changbai Mountain area of forest fixed CO_2 value. Yanbian University (Natural Science Edition), 2002 (2).

[9] Kang W, X J. Economic evaluation of the effectiveness of forest public forests in Hunan Province III clean air benefits. Central South Forestry University, 2002 (1).

[10] Ma W, Y J. Shaoyang City forest ecosystems purify the atmosphere of the functional value estimate. Central South University of Forestry Science and Technology, 2010 (2).

[11] Guo Q, H J. Masson Lin Gongyi Economic Evaluation. Central South Forestry University, 2005 (3).

[12] Hu S, H J. The cloth Nature Reserve in Tibet ecosystem services valuation and management. Science Progress in Geography, 2010 (2).

[13] He H, W J. Limestone mining process, environmental impact and ecological protection. Acta Geologica Sichuan, 2009.

[14] Chen L, G J. GCPN mine ecological environment protection design. Chinese cement, 2006 (1).

[15] Tian Z, L J. Dalian City Forest Ecosystem Service Function Evaluation. Liaoning Normal University, 2010.

Load on the Atmosphere with Carrying out Explosive Works in a Quarry

V.I. Papichev, E.S. Chechneva

(*Institute of Comprehensive Exploitation of Mineral Resources (IPKON), Russian Academy of Sciences, Russia*)

ABSTRACT: Explosive works in the quarries are short-time but powerful source of impacts on the atmosphere. Significant amount of dust and gases is formed in the process of their production. Cloud of dust and gas after the blast spreads beyond quarry and pollutes the atmosphere of the adjacent territories but part of the long-lived gases dart out by air currents over considerable distances, causing a slowdown in the quality of atmospheric air.

Usually assessment of impact on of gases and dust is produced separately. A comprehensive index that would allowed them to take into account the cumulative impact on the atmosphere is missing.

As a result, the total withdrawal of it part by means of direct (event of withdrawal oxygen) and indirect (when introducing foreign substances) of consumption invited to use the amount of load on the atmospheric air as such an index.

Thus, the total amount of load consists of three components:

(1) directly from consumed resource, as a result of the direct withdrawal of part resource from the natural environment; (2) indirectly from the consumed resource, as a result of entering into a toxic foreign substance as result the temporary withdrawal of the resource from the environment for a period of destruction of the substance; (3) indirectly from consumed resource, as a result of entering into non-toxic foreign substance causing a temporary replacement of part of the resource from the environment for a period of destruction substances.

With the application of this index are performed comparative assessments load on the atmosphere of the main of technological processes on the iron ore quarries Russian of mining leases within the limits enterprises. The results of evaluations show that the main load on the atmosphere associated with the process of rock mass transportation vehicles technology, which accounts for 57% of the overall load on the explosive works in the quarry with 26% on piling and dusting of moldboard surface - 16%.

By manufacture explosive works a major component of emissions has the greatest impact on the atmosphere is dust, which accounts for 76.8% of the load from the explosions, gas accounts for 23.2%, including 22.9% for nitrogen oxides.

But if nitrogen oxides, the existence of which in the atmosphere is limited to a few days, outside the mining lease affect the atmosphere for a short time, the other gases, the bulk of which is accounted for oxides of carbon accumulating in the atmosphere will impact on the atmosphere of remote areas for a long time.

KEYWORDS: quarries; environment; load; explosive works

Explosive works in the quarries are short-time but powerful source of emissions to the atmosphere. Large quantities of dust and gases during carrying out of such works are produced. Cloud of dust and gases after the explosion spreads beyond the quarry and pollutes the atmosphere of the adjacent territories but part of the long-living gases is taking out by air flows over considerable distances, causing reduction of the quality of atmospheric air.

Rated estimations on statistic data about functioning of iron-ore quarries of Russia are testify that motor transport, explosive works and dumps are the main sources of emissions to the atmosphere. An average part of this sources in all mass of emissions to the atmosphere is more than 96 % (Tab.1).

Tab.1 Distribution of average mass of emissions to the atmosphere in iron-ore quarry on sources of emissions per 1 year

Sources of emissions	Average common mass of emissions per 1 year by the quarry/t	Including	
		Dust/t	Gases/t
Drilling	15	15	0
Explosive works	6683	763	5920
Excavation	500	500	0
Transportation	17441	400	17041

Corresponding author E-mail: ipkon_sovet@mail.ru

Continues Tab.1

Sources of emissions	Average common mass of emissions per 1 year by the quarry/t	Including	
		Dust/t	Gases/t
Dumps	5711	5711	0
In total	30350	7389	22961

Data showed in Tab.1 are testify that the main load on atmosphere is connected with a process of transportation of mining mass by technological motor transport. The part of this process in all mass of emissions is 56 %. The part of explosive works is 22 %. The part of dusting of a dump's surface is 18 %.

The part of dust in composition of emissions in all mass of emissions is near 26 %. Other part consists from emissions of toxical and non-toxical gases, and the part of last is much larger (more than 90 %). When get to atmosphere, dust and gases (both toxical and non-toxical) make negative impact on atmospheric air by changing its composition and in essence causing withdrawal of a part of its resource.

An estimate of influence of gases and dust on the atmosphere is usually carried out separately, and an influence of non-toxical gases and consumption of oxygen from the atmospheric air isn't considered in general. A complex indicator that would allow to consider the cumulative influence of all of emissions on the atmosphere was offered a few years ago [1,2]. But it had a number of shortcomings, and the main of shortcomings was an equating of non-toxical emissions getting to the environment to toxical with giving an indicator of toxicity to these emissions. The indicator of toxicity gived to these emissions was accepted by the equal to the indicator of toxicity of the least toxic substance in the list of substances, and it causes to overestimate of load from non-toxical substances.

For elimination of this shortcoming a new indicator of load is offered.

In general view formula of the indicator of load on the atmospheric air is expressed as follows:

$$\Pi^\tau = \frac{\sum_{i=1}^{\tau}\frac{M_i}{n_i \cdot \rho_i}}{R_t} + \frac{\sum\left(\frac{G_i^T}{\text{ПДК}_i} \cdot T_i\right)}{R_t} + \frac{\sum_{i=1}^{\tau}\left(\frac{G_i^H}{\rho_i} \cdot T_i\right)}{R_t}$$

where, M is mass of substance directly withdrawn from natural component; G_i^T is mass of toxical alien substance brought to the resource; n_i is a part of the withdrawn substance in natural composition of resource; G_i^H is mass of non-toxical alien substance brought to the resource; ρ is density of non-toxical substance withdrawn or brought to the resource; ПДК is maximum permissible concentration of toxical alien substance brought to the resource; T is an amendment on the time of existence of admixture on the estimated territory; R_t are stocks of the resource on the estimated territory in units of volume.

Thus, the general size of load consists of three parts:

(1) from directly consumed resource as a result of direct withdrawal of a part of resource from environment; (2) from indirectly consumed resource as a result of bringing toxical alien substance, causing temporary withdrawal of a part of resource from environment, in it; (3) from indirectly consumed resource as a result of bringing non-toxical alien substance, causing temporary substitution of a part of resource of environment jn a period of destruction of substance, in it.

An offered indicator allows to get sizes of load on resources of environment as a result of the greatest possible consumption of these resources during direct and indirect withdrawal of these resources from the volume of the resource accepted to an assessment. Indirect withdrawal includes calculation of withdrawal and replacement of a part of resource on a period of destruction of substances taking part in this process. In calculations of consumption by non-toxical substances value of density of the substance which allows to get a size of volume occupied by it in general volume of the estimated resource is used instead of ПДК (maximum permissible concentration).

With using of the offered indicator the comparative estimates of load on the atmosphere from the main objects and technological processes of iron-ore quarries of Russia in limits of land branches of the enterprises and in environment are executed. These estimates showed that load from such processes as drilling and excavation practically isn't extend out of limits of quarry. Load on the atmosphere of the adjacent territories appears because of functioning of dumps, transportation of mining mass and carrying out mass explosions.

The main sources of load in limits of territory of land branch of quarries at the expense of the huge emissions of dust are motor transport of quarries (more than 45%) and dumps (near 42%). With increasing of area of estimated territory and remoteness of sources of emissions from its limits and, respectively, with increasing of volumes of resources, a part of loads from transportation of mining mass constantly increases, and a part of loads from dumps decreases. A part of loads from explosive works rather close territories keeps on constant level and decreases only at a great distance (Tab.2).

Tab.2 Distribution of load on the atmosphere from the main sources of emissions

Sources of emissions	Area of estimated territory			
	Land branch	Out of limits of land branch		
		100 thousands of hectares	1 million of hectares	1000 millions of hectares
	Part from general load/%			
Explosive works	13.1	12.2	14.0	4.0
Transportation	45.2	58.0	84.5	95.5
Dumps	41.7	29.8	1.5	0.5
In total	100	100	100	100

Main cause of decreasing of a part of loads with removal from a quarry is reduction of load from emissions of dust as a result of sedimentation of parts of dust on the earth surface, and a cause of increasing of a part of motor transport is long time of existence of gases in the atmosphere.

Estimation of loads on the atmospheric air from emissions of dust shows that in the time of explosive works a part of dust in the limits of land branch fluctuates between 13 % and 92 % and average size is 52 %. A part of dust from dumps in the limits of land branch fluctuates between 49 % and 87 % and average size is 70 %. It shows that dust from explosive works extends at bigger distance than dust from dumps, and it is confirmed by calculation estimates of distances of sedimentation of dust in iron-ore quarries (Tab.3).

Tab.3 Remoteness of limits of sedimentation of dust from sources of emission

Sources of emissions of dust	Distance of sedimentation of dust from the source of emission/km		
	The minimum	Average	The maximum
Explosive works	2.8	50.6	118
Dumps	15.6	42.2	83.3

With increasing of the area of estimated territory and respectively with removal from a quarry the role of more long-living in the atmosphere components of emissions is increasing. It is confirmed by calculations of changing of load on the atmospheric air from explosive works in dependence from the area of estimated territory (Tab.4).

Tab.4 Distribution of load by components of emissions

Components of emissions	Land branch of a quarry		Area 100 thousands of hectares beyond the land branch	
	Load, parts of unit	general load/%	Load, parts of unit	general load/%
Dust	0.93	44.3	0.056	24.8
Gases	1.17	55.7	0.226	75.2
Including:				
CO	0.031	1.4	0.004	2.0
NO_x	1.136	53.8	0.166	70.8
Other gases	0.003	0.5	0.015	2.4
In total	2.12	100	0.226	100

On the area of the land branch load from emissions of gases exceeds load from emissions of dust by 1.3 times. With increasing of estimated territory to area 100 thousands of hectares the excess increases. At this territory load from emissions of gases exceeds load from emissions of dust already by 3 times. And the main role in load from gases plays nitrogen oxides of which falls near 95 % of general load from gases. Character of dependence of changing of load at emissions of main pollutants (dust and nitrogen oxides) from dimensions of estimated territory respectively with removal from a quarry is shown at Fig.1.

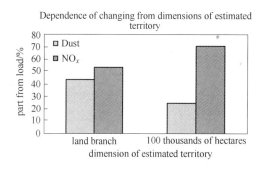

Fig.1 Character of dependence of changing of load at emissions of dust and nitrogen oxides from dimensions of estimated territory

The results show that the main components of emissions to the atmosphere from explosive works making negative impact on the atmosphere are gases, generally nitrogen oxides. The quantity of nitrogen oxides is a little more than the quantity of dust, but out of limits of land branch load from gases becomes prevailing.

With sedimentation of parts of dust from the atmosphere on soil and exit of a cloud of dust and gases out of limits of sedimentation only gases will render load on the atmosphere, and such load will constantly decrease in a process of their destruction. At a big distances from a quarry the role of each of pollutants will change: gases with more long time of existence in the atmosphere will become the most dangerous.

Using of offered approach allows to determine rational parameters of processes providing minimal negative impact on the atmospheric air in the limits of quarry and land branch of enterprise and out of these limits in specific technological conditions and geographical position of a developed quarry.

REFERENCES

[1] Chaplyigin N N, Galchenko Yu P, Papichev V I, Zhulkovskiy D V, Sabyanin G V, Proshlyakov A N. Ecological problems of geotechnologies: new ideas, methods and decisions. M.: OOO Izdatelstvo "Nauchtehlitizdat" , 2009: 320.

[2] Papichev V I. Character of changing of load on the atmosphere during carrying out explosive works. Mining information and analytical bulletin, No1, Separate release "Works of a scientific symposium" "Week of the miner-2010", 2010: 196~206.

Test and Study on the Low Dust Blasting Charge Structure

YANG Haitao[1,2], LIU Weizhou[1,2,3], ZHANG Xiliang[1,2,3], YI Haibao[1,3]

(1. Maanshan Kuangyuan Blasting Engineering Co., Ltd., Maanshan, Anhui, China;
2. State Key Laboratory of Safety and Health for Metal Mines, Maanshan, Anhui, China;
3. Sinosteel Maanshan Institute of Mining Research, Maanshan, Anhui, China)

ABSTRACT: The environment is seriously polluted and a high incidence of occupational disease such as silicosis is caused by blasting dust containing a large number of fine particles. Now a lot of research work has been done solving the problem of blasting dust, the some achievements on dust removal has been obtained. Due to the particularity of blasting dust, the blasting dust hasn't been completely eliminated. In this paper, the blasting dust test system is designed and established independently, according to the different position of water bag, the charge structure model tests are done respectively. Based on results of model tests, the best dust blasting charge structure is selected. Through the model test result, dust absorption effect is the best, adsorption rate of is 73.49% in using the radial water arrangement tests and dust adsorption rate is 67.47% in using orifice water arrangement tests. Based on model test in the laboratory and site test, a blasting effect is achieved. In using orifice way dust charging structure of water, dust absorption effect is obvious, dust adsorption rate is 70.27%~77.42% after 20 min ventilation.

KEYWORDS: blasting dust; charge structure; mist dust-reduction; experimental study

1 INTRODUCTION

Blasting dust contains large amounts of fine particles, seriously contaminating the work face, resulting in a high incidence of silicosis and other occupational diseases[1]. A lot of research work has been done by many blasting engineers to solve the problems of blasting dust and a certain controlling effects have been achieved. However, due to many characteristics such as fast blasting dust producing, dust holding capacity, high dispersion, particle size et al., blasting dust research is very difficult and has not completely eliminate yet. With the increasing requirements for blasting, how to reduce blasting dust pollution on the working surface and surrounding environment become one of the urgent technical problems to be solved.

Based on the mechanism analysis of blasting water mist, the model tests of different charging structure were carried out and the best blasting charge structure were selected.

2 MECHANISM ANALYSIS OF DUST REMOVAL BY WATER MIST AND THE LOW DUST BLASTING CHARGE STRUCTURE

There are two main methods to reduce dust. One is to increase the dust density, the other is to increase the dust particle size. Dust removal by water mist is based on these two basic principles. The dust density increases significantly after absorbing, which is conducive to settlement and condensation. The increasing of dust particle size prompts the dust settling.

Research shows that aggregation of any two small dust particles will make the sedimentation rate increase by three times[2]. The explosion energy makes the water become water mist and steam, increasing the contact area with dust. With the methods of collisions interception, diffusion an agglutination between droplets and dust particles, they condense into larger particles and settlement. Foreign dust measurement indicates that the dust reducing effect of water vapor is 8 times higher than water. Charging structure as shown in Fig.1.

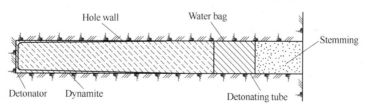

Fig.1 Schematic diagram of charging structure

Corresponding author E-mail: yht 03121 @ 163. com

3 CEMENT MORTAR MODEL EXPERIMENT

3.1 Experiment Materials and Instruments

The size of cement mortar model is 15.5cm × 15.5 cm × 15.5cm, 10kg weight, with a center blasting hole. The blast hole diameter is 20mm and depth is 105mm. The ratio of cement and sand (volume ratio) is 1:2, as shown in Fig.2.

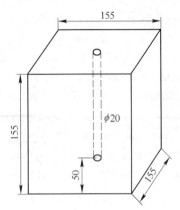

Fig.2 Cement mortar model

Blasting materials: Detonating cord, detonator.
Laboratory instruments: FCS-30 type dust sampler, electronic analytical balance, plastic straws, water bags, tape, etc.

3.2 Experiment and Analysis Results

3.2.1 Experiment Plan

The experiments were conducted in homemade dust test experimental device.

Three kinds of arrangements were selected, namely, upper water, bottom water and full hole water, to study the dust reducing effect of hydraulic charge structure, so that to determine the best charging structure. As shown in Fig.3.

3.2.2 Experiment Results Analysis

The sampling head was set into a tin to protect it from impact of blasting shockwave. The dust concentration of unit air volume is calculated by the membrane increment before and after dust sampling with FCS-30 dust sampler. The calculation formula is shown as followed.

$$R=[(m_2-m_1)/(Qt)]\times 1000 \quad (mg/m^3) \quad (1)$$

where R——total dust concentration, mg/m^3;
m_2——membrane quality after sampling, mg;
m_1——membrane quality before sampling, mg;
Q——sampling flow, L/min;
t——sampling time, min.

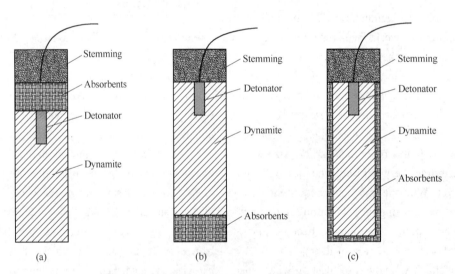

Fig.3 Schematic diagram of different charging structure
(a)upper water; (b)bottom water; (c)full hole water

Tab.1 Experimental results table

Name	Membrane quality before sampling /mg	Membrane quality after sampling /mg	Dust quality /mg	Time /min	Sampling flow /L·min^{-1}	Dust concentration/mg·m^{-3}	Dust reducing rate/%
Bottom water	68.7	79.8	11.1	2	10	0.56	32.53
Upper water	68.1	72.9	4.8	2	10	0.24	71.08
Full hole water	64.9	69.3	4.4	2	10	0.22	73.49
No water	65.8	82.4	16.6	2	10	0.83	Comparative Experiment

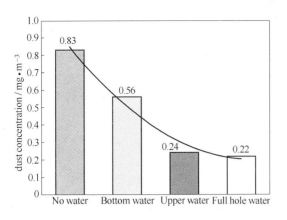

Fig.4 Dust concentration of different charging structure

The experimental results were shown in Tab.1 and Fig.4. The dust concentration of different charging structure was as followed. No water(0.83mg/m^3)＞Bottom water(0.56mg/m^3)＞Upper water(0.24mg/m^3)＞Full hole water(0.22mg/m^3). It illustrated that the dust absorption effect of bottom water was worst, with the dust reduction rate 32.53%. The dust reduction rates of upper water and full hole water respectively were 71.08% and 73.49%, being very close.

3.3 Conclusion of Mortar Model

Through analyzing the mortar model test results, upper water (that is low dust blasting charge structure) dust removal effect is greater than the charging structure of water at the bottom of the bore. The dust reduction rates of upper water and full hole water respectively were very close, but the site construction is more complex by the charging structure of hole full water. Experiments showed that the dust reducing effect of hydraulic charging structure was ideal, making the foundation for on-site verification testing.

4 FIELD TEST

To verify the dust reducing effect of low dust blasting charge structure, the on-site experiment was carried out in an excavating section of an iron ore mine. The water was filled in plastic bags, with length 0.4~0.5m, diameter 35mm, charging structure as shown in Fig.1.

Dust sampling work was conducted with 5min, 10min, 15min and 20min after blasting. Sampling time was 1min and sampling flow was 10L/min. Three representative test data were shown in Tab.2, dust concentration curve shown in Fig.5.

Tab.2 Representative test dust concentration table

Date	Time/min Dust concentration/mg·m^{-3}	5	10	15	20	Dust adsorption rate after ventilation 20min/%
July 25	Hydraulic	1.03	0.41	0.23	0.09	72.73
	Comparison	1.76	1.10	0.49	0.33	
August 1	Hydraulic	1.12	0.53	0.26	0.11	70.27
	Comparison	1.86	1.25	0.65	0.37	
August 5	Hydraulic	0.98	0.51	0.25	0.07	77.42
	Comparison	1.53	1.02	0.52	0.31	

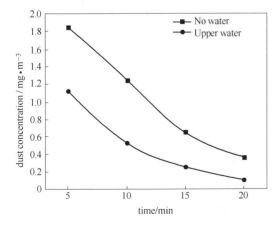

Fig.5 Dust concentration curve of August 5

The results showed that the dust concentration decreased with the prolonging of ventilation time. The dust spreading and concentration decreasing were obvious within the first 10min, then gradually leveled off. The filed test achieved a better dust reducing effect.

The industry experimental results showed that:

(1) The low dust blasting charging structure was reasonable;

(2) When the water bag length was 0.4~0.5m, the dust removal effect could be guaranteed.

(3) Blasting dust could be reduced in the low dust blasting charge structure, and the contrast test proved that the dust reducing effect was significant.

5 CONCLUSIONS

The study showed that low dust blasting charge structure was feasible and effective to reducing blasting dust. It could significantly lower blasting dust, with dust reduction rate 70.27%~77.42% after ventilating 20min. It cut down ventilation time, increased productivity and improved the working environment. The technology is not only applicable to underground mine, deep open pit mine, equally applicable to other non-mining blasting. The technology has broad application prospects.

REFERENCES

[1] Yang G G, Li H Y, Cheng X J. Mechanical analysis blasting dust particles in movement[J]. *Journal of Hebei Institute of Technology*,1996, 18(4): 1~5.

[2] Xue L, Yan S L. Application discussion of mist dust in demolition blasting[J]. *Mining Technology*, 2004,4(3): 65~67.

Vulnerability Assessment of Pressurized Pipeline by Surface Explosives

Yumin Li[1], Ettore Contestabile[2]

(1. 9132-0663 Quebec Inc., Montreal, Canada;
2. Canadian Explosives Research Laboratory, Ottawa, Canada)

ABSTRACT: Purposely or accidentally detonating explosives near pressurized gas or liquid pipelines may have severe consequences on them, ranging from permanent deformation to gas or liquid loss or even violent rupture. Subsequence, fire and vapor cloud explosions due to release of flammable gases or liquids may occur. This will result in the forced shutdown of energy supply and extensive economic and ecological damage. This paper presents a method for analyzing the effects of surface explosive detonation on the pressurized pipeline. Based on the dynamic criterion of pipeline yielding, the safe burial depth of the pressurized pipeline dependent on explosive loading and pipeline properties were evaluated to establish explosive loading - safe burial depth curves. Laboratory and field tests have been done to verify this estimation for pipeline safety. In case of rupture of the pressurized pipeline, the consequences, including the thermal hazard from a jet fire and a vapor cloud fire/explosion, are also simulated by using ALOHA code.

KEYWORDS: vulnerability assessment; pipeline; surface explosion; safety burial depth; fire and explosion; thermal hazard

1 INTRODUCTION

Pipelines are the main form of transportation for oil and gas across countries and continents, being the life-blood of economic growth for business around the world. Terrorist organizations have always been interested in targeting oil and gas facilities since striking pipelines, tankers, refineries and oil fields accomplishes two desired goals: undermining the stability of the regimes they are fighting, and economically weakening foreign powers with vested interests in their region. In the past decade alone, there have been many attacks against oil targets primarily in the Middle East, Africa and Latin America (Ashild & Brynjar, 2001). These attacks have never received much attention and have been treated as part of the industry's risk and the cost of doing business. However, after the Sept 11, 2000, attacks against transportation networks, military bases and government installations have become more difficult to execute due to the heightened security. By looking for a big bang to oil industry, the terrorists can deliver a blow to a country's oil dependent economy as well as global economy at large.

What makes striking pipeline of interest to terrorists is the ease of access and the resulting effective. Pipelines run over thousands of kilometers. A simple explosive device can puncture a pipeline and render it non-operational. While the economic and social disruption from a massive terrorist strike on a pipeline would likely be enormous, the incident in Alaska illustrates how even "minor" attacks on the pipeline system can have widespread downstream consequences. "A drunken local resident had pierced the 25-year-old pipeline with a bullet from a 0.308 caliber rifle. The single shot resulted in the release of more than 1.3 million liters of crude oil across two acres of tundra forest from the pipeline which carries one million gallons of oil per day. It also essentially shut down the pipeline for more than three days" (Parformak, 2004; Parformak, 2006).

To estimate the response of the buried pipeline to stress wave from blasting and surface explosion, some prediction equations and methods are found in the literature, using analytical methods, approximate numerical methods and measurements of pipeline strains (Weigand, 1994; Esparza, 185; ASCE, 2001; Rigas, 2009). Southwest Research Institute has conducted a series of tests for the buried pipelines. The following prediction equation has been obtained from blasting experiments and pipe strain measurements (Esparza, 1985; ASCE, 2001).

$$\sigma = 4.44 E \left(\frac{K_1 W_{\text{eff}}}{\sqrt{Et} R_s^{K_2}} \right)^{K_3} \quad (1)$$

where, σ is peak pipe stress, psi; E is pipe modulus of elasticity, psi; t is pipe wall thickness, inches; K_i are empirical coefficients, $K_1=1.0$, $K_2=2.5$, $K_3=0.77$ respectively for

---
Corresponding author E-mail: yuminli205@gmail.com

the point source; R_s is standoff distance, feet; W_{eff} is effective explosive weight, pounds, $W_{eff} = nW_{act}$; W_{act} is actual weight of explosive charge, pounds; n is factor to normalize explosive to ANFO (94/6).

As we expected, the breaching and cracking of pipeline are resulted from the extreme stress and big deformation of pipeline itself. The estimation of stress level of pipeline due to surface explosion is a key issue. Equation (1) indeed links stress of pipeline, mass of explosive, and the property of pipeline. However, the fact that property of soil will affect the propagation of stress wave is ignored in equation (1).

Rigas (Rigas, 2009) studied safety of natural gas transmission system in the vicinity of explosives by taking into account the explosion source, the ground and the structure characteristics. However, some parameters such as attenuation factors of ground waves with distance, stress intensification factor, soil-pipe coupling factor are not determined in his paper. This results in the difficulty to use his equation to estimate the safe distance of pipeline under surface explosion.

This paper addresses some issues of pipeline vulnerability by surface explosion. The results of the laboratory and field tests are also analyzed. The consequences of pipeline rupture are simulated by ALOHA code.

2 VULNERABILITY OF PIPELINE BY SURFACE EXPLOSIVE

2.1 Vulnerability of Pipeline above Ground or Underwater

It is extremely easy to severely damage above ground or underwater pipelines with explosives. A contact charge of plastic explosive, such as C4, will cut through even a filled oil pipeline without difficulty. The massive detonation pressures (approximate 20 GPa) produce a high pressure wave in the wall of pipe which totally overcomes the strength of the steel, punching a hole through the pipe wall as shown in Fig.1.

(a)

(b)

Fig.1 Explosively formed hole in a steel pipe (test in CERL chamber)

(a) setup; (b) explosive-formed hole

To assess the vulnerability of oil and gas pipelines to contact charges, the following empirical formula may be used to estimate the breach hole size in above-ground pipelines (Long et al, 1996).

$$Q = 0.1ct^2 B \quad (2)$$

where, Q is explosive mass (TNT equivalence), kg; t is thickness of pipeline, cm; B is breach length or diameter, m; c is pipeline strength factor, $c = 5$ and 7.7 for iron and steel, respectively.

Keil (Keil, 1961) has performed breaching tests on steel plates in underwater contact charge configurations. He ruled out the influence of depth of submergence on the damage potential of contact explosions and described the following relationship.

$$R = 0.0704\sqrt{\frac{Q}{t}} \quad (3)$$

where, R is radius of the hole produced, m; t is thickness of steel plate, m.

Equation (3) holds only above a certain charge mass since a minimum amount of explosive is required for making a hole in a plate of specified thickness. The critical charge mass Q_{cri} is given by,

$$Q_{cri} = 2.72t \quad (4)$$

Since the curvature of the pipe has little affect on the breaching capacity of contact charges, equations (3) and (4) can be used to assess the vulnerability of underwater pipes to contact charges and minimum charge mass to breach a pipeline.

Fig.2 shows calculated breach sizes in 12 mm thick steel pipe resulting from contact charges. With the same explosive mass, the damage to underwater pipes is always greater than that to aboveground pipes. A 1 kg TNT contact charge can easily cause more than a half meter diameter breach on steel pipes which wall thickness is 12 mm.

2.2 Tests of the Buried Pipeline by Surface Explosive

A relatively serious s effort is required to breach a buried pipeline. A series of tests were performed by Canadian Explosives Research Laboratory (CERL) to determine the charge mass required to rupture a pressurized gas pipeline buried in sand (Rosen, 2007). The 1 m diameter and 12 mm thick steel pipe was pressurized to 900 psi and buried to 1 m depth under loose soil. Explosive charges were placed at various locations on and above the ground over the pipeline. It was found that the pipelines are very ductile and tough, and the mass of explosive required to breach them is quite large. The result of an attack with an insufficient quantity of explosives is shown in Fig.3. Note that the pipe underwent significant deformation, it was not breached. The pressure in pipeline is still maintained. However, if the explosive mass is increased and a breach achieved, the pressure inside the pipe causes a catastrophic outwards rupture, as may be seen in Fig.4.

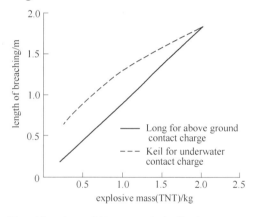

Fig.2　Breaching sizes of 12 mm steel pipeline by contact charges

Fig.3　Test result, damaged but not breached

Fig.4　Test result, catastrophic failure

Exact amounts of explosives for breaching and greatly deforming the buried pipeline can't be released due to confidential.

2.3 Theoretical Estimation of Soil Crater and Safe Distance of the Buried Pipeline

To estimation the crater depth in soil and concrete by surface explosion, Conwep code (Hyder, 1990) can be used. However, we still don't know how to link explosion crater depth and response of pipeline. In turn, to calculate the pipe stress caused by surface explosion and then to determine if the maximum stresses on the pipe exceeding allowable limits are suitable ways to evaluate the vulnerability of the pressurized pipelines.

From the equation (1), considering the explosive energy equivalency of surface charge, we can derive the critical burial depth of pipeline for a given deformation mode and a given explosive mass provided the failure criteria of pipeline are knew.

$$R_{cs} = \left[\frac{K_4 W_{\text{eff}}}{\left(\frac{Y_s - \sigma_h}{4.44E}\right)^{\frac{1}{K_6}} \sqrt{Et}} \right]^{\frac{1}{K_5}} \quad (5)$$

where, R_{cs} is critical burial depth by allowable stress criteria, feet; Y_s is allowable stress criteria of pipeline, psi; σ_h is hoop stress from internal pressure, psi.

$$\sigma_h = \frac{PD}{2t} \quad (6)$$

where, P is operating pressure of pipeline, psi; D is diameter of pipeline, feet. Here, the imperial system is used in order to keep accordance with equation (1). Typically, if maximum stress in pipeline wall is not beyond dynamic yield strength, we consider the pipeline is safe. Therefore, the critical burial depth derived from this criterion will be defined as safe burial depth. In another word, if

$$Y_s \leqslant KY_{ss} \quad (7)$$

Then, the critical burial depth derived from equation (5) is the safe burial depth. Where Y_{ss} is static yield strength of pipe, psi; K is dynamic increase factor of yield strength.

Fig.5 shows pipeline's safe burial depth vs. ANFO explosive mass by equation (5). The parameters of pipeline are $D=3.28$ feet, $t=1$ inch, $Y_{ss}=60000$ psi, $K=2$, $E=3\times10^7$ psf, $P=1000$ psi. For comparison, the crate depth of surface explosion calculated by Conwep is also shown in Fig.5. From Fig.5, it is clear that the safe burial depth is always bigger than the depth of crater for a given explosive mass. In other word, the pipeline with the buried depth less than the depth of crater is not safe. For a given buried depth, for example $D = 4$ feet, the calculated maximum explosive mass for pipeline safety is about 20 pounds with the dynamic yield. From our field tests, a 20-pound ANFO can't result in the permanent deformation or the rupture of the buried pipeline. Since unknown failure criteria corresponding to large deformation and rupture of pipeline, it is difficult to calculate critical burial depths corresponding to large deformation and rupture of pipeline. However, from our field tests, the detonation of a surface explosive charge really results in permanent deformation of pipeline when the burial depth of pipeline is less than the depth of crater produced by the explosive charge as shown in Fig.5.

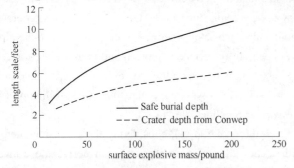

Fig.5　Safe distance of pipeline vs. surface explosive mass

3　CONSEQUENCES OF THE PRESSURIZED PIPELINE RUPTURE

When a high-pressure pipeline fails, immediate and rapid de-pressurization occurs over a matter of seconds, and is followed by relatively stable flow as the pipeline unloads due to the leak and continued pumping of gas and/or liquid from the pipeline. Flow may last for several hours depending on the location and topography of the pipeline and the time that it takes to turn off the valves. A leak or rupture of the pipeline would expose two major hazards.

3.1　Jet Fire and Thermal Hazard

The explosive attacks to oil and gas pipelines usually cause the immediate fire and explosion of pipelines. The damage of this jet-fire, to its vicinity can be serious due to thermal radiation. The thermal hazards can result in casualties and serious burns for nearby human beings. Radiation levels can also be sufficiently high to cause buildings to catch fire.

Techniques are available for estimating the jet flame length and the thermal radiation from an estimated quantity of gas and liquid released over time. ALOHA code (USEPA, 2006) has been used to calculate the maximum flame length and the distance to cause second degree burns to humans (5 kW/m^2 thermal radiation levels) from burning of pressurized natural gas leaking from a pipeline. In our calculation, the following inputs for ALOHA are used:

Site data:

location: Calgary, Canada;

building air exchanges per hour: 1.06 (unsheltered single storied).

Chemical data:

chemical name: methane; molecular weight: 16.04 g/mol;

TEEL-1: 15000 ppm;

TEEL-2: 25000 ppm; TEEL-3: 50000 ppm; LEL: 44000 ppm; UEL: 165000 ppm;

vapor pressure at ambient temperature: greater than 1 atm;

ambient saturation concentration: 1000000 ppm or 100.0%.

Atmospheric Data:

wind: 5 m/s from ESE at 3 m; ground roughness: open country; cloud cover: 5 tenths;

air temperature: 15°C; stability class: D, no inversion height; relative humidity: 50%.

Source Strength:

flammable gas is burning as it escapes from pipe; pipe diameter:100 cm;

pipe length: 10000 m, unbroken end of the pipe is closed off; pipe temperature: 15°C;

pipe roughness: smooth; hole area: 706.86 cm^2; pipe pressure: 1000 psi.

The typical outputs in ALOHA are summarized as following:

max flame length: 28 m;

burn duration: ALOHA limited the duration to 1 hour;

max burn rate: 49600 kg/min;

total amount burned: 357810 kg.

Threat Zone:

threat Modeled: Thermal radiation from jet fire;

red: 174 m——(10.0 kW/m^2 = potentially lethal within 60 s);

orange: 242 m——(5.0 kW/m^2 = 2nd degree burns within 60 s);

yellow: 375 m——(2.0 kW/m^2 = pain within 60 s).

Fig.6 and Fig.7 show the maximum flame lengths of natural gas and the distances to second degree burn with various sizes of the breached holes respectively. As shown in these figures, the larger the breached hole, the more serious the thermal hazards. For 1 m diameter, 10 km long natural gas pipeline with the inner pressure of 1000 psi, the complete rupture of pipeline may results in 115 m long jet of fire and the second degree burn to people within 550 m from pipeline.

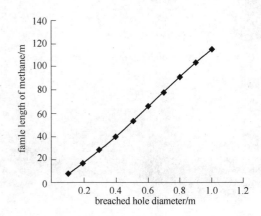

Fig.6　Calculated maximum flame length of natural gas pipeline

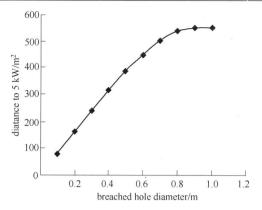

Fig.7　Calculated distance to second degree burns

3.2　Vapor Cloud Fire or Explosion

If a release of gas or an evaporation of flammable liquid does not ignite immediately, it will form a gas or vapor cloud, which will disperse over a large area. If the cloud subsequently encounters an ignition source, those portions of the cloud with certain combustible gas-to-air ratios, e.g. 5%~15% for natural gas will ignite. The cloud fire can then burn its way back to the point of the original leakage point. If the gas or vapor cloud reaches a location with some confinement and sufficient mixing with air, an explosion in the form of a deflagration with damaging overpressures can result. The distance over which such a release can disperse depends on the type of release and the prevailing weather conditions. Figure 8 shows the dispersed distance of natural gas to the low flammability limit (LFL) by ALOHA with the same inputs listed in 3.1. For a 1 m diameter and 10 km long natural gas pipeline pressurized to 1000 psi, the dispersed distance to LFL can reach to about a maximum of 1.6 km. Compared with the 550 m thermal hazard zone of jet fire, the potential hazard of vapor cloud fire is large. If the fuel involved is odorless like natural gas, people will pay less attention to its flammability and are least prepared to its deflagration. In case of having ignition sources, these vapor cloud fire or explosion incidents can occur all of a sudden and result in the most severe casualty. An extremely lethal incident in Nigeria on October 1998 is a typical example of vapor cloud fire and explosion, as more than 1000 people burned to death after a ruptured pipeline caught fire (Scotton, 2006). Most of the victims of the inferno had been trying to collect leaking oil when there was an explosion, apparently set off by a spark from either a cigarette or a motorbike engine.

4　CONCLUSIONS

The pipeline system is a key component of the many interconnected and interdependent critical infrastructure systems. In many cases, they are also the life-blood of cities and business around the world. The calculations and tests refered to in this paper show that the pipelines above ground and underwater are extremely easy to severely damage with explosives. A 1 kg TNT contact explosive charge can easily breach a half-meter diameter hole in a pipeline wall that is 12 mm thick. On other hand, a relatively serious effort is required to breach a buried pipeline. Therefore, if possible, new pipelines should be buried. Theoretical critical burial depth of the pipeline derived from existing empirical equations can be used but strongly depend on the allowable criteria. Compared to the results of field test, dynamic yield strength criterion predicts a conservative safe burial depth. The calculation results also shows that the pipeline is not safe and may be permanently deformed or breached by surface explosive when the burial depth of pipeline is less than the depth of crater produced by the charge. Due to unknown criteria for large deformation and rupture of pipelines, theoretical prediction of critical burial depth corresponding to the large deformation and rupture is not available now and will be our further work. In case of rupture of the pressurized pipeline, the consequences, including the thermal hazard from a jet fire or a vapor cloud fire/explosion, can be very serious up to distances as far as 1600 meters.

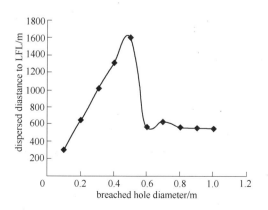

Fig.8　Dispersed distance to LFL of natural gas

REFERENCES

[1] Ashild K, Brynjar L. 2001. Terrorist and oil - an explosive mixture? A survey of terrorist and rebel attacks on petroleum infrastructure 1968-1999, *FFI/RAPPORT-2001/04031.*

[2] Parformak P W. 2006. Pipeline safety and security: federal program. *CRS Report for Congress,* Order Code RL33347.

[3] Parformak P W. 2004. Pipeline security: An overview of federal activities and current policy issues. *CRS Report for Congress,* Order Code RL31990.

[4] ASCE. 2001. Guideline for the design of buried steel pipe, *American Society of Civil Engineers.*

[5] Esparza E D. 1985. Trench effects on blast-induced pipeline stresses, *J. of Geotechnical Engineering.* 111(10):

1193~1210.

[6] Wiegand J E. 1994. Blasting effects and recommendations when blasting near pressurized buried pipeline, *Fifty high-tech seminar,* New Orleans, Louisiana, U.S.A.

[7] Rigas F. 2009. Safety of buried pressurized gas pipeline near explosion sources. In H. Alfada (ed.), *Proceeding of the 1st annual gas processing symp.* Elservier B.V.

[8] Long W, Ho Y. 1996. Blasting engineering. *Metallurgical Industry Press.*

[9] Keil A H. 1961. The response of ships to underwater explosives. *Trans. Soc. of Nav. Arch. Mar. Eng.* 69: 366~410.

[10] Rosen B. 2007. Full-scale pipeline vulnerability trials Phase 2. *CERL Report* 2007(15).

[11] USEPA. 2006. ALOHA User's Manual. *U.S. Environmental Protection Agency.*

[12] Hyde D W. 1990. ConWep. *U.S. Army Corps of Engineer.*

[13] ScottonG. 2006. Alberta oil an attractive target for terrorists. *http://www.canada.com.*

SVM Prediction Model of Tunnel Blasting Excavation for the Destruction of the Ancient Great Wall

LI Longfu [1, 2], CHEN Nengge [2], ZHANG Xiliang [1, 3], JIANG Dongping [1, 3]

(1.Maanshan Kuangyuan Blasting Engineering Co., Ltd., Maanshan, Anhui, China; 2.Maanshan Iron and Steel Group Mining Co., Ltd., Maanshan, Anhui, China; 3.State Key Laboratory of Safety and Health for Metal Mines, Maanshan, Anhui, China)

ABSTRACT: Taking the blasting excavation of a highway tunnel engineering in Shanxi Province as an example, using the principle of support vector machine (SVM) learning, with aperture, hole depth, hole spacing and row spacing, the maximal segment charge, total quantity and distance from an explosive source as the main factors influencing the blasting vibration, SVM model is built. To predict respectively the particle radial, tangential and vertical direction of the peak vibration velocity and frequency, the prediction results were compared with the measured values. The experimental results show that SVM prediction model predict the peak velocity and frequency of blasting vibration, it has fast convergence, high precision, small error characteristics. The model can be used to accurately forecast blasting vibration parameters, according to the forecast results can be better to take measures to protect the ancient Great Wall.

KEYWORDS: tunnel engineering; the ancient Great Wall; blast vibration; SVM

1 INTRODUCTION

Rock blasting seismic effect can cause building (structure) nearby, content in different degree of damage, has a great influence on its stability [1]. For example, building (structure) appear the structural instability, crack, even collapse.

In recent years due to the development of blasting technology and wide application, the problem of blasting seismic effect have become increasingly prominent, caused by civil disputes are also increasing, affect social stability. Therefore, strengthening the blasting technology research, especially the analysis of the blasting vibration related characteristic parameters, making to the building (structure) identification and prediction of building damage, is the key problem of engineering blasting field to be carefully studied and solved [2~4].

SVM is a statistical learning method based on structural risk minimization principle, having excellent learning and generalization ability to solve the small sample. It is based on the theory of statistical learning theory created by Vapnik. Compared with the traditional statistics, it shows unique advantages.

In solving the small sample, nonlinear and high dimensional pattern recognition problems [5~7]. Therefore, this article is based on the matlab platform to build the SVM algorithm near the ancient Great Wall in the tunnel excavation blasting of blasting vibration prediction model.

2 BRIEF INTRODUCTION OF THE PRINCIPLE OF SVM

The basic train of thought for SVM is to transform the input space to a high-dimensional space by nonlinear transformation, then finding the optimal linear classification in this new space, and the nonlinear transformation is defined by the appropriate inner product function. SVM for classification of function form is similar to a neural network, the output is the linear combination of several intermediary nodes, and each intermediate node corresponding to the input sample and a support vector inner product [3].

The idea of SVM can be well applied to the function fitting. Considering the problem of $F(x) = wx + b$, fitting data $\{x_i, y_i\}$, $i = 1,...,n$, $x_i \in R^d, y_i \in R$, the assumption that the training data is available for all linear function without error in precision ε of fitting, namely:

$$y_i - \varepsilon \leqslant wx_i + b \leqslant y_i + \varepsilon \quad i = 1,..., n \quad (1)$$

Similar to maximize classification interval in the optimal classification plane, this control method is to make the regression function on the complexity of function sets the flattest, equivalent to minimizing $\|\varepsilon\|^2/2$. Considering to allow the fitting error conditions, the relaxation factor $\xi_i^+ \geqslant 0$ and $\xi_i^- \geqslant 0$, then the condition becomes to:

$$y_i - \varepsilon - \xi_i^+ \leqslant wx_i + b \leqslant y_i + \varepsilon - \xi_i^- \quad (2)$$

The objective function is:

Corresponding author E-mail: yht 03121 @ 163. com

$$\varphi(w,\xi) = \frac{1}{2}\|\varepsilon\|^2 + C\left[\sum_{i=1}^{n}\xi_i^+ + \xi_i^-\right] \quad (3)$$

Using the optimized method can get the problem of the dual problem, in the condition of:

$$\sum_{i=1}^{n}(a_i^+ - a_i^-) = 0\ ;\ 0 \leqslant a_i^+,\ a_i^- \leqslant C \quad (4)$$

Under the constraints, the Lagrange takes on multiplier a_i^+ and a_i^- to maximize the objective function:

$$Q(a_i^+, a_i^-) = -\varepsilon\sum_{i=1}^{n}(a_i^+ + a_i^-) + \sum_{i=1}^{n}y_i(a_i^+ - a_i^-) - \frac{1}{2}\sum_{i,i=1}^{n}y_i(a_i^+ - a_i^-)(a_i^+ - a_i^-)x_i y_i \quad (5)$$

Then the regression function is:

$$F(x) = w^*x + b^* = \sum_{i=1}^{n}(a_i^{+*} - a_i^{-*})x_i x + b^* \quad (6)$$

Only a small part of a_i^+ and a_i^- is not zero, its corresponding sample generally in the function changing position is support vector. And here only involves the inner product operation, as long as using the kernel functions $K(x_i, y_i)$ instead of the above two type of inner product operation can realize nonlinear function fitting, then the maximum objective function becomes to:

$$Q(a_i^+, a_i^-) = -\varepsilon\sum_{i=1}^{n}(a_i^+ + a_i^-) + \sum_{i=1}^{n}y_i(a_i^+ - a_i^-) - \frac{1}{2}\sum_{i,i=1}^{n}y_i(a_i^+ - a_i^-)(a_i^+ - a_i^-)Kx_i y_i \quad (7)$$

And the Nonlinear regression function is:

$$F(x) = w^*x + b^* = \sum_{i=1}^{n}(a_i^{+*} - a_i^{-*})Kx_i x \quad (8)$$

The training data is fitted by using the principle. In this process, it does not need specific regression function expression, as long as selecting the proper kernel function and then determining the error penalty parameter C, kernel parameter σ and precision ε. Namely, it establishes the forecast model under certain conditions.

3 ENGINEERING APPLICATION OF BLASTION VIBRATION PARAMETERS PREDICTION MODEL

3.1 Project Summary

Tunnel of an expressway in Shanxi Province is located in Jin Meng at the junction of two provinces through the ancient Great Wall, it has flat terrain and loess slope on both sides and sparse vegetation. The ancient Great Wall is Shanxi provincial key protected cultural relics, which is the military defense project in the northern area during the Ming Dynasty which has more than 600 years of history. But most of it only is soil ridge due to the long wind and rain erosion, sand silt diffuse and man-made destruction, so it is particularly important to protect the ancient Great Wall.

Tunnel adopts step construction. Because regional lithology is the strong-weathered limestone, limestone and rock mass integrity is poor, the excavation is used by drilling and blasting method. Blasting task is heavy and frequent. Due to frequent and continuous pressure and torsion effect of blasting seismic wave, the ancient Great Wall is very easy to form crack. Therefore, it should fix up measuring point on the base of the Great Wall corner to study the vibration response in this blasting construction.

3.2 The Influence Factors of Blasting Vibration Parameters

According to the study of Li Hongtao and Xu Hongtao[8,9] influence factors of blasting vibration in addition to geological terrain conditions are mainly its peak of blasting vibration velocity and the main vibration frequency, and single maximum loading, height, distance, number of sections, the total amount of blasting, the first three factors play a main role. In order to improve the training speed of the model, the characteristics of seismic wave in the process of blasting parameter prediction factors affecting smaller can be ignored. Combining with the current situation of tunnel excavation, emulsion explosive, detonator, the charging structure and detonating in blasting process are the same, therefore, it does not consider their influence. Because of the fixed monitoring points and similar engineering geological conditions of blasting excavation of the section, the elevation difference and the influence of the engineering geological conditions are also ignored. It finally choices selection aperture, hole depth, hole spacing and row spacing, single segment from an explosive source, total quantity and the largest quantity of seven factors as input parameters of the model. The horizontal radial (CH1), horizontal tangential (CH2), vertical (CH3) in 3 directions are forecast.

3.3 Model Study and the Choice of Test Sample

SVM learning training data is collected during the normal tunnel excavation blasting; there are heading (step) excavation blasting and the inverted arch (next step) on both sides of excavation blasting. Three times of blasting is a cycle, several times in the middle of the large blasting does not make the records. Using 2# rock emulsion explosive, cartridge diameter 32mm, the hole diameter 42mm, the total 30 sets of blasting field monitoring records should be collected. The first 20 sets of data as training samples and after 10 sets of data as the test samples were analyzed to establish the relevant model. Among them, D is the aperture; h is the hole depth, a is the pitch; b is row spacing; q is the largest single segment dose; Q is the total dose; R is a blast from the source, v is blasting vibration velocity, f is the vibration frequency. See Tab.1.

Tab.1 Blasting related training parameters

No.	1	2	3	4	5	6	7	8	9	10	11	12	13	14	15
D/mm	40	40	40	40	40	40	40	40	40	40	40	40	40	40	40
h/m	3.2	2.5	2.5	3.2	2.6	2.6	3.1	2.5	2.5	3.2	2.7	2.7	3.1	2.7	2.7
a/m	1.4	1.2	1.2	1.4	1.2	1.2	1.4	1.2	1.2	1.4	1.2	1.2	1.4	1.2	1.2
b/m	1.4	1.2	1.2	1.4	1.2	1.2	1.4	1.2	1.2	1.4	1.2	1.2	1.4	1.2	1.2
q/kg	27	11	15	22	10	10.3	19	9.6	10.2	18.5	11.6	11.4	17.5	11.6	10.8
Q/kg	102	44.6	47	97.2	43	41.8	99.2	49	47.8	100.2	51.4	50.8	100.5	51.4	52
R/m	151	168	168	148	165	165	146	162	162	143	159	159	140	159	159
No.	16	17	18	19	20	21	22	23	24	25	26	27	28	29	30
D/mm	40	40	40	40	40	40	40	40	40	40	40	40	40	40	40
h/m	3.2	2.5	2.5	3.0	2.4	2.4	3.1	2.4	2.4	3.0	2.6	2.6	3.2	2.5	2.5
a/m	1.4	1.2	1.2	1.4	1.2	1.2	1.4	1.2	1.2	1.4	1.2	1.2	1.4	1.2	1.2
b/m	1.4	1.2	1.2	1.4	1.2	1.2	1.4	1.2	1.2	1.4	1.2	1.2	1.4	1.2	1.2
q/kg	17.6	11.2	10.9	15.8	9.2	9.5	16.5	10.1	9.9	17.2	10.5	11.2	16.6	9.2	9.4
Q/kg	102.3	51.6	52.2	98.7	48.1	48.5	103.2	49.5	48.8	100.5	51.4	49.7	102.2	52.4	53.5
R/m	137	157	157	135	155	155	132	153	153	129	151	152	126	149	149

3.4 Model Parameter Settings

According to the result of the research literature [10] of Yang Chengxiang et al, compared to other kernel function, Gauss kernel function has obvious advantage in dealing with complex model and the calculation accuracy. Therefore, Gauss radial basis kernel function is chosen as the kernel function of support vector regression. The method of adjusting the parameters is used to select parameter C, ε and kernel function parameters σ. According to the results of the training, the value of parameter σ, C and ε is shown in Tab.2, the ability of learning and promoting the model is best under the parameter.

Tab.2 Model related parameter of training result

Model parameter	C	ε	σ
Vibration speed of training result	48.5	0.00025	1.48
Vibration frequency of training result	115.6	0.00025	0.72

3.5 Comparison and Analysis

The trained model is used to forecast the 10 groups after blasting data (a maximum of three test direction), which are compared with the measured results to test precision of the model. v_r and v_p is respectively measured and predictive vibration velocity, f_r and f_p is respectively measured and predictive vibration frequency, e_r is the relative error. Comparison results are shown in Tab.3 and Tab.4.

Tab.3 Comparison of measured and predictive blasting vibration velocity

No.	The measured values v_r/cm·s^{-1}			The predictive values v_p/cm·s^{-1}			The relative error e_r/%		
	CH1	CH2	CH3	CH1	CH2	CH3	CH1	CH2	CH3
21	0.396	0.212	0.225	0.539	0.386	0.381	14.320	17.330	15.620
22	0.322	0.222	0.280	0.451	0.341	0.392	12.890	11.880	11.220
23	0.392	0.268	0.202	0.444	0.392	0.282	5.220	12.430	8.000
24	0.427	0.238	0.285	0.473	0.394	0.420	4.630	15.670	13.550
25	0.412	0.255	0.216	0.588	0.488	0.417	17.600	23.350	20.130
26	0.386	0.253	0.258	0.526	0.336	0.400	13.990	8.240	14.210
27	0.378	0.304	0.282	0.543	0.458	0.342	16.450	15.310	6.000
28	0.399	0.322	0.275	0.481	0.449	0.410	8.200	12.770	13.490
29	0.437	0.323	0.256	0.573	0.408	0.378	13.550	8.500	12.180
30	0.508	0.272	0.263	0.650	0.383	0.441	14.120	11.160	17.780

Tab.4 Comparison of measured and predictive blasting vibration frequency

No.	The measured values f_t/Hz			The predictive values f_p/Hz			The relative error e_t/Hz		
	CH1	CH2	CH3	CH1	CH2	CH3	CH1	CH2	CH3
21	21.243	36.056	19.415	21.396	36.173	19.515	15.320	11.680	10.000
22	20.446	35.250	18.575	20.568	35.418	18.648	12.140	16.770	7.320
23	20.084	36.305	19.277	20.228	36.429	19.426	14.370	12.370	14.870
24	20.291	35.728	19.584	20.384	35.778	19.624	9.220	5.000	4.000
25	20.877	36.204	19.211	21.033	36.358	19.336	15.630	15.380	12.510
26	20.757	37.323	20.444	21.000	37.441	20.548	24.280	11.770	10.450
27	22.829	38.682	19.611	22.930	38.779	19.759	10.150	9.730	14.790
28	22.686	38.261	20.666	22.861	38.406	20.766	17.540	14.550	10.020
29	21.537	38.522	20.782	21.666	38.619	20.935	12.850	9.740	15.250
30	22.980	38.852	21.045	23.132	38.989	21.196	15.270	13.760	15.120

The Tab.3 shows that in the prediction of blasting vibration velocity peak the average relative error (horizontal radial and tangential horizontal and vertical direction) of SVM prediction was 12.10%, 13.66%, 13.22%.The Tab.4 shows that in the prediction of blasting vibration frequency the average relative error (horizontal radial and tangential horizontal and vertical direction) of SVM prediction was 14.68%, 12.08%, 11.43%.

According to error analysis result, SVM prediction model has higher accuracy in prediction of blasting vibration velocity and vibration frequency, the average error is about 13%. It can use the model to forecast the peak vibration velocity and frequency in subsequent blasting construction Combined with blasting vibration safety control standards in "Blasting Safety Regulations" (GB 6733—2003)[11], optimizing blasting parameters and detonating network, In the premise of ensuring the safety of the ancient Great Wall to reduce the maximum detonation quantity of a same section, it could control the blasting vibration effect and play a good role in guiding the blasting construction of near the ancient Great Wall.

4 CONCLUSIONS

(1) By the blasting vibration data measured near the ancient Great Wall to establish SVM prediction model based on matlab platform, the measured data is used to test the model. According to error analysis result, SVM prediction model has higher accuracy in prediction of blasting vibration velocity and vibration frequency, the average error is about 13%.

(2) In engineering practice, the influence factors of blasting vibration are more and often more complex, therefore, several main affecting factors are selected as input parameters. Application of SVM principle to build the model to forecast blasting vibration velocity peak value and frequency of a maximum of three test direction, forecast results have the smaller relative error and the higher precision.

(3) It can use the model to forecast the peak vibration velocity and frequency in subsequent blasting construction Combined with blasting vibration safety control standards in "Blasting Safety Regulations" (GB 6733—2003)[11], optimizing blasting parameters and detonating network, In the premise of ensuring the safety of the ancient Great Wall to reduce the maximum detonation quantity of a same section, it could control the blasting vibration effect and play a good role in guiding the blasting construction of near the ancient Great Wall.

REFERENCES

[1] Ding K, Fang X, Lu F D, Li D. Forecasting Model of Blasting Vibration Acceleration Peak value based on SVM[J]. Journal of detection and control, 2010, 32 (4): 38~47.

[2] Zong Q, Wang H B. Zhou S B. RESEARCH ON MONITORING AND CONTROLLING TECHNIQUES CONSIDERING EFFECTS OF SEISMIC SHOCK[J]. Chinese Journal of Rock Mechanics and Engineering, 2008, 27 (5): 938~945.

[3] Zhang H L, Wang S L, Li Y H, Ying X T. Analysis and Forecast of Basting Vibration Effect Based on Support Vector Machine[J]. MINING R & D, 2007, 27 (4): 57~59.

[4] Yan Z X, Wang Y H, Jiang P, Wang H Y. STUDY ON MEASUREMENT OF BLAST-INDUCED SEISM AND BUILDING SAFETY CRITERIA[J]. Chinese Journal of Rock Mechanics and Engineering, 2003, 22 (11): 1907.

[5] Li J M, Zhang B, Lin F Z. Training algorithms for support vector machines. J T qinghua Univ (Sci & Tech), lancet, 2003(1): 120~124.

[6] Guo W M, Hong H, Ying X Z .Assessment of structure damage to blasting induced ground motions[J]. Engineering Structures, 2001, 22(1): 1378~1389.

[7] Zhao H B, Feng X T, Yin S D. Classification of engineering rock based on support vector machine[J]. Rock and Soil Mechanics, 2002, 23 (6): 698~701.

[8] Li H T, Shu D Q. Influential factors analysis of blasting vibration attenuation law[J]. Journal of Wuhan University of Hydraulic and Electric Engineering, 2005, 38 (1): 79~82.

[9] Xu H T, Lu W B. Advance on Safety Criteria for Blasting Vibration[J]. Blasting, 2002, 12 (1): 8~10.

[10] Pan Y Z, Zhang Y P, Wang Q, Tang L Y. Study on Support Vector Machines Model for Prediction of Rock Fragment Size of Bench Blasting[J]. MINING R & D, 2010, 30 (5): 97~99.

[11] National standard compilation group. GB 6733—2003 Blasting safety regulations [S]. Beijing: China Standard Press, 2004.

Study on Control Measures for Blast-Induced Ground Vibration in Open Pit Metal Mine by Means of Numerical Simulation

Kento Fukui[1], Hirokuni Inoue[1], Takashi Sasaoka[1], Hideki Shimada[1], Akihiro Hamanaka[1], Kikuo Matsui[1], Shiro Kubota[2], Tei Saburi[2]

(1.Kyushu University, Japan; 2.National Institute of Advanced Industrial Science and Technology, Japan)

ABSTRACT: In modern mining industry, the blasting technique is the rock excavation and fragmentation techniques most widely adopted due to its economical and efficient aspects. However, compared with other methods, use of explosives in blasting is always limited by law as it may have a severe impact on the surrounding environment such as dust, noise and vibration. Especially, a ground vibration induced by blasting has to be paid much attention in the mining operation. The ground vibration level is regulated by three parameters: duration, amplitude (peak particle velocity/*PPV*) and frequency. Current study focuses on two parameters suggested by USBM for structure, i.e. amplitude and frequency. Hence, the control of *PPV* and dominant frequency are very important in order to design an appropriate blasting standard and minimize its environmental impacts.

This paper discusses the propagation characteristics of blast-induced ground vibration in the open pit metal mine and the applicability of the numerical simulation to predict *PPV* by using 2D analysis code Ls-dyna based on the results of field experiments. Then, the measures for reduction and/or control of blast-induced ground vibration are discussed by means of numerical simulation in this study.

KEYWORDS: blast-induced ground vibration; peak particle velocity (*PPV*); Scaled distance; trench; vibration reduction wall; numerical simulation

1 INTRODUCTION

A blasting technique is widely adopted not only in mining operation but also in civil engineering works such as tunnel, subway, highways and damsdue to its economical and efficient aspects (Reza, N. 2012). However, the application of blasting technique is limited by law because it may have a serious impact on surrounding environment such as ground vibration, noise and dust. Especially, blast-induced ground vibration may cause damage to the surroundingbuildings. Thus, mining companies which apply the blasting technique have to control the blast-induced ground vibration in order to minimize and eliminate the damage to nearby structures (Charles H. D. 1985). As many researchers have already studied on blasting techniques and their impacts on the surrounding environment, several equations and methods for prediction of blast-induced ground vibration have been proposed so far. Most of them adopt the Peak Particle Velocity for evaluating the degree of blast-induced ground vibration. However, the general prediction equation and/or method have not been developed yet because the different rock parameters and geologicalconditions have an obvious impact on the propagation of blast-induced ground vibration.

In the K mine, it is necessary to control the blast-induced ground vibration which may causes damage to the surrounding buildings during recent years. Therefore, the propagation behavior of blast-induced ground vibration has to be considered, which aims at developing of suitable blasting standards and vibration-proofing method of minimizing its impact on the surrounding environment.

From these points of view, a series of field experiments and a 2D numerical simulation by using Ls-dyna have been conducted in this study in order to comprehend the propagation behavior of blast-induced ground vibration in this mine. Then,the effect ofvibration-proofing trench and the decrease in quantity of chargeon blast-induced ground vibration were studied by means of the 2D numerical simulation in order to discuss the most appropriate vibration-proofing method in this mine.

2 FIELD EXPERIMENTS

2.1 Outline of Field Measurements

Blast-induced ground vibration is affected by many factors such ascharge weight, number of free faces, distance from

Corresponding author E-mail: sasaoka@mine.kyushu-u.ac.jp

the blast source, and geological conditions. Hence, it is very difficult to develop a common blasting prediction equations or methods for different locations/geological conditions. Thus, it is necessary to develop the specific prediction formula and propagation behavior of blast-induced ground vibration in each mine site based on the results of field measurements.

A series of field blasting tests were conducted at K mine located in southern part of Kyushu Island, Japan. In this test, Ammonium Nitrate Fuel Oil (ANFO) was used as an explosive. Three-axial accelerometer was used as an instrument to monitor a blast-induced ground vibration. The measurement threshold is ±15 G (where 1 G is equal to 9.80665 m/s^2), frequency response is 3~5000 Hz, signal response is 100 mV/G, and signal gain is alternative 10 and 100. Additionally, the SA-611 amplifier and the LX-100 data recorder were also used. This digital recorder can record the low frequency signal and analyze waveforms owing to the A/D converter. Layout of field experiment is illustrated in Fig.1 and the blasting design is listed in Tab. 1. In these experiments, the Cartesian coordinate system was defined with the x axis parallel to the bench face, the y axis perpendicular to the bench face, and the z axis perpendicular to ground surface. The vibration acceleration sensors were installed along the perpendicular to the ground surface. The distances from the row of blast holes to these three sensors were 15 m, 25 m, 30 m, 40 m and 300 m, respectively.

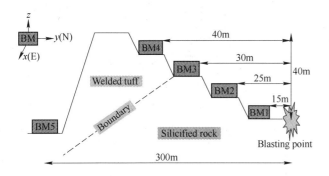

Fig.1 Layout of field experiments

Tab.1 Blasting design

Spacing/m	1.5	Drilling angle/(°)	80
Burden/m	2	Delay time/s	0.25
Powder factor/g·t^{-1}	254	Bench height/m	10
Hole diameter/mm	76	Drilling depth/m	13
Bench angle/(°)	80	Number of blast holes	11

2.2 Results of Field Experiments

In this paper, the cubic root, $R/W^{1/3}$ was used as a scaled distance to predict the PPV. Fig.2 shows the relationship between three dimensional combination PPV and scaled distance $R/W^{1/3}$ (R: Distance from blasting hole to observation point (m); W: Weight of charge (kg)) in this field experiment.

Based on this result, it is obvious that the PPV is uniformly damped with increasing scaled distance, and the prediction equations for PPV in this mine site can be derived by data fitting. Also, it can be seen that the propagation behavior of blast-induced ground vibration is different depending on the type of rocks. The prediction equation of Welded tuff and Silicified rock can be respectively expressed as Equation (1), (2).

$$PPV=7.21(R/W^{1/3})^{-1.62} \quad (1)$$
$$PPV=4.79(R/W^{1/3})^{-1.20} \quad (2)$$

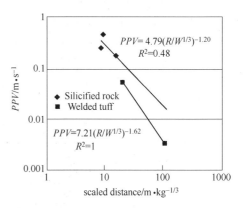

Fig.2 Field experiment result

3 NUMERICAL SIMULATION

3.1 Outline of Numerical Simulation

The numerical simulation was conducted by means of two dimensional Finite Element Method (FEM) Code; Ls-dyna in order to know the propagation behavior of blast-induced ground vibration and develop the simulation model for prediction and evaluation of control measures for blast-induced ground vibration in this mine. The monitoring data obtained from a series of field experiments was used in this simulation. Fig.3 shows the simulation model. 2D elastic model was applied in this simulation.

Pressure wave induced by blasting around the epicenter can be expressed as the following function (3)(Wilbur I. D. 1953).

$$P(t) = P_0 \xi \{\exp(-At) - \exp(-Bt)\} \quad (3)$$

where P_0——maximum pressure, MPa;
ξ, A, B——constants.

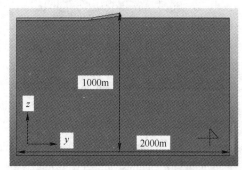
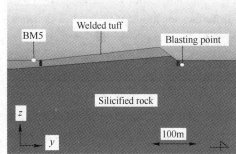

Fig.3　2D simulation model

The time of reaching peak pressure is inferred 3 msec, and $A=250$, $B=350$ are determined by the detonation velocity of ANFO and length of explosive charge. Fig.4 shows the pressure variation with time at blasting hole. The damping factor of blast-induced ground vibration in the rock mass is determined by the minimum frequency in this simulation: it can be estimated as 3Hz based on the results of the field experiments. Tab. 2 provides the mechanical properties of rocks obtained from the laboratory test.

Tab.2　Mechanical property of rock

	Silicified rock	Welded tuff
Density /10^3kg·m^{-3}	2.56	2.39
Poisson's ratio	0.25	0.28
Young's modulus /GPa	74.0	17.7
Uniaxial compressive strength /MPa	163	55.8
Tensile strength /MPa	21.1	4.92

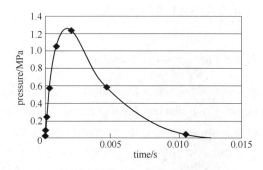

Fig.4　Relationship between pressure and the elapsed time at a blast source

Fig.5 shows the contour of stress wave propagation. From this contour, it can be seen that the damping of blast-induced stress wave of Welded tuffs is larger than that of Silicified rocks. Moreover it can be considered that blast-induced stress wave to BM5 is propagated from Silicified rocks to observation point outside of mine. This is because the magnitude of elastic wave velocity of Silicified rocks is the larger than that of Welded tuffs due to the difference of their Young's modulus (Japan Explosive Society. 2002).

Fig.6 shows the results of numerical simulation and field experiments. As compared both data, it can be seen that the result of numerical simulation gives good agreement with the value obtained from field experiments. Therefore, it can be said that the simulation model and method used in this study can simulate the propagation behavior and value of blast-induced vibration around the targeted area in this research quantitatively.

From the results of a series of field experiments and simulations, the wave propagation of blast-induced ground vibration can be predicted by means of this numerical simulation and analysis model. Thus, this simulation method can beused for the development of an appropriate vibration-proofing method in this mine as follows.

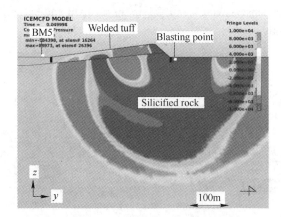

Fig.5　Contoursof stress wave propagation

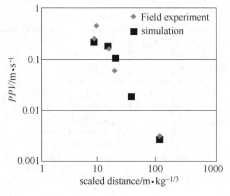

Fig.6　Comparison simulation results and field experiment

3.2 Control Measures of Blast-Induced Ground Vibration

In order to reduce the blast-induced ground vibration, there are several measures at different locations or points such as the blasting point, the propagation path of blast-induced ground vibration,and the vibration receiving point(K. Narisawa & K. Kajitani. 2006).In this study, the effects of the vibration-proofing trench constructed in the propagation path and the reduction of explosive on blast-induced ground vibration werestudied by means of the 2D numerical simulation.

Here,in order to evaluatethe effects of both measureson blast-induced ground vibration,the reduction rate of that was used. It can be expressed as Equation (4).

$$\text{Re}\,ducing\ ratio = (PPV_{after} - PPV_{before}) / PPV_{before} \quad (4)$$

3.2.1 Effect of Vibration-Proofing Trench on the Propagation of Blast-Induced Ground Vibration

There are many types of wave barrier methodssuch as stiff concrete walls and piles and they work as a flexible gas cushions. Among them, an open trench is the most common method in practical application since it presents an effective and low cost isolation method (M. Adam. 2005).

From the simulation results of stress wave propagation, the blast-induced ground vibration propagatesmainly in Silicified rocks. In order to control or reduce impact on the surrounding environment, an open trench shouldbe constructed in Silicified rocks. Thus, the effects of trench constructed near the blasting hole on the propagation of blast-induced ground vibration are discussed under different conditions in this section.Tab.3 shows the input parameters of an open trench.

Tab.3 Input parameters of open trench

Density/kg·m^{-3}	2.00×10^3
Poisson's ratio	0.35
Young's modulus/GPa	7.40×10^{-3}

Fig.7 shows the layout of simulation models. In this simulation,the effects of trench depth and width on the magnitude of blast-induced ground vibrationare discussed. Figs.8 and 9 show the effects of the trench depth and width on the magnitude of blast-induced ground vibration at each observation points, respectively.

It can be recognized from Fig. 8 that the trench depth has an obvious impact on blast-induced ground vibration. Especially, the vibration reduction effect is increased obviously when the trench depth is larger than the depth of blast hole.It can be also recognized that the reduction ratio increases with decreasing the distance between the blast hole and observation points. This result can be explained by the difference of travel distance of the blast waves around the trench. On the other hand, from Fig.9, the trench width has a small effect on the reduction of blast-induced ground vibration because of small difference of the wave travel distance.

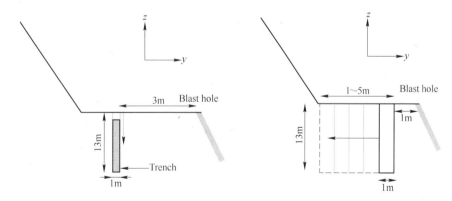

Fig.7 Layout of the simulation model (left : depth , right : width)

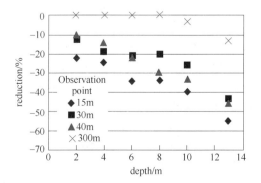

Fig.8 Relationship between reduction ratio and trench depth

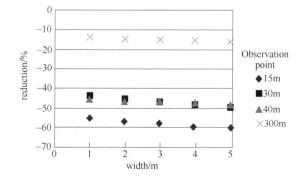

Fig.9 Relationship between reduction ratio and trench width

From these results, it can be said that the trench depth have to be considered for the effective vibration reduction and it should be larger than the depth of blast hole.However, if the blast-induced ground vibration at neayby buildings is controlled to be less than the self-regulated value, which *PPV* is less than 0.0002m/s, more than 20% of reducing ratio is needed in this condition. Hence, the application of another measures also has to be considered.

3.2.2 Effect of the Amount of Explosive on the Propagation of Blast-Induced Ground Vibration

Pressure wave of equation (3) is changed as the amount of explosive decreases. Thus, the effects of the amount of explosive on the propagation of blast-induced ground vibration are discussed in this section. Fig.10 shows the relationship betweenthe reduction of *PPV* for observation points and the weight of charge.

From these results, it can be said that the blast-induced ground vibrationdecreases with decreasing the weight of charge dramatically. Moreover, the vibration reduction effect can be recognized at the farthest observation pointwhich is 300 m far from blasting hole (BM5)obviously. When the amount of charge is less than 10 kg, the reducing ratio is up to 20% and the blast-induced ground vibration at BM5 can be controlled below the limiting value, *PPV*=0.002 m/s defined by *K* mine.However, the efficiency of blasting operation decreases with decreasing the amount of explosive dramatically. Hence, both the vibration-proofing trench and the reduction in the amount of charge should be applied depending on the conditions.

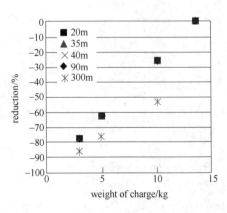

Fig.10   Relationship between the reduction ratio and the weight of charge

The distance from blasting point to observation point where the blast-induced ground vibrationreaches limiting value for nearby buildings is studied.Fig.11 shows the relationship between the weight of charge and the distance from blasting point to the observation point where the *PPV* is 0.002 m/s.From these results,as the current amount of charge is 13.8 kg in this mine, when the distance from the blasting point to the observation pointis smaller than 328 m, the *PPV* at observation point becomes to be larger than 0.002 m/s. Thus, any control measuressuch as the installation of trench and/or the reducing the amount of charge have to be conducted when the distance from the blasting point and nearby buildings is less than 328 m.

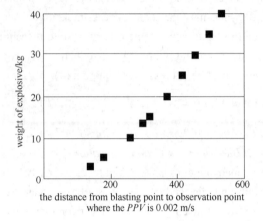

Fig.11   Relationship between the weight of charge and the distance from blasting point to observation point where the *PPV* is 0.002 m/s

4   CONCLUSIONS

In this research, the propagation behavior of blast-induced ground vibration in K mine have been made clear and the analysis model for prediction of it has been developed. Moreover, from the results of a series of numerical simulations, both aconstruction of open trench and reducing the amount of explosive can control the propagation of blast-induced ground vibration effectively. However, in order to develop the guidelines and the appropriate designs for two measures, more detailed study have to be conducted.

REFERENCES

[1] Reza N. Evaluation of blast induced ground vibration for minimizing negative effects on surrounding structures[J]. *Soil Dynamics and Earthquake Engineering*, 2012, 43: 133～138.

[2] Charles H D. *Blast Vibration Monitoring and Control*, Prentice-Hall International Series in Civil Engineering and Engineering Mechanics,1985.

[3] Mogi G. Study on control of local vibration by using delay blasting[J] *Journal of Japan Explosive society*, 1999, 60(5): 233～238 (in Japanese).

[4] Wilbur I D. Strain-wave shapes in rock near explosions Geophysics, 1953, 18(2): 310～323.

[5] Japan Explosive Society. *Collection of blasting cases*, 2002: 1~13(in Japanese)Japan.

[6] Narisawa K, Kajitani K. The Progressive Action for Mitigating the Blast Vibration Suited to Condition of Slicing the Deposit Down at Shuhou Mine to the Local Residents[J]. *The Mining and Materials Processing Institute of Japan*, 2006, 122: 613~317.

[7] Adam M. Reduction of train-induced building vibrations by using open and filled trenches[J]. *Computers and Structures*, 2005, 83: 11~24.

Analysis of Vibration Effects of Shallow Highway Tunnel under Blasting Seismic Waves

YAO Jinjie[1], LUO Zhenhua[1], XIANG Aiguo[2], ZHAO Rundong[1], XIONG Yazhou[1]

(1. College of Hydraulic &Environmental Engineering of China Three Gorges University, Yichang, Hubei, China;
2. Daxing Blasting LLC, Yichang, Hubei, China)

ABSTRACT: According to blasting excavation instance on the top of shallow highway tunnel in Pingyikou Industrial Park 270 platform Project of field, analyzing the effect of blasting vibration to the shallow highway tunnel. In order to ensure the safety of the tunnel, monitoring in real-time vibration velocity of tunnel blasting by using TC-4850 vibration meter. And for safety reasons, determine the security of the structure referring to the allowed maximum safety velocity of the general brick and large blocks of non-seismic building, and give improving advices to the follow-up blasting design and construction.
KEYWORDS: project of field; shallow highway tunnel; monitoring; vibration velocity

1 INTRODUCTION

With the development of construction projects, in order to accelerate the construction schedule and reduce costs, blasting have been widely used in various engineering fields, which also requires blasting need to be constructed [1]in the vicinity of the existing facilities. Now blasting operations will happened above an existing traffic tunnel for shallow tunnels. Because of affecting by blasting vibration, monitoring the effects of vibration has been needed to judge the result in order to ensure the safety of the tunnel construction and to provide the necessary basis for the subsequent construction.

2 PROJECT OVERVIEW, BLASTING AREA TOPOGRAPHY, GEOLOGICAL CONDITIONS, ENVIRONMENTAL AND TECHNICAL REQUIREMENTS

2.1 Project Overview

Pingyikou project is located on the top of industrial silicon traffic tunnel, the west of the 240 platform, belongs to phase II of Pingyikou project. Needed blasting about 5 cubic meters, about 18 tons planned dosage.

2.2 Blasting Terrain, Geological Conditions

The mountain which needs for blasting is continental sedimentary rocks.North-south trending ridge;layered structure;oblique structure (ride west to east, dip about 75°). Lithology of sandstone (gray and yellow sandstone, purple sandstone alternating);strongly weathered part has been stripped in the early construction; the remaining part is weathered, weakly weathered layer, developed joints.

2.3 Environmental and Technical Requirements

The mountain which needs for blasting located in 100 meters north of Xingshan County power company.10 meters away from equipment and facilities of the new coating plant, the height is just 24 meters between ▽270 platform and the top of the tunnel. Traffic tunnel of industrial silicon crosses aslant from the bottom of the blast zone.Critical Area environment is complex.After considering,power lines are not migrated in the construction,the main power line of Xingshan County cannot be stopped.Ensure the safety of the traffic tunnel, high-voltage power lines, paint factory equipment and facilities in the blasting. Thus, safe blasting needs a higher and more stringent requirement because of environmental conditions.

3 SHALLOW TUNNEL BLASTING VIBRATION MONITORING OF PINGYIKOU ▽270 PLATFORM

3.1 Measuring Point Scheme

The international community in general uses round particle vibration velocity to measure the impact of vibration on the surrounding buildings [1, 2]. This monitoring uses TC-4850 blasting vibrometer to measure the three-way vibration velocity and the main vibration frequency. Get vibration velocity and the main vibration frequency by synthesizing

---

Corresponding author E-mail: 779662311@qq.com

three vectors. The foundation of the building is the architectural structure of the carrier member, vibration waves spread through the upper part of the building's structural foundation, the vibration of the basis can be a true reflection of buildings[3], so for this vibration monitoring, vibration measurement points arranged in the central axis of the tunnel floor, the sensor is fixed with plaster.

3.2 Test Results

Tunnel axis horizontal distance measured focal distance of 38m, blasting tunnel in the ground plane of the axis distance 32m. The test results are:

Tab.1 Measured data

Blasting Number	Distance /m	The maximum single dose response/kg	The maximum vibration velocity/cm·s^{-1}	Main vibration frequency/Hz
1	50	28	0.91	62.500
2	50	25	0.71	18.519
3	50	18	0.46	96.238
4	50	35	1.92	28.57

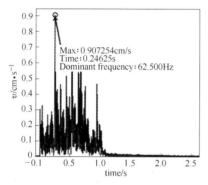

Fig.1 The first blast vibration vector synthesis figure

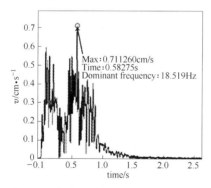

Fig.2 The second blast vibration vector synthesis figure

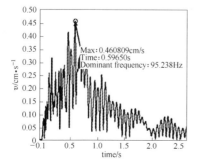

Fig.3 The third blast vibration vector synthesis figure

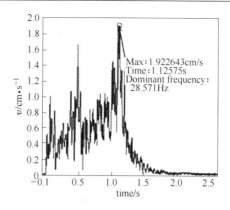

Fig.4 The fourth blast vibration vector synthesis figure

3.3 Analysis of Monitoring Results

According to Tab.1, the measured data, combined with the division of the protected object categories of "Allowing blasting vibration safety standards", allowing safe traffic tunnel velocity 15~30 cm/s, for safety reasons, here referring to the general brick, large non-seismic vibration safety standard building blocks to allow the maximum $v = 2.8$ cm/s[4]. According to Savannah Road Minkowski formula [5]

$$v = K\left(\frac{R}{\sqrt[3]{Q}}\right)^{-\alpha} \quad (1)$$

where, v is the speed of vibration monitoring points, cm/s; K and α are bursting point to the terrain, geological conditions point correlation between the vibration attenuation coefficient and exponent; Q is the largest single segment homogeneous dose explosion, kg; R is the distance from the center of the explosion source, m. In this calculation, according to the Pingyikou blasting practice, $K = 200$, $\alpha = 1.6$. Return by the TC-4850 Savannah Road Minkowski analysis, forecasting blasting vibration velocity in the following table.

Tab.2 Prediction vibration velocity

Blasting Number	The maximum single dose response/kg	R/m	K	α	The maximum vibration velocity/cm·s^{-1}
1	28	50	200	1.6	2.26
2	25	50	200	1.6	2.13
3	18	50	200	1.6	1.79
4	35	50	200	1.6	2.55

Thus, although there is the relatively large gap between measured data and forecast data, but they all met allowed blasting vibration velocity in "Blasting Safety Regulations". You can learn tunnel blasting process is safe.

For $K=200$, $\alpha=1.6$, $R=50$ m, when control blasting vibration velocity v is 2.8 cm/s, calculated $Q = 41.77$ kg by Savannah Road Minkowski regression analysis, in the four times, the maximum single loud blast dose were less than 41.77 kg, experimental results are consistent with predictions.

4 CONCLUSIONS

(1) The maximum measured vibration velocity and main vibration frequencies are within the scope of "Blasting Safety Regulations" and will not cause harm to the tunnel.

(2) It is simple and effective to predict blasting vibration velocity and synthesize three vector to measured data by using of TC-4850.

(3) Blasting design is effective. For this project with the complex characteristics of the environment, follow the principle of multi punch and less charge, use shallow hole bench blasting construction program, multi detonating technology, and non-electric extension initiation network outside the hole, in order to avoid hops, reduce the possibilities of seismic wave superposition.

(4) The lack of testing is just a monitoring point, and multiple measuring points can be arranged in the follow-up tests, to get more data, and then get blasting seismic wave propagation in the region. According to various forms of blasting to analyze its characteristics and its variation of stress-induced breaks, to grasp the reasons for the destruction damage and damage processes [6], to provide experience for future design and construction. While only the largest particle vibration velocity was considered as a safety criterion, main vibration frequency and duration of action of seismic waves was considered in this project [7~10].

REFERENCES

[1] Wang X Z, Wang Z L, Long Y, etc. Blasting vibration-rise buildings collapse digital model[J]. *Explosion and Shock*, 2002, 22 (2): 188~192.

[2] Chen J, Song H W, Shen Z Y, Zhou Y. The practice of tunnel blasting vibration control under neighborhoods[J]. *China Three Gorges University (Natural Science Edition)*, 2001, 33 (1): 55~57.

[3] Wang D A. 79 m high chimney demolition blasting vibration and vibration slump touchdown comparative analysis[J]. *Blasting*, 2006, 12 (1): 86~89.

[4] People's Republic of China national standard. GB 6722—2003 *Blasting Safety Regulations*[S]. Beijing: China Standard Press, 2004.

[5] Wu L, Rao X A, Huang C B. Methods of prediction and control of blasting vibration analysis[J]. *Hydrogeology and Engineering Geology (Supplement)*, 2004: 136~140.

[6] Wang Z S. Rock blasting destruction causes foothold research analysis [J]. *China three Gorges University (Natural Science Edition)* 2001, 23, (2):120~123.

[7] Zhang Y P. HHT blasting vibration signal analysis and application of . Changsha: Central South University, 2006.

[8] Zhong G S. Applied basic research of blasting vibration analysis based on wavelet transform . Changsha: Central South University, 2006.

[9] Li H T. Study based on seismic effect of blasting energy principle [D]. Wuhan: Wuhan University, 2007.

[10] Zhao M S, Liang K S, Cao Y, etc. Explore of safety standards for building (structure) under the action of blasting seismic[J]. *blasting*, 2008, 25 (4): 24~27.

The Various Reasons and the Preventive Measures of Misfire in the Nonel Detonation Circuit

LI Shuming [1], LUO Wei [2], ZHANG Fuyang [3]

(1. Shenzhen Huahai Blasting Engineering Co., Ltd., Shenzhen, Guangdong, China;
2. Shenzhen Dijian Engineering Co., Ltd., Shenzhen, Guangdong, China;
3. School of Chemical Engineering, Nanjing University of Science and Technology, Nanjing, Jiangsu, China)

ABSTRACT: This paper analyzes the various reasons of misfire in the nonel detonation circuit, and put forward the preventive measures of misfire. A lot of practice has proved that the preventive measures can greatly improve the reliability of nonel detonation circuit by taking these measures, such as making nonel tube symmetrical and tied firmly around detonator, protecting nonel tube connection node and processing waterproof, but also ensure blasting safety and create significant economic benefits.

KEYWORDS: blasting safety; non electric initiating circuit; misfire; preventive measures

1 PREFACE

Compared to the electric detonator, the nonel detonator have two advantages(Wang Z H et al, 2009; WU C S et al, 2013), first is anti stray current, second is infinite setting delay time. Nonel Detonator also has some disadvantages, such as its blasting circuit can not detect by any measuring apparatus, but electric blasting circuit can be detect by special apparatus. Compared with ordinary electric detonator, its cost is very high in the shallow hole blasting, and the operation is complex, also easily misfire by misoperation(BAI X C, 2006).

The nonel detonation circuit blind shot mainly have 3 type phenomena as follows: nonel tube did not detonated, nonel tube detonation propagation interrupt, nonel tube detonator misfire.

2 MISFIRE PHENOMENON AND ANALYSIS

2.1 Nonel Tube Did not Detonated

Nonel tube did not detonated mainly by nonel tube binding asymmetry or percussion detonator connection node protection is not good.

2.1.1 Twining Asymmetry

One project of the pile well blasting construction in Shenzhen, the style of one electric detonator binding on a piece of nonel tube was used in initiating circuit, and as shown in Fig.1, the statistics detonation result indicated that, the nonel tube undetonation rate reach to 10%.

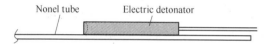

Fig.1 Sketch map of nonel twining on one side

The reason for this phenomenon may be: only a piece of nonel tube blind on the side of detonating detonator, and the other side of the detonating detonator in the free state, when detonating detonator explosion, the detonation wave in the free side to can not forming counterforce which provide to the nonel tube on other side of the detonating detonator, result in the phenomenon of nonel tube did not detonated.

2.1.2 Connection Node Protection

The parallel or series nonel tube circuit connection node detonated by electric detonator, if the connection node is not binding tape strictly and do not stack sandbags, it caused the blasting energy of detonating detonator unconcentrate, hence the nonel tube unexploded.

2.2 The Nonel Detonation Interrupt

The reasons of detonation propagate interrupt of nonel tube is: one is the problem of the quality of the products, such as coating explosive on the nonel tube wall unevenness, coating explosive break off or coating explosive expired. The other is the operating problems, such as the nonel tube initiating circuit did not take protective measures or the nonel tube in circuit did not straighten, leading to the blasting damage by detonating detonator. In addition, the nonel tube

---

Corresponding author E-mail: 13902452791@163.com

knotting and the nonel tube wall damage would caused the nonel tube detonation propagate interrupt.

2.3 Nonel Tube Detonator Misfire

On the condition of water hole, sometimes nonel can normally detonation, but the detonator misfire. Since the nonel tube detonator did not waterproof processing, misfire increased significantly when in the rainy season, according to take out the detonate charge in the misfire, found that the majority reason of charge misfire is detonator misfire. That's because of the seals between nonel and detonator shell is not tight, water enter into detonator result in misfire.

3 PREVENTIVE MEASURES

3.1 Nonel Tube Symmetrical and Tied Firmly Around Detonator

Nonel tube detonator connection with initiating detonator, the nonel should be symmetrical around the initiating detonator, and twine electrical tape more than 4 layers. If use single nonel tube, take other 5 to 6 root scrap nonel binding around to initiating detonator uniformly and symmetrically, or using single nonel tube folded back and forth repeatedly, then binding symmetrically around the initiating detonator.

3.2 Nonel Tube Connection Node Protection

The connection node protection is very important, because of the detonator blasting formed high-speed jet during normal initiation can cut off the nonel tube. The concreteness method is: After completion of binding detonating detonator with nonel tube, in the direction with shaped aperture using electrical tape twining two layers into braided shape, its length should be greater than 15cm, then the connection node is placed in the flat and no stones position which covered with sand bags, and as shown in Fig.2, the sand bag cover range including the detonator and the braided shape parts of nonel, weight of sand bag about 10kg. In addition, detonation propagate nonel tube should be straightened, and the distance between initiating detonator and nonel should reach to a certain safety value to protect the initiating circuit is not broken. At the same time, the sand bags can restraine explosion energy of the initiating detonator, make the initiating detonator effectively detonating nonel.

3.3 Waterproof Processing

The nonel tube detonators using in water hole blasting should had waterproof processing. First using electrical waterproof tape wrapped on the detonator crimping parts a layer tightly, then wrapped a layer of electrical insulation tape around the outside of electrical waterproof tape to ensure the waterproof tape do not loose.

Fig.2 Sand bags protected circuit node picture

4 ENGINEERING EXAMPLES

Because of appear the phenomenon of nonel tube did not detonated using reverse initiation connection circuit in the early times, recently, we using normal connection achieve at good effect. The blasting construction of quarry group comprehensive treatment project in Buji Shenzhen city was continuing exploiting more than ten years, the distance between blasting face and residential areas is more and more come near, the explosive charge weight of single hole is very small, nonel initiating circuit needs to use the out hole delay. The detonators connected in out hole relay circuit are use series connection (commonly known as out hole relay), each holes have a connection points, out hole delay using two detonator. General every blasting detonating 30~35 holes, and the total number of connection point is 30~35 in once blasting, the longest series node is up to 15 (such as the rear row hole, row hole delay nodes and inter hole delay nodes); the maximum numbers of holes exceed 60, and the longest series more than 20 (shown in Fig.3).In these years, according the above method, carry out more than million nonel tube detonator applied to initiating circuit, also applied on water holes blasting thousands times, almost never appears misfire.

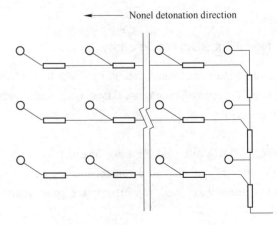

Fig.3 Nonel detonators initiating circuit of out hole relay

5 CONCLUSIONS

Practice has proved, the prevention of nonel initiating circuit misfire measures greatly improves the reliability of nonel initiating circuit, for using nonel initiating circuit provides a reference for the select of prevent misfire measures.

REFERENCES

[1] Wang Z H, Zhou M Y, Zhao G C, Tian X B. Application of high precision fuse blasting cap and hole-by-hole detonation technique[J]. Metal Mine, 2009, 395(5): 43～45.

[2] WU C S, CHEN Z J, HE C J. Discussion on safe usage of nonel-detonators, 2013, 30(1): 93～95.

[3] BAI X C. Application of the plastic cartridge igniter V detonation detwork in the deep-hole demolition[J]. *Blasting,* 2006, 23(2): 53～56.

Using Sublevel Pre-splitting Blasting Effect to Reduce the Seismic Wave Velocity

SUN Jing[1], YAN Jianhua[2]

(1. Xinjiang Nonferrous Metallurgical Design and Research Institute Co., Ltd., Urumqi, Xinjiang, China;
2. Xinjiang Jinhaikun Blasting Engineering Co., Ltd., Urumqi, Xinjiang, China)

ABSTRACT: Through the measured blasting seismic wave data on the construction site, key paragraph in blasting, the time differential, application sublevel pre-splitting blasting, reducing the peak vibration superposition effect, safe construction task smoothly.

KEYWORDS: blasting seismic wave; pre-splitting blasting; superimposition

1 GENERAL SITUATION OF ENGINEERING

Spider mountain in urumqi expansion of outer ring expressway reconstruction A2 blocks the whole circuit is designed to separate, the Z1 line length of 1055 meters, the tunnel length 370 meters,Z2 line length of 1055 meters, the tunnel length 370 meters.Cover for quaternary pleistocene series diluvium western group, grey pebbles, the top of the hill in the middle of 18 to 23 m thick.

Outward gradually thinning, the skeleton particles most continuous contact, arranged disorderly, average particle size 5~10 cm, solid state, serious local cementation, sloping fields with loess soil powder, thickness of about 1 to 2 m, bedrock under the main group for Jurassic qi ancient group kara argillaceous siltstone, lithic sandstone and mudstone, argillaceous structure, the silt structure, layered structure.

Bedrock under the main group for Jurassic qi ancient group kara argillaceous siltstone, lithic sandstone and mudstone, argillaceous structure, the silt structure, layered structure.

Broken rock mass integrity is broken, poor stability, level of surrounding rock grade V, surface water and groundwater is not development.

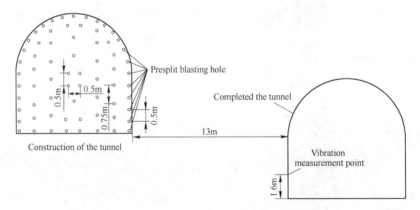

Fig.1 Tunnel related sketch map

2 THE BLASTING CONSTRUCTION TECHNOLOGY

This blasting engineering using the step-by-step excavation method, blasting construction technology for drilling—charging—fill—alerts—attachment—warning—initiation, specific parameter selection are as follows.

2.1 The Selection of Tunnel Excavation Blasting Parameters

2.1.1 The Hole Arrangement

Hole arrangement according to the following parameters:
Aperture 40 mm diameter, cut hole for V-cut, the rest for the vertical hole drilling.

Prespitting hole depth $L_y = 1.1$m.

Cut hole depth $L_t = 1.1$m.

Contour(perimeter)hole, trim hole depth $L_2=0.9$m.

Cut hole, auxiliary hole spacing $a_1=0.6$~0.8m, row-spacing $b_1=0.5$~0.7m.

Trim hole depth spacing $a_2=0.5$m, row-spacing $b_2=0.4$~0.5m.

2.1.2 Blasting Equipment

Hole within 2# emulsification rock ammonium nitrate explosives, non-electric detonator detonating in milliseconds.

Corresponding author E-mail: 2953023623@qq.com

2.1.3 Specific Charge

Select specific charge $k_b = 0.2$ kg/m, $k_b = 0.15$ kg/m, $k = 0.5$ kg/m³, the calculation a single charge $q_y = 0.2$ kg.

$q_t = 0.3$ kg, and $q_f = 0.2$ kg, $q_y = 0.15$ kg.

2.1.4 With Period of Initiating Maximum Dose at a Time

The main blast hole maximum dose with a period of initiating explosive $Q_1 = 6q_f = 1.2$ kg.

With presplit in a period of detonation dose $Q_2 = 4q_y = 0.8$ kg.

2.2 Detonator Piecewise Division

Pre-split blasting using 3 and 5 of segment detonator, main blasting area using 6,7,8,9,10,11,12,13,14,15 segment detonator.

2.3 The Blasting Network and Blasting Method

Blasting network using cluster mode, use MFB200 type initiating.

Near the building or the original tunnel on peripheral hole first initiation, forming a crack, and then cut slots, auxiliary hole, peripheral hole blasting in turn.

Pre-splitting fissure a shock absorption effect, this effect is generally can reach 50%.

2.4 The Blasting Safety to Check Calculate

2.4.1 Blasting Fly Rocks Farthest Distance
$$L_f = k_f d = 64 \text{m}$$
where k_f——value range of 15~16;

d——borehole diameter, cm.

In the tunnel with the curtain shade, and in the tunnel with pile retaining wall, retaining wall must be right tunnel, It's the same height and tunnel excavation area, prevent slungshot fly out of the alert area.

2.4.2 Blasting Vibration Velocity

(1) Main blasting area vibration velocity $v = k'k(Q_1^{1/3}/R)^\alpha = 1.3$ cm/s, pre-splitting fissure after shock blasting vibration velocity of 0.65 cm/s, less than 1 cm/s design requirement of vibration speed, has been built tunnel safety.

where k——factor, is related with the nature of rock, blasting method and conditions, $k = 200$;
k'——correction coefficient, $k' = 0.6$;
Q_1——single biggest controll blasting, $Q_1 = 1.2$ kg;
R——for vibration measurement points to the blasting center distance, $R = 13$ m;
α——seismic attenuation index, $\alpha = 1.8$.

(2) Primary area blasting vibration velocity $v = k'k(Q_2^{1/3}/R)^\alpha = 1.0$ cm/s, less than 1 cm/s design requirement of vibration speed, has been built tunnel safety.

3 VIBRATION DATA

3.1 Before the Adjustment

1.28 Z2 line K1 + 1162.9 merge
　Recording time 2013.01.28 12:36:45
　Record length 2.0000 s
　Record the rate of 8000,sps

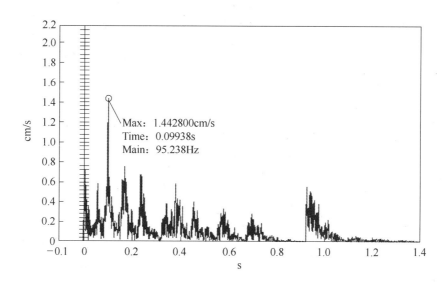

Fig.2　Before the adjustment

3.2 Plan Adjusted

1.29 Z2 line K1 + 1162.1 merge

　Recording time 2013.01.29 12:49:07
　Record length　2.0000 s
　Record the rate of　8000,sps

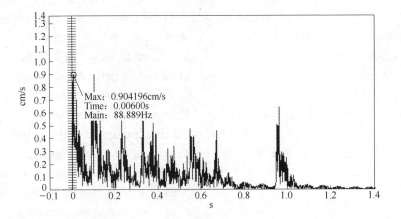

Fig.3 Plan adjusted

4 PLAN ADJUSTMENT

According to the design scheme of four tunnel face blasting construction there is a face of blasting vibration design of vibration velocity over the design to 1.44 cm/s.

From the blasting vibration on the drawing as you can see, 0.099 seconds after blasting, blasting vibration velocity peak 1.44 cm/s, so we conclude that more than standard from presplit blasting vibration speed, it is due to the 3 and 5 paragraphs caused by blasting vibration superposition. We adjusted the amount of pre-splitting blasting paragraphs, will advance pre-split hole in 3, 5, 6 three paragraphs initiating, 3 and 5 each for the 2 hole, 6 paragraph for the 4 hole, is the main blast hole 7~15 paragraphs.

After adjustment scheme, blasting vibration data as shown, the biggest vibration velocity of 0.9 cm/s, comply with the design requirements.

5 SUMMARY

Adopting sublevel pre-splitting blasting in the practical engineering, in order to increase the presplit blasting paragraph time differential, that can avoid the seismic wave superposition and reduce the seismic effect, achieve the goal of safe To complete the construction task.

REFERENCES

[1] Meng jifu. Blast Vibration Monitoring Techniques[M]. Beijing: Metallurgical Industry Press, 1992.

[2] Liu Dianzhong. Engineering Blasting Applied Manual[M]. Beijing: Metallurgical Industry Press, 1999.

[3] Yu Yalun. Theories and Techniques of Engineering Blasting[M]. Beijing: Metallurgical Industry Press, 2004.